# BRISTOL BAY ALASKA

## NATURAL RESOURCES OF THE AQUATIC AND TERRESTRIAL ECOSYSTEMS

EDITED BY

### CAROL ANN WOODY, PhD

J.ROSS
PUBLISHING

# CONTENTS

# FOREWORD

The Bristol Bay Partnership is pleased to support the publication of this important scientific reference book, the very first to focus on the diverse natural resources of Bristol Bay. These abundant natural resources have sustained people, cultures, economies, and communities across Bristol Bay, the nation, and the world for many generations. It is everyone's responsibility to ensure that our decisions and actions sustain these resources for many generations to come. Our residents rely on these natural resources for their economic and food security, as well as to continue traditional lifestyles and cultural practices. Yet, while these resources are fundamental to the region's culture and communities, we've always shared our abundance with people and firms around the globe. For example, our world-class commercial fisheries have sustainably and reliably supported international businesses for over 125 years.

Sustaining the region's renewable natural resources is very important to the Bristol Bay Partnership. We are the chief executives of the major regional organizations: Bristol Bay Native Association, Bristol Bay Native Corporation, Bristol Bay Housing Authority, Bristol Bay Area Health Corporation, and Bristol Bay Economic Development Corporation. Together, our organizations provide essential and critical services to 31 tribal communities in the Bristol Bay region of southwest Alaska. Our region covers approximately 44,000 square miles—about the size of the state of Ohio. Our natural resources have a local, regional, national, and global reach. They support traditional subsistence practices that continue to be handed down from one generation to the next, and provide economic benefits to guides, outfitters, and visitors.

The Bristol Bay Partnership is united in its dedication to sustaining our region's economy and lifestyles. We know that our waters, fish, wildlife, and migratory resources—as well as our cultural ways—can be dangerously impacted by proposed mineral, oil, and gas development on our homelands. The Partnership is unanimously against resource development that threatens the sustainability of our renewable natural resources. We want our natural resources to continue to be enjoyed and shared by millions for generations to come.

That is why this book is so important—it is the very first of its kind to compile, document, and share the science regarding the bountiful natural resources and ecosystems that surround Bristol Bay. It is written by scientists and researchers who are experts in their fields and who have dedicated themselves to understanding the region. It is a *must-read* for policymakers, resource managers, scientists, students, visitors, and businesses that want to know more about our lands and our natural resources. Most importantly, it provides a foundational *state of our understanding* that this and future generations can use to encourage wise and sustainable management of Bristol Bay's natural resources and the livelihoods (both cultural and economic) that depend on them.

We hope you enjoy the educational journey this book provides through our region and that it inspires you to come experience all the region has to offer.

With warm regards,

The Bristol Bay Partnership

# INTRODUCTION

Bristol Bay, Alaska, and its surrounding watershed, is one of the most pristine and ecologically significant places on earth. It is a wild, fascinating region that is bordered by the tall, fuming volcanoes of the Alaska Range and the rich marine habitats of Bristol Bay. No interstate traverses this region, although it is quite large—around 44,000 square miles. It is home to Native Alaskans practicing ancient ways, tough fishing families, and eclectic immigrants. Iconic creatures such as bear, moose, caribou, and eagles are relatively common, while a rare population of freshwater seals inhabits Iliamna Lake. The rich bay provides important breeding, feeding, and migration areas for Pacific walrus, sea lions, seals, dolphins, porpoises, and whales, including the endangered North Pacific right, humpback, fin, and sperm whales. Tens of millions of seabirds, shorebirds, and waterfowl migrate annually to the bay from distant regions (e.g., New Zealand) to forage, molt, rest, or breed. This rich quilt of life is directly or indirectly connected to salmon, specifically sockeye salmon (*Onchorhynchus nerka*)—the keystone species that holds the region together.

Bristol Bay is the most extraordinary, valuable, wild sockeye salmon ecosystem left on the planet. More than 56,000,000 wild sockeye salmon swarmed through Bristol Bay in 2017 on their way to spawn in the region's sparkling lakes, rivers, and streams. On this journey, millions of salmon were caught—filling fishers' nets and literally sinking some boats with their bounty. Native Alaskans fished and put up their harvest following ancient traditions, while commercial fishers set their nets for the 133rd year. These wild salmon are sustainably caught, and they fuel the people, the economy, and the ecosystems of Bristol Bay. Such salmon runs were once common across the Pacific Northwest, but now most populations in the Lower 48[1] are at less than 10% of their former abundance—due primarily to habitat degradation and alteration. The remaining wild population biodiversity is further threatened by climate change and genetic contamination from hatchery stocks. And yet, Bristol Bay is still intact and wildly productive.

As a young fish biologist, I spent a decade attempting to restore damaged fish habitats before my unforgettable introduction to Bristol Bay. I flew in a Cessna 206 prop plane from Anchorage to Iliamna, a small community of just over 100 people—my camera glued to the window the entire flight. Three massive, glacial-capped volcanoes higher than 10,000 feet dominated the view—Mount Spur (which erupted that summer), Mount Redoubt, and Mount Iliamna. We cut through a narrow pass of jagged glacier-covered peaks, then emerged over Lake Iliamna, Alaska's largest lake, and one of the world's largest sockeye salmon nurseries. Flying along the shoreline, I watched sinuous red schools of salmon swimming along the gin clear beaches and swarming by the thousands at river mouths. I was thrilled to see the rare population of freshwater Iliamna seals chasing salmon and basking on islands in the sand. As we flew to the small airstrip, passing over many free running rivers and streams, I realized I had arrived at the largest intact sockeye salmon ecosystem left in the world. It was a fish biologist's dream! Here lies an ecosystem that produces half of the world's sockeye salmon for human consumption—without hatcheries or human interference. Little did I realize that I would spend half of my career dedicated to the scientific research and conservation of the salmon of Bristol Bay.

The Bristol Bay region contains incomparable aquatic and terrestrial ecosystems. These ecosystems are all interconnected and fueled, in part, by the nutrients that salmon annually transport from the sea. Bristol Bay harbors many treasures besides the obvious renewable salmon, fish, and wildlife—it also contains oil, gas, and mineral resources. The debate over development of these nonrenewable resources creates a constant tension, often erupting into controversy, regarding what is appropriate relative to salmon conservation. Every large development proposal prompts a flurry of scientific studies.

Most of the scientific information on Bristol Bay resources is scattered in government documents, scientific publications, or lost in some dusty file cabinet. Consequently, my inspiration for this volume came from the fact that there was no definitive work on the natural resources within Bristol Bay and their incredible importance to the ecosystems, cultures, and economy of the region.

This book is organized into six sections—starting with the indigenous people who have resided in the region for millennia and the land itself, including vegetation, flora, parks, and refuges. Next, we discuss key wildlife species that can be found in the terrestrial environment. Sections three and four include marine wildlife and fisheries, while section five

---

[1]In Alaska, *Lower 48* is the common way of referring to the 48 states that make up the contiguous United States.

provides extensive information on the freshwater ecosystems. Section six explores the nonbiological resources such as oil, gas, mineral, and renewable energy resources of the region. Each chapter was written by scientific experts. The reader should keep in mind that Bristol Bay is a huge region and relatively few studies exist for many species across this amazing but sometimes harsh region. Subsequently, the scientists who conducted their research and wrote their respective chapters also relied on any and all available regional information as well as general life history information for Alaska and North America. In addition these authors provided recommended research to further the understanding of each topic.

My intention is for this book to be extremely helpful to policy makers, scientists, resource managers, students, and others who may find themselves involved in decisions that affect Bristol Bay. All future decisions affecting this region are of paramount importance as they may ultimately impact the iconic salmon fisheries, which will in turn affect the people, their cultures, all other wildlife and fisheries, and the regional economy. Hopefully this book will also spur interest and excitement in young scientists and inspire them to focus their intelligence and energy to help us better understand and conserve this beautiful, bountiful place. Lastly, I hope that this book can help identify needed research to establish robust, defensible scientific baselines and ensure that the region's renewable resources are well managed and endure for future generations to experience and enjoy.

Carol Ann Woody

# ABOUT THE EDITOR

CAROL ANN WOODY has been adventuring, researching, teaching, and living in Alaska since 1988. She became fascinated with Bristol Bay in 1993 when, clad in a leaky dry suit, she spent a chilly summer floating around North America's largest sockeye salmon nursery (Iliamna Lake) surrounded by ruby red spawning sockeye, studying their behavior for the University of Washington (UW). She went on to earn her Ph.D. in Fisheries Science from UW. She also holds an M.S. in biology from the University of Wisconsin and a B.S. in Wildlife and Fisheries Management from Utah State University.

Carol Ann served almost 20 years as a federal scientist for the U.S. Fish and Wildlife Service, the U.S. Forest Service, the U.S. Geological Survey (USGS), and, most recently, the National Park Service. During her four years on the 17 million-acre Tongass National Forest in southeast Alaska, she and her teams earned multiple awards for exceptional productivity and meritorious service. While a fisheries research biologist with USGS, she earned multiple awards for exceptional research productivity in population status and trends, ecology, genetics, and evolution. She also received a meritorious service award for acting as the nation's USGS Director of Fisheries.

During a decade of research in Bristol Bay with USGS, Carol Ann recruited and mentored a diverse cadre of graduate students, local interns, and volunteers, many of whom continue to serve in fisheries and environmental science in Alaska. When industrial mining was proposed for sensitive headwater regions of the Kvichak and Nushagak watersheds, she left USGS and started Fisheries Research and Consulting. Collaborating with a team of talented scientists, she helped increase publicly available scientific information for the proposed mine region, provided technical review and translation of mine proponent studies and claims for the public, and educated the public and decision makers on potential risks to Bristol Bay fisheries from industrial copper mining. Later, she joined the Center for Science in Public Participation (CSP2) and focused on a diversity of resource extraction and fisheries issues across Alaska and the nation.

Carol Ann has academic affiliations with the University of Alaska, University of Idaho, and University of Montana, mentoring graduate and undergraduate students and sparking interest in fisheries and environmental science through courses and research internships. In 2016 and 2017, the Alaska Native Science & Engineering Program honored her for creating systematic change in the hiring patterns of Indigenous Americans in the fields of science, technology, engineering, and mathematics.

Embracing the responsibility of a scientist to translate and communicate scientific findings and their implications to all interested parties—peers, resource managers, tribes, decision managers, policy makers, stakeholders, and the public—has led Carol Ann to become an expert fisheries advisor to diverse groups including indigenous tribes, the World Wildlife Fund, Patagonia Inc.'s Wild Salmon Advisory Team, and Alaska Governor Walker's Fisheries Transition Team. She is a Past President of the Alaska Chapter of the American Fisheries Society (AFS) and served on the Western Division Environmental Concerns Committee and Endangered Species Committee of AFS. She is the first woman to receive one of the American Fishery Society's highest honors—the President's Fishery Conservation Award (2017).

She has published more than 20 peer reviewed articles; numerous technical reports; edited two books; prepared policy and law articles, including an amicus brief for the U.S. Supreme Court; served as an expert witness for various court cases; been an invited speaker on three continents; given innumerable talks; and led discussion panels and workshops on Bristol Bay fisheries. Carol Ann is currently the Regional Fishery Biologist for the National Park Service in Anchorage, Alaska.

# ACKNOWLEDGEMENTS

Conducting scientific research in the wilds of the Bristol Bay region is not an easy task. Most field work is conducted in the summer months due to the limited light and weather the remainder of the year. The lack of roads requires that work be accomplished using small planes, boats, or helicopters from remote camps that are far from home, internet access, and any medical facility. To these hardy scientists, I extend my sincere gratitude for graciously contributing their time and energy in researching and writing their respective chapters. This book would not have been possible without their dedication, generosity, and expertise. In addition, my sincere thanks to Dr. David Chambers of the Center for Science in Public Participation for his unwavering support. The people of Bristol Bay are truly remarkable, and I thank them for their continued efforts to conserve the extraordinary Bristol Bay salmon fishery as well as their imaginative work to improve and diversify fishery products. Finally, I am very grateful to my publisher Gwen Eyeington, of J. Ross Publishing, who originally talked me, a rather reticent editor, into taking on this project, which four years ago was merely an idea. Without her continued support, encouragement, and enthusiasm, this book would not exist.

Carol Ann Woody

At J. Ross Publishing we are committed to providing today's professional with practical, hands-on tools that enhance the learning experience and give readers an opportunity to apply what they have learned. That is why we offer free ancillary materials available for download on this book and all participating Web Added Value™ publications. These online resources may include interactive versions of material that appears in the book or supplemental templates, worksheets, models, plans, case studies, proposals, spreadsheets and assessment tools, among other things. Whenever you see the WAV™ symbol in any of our publications, it means bonus materials accompany the book and are available from the Web Added Value Download Resource Center at www.jrosspub.com.

Downloads for *Bristol Bay Alaska: Natural Resources of the Aquatic and Terrestrial Ecosystems* include material on North Aleutian Basin oil and gas potential.

# SECTION I

## THE PEOPLE AND THE LAND OF BRISTOL BAY, ALASKA

# THE INDIGENOUS SALMON CULTURES OF THE BRISTOL BAY WATERSHEDS

**Alan S. Boraas and Catherine H. Knott**
**Kenai Peninsula College**

*. . . Salmon more or less defines this area. It defines who we are. When you look at our art, you will see salmon . . . It is who we are. When you listen to the stories and take a steam, even in the middle of winter, people talk about salmon. It is in our stories; it is in our art. It is who we are; it defines us.* **M-61, 9/16/11**[1]

## INTRODUCTION

Before colonization by European, American, and Asian nations, several dozen salmon cultures subsisted around the North Pacific and North Atlantic rim. Indigenous cultures such as the Sami of Fennoscandia, Micmac and Abnaki of northeastern North America, the Ainu and Nvenk of the northwest Pacific, and cultures of the northeast Pacific from California to Alaska relied on salmon as their primary subsistence species. For each culture the breadth of foods hunted or gathered was important, but the abundance, reliability, and nutritional value of salmon made it their keystone subsistence species. In most cases salmon were harvested for thousands of years, shaping each culture such that mythological stories depict the people and the salmon as one inseparable entity. With colonization, wild salmon populations suffered drastic declines, particularly in recent years (Colombi and Brooks 2012). The building of hydroelectric and irrigation dams, mining, agricultural development, and urbanization caused salmon habitat destruction and degradation which, along with high-seas over-fishing, were of such magnitude that almost none of the indigenous cultures can currently harvest wild salmon in numbers anywhere close to that of earlier times.

For most, the traditional first salmon ceremony that marked world renewal and an affirmation of the identity of people with salmon is nothing more than a symbolic expression of what once was. Everywhere, that is, except Alaska where wild salmon continue to thrive, particularly in the watersheds of Bristol Bay. Here indigenous cultures of the Yup'ik and Dena'ina have been harvesting wild resources for at least 12,000 years and have intensively caught salmon for at least 4,000 years. This immense time depth has shaped all aspects of the cultures, including social and spiritual dimensions. Because wild salmon still return in numbers approximating, perhaps even exceeding, those of prehistory (e.g., 45–63 million sockeye salmon were predicted to return to Bristol Bay in 2015),[2] the indigenous cultures maintain a continuum with the past that is rare in North America. In few places in the world do the same wild foods that their ancestors ate dominate the diet and configure the culture as they do today in Bristol Bay watersheds. The indigenous cultures of the Bristol Bay watersheds have made a successful transition from prehistory to the present based on the same keystone species as their ancestors—salmon.

---

[1] These interviews followed University of Alaska Anchorage Institutional Review Board guidelines. Because of the sensitive nature of the information regarding controversy over potential development of the Pebble Mine, it was determined that interviewees could be criticized, harassed, or otherwise adversely affected. Hence the names of the interviewees are not used and are designated M (male) or F (female), a random number, and the date of the interview.

[2] Alaska Department of Fish and Game Commercial Fisheries Division News Release. 11/13/2014. Anchorage.

**Figure 1.1**    Salmon art on the wall of the Sam Fox Museum at Dillingham (Curyung). September 2011. *Photo credit: Alan Boraas.*

In 2011 we were asked by the United States Environmental Protection Agency (USEPA) to undertake a characterization of the indigenous cultures of the Nushagak and Kvichak River watersheds to help in their assessment to "characterize the biological and mineral resources of the Bristol Bay watershed, increase understanding of the potential impacts of large-scale mining . . . on the region's fish resources, and inform future decisions . . . related to protecting and maintaining the chemical, physical, and biological integrity of the watershed" (USEPA Region 10). To best translate the indigenous importance of salmon and clean water to the USEPA, we undertook qualitative, semi-structured interviews of over 50 Elders and culture-bearers in the tribal centers or homes of the people in their villages. In addition, we contextualized the interviews, relying on anthropological and related literature as well as observations from our visits. The result was: "Traditional Ecological Knowledge and Characterization of the Indigenous Cultures of the Nushagak and Kvichak Drainages, Alaska" (Boraas and Knott 2013). This chapter expands the perspectives of that document to the villages of the Bristol Bay area. For both the USEPA study and this chapter we want to thank the Elders and culture bearers who shared their knowledge and deeply personal and culturally sensitive attitudes and beliefs. We have tried to depict those perspectives as accurately as possible. To give voice to those viewpoints, we have included numerous direct quotations from the Elder and culture bearer interviews, which are presented here as *Voices of the People.* Consensus of cultural practices or beliefs is cited as *Interviews 2011.*

Indigenous villages of the Bristol Bay watershed are shaped by a very high degree of reliance on wild salmon, both today and historically. However, they are by no means uniform (Table 1.1 and Figure 1.2). Most of the occupants of eleven of the twenty-four towns and villages in the Bristol Bay watershed are from a Yup'ik-speaking cultural heritage. These consist of villages along the Nushagak, Kvichak, Wood, Naknek, and Igushik Rivers as well as three coastal villages on smaller rivers. Four of the villages in the Kvichak River drainage are primarily Dena'ina with close associations with the Northern Déné culture that extends from Alaska to Hudson Bay. Linguistically the Dena'ina represent an entirely different language family from the Yup'ik and Alutiiq speakers with different pre-Orthodox spiritual practices and a different

**Table 1.1** Census and subsistence data for villages and towns of the Bristol Bay watersheds

| Watershed | Community | Native Name | 2010 Census | Percent Alaska Native | Ethnic Majority | Per Capita Wild Food Harvest (includes salmon) In pounds | Per Capita Wild Salmon Harvest In pounds |
|---|---|---|---|---|---|---|---|
| Nushagak River | *Dillingham | Curyung | 2378 | 55.9 | Yup'ik | 242 | 141 |
| | *Ekwok | Iquaq | 115 | 90.4 | Yup'ik | 797 | 456 |
| | *Koliganek | Qalirneq | 209 | 95.7 | Yup'ik | 899 | 565 |
| | *New Stuyahok | Cetuyaraq | 510 | 93.5 | Yup'ik | 389 | 188 |
| Kvichak River | *Igiugig | Igyaraq | 50 | 40.0 | Yup'ik | 542 | 205 |
| | *Iliamna | Iliamna | 109 | 54.1 | Dena'ina | 469 | 370 |
| | *Kokhanok | Qarr'unaq | 170 | 80.0 | Dena'ina | 680 | 513 |
| | *Levelock | Liivlek | 69 | 84.1 | Yup'ik | 527 | 152 |
| | *Newhalen | Nuuriileng | 190 | 80.0 | Yup'ik | 692 | 502 |
| | *Nondalton | Nundaltin | 164 | 63.4 | Dena'ina | 358 | 219 |
| | *Pedro Bay | NA | 42 | 66.7 | Dena'ina | 306 | 250 |
| | Port Alsworth | NA | 159 | 21.4 | Non-Native | 133 | 89 |
| Wood River | *Aleknagik | Alaqnaqiq | 219 | 81.9 | Yup'ik | 296 | 143 |
| Igushik River | *Manokotak | Manuquutaq | 442 | 95.7 | Yup'ik | 298 | 135 |
| Naknek River | Naknek | Nakniq | 544 | 30.3 | Non-Native | 264 | 177 |
| | *South Naknek | Qinuyang | 79 | 82.2 | Yup'ik | 267 | 200 |
| | King Salmon | NA | 374 | 27.8 | Non-Native | 313 | 256 |
| Coastal | *Clark's Point | Saguyaq | 62 | 88.7 | Yup'ik | 1210 | 637 |
| | Egegik | Igyagiiq | 109 | 39.4 | Non-Native | 384 | 94 |
| | Pilot Point | Agisaq | 56 | 80.3 | Alutiiq | 384 | 35 |
| | Ugsahik | Ugaasaq | 10 | 70.0 | Alutiiq | 814 | 320 |
| | Port Heiden | Masrriq | 100 | 85.0 | Alutiiq | 408 | 85 |
| | *Twin Hills | Ingricuar | 74 | 94.6 | Yup'ik | 499 | 172 |
| | *Togiak | Tuyuryaq | 817 | 78.0 | Yup'ik | 304 | 106 |
| | Total Population | | 7051 | | | | |

*This chapter's cultural description focuses on the Yup'ik and Dena'ina villages indicated with an asterisk (*). Per Capita salmon data does not include fish retained from commercial harvest. Data from U.S. Census, Alaska; Alaska Community and Regional Affairs Database; Native Place Names from Indigenous Peoples and Languages of Alaska, Gary Holton Alaska Native Language Center, 2011; Subsistence Data from Fall et al. 2009 and Alaska Division of Fish and Game, Subsistence Division On-line Harvest Database (https://www.adfg.alaska .gov/sb/CSIS/index.cfm?ADFG=harvInfo.harvestCommSelComm).

oral tradition. These seventeen Yup'ik and Dena'ina villages with up to 95% Alaskan Native populations form the hub of the salmon cultures described in this chapter and are listed in Table 1.1 where they are identified with an asterisk.

Three of the outer Bristol Bay villages are primarily of Alutiiq heritage with close ties to the Alaskan Peninsula where marine mammals are a significant subsistence food. Because of the differences in subsistence and cultural heritage, and because we did not do research there, the Aluittq villages are not included in the following discussions. Within the watershed, King Salmon, Egegik (Igyagiiq), Naknek (Nakniq), and Port Alsworth are occupied primarily by non-Native migrants from outside of Alaska who do not share the cultural practices described in this chapter and are not included in this cultural characterization. The first three, like Dillingham (Curyung), are ports and fish processing centers for the Bristol Bay commercial salmon harvest, while Port Alsworth is headquarters for Lake Clark National Park and Preserve, as well as a center for evangelical missionary activities.

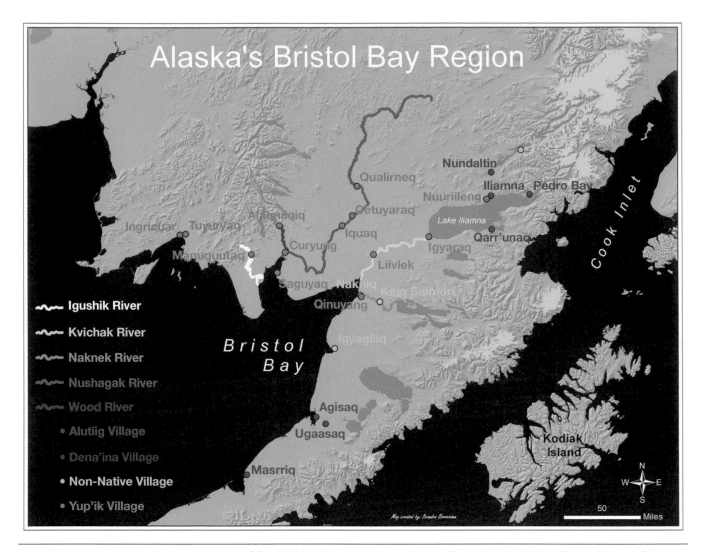

**Figure 1.2**   Primary Watersheds and Villages of the Bristol Bay Region. *Map credit: Brandon Bornemann.* See Table 1.1 for English Names. Native Place Names from Indigenous Peoples and Languages of Alaska *(map), Gary Holton Alaska Native Language Center, 2011.*

Dillingham (Curyung, population 2,378), the hub city of the region, has all the characteristics of a small town (hospital, airport, stores, hotels, etc.), and is homeport for much of the Bristol Bay commercial fishery—the largest wild salmon fishery in the world. It is 56% Alaska Native, mostly Yup'ik, and is home to the Bristol Bay Native Association and Curyung Tribe.

Whether Yup'ik or Dena'ina, indigenous Alaskan's of the Bristol Bay watersheds revere taking wild salmon; practice, or know about, the meaning of sharing wild resources; practice, or know about, the first salmon ceremony; believe, or have close relatives who believe, that wild creatures such as salmon have a soul and a will; and believe, or have close relatives who believe, that clean water is sacred. Everyone we talked to in the 2011 interviews believes that wild salmon are vital to their health and most associate salmon with cultural survival.

## PREHISTORY AND HISTORY

### Voices of the People

> *Salmon and fresh water have been the lifeline of the people here for thousands of years. If you look at the water, that is why fish and game has survived so well here, because we have such clean water.* **M-62, 9/16/11**

*So the importance of this resource, specifically salmon, has a major impact on my people here. That's the reason why we live here. We have sockeye salmon until March, when everyplace else has no more. That's why my ancestors fought over this region . . . The reason why they've been here for so long is it's a healthy environment, and we have been kind of watching over it all these years.* **M-33, 8/18/11**

*When you look at the map and where the old villages were, they were there because of the salmon. . . . all those villages. Site selection of those communities was very important and it was because of the production of subsistence foods at each of those sites processed. Most of those produced salmon in addition to [other foods].* **M-61, 9/16/11**

[If the salmon were to be impacted], *it would stop 10,000 years' plus of tradition, culturally and spiritually, for my people; not only my people, all the other communities and villages in this region would go away. We would cease to exist.* **M-33, 8/18/11**

People of the Paleoarctic tradition, indigenous colonizers with close ties to Siberia (see Dumond 2005), occupied the Bristol Bay watersheds as early as 12,000 years ago (Alaska Historic Resources Survey database, accessed July 1, 2011). Subsequently, archaeological cultures of the Northern Archaic and Ocean Bay traditions occupied the area, but current archaeological evidence indicates none of these groups practiced intensive salmon fishing.

In Southwest Alaska, salmon fishing first appears with the Arctic Small Tool tradition (ASTt) named for their small distinctive lithic artifacts where they are part of the Brooks River Gravels phase. These Bristol Bay watershed ASTt sites date from ca. 1800 BC to 1100 BC (Dumond 2005; Holmes and McMahan 1996). The houses at these sites were permanent structures, generally measuring four meters on a side, indicative of sedentary or semisedentary people and are located adjacent to salmon spawning streams such as a site near Igiugig (Igyaraq) (Holmes and McMahan 1996) (Figure 1.3). Only a small number of salmon bones have been found in the limited excavations (it is rare to find salmon bones in any Southwest or Southcentral Alaskan archaeological site). This is likely due to the antiquity of Dena'ina and Yup'ik practices that returned salmon carcasses to the water after they have been cleaned of meat, in effect, ritualizing

**Figure 1.3**    Kvichak River at Igiugig (Igyaraq) Village, Iliamna Lake in the distance. May 2011. *Photo credit: Alan Boraas.*

the functional importance of marine derived nutrients in northern aquatic ecosystems. The evidence shows indigenous people in Southwest Alaska have been subsisting on salmon for at least 4000 years.

Dena'ina involvement in salmon subsistence does not appear in the archaeological record until about AD 1000. At that point Dena'ina sites exploded on the landscape, particularly where salmon could be caught with weirs (obstructions used to corral and catch fish), possibly in response to increased salmon runs coincident with climatic changes of the Medieval Climatic Anomaly (Medieval Warm Period) (Boraas 2015). What appears in this subsistence revolution are underground cold storage pits called *elnen t'uh* by the Dena'ina (Figure 1.4). These complex salmon storage units were about a meter or more deep and up to two meters in diameter. The pit was lined with birch bark and the seams sewn with spruce root and sealed with partially dried salmon eggs to make a waterproof and insect resistant storage unit. Moss provided insulation and, after a second birch bark layer, the lined pit was filled with frozen salmon, capped with moss, and gummed birch bark on the sewn seams. This would have been done at freeze-up in the fall and the unit would remain frozen throughout the winter despite periodic warm weather above 0°C thus solving the problem of preserving, on a large scale, summer and fall-caught salmon for winter and early spring consumption. To fill these freezers, intensive fishing was done by fish weirs targeting late run coho salmon, and sockeye salmon if coho salmon were not present (Osgood 1976). At the same time, large log houses appear in the archaeological record indicating a shift from nomadism to sedentism (Osgood 1976).

The ability to store salmon, in turn, gave rise to complexity of social organization. To organize labor for intensive salmon fishing and to form alliances between villages, a complex social system evolved. The married men of a village were members of the same matrilineal clan and their wives and children were members of a different clan (Osgood 1976). The Dena'ina called this group the *nakilaqa* (Kari 2007) (*ukilqa* in Osgood 1976) or clan helpers. The clan helpers recognized a chief, called a *qeshqa*; in the Iliamna area the position was related to being a family head (Osgood 1976, Fall 1987). The *qeshqa* had numerous characteristics, among them wisdom, experience, and generosity. He or she had three primary duties: first, to arbitrate and resolve disputes; second, to care for the elderly and orphaned; and third, to assure the survival of the clan helpers through the equitable distribution of food. Regarding the latter, the *qeshqa* controlled the foods that were gathered, processed, and stored by the clan helpers and had authority to redistribute salmon stored in

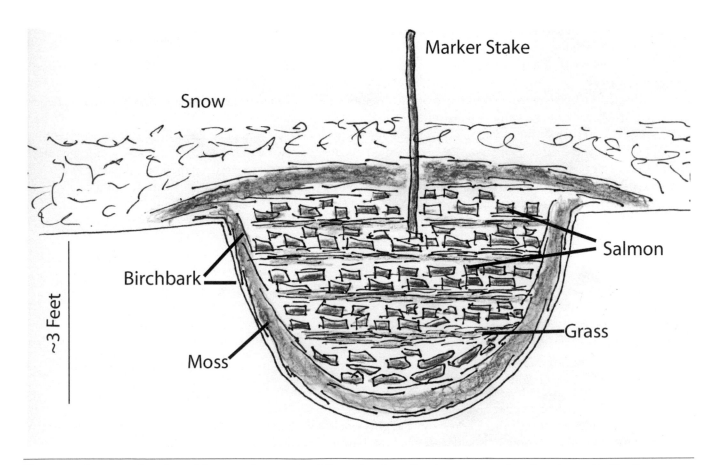

**Figure 1.4**    Precontact Dena'ina cold storage pit, preserving fall-caught salmon for winter consumption. *Drawing credit: Alan Boraas.*

underground cold storage pits back to people throughout the winter on an as-needed basis. This system provided food security based on the ability to over-produce stored salmon through cold storage. Each *qeshqa* had a partner in a distant village, called a *slocin*. If bacteria or a rogue wolverine affected one village's food supply, the *qeshqa* could request aid from his partner, who would divert some of his village's food resources to the needy village. The second *qeshqa* would be willing to do this because, at some point, his village might be short of food, and the partner he helped would return the favor. The socio-political system institutionalized survival based on the ability to over-produce and store frozen salmon.

Into the twentieth century Dena'ina practiced a ritual that involved sending the spirit of the animal to the *reincarnation place*. Land animal bones were burned in the fire and water animal bones, like salmon, were returned to the water. These practices ritualized ecology and were said to bring the animal back to be hunted or fished again (Boraas and Peter 1996) and is a form of *sacred ecology* (Berkes 1999) encoding ecological practices in spirituality: concepts that permeate Dena'ina and Yup'ik culture.

Salmon also was the staple of the early contact Yup'ik in the Bristol Bay watersheds (VanStone 1967). Although comparable ethnographies were not done, salmon were almost certainly caught using techniques similar to those of the middle Kuskokwim River just to the north. There the Yup'ik used woven willow bark set nets or drift nets in the river with notched stones tied off at the bottom to keep the net perpendicular to the water flow. Long handled dip nets and basket trap nets were also used (Oswalt and VanStone 1967). Salmon were dried on racks and stored in aboveground caches on stilts. The Yup'ik also stored salmon in underground, birch bark lined pits, some of it for fermented fish, and others frozen for winter consumption (Oswalt and VanStone 1967, Oswalt and Neely 1996).

Largely because of reliance on salmon and organization for intensive salmon fishing, the Yup'ik historically were organized in extended families of up to thirty people settled in permanent and semipermanent villages. Many of the villages contain a *qasgiq* or men's house, and were relatively small, averaging five to six houses per village in the 12 precontact villages for which there is house data (Alaska Historic Resources Survey database, VanStone 1967, Oswalt and Neely 1996).

In the summer, the people harvested salmon by net fishing, and processed the catch at fish camps where women managed the processing (Oswalt and Neely 1996). Family groups might put up as many as 5,000 fish, including fish for their dogs (pers. comm., Lena Andre, Dillingham, July 2011). The Yup'ik traveled to different subsistence sites either overland, by foot or dogsled, or on the water, in vessels that ranged from small skin kayaks and bark canoes, to larger skin boats (Oswalt and Neely 1996).

As during precontact times, after contact the men and boys who were older than seven lived in the *qasgiq*, the large communal men's houses, while wives, daughters, and young sons lived in a smaller individual family house called an *ena*, both built from sod and wood (VanStone 1967). During winter, the community came together for dances and storytelling, but otherwise, men and women stayed in separate dwellings and worked to do gender-specific chores. Men made and repaired the tools for hunting, while women sewed clothes, including waterproof rain gear.

The *qasgiq* continued as the communal sweat bath for the men. They would open the central smoke hole, feed the fire until the heat was intense, and then bathe. Men would put a wad of wood shavings in their mouth as a respirator to breathe the hot air (Oswalt and VanStone 1967). Men sat in the sweat house in the order of their social status. The *nukalpiaq*, or good provider, held a high social position and contributed wood for the communal sweat bath; he also played an important role in midwinter ceremonial distributions of food (Fienup-Riordan 1994).

Traditional Yup'ik festivals included the Bladder Festival (*Nakaciq*) which marked the opening of the winter ceremonial season; the yearly and ten-year Feast of the Dead (*Elriq*) recognized the namesakes of the dead; *Kelek* to *interact with and influence the spirits of the animals*; the Messenger Feast (*Kevgiq*) performed between villages; and *Petugtaq*, a ceremony between men and women (Fienup-Riordan 1994). Food exchanges played an important part in these festivals but the underlying theme was interaction with natural spirits, particularly those of food animals, including the salmon that sustained them (VanStone 1967).

At the turn of the nineteenth century, the extended family, stretching over several generations, still formed the basis of Yup'ik villages (Fienup-Riordan 1994). Winter villages could be just one family, but ranged up to 150 to 300 people in some places. The practice of men and women living separately continued. Winter villages had one or more *qasgiq*, where men and boys over the age of six or seven lived and worked together, telling stories, making tools, and preparing for subsistence activities as in earlier times. In the *ena*, women, girls, and the youngest boys lived in groups of up to a dozen, and the women taught the girls how to sew and cook. They cooked the meals there, either in the entryway, or in a central fireplace. Each winter, for three to six weeks, boys and girls would switch homes, and the men would teach girls survival and hunting skills, while the women would teach the boys how to sew and cook (Fienup-Riordan 1990).

Historically, the interior villages of the Bristol Bay watersheds were largely by-passed by nineteenth and twentieth century economic development. A Russian fur-trading post was built near Pedro Bay on Lake Iliamna in the 1790s but was abandoned shortly after because of attacks by local Dena'ina incensed by Russian abuse of their children (Boraas and Leggett 2013). A Russian fur-trading post, Alexandrovski Redoubt, was established in 1818 in Yup'ik territory on the coast near the mouth of the Nushagak River (VanStone 1967). While Russian population and economic influence was minimal, Russian Orthodoxy had a significant effect particularly after Fr. Veniaminov's arrival in 1829 (VanStone 1967). Veniaminov was flexible in his approach to the Yup'ik and their traditional religion and numerous conversions were registered in church documents. Veniaminov noted that *the Nushagak River was for them [Yup'ik] the River Jordan* (cited in Barsukov, 1887–1888), an observation which resonates even today through the *Great Blessing of the Water* (described later on).

European diseases, to which the Yup'ik, Dena'ina, and other Alaska Native populations had no immunity, came with the colonizers. The first epidemic known to have occurred in the Nushagak River region was before 1832 (VanStone 1967). The 1838–1839 smallpox epidemic caused hundreds of deaths among the Yup'ik and Dena'ina. Subsequent years, through the early twentieth century, brought more death and illness to the region, particularly in coastal areas where Western interaction was highest (VanStone 1967). VanStone notes the loss of population (especially Elders), the disruption of families, the number of orphans, and subsequent rearrangements of the social order created a social and cultural upheaval that the coastal Yup'ik struggled to overcome for decades.

After the Alaska purchase, Bristol Bay became an important commercial salmon fishery and remains so today. The first salmon cannery, The Arctic Packing Company, began operation in 1884 at the village of Kanulik at the mouth of the Nushagak River, and the fourth cannery, built at Clark's Point in 1888, is now the oldest surviving cannery in the region (Troll 2011). In the late nineteenth century, unregulated Bristol Bay canneries regularly blocked the mouths of spawning rivers to harvest salmon; consequently, there were years when few salmon escaped, creating extreme hardship for the

**Figure 1.5**    Nondalton (Nundaltin) Village. September 2011. *Photo credit: Alan Boraas.*

upriver Dena'ina and Yup'ik subsistence communities (Gaul 2007). Starting in the early 1900s, men from the inland villages traveled to the coast to work seasonally in the commercial fishery, as many still do today.

Small scale gold mining occurred on the Mulchatna River during the nineteenth and early twentieth centuries, but it was not economically successful (VanStone 1967). In 1902 a copper mine was staked about nine and a half miles from Cottonwood Bay on Cook Inlet toward Lake Iliamna (DeArmond, no date), but the deposit proved unprofitable and by 1909 the mine was abandoned.

# MYTHOLOGY AND LANGUAGE

## Voices of the People

*Talk Native, no English. . . . They talk Native* [Yup'ik] *better* [than English]. [In reference to Elder interviews in Yup'ik] **M-25, 5/18/11**

*When we first went to school they took our dialect away from us and told us to speak English only. If we spoke our Native tongue we would get hit by the teacher, which isn't right. Now they call it abuse. Anyways, none of us speak our Native tongue* [Dena'ina] *because of that. My mom didn't speak English . . .* **F-48, 8/20/11**

A Dena'ina origin story (Osgood 1976, Kari and Fall 2003) tells of a young woman who was admonished not to go near the fish weir, but, being strong-willed, she does. There she meets a king salmon, they talk, and she slips into the water and swims away with him. Her parents and the village are distraught when they cannot find her. Several years later her father, a chief or *Qeshqa*, is at the same weir collecting king salmon and sees a small one among the fish he has laid out on the grass. Upon further examination, he sees in it the face of his grandson. These and other stories allegorically portray the inescapable—the people and the salmon are so intertwined they are one.

Language, like mythology, is intimately tied to cultural identity, and Yup'ik and Dena'ina have evolved as languages of place for their respective areas over thousands of years. The intersection of landscape, subsistence, social relations, and spirituality are reflected in both languages. Given their cultural importance, it is not surprising that both Yup'ik and Dena'ina have numerous highly detailed terms involving salmon. Streams are also intimately tied to Yup'ik and Dena'ina psyche and their languages mirror that connection through principles like the upstream/downstream nature of the directional system.

In the Yup'ik language there are at least 36 words for salmon (Jacobson 1984). The nuanced and highly specific meaning of some of the terms reflects a deep and long relationship with the fish. For example, the Yup'ik word *kiarnaq* means 'unsalted strip or fillet of fish flesh without skin, cut from along the backbone and hung to dry' and *aciirturtet* means 'king salmon swimming under smelt' and refers to the fact that both king salmon and smelt are among the first anadromous fish to return to the rivers in the spring, often running one above the other, and would be highly prized as the first fresh fish of the year.

Dena'ina also has 36 words for salmon (Kari 2007) (the number is a coincidence and the words do not all refer to the same concepts). As with Yup'ik, the words reflect subtle meanings or culturally significant concepts. The Dena'ina named a general category of animal or plant by the name of its most culturally important representative. For example, the name for animal is *ggagga*, meaning brown bear, and the name for tree is *ch'wala*, or white spruce. Not surprisingly, the name for fish is the name for salmon, *łiq'a*. The language also has culturally significant terms such as *veghutna qilin*, a nickname for salmon that means 'it exists for people.'

The spirituality of water is also embedded in the language. The Dena'ina have numerous terms for streams (Kari 2007), among those the primary word for 'water' is of special note. 'Water' *viniłni* (Inland dialect) is unique among other Déné languages and Dena'ina linguist James Kari considers it to be esoterogenic, meaning a special word that reflects special importance or sacredness (pers. comm., James Kari, December 2011). Dena'ina Elders Clare Swan and Alexandra Lindgren (2011) state that *the Dena'ina word for water was held sacred* and by implication, the water was sacred. The word *viniłni* and its sacred connotations is reflected today in the Orthodox Great Blessing of the Water ceremony described below in which river water is annually baptized and made holy. In Yup'ik, baptized river water is called *malishok* and has special curative powers.

During the early part of the twentieth century policies of the U.S. Department of Education tried to eradicate indigenous languages by corporal punishment of students speaking anything other than English in school. Both Yup'ik and Dena'ina children were affected by this cruel and counter-productive policy. Today Central Yup'ik is a reasonably healthy language; about 42% of the Yup'ik of the Bristol Bay watersheds speak their language (Krauss 2007). Dena'ina, however, is moribund and the few remaining speakers are in old age. Efforts are being made to revitalize the language.

# SALMON AND SUBSISTENCE

## Voices of the People

*It may be different, the way we gather it nowadays, but it's the same end product. It's the same.* **F-69, 9/18/11**

*We would starve if we don't have fish or salmon. In this area we have lived with fish all our lives, from generation to generation. The people that stayed before us and kids that are behind us will be living on fish. Salmon is very important . . . Without fish we are very poor; we have no food to eat. With fish we are very rich; our stomach is full. That's the way I look at it.* **F-48, 8/20/11**

*My Auntie would say, "Don't forget how to live off the land," and I'd think, "Oh, we could just go to the store and have microwave stuff." She said, "One day in this world something's going to happen where you guys are going to rely on living off the land." . . . We can't just forget our ways; how to live off the land, because one day there's going to be something that happens in the world.* **F-32, 8/18/11**

*If you get out in these outlying villages, about 80–90% of what they eat is what they gather from their front yards. I was in Igiugig this spring . . . Do you know how much a can of SPAM is in Igiugig? Eight dollars for a can of SPAM!* **M-60, 9/16/11**

Subsistence is not a return to practices of earlier centuries but is a continuum from the past to the present that now employs modern technology but retains traditional perspectives. Nylon nets have replaced spruce-root or willow-bark nets; aluminum skiffs and four-stroke motors have replaced kayaks or canoes; and freezers have replaced underground cold storage pits. Moreover, subsistence activities follow management practices formulated by the Alaska Department of Fish and Game (ADF&G), dictating limits and seasons. In Southwest Alaska, subsistence is a fundamental nonmonetized

**Figure 1.6**    Subsistence Skiffs at New Stuyahok (Cetuyaraq) Village Landing, Nushagak River. May 2011. *Photo credit: Alan Boraas.*

economic activity that forms the basis of cultural life. Most of the protein in a high protein diet comes from subsistence harvesting. Moreover, cultural and personal identity largely revolves around subsistence. Fall et al. (2009) wrote of the Bristol Bay villages, "At the beginning of the 21st century, subsistence activities and values remain a cornerstone of area residents' way of life, a link to the traditions of the past, and one of their bases for survival and prosperity." Everyone who was interviewed felt that subsistence is *life* and the foundation of culture for the Bristol Bay watershed villages.

In urban Alaska subsistence is frequently marginalized as welfare subsistence—the perception that subsistence should be available only to the poor who don't have access to a supermarket (Thornton 1998). Consequently, many indigenous Alaskans feel the term *subsistence* misconstrues what it means to fish, hunt, and gather in the manner of their ancestors because ADF&G data emphasizing harvest and consumption ignores the holistic social and spiritual elements of tradition. In 1983 the Inuit Circumpolar Conference and the World Council of Indigenous Peoples sponsored the Alaska Native Review Commission to conduct hearings in rural Alaska aimed, in part, to help the non-Native community understand the importance of subsistence to Alaskan Natives. In the commission's final report, Thomas Berger (1983:51) summarized subsistence as follows and is the definition of subsistence used in this chapter:

> The traditional economy is based on subsistence activities that require special skills and a complex understanding of the local environment that enables the people to live directly from the land. It also involves cultural values and attitudes: mutual respect, sharing, resourcefulness, and an understanding that is both conscious and mystical of the intricate interrelationships that link humans, animals, and the environment. To this array of activities and deeply embedded values, we attach the word *subsistence*, recognizing that no one word can adequately encompass all these related concepts.

Table 1.1 indicates the importance of subsistence resources in each village on a per-capita basis. While all subsistence foods are important—particularly for the physical and emotional benefits derived from a varied diet—salmon is, by far, the most important food, generally constituting well over 50% of the subsistence diet for Yup'ik and Dena'ina villages. The data indicates as much as 1210 pounds of dressed meat is harvested per capita (Clark's Point) and an average of 491 pounds of meat per capita is harvested per Yup'ik and Dena'ina village, of which 297 pounds per capita is salmon. By comparison, Americans consumed an average of 198.6 pounds of red meat, chicken, and fish per year per capita in 2012 (U.S. Department of Agriculture "Economic Research Service"). The significant per-capita difference is partially attributable to the fact that the subsistence data is pounds per capita harvested, not pounds per capita consumed. A substantial amount of subsistence-harvested food is shared, which partially accounts for such high numbers of per-capita harvest. Nevertheless, the numbers are high because the people eat a great amount of wild food; subsistence foods, particularly salmon, are the staple of the culture.

Berger noted that subsistence is an interplay of the time, effort, and skill needed to catch, process, and store subsistence foods and part-time wage employment necessary to support the means of subsistence: boats, motors, fuel, etc. (Berger 1985). Today, interviewees reiterate this finding and indicate that, for those fully engaged in it, subsistence is a full-time job; however, it is necessary to supplement subsistence with cash from part-time wage labor or commercial fishing to defray the costs of subsistence activities. With gasoline costs in the $6 per gallon range (summer 2011), trips to fish camps and other subsistence areas are expensive. Skiffs, four-stroke engines, ATVs, snow machines, guns, ammunition, fishing gear, among other expenses, also add to the subsistence investment.

Interviewees indicate that to meet these costs, many families have commercial fishing permits and fish the sockeye run in Bristol Bay during late June and into mid-July or engage in other forms of part-time employment. These forms of employment, with their short duration and/or seasonal nature, are ideally suited to provide another ingredient critical to a subsistence lifestyle—time to engage in subsistence activities. Thomas Lonner (cited in Lowe 2007) indicates that in Bristol Bay villages, the cash obtained from working in the commercial fishery (also other part-time employment, Native corporation dividends, and social welfare payments, among other sources) is considered an investment in subsistence. For most, subsistence is the prime activity and is not merely a supplement to wage labor.

Several attempts have been made to measure subsistence economically by monetizing wild food resources. Fall et al. (2009) measured the economic importance of subsistence by calculating the cost of replacing wild foods obtained from hunting, fishing, and gathering with similar foods obtained in a market. Their data indicate food purchased in the store ranged from $1,467 to $2,622 per capita in the villages of ADF&G's Bristol Bay management area. The people in their study area consumed 304 pounds of wild salmon and 203 pounds per capita of other wild foods. Fall et al. (2009) concluded that it would have cost an additional $7000 per capita if the wild foods were to have been replaced by store-bought foods. While monetizing subsistence gives a measure of its importance to the economy, these values do not

**Figure 1.7**   Ekwok (Iquaq) Village on the Nushagak River. *Photo credit: Alan Boraas.*

reflect the fact that the people of the region almost unanimously reject replacing their traditional subsistence foods with imported products.

Villages in the Bristol Bay watersheds have high unemployment rates (from 14% in Igiugig to 37% in Newhalen—computed from 2012 Alaska Division of Regional Affairs Community Database), and high rates of people below the poverty level (17.1%, 2010 U.S. Census). These numbers are high, but labor statistics do not identify subsistence as an employment category—and wealth, and conversely poverty, is defined by income, not in terms of subsistence measures of wealth such as a freezer full of fish as described below. Subsistence fishers, hunters, and gatherers do not consider themselves to be unemployed or in poverty. Those who choose the subsistence lifestyle work long hours, utilizing considerable skill to provide food for themselves and their families and in interviews described subsistence as a full-time occupation. Subsistence is dictated by the seasons, is time-consuming, and, for many Bristol Bay village residents, is one's work obligation. Eight-to-five or two-weeks-on/two-weeks-off types of employment in the cash economy severely limit the time necessary to obtain and process food for a family for a year and makes full-time subsistence impossible.

# NUTRITION

## Voices of the People

> *Wild salmon is more important for us, or wild fish. I don't believe in farmed fish because wild fish is better for all our health. It has natural oils and we don't paint it with artificial paint like the farmed fish you get. You can sell your farmed fish all you want, but wild salmon is more important to us.* **F-48, 8/20/11**

*You know, it's got that one oil in it that is a cancer-fighting oil, and it's really good.* **F-38, 8/18/11**

*I've seen kids teethe on smoked salmon strips. They're hard. They get all fishy and smelly, but man, they just chew. It's better than the rubber toy.* **F-38, 8/18/11**

*I definitely limit my child; you know, the fast foods, we eat it once a week . . .* [They eat] *moose meat, the fish . . . berries, and wild plants . . . We want to give to our children the fish and we want to keep the water clean for them. It was a gift to us from our ancestors, which will then be given to our children.* **F-69, 9/18/11**

*To me, I think eating salmon has sustained our way of life. I think by eating a lot of salmon, we are a healthy, healthy Dena'ina. I always tell children there at potlatches or wherever; I say that, "If you eat this piece of fish you're going to be a smart Dena'ina woman, you might be able to be a lawyer or a doctor."* **F-32, 8/18/11**

*We can't live without salmon. We'll be missing something.* **F-27, 8/17/11**

*We grew up with it. We need it. If we don't have it, we miss it. I can't see anybody that lives around here without it.* **F-30, 8/17/11c.**

The dietary habits of Yup'ik and Dena'ina living in the villages of the Bristol Bay region show regular dependence on several species of wild salmon, which they usually consume daily—often several times a day, as the interviews attest. Yup'ik and Dena'ina primarily prepare and eat three species of Pacific wild salmon (sockeye, coho, and Chinook salmon) in many different ways, including fresh, salted, pickled, canned, dried, and smoked. Salmon and other traditional wild foods comprise a large part of the villagers' daily diet throughout their lives, beginning as soon as they are old enough to eat solid food (Interviews 2011). In addition to salmon, villagers also regularly consume other wild fish species, and a wide variety of mammals, bird, shellfish (locally), and plant foods. The people also eat Western store-bought foods, but

**Figure 1.8** Salmon Drying at Koliganek (Qalirneq). September 2011. *Photo credit: Alan Boraas.*

adults, at least, do not prefer it (Interviews 2011). The Yup'ik and Dena'ina consider their traditional foods to be healthy and satisfying in addition to providing strength, warmth, and energy in ways that processed store-bought food does not (Hopkins 2007).

Yup'ik and Dena'ina dependence on subsistence foods has the additional health benefit of providing opportunities and incentive for physical fitness, since engaging in subsistence harvesting improves fitness—and fitness, in turn, enhances the efficiency of subsistence harvesting. Subsistence hunting, fishing, and gathering demand both stamina to endure long periods of physical activity and strength to handle meat, large quantities of fish, and heavy fishing gear. Processing large amounts of subsistence foods is equally demanding. Hopkins (2007) quotes from a Yup'ik woman, "I think today most of the women are healthy from activity—physical activities. When they go berry picking, they're working using their body—everything. When we are cutting fish, we are using everything, our muscles, lifting things."

Beyond the Yup'ik and Dena'ina personal conceptions and cultural knowledge about the importance of wild foods in their diets, many studies also confirm the remarkable health benefits of omega-3 fatty acids and other nutrients found in high percentages in wild salmon, and the combination of salmon, wild greens, blueberries, and other berries for preventive health. O'Brian et al. (2009, 2011) found that salmon-rich diets are significant in the prevention of chronic diseases, including cardiovascular diseases and type 2 diabetes. In a cohort study of Yup'ik from the Yukon-Kuskokwim area, Boyer et al. (2007) of the Center for Alaska Native Health Research (CANHR) found that metabolic syndrome is less common in salmon-consuming populations relative to others, occurring at a rate of 14.7% in the study population, compared to 23.9% in the general U.S. adult population. The study population also had significantly higher high-density lipoprotein cholesterol levels and lower triglyceride levels than the general U.S. adult population. In related studies, Makhoul et al. (2010) and the Fred Hutchinson Cancer Research Center (2011), in collaboration with the CANHR, found that Yup'iks consume 20 times more omega-3 fatty acids from fish than the average American and display a much lower risk of obesity-related disease despite having similar rates of being overweight and obesity. According to Bersamin et al. (2007,

**Figure 1.9**   Subsistence Skiffs at Koliganak (Qalirneq) Village Landing, Nushagak River. September 2011. *Photo credit: Alan Boraas.*

also Bersamin et al. 2008), "Diets emphasizing traditional Alaskan Native foods were associated with a fatty acid profile promoting greater cardiovascular health than diets emphasizing Western foods." A study by Adler et al. (1994) regarding the benefit of salmon and seal oil consumption concluded these wild foods played a significant role in combating diabetes among Yup'ik and Athabascan Native Alaskans.

Compounds derived from their subsistence diet, including omega-3 fatty acids from wild salmon consumption, may also benefit mental health in Yup'ik populations. Lesperance et al. (2010) report that omega-3 fatty acids can help prevent depression. Nemets et al. (2006) showed greater improvement in symptoms for patients with chronic depression who consumed omega-3 fatty acids with their medication compared to those receiving only a placebo with their medication. (See the section *Behavioral and Mental Health* later on in this chapter for additional information concerning the behavioral and mental aspects of a subsistence lifestyle.)

The health benefits of wild salmon consumption may have an epigenetic explanation. Among southwest Alaskan Yup'iks, Aslibekyan et al. (2014) identified DNA methylated genomic regions that facilitate the metabolism of omega-3 fatty acids that they hypothesize would give nutrigenomic protection from chronic diseases such as those listed previously. To the extent that it can be demonstrated that these epigenic factors are passed from generation to generation, it can be said that Yup'iks are evolved to metabolize salmon.

Health benefits, particularly the long-term fit between the human and fish populations, depend upon maintaining the local wild salmon for subsistence fishing. While it would be easy to assume that any salmon would provide a similar quantity and quality of nutrients, Sincan (2011) showed that farmed salmon that were fed a typical farmed salmon diet, did not have the omega-3 fatty acids in beneficial quantities, in contrast to the wild salmon, which did.

When asked how often they ate salmon, respondents frequently said "all the time," and when asked if they needed salmon to be healthy, all who responded said "yes" (Interviews 2011). As noted before, it is possible the metabolic aspects of salmon consumption may have an evolutionary component occurring in the 4000 years the people have been utilizing wild salmon. The loss of the local wild salmon as a large component of the Yup'ik and Dena'ina diet would result in risks to the physical and psychological health of the population, including greater risks of cardiovascular disease, type 2 diabetes, and depression.

**Figure 1.10**   Children's art on storage unit at Koliganak (Qalirneq). September 2011. *Photo credit: Alan Boraas.*

# FAMILY AND SOCIETY

## Voices of the People

*You are a very rich person if you share. If you don't share, you are nobody.* **F-46, 8/20/11**

*Yeah, we always share. Holidays, we share, and if somebody passes away, after burial we have a potlatch; we share. We share with people; that is the way we are brought up.* **F-41, 8/19/11**

*We share with people here and in Anchorage. . . . I like to go fishing, so if we run out of freezer space, I will ask people* [who can't fish in the village, e.g., Elders] *if they want fish, then I'll go out and catch some fish if they want.* **M-70, 9/18/11**

*I feel good, proud* [to share]. *And when our friends give us back, way proud.* **M-60, 9/16/11**

[My wife] *and I have been doing it for thirty some years, doing the fish camp, and putting up fish for the winter. When the kids were small, we were down there for them too, and hopefully, they will have a family, too, and carry on the tradition.* **M-33, 8/18/11**

*We catch moose and caribou and give it away; it ensures good luck back. Even beaver, you give the whole beaver away after you skin it. After you skin the beaver, you give it away; give the whole beaver away. That animal that you give away . . . give*[s] *you back in return good luck.* **M-54, 8/20/11**

*The parents, their sisters, their aunties, their grandparents, their great-grandparents—everybody is there* [at fish camp], *you know, telling them* [the children] *how to do this . . . Everybody does it at their own camps . . . . Everybody is living in different fish camps, so all these families that are together, that's how they taught the younger kids.* **F-28, 8/17/11**

*Salmon is one thing. They make you feel rich, because you have something to eat all winter. Smoked salmon, sun-dried spawned-out fish, all of those make you feel good, because you grew up with it; it is in your body.* **M-53, 8/20/11**

*As long as we have a lot of fish and meat and stuff, we are wealthy. We don't believe in . . . having lots of money. The wealth to us is having more fish put away for the winter, and meat; that's our wealth.* **F-27, 8/17/11**

The Yup'ik and Dena'ina cultures center on belonging to a community; sharing food is a means of creating and maintaining the living bonds of relationship. The practice of sharing is elemental in both indigenous and other cultures both from a material and a social standpoint (Counihan 1999). The Yup'ik and Dena'ina of Bristol Bay, as traditional cultures, continue these practices through harvesting, preserving, and preparing food together and sharing food through traditional practices and ritual celebrations. Sharing remains a fundamental institution within Yup'ik and Dena'ina cultures today (Interviews 2011). Among the Yup'ik, *elaqyaq* means "those of the same stomach" and refers both to sharing food and being biologically related. Oscar Kawagley noted a similar linguistic reference: "The Yupiaq [Yup'ik] term for relatives is associated with the word for viscera, with connotations of deeply interconnected feelings" (Kawagley 2006). Villagers do not consider sharing to be an obligation, but a way of life. Interviewees universally indicated that giving or receiving salmon or other subsistence foods makes them feel good. The altruism of sharing Native food expresses social solidarity between the participants. Almost universally, Yup'ik and Dena'ina seem to have small jars of salmon available for favored visitors to take with them.

Villagers particularly recognize some Elders who cannot participate in the rigors of subsistence harvesting as people with whom to share salmon and other subsistence foods. The informal first salmon sharing, for instance, always includes Elders. Sharing salmon and other subsistence foods with family living in Anchorage or even farther away is an important bond with home, family, and place. Interviewees consistently talked about how much they appreciated a gift of canned or jarred salmon from home when they were away from the village. They also talked about how important it is for them to send a part of the place to family and friends living away from Bristol Bay.

The Dena'ina believe that tangible items can take on aspects of the owner. This personification is called *beggesha* if the aspects are positive and *beggesh* if negative (Boraas and Peter 2008). Artifacts or places can have *beggesha* or *beggesh* depending on events associated with them. A place, something someone made, such as a birch bark basket, or salmon someone prepared take on *beggesha*. The term does not easily translate into English, so today people talk about giving *love* when giving a gift of something they made or prepared. Conversely, one receives *love* when receiving a similar gift. This perspective is one of the reasons that Alaska Native foods, especially salmon, are served at gatherings such as potlucks and potlatches. Preparing and giving food is a tangible act of love. Recipients appreciate non-Native foods, but they are not from the place, were not made by the giver and, consequently, are not the same expression of love when gifted. Athabascan Elder, the late Reverend Peter John (1996) expresses love this way, "True love is something that you never see . . . By gathering to share food, songs, and speeches, love grows among the people."

Writing of subsistence in general, including fish camp, Yup'ik Elder and scholar Mary C. Pete (1993:10) wrote:

> For many Yup'iks, subsistence activities teach children much more than hunting and fishing: they convey respect and proper conduct toward the land and water and animals and other humans; they promote satisfaction from hard work and contribution to the kin group. For many Yup'iks, subsistence goes beyond mere economy—it is a vital way of life and a source of pride and identity.

Some of the villages harvest salmon either at or very near town, and fish camp may be only a short trip to a traditional fishing locality where they may or may not camp out (Fall et al. 2010). Many villagers, however, travel to a traditional place, set up camp, and live for several weeks catching and putting up salmon. Villagers from Kokhanok, for example, travel to fish camp on Gibraltar Lake, residents of New Stuyahok, Ekwok, and Koliganek stay at various camps downstream on the Nushagak River, primarily at Lewis Point (Nunaurluq), and villagers from Nondalton go to nearby camps on Sixmile Lake, the Newhalen River, or up to Lake Clark. Generally, the fish camp consists of an extended family, with three or more generations, but close friends may also participate (Interviews 2011, Fall et al. 2010).

Families view fish camp as a good time when they can renew bonds of togetherness by engaging in the physical work of catching and processing salmon—multigenerational, meaningful work. Family members who don't live in the villages often schedule vacation time to return home to fish camp, not just for the salmon, but also for family. Sharing in vigorous, meaningful work creates cross-generational bonds between children, their parents, aunts, uncles, and grandparents (Interviews 2011). Fish camp is a time when children and teens learn not only the practice of how to catch, clean, and process fish properly, but the values that are an integral part of harvesting salmon and interacting with nature (Interviews 2011, Fall et al. 2010).

**Figure 1.11** An active fish camp. *Photo credit: Michael Melford.*

*I used to live in Portage where there is no clinic. That is the only thing I could give my kids* [holy water, when they were sick]. *You know, pray upon them and let them make the sign of the cross and let them have a taste of the holy water.* **F-72, 9/19/11**

*That holy water is strong. . . . A long, long time ago, before I become a lady, we were upriver with my mom and dad. . . . In nighttime, I guess I almost go* [die] *you know. But my dad, he prayed for me. If you're really true, praying really hard, I guess He'll answer you. . . . I was going to go to Big Church* [heaven], *and my dad said 'you can't go to Big Church.' When he tell me that, I told him holy water—I call Native way, malishok, holy water, malishok* [Yup'ik]— *'give me holy water to drink.' He did, my dad, he did. A little bit you know. I opened my mouth, I swallowed, . . . I closed my eyes, pretty soon I come through . . . . Almost going to that Big Church.* **F-66, 9/18/11**

*I think that, if you treat animals disrespectful, that they are not going to show up again.* **F-32, 8/18/11**

*The first salmon, it's still tradition to share with everybody. You do say a prayer.* **F-47, 8/20/11**

Traditional Native American cultures, including Yup'ik and Dena'ina, approach the land, the water, the plants, the animals, and themselves through the lens of spirituality. Roy Rappaport has written that all cultures have one central concept that guides the understanding of everything else. He called these *ultimate sacred postulates*[4]. Rapport (1999) writes of ultimate sacred postulates: "They sanctify, which is to say certify, the entire system of understanding in accordance with which people conduct their lives . . . It becomes something like an assertion, statement, description or report of the way the world in fact is." For the traditional Yup'ik and Dena'ina the ultimate postulate is the concept that everything has an animating spirit or soul: people, animals, plants, and the landscape, and are to be understood in terms of that spirit.

The legitimacy of Native spirituality has largely gone unrecognized by historians and social scientists constructing the master narrative of history disempowering identity and robbing history of an indigenous perspective. Moreover, scientists unilaterally applying the primacy of science and its methods to the understanding of nature have marginalized Native spiritual perspectives of the natural world.

Most residents of the interior villages of the Bristol Bay drainage are Russian Orthodox Christians or were brought up as Russian Orthodox, and the Orthodox Church, along with the public school and the tribal structure, is among the dominant institutions in the villages. Many of the villages have a resident indigenous priest or priests; for others, clergy visit periodically, on a scheduled basis. In some villages, Protestant churches have formed but none, except Dillingham, have a dedicated church structure.

Beliefs concerning streams and salmon in those villages where Orthodoxy is the dominant religion, involve expressing traditional beliefs (ultimate postulates) through Russian Orthodox practice. Dena'ina writer Peter Kalifornsky (1991) described his great-great-grandfather's nineteenth century message to the Dena'ina people after his conversion to Orthodoxy: "Keep on respecting the old beliefs, but there is God to be believed in; that is first of all things on earth." Russian Orthodoxy itself has a syncretic tradition of melding Middle Eastern-derived Christianity with spirituality influenced by the northern environment. Billington (1970) points out that, though Orthodoxy moved north from Greece and Asia Minor into Russia in the ninth century A.D., its long history in the northern forest has shaped the belief system to interpret and interact with aspects of the subarctic taiga. Billington writes, "God came to man not just through the icons and holy men of the church but also through the spirit-hosts of mountains, rivers, and above all, the forests" (Billington 1970).

Consequently, many Russian Orthodox rituals interface with nature through spirituality. The mystical aspects of Orthodoxy fit well with traditional Dena'ina and Yup'ik beliefs, many of which relate to interacting with the landscape on which their survival depended (Boraas, 2013) and are an extension of the sacred ecology of their ancestors (Birkes 1999). For the Dena'ina and Yup'ik living in the Bristol Bay watershed, beliefs regarding pure water and the return of the salmon raise to the sacred what is most important in their lives and ritually express the meaning of life as people of the salmon.

An example is the Orthodox *Great Blessing of Water* that takes place during the Feast of Theophany, a major event in the Orthodox Church calendar and is celebrated on January 6 in the Julian calendar, the calendar of Orthodoxy (January 19 in the Gregorian calendar). While all church rituals are important, Theophany is the third most important church ritual after Christmas and Easter to the Orthodox of the Bristol Bay area (pers. comm., Fr. Alexie Askoak, St. Sergius Russian Orthodox Church, New Stuyahok, January 19, 2012). According to Fr. Askoak, Orthodox baptism both redeems sin

---

[4]The term "ultimate postulates" might be better to avoid confusion with Christian or other religions principles and to recognize not all are sacred. Principles of science, for example, are an explanatory worldview that does not necessarily recognize the sacred.

and brings the Holy Spirit to the recipient. Orthodoxy transfers the baptismal ceremony from infants to one of God's most important creations, water, removing sin in the form of human-caused pollution and bringing the Holy Spirit to the water; one of the most important creations to the people since salmon and related wild foods are dependent on clean water.

The two-day ritual is a liminal event with believers moving into a deep spiritual state. An evening church service is held on the eve of Theophany in preparation for the blessing the next day (Figure 1.12). The next morning a communion service is held involving forgiveness of sins and, as the sun rises, the people, led by the priests, go out onto the frozen river where an Orthodox cross has been cut into the ice and a small hole has been made to withdraw holy water (Figure 1.13). There a baptism service is held and, at the moment in the service when the priest dips the cross through the hole in the ice into the water for the third time, God is believed to sanctify the river, making it holy. According to Father Michael Oleksa the Great Blessing of the Water is done to *reaffirm the Church's belief that the natural world is sacred and needs to be treated with care and reverence* (Orthodox Church in America, no date). For Orthodox Yup'ik and Dena'ina that sacred world includes the water and the salmon.

Holy water from the sanctified rivers is believed to have curative powers for both physical and mental illness and is drunk or put on the affected body part for healing purposes (Fr. Alexi Askoak, pers. comm., January 19, 2012). From a secular standpoint, the question is not whether holy water has healing efficacy or whether the water is actually purified, but how the Great Blessing of the Water ceremony and holy water reflect values of the people. By elevating water to sacred status, the people of the villages define core values. As described previously, the Dena'ina word for water, *vinłni*, has sacred overtones and water, itself, is sacred. Since the word predates Christianity in Southwest Alaska, we can assume sacred water has long been a part of the salmon cultures of the Bristol Bay watersheds because the people recognize that clean water and wild salmon are fundamental to life itself. The Great Blessing of the Water ceremony is an extension of

**Figure 1.12** Theophany Eve Orthodox service, St. Sergius Church at New Stuyahok (Cetuyaraq). January 2012. *Photo credit: Alan Boraas.*

**Figure 1.13**  The Great Blessing of the Water on the Nushagak River at New Stuyahok (Cetuyaraq) conducted by Father Alexie Askoak. January 2012. *Photo credit: Alan Boraas.*

that very old concept, rendering in Christianity the belief that water and salmon are sacred to life and culture. Through the liturgy of baptism, the ceremony becomes a form of world renewal ceremony, reestablishing a sacred order.

Water and salmon play additional roles in modern Orthodoxy as derived, in part, from traditional subarctic spiritual practices. Describing traditional Dena'ina beliefs, Kalifornsky (who was also a devout Orthodox Christian) writes (1991) that, after putting out his net, "*Ouq'a shegh dighelagh*" or "a fish swam to me," indicating that the spirit of the salmon had a will and would allow itself to be taken for food if the net-tender had the correct attitude. Today, all interviewees who commented on it believe that salmon have a spirit or soul and that soul is a creation of God.

Interviewees also believe in treating all animals, including salmon, with respect. Several modern practices reflect this belief—for example, using the entirety of a fish for food, except the bones and entrails, which villagers return to the water along with the bones that remain after consumption. To not use all of the edible parts of a salmon is considered to be abuse (interviewees). Another example, interviewees report, is never allowing fish or meat to spoil. Interviewees repeatedly stressed the importance of giving respect to salmon and all subsistence animals. This attitude echoes the precontact beliefs that animals have a will and, if not treated properly, will not allow themselves to be taken for food, leading to dire consequences for the people (Boraas and Peter 1996).

The First Salmon Ceremony is a world renewal ceremony, which, like other world renewal ceremonies, recognizes the cyclical onset of the most important yearly event in the culture. The First Salmon Ceremony was described by ethnographer Cornelius Osgood (1976) and was practiced in precontact times and is based on a mythical story that merges people and salmon as described earlier. Because of the importance of salmon in the lives of the Bristol Bay villagers, interviewees report that they continue to mark the return of salmon in the spring by a special observance. The actual practice varies, but involves a prayer of thanks to God for the return of the salmon and sharing the first salmon caught in the spring with Elders and others in the community. Typically, according to interviews, each receives a small piece, and there is a general feeling of happiness that the salmon have returned and the cycle of the seasons has begun again and nature will provide the people with sustenance. In some places the First Salmon Ceremony takes place at fish camp, where extended families and others present share the first salmon they catch with one another, including the Elders. In at least one village, New Stuyahok, the ceremony includes sharing the first salmon with *the underground*, by placing a small piece of it under the forest mat at the cemetery, symbolically sharing salmon with the ancestors buried there.

# FREEDOM

## Voices of the People

> *It's free, it's free and peaceful here, and we can get fish* . . . **F-27, 8/17/11**

Freedom, more specifically cultural freedom, is a vital concept among the subsistence villages of the Bristol Bay watersheds. Though hunting and fishing require abiding with ADF&G regulations, most villagers see those activities as involving a degree of freedom that does not often occur in nonsubsistence work settings. As described in many interviews, with subsistence as your job, you don't have to punch a clock, you only follow nature's clock; you don't have a boss, you are your own boss, and you either suffer the consequences if you do not perform well or reap the benefits if you do. During our May 2011 visit to one village on the Nushagak River, two young men in their early twenties left on a 17-day goose hunting trip, upriver into the Mulchatna area—one of the most remote places in North America at any time of year, but virtually deserted in spring when snow is still present. They were on their own, and apparently all who were connected to the endeavor embraced that freedom. As they left, for example, the mother of one of the boys simply said, "Be careful," just as a parent living on Alaska's road system might say to a son embarking on a trip to Anchorage. This view comes from villagers having knowledge of, and ranging over, a vast territory, and having, even at a relatively young age, the skills and focus to deal with the harsh reality of being on your own in a natural area days from help—where a small mistake can be deadly. Cultural practices and cultural values are essential to a good outcome.

Freedom does not mean *free to do whatever one wants*. Freedom means to be free to express your traditional cultural practices, with all of its opportunities and restraints, on your traditional lands. One of the interview questions we asked was, "How do people define wealth in this village?" No one said money or aspects of materialism that are common in Western society. Most said a freezer (or freezers) full of fish, some said a big family, many said both. One young man said *freedom*. By that he meant cultural freedom. Among other Yup'ik and Dena'ina interviewees, cultural freedom proved to be at the heart of living in the salmon culture villages of the Bristol Bay watersheds.

# SUMMARY—SURVIVANCE OF PLACE

Gerald Vizenor (2008), an Ojibwa (*Anishinaabe*), along with other Native American scholars, has developed the concept of *survivance* to express traditional viewpoints. Survivance contextualizes the significance of the historically situated salmon cultures of the Bristol Bay watersheds. As used by Vizenor, survivance combines the term *survival* with the suffix *ance* to literally mean *the action of surviving*—emphasizing that Native American survival is a dynamic process. Survivance is not merely survival, but survival with dignity, worthiness, and control of history—and involves what Vizenor (2008) describes as "the heritable right of succession . . . in the course of international declarations of human rights." Vizenor et al. (2008) reject the themes of many historians, sociologists, anthropologists, and other scholars to define Native Americans by a narrative of despair, victimhood, alienation, and annihilation.

Of course, Dena'ina and Yup'ik tribal institutions and some individuals continue to grapple with forced adaptations often not of their making, what Vizenor calls *cultural schizophrenia*. But the salmon cultures have an important place in the Alaskan and American story. Theirs is a meaningful survivance in which the people identify with being Dena'ina or Yup'ik based on the continued ability to harvest wild salmon and other wild foods following a time-honored, seasonal round, to practice sharing as a way to define group identity, and to pay homage to traditional interaction with the forces of nature through rituals of place in the performance of spiritual ecology. Because of abundant wild salmon caught through the tradition of subsistence, Yup'ik and Dena'ina survivance looks much different from the way it does among most other traditional salmon cultures that can no longer rely on wild, nonfarmed, nonhatchery-reared, nongenetically modified salmon as the keystone species of their culture.

In addition to the will of the people to make it so, two factors have made salmon survivance possible. As discussed elsewhere in this volume, the largely unaltered Bristol Bay watersheds are ideal salmon-rearing habitat providing an

**Figure 2.5**   Open spruce woodland and low shrub-lichen tundra along Lower Tazimina Lake in the Lime Hills Ecoregion. *Photo credit: Mike Fleming.*

to gently rolling lowlands, but isolated hills and rounded, low-relief mountains occur. The lowlands are bounded by the mountain toeslopes of the Ahklun Mountains and Aleutian Range. Large lakes, filling mountain valleys in the Ahklun (Wood-Tikchik Lakes) and Alaska Ranges (Ilamna, Naknek, and Becharof Lakes), are contained by terminal moraines. The low-gradient, meandering rivers that drain the central lowlands are bordered by narrow floodplains and stream terraces. Numerous small lakes, ponds, stream channels, and wetlands occur across the Bristol Bay lowlands.

The underlying geology of the areas is sedimentary with occasional volcanics. As the entire region was glaciated in the Pleistocene, extensive glacial till and outwash comprise most of the surficial deposits; elsewhere in the ecoregion alluvial and marine sediment predominates. Layers of volcanic ash and loess can be found in much of the area. Permafrost is discontinuous, occurring primarily in finer-textured deposits.

Exposed alpine ridges are occupied by a mosaic of barren ground, dwarf shrub, and lichen. Well-drained uplands are dominated by low and dwarf scrub, ericaceous shrubs, and high lichen and moss cover. Moderately drained areas are largely comprised of dwarf scrub that is composed of dwarf birch, ericaceous shrubs, tussock-forming sedges, and deep moss. Wet herbaceous communities comprise the poorly drained lowlands, fens, and lake margins (Figure 2.6). Salt-tolerant sedge meadow communities occupy the coast in estuaries and narrow beach margins. The region supports few trees; however, small stands of balsam poplar and white spruce with a tall and low shrub understory line the floodplains of major rivers. Figure 2.7 illustrates the network of landcover types in the Bristol Bay lowlands with woodlands and forests associated with the margins of Naknek Lake and Naknek River and smaller floodplains, low shrub and tundra habitats on hills and moraines, and wetlands primarily associated with the lowlands adjacent to Kvichak Bay and ponds and lakes scattered throughout the region.

## Alaska Range Ecoregion

The northeastern corner of the Bristol Bay watershed extends a short distance into the Alaska Range Ecoregion, an ecoregion that is characterized by high jagged mountains that form a broad 600-mile arc from the Aleutian Range to the Wrangell Mountains. The massive mountains are a complex series of accreted terrains with high ice-capped peaks. The bedrock lithology includes a wide diversity of metamorphic and volcanic-derived rocks with smaller contributions of sedimentary rocks. The range intercepts moisture-laden air from the Gulf of Alaska, fueling the numerous glaciers that flow from the peaks and small icefields. Glacial rivers cut through the mountains and deposit unconsolidated sediments in the valleys. The climate is cold and harsh. Drier conditions prevail on the northern and western sides of the Alaska Range Ecoregion. Much of the ecoregion is either barren or ice-covered with alpine tundra occurring at mid-elevations with more stable slopes. Tall shrub communities occupy lower mountain side slopes. The lower valleys in the Alaska Range support a diversity of shrub and forest communities.

Within the Bristol Bay watershed the Alaska Range Ecoregion occurs from the eastern edge of Iliamna Lake north to Turquoise Lake, and east up the Tlikakila River and Pile River drainages. The high rugged Neacola Mountains dominate the northern section where the watershed overlaps the Alaska Range Ecoregion—and the Chigmit Mountains form the

**Figure 2.6** Bristol Bay lowlands in the upper King Salmon River drainage. Dwarf ericaceous shrub tundra is present in the foreground; young alder shrubs are establishing in the herbaceous graminoid tundra in the mid-ground. In the background, a mosaic of low willow shrub tundra and wet sedge tundra is visible. *Photo credit: Matthew Carlson.*

**Figure 2.7**   Close-up of the Bristol Bay Lowland Ecoregion, showing the diversity of vegetation types (landcover classes), indicated by different colors in the watershed. Green colors denote forests and woodlands; browns and grey represent shrub habitats; pink denotes tussock tundra habitats; blue, purple, and yellow denote wet, mesic, and aquatic herbaceous habitats, respectively.

southern portion of the ecoregion. The majority of the landscape here is glaciated, barren, or composed of isolated alpine tundra vegetation (Figure 2.8). The more heavily vegetated areas are confined to the narrow, low elevation drainages.

## Alaska Peninsula Ecoregion

The Alaska Peninsula Ecoregion is part of the Aleutian Meadows group, which extends from the peninsula down the length of the island chain. Here, the narrow Aleutian Range separates the Gulf of Alaska from the Bering Sea. The cool maritime climate is characterized by low temperatures, short summers, abundant rainfall, strong winds, fog, persistent low clouds, and salt spray. Maritime influence precludes the development of permafrost, while wind, wet soils, and low summer temperatures largely restrict the growth of trees. This is one of the most seismically active regions in North America—subject to both earthquakes and volcanic activity. Volcanoes, many of which are active, dominate the landscape. The Aleutian Range is comprised of rugged mountains deeply dissected by narrow, high-gradient valleys which discharge to broad, braided alluvial fans and floodplains. At lower elevations toward the Bristol Bay lowlands, valley bottoms are occupied by narrow floodplains, stream terraces, and shallow basins punctuated by small lakes and interconnected wetlands, which transition to salt marshes toward the coast.

Mountains derived from sedimentary rocks extend down the peninsula with intermittent volcanic cinder cones along the spine of the range. While this ecoregion was glaciated in the Pleistocene and relict glaciers and ice cap the higher elevation peaks and side slopes of the Aleutian Range, permafrost is no longer present. Numerous large lakes, such as Iliamna, Naknek, and Becharof, which formed behind terminal glacial moraines, extend out from the mountain foothills

**Figure 2.8** Alaska Range Ecoregion, Chigmik Mountains, dwarf-shrub lichen graminoid tundra on an ice-free mountain surrounded by glaciers (nunatak). *Photo credit: Matthew Carlson.*

northwest of the Aleutian Range. Glacial deposits across much of the area have eroded or have been overlain by water- and gravity-driven sediment. Volcanic rubble, ash, and cinder mantle calderas on the lower Alaska Peninsula.

The vegetation of the Alaska Peninsula Ecoregion is largely devoid of trees with dwarf shrub (crowberry dwarf shrub and ericaceous heath and dryas-lichen) communities dominating high-elevation, exposed alpine areas (Figure 2.9). Alder and willow shrub is common along mountain slopes and foothills with low-elevation or otherwise protected areas supporting mesic graminoid (bluejoint grass)—herbaceous communities. Salt-tolerant grasses, forbs, and sedges are common along the coast.

**Figure 2.9**    Dwarf ericaceous shrub vegetation on hummocks at a mountain toeslope on the Alaska Peninsula. *Photo credit: Keith Boggs.*

# ECOLOGICAL DRIVERS

The composition, structure, and function of vegetation are ultimately directed by landscape-scale factors. Substrate, temperature, precipitation, and disturbance directly influence the type and location of plant species that are able to establish and thrive in a given area. The ecological drivers in the Bristol Bay watershed include a diversity of geologic and climatic conditions, as well as disturbance types that are observed across other regions of Alaska.

## Geology

The geology of the region is representative of the many forces active in Alaska and in the Bristol Bay watershed, creating a diversity of landforms and bedrock types on which the vegetation is mantled. The Bristol Bay lowlands are flanked by the Togiak Terrane in the northwest and the Peninsular Terrane in the southeast, which were accreted to the growing margin of the North American Plate in the late Jurassic and early Cretaceous (Plafker and Berg 1994). The predominately volcanic and volcaniclastic Togiak Terrane comprises the Akhlun and Kilbuck Mountains, whereas the igneous and sedimentary rocks of the Peninsular Terrane form the foundation of the Alaska Peninsula (Wilson et al. 2015). The Bristol Bay lowlands are occupied by a former shallow marine embayment, overlapped by assemblages of geologically-young deposits shed from the adjacent terrains (Decker et al. 1994). Granitic plutons in the Taylor Mountains, at the head of the Bristol Bay lowlands, were emplaced within this terrain during late-Cretaceous subduction-related magmatism (Hudson et al. 2010). Younger Cenozoic igneous rocks intruded into the Peninsular Terrane are part of the Aleutian volcanic arc, which has been active for at least the last 40 million years (Moll-Stalcup 1994). The modern Aleutian volcanic arc, which runs down the length of the Alaska Peninsula, has formed due to northwest-directed subduction of the Pacific Oceanic Plate beneath the continental crust of western Alaska.

The terrains comprising the Bristol Bay watershed have a range of lithologies, from more acidic granodiorites to basic limestone and other calcareous rocks. Substrate acidity has a strong influence on plant community composition and structure with many plant species restricted either to acid or basic conditions (Kruckeberg 2002). Limestone substrates at moderate to high elevations elsewhere in Alaska are typically associated with an open *Dryas* tundra, often with unusual or rare plant species present, while adjacent acidic substrates typically support a lusher lichen and herb-dominated tundra (Juday 1989). Restricted limestone bodies have been described near Becharof Lake and along the Bruin Bay Fault to the north (Blean 1976). Volcanic ash is an important substrate in much of the eastern portion of the watershed, where large volumes were deposited during the eruption of Novarupta in 1912 and previous eruptions from other volcanic centers. Much of the previously existing vegetation in Katmai National Park and Preserve was buried in the pyroclastic ash and in some areas—such as the Valley of Ten Thousand Smokes—this welded ash remains largely unvegetated (Griggs 1934, Boggs et al. 2003). In some parts of the eastern lowlands, an impermeable welded ash layer impedes drainage, allowing wet meadows to develop above the layer. Weakly-vegetated pyroclastic lava fields are found intermittently along the Aleutian Range. Extensive pyroclastic lava fields flanking the Aniakchak caldera are colonized by an unusual liverwort community. This nonvascular plant community has a well-developed cryptogamic crust layer, composed primarily of the liverwort species *Anthelia juratzkana* and *Gymnomitrion corallioides* (Boucher et al. 2012) (Figure 2.10). These pyroclastic flows are deep and exceedingly well-drained, creating difficult conditions for vascular plant establishment; however, the cool misty summer conditions that characterize alpine environments on the Alaska Peninsula provide adequate surface moisture for the establishment of these unique nonvascular communities.

Vegetation communities are also strongly influenced by soil development. Areas with coarse, mineral material are typically well-drained and possess plant communities quite distinct from those supported by more mature soils. For example, well-drained floodplain communities with little soil development may be composed of largely *Racomitrium* moss and *Stereocaulon* lichen, tall willow, alder, or balsam poplar woodlands; whereas areas with fine substrates and high organic content are more likely to support rich forb-graminoid meadows or sedge wetlands.

Glaciation played an important role in shaping the landscape and vegetation of the Bristol Bay watershed. In the Pleistocene, nearly the entire region was covered by ice sheets that originated in the Alaska Range. The only area that was not glaciated at this time was the Kilbuck Mountains that were peripherally contiguous with the large Bering Land Shelf, connecting Siberia with Alaska. The current vegetation represents those species that have successfully colonized

**Figure 2.10** *Anthelia juratzkana–Gymnomitrion corallioides* Plant Association on a side slope of the Aniakchak Volcano, Alaska Peninsula. *Photo credit: Tina Boucher.*

the post-glacial landscape. Owing to this widespread glaciation, much of the low relief landscape is mantled by fluviogla-cial deposits, such as outwash plains, braided streams, eskers, kettles, and kames. This results in a diversity of landforms, which in part direct the formation of vegetation communities. Time since deglaciation also shapes the distribution of spe-cies on the landscape, including the southern migration of white spruce into the Bristol Bay region (Brubaker et al. 2001).

## Climate

Climate has long been recognized as one of the primary forces shaping plant communities (e.g., von Humboldt and Bon-pland 1807, Billings 1952). In general, the climate of the Bristol Bay watershed is cool and moist with substantial maritime influences from the Gulf of Alaska and Bering Sea in the southern and western portions of the region, respectively. Cli-mate is distinctly continental to the north and eastern edges of the watershed with cold winters and short, but relatively warm summers. Lower annual temperature variability, cool summers, more moderate winters, and greater precipitation are typical toward the coastal areas with strong winds, fog, persistent low clouds, and salt-spray particularly prevalent along the Alaska Peninsula. Topography also has effects on the regional climate. Elevations range from sea-level to nearly 7,000 feet in the Aleutian Range and Neacola Mountains within the Bristol Bay watershed and as air temperatures de-cline with elevation, zones of vegetation are evident in many areas. Elevational tree line occurs at approximately 1,500 to 2,000 feet and tall shrubs reach their altitudinal limit at approximately 2,500 feet across much of the eastern portion of the watershed. Orographic effects are also evident in patterns of precipitation with greater rain and snowfall along the southeastern side of the Aleutian Range and in the Ahklun Mountains.

Low summer temperatures and wet soils may be partially responsible for the low prevalence of forests in the water-shed. Forests and woodlands are restricted in the Bristol Bay lowlands, and are typically restricted to well-drained flood-plains and areas with mean summer temperatures of at least 12°C (Figures 2.11A and 2.11B.). The northern limit of the

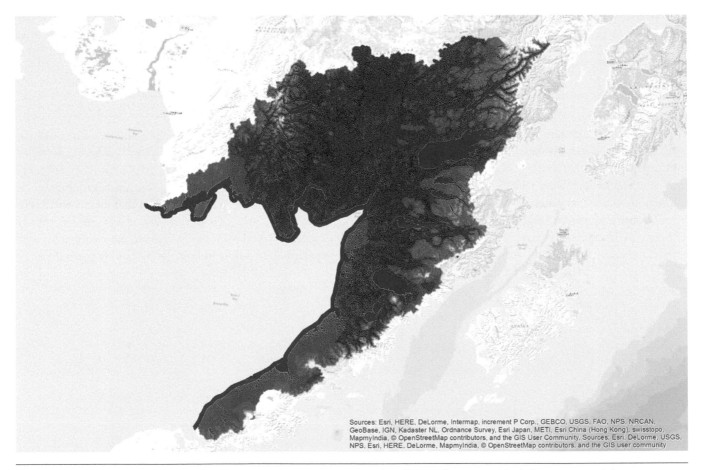

**Figure 2.11A** Forested areas in relation to summer temperatures in the (2000) Bristol Bay Watershed showing mean July temperatures (yellow to orange) (from SNAP 2016) and forested areas (green). Areas with 11°–12°C mean July temperature are shown as yellow-orange and areas of 12°–13°C are shown in orange.

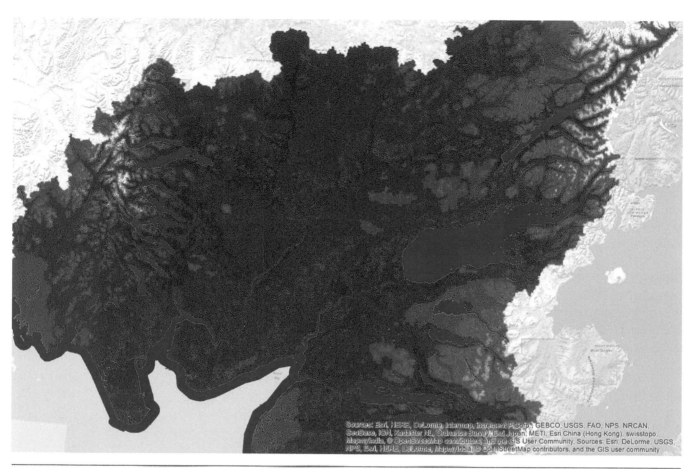

**Figure 2.11B** Close-up of Bristol Bay Watershed showing mean current mean July temperatures (yellow to orange) and forested areas (green). Areas with 11°–12°C mean July temperature are shown as yellow-orange and areas of 12°–13°C are shown in orange.

boreal forest more generally occurs approximately at the 12°C mean July isotherm (Walker 2000) and tree-line advance is strongly linked to cumulative summer warmth (Breen 2014). While specific environmental factors that are responsible for the position of tree line in Alaska are quite variable among regions (see Sveinbjörnsson et al. 2010), summer temperatures are highly correlated with the presence of trees in the Bristol Bay watershed. Based on projections for climates in the region, in approximately 50 years the majority of the landscape will be above 12°C, suggesting that forests will likely become more prevalent (Figure 2.12; and see Beck et al. 2011, Juday et al. 2015). However, even if summer warmth is associated with a growth threshold for spruce in this region, we expect a delayed response by trees to increasing temperatures (Dial et al. 2016), as the establishment of less cold-tolerant species will be slow in areas underlain by permafrost.

The Bristol Bay Watershed is situated at the southern edge of the discontinuous permafrost zone with permafrost present in isolated patches with mean annual temperatures below freezing. Ice-rich permafrost develops in fine substrates with adequate moisture and can create unique substrate features that determine the vegetation composition. For example, tussock tundra (dominated by tussock cottongrass and Bigelow's sedge) develops over shallow permafrost whereas ericaceous dwarf shrub-lichen communities develop on raised permafrost plateaus characterized by a greater depth to permafrost. The depth of the soil portion that thaws each summer (active layer) may limit the growth of more deeply rooted plants, however depth of thaw is generally deep within the discontinuous permafrost zone. Permafrost can act as an impermeable barrier, creating areas of perched groundwater and allowing wetlands to develop in areas where they would otherwise not occur. Additionally, seasonal and daily cycles of freezing and thawing can create substrate disturbance at multiple scales, such as permafrost plateaus, hummocks, and frost boils. Some plant species—such as many primroses, marsh fleabane, koenigia, and others—are largely restricted to recently exposed, wet mineral soils, such as frost boils, and thus represent small and ephemeral plant communities that are eventually outcompeted by neighboring vegetation. Frost boils also represent an important microhabitat for establishment of expanding spruce in tussock tundra habitats (Sullivan and Sveinbjörnsson 2010). The degradation of ice-rich permafrost can result in irregular subsidence of the land, resulting in a landscape mosaic of small valleys, pits, ponds, and raised benches that is common in the Bristol Bay lowlands. Where

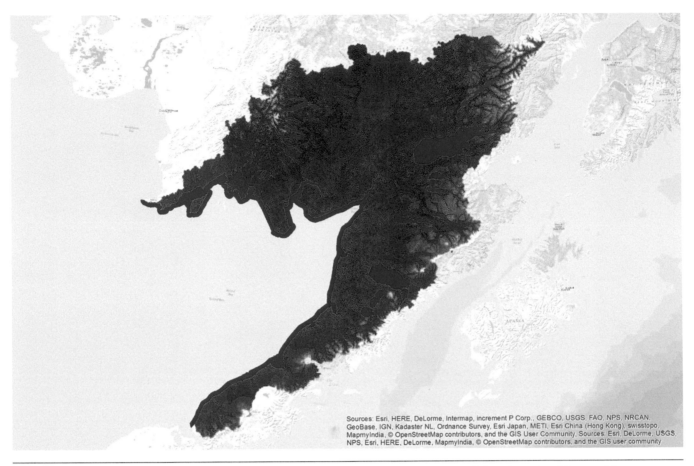

**Figure 2.12**   Forested areas (green) in relation to projected 2060 summer temperatures. Areas below the 11°–12°C mean July temperature threshold, shown as yellow-orange, are restricted to higher elevations in the watershed.

permafrost loss results in paludification, vegetation can undergo dramatic change (see Jorgenson et al. 2001, Jorgenson and Ely 2001). Thermokarst may cause mesic habitats such as those vegetated by ericaceous dwarf shrubs to subside below the water table, thus initiating transformation to graminoid-dominated wetlands and ponds. In low-gradient, coastal areas of western Alaska, storm surges can move tidewaters many miles inland, resulting in the thermal degradation of permafrost and the conversion of plateaus to thermokarst trenches (Terenzi et al. 2014). Permafrost may aggrade in areas where deep peat mosses insulate the soil from solar radiation. The current rates of permafrost loss are expected to accelerate with increasing mean annual temperatures (see Jorgenson and Ely 2001), and the entire Bristol Bay watershed is projected to lose all permafrost in 100 years (Marchenko et al. 2008).

## Disturbance

Disturbances such as volcanism, wildland fire, insect outbreaks, saltwater inundation, and erosion affect plant community structure and composition at multiple scales in the Bristol Bay region. The magnitude of the disturbance greatly influences the pattern of community change through time (ecological succession) with severe disturbances reducing habitats to unvegetated, mineral substrates requiring much longer periods for the development of plant communities. Such disturbances (e.g., glaciation, volcanism) may also cover large spatial areas, resulting in a smaller pool of species for colonization of the newly exposed terrain due to dispersal limitations. Disturbances that leave the soil largely intact (e.g., low-intensity wildfire, defoliating insect outbreak) are more quick to recover, but may not return to the predisturbance community. The succession of plant communities following particular disturbances often follows a similar pattern. For example, newly exposed floodplain alluvial surfaces in the Lime Hills Ecoregion are generally colonized within the first few years by forbs and willows. The willows may form dense thickets that overtop the forb layer. Willows may then be replaced by alders, and/or balsam poplar, which are eventually replaced by white spruce (Drury 1956, Viereck

1970, Van Cleve et al. 1983, Viereck et al. 1993, Mann et al. 1995, Yarie et al. 1998, Boggs et al. 2003, Hollingsworth et al. 2010). The introduction of marine water and sediment to terrestrial and freshwater systems by coastal storm surges can both degrade permafrost and weaken or kill resident species, thereby affecting the conversion of tundra and ponds to more saline types. The increasing magnitude and frequency of storms, exacerbated by sea-level rise, further extends the reach of storm-induced flooding across the Yukon-Kuskokwim Delta (Jorgenson and Ely 2001, Terenzi et al. 2014). Wildfire is a primary driver of boreal plant community succession (Viereck 1973), but is of more limited influence in the Bristol Bay watershed with the more interior regions most frequently subjected to the impacts of fire. However, we expect fires to become a more influential disturbance as the climate warms and forests expand westward.

# DESCRIPTION AND PATTERNS OF VEGETATION TYPES

The classification of vegetation seeks to describe the continuum of species occurring across a landscape; specifically, to organize vegetation into discrete, observable classes. While these descriptions provide a common language that enables us to communicate ecological concepts more precisely and consistently, it also requires generalization and arbitrary boundaries that are often based on the presence or absence of a given species within a community. Vegetation types may be described hierarchically. Plant associations, or communities, represent the finest scale vegetation type and are present on the landscape at the scale of a few meters. Progressively coarser-scaled vegetation units are classified by the inclusion of physiognomy and biogeography. For example, on a mountain ridge, one may be surrounded by crowberry, bog blueberry, dwarf birch, alpine sweetgrass, sedges, mosses, and lichens. This combination of species, as well as their heights and relative abundance, would constitute a 'crowberry-dwarf shrub tundra' community. Nearby along the slope of the mountain, one may find nearly complete cover of alders with very few understory species; a community described as a 'tall alder shrub'. These dwarf and tall shrub communities occur over a large geographic area, but just like all fine-scale vegetation types, they tend to be restricted to areas with particular climate, topography, soil, and disturbance patterns. Figure 2.12 shows mapped landcover classes between Kvichak Bay to Naknek Lake, illustrating the mosaic of forest and woodland, tundra, and wetland habitats.

The transition from one plant community to another may be very gradual (e.g., the transition from open spruce forest to woodland can be difficult to delineate). Vegetation communities may also transition at rather fine spatial scales. Further, vegetation communities are in a constant state of flux. A recently burned area may be mapped as a low shrub community following the disturbance, but will likely develop into a deciduous forest (at least for much of boreal Alaska) if it is mapped some decades after the disturbance.

## Shrub Communities

The Bristol Bay watershed is distinctively shrubby with some form of shrub vegetation covering half of the total area in the watershed. The majority of the shrub vegetation is dwarf shrub and low shrub-lichen-dominated communities (35%). Dwarf shrub communities (dominated shrubs less than 8 inches in height) are common throughout the hills and mountains of the region, and are found on a range of landforms including side slopes, late-lying snowbeds, ridges, and summits. Soils associated with dwarf shrub communities typically are poorly developed with low organic content. Common species associated with dwarf shrub communities include crowberry, bog blueberry, dwarf birch, mountain avens species, alpine azalea, Kamchatka rhododendron, dwarf willow species, alpine sweetgrass, and sedge species. Lichens and moss species often comprise a significant component to the dwarf shrub community.

Low shrub communities (dominated by intermediate-sized shrubs between 8 and 50 inches) cover 16% of the Bristol Bay watershed. This vegetation class is comprised of two elements: a low willow community and a low dwarf birch and ericaceous shrub community. The low willow community is common on wet to mesic mountain slopes, hillslopes, flats, and adjacent to streams and along water tracks. Soils can range from mesic to wet and mineral to organic peat. Diamond willow and grayleaf willow typically are dominant, often co-occurring with alder, dwarf birch, and blueberry. Low shrub-tussock tundra vegetation is uncommon in the region, covering just 2% of the area, and associated with permafrost-dominated areas in the Bristol Bay lowlands. Here, tussock cottongrass and Bigelow's sedge are the common tussock-forming sedges. The low dwarf birch and ericaceous shrub community is associated with mesic mountain slopes and hillslopes, as well as flats, with a well-decomposed organic soil layer. Species associated with this community include dwarf birch, Kamchatka rhododendron, Barclay's willow, and diamond willow. Low shrub communities here are often associated with permafrost.

Tall shrub communities comprise 13% of the area. This community includes shrubs over 50 inches tall. Alder-willow shrub communities are abundant along mesic mountain side slopes and foothills. In many cases, thick bands of alder extend from the toeslopes of mountains to over 2,000 feet in elevation. Alder and tall willows commonly occur along riparian areas in the Bristol Bay lowlands and Aleutian Range ecosystems where feltleaf willow and diamond willow are present. Alders are ecologically important as a species that have symbiotic *Frankia* bacteria in root nodules that are able to fix atmospheric nitrogen into a form readily available for plants. The abundance of alder in watersheds is correlated with stream productivity and influences the abundance of stream invertebrates and overall stream productivity (Schaftel et al. 2012, Callahan et al. 2017). Tall shrub communities have been recognized to have been increasing in distribution and abundance in the Bristol Bay Watershed in the last century (Robinette 2014).

## Herbaceous Communities

Nonwoody (herbaceous) dominated vegetation types comprise roughly 14% of the watershed. The herbaceous vegetation is a mixture of tidal marshes, wetlands, freshwater marshes, and mesic grasslands and forb meadows. Herbaceous vegetation is more common along the coast, as well as in the Bristol Bay lowlands in poorly drained areas and valley floors and floodplains in the upper Aleutian Peninsula Ecoregion. Herbaceous, freshwater wetlands are the most commonly encountered type in the region and are comprised of finer-scale wet sedge, sedge/sphagnum peatland, and herbaceous peatland communities. The wet sedge community occurs on wet sites, often with visible surface water in valley bottoms, basins, water tracks. Soils are highly organic and range from acidic to nonacidic and are typically underlain by permafrost. Water sedge and tall cottongrass are dominant species in this community. In closed bogs and nutrient poor fens with thick peat deposits, the sedge/sphagnum peatland community can be found. Peat-forming sedges such as tufted bulrush and other sedge species are common along with high cover of sphagnum mosses. A low abundance of dwarf and low shrubs such as small cranberry, bog rosemary, bog blueberry, and crowberry may be present in the sedge/sphagnum peatland community. The herbaceous peatland community occurs in shallow depressions and basins, pond margins, and thermokarst pits. Herbaceous peatlands are nutrient-rich peatlands that have a thick peat layer that may be floating or submerged. Standing water is usually present (Viereck et al. 1992). Purple marshlocks, bog-bean, water horsetail, tall cottongrass, and water sedge are common. A low abundance of shrubs, such as sweetgale and dwarf birch may be present, as well as submerged aquatic species, such as common mare's tail, pondweed species, burreed species, and aquatic mosses.

## Forest Communities

Forests and woodlands comprise just 12% of the watershed area. Deciduous forests, comprised primarily of Alaska paper birch, Kenai birch, and balsam poplar are common on mesic upland terrains on glacial till, loess, and colluvium derived soils. Balsam poplar forests are restricted to well-drained floodplains and are scattered widely throughout the Bristol Bay watershed. In some cases, birch co-dominates with balsam poplar. Small, isolated stands of quaking aspen can be found in central and northern portions of the watershed on relatively dry sites, especially on south-facing slopes.

White and black spruce forest and woodland communities are uncommon in the Bristol Bay watershed (6% of the area). Forests are defined by a foliar cover of trees greater than 25%, and woodlands with less than 25% cover (Viereck et al. 1992). The larger expanses of spruce forests and woodlands are restricted to the upper Mulchatna and Nushagak rivers, eastern Wood-Tikchik Lakes, and low elevation valleys from Naknek Lake north to Lake Clark and are associated with areas with warmer summers in the watershed. In upland areas, white spruce forest communities are more prevalent, occurring on well-drained rolling hills, old river terraces, and mountain side slopes. Black spruce and Alaska paper birch may co-occur with white spruce in these communities, along with an understory of prickly rose, mountain cranberry, dwarf birch, twinflower, wintergreen, tall bluebells, and stairstep moss. Black spruce forest communities in this area are found on well-drained to moderately well-drained alluvial fans and river terraces, as well as uplands. Little to no peat development is associated with black spruce forest communities in Bristol Bay; however, an organic layer of nonsphagnum mosses may be present. Understory species associated with this community include dwarf birch, prickly rose, bog blueberry, crowberry, bluejoint grass, horsetail species, stairstep moss, and reindeer lichen. Sitka spruce forest and woodlands barely reach the very northwestern portion of the Bristol Bay watershed (Boggs et al. 2003).

# RECOMMENDED RESEARCH

The composition and distribution patterns of vegetation have substantial influence on the region's ecology. As the Bristol Bay watershed is faced with continued changes to many of the underlying factors that influence vegetation communities, we can expect broad-scale changes in accompanying ecological processes such carbon balance and primary production, nutrient cycling, disturbance regimes, and changes in distribution and abundance of species. Many of the vegetation communities are at their distributional limits and at perceived tipping points. Small increases in summer warmth, for example, are expected to facilitate forest expansion in the region. Thus, we recognize a need to understand how vegetation communities are currently changing and how these changes are linked to other abiotic and biotic drivers. Notably, determining the rate, pattern, and nature of spruce forest expansion in the region is recognized as a research goal. Additionally, shrubs are expanding across many tundra and graminoid-forb communities, but increased effort would be beneficial to explore the context for this expansion and predict landscape-level outcomes. Understanding the dynamics between transitional shrub and forest communities would be very beneficial. In particular, it is not clear how resilient shrub communities are to the establishment of spruce, and which circumstances facilitate spruce establishment. The longer-term response of vegetation communities to permafrost loss is poorly understood yet important, as permafrost is expected to be absent from the area in the coming century. Additionally, the dynamics of estuaries and low-gradient coastal ecosystems is not well-studied in the Bristol Bay watershed. Similar habitats in Kuskokwim Bay are recognized to be a dynamic interplay of sedimentation, erosion, and increasingly frequent storm surges. These factors have pronounced impacts on the vegetation communities—and very small changes in elevation can result in dramatically different vegetation, which in turn influences shorebirds and waterfowl, as well as other species (Jorgenson and Ely 2001). Little effort has been applied toward understanding the importance of interactions among plants and other organisms in generating vegetation patterns in this region. Herbivorous insects and vertebrates can influence competitive outcomes among plants. The role of pollinators in plant reproduction, establishment tundra, and boreal habitats is very poorly understood. Furthermore, the role of mycorrhizal fungi, as well as pathogenic fungi, may be very important in the structuring of vegetation communities, but is not well-studied in transitional boreal forests.

# SUMMARY

The ecological drivers at play in the Bristol Bay watershed produce a diversity of ecoregions and vegetation types. Lithologically-distinct accreted terrains have been extensively reworked by Holocene-epoch glaciations and more recently modified by tectonic, volcanic, and fluvial activity. These episodic, catastrophic events coupled with more frequent, yet less severe disturbances such as wildland fire and insect outbreak continue to reshape the relationships between ecological process and vegetation patterns. Distinctly mountainous ecoregions flank the broad Bristol Bay lowlands. Ecoregions include those with affinities to subarctic, interior-boreal, and maritime climates, while vegetation types are characteristic of arctic and alpine tundra, boreal forest, coastal beaches and estuaries, and Aleutian meadows, as well as a range of freshwater wetland types. While many of the state's characteristic vegetation communities can be found within the watershed, the proportion of the landscape that is occupied by particular communities is distinct from many other regions. For example, spruce forests are rather restricted. Permafrost is discontinuous in much of the region and largely absent from the southern portion of the watershed on the Alaska Peninsula. Glaciation in the eastern margins of the watershed and erosion on river floodplains and mountain slopes can result in barren substrates that take a considerable length of time to be recolonized; wildfire and defoliating insect outbreaks can also cause substantial alterations in the vegetation.

The vegetation of Bristol Bay is notably shrubby. Willow and alder-dominated communities occur throughout the watershed and often extend over elevational gradients much broader than in other areas of the state. Emergent freshwater wetlands are particularly common in the Bristol Bay lowlands, and otherwise are found intermittently across the landscape. Forested habitats are restricted in the region with most spruce forests found at low elevations in the eastern portion of the watershed and balsam poplar stands along low elevation floodplains throughout the watershed.

The primary drivers influencing the composition and distribution of plant communities, such as average annual temperature and precipitation, summer warmth, and sea-level rise, are undergoing continued changes in this region. Many of the vegetation communities are at their distributional limits and thus exist at perceived tipping points. Summer temperatures are correlated with the occurrence of spruce forests, and much of the watershed is expected to shift from unsuitable to suitable climatic conditions for boreal forest in the coming decades. If boreal forests do expand in the region, it will likely be at the expense of the shrub habitats that are so characteristic of Bristol Bay. This would likely result in broader ecological ramifications, such as increased frequency and intensity of wildfires, and potential reductions in nitrogen inputs to soils and stream from alder. In recent decades, however, shrubs such as alder are noted to be expanding in tundra habitats, particularly in the western portion in this region. Additionally, as the climate continues to warm, areas of permafrost will likely be lost, resulting in shifts in plant communities (e.g., Jorgenson et al. 2001). Last, climate change and associated sea-level rise is expected to result in more frequent and extensive storm surges across the Bristol Bay lowlands, resulting in shifts from freshwater aquatic and terrestrial habitats to more saline types with the most significant effects predicted in areas just off of ice-rich permafrost, as well as potential impacts to wildlife (Terenzi et al. 2014). Indeed, some of the most dramatic changes in vegetation for the state are predicted to occur in the Bristol Bay watershed in the next century, with some areas expected to have temperature and precipitation patterns that have no current analogies in Alaska (Murphy et al. 2010). In this way, the Bristol Bay watershed provides an important bellweather for biotic response to climate-induced ecological change in Alaska.

# REFERENCES

Ackerly, D.D., S.R. Loarie, W.K. Cornwell, S.B. Weiss, H. Hamilton, R. Branciforte, and N. J. B. Kraft. 2010. The geography of climate change: implications for conservation biogeography. Diversity and Distributions 16:476–487.

Amiro, B.D., A.L. Orchansky, A.G. Barr, T.A. Black, S.D. Chambers, F.S. Chapin III, M.L. Goulden, M. Litvak, H.P. Liu, J.H. McCaughey, A. McMillan, J.T. Randerson. 2006. The effect of post-fire stand age on the boreal forest energy balance. Agricultural and Forest Meteorology. 140:41–50.

Baldocchi D., F.M. Kelliher, T.A. Black, and P.G. Jarvis. 2000. Climate and vegetation controls on boreal zone energy exchange. Global Change Biology, 6 (Suppl. 1), 69–83.

Beck, P.S.A., G.P. Juday, C. Alix, V.A. Barber, S.E. Winslow, E.E. Sousa, P. Heiser, J.D. Herriges, S.J. Goetz. 2011. Changes in forest productivity across Alaska consistent with biome shift. Ecology Letters 14:373–379.

Billings, D. 1952. The environmental complex in relation to plant growth and distribution. Quarterly Review of Biology. 27:251–265.

Blean, K.M., 1977. The United States Geologic Survey in Alaska: Accomplishments During 1976. Geologic Survey Circular 751-B. Pp. 428.

Boucher, T.V., K. Boggs, B. Koltun, T.T. Kuo, J. McGrath, and C. Lindsay. 2012. Plant associations, vegetation succession, and earth cover classes: Aniakchak National Monument and Preserve. Natural Resource Technical Report NPS/ANIA/NRTR—2012/557. National Park Service, Fort Collins, Colorado. Pp. 241.

Breen, A.L. 2014. Balsam poplar (*Populus balsamifera* L.) communities on the Arctic Slope of Alaska. Phytocoenologia 44:1–24.

Boggs, K.W., S.C. Klein, J.E. Grunblatt, and B. Koltun. 2003. Landcover classes, ecoregions and plant associations of Katmai National Park and Preserve. Natural Resource Technical Report NPS/KATM/NRTR—2003/001. National Park Service, Fort Collins, Colorado. 274 p.

Brubaker, L.B., P.M. Anderson, and F.S. Hu. 2001. Vegetation ecotone dynamics in Southwest Alaska during the Late Quaternary. Quaternary Science Reviews 20:175–188.

Bryant, J.P. 1987. Feltleaf willow-snowshoe hare interactions: plant carbon/nutrient balance and floodplain succession. Ecology 68:1319–1327.

Callahan, M.K., D.F. Whigham, M.C. Rains, K.C. Rains, R.S. King, C.M. Walker, J.R. Maurer, and S.J. Baird. 2017. Nitrogen subsidies from hillslope alder stands to streamside wetlands and headwater streams, Kenai Peninsula, Alaska. Journal of the American Water Resources Association (JAWRA) 53(2):478–492, DOI: 10.1111/1752-1688.12508.

Consortium of Pacific Northwest Herbaria. 2016. Specimen Data. Managed by the University of Washington Herbarium, Burke Museum of Natural History and Culture University of Washington. Accessed [26 May 2016]. http://www.pnwherbaria.org/data/results.php?DisplayAs=WebPage&ExcludeCultivated=Y&GroupBy=ungrouped&SortBy=Year&SortOrder=DESC&SearchAllHerbaria=Y&QueryCount=1&IncludeSynonyms1=Y&Genus1=picea&Species1=sitchensis&Zoom=4&Lat=55&Lng=-135&PolygonCount=0.

Devotta, D. A. 2008. The influence of Alnus viridis on the nutrient availability and productivity of sub-arctic lakes in southwestern Alaska. Masters Thesis, University of Illinois at Urbana- Champaign, Urbana, Illinois, US. Pp. 49.

Dial, R.J., T.S. Smeltz, P.F. Sullivan, C.L. Rinas, K. Timm, J.E. Geck, S.C. Tobin, T.S. Golden, and E.C. Berg. 2016. Shrubline but not treeline advance matches climate velocity in montane ecosystems of South-central Alaska. Global Change Biology 22:1841–1856.

Dorrepaal, E., Toet, S., van Logtestijn, R., S.P., Swart, E., van de Weg, M.J., Callaghan, T.V., and Aerts, R. 2009. Carbon respiration from subsurface peat accelerated by climate warming in the subarctic. Nature 460: 616–619A.

Drury, W. 1956. Bog flats and physiographic processes in the upper Kuskokwim River region, Alaska. Contributions from the Gray Herbarium 178. Pp. 130.

Ehrlich, P.R., and Raven, P.H., and 1964. Butterflies and plants: a study in coevolution. Evolution. 18:586–608.

Gorham, E. 1991. Northern peatlands: role in the carbon cycle and probable responses to climatic warming. Ecological Applications 1:182–195.

Griggs, R.F. 1918. The Valley of Ten Thousand Smokes: An Account of the Discovery and Exploration of the Most Wonderful Volcanic Region in the World. National Geographic. Pp. 115–169.

Griggs, R.F. 1934. The Edge of the forest in Alaska and the reasons for its position. Ecology 15:80–96.

Jorgenson, M.T. and Ely, C.R. 2001. Topography and flooding of coastal ecosystems on the Yukon-Kuskokwim Delta, Alaska: implications for sea-level rise. Journal of Coastal Research 17:124–136.

Jorgenson, M.T., C.H. Racine, J.C. Walters, and T.E. Osterkamp. 2001. Permafrost degradation and ecological changes associated with a warming climate in central Alaska. Climatic Change 48:551–579.

Jorgenson, M.T., G.V. Frost, W.E. Lentz, and A.J. Bennett. 2006. Photographic monitoring of landscape change in the southwest Alaska network of national parklands. Report No. NPS/AKRSWAN/NRTR-2006/03. ABR, Inc.–Environmental Research & Services, Fairbanks, Alaska.

Juday, G.P. 1989. Alaska Research Natural Areas. 2: Limestone Jags. US Forest Service. General Technical Report: PNW-GTR-237.

Juday, G.P., C. Alix, C., and T.A. Grant. 2015. Spatial coherence and change of opposite white spruce temperature sensitivities on floodplains in Alaska confirms early-stage boreal biome shift. Forest Ecology and Management 350:46–61.

Hollingsworth, T.N., A.H. Lloyd, D.R. Nossov, R.W. Ruess, B.A. Charlton, and K. Kielland. 2010. Twenty-five years of vegetation change along a putative successional chronosequence on the Tanana River, Alaska. Canadian Journal of Forest Research, 40: 1273–1287. 10.1139/X10-094.

Hocking, M.D., and J.D. Reynolds. 2011. Impacts of salmon on riparian plant diversity. Science 331.6024:1609–1612.

Hu, F. S., B.P. Finney, and L.B. Brubaker. 2001. Effects of Holocene Alnus expansion on aquatic productivity, nitrogen cycling, and soil development in southwestern Alaska. Ecosystems 4:358–368.

Hudson, T.L., M.L. Miller, E.P. Klimasauskas, and P.W. Layer. 2010, Reconnaissance study of the Taylor Mountains pluton, southwestern Alaska, in Dumoulin, J.A., and Galloway, J.P., eds., Studies by the U.S. Geological Survey in Alaska, 2008–2009: U.S. Geological Survey Professional Paper 1776–A, 13 p.

Krukeberg. A.R. 2002. Geology and Plant Life: the Effects of Landforms and Rock Types on Plants. University of Washington Press, Seattle, Washington. Pp. 362.

Lisi, P.J., and D.E. Schindler. 2011. Spatial variation in the timing of marine subsidies influence riparian phenology through a plant-pollinator mutualism. Ecosphere 2:1–14. DOI: 10.1890/ES11-00173.1

Mann, D.H., C.L. Fastie, E.L. Rowland, and N.H. Bigelow. 1995. Spruce succession, disturbance, and geomorphology on the Tanana River floodplain, Alaska. Ecoscience. 2(2): 184–199.

Marchenko, S., V. Romanovsky, and G. Tipenko. 2008. Numerical Modeling of Spatial Permafrost Dynamics in Alaska, Ninth International Conference on Permafrost, Fairbanks, Alaska, Institute of Northern Engineering UAF, Fairbanks, pp. 1125–1130.

McFadden, J, G. Liston, M. Sturm, R.A. Pielke, Sr., F.S. Chapin, III. 2001. Interactions of shrubs and snow in arctic tundra: measurements and models. In Soil-Vegetation-Atmosphere Transfer Schemes and Large-Scale Hydrological Models (Proceedings of a symposium held during the Sixth IAHS Scientific Assembly at Maastricht, The Netherlands, July 2001). IAHS Publ. no. 270, 2001.

Moll-Stalcup, E.J., 1994, Late Cretaceous and Cenozoic magmatism in mainland Alaska, in Plafker, George, and Berg, H.C., eds., The geology of Alaska: Geological Society of America, DNAG Series, v. G-1, Pp. 589–619.

Moore, J.P., D.M. Moore, M. Clark, D.R. Kautz, D. Mulligan, M. Mungoven, D.K. Swanson, and D.V. Patten. 2004. Land resource regions and major land resource areas of Alaska. Kautz, D. R., and P. Taber, eds. United States Department of Agriculture–Natural Resources Conservation Service Alaska. Palmer, Alaska.

Murphy, K., F. Huettmann, N. Fresco, and J. Morton. 2010. Connecting Alaska landscapes into the future: results from an interagency climate modeling, land management and conservation project. U.S. Fish and Wildlife Service. 96 Pp.

Nowacki, G., P. Spencer, M. Fleming, T. Brock, and T. Jorgenson. Ecoregions of Alaska: 2001. U.S. Geological Survey Open-File Report 02-297 (map).

Philip, K.W., Ferris, C.D., 2016. Butterflies of Alaska: A Field Guide. 2nd ed. Alaska Entomological Society. 110 pp. ISBN 978-1-945170-60-7.

Plafker, G. and H.C. Berg, 1994. Overview of the geology and tectonic evolution of Alaska. The Geology of North America. G-1. Pp. 989–1021.

Robinette, J. 2014. Assessing the vulnerability of Western Alaska ecosystems and subsistence resources to non-native plant invasion. Committee for Noxious and Invasive Plant Management, Annual Meeting. Poster Presentation. Accessed [26 May 2016] https://westernalaskalcc.org/projects/Lists/Project%20Products/Attachments/105/Assessing%20the%20vulnerability%20of%20%20Western%20Alaska%20ecosystems%20Poster.pdf

Roland, C.A., J.H. Schmidt, E.F. Nicklen. 2013. Landscape-scale patterns in tree occupancy and abundance in subarctic Alaska. Ecological Monographs 83:19–48.

Ruess, W. R., M. D. Anderson, J. S. Mitchell, and J. W. McFarland. 2006. Effects of defoliation on growth and N fixation in Alnus tenuifolia: Consequences for changing disturbance regimes at high latitudes. Ecoscience 13:404–412.

Schaftel, R.S., R.S. King, and J.A. Back. 2012. Alder cover drives nitrogen availability in Kenai lowland headwater streams, Alaska. Biochemistry. 107:135–148.

SNAP. 2016. Scenarios Network for Alaska and Arctic Planning. University of Alaska Fairbanks. Accessed [26 May 2016] https://www.snap.uaf.edu/.

Soja. A.J., N.M. Tchebakova, N.H.F. French, M.D. Flannigan, H. H. Shugart, B.J. Stocks, A.I. Sukhinin, E.I. Parfenova, F.S. Chapin III, and P.W. Stackhouse Jr. 2007. Climate-induced boreal forest change: Predictions versus current observations. Global and Planetary Change. 56:274–296.

Sullivan, P.F., and B. Sveinbjörnsson B. 2010. Microtopographic control of treeline advance in Noatak National Preserve, northwest Alaska. Ecosytems. 13:275–285.

Sveinbjörnsson B., M. Smith, T. Traustason, R.W. Ruess, P.F. Sullivan. 2010. Variation in carbohydrate source–sink relations of forest and treeline white spruce in southern, interior and northern Alaska. Oecologia. 163:833–843.

Tape K.D., D.D Gustine,R.W. Ruess, L.G. Adams, and J.A. Clark. 2016. Range Expansion of Moose in Arctic Alaska Linked to Warming and Increased Shrub Habitat. PLoS ONE 11(4): e0152636. doi:10.1371/journal.pone.0152636.

Terenzi, J., M.T. Jorgenson, C.R. Ely, and N. Giguère. 2014. Storm-surge flooding on the Yukon-Kuskokwim Delta, Alaska. Arctic 67:360–374.

Van Cleve, K., L. Oliver, R. Schlentner, L. Viereck, and C. Dyrness. 1983. Productivity and nutrient cycling in taiga forest ecosystems. Canadian Journal of Forest Research 13:747–766.

Viereck, L. 1970. Forest succession and soil development adjacent to the Chena River in interior Alaska. Arctic and Alpine Research 2:1–26.

Viereck, L.A. 1973. Wildfire in the taiga of Alaska. Quaternary Research. 3:465–495.

Viereck, L.A., C.T. Dyrness, A.R. Batten, and K.J. Wenzlick. 1992. The Alaska vegetation classification. Portland, OR, USA: US Department of Agriculture, Forest Service, Pacific Northwest Research Station. Pp. 278.

Viereck, L.A., C.T. Dyrness, and M. Foote. 1993. An overview of the vegetation and soils of the floodplain ecosystems of the Tanana River, interior Alaska. Canadian Journal of Forest Research 23:889–898.

von Humboldt, A. and A. Bonpland. 1807. Essay on the Geography of Plants. S. T. Jackson (ed.), Translated by S. Romanowski (2009). University of Chicago Press, Chigaco, Illinois. Pp. 296.

Walker, D.A., M.D. Walker, K.R. Everett, and P.J. Webber. 1985. Pingos of the Prudhoe Bay Region, Alaska. Arctic and Alpine Research (Institute of Arctic and Alpine Research, University of Colorado) 17:323.

Wilson, F.H., R.L. Detterman, and G.D. DuBois. 2015, Geologic framework of the Alaska Peninsula, southwest Alaska, and the Alaska Peninsula terrane: U.S. Geological Survey Bulletin 1969–B, scale 1:500,000, 2 plates, 34 p.

Yarie, J., L. Viereck, K. Van Cleve, and P. Adams. 1998. Flooding and ecosystem dynamics along the Tanana River. BioScience 48: 690–695.

# 3

# FLORA OF THE BRISTOL BAY WATERSHED

Matthew L. Carlson and Brian Heitz
Alaska Center for Conservation Science, University of Alaska Anchorage

## INTRODUCTION

Southwestern Alaska is a floristically diverse region noted as an area with a large number of geographically restricted species relative to other high latitude regions (Orme et al. 2005) and an area that has served as a source of species radiations (Hultén 1937). Strong climatic gradients and pronounced topographic and geologic diversity all contribute to a wide range of habitats that promote the high regional plant diversity. High elevation and cold and barren areas in the Aleutian Range and Ahklun Mountains, for example, represent important habitats for alpine species. Some of these species, such as ciliate saxifrage (*Saxifraga eschscholtzii*), are restricted to calcareous substrates like limestone, and some species, such as many in the heath family, are associated with acidic substrates. Likewise, particular plant species are associated with tidal marshes, freshwater wetlands, forests, and floodplains—all habitats that occur in the region. Substrates underlain with permafrost represent an important habitat for many cold- and moisture-tolerant plants that typically have more arctic or boreal distributions. A diversity of natural disturbances also promotes floristic richness of the area. Notable disturbances include riparian flooding along with ice-scour, avalanches, wave action, insect outbreaks, wildfire, and periglacial action. These disturbances create a mosaic of various-aged terrains and provide habitats for early colonizing to late-seral species.

Hultén (1937) described southwestern Alaska as an important epicenter of boreal and arctic plant radiations. Much of the region, however, was overlain by glaciers as recently as 11,000 years ago. Areas particularly affected by glaciation include the Alaska Peninsula, Alaska Range, and Aklun Mountains (Manley and Kaufmann 2002). Thus, contemporary species in the area presumably represent more recent migrants from areas that are now submerged to the west on the Beringian subcontinent that remained ice-free during maximum glaciation and the Wisconsin glaciation (Carlson et al. 2013). Additionally, post-glaciation migrations of once isolated floras converge in the Bristol Bay region (Carlson et al. 2013). Numerous Pacific-coastal species (Vancouverian, *sensu* Cronquist 1982) such as Sitka spruce (*Picea sitchensis*), devil's club (*Oplopanax horridus*), and false blueberry (*Menziesia ferruginea*) reach their western range limit in the region. Primarily arctic taxa, such as Fisher's tundragrass (*Dupontia fisheri*) and the arctic lousewort (*Pedicularis arctoeuropaea*), extend down the western Bering Sea into the Bristol Bay area. Species associated with the Aleutian Flora reach Bristol Bay, including Aleutian chickweed (*Cerastium aleuticum*), oriental popcornflower (*Plagiobothrys orientalis*), and Kamchatka rhododendron (*Therorhodion camtschaticum*). A number of species with more Aleutian distributions extend eastward beyond Bristol Bay and the Alaska Peninsula. Species associated with the Boreal Forest Flora are well represented in Bristol Bay with many species reaching their southwesterly range limit in North America such as aspen (*Populus tremuloides*), muskroot (*Adoxa moschatelliana*), and black spruce (*Picea mariana*).

The Bristol Bay watershed also harbors species with distributions that span the north Pacific, as well as those that occur broadly across the northern hemisphere. Species such as the wedgeleaf primrose (*Primula cuneifolia* ssp. *saxifragifolia*, Figure 3.1), which can be abundant in moist alpine areas of Bristol Bay, is primarily north Pacific in its distribution, ranging from Japan, north and east through the Aleutians, southern Bering Sea coast, and Gulf of Alaska coast. More broadly distributed plants bridging eastern Asia and northwestern North America include narcissus anemone (*Anemone narcissiflora* var. *monantha*, Figure 3.2). Last, species such as felwort (*Swertia perennis*, Figure 3.3) typify plants of the region that are both common in Alaska and have ranges that span widely across the northern hemisphere, although often with extensive gaps in their distributions.

**Figure 3.1**　Wedgeleaf primrose (*Primula cuneifolia* ssp. *saxifragifolia*), an alpine plant, ranging from Japan, east through the Aleutians and northern Gulf of Alaska coast. *Photo credit: Jan Anderson.*

**Figure 3.2**　Narcissus anemone (*Anemone narcissiflora* var. *monantha*), a common plant of meadows from eastern Asia to British Columbia. *Photo credit: Jan Anderson.*

**Figure 3.3** Felwort (*Swertia perennis*), a plant common to wetlands in the region with a distribution ranging from Europe through Asia to North America. *Photo Credit: Jan Anderson.*

A contingent of narrowly endemic Bristol Bay taxa have distributions restricted to either side of the the Bering Strait. These species include *Rumex beringensis, Thalictrum minus*, and *Primula tschuktschorum*. Species such as these are found in the Russian Far East, Bering Sea Islands, and western Alaska. Disjunct populations of *Rumex beringensis* are also known from the Wrangell Mountains on the Alaska-Yukon border. For rare species, we do not include common names, as these names are not in regular usage (Murray and Lipkin 1987), however, Table 3.1 includes common names that were generated by the United States Department of Agriculture (USDA) PLANTS Database.

**Table 3.1** Rare plant species of the Bristol Bay watershed and subnational and global conservation ranks

| Family | Scientific name | USDA common name | Federal rank | S-rank | G-rank |
|---|---|---|---|---|---|
| Ophioglossaceae | *Botrychium alaskense* Wagner and Grant | Alaska moonwort | | S3 | G4 |
| Ophioglossaceae | *Botrychium ascendens* W. H. Wagner | triangleglobe moonwort | Sen[1] | S2S3 | G3 |
| Ophioglossaceae | *Botrychium pedunculosum* W. H. Wagner | stalked moonwort | | S1 | G2G3 |
| Ophioglossaceae | *Botrychium virginianum* (L.) Swartz | rattlesnake fern | | S3 | G5 |
| Cyperaceae | *Carex atherodes* Spreng. | wheat sedge | | S3S4 | G5 |
| Cyperaceae | *Carex phaeocephala* Piper | dunhead sedge | | S3 | G4 |
| Poaceae | *Catabrosa aquatica* (L.) P. Beauv. | water whorlgrass | | S1S2 | G5 |
| Ceratophyllaceae | *Ceratophyllum demersum* L. | coon's tail | | S3S4 | G5 |
| Crassulaceae | *Crassula aquatica* (L.) Schönland | water pygmyweed | | S1S2 | G5 |
| Brassicaceae | *Draba chamissonis* G. Don | Cape Thompson draba | | S1Q | GNR |

| | | | | | |
|---|---|---|---|---|---|
| Brassicaceae | *Draba incerta* Payson | Yellowstone draba | | S3 | G5 |
| Brassicaceae | *Draba macounii* O. E. Schulz | Macoun's draba | | S3 | G3G4 |
| Rosaceae | *Geum aleppicum* ssp. *strictum* (Aiton) R. T. Clausen | yellow avens | | S3 | G5T5 |
| Poaceae | *Glyceria pulchella* (Nash) K. Schum. | valley mannagrass | | S3S4 | G5 |
| Isoetaceae | *Isoetes occidentalis* L. F. Hend. | MacKenzie western quillwort | | S3S4 | G4G5 |
| Plantaginaceae | *Limosella aquatica* L. | water mudwort | | S3 | G5 |
| Saxifragaceae | *Micranthes charlottae comb. nov.* (Calder & Savile) | heartleaf saxifrage | | S2 | GNR |
| Saxifragaceae | *Micranthes porsildiana* (Calder & Savile) Elven & D. F. Murray | Porsild's saxifrage | Watch | S2 | G4 |
| Caryophyllaceae | *Minuartia dawsonensis* (Britt.) House | rock stitchwort | | S3S4 | G5 |
| Boraginaceae | *Plagiobothrys orientalis* (L.) I. M. Johnst. | Oriental popcornflower | Watch | S3 | G3G4 |
| Polygonaceae | *Polygonum fowleri* ssp. *fowleri* | Fowler's knotweed | | S3S4 | G5TNR |
| Potamogetonaceae | *Potamogeton robbinsii* Oakes | Robbin's pondweed | | S2 | G5 |
| Potamogetonaceae | *Potamogeton subsibiricus* Hagstr. | Yenisei River pondweed | | S3S4 | G3G4 |
| Primulaceae | *Primula tschuktschorum* Kjellman | Chukchi primrose | Sen[1] | S3 | G2G3 |
| Ranunculaceae | *Ranunculus pacificus* (Hultén) L. D. Benson | Pacific buttercup | | S3S4 | G3 |
| Hydrophyllaceae | *Romanzoffia unalaschcensis* Cham. | Alaska mistmaiden | Sen[2] | S3S4 | G3 |
| Polygonaceae | *Rumex beringensis* V. V. Petrovsky | Bering Sea dock | | S3 | G3 |
| Saxifragaceae | *Saxifraga adscendens* ssp. *oregonensis* (Raf.) Bacig. | small saxifrage | | S2S3 | G5T4T5 |
| Brassicaceae | *Smelowskia pyriformis* W. H. Drury & Rollins | pearshaped smelowskia | Sen[1] | S3 | G2 |
| Ranunculaceae | *Thalictrum minus* ssp. *kemense* (Fr.) Cajander | lesser meadow-rue | | S2 | GNR |
| Violaceae | *Viola selkirkii* Pursh ex Goldie | Selkirk's violet | | S3S4 | G5? |
| Potamogetonaceae | *Zannichellia palustris* ssp. *palustris* L. | horned pondweed | | S3S4 | G5 |

Critically imperiled taxa at the state or global level = 1, imperiled = 2, vulnerable = 3, secure but of long-term concern = 4, secure = 5. Range ranks (e.g., S2S3) are given for taxa with insufficient information to assign a specific rarity level. Species not ranked = NR; taxonomically questionable = Q, and uncertain ranks = ? (see AKNHP 2015, NatureServe 2015 for greater discussion of ranks). Sen[1] = BLM Sensitive Species, Sen[2] = USFS Sensitive Species, Watch = BLM Watch-listed species. Common names were derived from the USDA PLANTS Database (http://plants.usda.gov/java/).

Florisitic surveys in the region have been intermittent with most surveys concentrated on U.S. National Park Service, U.S. Fish and Wildlife Refuge, and state park lands during the last 30 years (Figure 3.4). Lake Clark and Katmai National Park and Preserve, Alagnak Wild and Scenic River, and Aniakchak National Preserve have all had a diversity of plant ecological studies that generated floristic information (Griggs 1936, Cahalane 1959, Young and Racine 1978) as well as directed floristic inventories, typically focused on vascular plants (Lipkin 2002, Carlson and Lipkin 2003, Lipkin 2005, Carlson et al. 2008). Undersampled areas are concentrated in state and Native lands in the Nushugak and Mulchatna river

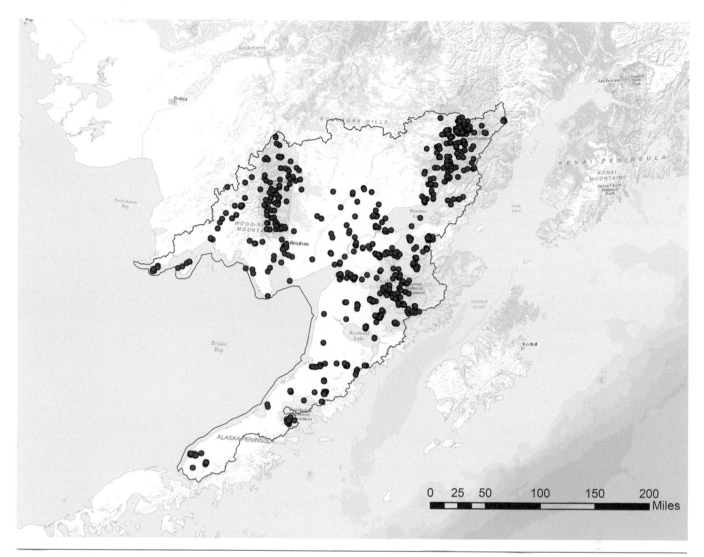

**Figure 3.4** Vascular and nonvascular plant collection locations in the Bristol Bay watershed; collection data were derived from the Consortium of Pacific Northwest Herbaria (2015).

lowlands and the Alaska Peninsula lowlands. More recent floristic work in the parks has collected more data on bryophytes and lichens (notably by W. Schofield and J. Walton, Consortium of Pacific Northwest Herbaria 2015), however overall survey efforts for nonvascular plants and lichens are very limited geographically. Targeted floristic inventories by lichenologists elsewhere in the state have revealed exceptionally high biodiversity and a number of species new to science (Spribille et al. 2010, Nelson et al. 2011). For example, in just a 53 km² area in southeastern Alaska, over 750 species of lichens were documented, the highest number of lichen species ever reported per unit area—including six species new to science (Spribille et al. 2010).

Collection records from the Consortium of Pacific Northwest Herbaria (2015), which include data from 38 regional herbarium collections, indicate the Bristol Bay watershed harbors 708 species of vascular plants. Approximately 40% of the total vascular plant diversity of Alaska is found within the Bristol Bay watershed despite occupying a small area (6% of the state). Thus, Bristol Bay can be viewed as distinctly floristically rich. All recorded species in the region have distributions that extend beyond the watershed; no plant species are known to be unique to Bristol Bay and no species are federally listed. However, there are 32 species considered to be rare at the state and/or global level and five of these species are listed as Species of Concern or Watch-listed species by the Bureau of Land Management, and one species is listed as Sensitive by the U.S. Forest Service (Table 3.1). Known populations of rare vascular plants are most concentrated in the Ahklun Mountains and Alaskan to Aleutian ranges (Figure 3.5).

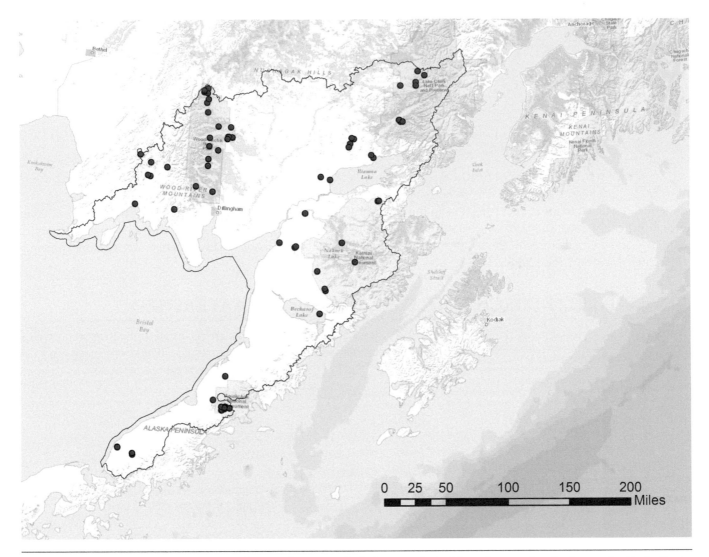

**Figure 3.5**   Rare plant collection locations in the Bristol Bay watershed; data were derived from the rare plant database at the Alaska Center for Conservation Science, University of Alaska Anchorage (2015).

*Smelowskia pyriformis* is the most regionally restricted species, with populations scattered in the Ahklun, Kuskokwim, Taylor, and Revelation mountains and extending northeast out of the Bristol Bay watershed into the western Alaska Range. *Smelowskia pyrifromis* is a perennial herb in the mustard family that is found on unstable rubble and talus slopes in both calcareous, shale, and mudstone substrates. There are only 17 known locations of *S. pyriformis* and approximately a third occur in the Bristol Bay watershed. *Botrychium pedunculosum* is a fern that is only known from two locations in the state, one outside of Aniakchak Crater, and just 20 small populations globally, primarily concentrated in the northern Rockies (Lipkin 2005). Three additional rare *Botrychium* species have been documented in Bristol Bay, including *B. alaskense, B. ascendens,* and *B. virginianum.* These species are all associated with herbaceous meadow habitats. *Botrychium* species are generally crypic and difficult to identify, generally overlooked, and may be more common.

A large proportion of the regionally rare species are aquatic or wetland-obligates, reflecting the ubiquity of these habitats in the region. The rare aquatic and wetland species include: *Carex atherodes, Catabrosa aquatica, Ceratophyllum demersum, Crassula aquatica, Glyceria pulchella, Isoetes occidentalis, Limosella aquatica, Potamogeton robbinsii, Potamogeton subsibiricum,* and *Zannichellia palustris* ssp. *palustris.* These aquatic and wetland species are uncommon in Alaska, typically with populations widely separated, however, these species are considered secure at the global level. The apparent rarity of these aquatic and wetland species statewide may be in part due to inadequate sampling of appropriate habitats (Cook and Roland 2002).

Alpine rocky outcrop and talus slope-associated rare species constitutes another large element of the rare flora of Bristol Bay. These species include: *Carex phaeocephala, Draba chamissonis, D. incerta, D. macounii, Saxifraga adscendens* ssp. *oregonensis,* and *Smelowskia pyriformis. Draba chamissonis* is a species primarily restricted to the Russian Far East and is known from just two locations in the state (between Aleknagik and Dillingham and near Naknek Lake). Some botanists suggest this plant is a form of the more widespread *D. nivalis* (Al-Shebhaz et al. 2010). *Carex phaeocephala, Draba incerta, Draba macounii,* and *Saxifraga adscendens* ssp. *oregonensis* are uncommon in the state, but are more common throughout the northern Rocky Mountains.

Other rare plants typically associated with moist to mesic, mid-elevation habitats include *Primula tschuktschorum,* a primrose of the Bering Strait region that reaches its southeastern range limit in Bristol Bay. *Plagiobothrys orientalis* and *Thalictrum minus* are species that range from the Russian Far East and Eurasia, respectively, that extend eastward through the Aleutian Islands and southwestern Alaska (Nawrocki et al. 2013). *Romanzoffia unalaschcensis* is an Alaskan endemic, restricted to moist rocky habitats in the Gulf of Alaska, which barely spills over into the Bristol Bay watershed near Aniakchak Crater. *Rumex beringensis* straddles the Russian Far East and western Alaska with 33 known Alaska sites, as well as a few occurrences in the southwestern Yukon, and is typically associated with moist, wet volcanic, or sandy substrates (Nawrocki et al. 2013).

## BRYOPHYTE AND LICHEN IN THE BRISTOL BAY WATERSHED

A total of 370 species of mosses and liverworts, and at least 160 species of lichens are recorded in the Bristol Bay watershed. Floristic surveys of bryophytes, and lichens in particular, are very limited and the true diversity of species in these groups is likely considerably higher in the Bristol Bay watershed. The majority of collections have occurred in U.S. Fish and Wildlife Refuges (notably by W. B. Schofield from University of British Columbia in collaboration with S. S. Talbot) and Lake Clark National Park and Preserve (J. Walton).

The bryophyte flora known from Bristol Bay includes 63 families and includes numerous wetland-associates (e.g., *Sphagnum* spp.), species from mesic forests and often shaded habitats (e.g., *Hylocomium splendens, Ptilium cristacastrensis,* and *Plagiogmnium* spp., Figure 3.6), and species from rocky habitats or mineral substrates (e.g., *Grimmia longirostris, Rhacomitrium canescens,* and *Bartramia pomiformis,* Figure 3.7). The most commonly collected species are *Andreaea rupestris, Dicranum scoparium, Sanionia uncinata, Sphagnum magellanicum,* and *Tetraplodon mnioides.* Eight

**Figure 3.6**   Species of *Plagiomnium* moss, from moist, shaded forests. *Photo credit: Brian Heitz.*

**Figure 3.7**    *Bartramia pomiformis* moss, from humid cliffs and shelves. *Photo credit: Brian Heitz.*

rare-to-uncommon bryophyte species are known from the region including: *Haplodontium macrocarpum, Pleuroziopsis ruthenica, Schistostega pennata, Sphagnum aongstroemii, Sphagnum balticum, Sphagnum orientale, Splachnum luteum,* and *Warnstorfia pseudostraminea* (Heitz, in prep.). Most of these species have rather spotty distributions across much of Alaska, but are associated with uncommon substrates or habitats. *Splachnum luteum* (yellow moosedung moss) is found on old moose dung, as well as soil and rotten wood; *Schistostega pennata* (goblin gold moss) is found on mineral soil, typically in shaded habitats such as caves, crevices, and upturned tree root wads; and *Haplodontium macrocarpum* (Porsild's bryum) is typically associated with wet limestone cliffs. *Haplodontium macrocarpum* is listed as a threatened species in Canada (Environment Canada 2014).

Despite a paucity of collections, the lichen flora of Bristol Bay is represented by a range of life forms and associated habitats, and includes two rare species: *Pseudocyphellaria perpetua* and *Hypogymnia pulverata* (Figure 3.8). *Pseudocyphellaria perpetua* was described recently (Miadlikowska et al. 2002) and is only known from approximately ten collections world-wide, from the Russian Far East to the Pacific Northwest. This species was collected in partial shade in a white spruce forest in Lake Clark National Park and Preserve in 2012. This is the only known collection in Alaska and it is disjunct from other North American populations in coastal Washington and Oregon by thousands of miles. *Hypogymnia pulverata* is widespread in Asia, but is only known to be from three widely separated locations in North America (Quebec, Oregon, and Alaska) (Nelson et al. 2011). In the Bristol Bay region it was collected from twigs and branches of white spruce in partial shade in Katmai National Park (Nelson et al. 2011). It should be emphasized that directed survey efforts by lichenologists elsewhere in Alaska uncovered numerous important finds, such as *Erioderma pedicellata* (Figure 3.9), an International Union for Conservation of Nature (IUCN)-listed species, previously known only from eastern Canada in North America (Nelson et al. 2009, Nelson et al. 2011); more thorough efforts in Bristol Bay are likely to reveal additional important finds.

**Figure 3.8** The rare *Hypogymnia pulverata* lichen that was found growing in the Bristol Bay watershed in Katmai National Park on white spruce branches. *Photo credit: Peter Nelson.*

**Figure 3.9** The rare and endangered *Erioderma pedicellata* lichen that was recently discovered in Alaska and found growing south of the Alaska Range, just east of the Bristol Bay watershed. *Photo credit: Peter Nelson.*

# CONSERVATION CONTEXT

While our perspective on the flora of Bristol Bay is limited by inadequate baseline data to perceive trends and gauge vulnerabilities, the region does not have the large-scale habitat conversions that are most directly linked to population declines and species loss elsewhere. Additionally, most of the rarest plant species are associated with largely inaccessible habitats or remote areas, such as high elevation scree slopes. The region is expected to face continued climate change however, which is anticipated to result in substantial shifts of dominant plant communities (see Murphy et al. 2010). Large-scale climate-driven changes are likely to result in novel species assemblages and ecological contexts that create greater uncertainly in the long-term viability of the region's vulnerable flora. Climate modeling efforts of rare plants from southwestern Alaska suggested that suitable habitats are likely to shift to the northeast, but not decline in overall spatial extent (Carlson and Cortés-Burns 2013). In general, the high topographic and climatic variability of the Bristol Bay watershed will likely buffer climatic effects, such that suitable habitats can be tracked by species over relatively short spatial scales (see Ackerly et al. 2010). One threat to the regional ecology, which may interact with both changes in land use and a warming climate, is the introduction and establishment of nonnative plant species.

Currently, 44 nonnative plant species have been documented in the Bristol Bay watershed (AKEPIC 2015, Pacific Northwest Consortium of Herbaria 2015). Common dandelion (*Taraxacum officinale*) is the most frequently encountered (890 recorded occurrences) nonnative plant, followed by the disturbance specialists: pineapple weed (*Matricaria discoidea*; 623 records), shepherd's purse (*Capsella bursa-pastoris*; 631 records), common sheep sorrel (*Rumex acetocella*; 205 records), and annual bluegrass (*Poa annua*; 158 records). Numerous records are also present for narrow-leaved hawksbeard (*Crepis tectorum*; 95 records) and fall dandelion (*Leontodon autumnalis*; 208 records), two species that are not uncommonly observed off of the immediate human footprint. Most of the nonnative species perceived to have large ecological impacts are restricted to the population centers and roadsides of Dillingham, King Salmon, Naknek, Port Alsworth, and Brooks Camp. The species with the highest invasiveness ranks (see Carlson et al. 2008) are reed canarygrass (*Phalaris arundinacea*), which is only known from two locations in Dillingham; orange hawkweed (*Hieracium aurantiacum*), which is restricted to Dillingham; European bird cherry (*Prunus padus*), known from a single site in King Salmon; Siberian peashrub (*Caragana arborescens*), known from King Salmon; and bird vetch (*Vicia cracca*), known from both King Salmon and Brooks Camp. At the landscape level, the watershed is not greatly impacted by nonnative plant species, however, the number of infestations and diversity of nonnative species present in the region continues to rise, which increases the probability of measurable impacts to native species and regional ecology. For example, invasive waterweed (*Elodea* spp.) is rapidly spreading in southcentral Alaska, potentially facilitated by floatplane traffic; it has reached high densities in some shallow lakes, posing both ecological and economic threats (Schwörer 2012). The large number of shallow lakes, ponds, and low gradient streams in the Bristol Bay watershed, coupled with the high volume of floatplane traffic pose a particular threat to the region by this species.

# RECOMMENDED RESEARCH

The Bristol Bay watershed is recognized to be a floristically rich area, despite the limited baseline data available. Targeted floristic surveys would help clarify species distributions and assist in the understanding of the importance of the area as a mixing zone of previously isolated species groups. This approach would be particularly effective when coupled with a population genetic approach. Areas that are poorly surveyed (especially state and Native lands), habitats that are underrepresented (aquatic, alpine), and taxonomic groups (especially bryophytes and lichens), should be targeted. Additionally, developing a strong baseline will be integral in gauging the responses of individual plant species, as well as vegetation communities in the face of climate change. Finer-scale case studies on population viability for species of particular conservation need (e.g., *Smelowskia pyriformis*) and a more thorough investigation into the ecology of the more at-risk species, particularly *Pseudocyphellaria perpetua* and *Hypogymnia pulverata*, is needed. This effort will be critical in identifying other regions with similar habitats that may harbor additional populations, allowing a more informed understanding of the species' rarity. Last, efforts should be underway to understand the ecological effects of invasive species establishment and to conduct a landscape-level vulnerability analysis to best target areas and resources at greatest risk (see Carlson et al. 2014, Carlson et al. 2015).

# SUMMARY

The Bristol Bay watershed is a diverse floristic region, harboring a large number of vascular plant species for the area encompassed. The floristic diversity is largely a function of high variability of habitats, climates, and natural disturbance regimes, as well as representing a zone of convergence among floras that were once largely isolated by glaciation. The region is a mixing zone of species coming from the Arctic, Aleutian, boreal, and Pacific coastal floras. While there is high species richness, there are no known plant species that are restricted strictly to this watershed. This is not unexpected, owing to its recent glacial history. One alpine species (*Smelowskia pyriformis*) is limited to mountain slopes in this watershed and extending to the western Alaska Range. In general, the rare plants of this region have few, but widely dispersed populations in the state and adjacent territories. The rare plants are well-represented in wetland habitats, alpine slopes and rocky outcrops, and mesic meadows. Survey intensity is low throughout the region, particularly outside of federally managed lands.

Mosses, liverworts, and lichen data is especially sparse. A relatively small number of byrophytes and lichen species have been collected, and most species are common in forests, wetlands, and open-mineral substrates in the state, however, two regionally rare lichen species are known from the watershed: *Pseudocyphellaria perpetua* and *Hypogymnia pulverata*. These two species are associated with white spruce forests and were found growing on twigs and branches, a habitat that is certainly not uncommon in the region. The reason for the rarity of these two lichens is not well understood.

While we know very little about the population trends, basic distributions, and ecology of the flora of the Bristol Bay watershed, we expect that most species are relatively secure despite their rarity. Anthropogenically driven habitat alteration is isolated to areas around communities and roads. The impact of climate change on the flora is difficult to predict, however, the region is expected to experience some of the largest climate change-induced effects in the state. These changes are likely to be responsible for alterations in community composition and a shift in the location of suitable habitat of most rare plants. Suitable habitats for many of the species are expected to shift northward and eastward. The high topographic complexity of the watershed will likely offer habitat refugia for many species, where suitable habitat tracking can occur over relatively fine geographic scales. The establishment and continued spread of nonnative invasive plant species is cause for concern. Most nonnative populations are centered in areas of highest human use, however, the large number of people who access more remote areas of the watershed (for example Brooks Camp, Alagnak River, etc.) may unwittingly assist in the establishment of unwanted species.

# REFERENCES

Ackerly, D.D., S.R. Loarie, W.K. Cornwell, S.B. Weiss, H. Hamilton, R. Branciforte, and N.J.B. Kraft. 2010. The geography of climate change: implications for conservation biogeography. Diversity and Distributions 16:476–487.

AKNHP 2015. Rank Definitions. Alaska Natural Heritage Program, University of Alaska Anchorage website. http://aknhp.uaa.alaska.edu/botany/rare-plant-species-information/rank-definitions/ Accessed 6/15/2015.

Al-Shehbaz, I.A., M.D. Windham, R. Elven. 2010. *Draba* Linneaus. In: Flora of North America, Vol. 7.

Cahalane, V.H. 1959. A biological survey of Katmai National Monument. Smithsonian Misc. Coll. 138 (5).

Carlson, M.L., and H. Cortés-Burns. 2013. Rare Vascular Plant Distributions in Alaska: Evaluating Patterns of Habitat Suitability in the Face of Climate Change. Conference Proceedings from Conserving Plant Biodiversity in a Changing World: A View from Northwestern North America, W. Gibble, J. Combs, and S. H. Reichard (eds.). Pp. 1–18.

Carlson, M.L., I.V. Lapina, M. Shephard, J.S. Conn, R. Densmore, P. Spencer, J. Heys, J. Riley, and J. Nielsen. 2008. Invasiveness ranking system for non-native plants of Alaska. USDA, Forest Service, Gen. Tech. Rep. R10, R10-TP-143. 218 Pp.

Carlson, M.L., and R. Lipkin. 2003. Alagnak Wild River & Katmai National Park vascular plant inventory, annual technical report. Natural Resources Tech. Report NPS/AKR/SWAN/NRTR-2003/01, National Park Service, Anchorage, AK. Website (http://science.nature.nps.gov/im/units/swan/Libraries/Reports/Inventories/CarlsonM_2003_ALAG-KATM_VascularPlant2002AnnReprt_568945.pdf).

Carlson, M.L., R. Lipkin and J.A. Michaelson. 2005. Southwest Alaska Network vascular plant inventory, summary report. Natural Resources Tech. Report NPS/AKR/SWAN/NRTR-2005/07. National Park Service, Anchorage, AK. Website: http://science.nature.nps.gov/im/units/swan/Libraries/Reports/Inventories/CarlsonM_2005_SWAN_VascularPlantSummaryReprt_642393.pdf).

Carlson, M.L., M. Aisu, E.J. Trammell, and T. Nawrocki 2014. Invasive Species. In: Trammell, E.J., M.L. McTeague, and M.L. Carlson (eds.), Yukon River Lowlands—Kuskokwim Mountains—Lime Hills Rapid Ecoregional Assessment Technical Supplement. Prepared for the U.S.

Carlson, M.L., M. Aisu, E.J. Trammell, and T. Nawrocki. 2015. Biotic Change Agents: Invasive Species. *In*: Trammell, E.J., M.L. Carlson, N. Fresco, T. Gotthardt, M.L. McTeague, and D. Vadapalli, eds. North Slope Rapid Ecoregional Assessment. Prepared for the Bureau of Land Management, U.S. Department of the Interior. Anchorage, Alaska. Pp. D1–D20.

Cook, M.B., C.A. Roland. 2002. Notable vascular plants from Alaska in Wrangell-St. Elias National Park and Preserve with comments on the floristics. Canadian Field-Naturalist 116:192–304.

Cronquist, A. 1982. Map of the floristic provinces of North America. Brittonia 34:144–145.

Environment Canada. 2014. Recovery Strategy for the Porsild's Bryum (*Haplodontium macrocarpum*) in Canada [Proposed]. Species at Risk Act Recovery Strategy Series. Environment Canada, Ottawa. v + 38 pp.

Griggs, R.F. 1936. The vegetation of the Katmai District. Ecology 17: 380–417.

Hultén, E. 1937. Outline of the History of Arctic and Boreal Biota during the Quaternary Period. Bokförlags Aktiebolaget Thule, Stockholm, Sweden.

Lipkin, R. 2002. Lake Clark National Park and Preserve vascular plant inventory, annual technical report. Natural Resources Tech. Report NPS/AKR/SWAN/NRTR-2002/01, National Park Service, Anchorage, AK. Website http://science.nature.nps.gov/im/units/swan/Libraries/Reports/Inventories/LipkinR_2002_LACL_VascularPlant2001AnnReprt_568944. pdf).

Lipkin, R. 2005. Aniakchak National Monument and Preserve vascular plant inventory, final technical report. Natural Resources Tech. Report NPS/AKR/SWAN/NRTR-2005/06, National Park Service, Anchorage, AK. Website http://science.nature.nps.gov/im/units/swan/Libraries/Reports/Inventories/LipkinR_2005

Murphy, K., F. Huettmann, N. Fresco, J. Morton. 2010. Connecting Alaska Landscapes into the Future.

Murray, D. F., and R. Lipkin. 1987. Candidate Threatened and Endangered Plants of Alaska: with Comments on Other Rare Plants. University of Alaska Museum. 76 pp.

NatureServe 2015. Rank Definitions. NatureServe Explorer website. http://explorer.natureserve.org/nsranks.htm. Accessed 6/15/2015.

Nawrocki, T., J. Fulkerson, and M.L. Carlson. 2013. Alaska Rare Plant Field Guide. Alaska Natural Heritage Program, University of Alaska Anchorage. 350 pp.

Nelson, P.R., J. Walton, and C. Roland. 2009. *Erioderma pedicellatum* (Hue) P.M. Jørg, New to the United States and western North America, discovered in Denali National Park and Preserve and Denali State Park, Alaska. Evansia 26:19–23.

Nelson, P.R., J. Walton, H. Root, T. Spribille. 2011. *Hypogymnia pulverata* (Parmeliaceae) and *Collema leptaleum* (Collemataceae), two macrolichens new to Alaska. North American Fungi 6:1–8. ISSN 1937–786X. Available at: //www.pnwfungi.org/index.php/pnwfungi/article/view/1098:. Date accessed: 25 May. 2015. doi:10.2509/naf2011.006.007.

Miadlikowska, J., B. McCune, and F. Lutzoni. 2011. *Pseudocyphellaria perpetua*, a New Lichen from Western North America. The Bryologist 105: 1–10.

Orme, C.D.L., R.G. Davies, M. Burgess, F. Eigenbrod, N. Pickup, V.A. Olson, A.J. Webster, T-S. Ding, P.C. Rasmussen, R.S. Ridgely, A.J. Stattersfield, P.M. Bennett, T.M. Blackburn, K.J. Gaston, and I.P.F. Owens .2005. Global hotspots of species richness are not congruent with endemism or threat. Nature 436: 1016–1019.

Pacific Northwest Consortium of Herbaria. 2015. University of Washington Herbarium, Burke Museum, website. http://www.pnwherbaria.org/index.php. Accessed 5/1/2015.

Schwörer, T. 2012. Decisions under uncertainty. Presentation at the annual Alaska Invasive Species Conference (CNIPM). Available at: http://www.iser.uaa.alaska.edu/Publications/presentations/2012_10_31-DecisionsUnderUncertainty.pdf. Kodiak, Alaska.

Spribille, T., S. Pérez-Ortega, T. Tønsberg, and D. Schirokauer. 2010. Lichens and lichenicolous fungi of the Klondike Gold Rush National Historic Park, Alaska, in a global biodiversity context. The Bryologist 113: 439–515.

Young, S.B, and C.H. Racine. 1978. Ecosystems of the proposed Katmai western extension, Bristol Bay lowlands, Alaska. Contributions from the Center for Northern Studies 16. 94 pp.

# 4

# NATIONAL PARKS IN BRISTOL BAY, ALASKA

**Robert A. Winfree**
**National Park Service (retired)**

## INTRODUCTION

A flight from Anchorage across the Alaska Peninsula and around Bristol Bay crosses areas of remarkable natural beauty and visual contrasts. Scorched valleys rest uneasily below still-smoking volcanoes among seemingly endless snow-capped mountains and glaciers, rich forests and tidal marshes, lakes vibrating with fish and wildlife, and rivers flowing in all directions toward seawater. Katmai National Park and Preserve (KATM) and Lake Clark National Park and Preserve (LACL) each encompass about four million acres, including several privately owned parcels within park boundaries. Aniakchak National Monument and Preserve (ANIA) encompasses about 600,000 acres. About 90% of KATM and 65% of LACL are *wilderness*, a designation that focuses attention to preserving the natural, wild, and undeveloped values in those areas (Table 4.1). In addition to these national parks, preserves, and monuments, five designated stretches of the National Wild and Scenic Rivers System (Figure 4.1) have been established in the Bristol Bay region, and the Alagnak

**Table 4.1**  Parks by the numbers

| Park unit | Size | Estimated visitation (2014) |
|---|---|---|
| Katmai | Gross acreage: 4,093,067 <br> Katmai National Park acres: 3,674,368 <br> Katmai National Preserve acres: 418,699 <br> Katmai Wilderness acres: 3,384,358 <br> Privately owned acres: 21,226 | 30,896 |
| Lake Clark | Gross acreage: 4,030,130 <br> Lake Clark National Park acres: 2,619,836 <br> Lake Clark National Preserve acres: 1,410,294 <br> Lake Clark Wilderness acres: 2,619,550 <br> Private acres: 194,072 | 16,100 |
| Aniakchak | Gross acreage: 601,294 <br> Aniakchak National Monument acres: 137,176 <br> Aniakchcak National Preserve acres: 464,118 <br> Private acres: 162 | 134 |
| Alagnak Wild River | Gross acreage: 30,665 <br> Private acres: 1,982 | |

Wilderness acreage from Wilderness.net. Land acreage from landsnet.nps.gov. Visitation figures from the National Park Service (NPS) Stats Report Viewer at https://irma.nps.gov/Stats/

Wild River (ALAG), which flows partly through KATM, was designated as a distinct unit of the National Park System. Congress created the National Wild and Scenic Rivers System to preserve certain rivers having outstanding scenic, recreational, geologic, fish and wildlife, historic, cultural or other values in their natural free-flowing condition.

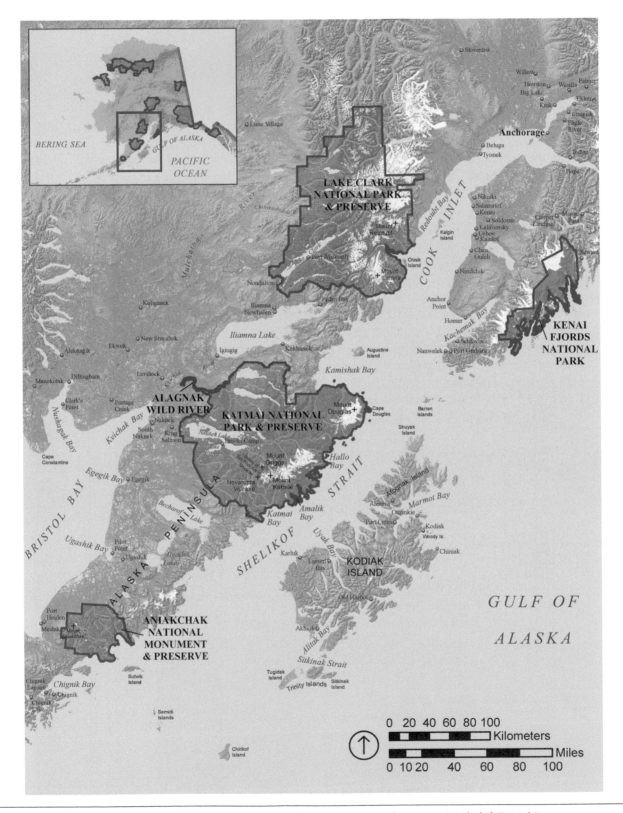

**Figure 4.1**   U.S. National Park Service (NPS) map of national parks, monuments, and preserves in Alaska's Bristol Bay area.

# KATMAI NATIONAL PARK AND PRESERVE

Katmai was initially established as a national monument by Woodrow Wilson through presidential proclamation in 1918 to protect volcanic features surrounding Mount Katmai and the Valley of Ten Thousand Smokes. Katmai's national park and wilderness designations were conferred in 1980 when Congress expanded the boundaries by adding the Katmai National Preserve. Katmai's purposes were also expanded to protect populations of brown bear and salmon, along with their habitats—and also other fish, wildlife, scenic, geological, cultural, and recreational features. Congress also ensured continued opportunities in the preserve for rural residents to continue engaging in a subsistence way of life that is consistent with fish and wildlife management in accordance with recognized scientific principles and with other purposes of the area. The primary difference between parks and preserves is that the 1980 enabling legislation (Alaska National Interest Lands Conservation Act, ANILCA) permits sport hunting in Alaska's preserves, but not in the parks.

The natural and cultural history of KATM is firmly rooted in plate tectonics and volcanology, as it lies along the northeast margin of the Ring of Fire (a region of high tectonic and volcanic activity caused by sea floor spreading and subduction that follows the approximate perimeter of the Pacific Ocean). There are at least 50 discrete volcanic vents within KATM, many of which are still active (Fierstein 2012a). The 1912 eruption of the Novarupta Volcano, which collapsed Mount Katmai (Figure 4.2), was the largest volcanic eruption on earth during the 20th century. Precipitation has collected in Mt Katmai's caldera over many years, forming a mineral-rich and visually distinctive lake. Novarupta's 1912 eruption blasted out three cubic miles of magma—three times the erupted volume from Vesuvius in 79 AD (that destroyed the ancient Roman cities of Pompeii and Herculaneum) and 30 times the volume from the eruption of Mount St. Helens in 1980 (Fierstein 2012b).

Pyroclastic deposits buried glaciers and filled streams and valleys with abrasive volcanic sediments creating a desolate wasteland of steaming cracks and fissures in an area that became known as the Valley of Ten Thousand Smokes (VTTS)

**Figure 4.2** A view inside Katmai Caldera from overhead with Mt. Griggs in the background. *Photo credit: NPS.*

(Figure 4.3). The eruption forced rapid evacuation and permanent displacement of several Native villages and exterminated wildlife within the immediate area. The losses of formerly productive salmon habitat were also evidenced by salmon catches, which declined sharply at Kodiak for several years thereafter (Schaaf 2012).

Robert Grigg's study following the 1912 eruption was financed and publicized by the National Geographic Society in several issues of their magazine and by a monograph, as well as by scores of photographers and filmmakers. Father Hubbard, a famed adventurer, explorer, Jesuit priest, and the highest-paid lecturer of his time, also made two trips to VTTS. Today, VTTS is one of the park's premier destinations—a place that provides people with an entirely unique opportunity to experience and explore, on their own terms, the aftermath of the most powerful volcanic eruption of the 20th century.

The 1912 eruption also buried sedimentary deposits of the 145 to 155 million year old Naknek Formation, which has again been exposed where the Ukak River cuts through Novarupta Volcano's pyroclastic deposits. A fossilized bone fragment recently found there in a Jurassic Age deltaic deposit suggests the potential for other dinosaurian remains to be awaiting discovery in KATM (Fiorillo et al. 2004).

Scientists with the NPS's Southwest Alaska Network (SWAN) Inventory and Monitoring are tracking a carefully selected set of physical, chemical, and biological elements and processes of park ecosystems as indicators or *vital signs* of park resource condition, stressors, and ecosystem elements having important human values. At the time of writing, SWAN has completed or is developing long-term monitoring protocols for 23 vital signs in five broad categories (Table 4.2) (Shephard 2010, NPSa). Monitoring data, along with seasonal and topical summaries, are regularly posted on the SWAN website (NPSa).

Many of Alaska's explorers, scientists, and tourists took advantage of new photographic technology to document their travels, and some of their collections have found their way into museums, archives, and library collections. Recent comparisons of historic and repeat photography at KATM demonstrate vegetation spreading from isolated areas that avoided searing volcanic impacts, and returning above areas still covered with deep ash layers. Trees that survived the initial eruption by chance or circumstance frequently experienced rapid growth for several years afterward—possibly a result of more light from canopy openings, reduced competition, and increased soil moisture retention (Miller et al. 2012). However, ecological recovery in the immediate impact zone is still limited, more than one hundred years after the eruption, probably due to a combination of physical and chemical factors (e.g., mobility, abrasiveness, sediment load, drainage patterns, nutrient limitations, toxicity) (Jorgenson et al. 2012).

**Figure 4.3**    Deep volcanic ash deposits from Novarupta's 1912 eruption still fills Katmai's VTTS. *Alaska Volcano Observatory/U.S. Geological Survey photo credit: Game McGimsey.*

**Table 4.2** National park vital signs in the Bristol Bay area

| Vital Signs Category | Vital Sign | Katmai | Lake Clark | Aniakchak | Alagnak |
|---|---|---|---|---|---|
| Freshwater Flow Systems | Surface Hydrology | X | X | X | X |
| | Freshwater Chemistry | X | X | X | X |
| | Resident Lake Fish | X | X | - | - |
| | Salmon | O | O | - | O |
| Landscape Dynamics and Terrestrial Vegetation | Glacier Extent | X | X | - | - |
| | Landscape Processes | X | X | X | X |
| | Repeat Photography | X | X | - | - |
| | Vegetation Composition and Structure | X | X | X | X |
| | Sensitive Vegetation Communities | X | X | - | - |
| | Insect Outbreaks | O | O | - | O |
| | Invasive Exotic Species | O | O | O | O |
| | Volcanic and Earthquake Activity | O | O | O | O |
| Marine Nearshore | Sea Otter | X | - | - | - |
| | Kelp & Eelgrass | X | X | - | - |
| | Marine Intertidal Invertebrates | X | X | - | - |
| | Marine Birds | X | X | - | - |
| | Black Oystercatcher | X | - | - | - |
| | Geomorphic Coastal Change | + | X | + | - |
| | Marine Water Chemistry | X | X | - | - |
| | Harbor Seal | O | O | - | - |
| Terrestrial Animals | Brown Bears | X | X | + | + |
| | Moose | X | X | - | - |
| | Caribou | O | O | - | - |
| | Bald Eagle | + | O | + | + |
| | Wolf | + | + | - | + |
| Weather, Climate & Air Quality | Weather and Climate | X | X | - | - |
| | Air Quality | X | O | O | - |
| Human Use | Consumptive Use | O | O | O | O |
| | Visitor Use | O | O | O | O |

SWAN vital sign monitoring for the selected park units are indicated by an "X". An "O" indicates that SWAN uses data from other monitoring programs. SWAN monitoring may be expanded in the future to include vital signs and park units marked with a "+". A hyphen "-" indicates a vital sign that is not monitored in the park unit indicated.

A recent remote sensing inventory by SWAN-funded scientists (Loso et al. 2014) counted nearly 300 glaciers within KATM. Several of the glaciers in the southwestern part of the park that were previously impacted by ash fallout from the 1912 eruption of Novarupta have advanced since the eruption, but most of KATM's glaciers are in retreat. The total count of glaciers has increased since the area was mapped from aerial photographs during the mid-20th century, but this larger number can be explained by the habit of large glaciers to divide into separate drainages as they grow thinner. The total area of KATM's glaciers has declined (−14%), as has their total volume (−20%), over the same time frame (Loso et al. 2014).

Archaeological deposits indicate human habitation in the KATM area for at least 7,000 years. A midden found on Mink Island within the Amalik Bay Archaeological District, off the park's eastern coast, contained layered deposits of discarded shells, bones, soil, and volcanic ash more than ten feet deep (Schaaf 2008). Contextual evidence and dating of faunal remains and artifacts, including finely crafted basalt blades, stone lamps, ground slate tools, bone needles, and pumice net floats, indicates successive waves of occupation over several millennia. Mink Island's inhabitants were well adapted to the sea. They lived in houses constructed of locally collected driftwood and used watercraft to harvest marine species, including: two species of porpoise, several seal species, sea otter, sea lion, walrus, whale, at least 28 species of shellfish, and multiple fish and bird species (Schaaf 2008). Evidence of terrestrial mammal use—such as bear, caribou, dog or wolf, and various small mammal species—is less common in this particular site (Schaaf 2008). Systematic cataloging, taxonomic identification and characterization of zoo-archaeological remains, including dating and site context, can provide unique insights into prehistoric adaptations by marine life and human users over long periods of climatic and ecological change. Despite the scientific potential of the resource, faunal remains from only about 5% of the southwest Alaska archaeological sites recorded in the Alaska Heritage Resources Survey have been analyzed (Etnier and Schaaf 2012). Mink Island, like many other coastal Alaska archaeological sites, has been severely affected by erosion. Following research and data collection, the NPS stabilized and sealed the Mink Island archaeological site with cobble-filled gabions. Resource managers regularly revisit the Amalik Bay area to monitor site condition.

About 4,500 years ago, falling water levels created two lakes—Lake Brooks and Naknek Lake—from one in the heart of KATM, with a very short stretch of river connecting them (Birkedal 1993). As former aquatic habitats turned into dry land, fish began using the Brooks River for passage between the two lakes, while animals used the isthmus and Brooks Falls, a shallow waterfall, for crossing the river. The remains of permanent human settlements found in this area date back at least 4,000 years. A complex of 19 recorded archaeological sites containing at least 900 house depressions has been discovered along the banks of the Brooks River. Stylized figures and designs found incised into rounded slate pebbles here suggest contact with Kodiak Island, where similar artifacts were discovered (Bundy et al. 2005, Vinson and Bundy 2003). Human use continued for thousands of years in the Brooks River area, with occasional temporary interruptions, until after Ruso-European contact when measles and influenza epidemics resulted in sharp contractions. The 1912 eruption of Novarupta made the area uninhabitable for several years and forced hasty and permanent abandonment of several Native villages with assistance from the U.S. government and other ships. Today, the NPS is cooperating with descendants to visit, relocate, study, and preserve the cultural history of these sites.

The Naknek Lake drainage provides critical freshwater habitat for spawning and early development of red or sockeye salmon (*Oncorhynchus nerka*), the cultural and economic foundation of the Bristol Bay region. A fishing lodge was established at Brooks Camp, where the Brooks River flows into Naknek Lake in 1950. Brown bears (*Ursus arctos*) were not common in the area then, but by the late 20th century brown bears markedly increased their presence during summer and fall salmon migration. Visitor use for fishing, bear viewing, and VTTS access also increased. KATM contains the full range of habitats required for all life stages of brown bears to survive and flourish, and it now supports the world's largest protected population of brown bears (Figure 4.4). Given naturally high fecundity and long brown bear life spans (up to 35 years), federal and state regulations allowed continued subsistence and sport hunting in Katmai National Preserve, consistent with legal and traditional practices and the preservation of healthy wildlife populations. Katmai's extensive vegetated mountain slopes provide bears with safe places to spend winter and give birth to cubs in the security of dens unaffected by human interference. Extensive coastal salt marshes provide critically important spring food supplies (plants, clams, and other marine life) before migrating salmon arrive to complete their own life cycles by spawning in streams and rivers throughout the region during summer and early fall. Several large and predictable brown bear feeding aggregations that occur along the coast also provide early summer viewing opportunities for intrepid visitors who arrive in small planes equipped with balloon tires (for landing on beaches) or floats (for landing on water) and occasionally by boats, including small cruise ships.

During the last few decades, bear viewing has become a very popular activity in Alaska and in July and September there may be no better place where large numbers of visitors can view and study large numbers of brown bears than Brooks Camp. Park employees instruct every arriving visitor on responsible bear viewing etiquette at KATM's *Bear School*. Roving park interpreters are also present throughout the day at popular viewpoints, interpreting natural and cultural resources, guiding visitors, and maintaining safe viewing distances between constantly moving anglers, photographers, and bears. The trails along the woods and waters around Brooks Camp vary somewhat in difficulty, but reaching the expertly-designed elevated walkways and bear viewing platforms requires only good walking shoes, good judgment, and a willingness and ability to use both.

In the early 1990s, the NPS engaged in planning to both manage visitor-resource interactions at Brooks Camp and mitigate impacts to wildlife and cultural sites (NPS 1996). Several elevated platforms, with connecting walkways and a floating bridge were constructed to facilitate safe visitor access and unobstructed resource use by bears (Figure 4.5). Natural erosion near the mouth of the Brooks River has exposed elements of a later phase of the archaeological complex (ca. A.D. 1450–1800) that apparently extends under facilities constructed during the second half of the 20th century.

**Figure 4.4** Brown bears gather in July and September to feed on fish collected below Brooks Falls in Katmai. *Photo credit: NPS.*

**Figure 4.5** Katmai's raised viewing platforms help to reduce surprise encounters between distracted visitors and bears. *Photo credit: NPS.*

Preservation of natural, archaeological, and ethnographic culture were all important considerations in the park's Development Concept Plan for the Brooks River area, which calls for shifting new and replacement administrative facilities to less environmentally sensitive locations across the river. NPS archaeologists and ethnographers are working closely with culturally affiliated Alaska Native groups through the Council of Katmai Descendants to develop and implement excavation plans.

About 99% of park visitation occurs during four months—June through September—when salmon are running and bears are congregating to feed along the rivers or coastline. Ease of access is one strong point for visiting Brooks Camp, where people can watch bears in their natural habitat, fly fish, canoe, and learn about volcanology, archaeology, and biology. During the summer season, the park concessioner provides regular float plane service from the community of King Salmon (where there is a commercial airport), and daily VTTS bus tours from Brooks Camp (including VTTS shuttles for backpackers). Three-quarters of all visitors are day users—visitors who arrive at Brooks Camp for several hours of bear viewing and then depart on the same day. However, there is an NPS campground and the concessionaire offers food services, along with lodge and cabin reservations for comfortable and memorable multiday visits. About 15% of the visitors to Brooks Camp stay in concession lodging and 10% utilize the park campground. KATM's wilderness areas also provide limitless opportunities for properly equipped, self-reliant, independent adventurers to find solitude and inspiration—but backcountry campers constituted less than 2% of the visitors in 2013.

# LAKE CLARK NATIONAL PARK AND PRESERVE

Today, LACL includes about 2.5 million acres of lands, submerged lands, and waters around Lake Clark that were first set aside by presidential proclamation in 1978 by President Carter. ANILCA added about 1.5 million acres in 1980. The U.S. Congress established LACL to protect the watershed necessary for perpetuation of the red or sockeye salmon fishery in Bristol Bay; to maintain unimpaired the scenic beauty and quality of portions of the Alaska and Aleutian Ranges; and to protect habitat for populations of fish and wildlife. Three of LACL's rivers—the Tlikakila, Mulchatna, and Chilikadrotna—hold National Wild River status. Congress also ensured the opportunity for rural residents to continue engaging in a subsistence way of life that is consistent with scientific principles of fish and wildlife management and the purposes of the area.

About 80% of Lake Clark's visitation occurs from June through September. Coastal use, including fishing and bear viewing, constitutes the vast majority. Dick Proenneke's legacy of self-reliance and wilderness stewardship draws about 20% of park visitors to visit his cabin site on Upper Twin Lake today (Figure 4.6). Traditional and subsistence uses include access, cabins, hunting, trapping and fishing for food, shelter, clothing, transportation, handicrafts, and trade. Sport hunting is permitted only in the preserve. Wilderness is *managed* in LACL to preserve natural, cultural, and ethnographic resources; scenic vistas; and opportunities for solitude, independent self-reliant recreation, scientific research, and education. Backcountry campers are estimated to make up less than 3% of the total visitation.

**Figure 4.6**    Dick Proenneke's wilderness cabin on Upper Twin Lake. *Photo credit: NPS.*

LACL is unique in that it is the only national park in Alaska that contains four biogeographic provinces: subarctic, boreal, maritime, and alpine. To adequately describe its landscape demands a vocabulary of superlatives and metaphors; jagged peaks uplifted at the juncture of the Alaskan and Aleutian Ranges by stupendous tectonic activity—still-steaming volcanoes flanked by hundreds of frozen alpine glaciers (Figure 4.7)—with thousands of waterfalls, sky-blue lakes, and wild free-flowing rivers filled with salmon that are returning to spawn after years in the Pacific Ocean. With vast natural landscapes, largely unaltered by human *improvements*, a full complement of wildlife native to the area, including (among others) caribou, Dall's sheep (Figure 4.8), brown/grizzly bears, bald eagles, and peregrine falcons can be found. With clean night skies devoid of artificial light, but filled with predominately natural sounds, LACL gives special meaning to the word

**Figure 4.7** Iliamna Volcano and Red Glacier. *Photo credit: NPS.*

**Figure 4.8** Dall sheep in mountains around Lake Clark. *Photo credit: NPS.*

*wilderness* for those who visit. These watersheds protect the headwaters for critical spawning and rearing habitat of the world's most productive red salmon fishery—a critical and traditional resource for the Athabascan Dena'ina people. Approximately 123 miles of LACL's borders and protects biologically productive Cook Inlet. While not immune to the effects of global climate and atmospheric processes, LACL comes as close as any place on earth to the definition of *pristine*.

Ten thousand years of vibrant, living Dena'ina culture is intricately woven through LACL's land, water, and resources. Archaeological and historic sites, as well as museum collections preserved by the park, document human activities from prehistory, through Ruso-European exploration in the 18th century, to present. This park also preserves the only known rock paintings in the U.S. that depict prehistoric whale-hunting rituals. The Kijik area, located between Kijik Lake and Lake Clark, is a documented cultural landscape; one of only three areas in Alaska conferred with both National Historic Landmark and an Archeological District status. More than a dozen archeological sites that are clustered here, collectively preserve the most complete, intact archaeological record from the previous 1,000 years of the inland Dena'ina Athabascan people. The Telaquana Trail (Figure 4.9), a historic trade route, connected families and provided access to other trade networks along Cook Inlet, as well as contact with missionaries, miners, trappers, and explorers. Repeated disease outbreaks introduced by outsiders in the early 20th century decimated Kijik village, with its Russian Orthodox mission on the scenic shore of Lake Clark. The measles outbreak during winter of 1901–1902 killed hundreds of people and ultimately caused survivors to abandon Kijik village. Today, descendants of those survivors can now be found beyond Lake Clark's outlet, living in Nondalton on the shores of Sixmile Lake.

Salmon are at the heart of this park's purpose and are central to its existence. Wild salmon provide the link between marine, freshwater, and land ecosystems. LACL protects critical headwater habitat for Bristol Bay—the world's most productive sockeye or red salmon fishery. Fresh salmon provide food that nourishes wildlife and humans, and also the marine-derived nutrients that are necessary for productive aquatic and terrestrial ecosystems. Fishing for red salmon provides the primary source of local employment and about three-quarters of subsistence users' diets. Salmon were traditionally harvested from June through August with nets, weirs, spears, and fish traps—and then preserved through slicing, drying, and smoking. Other native fish species supplemented the salmon-based diet, and were also used to feed dog teams, which were important for winter travel until recently. Historically, the most productive natural river system was the Kvichak, which includes Lake Iliamna and Lake Clark (Figure 4.10). However, the Kivchak's salmon runs declined markedly and failed to meet escapement goals for several years after 1995 (Woody 2003). When harvest restrictions were put in place, competition with commercial and sport fishing (including catch-and-release fishing) also became more of an issue (Stickman et al. 2005). Kvichack system sockeye runs have subsequently increased. Although they are still lower than many years in the last quarter of the 20th century, minimum escapement goals have been met or exceeded after 2004 (Morstad and Brazil 2012).

LACL, like KATM is part of the NPS Southwest Alaska Inventory and Monitoring Network, and monitors a similar suite of natural processes and resources (Table 4.2). A recent study (Loso et al. 2014) revealed more than 1700 individual

**Figure 4.9**    The Telaquana trail route begins at Telaquana Lake. *Photo credit: NPS.*

**Figure 4.10**   Fall on Lake Clark. *Photo credit: NPS.*

glaciers within LACL, about 200 more than were shown in mid-20th century (1956) maps. Although the authors attributed some part of the change to improved remote sensing capability for small glaciers, they noted that total glacier cover (−12%) and volume (−17%) declined over the same period. As for KATM, the rising number of small glaciers resulted mostly from the breaking of larger glaciers into smaller disconnected tributary glaciers.

## ANIAKCHAK NATIONAL MONUMENT AND PRESERVE

ANIA preserves the remote, rarely-visited, and awe-inspiring site of a massive prehistoric volcanic eruption on the Alaska Peninsula (Figure 4.11). Aniakchak was originally designated as a national natural landmark in 1967 and then as a national monument in 1978. The boundaries were expanded in 1980 through the addition of the preserve by ANILCA. ANIA's management purposes include protecting and maintaining in natural condition, Aniakchak Volcano's caldera and the landscape of associated volcanic features including lakes, rivers and streams, fish and wildlife. Study and interpretation of the park's resources and natural succession are also inherent to this area's purpose. Subsistence uses are permitted in ANIA by local residents and taking of fish and wildlife for sport purposes is permitted in the preserve. Aniakchak River, part of the National Wild River System, offers the unique opportunity of rafting free-flowing Class IV rapids from the center of a dormant Alaskan volcano to the Pacific Ocean (Figure 4.12). Recreational visits are uncommon, but well-prepared adventurers have explored ANIA by landing a float plane in Surprise Lake or by flying to a nearby village, hiking in and floating out, followed by more hiking.

Mount Aniakchak is the remains of an ancient stratovolcano that is underlain by Tertiary-age sedimentary and volcanic rocks, and sedimentary rocks dating back to the Cretaceous Period (Alaska Volcano Observatory n.d.). Similar Tertiary-age rock formations have been developed for oil and gas production in other locations in and around Cook Inlet, and there has been interest in determining the production potential of lands conveyed under the Alaska Native Claims Settlement Act.

Natural fossilized casts of footprints attributed to hadrosaurs (duck-billed dinosaurs) were found preserved in the older Cretaceous-age Chignik Formation in ANIA, along with fossilized remains of contemporaneous vegetation. Such evidence, when combined with information from other locations in Alaska, plausibly demonstrates that a land bridge between the continents of Asia and North America has existed more than once in the geologic history of Beringia (Fiorillo et al. 2004).

The eruption of Aniakchak Volcano, 3500 years ago, produced a six-mile-wide (10 km) crater (NPSb), buried large areas in deep debris flows and ash deposits, and caused a tsunami in Bristol Bay. The archaeological record of what happened to people living in the area at the time of the eruption is sparse and probably deeply buried, but Richard Vander-Hoek postulates that the resulting biological *dead zone* physically separated and culturally isolated people living in the

**Figure 4.11**    Aerial view of a volcanic vent in Aniakchak caldera. *Alaska Volcano Observatory/U.S. Geological Survey photo credit: Game McGimsey.*

**Figure 4.12**    Aniakchak Wild River offers the unique opportunity of rafting free-flowing Class IV rapids from the center of a dormant Alaskan volcano to the Pacific Ocean. Hamon et al. 2004. *Photo credit: NPS.*

western Aleutians from Bering Sea populations to the North for over 1,000 years, resulting in archaeological and linguistic differences between Aleut and Eskimo cultures (VanderHoek 2009).

Aniakchak Volcano collapsed upon itself following the eruption 3,500 years ago, creating a caldera that gradually filled with water. About 1,800 years ago, a catastrophic flood of water breached the caldera wall, cutting a channel to the Gulf of Alaska—the Aniakchak River (Hamon et al. 2004). ANIA represents a challenging—even hostile—environment for life. Volcanic eruptions have seared and buried land surfaces repeatedly over thousands of years, leaving in their wake abrasive volcanic sediments and anoxic and heavy metal-laden water drainages (Figure 4.13). Hubbard visited Aniakchak following the 1931 eruption and recorded first-hand descriptions of the event by residents of the Aleut village of Meshik 15 miles away (Ringsmuth 2007). Hubbard's report also provided graphic details of bird and wildlife mortality from the effects of the eruptions.

Surface vegetation remains sparse within the caldera, but today ANIA provides habitat for brown bears, moose, caribou, marine mammals, migratory and breeding bird species, and fish. Salmonid fish, including Dolly Varden (*Salvelinus malma*) and red salmon have successfully colonized Surprise Lake—the remnants of a large prehistoric lake that once filled the caldera (Figure 4.14). The multiyear life history and use of both fresh and marine habitats by anadromous Pacific salmon provide breeding populations some protections against extirpation by short-term catastrophic events. Genetic studies indicate Aniakchak's red salmon recolonized this area more than once, differentiated into multiple population segments, persisted through, and recovered again after the 1931 eruption (Hamon et al. 2004, Hamon and Pavey 2012).

We have much still to learn from studying the physical, biological, and cultural resources of Aniakchak and surrounding areas—about geological and biological processes and how humans and other species can recover from rapid and catastrophic environmental change. The examples already discussed suggest that what we learn at ANIA also has potential to greatly change the way we think about prehistoric, historic, and future environments in Alaska.

**Figure 4.13** An iron spring flowing into Surprise Lake. Hamon et al. 2004. *Photo credit: NPS.*

**Figure 4.14**    Surprise Lake, seen here from the rim of Aniakchak crater, is the remnant of a once much larger lake in Aniakchak's caldera. *Photo credit: NPS.*

## ALAGNAK WILD RIVER

The Alagnak River drains northwest from Katmai into Bristol Bay, via the Kvichak River. Alaska Natives lived along the Alagnak harvesting its rich salmon bounty for at least 2,300 years; cultural evidence in the Paleoarctic tradition suggests their occupation for perhaps 7,000–9,000 years (Deur et al. 2013). However, the 1919 Spanish influenza epidemic decimated populations in many Alagnak villages as it did throughout Alaska. Then during the 1960s, a combination of federal schooling requirements and other modern pressures resulted in people relocating to larger communities from traditional villages located along the Alagnak (Kedzie-Webb et al. 2006, Deur et al. 2013). In 1980, 67 miles of the Alagnak was federally protected and redesignated by ANILCA. Today, the ALAG, which is managed by KATM, is experiencing increasing, and sometimes competing uses including floating, camping, wildlife viewing, fishing, hunting, and subsistence. The free-flowing Alagnak protects habitat for five species of salmon and trophy-class rainbow trout, providing outstanding opportunities for world-class wilderness recreation. Sport fishing is supported by fly-in guide services and private lodges, while much of the land along the river's banks is privately owned and continues to be used for subsistence by residents of Levelock, Igiugig, Kokhanok, King Salmon, Naknek, and South Naknek.

## FUTURE CHALLENGES FOR NATIONAL PARKS IN THE BRISTOL BAY AREA

The NPS was created in the Organic Act of 1916 (just two years before establishment of Katmai National Monument), with an open-ended mission *to conserve the scenery and the natural and historic objects and the wild life therein and to provide for the enjoyment of the same in such manner and by such means as will leave them unimpaired for the enjoyment of future generations* (NPSc). The NPS strives to preserve the nation's natural and cultural heritage in perpetuity, but accomplishing such a lofty purpose requires hard work and eternal vigilance. Since its earliest days, the NPS has worked to protect human life and property while also preventing or mitigating negative impacts on park resources from human activities. It is not in the NPS mission (or practical ability) to prevent natural disturbance regimes from occurring (e.g., volcanic eruptions, storms, flooding, fire), but the agency does work to mitigate both human and natural impacts where feasible and appropriate. Using the previously mentioned case of KATM's Mink Island as an example, park employees sought first to preserve the ancient archaeological remains in situ, but also selectively stabilized eroding sites and implemented data recovery where neither was feasible.

Unique enabling legislation has sometimes produced differences among activities allowed in different parks. For example, in 1980 ANILCA protected rural subsistence as a traditional use among newly established and expanded sections of Alaska parks and monuments, and also provides opportunity for sport hunting in the new Alaska preserves. Subsistence harvest and gathering rights are protected by the NPS and managed cooperatively with other state and federal agencies as regulated activities in order to preserve healthy populations. ANILCA also authorized the Department of Interior to conduct mineral assessments through the Alaska Mineral Resource Assessment Program. Where there are valid preexisting claims, mining and drilling can sometimes be permitted under approved plans of operation. However, authorities that allow consumptive and extractive uses did not diminish the agency's responsibility to preserve and protect other resources and approved park uses from unacceptable impacts. Environmental compliance and permitting, informed by scientific and scholarly research, are important tools for accomplishing the parks' preservation mandates while respecting private property rights and other legal obligations.

As we have already seen, the four national park units in the Bristol Bay area (KATM, LACL, ANIA, and ALAG) exist in a dynamic environment of geological, environmental, cultural, and economic change. Winfree and Marcy (2005) interviewed 125 NPS managers and scientists to identify pressing science information needs, challenges, and opportunities in Alaska's national parks. Although their list of challenges is now over ten years old (Table 4.3), it remains pertinent to national parks in the Bristol Bay region and across Alaska: climate change; global and local contaminants; exotic species; increasing human use; and development within and surrounding the parks. Each of those five challenges relates in some way to environmental disturbance, and several of the challenges are interconnected (e.g., increasing human use and exotic species, development, and contaminants).

**Table 4.3** Resource management challenges identified by National Park Service managers and scientists in Alaska (Winfree and Marcy 2005)

| | |
|---|---|
| **Climate change** | Climate change is changing habitats, use of areas, accessibility, biotic communities, diseases, and causing other effects that will change the characteristics of parks as well as the type of management action required to maintain park values and mission. |
| **Global and local contaminants** | Long-range atmospheric and oceanic transport is bringing contaminants to parks with potential direct impacts on the viability of park resources, and the value of subsistence harvest. Local contaminants are being introduced through development of natural resources (e.g., mining) and use of park resources with industrial based transportation and activities (e.g., development, ATVs, boats, vehicles, hover craft). |
| **Exotic species** | Coupled with climate change and increasing use of parks, exotic species are increasingly transported to, and able to thrive, in areas where they did not exist previously. This has significant impacts on natural communities. |
| **Increasing human use** | As human population continues to expand exponentially and Alaska parks become an increasing target of visitor enjoyment and subsistence and hunter use, potential impacts on natural and cultural resources increase. |
| **Development within and surrounding parks** | Park ecosystems are directly linked to surrounding areas around park boundaries. Fragmentation, contamination, loss of habitat, and overuse are likely to increase. |

Climate change is certainly not unique to Alaska's national parks, but neither are parks immune to its effects. The rate and magnitude of climate change during the coming decades will be influenced by future greenhouse gas emissions and other uncertainties. However, some of the effects of climate change—such as warmer air temperatures, wasting glaciers, longer ice-free seasons, thawing permafrost, and changing tree lines—are already becoming evident in Alaska. Permafrost thaw, reduced sea ice coverage, and changing weather patterns are making historic and prehistoric cultural resources more vulnerable to loss across the north. Rapid coastal erosion, incomplete baseline inventories, and accessibility can make coastal archaeological sites especially challenging to protect. The NPS recently worked with the University of Alaska's Scenarios Network for Alaska and Arctic Planning, local area residents, and others to identify a range of hypothetical but plausible and relevant climate change scenarios, and to assess their implications for national parks and surrounding areas across Alaska, including parks in the Bristol Bay area (Winfree et al. 2014). Impacts to wildlife and fisheries habitat were deemed likely across a wide range of scenarios, including changes to the water cycle (e.g., more rain or snow events, changing glacial runoff), forests (e.g., drought, pests, fire), and marine ecosystems (e.g., acidification, fresh water, sediment runoff). Scenario planning is not prescriptive, but it does provide park managers and others with useful points of reference for assessing proposed projects, priorities, and activities through the windows of a changing world or *rehearsing the future*.

Alaska's national parks have coexisted for many years with nearby resource development, including mining, oil, and gas—but not always without negative effects. The 1989 Exxon Valdez Oil Spill (EVOS) is a prime example of the hazards associated with resource development. Although the oil tanker ran aground in Prince William Sound, the resulting spill spread widely across the water and with disastrous consequences for fish, wildlife, and coastal communities. Empirical observations and recent surveys document that—25 years later—oil from that massive spill is still lingering in and outside of Prince William Sound, including along the Katmai coast over 450 miles away (http://www.evostc.state.ak.us/index.cfm?FA=status.lingering). All it takes to still find oil in some of the spill-impacted areas is to flip over a beach cobble or drive a shovel into sediments below high tide. It is important to recognize the EVOS oil was neither produced nor spilled locally in Cook Inlet. It was pumped from Alaska's North Slope through the Trans-Alaska Pipeline System to the port at Valdez for storage and loading into tankers. As noted above, oil and gas have also been produced in and around Cook Inlet, very close to these parks, for nearly 50 years—and again not entirely without risk or incidents.

At least two of LACL's volcanoes show recent activity—Mt. Iliamna with its recurrent vapor emissions and Mt. Redoubt with no less than four recorded eruptions since 1900. Volcanic activity also affects the glaciers on and around Mt. Redoubt. Geologic research has discovered Holocene lahar deposits in the Crescent and Drift River drainages (Figure 4.15), along the Cook Inlet coastline between LACL and KATM. The Drift River Terminal, an 80-million-gallon capacity crude oil storage facility (seven times the spilled volume of EVOS), is located outside park boundaries below Mt. Redoubt. Protective dikes were raised and strengthened to mitigate against potential recurrences after the tank farm was flooded in 1989–1990, when eruptions from Redoubt Volcano triggered lahars in the Drift River drainage. When Redoubt threatened to erupt again in 2009, much of the oil was removed by tanker and most of the facility's employees evacuated before another lahar struck the dikes. There was extensive flooding, but the strengthened facilities prevented loss of life and the potential threat of another massive oil spill affecting Cook Inlet, Shelikof Strait, and the Gulf of Alaska was avoided. On-site oil storage was subsequently suspended for several years, and storage capacity was reduced by decommissioning several tanks. In 2012, new owners announced plans to reopen the tank farm. Cook Inlet oil and gas production is widely recognized in Alaska as a regionally important energy source, but after the events of 2009, suggestions were also put forward for building an undersea pipeline to the Kenai Peninsula to reduce the risk of an oil spill triggered by future volcanic eruptions. About 185,000 acres of ANIA, including subsurface oil and gas rights, have also been selected for conveyance to nonfederal ownership, under provisions of the Alaska Native Claims Settlement Act. Aniakchak Volcano, previously discussed relative to its eruptive history, has been dormant for about 80 years.

Around 2006, new prospects surfaced for large-scale copper/gold mining on the Alaska Peninsula. Support for new local employment opportunities and income—and concerns about compatibility of industrial scale mining in an area of intense tectonic and volcanic activity with traditionally important resources, values, and livelihoods—raised many questions about the magnitude of development and associated infrastructure (mine, processing plant, tailings pit, road, harbor, power generation/transmission). Although detailed plans had not been released nor were required permits applied for at the time of writing, there is still considerable debate about whether large-scale mining can be accomplished safely in and above anadromous fish habitat—and whether potential benefits outweigh potential costs. There appears to be little, if any, argument that any development as large as what was envisioned would permanently affect the character of the area in many ways that are not yet fully understood.

**Figure 4.15** Mt. Redoubt and lahar deposits in the upper Drift River Valley about two months after the July 2009 eruption. *Alaska Volcano Observatory/U.S. Geological Survey photo credit: Game McGimsey.*

## ACKNOWLEDGEMENTS

The author gratefully acknowledges the contributions of many park employees, scientists, and scholars who provided information through the Foundation Statements for Katmai and Alagnak (KATM 2009), Lake Clark (LACL 2009), and Aniakchak (ANIA 2009) and numerous articles contributed to the Alaska Park Science journal. Drafts were also peer reviewed by each park.

## REFERENCES

Alaska Volcano Observatory. Aniakchak peak description and information. Available at: http://www.avo.alaska.edu/volcanoes/volcinfo .php?volcname=Aniakchak (March 2015).

Aniakchak National Monument and Preserve (ANIA). 2009. Aniakchak National Monument and Preserve. Foundation Statement. National Park Service. U.S. Department of the Interior.

Birkedal, T. 1993. Ancient hunters in the Alaskan wilderness: Human predators and their role and effect on wildlife populations and the implications for resource management. Proceedings of the 7th conference on Research and Resource management in Parks and on Public Lands. George Wright Society. Hancock, Michigan. pp. 228–234.

Bundy, B.E., D.M. Vinson, and D.E. Dumond. 2005. Brooks River cutbank. An archeological data recovery project in Katmai National Park. University of Oregon Anthropological Papers No 64. Museum of Natural and Cultural History and Department of Anthropology. University of Oregon.

Deur, D., K. Evanoff, A. Hermann and A. Salmon. 2013. Collaborative research to assess visitor impacts on Alaska Native practices along Alagnak Wild River. Alaska Park Science. 12(1):32–37.

Etnier, M.A. and J. Schaaf. 2012. Using archeofaunas from Southwest Alaska to understand climate change. Alaska Park Science. 11(2):20–25.

Fierstein, J. 2012a. Katmai National Park volcanoes. Alaska Park Science 11(1):14–21.

Fierstein, J. 2012b. The great eruption of 1912. Alaska Park Science 11(1):6–13.

Fiorillo, A.R., R. Kucinski, and T R. Hamon. 2004. New frontiers, old fossils: Recent dinosaur discoveries in Alaska national parks. Alaska Park Science. 3(2):4–9.

Hamon, T.R., S.A. Pavey, J.L. Miller, and J.L. Nielsen. 2004. Aniakchak sockeye salmon in investigations. Alaska Park Science. 3(2): 35–37.

Hamon, T. and S. Pavey. 2012. Salmon in a volcanic landscape: How salmon survive and thrive on the Alaska Peninsula. Alaska Park Science. 11(2):16–19.

Jorgenson, M.T., G.V. Frost, A.J. Bennett, and A.E. Miller. 2012. Ecological recovery after the 1912 Katmai eruption as documented through repeat photography. Alaska Park Science. 11(1):67–73.

Kedzie-Webb, S., J. Schaaf, M.O. Crow, and J. Branson. 2006. Alagnak Wild River of the Alagnak Wild River, An Illustrated Guide to the Cultural history of the Alagnak Wild River. Available at: http://www.nps.gov/alag/learn/historyculture/upload/Alagnak%20Final_1-30.pdf (March 2015).

Katmai National Park and Preserve (KATM). 2009. Katmai National Park and Preserve. Foundation statement. National Park Service. U. S. Department of the Interior.

Lake Clark National Park and Preserve (LACL). 2009. Lake Clark National Park and Preserve. Foundation Statement. National Park Service. U. S. Department of the Interior.

Loso, M., A. Arendt, C. Larsen, J. Rich and N. Murphy. 2014. Alaskan national park glaciers – status and trends. Final report. Natural Resource Technical Report NPS/AKRO/NRTR-2014-922.Kedzie-Webb, S., J. Schaaf, M. O. Crow, and J. Branson. 2006. Alagnak Wild River: An illustrated guide to the cultural history of the Alagnak Wild River. Available at: http://www.nps.gov/alag/historyculture/upload/Alagnak%20Final_1-30.pdf (March 2015)

Miller, A.E., R.L. Sherriff, and E.E. Berg. 2012. Effect of the Novarupta (1912) eruption on forests of southcentral Alaska: Clues from the tree ring record. Alaska Park Science. 11(1):75–77.

Morstad, S. and C.E. Brazil. 2012. Kvichak River sockeye salmon stock status and action plan, 2012, a report to the Alaska Board of Fisheries. Alaska Department of Fish and Game. Divisions of Sport Fish and Commercial Fisheries. Juneau, Alaska. 19 pp.

National Park Service (a). Inventory & Monitoring (I&M), Southwest Alaska Network, Monitoring. Available at: http://science.nature.nps.gov/im/units/swan/monitor/index.cfm (February 2015).

National Park Service (c). Explore geology. Available at: http://www.nature.nps.gov/geology/parks/ania/ (April 2015).

National Park Service (b). History. Available at: http://www.nps.gov/aboutus/history.htm (February 2015).

National Park Service. 1996. Brooks River area, Katmai National Park and Preserve, Alaska final development concept plan, environmental impact statement (SuDoc I 1.98: B 79/3/FINAL).

Ringsmuth, K.J. 2007. Beyond the moon crater myth. A new history of the Aniakchak landscape. A historic resource study for Aniakchak National Monument and Preserve. U.S. Department of the Interior. Research/Resources Management Report AR/CRR-2207-63. Available at: http://www.nps.gov/parkhistory/online_books/ania/hrs/contents.htm (March 2015).

Schaaf, J M. 2008. Mink Island, Katmai National Park and Preserve, Pacific Coast of the Alaska Peninsula. Alaska Park Science. 7(2):34–39.

Schaaf, J. 2012. Witness: Firsthand accounts of the largest volcanic eruption in the Twentieth Century. Alaska Park Science. 11(1):53–59

Shephard, M. 2010. The Southwest Alaska Network. Alaska Park Science. 9(1):42–43.

Stickman, K., A. Balluta, M. McBurney, D. Young, and K. Gaul. 2005. K'ezdlagh: Nondalton traditional ecological knowledge of freshwater fish. Alaska Park Science. 4(2):27–31.

VanderHoek, R. 2009. The role of ecological barriers in the development of cultural boundaries during the later Holocene of the Central Alaska Peninsula. Ph.D. dissertation. University of Illinois, Urbana-Champaign. 413 pp.

Vinson, D.M. and B.E. Bundy. 2003. Working on the edge: the Brooks River cutbank archeological data recovery project. Alaska Park Science. Winter 2003.38–39.

Winfree, R.A. and S.K. M. Marcy. 2005. Science in Alaska national parks: Challenges and opportunities for the 21st century. Alaska Park Science. 4(2):47–51.

Winfree, R., B. Rice, N. Fresco, L. Krutikov, J. Morris, D. Callaway, D. Weeks, J. Mow, and N. Swanton. 2014. Climate change scenario planning for southwest Alaska parks: Aniakchak National Monument and Preserve, Kenai Fjords National Park, Lake Clark National Park and Preserve, Katmai National Park and Preserve, and Alagnak Wild River. Natural Resource Report NPS/AKSO/NRR—2014/832. National Park Service, Fort Collins, Colorado. Available at: http://www.nps.gov/akso/nature/climate/south-west.cfm . (February 2015).

Woody, C.A. 2003. Unlocking the secrets of Lake Clark sockeye salmon. Alaska Park Science. Summer 2003:32–37.

# 5

# NATIONAL WILDLIFE REFUGES
# OF BRISTOL BAY

Stephanie Kuhns, Bureau of Land Management
Joel H. Reynolds, Western Alaska Landscape Conservation Cooperative

## INTRODUCTION

The four National Wildlife Refuges (NWRs) in the Bristol Bay region are a testament to the richness, diversity, and importance of the region's fish, plants, and wildlife resources (Figure 5.1). These refuges—Togiak, Becharof, Alaska Peninsula, and Izembek—are among the 560+ units of the National Wildlife Refuge System (NWRS)—a national network of lands managed by the United States Fish and Wildlife Service (USFWS), part of the Department of the Interior, "for the conservation, management, and where appropriate, restoration of the fish, wildlife, and plant resources and their habitats within the United States for the benefit of present and future generations of Americans" [National Wildlife Refuge System Improvement Act (NWRSIA) 1997].

President Theodore Roosevelt established the first NWR in 1903: Pelican Island Federal Bird Reservation in Florida (Woods 2003). The current boundaries of the four NWRs in Bristol Bay emerged from the Alaska National Interest Lands Conservation Act of 1980 (ANILCA). The legislation redesignated Togiak and Izembek NWRs from existing conservation units, and in the process, greatly expanded the area now designated Togiak NWR, and created Becharof and Alaska Peninsula NWRs. These four NWRs constitute over 12% of the current total area of the sixteen NWRs in Alaska (Woods 2003) and approximately 6% of the total acreage of the NWRS (USFWS 2013).

While each NWR in Alaska has its own mandates set forth in ANILCA, all of them share four broad purposes (paraphrased): (i) to conserve fish and wildlife populations and habitats in their natural diversity; (ii) to fulfill international treaty obligations with respect to fish and wildlife and their habitats; (iii) to provide the opportunity for continued subsistence uses by local residents; and (iv) to ensure water quality and quantity within the refuge (ANILCA 1980). The legislation further clarifies that purposes (iii) and (iv) are subordinate to purposes (i) and (ii). The legislative language also identifies specific species or groups of animals in each refuge's first purpose. For the four refuges in the Bristol Bay region, all include mention of salmon or salmonids and migratory birds, all but Izembek include mention of marine mammals, and all but Togiak include brown bears—Togiak's refers to *large mammals* (Table 5.1).

Across the nation, the NWRs attracts over 40 million visitors each year who engage in wildlife viewing, hiking, photography, hunting, fishing, camping, kayaking or canoeing, and environmental education (USFWS 2007). In the Bristol Bay region, the four refuges are important in terms of both the opportunity they provide to harvest subsistence resources (many of the region's communities are on or near a refuge) and their contributions to the regional, state, and national economies. An economic analysis in 1997 showed that the NWRs in Bristol Bay generated approximately 3,200 jobs and $127,000,000 in personal income *in Alaska* (Goldsmith et al. 1998). These results are heavily influenced by the commercial fisheries that target the salmon that are born, nurtured, and later return to spawn in these conservation units (ibid), with important additional contributions from sport hunters, anglers, and recreational users. Subsistence activities on the refuges constituted approximately two-thirds of their *net use* value; the *existence value* of the refuges, however, dwarfed this in contribution to their estimated total *net economic value* in the range of $2.3 to 4.6 billion (ibid).

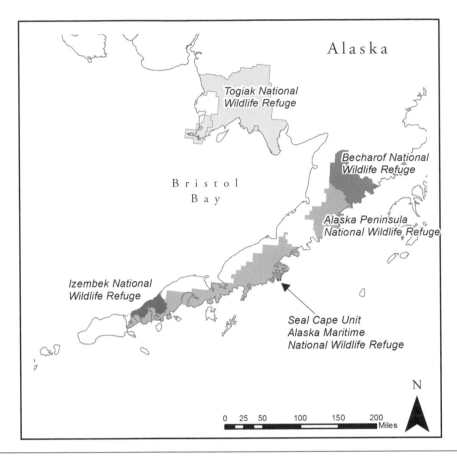

**Figure 5.1**    In total, the four NWRs in the Bristol Bay region—Togiak, Becharof, Alaska Peninsula, and Izembek—include over 3,723,100 ha (9,200,000 acres) of lands and waters for the conservation and management of the fish, wildlife, and plant resources and their habitats for the benefit of present and future generations of Americans. The 3,966 ha (9,800 acre) Seal Cape unit of the Alaska Maritime NWR, a headland south of Chignik, is managed by the Alaska Peninsula NWR. *Map credit: Ben Matheson, U.S. Fish and Wildlife Service (USFWS).*

**Table 5.1**    Fish and wildlife resources specifically identified in each refuge's establishment language in ANILCA. Each refuge's mandates include: (i) To conserve fish and wildlife populations and habitats in their natural diversity including, but not limited to, . . .

| Refuge | Fish and wildlife resources specifically identified in ANILCA |
|---|---|
| **Togiak** | salmonids, marine birds and mammals, migratory birds and large mammals (including their restoration to historic levels) |
| **Becharof** | brown bears, salmon, migratory birds, the Alaska Peninsula caribou herd and marine birds and mammals |
| **Alaska Peninsula** | brown bears, the Alaska Peninsula caribou herd, moose, sea otters and other marine mammals, shorebirds and other migratory birds, raptors, including bald eagles and peregrine falcons, and salmonids and other fish |
| **Izembek** | waterfowl, shorebirds and other migratory birds, brown bears, and salmonids |

     This chapter contains an overview of each of the four refuges, including select highlights of recent research and, where available, information on recent visitorship. While the refuges contain largely intact ecosystems with minimal development, relative to the NWRS units outside of Alaska, challenges remain. Foremost among the challenges is the task of managing these rich natural resources in the face of impacts of projected changes in the region's climate (Walsh 2012).

# TOGIAK

Togiak National Wildlife Refuge lies in the northwest portion of Bristol Bay (Figure 5.1). The area that is now incorporated as Togiak Refuge was originally public land managed by the Bureau of Land Management. In 1969, a portion of the land was designated as Cape Newenham National Wildlife Refuge. On February 11, 1980, the area was designated as Togiak National Wildlife Refuge under Section 204(c) of the Federal Lands Policy Management Act of 1976. In 1980—under ANILCA—the protected lands were expanded from 107,241 hectares to 1,902,022 hectares (265,000 acres to 4.7 million acres). The northern 930,777 hectares (2.3 million acres) on the refuge are designated wilderness (55% of the refuge area), the second largest contiguous wilderness within the National Wildlife Refuge System (USFWS 2013). Togiak NWR was established, in part, because of the area's importance to salmonids, marine birds and mammals, migratory birds, and large mammals (Table 5.1). Togiak is important economically for both commercial and subsistence fishing, as the region supports large salmon runs and has excellent habitat for rearing young salmon prior to migrating to the ocean. The refuge also offers outstanding opportunities for sport fishing, river floating trips, camping, and hunting. Nearby Wood-Tikchik State Park, managed by the state of Alaska, was established in 1978 so as to extend the protections for the important fish and wildlife resources. At 647,497 hectares (1.6 million acres), it is the largest state park in the United States.

## Landscape and Habitat

The landscape varies from rugged mountains to glacial valleys, sheer cliffs to sandy beaches—providing habitat for a plethora of species and inhabitants. Lakes, rivers, and streams provide excellent environments for a variety of fish species. The surrounding wetlands and marshes also provide key habitat and feeding grounds for many species of birds and small mammals. There are more than 500 plants or plant-like species (lichens are included in this list, but are actually a fungus and algae living together in a symbiotic relationship, rather than a plant) that have been identified within Togiak including eelgrass (in the lagoons), flowers, berry bushes, spruce, and cottonwood (USFWS 2009). The mid-elevations of Togiak NWR include forests of white spruce (*Picea glauca*), birch, and alder (Viereck and Little, Jr. 2007). Many of these plants serve as food for the various animals in the refuge.

## Geology

Togiak has a rich geologic history that includes volcanoes, glaciers, different types of erosion, and more. The features seen in the refuge today reflect the geologic functions of the past. Unlike the other Bristol Bay refuges, no active volcanoes are found within the boundaries of Togiak. There are, however, extinct volcanoes that created unique features that can still be seen. During a glacial period roughly 39,000 years ago, volcanic eruptions occurred under the ice, creating a steep, flat-topped volcano that forms when lava erupts under a sheet of ice, known as a *tuya* (Alaska Volcano Observatory 2014a). The Togiak tuya has an area of approximately 15 square kilometers (6 kilometers long by 2.5 kilometers wide), and its elongated shape indicates that it was formed in an actively flowing glacier. Togiak tuya is located 24 kilometers (15 miles) northeast of the village of Twin Hills, and is the only formation of its kind in Alaska.

Other unique features include hanging valleys and elongated lakes that were formed by ancient glacial movements carving deep gouges into the surrounding bedrock. The refuge has more than 500 lakes that are larger than 10 hectares (25 acres) in size, and many are more than 120 meters in depth.

Unfortunately, many of the glaciers are disappearing at a rapid pace. Recent research has shown that 10 of the 109 glaciers in the Akhlun Mountains (located in the northeast corner of the refuge) have disappeared completely, and the sizes of the remaining glaciers have decreased due to melting (Walsh et al. 2015). The water from these glaciers feeds rivers that provide important habitat for juvenile salmon.

## Human History

Archaeological evidence suggests that Togiak was inhabited by humans beginning nearly 2000 years ago. Some of the sites of original habitation have been continuously occupied, and are now the sites of present day villages (Alaska Native Heritage Center 2011a). Historically, there were three different groups of people living within the Togiak range in the areas of Nanvak Bay east to Cape Constantine, in the Nushagak Bay region, and from the Kuskokwim River south to Chagvan Bay.

The first known contact between historical peoples and Europeans was 1778, when Captain James Cook's expedition moored at Cape Newenham on July 16. Cook believed that he and his crew were the first outside contact the Natives had, as they did not have any foreign objects in their possession [Department of the Interior (a) n.d.]. There are a number of formerly occupied historic village sites on the refuge, including Kashiagamiut on the Togiak River and Kulukak on Kulukak Bay.

Today, the Native Alaskans in this region are collectively known as the Yup'ik and Cup'ik peoples, named after the two main dialects of the region (Alaska Native Heritage Center 2011a). Many of the residents maintain a balance of subsistence and modern lifestyles by hunting, fishing, and gathering local berries and other plants based on their seasonal availability, supplemented by store-bought items. Unfortunately, the changing arctic climate and the high costs of shipping goods to such remote locations are threatening this way of life.

An estimated 53,000 visitors came to Togiak Refuge in 2015 (Susana Henry, pers. comm., Togiak National Wildlife Refuge, Sept. 29, 2016). Togiak is very popular for guided sportfishing trips by motorboat or by raft, as well as for hunting moose, caribou, and brown bear. Residents of the nearby villages of Togiak, Manokotak, Quinhagak, Platinum, and Goodnews Bay are among the most common visitors. Among those people who visit for sportfishing, there are always a number of individuals from Northern Europe.

## Wildlife

The size of the refuge and the diverse habitats found there allow for a rich selection of wildlife to thrive. Including migratory birds, 201 avian species have been identified in Togiak, while an additional 15 species have been identified in the nearby town of Dillingham [Department of the Interior (a) n.d.]. Though there are far fewer species to be seen in the winter, some birds—such as bald eagles, boreal owls, willow ptarmigan, ravens, chickadees, and redpolls—are found year-round. In the spring and fall, hundreds of thousands of migratory birds, such as black brant and king eiders, can be found eating in the lagoons. During the summer, dozens of species of gulls, auklets, and other seabirds can be found along Togiak's extensive shoreline, especially at Cape Newenham and Cape Peirce.

Visitors to Togiak are less likely to see as many terrestrial animals as birds; not for lack of diversity, but for their elusive habits. Brown bears and wolves are top predators in Togiak, and are sustained by caribou, moose, and salmon, along with berries and other plants. Brown bears are common throughout the refuge and can often be seen near rivers and streams in conjunction with the salmon runs, while wolves are common in the northern region of Togiak, and not often seen by humans. Other common animals in the refuge include foxes (both red and arctic); river otters; several rodent species such as porcupines, hoary marmots, and beavers; little brown bats; mink; and two species of weasel.

The waters of Togiak also support an array of animal life. A total of 33 species of fish have been identified, among those being char, trout, whitefish, and salmon [Department of the Interior (a) n.d.]. The salmon fishery in Bristol Bay is unparalleled in its size and economic importance to Alaska (Goldsmith et al. 1998), and the sockeye salmon run in Togiak is especially plentiful. Salmon are natal spawners, meaning that they return to the freshwater stream where they hatched to lay their eggs and set the foundations for a new generation of fish. Each year, hundreds of thousands of fish swim upstream from the waters of Bristol Bay and the Bering Sea toward the streams where they first hatched, along the way providing food for bears, eagles, wolves, humans, and more. It is truly a sight to behold.

Outside of the freshwater habitats, 17 species of marine mammals have been identified in the waters of Bristol Bay in and around Togiak [Department of the Interior (a) n.d.]. Walrus can be seen resting on the shore during the ice-free months in particular locations (Figure 5.2). The locations that marine mammals such as walrus and sea lions use to rest are called *haulouts*, and it is not uncommon for thousands of males to haulout on the shores at the same time. Haulout locations may be on land or, during winter months, on sea ice. Walrus have been an important food source to Alaska Native people of the Bristol Bay and Arctic regions for generations, and also often fall prey to the killer whales of the region.

Other marine mammals common to Togiak include Steller sea lions, which since 1990 have been listed as *Threatened* under the Endangered Species Act. Steller sea lions have a varied diet, but it is most often comprised of such fish as walleye Pollock, Pacific herring, Pacific cod, and salmon (NOAA Fisheries n.d.). Fishermen often view sea lions as competition for precious resources, but a healthy population of predators (such as sea lions) indicates a thriving ecosystem.

The types of whales most often found in the region include gray whales, killer whales, and belugas. Gray whales are often found off the coast of Togiak while on their annual migrations to and from Arctic feeding areas to Southern calving areas. Killer whales can be found roaming the region in social groups known as *pods* that may include up to 30 individuals [Department of the Interior (a) n.d.]. These carnivorous whales have a diverse diet, ranging from fish and seabirds to seals, walrus, and other whales. Beluga whales are pure white in adulthood, but may be slate gray to pinkish-brown when

**Figure 5.2** A group of pacific walrus (*Odobenus rosmarus divergens*) rests on the coast of Cape Peirce, Togiak NWR. While Togiak NWR was once known for walrus haulouts, with thousands of walrus, currently haulouts on the refuge are more erratic and much smaller—with walrus numbers in the low 100s. Large haulouts are now beginning to occur at locations near Cape Greig on the Alaska Peninsula. *USFWS photo credit: Dave Keuhn.*

young. These whales also travel in pods of 2–25 individuals, though the average size is 10 (Sea World 2015). Native Alaskans hunt them for their meat, blubber, and oil; predation by other animals such as killer whales is rare, but not uncommon. Unlike other whales, belugas have flexible necks, which may help them as they forage for small bottom-dwelling prey such as crabs, worms, flounder, and octopus.

## Recent Research Findings

The refuge biological staff, in conjunction with the Alaska Department of Fish and Game (ADF&G), monitors habitat use and demography of two herds of caribou—the Nushagak Peninsula and the Mulchatna. In recent years, the Nushagak Peninsula caribou herd numbers have been estimated at over 1200 animals, while the Mulchatna herd has declined from highs of 200,000 in the 1990s to around 20,000 today. Caribou populations are known to fluctuate. The refuge staff also works cooperatively to monitor moose habitat use and demography. Moose populations on the refuge have been found to be gradually increasing. Long-term monitoring of stream, lake, and river water temperature takes place at over 20 locations. Water quantity is measured on two refuge rivers, while water quality is measured at the Salmon River downstream of historic and recently active platinum mines. There is an ongoing predation study of wolves, bears, and moose. Seabird nesting is monitored annually at Cape Peirce. In the summer of 2016, there were few common murre nests and even fewer successful nests, and the same could be said of black-legged kittiwakes and pelagic cormorants. The widespread nest failures (along with failure to nest at all) may be due to a lack of food fish for these birds—and that may be due to warming seawater temperatures (Doherty 2016).

# BECHAROF NATIONAL WILDLIFE REFUGE

Becharof National Wildlife Refuge was established as a National Wildlife Monument in 1978 by President Jimmy Carter. It was later designated as an NWR under ANILCA. One third of the refuge's 485,792 hectares (1,200,419 acres) are designated wilderness (161,874 hectares/400,000 acres). Becharof NWR was established, in part, because of the region's abundant brown bears, salmon, migratory birds, the Alaska Peninsula caribou herd, and marine birds and mammals (Table 5.1). The refuge offers visitors an array of recreational activities, including spectacular wildlife viewing, camping, hunting, environmental education, and more.

## Landscape and Habitat

The diverse geography and habitats of Becharof NWR offer visitors many stunningly scenic views. On the Pacific coast side, steep cliffs tower above sandy beaches, providing prime nesting grounds for many species of seabirds, including kittiwakes, gulls, and terns. Traveling west into the refuge, the glaciated mountains of the Aleutian Range rise 1,500 meters (4,835 feet) high above sea level. Within the Range and overlooking Becharof Lake is Mount Peulik—a stratovolcano still active today (Figure 5.3). Stratovolcanoes have very steep profiles, and are characterized by explosive eruptions with viscous lava with high silica content.

Covering approximately 121,000 hectares (300,000 acres), glacial Becharof Lake is the largest lake in the NWRS. It is the second largest lake in the state of Alaska—Lake Illiamna is more than double the size, covering just over 263,045 hectares (650,000 acres). Its depth reaches 183 meters (600 feet). It is fed by two major rivers, Egegik and King Salmon, and numerous tributaries (Alaska Department of Natural Resources 2005).

A large portion of the refuge is crisscrossed by streams, dotted with ponds and lakes, and covered in tundra wetlands. Caribou, moose, marmots, ground squirrels, and wolves are all common in these areas. Seen less frequently, but not considered uncommon, are wolverines and red foxes. The region also serves as an important stop for migratory birds such as geese and swans as they travel north in the spring to nesting grounds, and south in the fall to overwintering areas.

**Figure 5.3**    The active stratovolcano Mount Peulik rises above the windswept landscape of Becharof NWR. *Photo credit: USFWS.*

## Geology

Mount Peulik and the nearby geologic features provide geologists and other visitors alike with exciting opportunities for research, photography, and more. Standing at 1,474 meters (4,835 feet) tall, Mount Peulik is a stratovolcano with eruptions recorded in 1814 and 1845. Of the fourteen volcanos on Alaska Peninsula and Becharof Refuges, nine have erupted in the last century. Among the more recent geologic activity was the 1977 formation of the Ukinrek Maars over a ten-day period. Maars are volcanic craters formed by underground volcanic explosions when magma contacts groundwater; the extreme temperature change often results in a violent steam explosion. Maars typically fill with the groundwater that helped to form them, forming shallow lakes. There are many maars across Alaska, with the largest located on the Seward Peninsula.

Within Becharof Lake are the Gas Rocks—geologic features that emit a steady stream of carbon dioxide. Nearby the Gas Rocks, on the south shore of the lake, are two thermal hot springs. West Ukinek Spring has a temperature of about 178°F, and Gas Rocks Hot Spring has a temperature of about 127°F. Both are too far from inhabited locations to be easily utilized (Alaska Department of Natural Resources 2005).

## Human History

Evidence of a rich natural history has been found in the fossilized remains of Jurassic-era dinosaurs and plants. More recently, the lands within the refuge were home to a variety of indigenous peoples. Though Russian fur traders made contact with Native Alaskans in the area as early as the mid-eighteenth century, anthropologists and historians agree that the central Alaska Peninsula, including Becharof NWR, had been inhabited for approximately 7,000 years prior (McClenahan 2004).

These early peoples subsisted on the bounty that the area had to offer, with salmon, caribou, bird eggs, wild vegetables and berries comprising the majority of their diet (McClenahan 2004). Caribou bones were discovered during archaeological excavations of one of the earliest known prehistoric sites on the peninsula, which date back 9,000 years (Henn 1978). In historic times, the region was home to peoples of the Unangax and Alutiiq/Sugpiag cultures.

More recently, residents of small villages such as Kanataq continued a mixed subsistence and cash economy through the 1940s and 1950s. Residents tended to follow a seasonal subsistence pattern—hunting seals in the spring, gathering wild vegetables and fishing through the summer and fall, and hunting for moose and caribou in the fall. Cash income often came from cannery or commercial fishing work, and was supplemented by fur trapping in the winter. Kanataq's population declined during the late 1950s until the U.S. Postal Service stopped delivering mail there; eventually the village was completely depopulated (McClenahan, 2004).

Today, there are no permanent residents at the Becharof NWR. Visitors may camp, canoe, or hike, among many other activities. Hiking the Kanatak Trail has become an increasingly popular activity for visitors. This recently appointed National Historic Trail connects the former villages of Kanataq and Marraatuq, at Becharof Lake.

## Fish and Wildlife

Becharof NWR supports a broad array of fish and wildlife, from dozens of species of fish to mammals (both large and small) to nearly 200 species of birds, many of which are year-round residents of the refuge. The refuge surrounds Becharof Lake, the site of the world's second largest run of sockeye salmon (Woods 2003); young sockeye salmon rear in the lake one to two years prior to migrating out to marine feeding grounds (see Figure 5.4). Four other species of Pacific salmon (Chinook, chum, coho, pink) also spawn in the streams, rivers, and lakes in the refuge. Other notable fish found within the refuge include rainbow trout, Dolly Varden, burbot, Arctic char, Arctic grayling, and northern pike. Fish—especially salmon—fill an important role within the refuge, providing food for other animals, and returning nutrients to the ecosystem when they die and decompose.

Thanks, in part, to the abundant fish resources, the refuge also hosts a number of terrestrial mammal species, including river otters and beavers, caribou, moose (relative newcomers to the Alaska Peninsula), rodents, hares, wolverines, foxes, wolves, and—on occasion—bats. One of the largest concentrations of brown bears in Alaska can be found congregating at the refuge's streams and rivers when the salmon are returning to spawn (Woods 2003), making bear viewing a spectacular and popular activity for human visitors. Becharof and the nearby Alaska Peninsula NWR are estimated to be home to nearly 3000 brown bears (Woods 2003).

Over 200 bird species have had verified sightings at Becharof, with 45 of those species being year-round residents. Many others are migratory species that use the refuge as an important staging area during spring and fall migrations.

**Figure 5.4**    Sockeye salmon (*Oncorhyncus nerka*) spawn in Becharof Stream, Becharof NWR. *Photo credit: USFWS.*

The refuge's habitat, including the coastal zones, provides moderate to good breeding habitat and ample food for nesting seabirds, shorebirds, and waterfowl. Commonly seen birds include many species of geese and ducks, of loons and grebes, ptarmigan, cormorants, and several species of raptors, including bald eagles (USFWS 2010).

# ALASKA PENINSULA NATIONAL WILDLIFE REFUGE

The Alaska Peninsula National Wildlife Refuge lies in the southern portion of Bristol Bay, spanning approximately 540 km (330 miles) from the southwestern tip of the peninsula at False Pass to the border of Becharof National Wildlife Refuge, just west of Becharof Lake (Figure 5.1). The area was designated a NWR under ANILCA in 1980 after being created as a National Monument in 1978. None of the refuge's 1,450,760 hectares (3,584,906.7 acres) are designated wilderness. Alaska Peninsula NWR was established, in part, because of the area's importance to large mammals (including brown bears, caribou, and moose), marine mammals, migratory birds (including shorebirds), and raptors, along with salmonids and other fish (Table 5.1). The refuge is important economically for commercial and subsistence fishing, especially for the salmon that spawn and rear in the region's many lakes and rivers, as well as for subsistence and sport hunting and fishing, birdwatching, and other nonconsumptive uses. A study found that in 2011 Alaska Peninsula and Becharof NWRs received over 8,000 total visits from residents and nonresidents, generating an associated total local economic impact of over $1,200,000 and an additional total economic value beyond those impacts of approximately $319,000 (Carver and Caudill 2013).

In 1983, it was decided to manage the Ugashik and Chignik units of the Alaska Peninsula NWR, the Seal Cape area of Alaska Maritime NWR, and Becharof NWR as a *complex* because these areas share resources and resource issues. Izembek NWR, headquartered in Cold Bay, manages the Pavlof and North Creek units of Alaska Peninsula NWR.

## Landscape and Habitat

The refuge is considered the most scenically diverse of the Bristol Bay refuges (USFWS 2006). The Aleutian Mountain Range's massive volcanoes form the spine of the refuge—dividing the lakes, wetlands, and braided rivers of the Bristol Bay coastal plain from the rocky shores of the Pacific. The Pacific side is warmer and wetter than the Bristol Bay side, reflected in different vegetation communities. Vegetation communities also change across a gradient of elevation, shifting from alpine and tundra to lowland meadows, the limits of the boreal's northern spruce, and patchy stands of balsam cottonwood at the edge of their range.

The lakes, rivers, and streams provide excellent habitat for a diversity of fish species while the surrounding wetlands and marshes are key habitat for shorebirds, waterfowl, and other migratory birds. The cliffs along the Pacific coast provide sites for seabird nesting while the bays and lagoons host flocks of wintering emperor geese and Steller's eiders. The rolling tundra of the peninsula's northern side support the Northern Alaska Peninsula Caribou herd and other large mammals.

## Geology

The region reflects the violent geologic history of volcanoes, the erosive power of the North Pacific, and glacial influences—forces that continue to shape the landscape. Underneath the Peninsula the North Pacific plate subsides under a western portion of the North American plate, creating the region's many volcanoes. Mt. Veniaminof (2,506 m; 8,225 ft), a massive, smoldering volcano with a base almost 22 miles wide, dominates the southwestern end of the refuge; it was designated a National Natural Landmark by President Carter. The volcano's summit is the largest ice-filled caldera in North America, formed approximately 4,200 years ago. The glacier filling the caldera is the only one on the continent with a volcanic vent in its center. The volcano remains active, affecting residents of nearby villages when it releases gases and ash [Department of the Interior (c)]. Mt. Chiginagak (2,221 m; 7,287 feet), a stratovolcano, emits a nearly constant steam plume.

This geology underlies the rugged Pacific coast's steep rocky cliffs, deep-water bays, and long beaches, as well as the meandering glacial rivers of the Bristol Bay coastal plain. The sediment-rich rivers flow from the alpine glaciers through the lowlands to the bay's coast—continually renewing the broad outwash plains and, through their movement, limiting the growth of vegetation. The melting of ice 10,000 years ago formed the two Ugashik lakes on the coastal plain, an important area of human habitation.

## Human History

People have camped, hunted, and fished on these lands for millennia—stone tools, flakes, and other evidence of their presence have been found from over 9,000 years ago (USFWS 2005). The region's location and geography have long made it a crossroads for human cultures—an area where people from different cultures and ancestry settled to live side by side and enjoy the abundant resources. The local population includes Alaska Native people of Aleut, Alutiiq, Athabascan Indian, and Yup'ik Eskimo heritage, as well as more recent cultural additions, and some of the region's villages have been inhabited for millennia. The low mountain passes between the Ugashik lakes and Wide Bay on the Pacific, for example, provide a ready route for humans, as well as bears and others, to access rich resources of both salmon and coastal foods.

Many local residents engage in subsistence practices—hunting, fishing, trapping, and gathering wild plants and berries—supported by the refuge's lands and wildlife. As mentioned before, the refuge also is an important destination for recreational opportunities by people from outside the region.

## Wildlife

From high, cold alpine to low, swampy wetlands, the refuge's diversity of habitat supports a diversity of wildlife. Moose, a newcomer to the peninsula first observed in the early 1900s, spend summers in the low, flat wetlands of the coastal plain, feeding on emergent vegetation, then move up to higher, brushy zones providing winter forage. The Northern Alaska Peninsula caribou herd (Figure 5.5)—which is actually two large herds that intermingle at times—sometimes migrates up to 200 miles from its wintering range, which can extend to just below the Naknek River, to spring calving grounds between Ugashik and Port Moller (USFWS 2005). The herd size has varied from 2,000 to 20,000 animals in the last 60+ years. Brown bears occur at some of their highest known densities on the refuge, ranging from mountains to coast to feed on salmon, berries, sedges, ground squirrels, caribou, moose, and even dead marine mammals. Other mammals known to inhabit the refuge include gray wolves, coyotes, red fox, wolverines, river otters, mink, short-tailed weasels, and numerous other small mammal species (ibid). Sea otters, once abundant along the peninsula and now listed as *Threatened* under the Endangered Species Act, appear to be increasing [Department of the Interior (b)]. Other marine mammals found on the refuge include Stellar sea lions and harbor seals.

Birds from the North American, Asian, and Pacific migration flyways use the refuge's diverse habitats (ibid), with over 225 species identified on the refuge at some point in their life cycle (USFWS 2005), including waterbirds like tundra swans and dabbling ducks; tundra birds like Lapland longspur and savannah sparrow; boreal species like woodpeckers,

**Figure 5.5**    Caribou (*Rangifer tarandus granti*) near Ugashik Lakes, Alaska Peninsula NWR. *Photo credit: Robert Dreeszen.*

goshawks and pine grosbeaks; raptors like Northern harriers, parasitic jaegers, bald eagles, gyrfalcons, and peregrine falcons; songbirds, shorebirds, sandhill cranes, American dippers, rock ptarmigan, etc. At least 140 species are known to raise their young on Alaska Peninsula or Becharof Refuges (ibid). An isolated subspecies of marbled godwits that nests near the Ugashik River have been using the Ugashik Bay area from the Pleistocene Ice Age [Department of the Interior (b)]. Additionally, the cliffs on the Pacific coast support colonies of breeding seabirds, including murres and puffins, while the nearshore supports seaducks and shorebirds like the black oystercatcher and rock sandpiper (ibid).

The equally diverse freshwater habitats, formed by lakes, rivers, and wetlands emptying to both the Bristol Bay and Pacific sides of the Peninsula, support an exceptional abundance of fish (ibid). The state's largest grayling was caught on the refuge, and anglers are drawn from across the globe in quest of record length rainbow trout, Dolly Varden char, ocean-going steelhead, lake trout, and all five species of Pacific salmon.

## IZEMBEK NATIONAL WILDLIFE REFUGE

Izembek NWR is located near the tip of the Alaska Peninsula between the Bering Sea and the Gulf of Alaska. It was first established as a National Wildlife Range in 1960; twelve years later, in 1972, Izembek Lagoon and the associated state-owned tidal lands were designated as a state game refuge (ADF&G 2016). With the passing of ANILCA, the National Wildlife Range was converted to a NWR and 121,405 hectares (300,000 acres) within the refuge were designated wilderness. Izembek NWR today totals 125,893 hectares (311,088 acres), making it the smallest of Alaska's NWRs (USFWS 2013). Izembek NWR was established, in part, because of the area's importance to waterfowl, shorebirds and other migratory birds, brown bears, and salmonids (Table 5.1). The refuge supports a plethora of plants and wildlife specifically adapted to the challenging location; the refuge experiences an average wind speed of 29 kilometers per hour (18 miles per hour), and nearly one meter (three feet) of precipitation annually.

### Landscape and Habitat

Izembek NWR contains a wide variety of geographical features. The landscape varies from treeless tundra dotted with freshwater ponds and lakes to rugged mountains over 2,743 meters (9000 feet) high. Three of the most distinctive geographical features are Izembek Lagoon, Pavlof Volcano, and the Aghileen Pinnacles.

Izembek Lagoon is approximately 38,850 square hectares (96,000 square acres) in size, and is protected from the harsh storms that are common in the Bering Sea by a string of barrier islands. It contains the largest eelgrass bed in the

United States, which is one of the world's largest, providing a crucial food source for many species of migratory waterfowl. The eelgrass, and the lagoon itself, support a myriad of birds and aquatic species and because of this, the lagoon was recognized in 1986 as a *Wetland of International Importance* by the RAMSAR Convention, the first site in the U.S. to be designated as such. Additionally, in 2001, both Izembek and Moffet Lagoons were designated by the American Bird Conservancy as *Globally Important Bird Areas* (The American Bird Conservancy in association with the Nature Conservancy, 2003). The area is a vital stopover for many migratory bird species that are making transoceanic flights and also provides habitat for sea otters, Steller sea lions, and harbor seals. Similarly, several species of fish use the lagoon for spawning or rearing, and five species of Pacific salmon (Chinook, chum, coho, pink, and sockeye) migrate through on their way to and from the ocean and freshwater spawning grounds (RAMSAR 2014).

As is common across the peninsula, the tundra is characterized by an abundance of waterbodies of many different sizes (Figure 5.6). Streams and rivers flow into the lagoons that buffer the shoreline from the Bering Sea. Plants here typically grow close to the ground, as winds and storms can be violent, and the soils are shallow and acidic. Fruit-bearing bushes such as crowberry and many types of grasses are common. There are no maintained trails throughout the refuge, and the terrain can often be rugged.

## Geology

Izembek, like most of the Alaska Peninsula, contains many spectacular geologic features. The terrain that can be seen in the refuge today was formed by glaciers that covered the land centuries ago. At present, there are still some glaciers and snowfields; they appear in stark contrast to the towering mountains and volcanoes of the refuge. The spectacular Aghileen Pinnacles are volcanic rock that has been eroded into fortress-like spires by centuries of glacial activity (Figure 5.7). The pinnacles form a portion of the boundary between Izembek and the Alaska Peninsula National Wildlife Refuge to the north.

**Figure 5.6** Izembek NWR, like much of the Alaska Peninsula landscape, is punctuated by numerous wetlands and water bodies. *Photo credit: USFWS.*

**Figure 5.7**    The Aghileen Pinnacles above a glacial valley on Izembek NWR. *Photo credit: USFWS.*

Volcanos, both dormant and active, occur along the entire Alaska Peninsula. Two of the more recently active volcanoes are located within the boundaries of Izembek: Pavlof and Shishaldin. In mid-November of 2014, a ground observer in the nearby community of Cold Bay reported a small ash cloud from the summit of Pavlof Volcano. Over the next several days, seismic tremors were recorded, and lava and ash plumed from the volcano. At its highest point, the ash cloud reached more than 9,144 meters (30,000 feet) above sea level (Alaska Volcano Observatory 2014b).

Shishaldin Volcano is a stratovolcano that rises more than 2,743 meters (9000 feet) above sea level, with the upper portions perennially covered in snow and ice. Beginning in January of 2014, and continuing through the time of the writing of this book (mid-2015), low levels of seismic activity have been recorded. Increased surface temperatures at the summit have been recorded by satellite, as well as increased steam emissions and some steam and ash plumes. Though there was satellite evidence of low-level eruptive activity, it is unclear if any actual lava eruptions occurred (Alaska Volcano Observatory 2014c).

## Human History

Humans have inhabited the Alaska Peninsula for thousands of years. Archaeological evidence suggests that the area now incorporated as Izembek has been inhabited for at least the last thousand years, though there is debate as to whether it was occupied seasonally or year-round. Settlements dating to 900–1100 C.E. have been excavated in the Izembek Lagoon area (Maschner, 2004).

Beginning in the eighteenth century, Russian explorers and settlers influenced the region (Alaska Native Heritage Center 2011b). Russian vocabulary and practices were incorporated with traditional native customs, and the Russian influence can still be seen today in the prevalence of the Russian Orthodox Church in many of the villages on the peninsula. The nearby town of Cold Bay, Alaska, was an important Air Force base during the Aleutian Campaigns of World War II, at one point hosting over 20,000 troops (Wikipedia).

## Wildlife

While Izembek is known for its ecological diversity, it is most notable for the stunning numbers of birds that utilize the refuge's rich resources, either as part of a migratory route or as year-round residents. Izembek Lagoon holds extraordinary importance for migratory bird species that travel from northern summer feeding grounds to southern winter breeding grounds. The entire Pacific population—approximately 150,000 birds—of black brant, a small sea goose, descends on the lagoon in the fall, feeding on the nutrient-rich eel grass, as does the entire world's population of emperor geese. It is also an important molting and staging area for many other migratory bird species (including over 30 species of shorebirds). One such bird is the Steller's eider, which was designated as a *Threatened* species in 1997 under the Endangered Species

Izembek NWR is an important stop on the fall and spring migration routes for many migratory bird species, including the pacific black brant, a species of small goose. Approximately 150,000 black brant—the majority of the world's population—stage at Izembek Lagoon during their fall migration. They spend up to eight weeks feasting on the eelgrass (*Zostera marina*) of the lagoon before continuing on their path to overwintering grounds in Baja California and mainland Mexico (Reed, Stehn, & Ward 1989).

Pacific black brant (*Branta bernicla nigricans*) take flight at Izembek NWR. *USFWS photo credit: Ryan Hagerty.*

Act. The eiders depend upon the invertebrates that feed on eelgrass in the lagoons for nourishment. In addition to the brant and eiders, more than 150,000 ducks and 50,000 geese stop to stage in Izembek in the fall as they migrate to warmer southern locations for the winter. Another unique bird found within Izembek is the tundra swan, which is essentially the only nonmigratory population in North America [Department of the Interior (b) n.d.]. It would not be uncommon to see more than 300,000 birds at Izembek in the fall.

In addition to the many bird species found in Izembek, the refuge supports myriad other wildlife, from large terrestrial mammals such as caribou (the Southern Alaska Peninsula Caribou Herd numbers around 5000 animals) [Department of the Interior (b) n.d.], wolves, wolverines, foxes, and brown bears; to diverse aquatic species, including five species of Pacific salmon, harbor seals, Steller's sea lions, and sea otters. The salmon resources are so rich that the Joshua Green River Valley located within the refuge supports some of the world's highest densities of brown bears in late August: nearly one bear per square mile! Gray and Minke whales can be seen on their annual migrations. These whales occasionally stray into the lagoon itself.

For several years, the residents of the town of King Cove have asked for an approximately 20 mile long road to be constructed through Izembek NWR, including Congressionally designated wilderness, to better connect them to the airport at the community of Cold Bay. Citing the need of King Cove residents for quicker access to medical facilities via the Cold Bay airport, the state of Alaska agreed with the need, and requested permission from the Department of the Interior to have the road built. After heated debate and much research into the environmental impacts of such a road, Secretary of the Interior Sally Jewell announced on December 23, 2013, that the Department of the Interior had decided the road posed too great a threat to the important wildlife in Izembek, and the construction of the road would not be approved. The decision is still being debated.

## Challenges and Research Needs

All the refuges have active, collaborative monitoring programs with strong emphasis on priority subsistence species such as moose and caribou, as well as continuing inventorying efforts to better document and characterize the resources under their management (see, for example, USFWS 2015). Like the rest of Alaska, refuge staff and managers face the challenge of preparing for and responding to the impacts of climate change on the lands, habitats, and species they manage. Current projections suggest one of the greatest changes will be a shift in winter temperatures from below or at freezing to above (Walsh 2012)—a fundamental change with far-reaching consequences that the science community and other stakeholders are working to identify (Reynolds and Wiggins 2012) and prepare for (Western Alaska LCC 2016).

# ACKNOWLEDGEMENTS

This chapter greatly benefited from the advice and assistance of the managers and staff of Togiak, Becharof, Alaska Peninsula, and Izembek National Wildlife Refuges—especially managers Susanna Henry (Togiak) and Susan Alexander (Alaska Peninsula/Becharof). Thank you for your time and all you do to maintain these important places and populations for future generations. The findings and conclusions in this article are those of the authors and do not necessarily represent the views of the U.S. Fish and Wildlife Service.

# REFERENCES

Alaska Department of Fish and Game. 2016. Izembek State Game Refuge Land Status Map. http://www.adfg.alaska.gov/static/lands/protectedareas/_land_status_maps/izembekls.pdf, accessed 2016 Sep 29.

Alaska Department of Natural Resources. 2005. *Bristol Bay Area Plan For State Lands.* Juneau: Alaska Department of Natural Resources Division of Mining, Land & Water Resource Assessment & Development Section.

ANILCA. 1980. Alaska National Interest Lands Conservation Act of 1980 (Public Law 96-487). Signed Dec 2, 1980.

Alaska Native Heritage Center. 2011a. *Yup'ik and Cup'ik Cultures of Alaska.* http://www.alaskanative.net/en/main-nav/education-and-programs/cultures-of-alaska/yupik-and-cupik/, accessed 2015 May 13.

Alaska Native Heritage Center. 2011b. *Unangax & Alutiiq (Sugpiaq) Cultures of Alaska.* http://www.alaskanative.net/en/main-nav/education-and-programs/cultures-of-alaska/unangax-and-alutiiq/, accessed 2015 May 13.

Alaska Volcano Observatory. 2014a. *Togiak volcanics description and information.* https://www.avo.alaska.edu/volcanoes/volcinfo.php?volcname=Togiak+volcanics, accessed 2015 May 25.

Alaska Volcano Observatory. 2014b. *Pavlof Volcano description and information.* https://www.avo.alaska.edu/volcanoes/volcinfo.php?volcname=Pavlof, accessed 2015 April 23.

Alaska Volcano Observatory. 2014c. *Shishaldin Volcano description and information.* https://www.avo.alaska.edu/volcanoes/volcinfo.php?volcname=shishaldin, accessed 2015 April 19.

Carver, E. and J. Caudill. 2013. Banking on Nature: the economic benefits to local communities of national wildlife refuge visitation. U.S. Fish and Wildlife Service Division of Economics, Washington, D.C. https://www.fws.gov/refuges/about/refugereports/pdfs/BankingOnNature2013.pdf, accessed 2016 Nov 5.

Department of the Interior (a). (n.d.). *Togiak National Wildlife Refuge.* http://www.fws.gov/refuge/togiak/, accessed 2015 May 29.

Department of the Interior (b). (n.d.). *Alaska Peninsula National Wildlife Refuge.* https://www.fws.gov/refuge/Alaska_Peninsula/About.html, accessed 2016 Nov 5.

Department of the Interior (c). (n.d.). *Izembek National Wildlife Refuge.* http://www.fws.gov/refuge/izembek/, accessed 2015 April 27.

Doherty, S. 2016. Alaska Fish & Wildlife News, April 2016. Common murre update: growing awareness of sea bird die-off thanks to citizen reporting. http://www.adfg.alaska.gov/index.cfm?adfg=wildlifenews.view_article&articles_id=770, accessed 2016 Nov 5.

Goldsmith, O.S., A. Hill, T. Hull, M. Markowski, R. Unsworth. 1998. Economic assessment of Bristol Bay Area National Wildlife Refuges: Alaska Peninsula/Becharof, Izembek, Togiak. http://www.iser.uaa.alaska.edu/Publications/EconAssessment_Bristol_Bay.pdf, accessed 2016 Sep 5.

Henn, W. 1978. *Archaeology on the Alaska Peninsula: the Ugashik drainage, 1973–1975.* Eugene: Dept. of Anthropology, University of Oregon.

Maschner, H. D. 2004. Traditions Past and Present: Allen McCartney and the Izembek Phase of the Western Alaska Peninsula. *Arctic Anthropology,* 98–111.

McClenahan, P. 2004. Historic Kanataq: One Central Alaska Peninsula Community's Use of Subsistence Resources and Places. *Arctic Anthropology,* 55–69.

NWRSIA. 1997. National Wildlife Refuges System Improvement Act of 1997 (Public Law 105–57). Signed Oct. 9, 1997.

National Wildlife Refuge Association. (n.d.). *Protecting Izembek National Wildlife Refuge.* http://refugeassociation.org/advocacy/refuge-issues/izembek/, accessed 2015 May 13.

NOAA Fisheries. (n.d.). *Steller Sea Lions Biology.* http://www.afsc.noaa.gov/nmml/alaska/sslhome/biology.php, accessed 2015 May 16.

RAMSAR. 2014. *Izembek Lagoon National Wildlife Refuge.* http://www.ramsar.org/izembek-lagoon-national-wildlife-refuge, accessed 2015 Apr 21.

Reed, A., R. Stehn, and D. Ward. 1989. Autumn Use of Izembek Lagoon, Alaska, by Brant from Different Breeding Areas. *The Journal of Wildlife Management,* 720–725.

Reynolds, J.H. and H.V. Wiggins, eds. 2012. *Shared Science Needs: Report from the Western Alaska Landscape Conservation Cooperative Science Workshop. Western Alaska Landscape Conservation Cooperative Anchorage, AK, 142 pp.* https://westernalaskalcc.org/science/Shared%20Documents/walcc_final_report_web_27june12.pdf, accessed 2016 Nov 5.

Sea World. 2015. *Beluga Whales Animal InfoBook.* http://seaworld.org/en/animal-info/animal-infobooks/beluga-whales/scientific-classification/, accessed 2015 June 1.

The American Bird Conservancy in association with the Nature Conservancy. 2003. *The American Bird Conservancy guide to the 500 most important bird areas in the United States: Key sites for birds and birding in all 50 states.* New York: Random House.

U.S. Fish and Wildlife Service. 2005. Revised comprehensive conservation plan and environmental impact statement: Alaska Peninsula and Becharof National Wildlife Refuges (Oct 2005).

U.S. Fish and Wildlife Service. 2006. Revised comprehensive conservation plan: Alaska Peninsula and Becharof National Wildlife Refuges. https://catalog.data.gov/dataset/revised-comprehensive-conservation-plan-alaska-peninsula-and-becharof-national-wildlife-re, accessed 2016 Nov 5.

U.S. Fish and Wildlife Service. 2007. America's National Wildlife Refuges: factsheet. Last updated 31 July 2007. https://www.fws.gov/refuges/pdfs/factsheets/FactSheetAmNationalWild.pdf, accessed 2016 Nov 5.

U.S. Fish and Wildlife Service. 2009. Comprehensive Conservation Plan: Togiak National Wildlife Refuge (Sept. 2009). Appendix F: Togiak National Wildlife Refuge Species Lists.

U.S. Fish & Wildlife Service. 2010. Alaska Peninsula and Becharof National Wildlife Refuge Bird List [Brochure]. King Salmon, Alaska: USFWS.

U.S. Fish and Wildlife Service. 2013. Statistical Data Tables for lands under control of the Fish and Wildlife Service (as of 9/30/2013). https://www.fws.gov/refuges/land/PDF/2013_Annual_Report_of_Lands_Data_Tables.pdf, accessed 2016 Sep 5.

U.S. Fish and Wildlife Service. 2015. Science on Alaska Peninsula Refuge. https://www.fws.gov/refuge/Alaska_Peninsula/what_we_do/science.html, accessed 2016 Nov 5.

Viereck, L.A., and E.L. Little, Jr. 2007. Alaska Trees and Shrubs (2nd ed). University of Alaska Press, Fairbanks, AK.

Walsh, J. 2012. Climate change projections and uncertainties for Alaska. Chapter 3 *in* Reynolds, J.H. and Wiggins, H.V. eds. 2012. *Shared Science Needs: Report from the Western Alaska Landscape Conservation Cooperative Science Workshop. Western Alaska Landscape Conservation Cooperative Anchorage, AK, 142 pp.* https://westernalaskalcc.org/science/Shared%20Documents/walcc_final_report_web_27june12.pdf, accessed 2016 Nov 5.

Walsh, P., D. Kaufman, T. McDaniel, and J. Chowdhry Beeman. 2015. Historical retreat of alpine glaciers in the Ahklun Mountains, Western Alaska. Journal of Fish and Wildlife Management, 6 (1), 255–263. doi: http://dx.doi.org/10.3996/012014-JFWM-008.

Western Alaska Landscape Conservation Cooperative. 2016. Bristol Bay Coastal Resilience and Adaptation Workshop. https://westernalaskalcc.org:SitePages:Bristol%20Bay%20Coastal%20Resilience%20Workshop.aspx, accessed 2016 Nov 5.

Woods, B. 2003. *Alaska's National Wildlfie Refuges.* Alaska Geographic, Anchorage, AK.

# 6

# WOOD-TIKCHIK STATE PARK

Tim Troll, Bristol Bay Heritage Land Trust
Daniel E. Schindler, University of Washington

*There is one region in particular that, if not settled by home seekers, should be reserved from injury, and that is the Wood River lake country. Surely the region around the Wood River lakes is the Switzerland of Alaska, and the beautiful banks should not be robbed of the timber that has been growing so long and that can be secured in regions less noted for beautiful scenery. If the Government makes any park reserves for Alaska, surely the Wood River and its lakes should be set apart as such.* **Dr. Joseph H. Romig, Report to the Governor of Alaska, 1905**

Dr. Joseph Romig, one of Alaska's esteemed pioneers, came to Bristol Bay in 1904 to establish a hospital and government school on the banks of the Nushagak River. In one of his annual reports to the territorial Governor we find one of the earliest references to the beauty of a region we now know as the Wood-Tikchiks. Dr. Romig recommended the area be made into a park "if not settled by home seekers." Dr. Romig's recommendation languished for seventy-three years until Alaska became a state and Governor Jay Hammond signed legislation creating the Wood-Tikchik State Park (WTSP) in 1978—legislation promoted in part to prevent the area from being settled by home seekers.

Little has changed in the Wood-Tikchiks since Dr. Romig's time. Indeed, little has changed since Russian explorers ventured through the region in kayaks in 1829 and first described the lakes, the animals, and people living in this "Switzerland of Alaska." Certainly, remoteness accounts in part for the pristine nature of the region today, but the timely creation of the WTSP and the vigilant efforts of the state and local residents to protect its integrity have also been a reason people still enjoy the benefits of this region today.

Here we provide a brief overview of the ecology of the WTSP and a summary of its history over the last century that led to the establishment as a state park. We also discuss some of the ongoing challenges with protecting its ecological integrity while maintaining use of the land for a wide variety of subsistence, recreational, and scientific activities.

## THE ECOLOGICAL SETTING OF THE WOOD-TIKCHIK STATE PARK

The WTSP is situated on the northwest side of Bristol Bay, covering most of the watersheds of thirteen large lakes draining to the Nushagak River that enters Bristol Bay near the town of Dillingham. The southern half of the park covers the Wood River basin, which is a major tributary of the Nushagak, joining the main river a few miles upstream from tidal waters. The northern half of the park is associated with the Tikchik Lakes, which drain via the Nuyakuk River to the Nushagak in the northwest region of its watershed (Figure 6.1).

The WTSP straddles the Ahklun Mountains to the west (Figure 6.2) and the Nushagak/Bristol Bay lowlands to the east (Figure 6.3). The rugged Ahklun Mountains—extending to as high as 1200 meters above sea level—are a prominent feature of the park, as are the deeply carved valleys formed by repeated Pleistocene glacial advances (15,000–25,000 years ago, Briner and Kaufman 2008). Large fjord-like finger lakes occupy the glacially-carved valleys. The eastern ends of the lakes are dammed by terminal glacial moraines overlying bedrock, and the eastern region of the park is overlain with gravels from glacial outwash. While this region was heavily glaciated during the Pleistocene Epoch, only small remnant glaciers currently remain in the northwest mountains of the WTSP.

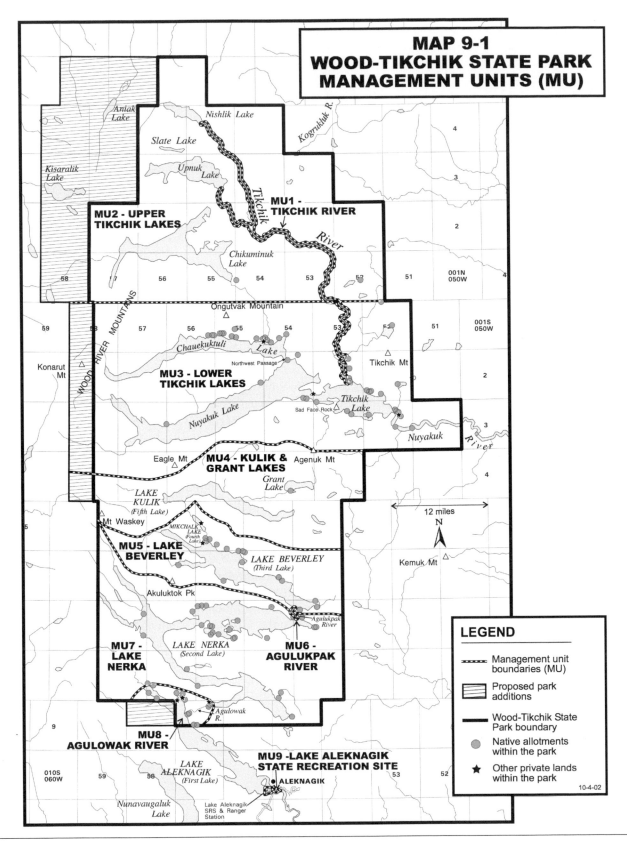

**Figure 6.1**    Map of Wood-Tikchik State Park in southwest Alaska, north of Dillingham. Map shows the park boundaries, different land use designations, and the locations of park inholdings. *Map Credit: Alaska Department of Natural Resources—Division of Parks and Outdoor Recreation.*

**Figure 6.2**  Lake Beverley, often called *Third Lake*, within the Wood River basin. The view is looking to the west into the Silverhorn Arm. *Photo credit: J.B. Armstrong.*

The long axes of the large lakes are oriented in an east-west direction with the lower six lakes (Grant, Kulik, Beverly, Nerka, Little Togiak and Aleknagik) draining south through the Wood River to the Nushagak River. The northern component of the WTSP is dominated by the Tikchik Lakes (from north to south: Nishlik, Slate, Upnuk, Chikuminuk, Chauekuktuli, Nuyakuk and Tikchik). The Allen River flows out of Chikuminik Lake into Lake Chauekuktuli, which then connects to Nuyakuk Lake. Nishlik and Upnuk lakes flow into the Tikchik River, which flows south to Tikchik Lake. From Tikchik Lake the entire system drains into the Nuyakuk River, which is a major tributary of the Nushagak River.

The dominant vegetation varies considerably across the WTSP. Within the Wood River basin, valleys are typically dominated by white spruce (*Picea glauca*) and while alder (*Alnus* spp.) provides dense scrub coverage on steep, well-drained slopes (Bartz and Naiman 2005). Riparian areas are dominated by willow (*Salix* spp.) in areas with poorly drained soils. Small stands of balsam poplar (*Populus balsamifera*) colonize coarser substrates found in the deltas of steeper streams throughout the park, particularly in the western side of the WTSP. Paper birch (*Betula papyrifera*) is also locally abundant throughout the basin. Sphagnum-dominated wetlands dominate the poorly drained soils of the shallow-sloping watersheds of streams in the eastern edge of the river basin. The northern region of the WTSP around the Tikchik lakes is distinctly less treed than the Wood River basin and grades into wet tundra. At elevations above about 400 m alpine vegetation communities dominate the landscape.

The WTSP is habitat to essentially all the wildlife resident in southwest Alaska. Brown bears are very common, while black bears are observed occasionally. Fur-bearing mammals are very common, particularly mustelids such as wolverines and otters. Wolves also inhabit the entire basin, though are not found at particularly high densities (P. Walsh, Togiak National Wildlife Refuge, pers. comm.). Moose are very common throughout the Wood River basin while caribou are somewhat rare. Caribou are more common in the upper regions of the park around the Tikchik lakes.

**Figure 6.3**    The north arm of Lake Nerka, often called *Second Lake*, within the Wood River basin. The view is looking to the east out toward the Nushagak Lowlands. *Photo credit: D.E. Schindler.*

Aquatic ecosystems are a dominant feature of the WTSP landscape. The Wood River basin is a very productive spawning and nursery ground for sockeye salmon (*Oncorhynchus nerka*) that spawn in habitat ranging from small streams to the largest rivers and lake beaches (Marriott 1964, Burgner et al. 1969) (Figure 6.4). Since 1963, the Wood River basin has produced more than 3.5 million sockeye salmon annually. Since 2000 this number has increased to more than 6.0 million annually. The Wood River also supports small populations of other species of anadromous Pacific salmon (Pess et al. 2015). In the case of coho salmon (*O. kisutch*), these appear to be self-sustaining populations though their genetic population structure has not been studied. Pink salmon (*O. gorbusca*) can be quite abundant in the large river in some years, and are often found in streams spawning alongside sockeye salmon. However, it is also not clear if these are self-sustaining populations. While small numbers of chum salmon (*O. keta*) and Chinook salmon, (*O. tshawytscha*) inhabit the waters of the Wood River, these fish appear to be strays from the main Nushagak River (Lin et al. 2011).

The Tikchik lakes system also provide habitat for sockeye salmon that spawn in streams and rivers, and on lake beaches of Tikchik Lake and Nuyakuk Lake (Nelson 1963). Notable sockeye salmon spawning areas include the Tikchik River, the river between lakes Chauekuktuli and Nuyakuk, and a series of streams flowing into Tikchik Lake. The lake beaches on Lake Chauekuktuli near the outlet of the Allen River can also be a particularly important spawning site for sockeye salmon (Nelson 1963). The Nushagak River, including production from the Tikchik Lakes, has produced about 1.4 million sockeye salmon since 1963 and about 1.8 million since 2000 (data from the Alaska Department of Fish and Game). Thus, the habitat protected by the WTSP produces a substantial component of the salmon that contribute to the lucrative Bristol Bay fisheries.

Like in all rivers draining to Bristol Bay, anadromous fish provide an invaluable food source for aquatic and terrestrial predators (Schindler et al. 2003). The importance of salmon resources to the resident fish community is mostly provided in the form of their eggs released during spawning, however, resident fish will also feed heavily on decomposing adult

**Figure 6.4** Sockeye salmon spawning in stream draining to Lake Nerka. *Photo credit: J.B. Armstrong.*

carcasses and on dipteran larvae (maggots) that colonize stranded carcasses and can be washed into streams from adjacent riparian habitat (Scheuerell et al. 2007). Terrestrial consumers ranging from song birds, gulls, and small mammals to bears all feed heavily on salmon carcasses during the salmon spawning season. Larger predators such as bears can catch live adult salmon while the fish are actively spawning, but this is mostly constrained to shallow water habitats in small streams.

The marine-derived resources provided by spawning anadromous fish are remarkably energy- and nutrient-rich, but are only accessible to aquatic and terrestrial consumers during the summer and fall spawning seasons. Sockeye salmon provide the majority of this marine resource subsidy to consumers in the watersheds.

One interesting but underappreciated dimension of the value of sockeye salmon to watershed consumers is provided by their life-history diversity. Watershed consumers in both aquatic and terrestrial habitats can only access salmon while they are actively spawning or shortly thereafter before their spawned-out carcasses decompose. Salmon populations are only present in a single site for 2–5 weeks, thereby placing substantial constraints on the duration of time that consumers can benefit from this trophic subsidy. However, there is substantial variation in spawn timing among the many populations of sockeye salmon that inhabit the WTSP, thereby extending the salmon feeding season for up to three months for consumers that can track their changing availability across the park (Schindler et al. 2013, Ruff et al. 2011, see Chapter 21 by Schindler, Seeb, and Seeb for more detail).

The community of nonanadromous fish is quite diverse throughout the WTSP (Schwanke 2013, Schindler, pers. obs.). Rainbow trout, Arctic grayling, Arctic char, and Dolly Varden char are widely distributed in the rivers and streams throughout the WTSP. Streams draining the west side of the park are typically very cold due to high inputs of snowmelt to streamflow and generally do not support notable populations of these resident species. Streams draining flatter watersheds down the middle and to the east of the park are generally productive habitat for these species. The rivers connecting the large lakes in the park support remarkably productive fisheries, particularly for rainbow trout, char, and Arctic grayling.

The resident fish species inhabiting the lakes are somewhat distinct from the stream-dwelling species. Northern pike are widely distributed throughout WTSP, but mostly constrained to small lakes and in the shallow (warmer) bays of the large lakes. It is not common to see pike in the main basins of the lakes. Lake trout and least ciscoe are absent from the Wood River basin but are common in Tikchik Lake and higher in that drainage. The absence of lake trout and ciscoes from the Wood River lakes remains somewhat of a mystery, given their presence in the Tikchik lakes and in a few small lakes to the east of the park within the Togiak National Wildlife Refuge—suggesting a complex glacial history that is currently not very well understood.

Other members of the resident fish community include: slimy sculpins and coastrange sculpins (which are widely distributed in the lakes and streams), burbot, long-nosed suckers, Alaska brook lamprey, humpback whitefish, round whitefish, Alaska blackfish, threespine and ninespine sticklebacks, and pond smelt. Starry flounder, an estuarine species, is found occasionally near the outlet of Lake Aleknagik, just outside the park boundary.

The zooplankton community within the Wood River lakes has been studied extensively and is dominated by a mixed community of copepods, cladocerans, and a wide variety of rotifers. The dominant calanoid copepods found in the large lakes are *Eudiaptomus gracilis*, *Eurytemora yukonensis*, and *Leptodiaptomus pribilofensis*. The dominant cyclopoid species are *Cyclops columbianus* and *Acanthocyclops brevisinious*. The dominant cladocerans are *Eubosmina longispina*, *Daphnia longiremis*, and *Holopedium gibberum* (Carter and Schindler 2012). The invertebrate communities in streams are not particularly well described, but are dominated by caddisflies, mayflies, and stone flies—all of which provide a forage base for resident fish during the periods of the year when the resources provided by spawning salmon are not available.

# GENESIS OF PROTECTION FOR THE SALMON OF THE WOOD RIVER SYSTEM

The availability of salmon, particularly sockeye salmon, as a reliable food source enabled formerly nomadic Alaska Native hunters to establish relatively permanent settlements around the fringes and even within the boundaries of today's WTSP. For thousands of years salmon were the exclusive domain of these early Alaska Native inhabitants and the bears, eagles, trout, and other creatures that fed upon them. That all changed after the purchase of Alaska by the U.S. from Russia and the discovery by American entrepreneurs of the wealth of salmon that could be harvested from Alaska's waters. Those waters included the Wood River. In 1884 the first salmon cannery in Bristol Bay went into operation not far from the mouth of the Wood River. Within four years there were four canneries exploiting the Wood River sockeye salmon run, and nine by the time Dr. Romig arrived in the region in 1904.

To harvest salmon in these early years, the canneries of Bristol Bay used small salmon traps and gillnets deployed from a fleet of sailboats. Not long after the first four canneries were operating, the owner of one, P.H. Johnson, convinced the other three that it would be more economical for all of them to share in the harvest of salmon running up the Wood River by cooperating in the construction of one large salmon trap across the river, just below its outlet from the lakes. Rumors of this trap reached officials of the U.S. Fish Commission, a division of the Treasury Department that was charged with enforcing the provisions of a new law enacted by Congress in 1889 to protect salmon in Alaska. The law prohibited the erection of barriers in Alaska's rivers that have *the purpose or result of preventing or impeding the ascent of salmon or other anadromous species to their spawning grounds*. Although little was known about the details of the life cycle of salmon in 1889, one fact was understood—salmon needed to return to their spawning grounds in sufficient numbers to survive as a species.

In 1890 the Revenue cutter *Albatross* was dispatched to Wood River to investigate the salmon trap used by the canneries. Upon arrival the commanding officer of the cutter, Lt. Commander Z.L. Tanner, contacted Johnson and along with John W. Clark, the resident trader at Nushagak (and namesake for Lake Clark), visited the site of the trap. According to Tanner's report the wings of the trap extended out from each bank of the river, stopping short of connecting; a 100-foot gap was left in the middle of the river to allow for the free passage of salmon. The report states that Clark insisted on the gap and gave assurances the gap would never be closed because the Natives who lived in the region were his customers and they would starve if the salmon were blocked from going up the Wood River. Johnson affirmed it was never the intent of the cannery to completely block off the Wood River.

Lt. Commander Tanner departed on the *Albatross* without noting any violations of the law or issuing any citations. In his report, however, he noted uncomfortably that, "While a 100-foot channel will serve for the ascent of salmon, a complete barricade of the stream can be accomplished with a net of that length, 12 to 15 feet in depth. Whether this simple appliance will be used depends, in the absence of a government inspector, upon the canneries themselves."

The trap on the Wood River was apparently not long-lived. Traps in general were difficult to manage in Bristol Bay because the extreme tide amplitude stranded salmon in the mud at low tide. Traps were totally abandoned in the 1920's in deference to the more efficient sailing gillnetters. The 1890 visit from the *Albatross*, however, was the beginning of a growing government presence in Bristol Bay as the commercial fishery grew and expanded to other rivers in the region. Although the geography of the Wood-Tikchik lakes remained relatively unknown until mapped in the 1930s by the U.S. Geological Survey, the importance of the lakes for spawning salmon was clear. Eventually federal government representatives were annually present during the commercial fishing season to monitor the fishery and enforce regulations. A counting station was also established on the Wood River at the outlet of Lake Aleknagik. In 1946, the Fisheries Research Institute from the University of Washington initiated ecological research throughout the region to develop scientifically based strategies for managing salmon in the region (Koo 1962, Burgner 2009). This work included establishing permanent, seasonally operated field camps at the outlet of Lake Aleknagik and on Lake Nerka that continue to support a wide variety of ecological research throughout the region, but particularly within the Wood River basin.

## CREATION OF THE WOOD-TIKCHIK STATE PARK

Not long after statehood, a field analysis of the recreational potential of the area was conducted by the National Park Service which recommended the area for national park consideration. However, state selection of the area under the Statehood Act and subsequent proposals to create a state park effectively thwarted any effort to set land in the Wood-Tikchik lakes aside for a federal park. State officials were concerned that federal park status would diminish the likelihood for potential mineral exploration, hydroelectric development, commercial fresh-water fisheries, and cabin sites. In 1970, however, the Alaska Division of Land inventoried the resources of the Wood-Tikchik lakes and proposed the area for the establishment of a state park. Finally, in 1978, the state legislature designated Wood-Tikchik State Park as part of the Alaska State Park System (AS 41.21.160). WTSP is now a 1.6 million acre wild area composed of mountains, forests, tundra, lakes, and rivers. The park is named for the two major river systems it protects—the Wood River and the Tikchik River. WTSP is the largest and most remote state park in the nation, represents almost half the acreage in the entire Alaska State Park System and 15% of all state parks land in the United States. WTSP is managed by the Division of Parks and Outdoor Recreation (DPOR) within the Alaska Department of Natural Resources (ADNR).

With the creation of WTSP the loop finally closed on establishing legal protection for the salmon of the Wood River system, the difference being that protection was established by state, and not federal legislation. A primary reason recited in the enabling legislation for WTSP was to protect its salmon-rich habitat. Another impetus for the creation of the park in 1978 was a concern among local residents that if a park was not created, the subsistence use of the region by local residents could be undermined by the 1978 Homestead Initiative that could have opened up state lands in the Wood-Tikchik region to settlement. New settlement was foreclosed by the creation of WTSP.

## MANAGEMENT OBJECTIVES OF THE WOOD-TIKCHIK STATE PARK

The enabling legislation for the Wood-Tikchik State Park is found at AS 41.21.160–41.21.167. This legislation sets forth the primary management objectives for WTSP: (1) protect and conserve the area's fish and wildlife populations and breeding systems; (2) provide for the continued use of the area for traditional subsistence and recreational purposes; and (3) protect the area's recreational and scenic resources. Regulations applying specifically to the park are found in Alaska's administrative regulations at 11 AAC 20.30–11 AAC 20.988. Management of WTSP lands and waters is delegated to ADNR with responsibility for fish and wildlife management assigned to the Alaska Department of Fish & Game (ADF&G).

In deference to the concern of local residents for continued use of the park for traditional subsistence purposes, the legislation created the Wood-Tikchik State Park Management Council with designated seats for local participation. ADNR management of the park was subordinated to the council. The seven members of the council are appointed by the governor and represent the interests of the village councils of Koliganek, New Stuyahok, and Aleknagik; the City Council of Dillingham; Bristol Bay Native Association; ADF&G; and ADNR. The management council was charged with developing a management plan for the park. The Wood-Tikchik State Park Management Plan (Plan) was finalized and adopted in October of 2002. A number of provisions of the Plan were adopted by reference under 11 AAC 20.365 and have the force and effect of law.

The Plan supplemented the three primary management objectives for WTSP with six management policies: (1) providing only those facilities which are necessary to serve existing uses or which mitigate against environmental degradation, as opposed to those which attract new visitation; (2) promoting the park only in regard to its natural and wilderness values; (3) authorizing commercial enterprises in the park through the Alaska State Parks' permit and concession procedures; (4) recognizing valid private property rights inside the park while negotiating with owners to protect public access and other park values; (5) avoiding potential conflicts between recreational and subsistence users of the park; and (6) maintaining the park's natural character by minimizing the numbers and types of management facilities inside its boundaries.

# IMPLEMENTING THE WOOD-TIKCHIK STATE PARK MANAGEMENT PLAN

The Plan establishes park-wide management practices and policies to control such matters as: the development of facilities within the park, visitor use of the park, levels of commercial use, limits on visitor and commercial use of individual rivers and lakes, and the use of motorized aircraft, watercraft, and snowmachines. Many of these policies are designed to avoid crowding and minimize conflicts between park users.

Particularly important provisions of the Plan are the creation of three general classifications of use within the park: *wilderness, natural* and *recreational development*. These classifications are designed to preserve particular kinds of experiences along a spectrum of human impact—from little human impact or expected human encounters in *wilderness* areas to more frequent and expected human impact in *natural* areas and unavoidable and thus accommodated human impact in *recreational development* areas. Within each general classification area, the Plan establishes management units that contain separate sets of management guidelines for that unit (see Figure 6.1).

The Plan designates the upper Tikchik Lakes and Kulik/Grant Lakes as *wilderness*; most of the remainder of the park is designated *natural*; and two limited areas are designed for *recreational development*. Within the upper Tikchik Lake *wilderness* area there are two management units. The Kulik/Grant Lakes *wilderness* area is its own management unit. Areas designated as *natural* contain four management units. To date there are two units designated for *recreational development*—the Agulowak River and Northern Lake Aleknagik unit and the Lake Aleknagik State Recreation Site unit.

# CONSERVATION ISSUES IN THE WOOD-TIKCHIK STATE PARK

The creation of the Wood-Tikchik State Park is unquestionably a milestone in the history of conservation in Alaska. It represents the first example of a government action in Alaska to specifically protect salmon habitat at a landscape scale. The salmon that successfully return to the Wood River and Tikchik Lake systems are relatively assured that their natal streams and spawning grounds will never be deliberately spoiled by humans. Only an act by the Alaska Legislature amending the enabling legislation along with the approval of the governor can modify that assurance. While it is unlikely both houses of the Alaska legislature and the governor would agree to allow a change to the enabling legislation of the Park, it can—and almost did—happen.

In 2014, a bill to amend the enabling legislation to allow for the possible construction of a hydroelectric dam on Chikuminuk Lake in the Upper Tikchik Lakes *wilderness* area (Figure 6.5) was urged by the potential developer of the dam and introduced and promoted by an urban legislator. The bill was introduced to counter the actions of local residents who challenged and defeated, through an administrative appeal, the issuance of a permit to the potential developer to allow helicopter access and drilling—activities not generally allowed in a *wilderness* area. Through some last minute maneuvers at the end of the legislative session, the bill was attached to a larger piece of legislation that fell short by one vote of being submitted to a supportive governor for his signature. The experience strongly suggests the conservation protections afforded by WTSP, while fairly solid, are not immune from politics.

The enabling legislation does allow for the possibility of hydroelectric dams at two specific locations—Elva Lake and Grant Lake. Both sites have since been determined to be economically unfeasible by studies conducted for the Nushagak Electric and Telephone Cooperative.

# A LANDMARK CONSERVATION INITIATIVE WITH ALEKNAGIK NATIVES LTD

The most challenging conservation problem for WTSP is dealing with the many privately owned inholdings within the park that are not subject to park rules and regulations. Roughly 33,000 acres within the park are privately owned. While development on these inholdings may not compromise the long-term salmon productivity of the Wood-Tikchik ecosystem, development could lead to localized habitat degradation, increased user conflicts, and could generally undermine the ability of the DPOR to effectively manage the park for the preservation of its natural characteristics and wilderness.

The largest private landowner in WTSP is Aleknagik Natives, Ltd. (ANL), the village corporation formed for the Alaska Native residents of the village of Aleknagik pursuant to the provisions of the Alaska Native Claims Settlement Act of 1971 (ANCSA). ANL selected and received title from the federal government to approximately 21,000 acres within WTSP. These lands lie along the Agulowak River (Figure 6.6), an important migratory corridor for salmon at the northwest end

**Figure 6.5** The lower reaches of the Allen River, flowing to Chikuminuk Lake. *Photo credit: D.E. Schindler.*

**Figure 6.6** Aerial photo of the Agulowak River flowing downstream to Lake Aleknagik. *Photo credit: J.B. Armstrong.*

of Lake Aleknagik and the east end of Lake Nerka. The regional corporation for Bristol Bay—Bristol Bay Native Corporation (BBNC)—received title to the subsurface beneath all ANL lands.

In 2010, ANL conveyed a conservation easement over the property to the state—and BBNC relinquished its subsurface rights. Under the terms of the conservation easement, ANL retains ownership of the land but forgoes development on the land in perpetuity and grants limited public access to the property for recreation. Funds to purchase the conservation easement and the subsurface rights were raised by The Conservation Fund, a national conservation organization, with assistance from the local Bristol Bay Heritage Land Trust.

## THE CHALLENGE OF PRIVATE INHOLDINGS

Most of the remaining private land within WTSP, some 8000 acres, is held by Alaska Natives as allotments under the Alaska Native Allotment Act of 1906. There were approximately 100 allotments ranging in size from 30 to 160 acres scattered throughout WTSP at the time of its creation. Most of these allotments were unapproved and pending adjudication by the Bureau of Land Management (BLM). Their ownership status was in suspension as a result of a complex series of tradeoffs that accompanied the passage of ANCSA in 1971. Language in ANCSA extinguished the 1906 Alaska Native Allotment Act limiting the future access of individual Alaska Natives to the land used for traditional activities to the claim procedures established in ANCSA. This provision prompted a last minute flurry of allotment applications statewide, including those in WTSP, in advance of the effective date of ANCSA. The result was thousands of Native allotment applications pending on December 18, 1971 when ANCSA became law.

To clear the backlog of pending allotment applications and eliminate a potential century or more of adjudications, Congress, with some exceptions, approved all allotment applications that were pending on the date the Alaska National Interest Land Conservation Act was passed in 1980. Most of the allotment claims within WTSP fell within that exception because they were claims that conflicted with land already selected by the state in the early 1960s and tentatively approved by BLM for conveyance. As such, most of the allotment claims within WTSP remained subject to the adjudication process of the 1906 Alaska Native Allotment Act—a process that placed the burden on the allotment applicant to prove exclusive use and occupancy of the claimed land for a period of five years before the land had been withdrawn for conveyance to the state.

Nothing had been done to resolve the uncertainty of title that surrounded these pending allotment claims in WTSP when its first park ranger, Dan Hourihan, arrived for duty in 1984. In addition to his single-handed responsibility to monitor activity in the 1.6 million-acre park, Hourihan made it a priority to find a way to move the allotment issue from conflict to compromise. Finally in the late 1990s, after years of community meetings and delicate negotiations, a compromise emerged. Applicants who wished to move their claim to unoccupied state lands outside the boundaries of WTSP could do so without the burden to prove use and occupancy. Twenty-seven applicants chose this acre-for-acre exchange.

For those applicants who wished to retain allotments within WTSP, the state offered to accept their claims if each agreed to accept conservation easements on their respective allotments. These easements imposed three levels of restrictions that varied according to the state's assessment of the strength of the original claim, the age of the applicant, and the location of the claim. Tier 1 allotments have no development restrictions and are subject only to a twenty-five-foot pedestrian access easement along shorelines; Tier 2 allotments have the pedestrian access easement and also restrictions that limit subdivision to ten-acre lots with no more than one five-acre commercial development site; and Tier 3 allotments have restrictions similar to Tier 2 except that no commercial development is allowed. Most allotments in WTSP ended up in the Tier 2 category.

While the negotiated tier system removed much uncertainty around land ownership in WTSP and reduced the likely impact of future habitat degradation, it did not address the likelihood that future commercial lodge and private land development could increase user conflicts and reduce the wilderness and natural characteristics of the park that its creators sought to preserve. To help address this lingering problem, the Alaska Native village corporation of Choggiung Ltd. fostered the creation of the Bristol Bay Heritage Land Trust. Since its incorporation in 2000, the land trust has raised funds to protect through acquisition or conservation easement several allotment parcels within WTSP, and has identified others that The Conservation Fund has been able to acquire and transfer to state ownership.

There are nine other patented parcels within WTSP that pre-date the establishment of the park. These sites—comprising 74 acres—were conveyed under the BLM's Trade & Manufacturing site or Headquarters site programs. Seven of the eight lodges within WTSP are located on these lands.

# ZONING

Finally, Alaska law does permit state parks to adopt and enforce zoning regulations to control some uses of private land within park boundaries. The Division of Parks and Outdoor Recreation has implemented some zoning regulations in WTSP. These regulations apply to private lands within the park, except for some Native allotments. The zoning regulations cover such matters as subdivision, building heights, fuel storage, and building color. The zoning restrictions also do not apply to the lands owned by ANL. However, the terms of the conservation easement conveyed by ANL to the state make the application of zoning restrictions to these lands unnecessary.

# SUMMARY

The WTSP represents a unique component of the natural heritage of western Alaska, and in fact, the world. It is the largest state park in the U.S.—providing a diverse landscape of productive habitat for a stunning range of terrestrial and aquatic wildlife. This ecosystem also provides a wide range of ecosystem services to people of the region by supporting commercial, sport, and subsistence fisheries, subsistence and sport hunting, berry-picking, and many other outdoors activities with immense intrinsic and economic value. The importance of the park to science is also invaluable as it provides one of the few long-term opportunities to understand how undeveloped river systems respond to global change in the absence of most direct human impacts. While the ecological integrity of the WTSP is protected by its status as an Alaska state park, conservation challenges remain—including sustainable management of wildlife and fishery harvests and the ongoing management of inholdings throughout the park.

# REFERENCES

Bartz, K.K., and R.J. Naiman. 2005. Effects of salmon-borne nutrients on riparian soils and vegetation in southwest Alaska. Ecosystems 8:529–545.

Branson, John. 2007. The Canneries, Cabin and Caches of Bristol Bay, Alaska. U.S. Department of the Interior, National Park Service, Lake Clark National Park and Preserve.

Briner, J.P., and D.S Kaufman. 2008. Late Pleistocene mountain glaciation in Alaska: key chronologies. Journal of Quaternary Science 23:659–670.

Bulletin of the United States Fish Commission, Government Printing Office 1894, Vol. XII for 1892.

Burgner, R.L. C.J. DiCostanzo, R.J. Ellis, G.Y. Harry, Jr., W.L. Hartman, O.E. Kerns, Jr., O.A. Mathisen, and W.F. Royce. 1969. Biological studies and estimates of optimum escapements of sockeye salmon in the major river systems in southwestern Alaska. Fisheries Bulletin U.S. 67:405–459.

Burgner, R.L. 2009. My career with Fisheries Research Institute University of Washington. University of Washington School of Aquatic and Fishery Sciences, Seattle, WA.

Carter, J.L., and D.E. Schindler. 2012. Responses of zooplankton populations to four decades of climate warming in lakes of southwestern Alaska. Ecosystems 15:1010–1026.

Koo, T.S.Y. (editor) 1962. Studies of Alaska Red Salmon. University of Washington Press, Seattle, WA.

Lin, J.E., R. Hilborn, T.P Quinn, and L. Hauser. 2011. Self-sustaining populations, population sinks or aggregates of strays: chum (*Oncorhynchus keta*) and Chinook salmon (*O. tshawytscha*) in the Wood River system, Alaska. Molecular Ecology 20: 4926–4937.

Marriott, R.A. 1964. Stream catalog of the Wood River lake system, Bristol Bay, Alaska. Special Scientific Report, Fisheries 494, US Fish and Wildlife Service, Washington D.C.

Nelson, M.L. 1965. Red salmon spawning ground surveys in the Nushagak and Togiak Districts, Bristol Bay, 1963. Alaska Department of Fish and Game, Division of Commercial Fisheries, Informational Leaflet 61, Juneau.

Pess, G.R., T.P. Quinn, D.E. Schindler, and M.C. Liermann. 2014. Freshwater habitat associations between pink (*Oncorhynchus gorbuscha*), chum (*O. keta*) and Chinook salmon (*O. tshawytscha*) in a watershed dominated by sockeye salmon (*O. nerka*) abundance. Ecology of Freshwater Fish 23:360–372.

Ruff, C.P., D.E. Schindler, J.B. Armstrong, K.T. Bentley, G. Brooks, G.W. Holtgrieve, M. McLaughlin, J. Seeb, and C. Torgersen. 2011. Temperature-associated population diversity in salmon confers benefits to mobile consumers. Ecology 92:2073–2084.

Scheuerell, M.D., J.W. Moore, D.E. Schindler, and C.J. Harvey. 2007. Varying effects of anadromous sockeye salmon on the trophic ecology of two species of resident salmonids in southwest Alaska. Freshwater Biology 52:1944–1956.

Schindler, D.E., M.D. Scheuerell, J.W. Moore, S.M. Gende, T.B. Francis, and W.J Palen. 2003. Pacific salmon and the ecology of coastal ecosystems. Frontiers in Ecology and the Environment 1:31–37.

Schindler, D.E., J.B. Armstrong, K.T. Bentley, K. Jankowski, P.J. Lisi, and L.X. Payne. 2013. Riding the crimson tide: mobile terrestrial consumers track phenological variation in spawning of an anadomous fish. Biology Letters 9: 20130048. http://dx.doi.org/10.1098/rsbl.2013.0048.

Schwanke, C. J. 2013. Tikchik Lake system lake trout assessment, Bristol Bay Management Areas, 2005–2006. Alaska Department of
    Fish and Game, Fishery Data Series No. 13–18, Anchorage.
Sherwonit, Bill. 2003. Wood-Tikchik, Alaska's Largest State Park, Photographs by Robert Glenn Ketchum with Essay by Bill Sher-
    wonit, Aperture Foundation Inc. 2003.
Van Stone, James (Editor) 1988. Russian Exploration in Southwest Alaska: The Travel Journals of Petr Korsakovskiy (1818) and Ivan
    Ya. Vasilev (1829), University of Alaska Press, Vol. IV, The Rasmuson Library Historical Translation Series.
Wood-Tikchik State Park Management Plan, Alaska Department of Natural Resources, Division of Parks and Outdoor Recreation,
    October 2002, www.dnr.state.ak.us/parks/plans/woodt/woodtpln.htm.

# SECTION II

## WILDLIFE RESOURCES OF BRISTOL BAY: TERRESTRIAL SYSTEMS

# 7

# BROWN BEARS

Colleen A. Matt, C. A. Matt Consulting
Lowell H. Suring, Northern Ecologic L.L.C.

## INTRODUCTION

Bristol Bay is unique for its annual run of tens of millions of wild Pacific salmon, which in turn support high densities of North American brown bears. Combined with largely unaltered watersheds, the region offers unparalleled opportunities for humans to study, view, photograph, and hunt for brown bears. We use the term *brown bears* to refer to all North American bears of the classification *Ursus arctos*, although bears in interior parts of North America are traditionally referred to as grizzly bears, and those in coastal salmon-rich areas, as brown bears. This chapter characterizes brown bears that spend most of their lives in the Bristol Bay watersheds. Some of the general information comes from decades of research on this species in its entire cosmopolitan ranges. We note where information varies from that available for the Bristol Bay watershed. Because of their interconnected home ranges and close genetic relationships, brown bear habitat selection and behavior show a great degree of overlap wherever they occur (Schwartz et al. 2003).

## HABITAT

Brown bears are wide-ranging animals that use diverse plant and animal communities throughout the course of a year. Habitat is defined here as the location or environment where an organism is most likely to be found. Habitats provide the food, water, and cover that a species needs to survive. Biologists generally use the term *home range* to describe the area that an animal uses to carry out the normal activities of securing food, mating, and caring for young. Brown bears are generally solitary, food-maximizing individuals whose home ranges vary with the availability of their seasonal foods. When food is abundant, as during Bristol Bay salmon runs (see Figure 4.4 in this volume), home ranges of female bears may be smaller than at other times because of their ability to obtain sufficient energy to meet their nutritional needs. Conversely, home ranges may increase in order to take advantage of more widely dispersed food resources (McLoughlin et al. 1999).

On the Alaska Peninsula, brown bears emerge from their dens between early April and early June. Males tend to emerge before females. Females with cubs of the year are often the last to leave the den (Miller 1990). Brown bears often spend June to mid-August in lowland and coastal areas, although some bears in the Bristol Bay watershed probably do not include coastal plains within their home ranges. A study conducted on Admiralty Island in southeast Alaska found that brown bears did not necessarily use rich coastal habitats as part of their home ranges, although reasons for these differential patterns of habitat selection are not known (Schoen et al. 1986).

Brown bears typically spend July through mid-September near streams that support salmon runs (Schoen et al. 1986) (Figure 7.1). The Nushagak and Kvichak watersheds contain at least 13,335 linear kilometers (8,286 linear miles) of anadromous fish habitat, not including lakes (Johnson and Blanche 2011). As salmon begin to appear in the streams, bears move closer to them, sometimes congregating where shallow streams make preying on fish more efficient (Aumiller and Matt 1994). Studies of bear predation on salmon in a series of streams in the Wood River system of Bristol Bay demonstrated that the probability of a fish being killed by a bear increases with decreasing stream size (Quinn et al. 2001).

**Figure 7.1**     Subadult brown bear taking a break from hunting salmon. *Photo credit: Carol Ann Woody.*

Brown bears move to higher elevations in the fall, presumably to feed on berries and other food items (Collins et al. 2005). Some bears continue to feed on fish until October, especially in shallow streams where dead and dying sockeye salmon wash out of lakes after spawning (Fortin et al. 2007).

Brown bears typically dig their dens at higher elevations within their home ranges (Van Daele et al. 1990), on 20- to 50-degree slopes between 300 and 1,600 meters (328–1,750 yds) in elevation. The mean date for den entrance is October 14, and most bears enter their dens by early November. Brown bears spent an average of 201 days in winter dens in the Nelchina area in southcentral Alaska (Miller 1990).

## FOOD HABITS

Brown bears on the Alaska Peninsula travel within their home ranges to exploit seasonally available food sources (Glenn and Miller 1980). During late summer and fall, bears gain weight rapidly and store it primarily as fat. Peak body mass generally occurs in fall, just prior to hibernation (Hilderbrand et al. 2000). Bears must maximize weight gain prior to hibernation because they metabolize only fat and muscle during that time and must rely on those stored energy reserves for reproduction and survival.

Following den emergence in spring, bears commonly graze on early season herbaceous vegetation, such as cow parsnip (*Heracleum lanatum*), horsetails (*Equisetaceae*), lupine (*Lupinus* spp.), false hellebore (*Veratrum viride*) and grasses (Gramineae) (Alaska Department of Fish and Game (ADF&G) 1985). They also search for and scavenge winter-killed carrion, as well as moose and caribou calves (Glenn and Miller 1980). Bears with access to salt marshes commonly graze on sedges (*Carex* spp.), grasses (e.g., *Elymus* spp.), sea-coast angelica (*Angelica lucida*), and forbs (e.g., *Plantago* spp. and *Triglochin* spp.). Brown bears are also known to dig for soft-shelled clams (*Mya arenaria*) and Pacific razor clams (*Siliqua patula*) on intertidal beaches (Smith and Partridge 2004). Bears on the coast scavenge for dead marine life (Glenn and Miller 1980) and recent studies indicate coastal Katmai bears will kill and feed on seals, sea otter, and flat fishes such as flounder and sole (pers. comm., Dr. Grant Hilderbrand, USGS, Anchorage, May 3, 2016).

In July through October, brown bears move to streams to take advantage of the predictable runs of salmon. Bristol Bay rivers support five species of Pacific salmon (*Oncorhynchus* spp.) with sockeye salmon (*O. nerka*) being most prolific; total runs average 35 million fish during 1994–2014 (Elison et al. 2014). The abundance of salmon in Bristol Bay watersheds makes them a prime food source for brown bears because salmon are an excellent source of lipids.

Lipids obtained through the consumption of salmon can account for approximately 80% of the mass gained by bears, as shown by Hilderbrand et al. (1999b) on the Kenai Peninsula of Alaska. More than other factors, the accumulation of fat determines whether brown bears will hibernate successfully or, in the case of females, produce cubs (Farley and Robbins

1995). In addition, bigger, fatter adult females produce faster growing cubs that survive better than do cubs produced by smaller, leaner females (Ramsay and Stirling 1988). Larger, fatter males also receive an advantage from their size. Dominance in males is necessary to win breeding opportunities and defend estrus females; larger males tend to compete better for these opportunities than smaller males (Robbins et al. 2007).

The availability, age, and spawning status of salmon all played significant roles in consumption choices made by brown bears in Bristol Bay and Southeastern Alaska. (Gende et al. 2004). When salmon were abundant, bears preferentially ate fish parts with higher caloric content, such as roe and brains. In addition, bears consumed sexually mature (or ripe) fish before fish that had spawned-out. Some bears were even observed catching and releasing fish that had spawned-out and consuming only ripe fish with higher energy content (Gende et al. 2004).

Brown bears are also known to feed extensively on wild fruits, including crowberry (*Empetrum nigrum*), lowbush cranberry (*Vaccinium vitis-idaea*), and bog blueberry (*Vaccinium uliginosum*) (Fortin et al. 2007, Rode et al. 2006a, 2006b). The simultaneous ingestion of salmon and berries appears to increase the growth rate of bears, over that attained by ingesting one or the other alone (Robbins et al. 2007).

Fall foraging is especially important for brown bears. In fall, brown bears seek out available meat sources including salmon, ungulates, and rodents, as well as berries, in order to store as much fat as possible. The more efficiently bears forage, the more vital lipids they can store, thereby improving their ability to survive and reproduce (Robbins et al. 2007).

# BEHAVIOR

Brown bears have generalist life history strategies, extended periods of maternal care, and omnivorous diets. Generalists thrive in a wide variety of environmental conditions and use a variety of food resources. Brown bears must travel long distances in their six months of activity to procure the abundance and variety of food needed to flourish.

Movement patterns that define home ranges are influenced by important food resources, breeding, reproductive status, individual dominance status, security, and human disturbance (Schwartz et al. 2003). The larger ranges of adult males overlap several females (Schwartz et al. 2003). Female brown bears are generally faithful to their home ranges. Subadult females tend to stay close to or within the home range of their mothers. Subadult males tend to disperse longer distances (Glenn and Miller 1980). Brown bears searching for alternative foods outside their usual home ranges, as well as dispersing bears, often run into more conflicts with humans (increasing human-caused bear mortality) than bears staying within their home ranges (Schwartz et al. 2003).

Differences in home range size between study areas are attributed to differences in habitat quality and distribution (McLoughlin et al. 2000). The fact that brown bear home ranges respond to habitat quality is illustrated by the difference in home range size in various study areas. For example, in the Nelchina River basin of southcentral Alaska, adult female home ranges averaged 408 km$^2$ (158 mi$^2$), whereas those of adult males averaged 769 km$^2$ (297 mi$^2$) (Ballard et al. 1982). In contrast, Collins et al. (2005) estimated home range size for adult females as 356 km$^2$ (137 mi$^2$) in the southwest Kuskokwim Mountains, west of the Nushagak River watershed. On the relatively productive Alaska Peninsula, Glenn and Miller (1980) found an average seasonal range of 293 km$^2$ (113 mi$^2$) for adult females (n = 30) and 262 km$^2$ (101 mi$^2$) for adult males (n = 4), although the small sample size for adult males cautions against comparing these directly. Glenn and Miller (1980) pointed out that other data contradicted the apparent finding of smaller male than female seasonal ranges, e.g., the cumulative six-year movements of 13 adult males were greater than those of 49 females. In addition, they found that seasonal range movement of subadult males (744 km$^2$; 287 mi$^2$) was three times that of subadult females (249 km$^2$; 96 mi$^2$).

Brown bears hibernate in dens during winter and rely on stored energy reserves for survival. During hibernation, bears can remain dormant up to seven months without eating, drinking, defecating, or urinating (Schwartz et al. 2003). Generally, brown bears seek out remote, isolated areas and sites that will accumulate enough snow to insulate them from cold winter temperatures, often on steep slopes. Bears may prefer steeper slopes for denning sites, due to reduced potential for disturbance (Goldstein et al. 2010).

Female brown bears enter estrus beginning in late spring and, depending upon male availability, may continue breeding into August. Because males are rarely limiting, most breeding occurs during May and June. After eggs are fertilized development proceeds to the blastocyst stage and then halts. Embryo implantation is delayed until hibernation begins. It is possible for a litter to have multiple sires. Female brown bears that have successfully bred and have implanted embryos have an obligate denning requirement because the newborns are completely helpless at birth and remain so for several months. Most births occur in January and February after six to eight weeks of gestation. All maternal care from

**Figure 7.3**    Brown Bear Boar (male). *Photo credit: Carol Ann Woody.*

due to increased hunting pressure. Female survival rates are generally higher than males. In the middle Susitna study, Miller et al. (2003) estimated 90% annual survival for adult females in both 1985 and 1995. In their study of female survivorship in a hunted population in the southwest Kuskokwim Mountains, midway between Dillingham and Bethel, Alaska, Kovach et al. (2006) found mean annual survival estimates of 90.1 to 97.2% for radio collared females aged five years or older.

Because of the difficulty of observing them, causes of natural mortality are poorly known for brown bears. Although adult males are known to kill juveniles, and adults kill other adults, there are insufficient data to fully assess the effects of predation on younger bears by adult bears (Schwartz et al. 2003). Brown bears are exposed to more dangers during some life stages than others. Cubs are particularly vulnerable during their first year of life. In the middle Susitna study, survival for cubs of the year was estimated at 67% (Miller et al. 2003). Kovach et al. (2006) reported survival rates of 48.2 to 61.7% for cubs of the year. In the Kuskokwim Mountains study, Kovach et al. (2006) estimated survival rates of 71.3 to 83.8% for one- and two-year-old offspring, combined.

Dispersing subadults may be forced to choose marginal home ranges or areas near human habitation where mortality risks are high (Servheen 1996). Brown bears can harbor parasites and diseases that may contribute to mortality, and starvation has also been reported. However, there are no reported cases of parasites or diseases causing major die-offs within populations (Schwartz et al. 2003).

In most brown bear populations, human-caused mortality is higher than natural mortality among adults. The rate of human-caused mortality varies greatly in Alaska, where contact between bears and humans is a function of human population density, activities of both species, and hunting regulations. Servheen (1996) lists the following categories of human-caused mortality in order of increasing frequency:

- Direct human/bear confrontations (hikers, backpackers, photographers, hunters, etc.);
- Attraction of brown bears to improperly stored food and garbage associated with human habitations and other sources;
- Careless livestock husbandry, including the failure to properly dispose of dead livestock;
- Inadequate protection of livestock;
- Loss of bear habitat; and
- Hunting—both lawful and illegal.

## POPULATION DENSITIES

Alaska brown bear populations vary in density depending on the availability and distribution of food (Miller et al. 1997, Schwartz et al. 2003)—achieving highest densities where bears have access to multiple runs of Pacific salmon (Hilderbrand et al. 1999c, Miller et al. 1997). Based on a modified capture-mark-recapture method, Miller et al. (1997) estimated the density of all ages of brown bears along the Pacific Coast of the Alaska Peninsula at 551 bears per 1,000 km$^2$ (386 mi$^2$) the highest documented anywhere in North America. In contrast, brown bears on Alaska's north slope have the lowest estimated density at only 3.9 bears of all ages/1,000 km$^2$ (Reynolds 1976).

Population density estimates for most of Bristol Bay are not currently available, but recent aerial distance sampling surveys in some watersheds provide some data. Using the double observer aerial line-transect method, brown bear density in the Lake Clark National Park and Preserve portion of the Kvichak River watershed was estimated at 39 bears/1000 km$^2$ (E. Becker, ADF&G, pers. comm., 1999). Recent aerial line transects in the remainder of the inland Kvichak River watershed estimated 47.7 bears of all ages/1000 km$^2$ (Becker 2010). In the nearby northern Bristol Bay area (Togiak National Wildlife Refuge and Bureau of Land Management, Goodnews Block), an area including both coastal and inland habitats, brown bear population density was estimated as 40.4 bears/1,000 km$^2$ (Walsh et al. 2010).

As expected, spring surveys that include coastal habitat have higher population density estimates than those with only inland habitats. Researchers using the ADF&G aerial line-transect double-count survey method as in the previously mentioned studies, estimated brown bear density at 124 bears/1,000 km$^2$ in Katmai National Park (Game Management Unit (GMU) 9C), in an area that included both coastal and inland brown bear habitat. Along the coast of Lake Clark National Park and Preserve, (NPP) (GMU 9A), brown bear density was estimated at 150 bears/1,000 km$^2$ (Olson and Putera 2007) (Figure 7.4).

When inferring the geographic distribution of brown bears from density estimates, it should be noted that brown bears can move long distances. Bears that are counted in coastal areas in the spring may move inland and upstream in the summer and fall to take advantage of pre- and post-spawning salmon. It is common to see brown bears in interior Lake Clark NPP feeding on post-spawning salmon in September and October, and less commonly, in December (B. Mangipane, NPS, pers. comm., 9/27/11).

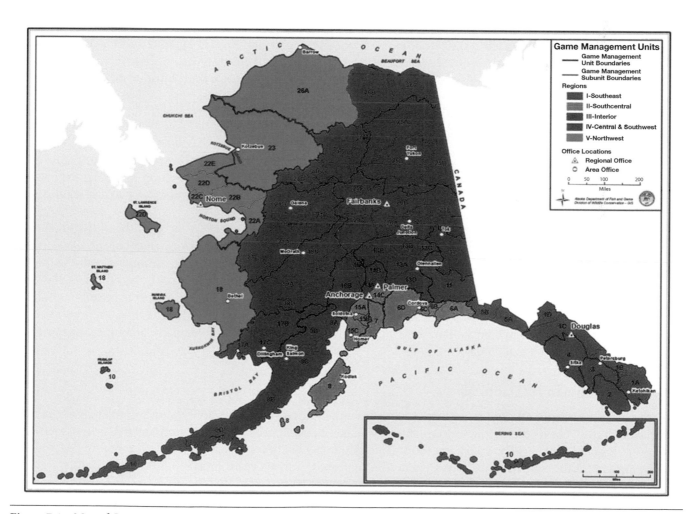

**Figure 7.4**   Map of Game Management Units (GMU) in Alaska. *Map courtesy of: Alaska Department of Fish and Game.*

# HUMAN USE/INTERACTION/MANAGEMENT

In Bristol Bay watersheds, humans value bears for sport hunting, subsistence hunting, and for viewing. Humans and bears interact in many other situations, including both residents and nonresidents visiting the area for purposes other than seeking encounters with brown bears.

## Sport Hunting for Brown Bears

The Alaska Board of Game has placed a high priority on maintaining a quality hunting experience for the large brown bears of the Alaska Peninsula in GMU 9, which includes the Kvichak River watershed (Figure 7.4). Due to the ease of access using aircraft and the high quality of bear trophies in the unit, an active guiding industry for the area developed during the 1960s. Nonresident sport hunters are required to use a guide for brown bear hunting throughout Alaska, and to have their harvest inspected and sealed by the ADF&G. The ADF&G management program strives to maintain stable guide and client numbers over time. As of 2007, approximately 75% of the GMU 9 brown bear harvest came from guided hunts (ADF&G 2009).

## Subsistence Hunting for Brown Bears

Federal and State subsistence hunting regulations differ. On federal lands within GMUs 9 and 17, residents must consult Federal Subsistence Management Program Regulations. For example, on federal lands in GMU 9B, there is a federal registration permit harvest quota of four females or 10 bears and the season closes as soon as the first quota is reached.

On nonfederal lands in GMUs 9b and 17, subsistence brown bear hunters must obtain a subsistence registration permit for bears to be taken for food. In addition to requiring a registration permit, the subsistence brown bear hunting regulations establish a hunting season of September 1 to May 31, limit take to one bear/regulatory year, and prohibit take of cubs and females with cubs (ADF&G 2011a). Salvage of the hide and skull is optional and the hide and skull need not be sealed, unless they are removed from the Western Alaska Brown Bear Management Area, in which case an ADF&G representative must seal them and their trophy value destroyed. All edible meat must be salvaged for human consumption.

## Bear Viewing

The Kvichak watershed contains specific destinations for recreational visitors to view brown bears. Lodges on Lake Clark, Kukakluk Lake, Nonvianuk Lake, and Battle Lake offer brown bear viewing in addition to fishing expeditions (see Figure 4.5 in this volume). Several guides and air taxis take brown bear viewers to Funnel Creek and Moraine Creek on day trips.

The 2006 National Survey of Fishing, Hunting, and Wildlife-Associated Recreation reported that 491,000 Alaskan residents and nonresidents participated in wildlife watching as a primary activity (U.S. Department of the Interior, Fish and Wildlife Service, and U.S. Department of Commerce, U.S. Census Bureau 2008). Bear viewing is now the leading recreational activity in Lake Clark NPP. Bear viewing altered spatiotemporal food resource use by brown bears; however, total resource use declined only when bears were exposed to 24-hour daily human activity. Energy expenditure, indexed as daily travel distances, was significantly higher when bears responded by altering spatial rather than temporal resource use. However, body weight and composition were unaffected as bears shifted their foraging to other locations or times (Rode et al. 2007, Tollefson et al. 2005).

## Other Bear-Human Interactions

In both GMU 9 and 17, both legally harvested bears and nonhunting mortalities are reported annually. Villages with open landfills attract bears during the spring, summer, and fall. Residential garbage, dog food, and fish drying racks also lure bears close to humans (Figure 7.5). Local residents commonly kill bears in defense of life or property near villages and fishing sites. Although reporting rates seem to have improved in recent years, most nonhunting mortalities of bears are reported either indirectly or not at all; as such, any conclusions based solely on harvest data or reported nonhunting mortalities should be viewed with caution (ADF&G 2009).

**Figure 7.5**   Subadult brown bear that was chewing on some old moose antlers under a lodge balcony in Lake Clark National Park. If coastal salmon runs fail and bears are hungry, or if bears associate humans with food, then bears can be destructive in their quest for food. *Photo credit: Carol Ann Woody.*

For example, during the 2007 regulatory year, ADF&G received 17 reports of bears killed by people in defense of life or property in GMU 9; however, wildlife managers estimated that the number of unreported brown bear killings in the unit might be over 50. During the same period in GMU 17, ADF&G received 5 reports of defense of life or property mortalities; however, wildlife managers assumed there were more unreported brown bear killings (ADF&G 2009).

## Other Recreational Users and Bear-Human Conflicts

Park managers analyzed 171 bear-human incidents over 24 years in Lake Clark NPP. They found that brown bears received food as a result of encounters with humans in 46% of the incidents, and that bears were killed in 23% of the incidents (Wilder et al. 2007). Managers were concerned about the large number of food-conditioning incidents, given that food-conditioned bears are responsible for the majority of human injuries from bears in national parks (Herrero 2002, Wilder et al. 2007). Food-conditioned bears have been found to be three to four times more likely to be killed by humans than nonfood-conditioned bears (Mattson et al. 1992). Wilder et al. (2007) also noted that casual bear photographers at private recreational cabins at Telequana Lake may have contributed to the high number of bear-human incidents because individuals repeatedly fed bears in this area to facilitate photography.

# RECOMMENDED RESEARCH

All aspects of the ecological relationships of brown bears in the Bristol Bay watershed lack adequate research and published results. Specific, local information currently available is limited to population estimates in portions of the Bristol Bay watershed (e.g., Walsh et al. 2010), a single study on home range and movements in the southwest Kuskokwim Mountains (Collins et al. 2005), a single study on seasonal distribution and movements in the central Alaska Peninsula (Glenn and Miller 1980), and a single study on foraging behavior near Lake Aleknagik (Gende et al. 2004). Other aspects of the natural history of brown bears in this area are only described through anecdotal observations of biologists working in this area. More comprehensive information on habitat use patterns and primary threats to brown bears in the Bristol Bay watershed is necessary to evaluate the effects of potential development in this area.

As indicated before, estimates of the population density of brown bears for most of the Bristol Bay watershed are not currently available. Recent work by Becker (2010) and Walsh et al. (2010) using aerial sampling surveys has provided some data on densities of brown bears in some areas of the watershed. Those efforts should be expanded to provide complete periodic (e.g., every three years) population estimates for brown bears within the watershed. Regular periodic population estimates would provide a basis upon which to evaluate the potential effects of development and other management actions in the watershed.

Understanding the habitat relationships of brown bears in the Bristol Bay watershed, particularly denning and foraging habitat, is crucial to understanding how these animals use the landscape. Habitat for a specific organism includes those areas that have a combination of resources and environmental conditions that promotes occupancy by the species and allows individuals to survive and reproduce (Morrison et al. 2006). Selection of den sites often involves a host of specific features ranging from local factors like the thermal environment to broader factors like proximity to spring foraging areas (Pigeon et al. 2014). Habitat models should be developed to provide a quantitative means of determining the potential of landscapes within the Bristol Bay watersheds as brown bear habitat (e.g., Suring et al. 2006). The results of these models may also be incorporated into analyses to describe movement patterns of brown bears and to identify potential movement corridors (Beier et al. 2007). These analyses would provide resource managers with information useful for planning habitat management, resource development, and infrastructure development (including transportation). Such planning may resolve conflicts regarding development of potential habitats by focusing management and mitigation efforts on habitat attributes and locations deemed critical to brown bears in the Bristol Bay watershed.

Salmon life histories are finely tuned to local environmental conditions. The potential for mineral extraction in the Bristol Bay watershed brings with it the potential for environmental contamination. Such contamination may lead to possible detrimental impacts to local environmental conditions critical to salmon populations in the Bristol Bay watershed. Anthropogenic climate change will also likely affect local environmental conditions for salmon populations in the Bristol Bay watershed. The associated progressive warming of the freshwater systems in the watershed has survival consequences for the populations of salmon in this area (Farrell et al. 2008). Reductions in salmon populations as a result of environmental contamination may lead to subsequent reductions in the brown bear population (Collins et al. 2005). Also, when salmon populations move or decline as a result of weather and climate-driven changes in hydrology, brown bear populations will also move or decline (Shanley et al. 2015). Brown bears in the Bristol Bay watershed likely rely significantly on the abundant salmon in this area. However, brown bears may be able to switch prey and depend more heavily on berries, ground squirrels (*Spermophilus parryii*), and caribou (*Rangifer tarandus*) if salmon populations decline. The relationship between brown bears and the salmon resource in the Bristol Bay watershed needs to be more fully examined so that the effects of potential reductions in salmon populations on the brown bear population can be more fully understood.

# SUMMARY

This chapter describes what is known about the habitat, food habits, behavior, interspecies interactions, mortality, productivity, and survivorship of the brown bears of Bristol Bay. References are also made to management and human-bear interactions. The brown bears of Bristol Bay are characterized by their large size and access to rich habitat, especially salmon. The lipids consumed from salmon can account for approximately 80% of the mass gained by salmon-eating bears and the accumulation of these fats often determines whether or not bears live through hibernation and whether females reproduce successfully. The relationship between anadromous salmon and brown bears of Bristol Bay is important to the entire ecosystem. Through their fish consumption, brown bears distribute the MDN, particularly nitrogen, in decaying salmon carcasses and through their excretion of wastes that are rich in salmon nutrients. The brown bears of Bristol Bay are also known for their tolerance of people and of other brown bears in proximity. They are among the most photographed of wildlife species in North America.

# REFERENCES

ADF&G. 1985. Alaska Habitat Management Guide, Southwest Region Volume 1: Fish and Wildlife Life Histories, Alaska Department of Fish and Game, Juneau, AK.

ADF&G. 2009. Brown Bear Management report of survey-inventory activities 1 July 2006–30 June 2008 Alaska Department of Fish and Game, Juneau, AK. pp. 336 pgs.

ADF&G. 2011a. 2011–2012 Alaska hunting regulations. No. 52, Alaska Department of Fish and Game, Juneau, AK.

Aumiller L.D., C.A. Matt. 1994. Management of the McNeil River State Game Sanctuary for viewing of Alaskan brown bears. Bears: Their Biology and Management 9:51–61.

Ballard W.B., S.D. Miller, T.H. Spraker. 1982. Home range, daily movements, and reproductive biology of brown bears in southcentral Alaska. Canadian Field-Naturalist 96:1–5.

Becker E. 2010. Preliminary final report on monitoring the brown bear population affected by development associated with the proposed Pebble mine project, Alaska Department of Fish and Game, Anchorage, AK.

Beier, P., D.R. Majka, and J. Jenness. 2007. Conceptual steps for designing wildlife corridors. Northern Arizona University, Flagstaff, Arizona, USA. Available at: www.corridordesign.org.

Collins G.H., S.D. Kovach, and M.T. Hinkes. 2005. Home range and movements of female brown bears in southwest Alaska. Ursus 16(2):181–189.

Elison, T., P. Salomone, T. Sands, M. Jones, C. Brazil, G. Buck, F. West, T. Krieg, and T. Lemons. 2015. 2014. Bristol Bay area annual management report. Alaska Department of Fish and Game, Fishery Management Report No. 15–24, Anchorage, AK.

Farrell, A.P., S.G. Hinch, S.J. Cooke, D.A. Patterson, G.T. Crossin, M. Lapointe, and M.T. Mathes. 2008. Pacific salmon in hot water: applying aerobic scope models and biotelemetry to predict the success of spawning migrations. Physiological and Biochemical Zoology 81:697–709.

Farley S.D. and C.T. Robbins. 1995. Lactation, hibernation, and mass dynamics of American black bears and grizzly bears. Can. J. Zool. 73:2216–2222.

Follman E.H. and J.L. Hechtel. 1990. Bears and pipeline construction in Alaska. Arctic 43:103–109.

Fortin J.K., S.D. Farley, K.D. Rode, C.T. Robbins. 2007. Dietary and spatial overlap amongst sympatric ursids relative to salmon use. Ursus 18(1):19–29.

Gende S.M., T.P. Quinn, R. Hilborn, A.P. Hendry, B. Dickerson. 2004. Brown bears selectively kill salmon with higher energy content but only in habitats that facilitate choice. Oikos 104:518–528.

Glenn L.P. and L.H. Miller. 1980. Seasonal movements of an Alaska Peninsula brown bear population. Bears: Their Biology and Management 4:307–312.

Goldstein M.I., A.J. Poe, L.H. Suring, R.M. Nielson, T.L. McDonald. 2010. Brown bear den habitat and winter recreation in southcentral Alaska. Journal of Wildlife Management 74:35–42.

Helfield J.M. 2001. Interactions of salmon, bear, and riparian vegetation in Alaska, (Doctoral dissertation, University of Washington). pp. 92.

Helfield J.M., R.J. Naiman. 2006. Keystone interactions: salmon and bear in riparian forests of Alaska. Ecosystems 9(2):167–180.

Herrero S. 2002. Bear attacks: their causes and avoidance. 2nd ed. Globe Pequot Press, Guilford, Connecticut.

Herrero S. and A. Higgins. 1999. Human injuries inflicted by bears in British Columbia: 1960–97. Ursus 11:209–218.

Herrero S. and A. Higgins. 2003. Human injuries inflicted by bears in Alberta: 1960–98. Ursus 14:44–54.

Herrero S., T. Smith, T.D. Debruyn, K. Gunther, and C.A. Matt. 2005. Brown bear habituation to people—safety, risks, and benefits. Wildlife Society Bulletin 33:362–373.

Hilderbrand G.V., S.D. Farley, and C.T. Robbins. 1998. Predicting the body condition of bears via two field methods. Journal of Wildlife Management 62:406–409.

Hilderbrand G.V., S.D. Farley, C.C. Schwartz, C.T. Robbins. 2004. Importance of salmon to wildlife: implications for integrated management. Ursus 15(1):1–9.

Hilderbrand G.V., T.A. Hanley, C.T. Robbins, and C.C. Schwartz. 1999a. Role of brown bears (*Ursus arctos*) in the flow of marine nitrogen into a terrestrial ecosystem. Oecologia 121:546–550.

Hilderbrand G.V., S.G. Jenkins, C.C. Schwartz, T.A. Hanley, and C.T. Robbins. 1999b. Effect of seasonal differences in dietary meat intake on changes in body mass and composition in wild and captive brown bears. Can. J. Zool. 77:1623–1630.

Hilderbrand G.V., C.C. Schwartz, C.T. Robbins, T.A. Hanley. 2000. Effect of hibernation and reproductive status on body mass and condition of coastal brown bears. Journal of Wildlife Management 64:178–183.

Hilderbrand G.V., C.C. Schwartz, C.T. Robbins, M.E. Jacoby, T.A. Hanley, S.M. Arthur, C. Servheen. 1999c. The importance of meat, particularly salmon, to body size, population productivity, and conservation of North American brown bears. Can. J. Zool. 77:132–138.

Holtgrieve, G.W. 2009. Linking species to ecosystems: effects of spawning salmon on aquatic ecosystem function in Bristol Bay, Alaska. (Doctoral dissertation,University of Washington). pp. 169.

Johnson J. and P. Blanche. 2011. Catalog of waters important for spawning, rearing, or migration of anadromous fishes—Southwestern Region, Effective June 1, 2011, Special Publication No. 11-08, Alaska Department of Fish and Game, Anchorage, AK.

much of this information is relevant to moose in Bristol Bay. We rely on the work of a distinguished group of moose biologists, which provides a near complete summary of moose biology and management (Franzmann 1998, Franzmann 2007) and also on a separate, condensed version of these data with more recent updates (Bowyer 2003).

## HABITAT

Both stable and transitory habitats are important in the evolution of moose (Geist 1971). Permanent habitats are those that persist through time without alteration in character or condition, such as riparian willow/poplar communities and high-elevation shrub/scrub communities that do not succeed to different kinds of vegetation (Peek 2007). Telfer (1984) characterized the full range of moose habitats as consisting of boreal forest, mixed forest, large delta floodplains, tundra subalpine shrubs, and stream valley/riparian zones. According to Peek (2007), boreal forest habitats are considered fire-controlled, and likely represent the primary environments in which moose evolved (Peterson 1955, Kelsall 1974). As noted by Peek (2007), delta floodplains are expected to have the highest density of moose, followed by shrub/shrub habitats, boreal forests, mixed forests, and stream valley/riparian zones. A Copper River Delta study supported the finding that large deltas and floodplains are the most productive of these five major habitat types for moose (MacCracken 1997). Boreal forest habitats are the least stable through time, whereas stream valley/riparian zones are the most stable. In Alaska and the Northwest Territories, climax tundra and subalpine shrub communities at higher latitudes and elevations are more stable in time and space than are alluvial and riverine habitats (Viereck 1980).

Transitory habitats of moose include boreal forests where fire creates successional shrub communities that provide extensive forage. Geist (1971) hypothesized that islands of permanent habitat found along watercourses and deltas and in the high elevation dwarf shrub communities serve as refugia where moose populations persist and from which they disperse into transient habitats created by fire. The frequency of fire in boreal forests is considered sufficient to promote adaptations favoring dispersal of yearling moose to newly created habitats (Peek 2007). The dominant land cover types in the study area consist of high elevation dwarf shrub, shrub/scrub and tundra habitats with lower elevation boreal forests (deciduous, evergreen, and mixed), and riparian habitats (woody wetlands) along watercourses (Table 8.1). All of these cover classes represent high-quality moose habitat.

Habitat selection by moose is influenced by availability of food, predator avoidance, and snow depth (Dussault 2005). Peek (2007) advanced the view that moose select habitat primarily on the basis of the most abundant and highest quality forage. Because these resources are unequally distributed in space and time, moose habitat may be considered as a series of patches of different types and sizes, with the value of each patch varying through the year. However, the total year-round value of a diverse habitat should be emphasized even if each part is only critical at one season or another. Further, sufficient size of both overall habitat and possibly each patch of any given habitat must be accessible to make an area suitable for occupation by moose. As a corollary, if a certain critically important community, such as shrub/scrub vegetation type, is unavailable in sufficient quantity, then the ability of an overall habitat to support moose may be reduced, even if it contains a highly diverse set of other plant communities.

The typical annual pattern of moose habitat selection includes open upland and aquatic areas that provide the best forage in early summer, followed by more closed canopy areas that provide the best forage as summer progresses and plant quality changes. In autumn, after the rut and into winter, moose intensify use of open areas with the highest biomass of dormant shrubs, which contain the remaining major source of palatable forage. Closed canopy areas are used in late winter when forage is naturally at its lowest value and quantity for the year. The nature of the cover used at this time will provide the best protection available from wind and cold, and may range from tall shrub communities to tall closed canopy conifer stands (Peek 2007). Metabolic activity in moose generally corresponds to this pattern, being highest in summer and lowest in winter (Regelin 1985). Alaskan moose generally do not use areas higher than 1,220 m in elevation (Ballard 1991). Also, in Alaska, because forage quantity and quality (nutritional value) in summer and winter can differ by orders of magnitude, winter habitat availability is often the ultimate limit on moose abundance (Stephenson 2006). Spatial heterogeneity of habitat on a relatively small scale ($\leq 34$ km$^2$) (13 mi$^2$) enhances habitat quality for moose (Maier 2005), probably because it enables moose to respond to rapidly changing conditions such as weather (Stephenson 2006).

Moose benefit from early successional stages of vegetation that provide woody browse (Schwartz 1992). A disturbance regime that provides persistent shrub communities, distributed in a diverse mosaic on the landscape, is essential to high moose density (Stephenson 2006). In Alaska, this disturbance can be provided by fire (Maier 2005, LeResche 1974), glacial outwash, or earthquakes (Stephenson 2006). On the Kenai Peninsula, forest succession following fire provided the most abundant forage for moose 20 years post-burn (Spencer 1953, Bangs 1985, Schwartz 1989, Weixelman 1998). Schwartz and Franzmann (1989) reported that, after a fire in 1969, moose abundance peaked about 15 years later, when

**Table 8.1**  Sub-basin land cover type composition relative to the National Land Cover Database* for Nushagak and Kvichak river sub-basins

| Land Cover Type | Kvichak River watershed | | | Nushagak River watershed | | | | | Total |
| --- | --- | --- | --- | --- | --- | --- | --- | --- | --- |
| | Lake Clark sub-basin | Lake Iliamna sub-basin | Subtotal | Mulchatna River sub-basin | Upper Nushagak River sub-basin | Lower Nushagak River sub-basin | Wood River sub-basin | Subtotal | |
| Size km² (mi²) | 9,148 (3,532) | 15,372 (5,935) | 24,519 (9,467) | 11,114 (4,291) | 13,017 (5,026) | 8,770 (3,386) | 3,541 (1,367) | 36,459 (14,077) | 60,961 (23,537) |
| | | | | *Percent of Watershed* | | | | | |
| Open-Water | 5.96 | 23.89 | 17.20 | 2.31 | 5.34 | 5.36 | 14.28 | 5.29 | |
| Perennial Ice & Snow | 6.51 | 0.38 | 2.67 | 0.39 | 1.61 | 0.00 | 1.32 | 0.82 | |
| Developed, Low Intensity | 0.02 | 0.01 | 0.01 | 0.00 | 0.00 | 0.00 | 0.00 | 0.00 | |
| Barren Land | 23.55 | 5.88 | 12.47 | 4.61 | 3.31 | 0.09 | .36 | 2.94 | |
| Deciduous Forest | 2.92 | 2.79 | 2.84 | 1.94 | 3.63 | 2.15 | 4.36 | 2.83 | |
| Evergreen Forest | 15.51 | 5.34 | 9.13 | 10.04 | 11.86 | 9.63 | 13.81 | 10.96 | |
| Mixed Forrest | 4.45 | 3.20 | 3.67 | 2.30 | 5.08 | 2.99 | 5.47 | 3.77 | |
| Dwarf Scrub | 10.85 | 22.57 | 18.19 | 24.50 | 12.77 | 20.06 | 5.26 | 17.24 | |
| Shrub Scrub | 28.27 | 31.79 | 30.66 | 51.08 | 54.97 | 44.86 | 44.12 | 50.23 | |
| Sedge Herbaceous | 0.01 | 1.67 | 1.05 | 0.48 | 0.37 | 3.86 | 0.13 | 1.22 | |
| Moss | 0.01 | 0.02 | 0.01 | 0.00 | 0.02 | 0.09 | 0.34 | 0.06 | |
| Woody Wetlands | 0.64 | 0.19 | 0.36 | 1.28 | 0.16 | 0.74 | 1.24 | 0.75 | |
| Emergent Herbaceous | 0.81 | 2.28 | 1.73 | 1.53 | 1.05 | 10.17 | 6.32 | 3.90 | |

*(http://alaska.usgs.gov/science/program.php?pid=23)

browse plants reached maximum productivity. The moose increase was attributed to high production and low mortality, with some initial shifting of home ranges from adjacent high-density populations. Where fire had been absent for 25 years, moose densities on the Kenai Peninsula were sufficiently high to cause the forage base to shift from a multispecies complex to a much less diverse community dominated primarily by white birch (Oldemeyer 1977). In the boreal forest, the optimum successional stages for moose are 11–30 years after burning (Kelsall 1977).

In Game Management Unit (GMU) 17 in northern Bristol Bay, moose habitat is enhanced primarily by the scouring of gravel bars and low-lying riparian areas by ice and water during the spring thaw (Woolington 2008) (for GMU map see Figure 7.4 of this volume, or refer to http://www.adfg.alaska.gov/index.cfm?adfg=huntingmaps.alaskamaps for a comprehensive listing of GMU maps in Alaska).

Willows and other plants quickly regenerate after bank scouring and subsequent deposit of river silt (Woolington 2004). This disturbance mechanism is particularly important for the Nushagak and Mulchatna Rivers and for the lower reaches of the major tributaries to those rivers (Woolington 2008). Major river systems with large riparian zones, such as those found in Bristol Bay, contain alluvial habitats that support an abundant moose population because they feature an abundance of nutritious food, primarily in the form of regenerating willow stands. Deciduous shrubs proliferate in these areas because of the annual influx of nutrients from waterways, sufficient soil moisture, and changing river channels. Wildfires caused by lightning also occur occasionally in GMU 17 (Woolington 2008), and provide disturbances that enhance moose habitat. Moose habitat has not been formally assessed for GMUs 17B and 17C. Much of GMU 17 is wet or alpine tundra, and moose are located mostly along the riparian areas (Woolington 2008).

In interior Alaska, female moose density was highest close to towns, at moderate elevations, in areas with the greatest amounts of riparian habitat, and in areas where fire had occurred 11 to 30 years earlier (Maier 2005). Moose avoided nonvegetated areas. Female moose preferred areas with plentiful diverse patches, indicating their need for habitat with both food and concealment. Maier et al. (2005) postulated that moose preferred to be near towns either because the human disturbance of vegetation provided high-quality food, or because predators such as wolves and grizzly bears tend to avoid human-inhabited areas.

On the Copper River Delta, tall, open alder-willow and low sweetgale-willow habitats were used most by moose, and use of tall, closed alder-willow habitat was intermediate (MacCracken 1997). Aquatic and woodland spruce habitats were used the least by Copper River Delta moose. Aquatic plants were used seasonally during the period from April through August, and were used primarily for foraging by both sexes (MacCracken 1993).

In northwest Alaska during March and April, moose used (quantified by percent of time) stands of felt-leaf willow (*Salix alaxensis*) (85.0%), followed by other willow (*Salix*) (6.4%), riparian areas (3.9%), gravel bars (2.5%), and upland areas (1.3%) (Gillingham 1992).

## WINTER HABITAT

Influence of snow on moose habitat use patterns has received considerable attention. Heavy snow during severe winters can be a limiting factor for moose populations. Deep snow can reduce browse availability by burying it, and travelling through deep snow requires increased energy expenditure (Ballard 1991).

Snow characteristics that have ecological significance for moose include temperature, density, hardness, and depth (Peek 1986). Because the temperature of snow fluctuates less than ambient temperatures (and never falls as low as the air temperature), snow provides insulation for moose against temperature extremes (Peek 2007). Density and hardness influence the ability of an animal to travel across or through snowpack. Under some conditions, snow density can be sufficient to support the mass of a wolf but not a moose. Under these circumstances wolf predation on moose tends to be high (Ballard 2007). For other cervids [e.g., mule deer (*Odocoileus hemionus*) and elk (*Cervus* spp.)] energy expenditure while moving through snow increases exponentially with increasing snowpack maturation through the winter. Hardness and density affect sinking depth, and snow level at front knee height has been suggested as a threshold parameter (Parker 1984). Applying the same principals and relationship to moose, snow depth beyond 50–60 cm (20–24 in) would result in a relatively large increase in energy expenditure for movement (MacCracken 1997). Snow depths of 70–100 cm (28–39 in) have been shown to limit the travel of moose (Des Meules 1964, Kelsall 1969, Kelsall 1971). When snow depths approach 97 cm (38 in), moose have been confined to areas where forest canopies are dense (Kelsall 1971).

The distribution of snow within the forest influences moose habitat use patterns. Snow depth is nearly always greater in open areas until late winter, when snow exposed directly to the sun melts more rapidly than snow protected by tree canopies. Snow falling on tree branches of fine-needled conifers, such as spruce, tends to be retained in the canopy and produces a lower snow depth immediately beneath the tree canopy in areas called tree wells (Pruitt 1959). This snow tends to produce a hard dense surface when it falls from the tree branches to the ground, which provides more support for moose traveling beneath the canopy (Peek 2007). In some geographic locations with deep winter snow, mature coniferous forests can provide zones of shallow snow accumulation that benefit moose survival (Balsom et al. 1996). In deep snow habitats where conifers are absent, such as in shrub/scrub tundra or riparian communities, moose use the best available microsites produced by combinations of shrub canopies and topography that reduce snow depths. However, the principal adaptation is simply to reduce energy expenditure (Peek 2007).

Severe winters have been associated with high rates of starvation among moose calves (Ballard 1991). In Quebec, cows with calves preferred habitats providing protection from predators, whereas solitary adult females preferred habitats with moderate food abundance, moderate protection from predators, and substantial shelter against deep snow (Dussault 2005). In Denali National Park and Preserve (DNPP) during the severe winter of 1986, large bulls were the only moose able to remain in the Jenny Creek unit, which had a higher forage biomass but deeper snow than other units (Miquelle 1992). Moose are very tolerant of cold temperatures, but are susceptible to heat stress. The upper critical temperature range for moose during winter is −5 to 0°C; in summer, upper thermal limits are 14 to 20°C (Renecker 1986).

## FOOD HABITS

Moose eat mainly leaves, twigs, and bark of woody plants (Schwartz 1992, Van Ballenberghe 1989). Renecker and Schwartz (2007) reviewed the diets of moose, listing more than 221 plant genera/species, with willow (*Salix*), birch (*Betula*), and alder (*Alnus*) predominating across North America. Daily use patterns of time and space explain how moose satisfy hunger, remain fit, avoid thermal stress, maintain security from predators, and reproduce. Many of an individual moose's life-cycle needs interact daily; thus, tradeoffs often occur, because many requirements are more critical at certain times than at others (Figure 8.2). The day-to-day needs of moose for food and thermoregulation are most often preempted in favor of other activities that accommodate reproduction. However, the survival instinct is satisfied most on a daily basis by optimizing food consumption while minimizing risk and effort. In this regard, a basic constraint for most moose is that food is abundant, yet of low quality (Renecker 2007).

Moose are ruminants, with a four-chambered stomach. In ruminants, browse, forbs, and grass are held in the large-chambered rumen until adequate nutrients are extracted from fibrous materials and plant particles are small enough to pass through to the omasum and abomasum. Based on feeding habits, specialization, and design of the digestive anatomy, ruminants are classified into three main groups (Hofmann 1973): browsers that eat mainly shrubs and trees; grazers that eat mainly grass; and mixed or intermediate feeders that eat a mixture of grass, forbs, and browse. The moose is a

**Figure 8.2**  Bull moose in velvet. A soft membrane—referred to as *velvet*—is a layer of skin that supplies the growing antlers with the nutrients needed to build the bone mass. *Photo credit: Carol Ann Woody.*

browser and has been classified as a seasonally adaptive concentrate selector (Hofmann 1989). Concentrate selectors have a relatively small ruminoreticular chamber and must search for high-quality foods that will pass rapidly through the digestive system. Moose select plant species and parts (twigs and foliage) high in cell-soluble sugars that ferment rapidly in the rumen. They generally avoid plants that are fibrous and require extensive breakdown in size before passage from the rumen. Moose have a relatively narrow muzzle, prehensile lips, and a tongue that allow them to select high-quality plant parts (Renecker 2007). Moose ferment mostly the soluble components of their food, and propel digesta rapidly through their digestive system (Schwartz 1992). Their digestive efficiency is regulated by forage selection, rumination, gut morphology, and mechanisms controlling the rate of passage of food (Schwartz 1992). Moose can ingest and process high-quality foods more rapidly, (e.g., aquatic plants eaten in summer), because both passage and digestion rates are increased (McArt 2009).

Plant species distributed within the range of moose respond to seasonality by growing during the short summers and entering a state of dormancy during the long winters. Plant nutrient quality varies seasonally. Plants begin their growth phase in early spring, long before actual green-up occurs. In general, spring and summer foods are one and a half to three times more nutritious than winter foods, depending on which constituent is examined (Schwartz 2007). Summer diets contain excess digestible energy and protein, whereas winter diets generally are insufficient to meet maintenance requirements (Renecker 1989, Schwartz 1987, Schwartz 1988).

Moose feeding habits and diets vary seasonally as a result of a complex interaction of internal physiological regulators and the external environment. There is an annual cycle of food selection and intake, fat metabolism, metabolic rate, and body mass dynamics that is not driven simply by food quality and availability (Schwartz 1992). The gastrocentric hypothesis predicts that large male moose will eat large amounts of low quality, fibrous foods, while smaller-bodied females will consume smaller amounts of higher-quality forage to meet the demands of reproduction and lactation (Oehlers 2011). Both sexes reduce food consumption and metabolic rates in winter when they operate at a net energy deficit by mobilizing fat reserves (Schwartz 1992, Miquelle 1992).

Protein and energy are considered the major limiting nutrients within the environment (Schwartz 1992). Summer protein intake is critical for lactating female moose (McArt 2009). Tannins have a negative influence on forage quality because they reduce the amount of protein available (Robbins 1987). In two areas of Alaska, browse quality differences were consistent with observed differences in moose reproductive success (McArt 2009). In recent years, the productivity of DNPP moose has been significantly higher than that of moose in the Nelchina Basin. A study of browse quality in the two areas found that, on average, nitrogen levels were 9% lower and tannin levels 15% higher in the Nelchina Basin than in DNPP, resulting in a digestible protein differential of 23%. McArt et al. (2009) concluded that the Nelchina moose population was nitrogen-limited. In both systems, browse quality declined significantly as summer progressed, with nitrogen levels decreasing and tannin levels increasing in all species of browse studied. In comparison with early-summer forage, digestible protein had decreased by an average of 35% by midsummer and 70% by late summer (McArt 2009).

High-quality summer forage, particularly near wetlands, allows nursing cows to regain body condition and calves to grow so they can better escape predators and survive their first winter (ADF&G 2011). During the spring-summer period, moose feed in aquatic habitats. In the Copper River Delta, aquatic habitats produced about four times more forage than terrestrial habitats, and the forage was more digestible (MacCracken 1993). Although some researchers have linked summer consumption of aquatic plants by moose to a craving for sodium (Jordan, 1987), data from the Copper River Delta did not support this hypothesis. Rather, these data suggested that moose selecting aquatic forage were switching from an energy-maximizing to a time-minimizing strategy (MacCracken 1993) because aquatic plants were high in water content. Although a moose can fill its rumen quickly, the relative quantity of dry matter consumed is less than when eating the same amount of terrestrial vegetation, such as leaves.

Moose also select bark as part of their diet seasonally, although it forms a relatively small part of the diet. Bark stripping occurs mostly in winter, when there are fewer twigs available due to snow cover (MacCracken 1997). In DNPP, female moose also stripped bark in aspen-spruce forests in May and June, coincident with birth and lactation (Miquelle 1989). Moose also avoid certain plant species because of low palatability due to their chemical defenses, such as black cottonwood (*Populus balsamifera trichocarpa*) on the Kenai Peninsula (Weixelman 1998) and white spruce (*Picea glauca*) both on the Kenai and in other parts of Alaska (Weixelman 1998).

In the Copper River Delta, moose consumed three different diets that varied among winter, spring/early summer, and late summer/fall (MacCracken 1997). Willow dominated all three diets; the differences were related to the amounts of sweetgale (*Myrica gale*), marsh five-finger (*Potentilla palustris*) and graminoids in the diet. Winter diets included sweetgale and alder (*Alnus* spp.), which are both nitrogen-fixers, leading to relatively higher protein content. Spring/early summer diets were most diverse, due to the increased use of emergent aquatic plants such as marsh five-finger. Late summer/fall diets were least diverse, consisting almost entirely of willow leaves and twigs (MacCracken 1997).

Moose diets in DNPP were also found to vary by season. In summer, seven species of willow constituted 81.5% of the diet (Van Ballenberghe 1989), and diamond-leaf willow (*Salix planifolia*) was eaten more than any other plant species (45.7%). Willow constituted 94% of DNPP moose diets in winter (Risenhoover 1989).

Moose can influence the composition and productivity of the terrestrial plant community through browsing (Bedard 1978). In DNPP, moose initiated positive feedback loops in their environment through browsing (Molvar 1993). Willows exhibited a growth response to moose herbivory; specifically, leaf area was significantly greater at a site with high moose density than at a site with low moose density. Annual biomass productivity per growing point on willow stems increased with rising browsing intensity on the plant as a whole because of release from apical dominance. Moose also increase nutrient cycling rates because their urine and feces transfer nutrients to soil. The organic content of soil can also be enhanced by moose, in turn benefitting microbiota such as decomposers (Molvar 1993). In interior Alaska, twigs regrowing from two-year-old willow stems that had been browsed by moose had larger diameters than those that had not been browsed in the previous year (Bowyer 2003). Browsing on felt-leaf willow had no effect on nitrogen content, digestibility, or tannin content, indicating that the willow had no tannin-mediated inducible defense system in response to herbivory (Bowyer 2003).

Marine-derived nutrients (MDNs) carried upstream by spawning salmon have implications for nutrient flow into riparian habitats, and are thought to enhance growth and productivity therein (Quinn 2009). Although it is plausible that MDNs contribute to increased plant productivity and thus benefit moose, evidence of this direct impact was not located in the scientific literature.

Moose density is often associated with food abundance (Eastman 1987, Joyal 1987, Oldemeyer 1987, Thompson 1987, Schwartz 1989). As reviewed by Renecker and Schwartz (2007), forage biomass varies with successional stage of forests. In Newfoundland, available woody biomass increased from about 200 kg/ha (179 lbs/ac) in two-year-old clear cuts to more than 2,000 kg/ha (1,786 lbs/ac) at eight years, at which time it peaked and subsequently declined gradually (Parker 1978). On the Kenai Peninsula, important browse species peaked about 15 years after fire (Spencer 1964). Oldemeyer and Reglin (1987) estimated the biomass of important browse species in successional stands on the Kenai National Wildlife Refuge (NWR); browse production measured at 3, 10, 30, and 90 years post-burn was 37, 1,399, 397, and 4 kg/ha (33, 1,249, 354, and 3.6 lb/ac), respectively (Oldemeyer 1987).

# BEHAVIOR

## Movements and Home Ranges

The ways in which moose use their environment both spatially and temporally are of great interest to resource managers (Hundertmark 2007). The dynamics of animal movements and distribution in space and time are integral to behavioral, ecological, genetic, and population processes. Thus, the attributes of the space occupied by individual animals, both annually and seasonally (home ranges), patterns of movement within home ranges, establishment of new home ranges by young moose, and colonization of new habitats (dispersal) and movements between seasonal ranges (migration) must all be considered in comprehensive management programs.

Moose home range sizes vary with the sex and age of the animal, season, habitat quality, and weather. Two studies from Alaska generated the largest estimates of home range size, although one of these (Grauvogel 1984) included migratory locations in the estimate of seasonal ranges, which can increase home range size significantly (Hundertmark 2007). Moose in south-central and northwest Alaska (Ballard 1991) had mean seasonal ranges > 92 km² (36 mi²). With the exception of home ranges of nonmigratory adults in the later study, total home range sizes exceeded 259 km² (100 mi²). In contrast, estimates of annual ranges for moose in northwest Minnesota were ≤ 3.6 km² (1.4 mi²) (Phillips 1973).

Seasonal ranges, when they exist, represent partitioning of the environment based on behavioral and energetic constraints. Migratory moose (those that use separate winter and summer ranges) use distinct seasonal ranges because they attempt to optimize their nutrient intake on summer range, but conditions on these ranges may preclude occupation during some or all winters. Moose that remain on the same range during winter and summer are termed nonmigratory, and remain because environmental conditions permit it. A third seasonal range, associated with mating, occurs in autumn, but many investigators define this as part of the summer range (Hundertmark 2007). Breeding areas for tundra moose are typically in open habitats where visibility is good. This is likely for behavioral purposes so bulls and cows can see each other as they display. It may also afford some protection from predators.

In several moose populations studied in Alaska, some individuals were nonmigratory residents, whereas others migrated seasonally. In the Copper River Delta, eight of 15 collared females were migratory, whereas two of five collared males were migratory (MacCracken 1997). Moose in the study area exhibited greater fidelity to summer range than to winter range (MacCracken 1997). Winter severity influenced yearly winter migratory behavior in the Copper River Delta (Stephenson 2006). Moose populations in south-central Alaska (GMU 13, consisting of the Nelchina and upper Susitna basins) also included both migratory and nonmigratory individuals (Ballard 1991). Migratory moose exhibited three seasonal periods of movement—autumn migration to wintering areas, spring migrations to calving areas or summer feeding grounds, and early fall migrations to rutting areas (Ballard 1991). In the Togiak River drainage of the northern Bristol Bay area (GMU 17A), some collared moose were resident, whereas others migrated seasonally (Woolington 2008). During a population estimation survey in February 1995, 29 moose were documented moving westward from the upper Sunshine Valley in GMU 17C (the Lower Nushagak watershed) into GMU 17A (Woolington 2008).

Cows with newborn calves restrict their movements for the first few weeks, after which they gradually expanded their home range to the approximate home range size of other adults (LeResche 1974). In one study, cow-calf pairs had smaller summer home ranges than did other moose, and calf movements increased exponentially with age during the first six weeks of life (Ballard 1980).

When comparing among the sexes, males are almost always found to occupy larger annual home ranges. In south-central Alaska, males had significantly larger home ranges than did females (Ballard 1991). In northwestern Alberta researchers found no difference between the sexes, but noted the tendency for bulls to occupy larger winter and spring home ranges (Lynch, 1984).

In several Alaskan moose populations, the timing of seasonal migration has been observed to vary significantly among individuals. In the Nelchina Basin (GMU 13), individual moose movements varied by month—both in the initiation and the duration of winter migration (Van Ballenberghe 1977). Snow depth was an important factor that influenced winter migratory behavior in that population. Cows with calves tended to migrate to wintering grounds earlier than did males and cows without calves (Van Ballenberghe 1977). During spring, the initiation of migration varied substantially between individuals, but all migrated quickly once they started moving (Van Ballenberghe 1977). Individual moose in GMU 13 initiated migration to wintering areas, ranging from mid-August to mid-February (Ballard 1991). Dates of spring migration ranged from March through mid-July; during some years, moose remained on the winter range for calving. Subsequent movement to the summer range in mid-summer seemed related to plant development (Ballard 1991).

Moose in various areas within Alaska migrated differing distances seasonally and had varied annual home range sizes (Ballard 1991, MacCracken, 1997, Gillingham 1992) (Table 8.2). Moose on the Seward Peninsula of northwest Alaska migrated up to 80 km (49.7 mi) seasonally (Gillingham 1992). In southcentral Alaska, the distance between winter and summer ranges of migratory moose averaged 48 km (29.8 mi), and varied from 16 to 93 km (9.9 to 57.8 mi) (Ballard 1991).

Moose use of seasonal home ranges varies little from year to year (Ballard 1991). In south-central Alaska GMU 13, only one of 101 radio-collared female adults dispersed from their home range during a 10-year study period (ibid). During the fall of 1978, that female moved 177 km (110 mi) from her previous location (ibid).

**Table 8.2** Mean (range) home range size (km$^2$) for selected Alaska moose populations

| Study area | Migratory status | Age/Sex | Mean home range size (km$^2$) | | | Reference |
| --- | --- | --- | --- | --- | --- | --- |
| | | | Total | Winter | Summer | |
| Kenai Peninsula | M | Adult/M | 137 (56–185) | | | (Bangs 1984) |
| | N | Adult/M | 52 (34–64) | | | |
| | N | Adult/F | 127 (25–440) | 63 (13–184) | 36 (2–152) | (Bangs 1980) |
| Seward Peninsula | M | Adult | 938 (236–1,932) | 311 (36–1,393) | 324 (41-1,323) | (Grauvogel 1984) |
| | N | Adult | 218 (91–350) | 98 (36–223) | 93 (44–150) | |
| | I | Adult | 339 (205–593) | 122 (21–334) | 210 (60–559) | |
| South-central | N | F | 290 (111–787) | 113 (10–430) | 103 (23–456) | (Ballard 1991) |
| | M | F | 427 (274–580) | 147 (15–375) | 263 (60–622) | |
| Southeast | N | Adult/F | 28 (9–51) | 11 (3–30) | 14 (2–30) | (Doerr 1983) |

Migratory status: M = migratory, N= nonmigratory, and I = intermediate. Data from Hundertmark (2007).

In northern Bristol Bay, some moose collared in GMU 17A beginning in 2000 moved westward within GMU 17A and into the southern part of GMU 18 (Woolington 2008). This is thought to be part of a continued westward expansion into previously unpopulated moose habitat (ibid).

## Sexual Segregation and Grouping Behaviors

Bowyer et al. (2003) provide a succinct discussion of sexual segregation in moose, and we paraphrase it here. Sexual segregation is the differential use of space by the sexes outside the mating season (Barboza 2000) and often includes differential use of habitats and forage. Sexual segregation is especially pronounced in moose and plays a crucial role in their ecology (Miller 1992, Miquelle 1992, Bowyer 2001). In Alaska, male and female moose select habitats differently, leading to their spatial segregation throughout most of the year (Oehlers 2011). Adult males select habitats with greater forage abundance and females select areas with more concealment cover during winter (Bowyer 2001, Miquelle 1992), whereas, cows with calves select denser cover and are more secretive than other age groups (Peek 2007). Females with calves remain solitary and prefer forested habitats, which provide better cover from predators. Miquelle et al. (1992) found that spatial segregation was most extreme during a deep-snow winter, when only large males could access forage at higher-elevation Jenny Creek due to their larger body size. Such differences in habitat use between the sexes have implications for surveying of moose populations, because it can affect the accuracy of sex and age ratio information obtained by direct observation (Peek 2007, Peterson 1955, Pimlott 1959, Bowyer 2002).

The effect of habitat enhancement on sexual segregation was studied in interior Alaska after willow stands had been mechanically crushed (Bowyer 2001). During winters following the treatment, males occurred predominantly on the more open, disturbed area. Adult females and young resolved the tradeoff between foraging on a greater abundance of food in the disturbed area and a reduced risk of predation in the mature stand by using older stands of untreated willow, where dense vegetation offered substantial concealment from wolves (Bowyer 2001).

The way that moose—either individually or in groups—partition their habitats and associate with other moose can be informative in determining the needs of various segments of the population (Hundertmark 2007). Moose have been referred to as "quasi-solitary," and large groups are uncommon (Houston 1968). The tendency of moose to lead a solitary life or to occur in groups depends on their age, sex, and reproductive status, and varies by season. Molvar and Bowyer (1994) note that Alaska moose are more gregarious than moose from Eurasia and suggest that the formation of social groups is a recently evolved adaptation in response to a relative abundance of predators and to relatively open terrain. In DNPP, larger groups ventured farther from cover but were less efficient at foraging due to inter-individual aggression (Molvar 1994).

Cows with calves are consistently the most solitary members of the population, probably because of predator avoidance (Hauge 1981, Miquelle 1992, Hundertmark 2007). Alaskan female moose with calves are nearly always solitary at the time of birth (Molvar 1994, Miquelle 1992). Females without calves are more likely than males to be solitary during early summer, but they become more gregarious as summer progresses (Miquelle 1992). In DNPP, during the summer, females without calves were seen alone only 23% of the time (ibid). During June-August, male moose in DNPP were consistently gregarious (ibid). When in a group, small males were more likely than large males to be in a group that included females at all times except during the rut and post-rut (ibid).

In south-central Alaska (GMU 13), calves separated from their mothers at an average age of 14 months (Ballard 1991). In that study, 33% of yearlings and a single two-year-old moose were observed in one to six temporary reassociations with their mothers after their original dispersal. Calves were more likely to reassociate with their mother if she was not caring for a new calf (Ballard, 1991).

## Mating and Maternal Behaviors

Moose in North America display two general patterns of mating behavior. In the taiga of Canada, moose have a serial mating system in which bulls search for cows in heat by traveling widely while calling and thrashing their antlers (Bubenik 2007). The bull digs shallow pit holes in which he urinates, but they are located randomly and seldom in the same spot in successive years. In order for all of the cows to be bred during the three-week mating season, the serial mating strategy requires the presence of a relatively high number of bulls. For moose living in tundra habitats, Bubenik (2007) concluded that due to differences in climatic conditions of the periglacial tundra, the tactic of serial mating was replaced by communal or harem mating. Among tundra moose, a prime bull settles in a mating area of about 10 km$^2$ (3.9 mi$^2$) toward the end of August. Rutting areas are used traditionally (Bubenik 2007). In early September, bulls begin scent-urinating on trails and in pit holes. Two prime-aged bulls may share a harem when it contains eight or more cows. During the eight to 10 days of breeding in the harem, a tundra bull can probably fertilize as many or more cows as a taiga bull does during the entire three-week rut because the tundra bull can mate with each female in his harem without traveling long distances to locate a new female.

Many mammals have evolved seasonal reproductive patterns that ensure adaptation to predictable annual changes in the environment. Moose exhibit marked seasonal changes in reproductive behavior that reflect adaptations to yearly fluctuations in food availability and that ensure favorable conditions for rearing young (Schwartz 2007). Thus, moose breed only during autumn. Breeding in the fall ensures that calves are born in spring when forage is high in nutrient quality and that cows have a high probability of producing enough milk to successfully raise their calves.

Day length may provide the clue to annual timing of the breeding season. The breeding season for moose is relatively short. Because it is difficult to determine the exact date of breeding under natural conditions, few studies provide detailed information. Researchers with the most robust data sets have concluded that moose exhibit a very well-defined breeding season, as judged by conception dates and the spread of observed breeding (Thomson 1991, Crichton 1992, Schwartz 1993). The mean date of breeding in British Columbia ranged from October 5–10, with a standard deviation of five days (Thomson 1991). The mean breeding date in Manitoba was September 29—and 93% of females were bred by October 12 (Crichton 1992). The mean breeding date in Alaska was October 5, with a range from September 28 to October 12. Annual variation in breeding dates in all studies was minor, suggesting that photoperiod, rather than weather influenced rut timing. Synchrony of the rut has also been observed in DNPP. Over a 12-year interval, rutting consistently occurred during the brief period from September 24 through October 5 (Van Ballenberghe 1993).

Moose cows across North America give birth during a relatively short period. The peak birthing period occurs from about May 15 through June 7 (Schwartz 2007). In DNPP, timing of birth in moose was consistent from year to year, despite variation in climate between years (Bowyer 1998). Birth timing exhibited *extreme synchrony* and Bowyer et al. (1998) hypothesized that moose were tracking long-term patterns of climate to time reproduction. Hence, there is concern that moose will be vulnerable to climate change independent of changes to vegetation (Bowyer 1998).

In DNPP, the primary drivers influencing birth site selection were microclimate, forage abundance and quality, and risk of predation (Bowyer 1999). Birth sites were not reused, and some females appeared to behave unpredictably, shortly before giving birth, perhaps in an attempt to thwart predators (ibid). Proximity to human development did not influence birth site selection. Moose preferred birth sites with abundant willow, high visibility (to detect predators), and a southeasterly exposure that would be warmer and drier. Bark stripping was common around birth sites, because females seldom traveled more than 100 m (328 feet) from their young and hence, rapidly depleted the birth site's forage (ibid) (Figure 8.3).

# ACTIVITY BUDGETS

Moose spend most of their active time foraging. Seasonal rates of forage intake tend to follow the cyclic nature of energy metabolism in moose (Regelin 1985), with higher rates of activity and intake in spring and summer and reduced rates during winter. Activity budgets tend to follow a similar pattern. Activity budgets have been studied for DNPP moose during winter (Risenhoover 1986) and spring/summer (Van Ballenberghe 1990). DNPP moose exhibited low activity levels from January through April, when they were active, on average, only 27.3% of the time (6.5 h/d) (Risenhoover 1986). Behaviors associated with resting and foraging comprised 99.3% of DNPP moose winter activity budgets. In contrast, Miquelle et al. (1992) found that during winter in DNPP, small males spent some of their active time engaged in social behavior. In both winter and early spring in DNPP, moose exhibited a polyphasic pattern, alternating between foraging and bedding, with about six cycles per 24 hours (Risenhoover 1986).

Following their relative inactivity in winter, DNPP moose increased their metabolic rate in April, as evidenced by the onset of antler development in males and increased mobility of cows (Risenhoover 1986). Activity increased during May to a peak in early June, then began to decline until mid-August (Van Ballenberghe 1990). DNPP moose were active 12.8 h/d at the peak, and activity declined to 9 h/d by late summer. In summer DNPP moose spent about equal amounts of time feeding, resting, and ruminating during each 24-hour period (Van Ballenberghe 1990). The differences between winter (Risenhoover 1986) and spring/summer (Van Ballenberghe 1990) activity budgets in DNPP, are summarized in Table 8.3.

On the Seward Peninsula of northwestern Alaska, moose winter activity budgets were 43.2% feeding, 42.8% bedding, 8.4% walking, 4.4% standing, and 1.2% other (Gillingham 1992). Walking time was far greater than reported

**Figure 8.3**  Cow moose and calf. *Photo credit: Carol Ann Woody.*

**Table 8.3**  Moose activity budgets in winter[1] and spring/summer[2] in Denali National Park and Preserve (averages)

| Activity Parameter | Winter | Spring/Summer |
|---|---|---|
| Total active time/day (hr) | 6.5 | 10.1 |
| Total resting time/day (hr) | 17.5 | 13.9 |
| # Activity bouts/day | 5.7 | 8.2 |
| Duration of activity bouts (min) | 68 | 73 |
| Duration of resting bouts (min) | 178 | 97 |
| Foraging time/day (hr) | 4.9 | 7.5 |
| Rumination time/day (hr) | 11.7 | 6.7 |

[1] *Winter* is January through April; data from Risenhoover 1986

[2] *Spring/Summer* is 1 May through 15 August; data from Van Ballenberghe and Miquelle (1990)

**Figure 8.4**    Adult moose face. *Photo credit: Carol Ann Woody.*

during winter in DNPP (where is was < 1%) (Risenhoover 1986). Gillingham and Klein (1992) attributed this difference to moose on the Seward Peninsula using the Kuzitrin River as a feeding and movement corridor during winter. The use of a narrow, linear feature, such as a river bottom, requires moose to travel farther up and down the river to obtain food, as opposed to feeding in a large, nonlinear area. At least two other differences between moose in DNPP and the Seward Peninsula are notable. First, DNPP is characterized by an abundance of predators (wolves and bears), whereas no predators use the Seward Peninsula in winter (Gillingham 1992). Second, moose activity on the Seward Peninsula was highly synchronized during mid-afternoon in late April, presumably due to heat stress (ibid). In contrast, there was no significant correlation between mean daily temperature and daily activity level in DNPP during winter (Risenhoover 1986).

On the Copper River Delta, the duration of inactive periods of moose was shortest on the west delta, which had the highest estimates of forage abundance and quality among the three areas studied (MacCracken 1997). This suggests that the relative duration of inactive bouts might be useful as an index of habitat quality for moose (Figure 8.4).

## INTERSPECIES INTERACTIONS

Boer (2007) provided an excellent review of the interspecific relationships between moose and other species. Interspecific interactions between moose and other species take on one or more of the following general forms: competition, parasite-mediated competition, predation, and commensalism (Boer 2007). Due to the diversity of habitats, species combinations, and abundance of sympatric species throughout the moose range, a variety of competitive mechanisms operate. Of the interspecific interactions possible, competition most obviously influences moose habitat use and distribution (Boer 2007). Throughout their North American range, moose compete with an array of other ungulate species. However, in Bristol Bay, caribou are the only other ungulate species abundant enough to compete with moose. Direct competition between moose and caribou appears limited and insignificant (Davis 1979). Food preferences of moose and caribou coincide to some degree, but the diet of caribou appears to be more specialized. In winter, caribou consume forbs and deciduous vegetation and lichens (Darby 1984, Servheen 1989). Moose primarily consume browse, but also use forbs and deciduous vegetation during summer (Dodds 1960, Eastman 1987).

As reviewed by Boer (2007), in multiprey systems, moose and caribou populations may influence each other indirectly. Increasing moose numbers in western and central portions of DNPP have resulted from increased availability of caribou as alternate prey for wolves (Singer 1985). In the eastern section of DNPP, migrating caribou were available as prey for only a brief period of time, and therefore, were not a particularly important factor of the area's prey base; moose populations have declined. Moose are the primary prey of wolves in other areas of Alaska as well (Gasaway 1983), although other authors have attributed an increase in moose numbers in northern Alaska to a preference by wolves for caribou (Coady 1980).

Interspecific population dynamics have been studied in several areas of Alaska with multiple predator and prey species. These relationships can be quite complex and can vary, based on both abiotic and biotic factors within the ecosystem. None of the interspecies studies reviewed here were conducted in Bristol Bay.

In Alaskan ecosystems with multiple predators, bears were responsible for more moose calf kills than were wolves. Black bears can be significant predators of moose calves (Franzmann 1980). Of 47 radio-collared neonatal calves on the Kenai Peninsula, black bears killed 34%, whereas brown bears and wolves each killed 6% (ibid). In the western Interior, near McGrath, black bears were also the dominant source of predation mortality of calves during six out of seven years studied; wolves and brown bears were secondary predators in that system (Keech 2011). In contrast, brown bears were the primary predators of moose calves in a south-central Alaska study, causing 73% of calf mortality (Ballard 1991). Brown bears were also the primary predator in east-central Alaska (GMU 20E), where 79–82% of radio-collared moose calves died by the age of eleven months (Gasaway 1992). In that study 52% of moose calves were killed by brown bears, 12–15% of calves by wolves, and 3% by black bears.

Several studies have compared the causes of calf mortality in the nutritionally unproductive 1947 burn and in the productive high-quality habitat of the 1969 burn on the Kenai Peninsula, Alaska (Franzmann 1986, Schwartz 1989, Schwartz 1991). Black bears killed 34 and 35% of the calves, respectively, whereas wolves and brown bears killed five and 13%, respectively. Total calf mortality from all causes ranged from 51–55%. Moose densities were four times greater in the 1969 burn area ($370/100 \text{ km}^2$) ($9.6/\text{mi}^2$) and the population was increasing, whereas the population in the 1947 burn was about $100/100 \text{ km}^2$ ($2.6/\text{mi}^2$) and declining. The investigators concluded that habitat quality had a significant impact on reproductive rate and population growth. The moose population in high-quality habitat (1969 fire area) was capable of sustaining this level of predation and continued to grow, whereas the population in poor habitat (1947 fire area) was not.

Wolves appeared to select moose calves in some areas and seasons in Alaska, but not in others. In south-central Alaska, moose calves were taken in proportion to their abundance during May through October (Ballard 1987). In contrast, during November through April, wolves preyed on moose calves selectively. During those winter months, calves were only 12–20% of the moose population, but they comprised 40% of kills by wolves (ibid). During autumn in northwest Alaska, wolves did not display selectivity for moose calves, which were killed in proportion to their abundance in the population (Ballard 1997). On the Kenai Peninsula, wolves killed mostly moose calves (47%), yearlings, and older adults (Peterson 1984). Half of moose adults killed by wolves during that study were > 12 years old. Wolf predation on moose calves was highest during the winter with deepest snow, and calves killed after January 1 were commonly malnourished, with a bone marrow fat content of ≤ 10%.

In east-central Alaska (GMU 20E), predation was the primary cause of nonhunting deaths for yearling and adult moose (Gasaway 1992). Of 46 nonhunting moose deaths during 1981–1987, 89% were killed by brown bears or wolves, 9% died from antler wounds or locked antlers, and 2% drowned. Peterson et al. (1984) examined the incidence of debilitating conditions among 109 wolf-killed adult moose on the Kenai Peninsula. They found that 20 had moderate or severe periodontitis, 14 had arthritis, 1 had a broken leg, and 1 had a leg wedged between trees. Of 40 wolf-killed adult moose assessed for bone marrow fat content, 4 had levels of ≤ 20%, indicating severe malnutrition (Peterson 1984).

Wolves in various regions in Alaska displayed different relative preferences for moose and caribou as prey. In south-central Alaska, moose were the primary prey of wolves, constituting 38% of observed kills, whereas caribou were the second most important prey at 21% (Ballard 1987). In northwest Alaska, caribou were the preferred prey of wolves (Ballard 1997). In January through April 1988, when caribou were abundant, 92% of observed ungulates killed by wolves were caribou. In contrast, in 1989 and 1990 when caribou were less abundant, they constituted 11% and 48% of observed ungulate kills, respectively (ibid).

Estimated kill rates for wolf packs on the Kenai Peninsula varied from one moose every 3.1 days to one moose every 21.4 days (Peterson 1984). The average kill interval in winter for Kenai wolf packs with more than two members was 4.7 days. In 38 wolf-moose encounters observed on the Kenai Peninsula, wolves succeeded in killing only two moose (ibid).

Ballard et al. (1997) speculated that the recent occurrence of moose (since the 1940s) in northwest Alaska has altered the historical migratory patterns of wolves in that area. There is evidence that wolves in northwest Alaska formerly migrated with the caribou herds, but now they do so only when alternate prey (moose) numbers are insufficient.

## MORTALITY, PRODUCTIVITY, AND SURVIVORSHIP

Understanding the dynamics of a population requires knowledge of how many individuals it contains, how fast it is increasing or decreasing, its rate of production of young, and its rate of loss through mortality (Van Ballenberghe 2007). Moose populations increase by the addition of calves born to the population each year and decrease by the loss of animals. Moose die from a variety of causes including hunting, predation, starvation, accident, drowning, vehicle collision, parasites, and disease. Mortalities are generally divided into two major categories: human-caused or natural. Moose populations are adaptable to artificially disturbed habitats, and therefore are often found in close proximity to roads, major highways, and railways, but this association can cause significant mortality (Child 2007). In populated areas of Alaska, large numbers of moose are killed each year by collisions with motor vehicles and trains (Bowyer 2003, Child 2007). During 1963–1990, 3,054 moose were killed on the Alaska Railroad, with losses ranging from seven to 725 annually (Modafferi 1991). In the severe winter of 1989–1990, deep snow caused many moose to travel on roads and railroads and fatalities exceeded the previous record by more than 100 animals. That winter, in the Willow-Talkeetna area, the number of railroad kills represented a 70% loss from the resident population (Schwartz 1991). Other sources of mortality include sport and subsistence hunting and poaching (Woolington 2004).

Prime-aged moose tend to have very high rates of survival because they are not as vulnerable to natural causes of mortality as younger (calves) or older age classes. Survival rates are generally estimated by radio collaring individuals and following them for some period of time (Van Ballenberghe 2007). Ballard et al. (1991) provided data on mean annual survival rates from a sample of radio-collared adult female moose during a 10-year period. From 25 to 80 moose per year were followed in a study area where hunting of cows was prohibited. Annual survival rates were estimated at 94.8%. Data from yearling females spanned four years with two to 22 individuals per year collared, and annual survival rates averaged 95.1%. Annual survival of yearling and adult males averaged 75.4 and 90.9% respectively, with hunting the major mortality factor. On the Kenai Peninsula, researchers followed 51 radio-collared females for six years—reporting a 92% annual survival rate (Bangs 1989). Survival of cows aged one to five years was estimated at 97%, and 84% for females aged 16 to 21 years. Hunting was not a significant cause of mortality of the study population. As reported by Van Ballenberghe and Ballard (2007), various other studies using radio-collared moose have reported annual survival rates of adults ranging from 75 to 94%, depending upon the extent of human hunting. In general, starvation and wolf predation during severe winters causes the greatest mortality in older age classes (Bowyer 2003, Ballard 1991); moose weakened from starvation are particularly vulnerable to wolf predation. Bull moose occasionally wound each other during the rut and die from these wounds (ADF&G 2011).

As reviewed by Van Ballenberghe and Ballard (2007), hunting is a major limiting factor of many moose populations throughout the world, and can reduce moose population density (Crete 1981). In Quebec, where natural mortality apparently was low, harvest rates as high as 25% were reported. Moose harvest rates ranging from 2 to 17% have also been reported for various other parts of North America (ibid). In some European environments, where severe winters, predation, and nutritional stress are absent, moose harvest as high as 50% of the winter population is sustainable (Cederlund 1991). In concert with other factors, including severe winters, high harvest rates have contributed to moose population declines in Alaska (Gasaway 1983). In addition, hunting can significantly reduce the number of bulls, perhaps sufficiently to reduce the level of first-estrus conception (Bishop 1974). When, due to heavy hunting pressure, there are fewer than ten bulls per 100 cows, some cows simply may not encounter a bull early in the mating season. Breeding early in the mating season means the rut, and therefore calving would be synchronous. A cow mating late in the breeding season will calve later in the spring. In nearly all areas where hunting is legal, harvest is managed under sustainable principals, so hunting mortality seldom results in unintended population declines (Timmerman 2007).

Both hunter numbers and moose harvest have increased in the Nushagak and Kvichak watersheds in recent decades. Responding to a 4-fold increase in moose hunters in GMU 17 from 1983 to 2006 (from 293 to 1,182), reported moose harvest tripled (from 127 to 380). In GMU 17B (the Upper Nushagak watershed), the reported harvests for the past five years that data were available varied from 113 to 183, with a mean annual harvest of 149 moose. In GMU 17C (the Togiak watershed), the five-year mean annual harvest was 224, with a range of 193 to 251 (Woolington 2008). Legal moose in GMUs 17B and 17C for Alaska residents are those with spike-fork antlers, antler spreads of no less than 50 inches (127 cm) or at with at least three brow tines on one antler. The largest antlers reported exceeded 69 inches (175 cm).

Juvenile moose tend to have lower survival rates than adults. Calves are typically the most vulnerable age class. Calf moose mortality can be divided into two general time periods when mortality is highest: from birth to about six months of age, and from about six months to one year of age (Van Ballenberghe 2007). These periods correspond roughly to particular vulnerabilities, specifically, to bear and wolf predation in the first period and hunting (in some areas), wolf predation, and winter starvation in the second period. According to Van Ballenberghe and Ballard (2007), neonatal mortality varies greatly, depending on several factors, most notably the extent of predation. Several studies of radio-collared moose calves have documented that predators may account for up to 79% of newborn deaths and that survival during the first eight weeks of life may be as low as 17% (Franzmann 1980, Ballard 1981, Ballard 1991, Larsen 1989, Osborne 1991, Gasaway 1992). Further losses during the first year of life may reduce annual survival to as low as 10% (Van Ballenberghe 2007). In south-central Alaska, Ballard et al. (1991) observed that brown bears caused the majority of natural death of calves younger than five months of age, whereas, on the Kenai Peninsula, Franzmann et al. (1980) documented that black bears were the greatest cause of moose calf mortality.

Moose breed in late September to early October (Van Ballenberghe 1993) and adult females give birth to one or two calves in late May to early June each year (Schwartz 2007, Testa 2000, Peterson 1955)—litters of three are rare, but not unheard of (Coady 1982). Production of moose calves is the result of a complex chain of biological processes including estrus cycles, rutting behavior, fertilization, gestation, prepartum events, and birth (Boer 1992, Van Ballenberghe 2007, Schwartz 2007). Fecundity, or productivity of individual moose, is related to sexual maturation and a broad array of ecological factors affecting food supply, forage quality, and weather that affect the physiological status of females. These factors influence ovulation, pregnancy rates, litter size, and fetal sex ratios. Ultimately, fecundity and subsequent survival of young determines recruitment rates and population trends.

Studies of reproductive tracts have shown that female moose do not ovulate in their first year of life and thus do not produce calves as yearlings. Cows may or may not breed in their second year, depending on body mass (Saether 1987). Most cows are sexually mature at around 28 months of age and females continue to breed to the end of their life (~18 years) (Schwartz 2007, ADF&G 2011). Twinning rates were shown to be strongly correlated with body condition of female moose (as influenced by habitat quality) in several diverse moose populations (Franzmann 1985). In an area known to contain abundant high-quality food resources on the Kenai Peninsula, up to 70% of cows with calves had twins the subsequent year. This contrasts to other populations, in which twinning rates as low as 5% were reported (Pimlott 1959, Houston 1968, Markgren 1969), but some of the estimates may have considered post-natal mortality. Twinning rates exceeding 40–50% are uncommon for moose populations strongly limited by nutrition. Twinning frequency is a good indicator of cow health condition and habitat quality (Dodge 2004). Calves that survive predation in the summer are weaned in August, but will remain with their mother until the next calf is born the following spring (Schwartz 2007), or for an additional year if no new calf is born (Testa 2004).

## POPULATION, SUBPOPULATIONS, AND GENETICS

The number of animals in a population (abundance) is only useful when the geographic boundaries of an area are well defined, because that allows biologists to estimate density (the number of individuals per unit area), which is a more useful parameter. Moose density (the number of individuals in a well-defined unit area) is an important indicator of population status and habitat quality over time. Gasaway et al. (1992) compared moose population densities over very large areas (: 2000 km$^2$) (772 mi$^2$) of generally continuous habitat across a broad area of Alaska and the Yukon Territories. They noted that smaller sites exhibited high variability in prey and predator densities and in habitat quality, making realistic comparisons more difficult. They focused their comparisons on the post-hunt, early winter season, thereby enhancing comparability. The mean density of moose from 20 populations was estimated at 0.15/km$^2$ (0.39/mi$^2$) (range 0.045–0.417/km$^2$) (0.017–0.16/mi$^2$) in areas where predation was thought to be a major limiting factor of moose. Densities of 16 other populations in the same area, where predation was not limiting, averaged 0.66 moose/km$^2$ (1/7/mi$^2$) (Van Ballenberghe 2007). Ballard et al. (1991) provided 29 moose density estimates from Alaska, including some populations studied by Gasaway et al. (1992); they ranged from 0.05 to 1.24/ km$^2$ (0.13–3.2/mi$^2$).

The ADF&G estimated the total population of moose in Alaska to be 175,000–200,000 animals (ADF&G 2011). The 2004 population estimate for the Nushagak and Kvichak watersheds was 8,100 to 9,500 moose (Woolington 2004, Butler 2004). This estimate was based on population data from the ADF&G GMUs 17B, 17C, 9B and less than half the area of GMU 9C, outside Katmai National Park.

Moose are relatively new inhabitants in the Bristol Bay area, possibly having migrated into the area from middle Kuskokwim River drainages during the last century (Woolington 2004). Moose were either not present or were rare in the northern Bristol Bay area until the turn of the twentieth century, and even then the moose population did not increase until three decades ago (Woolington 2004, Butler 2004). Suspected reasons for low moose populations in the Bristol Bay region are heavy hunting pressure, particularly on female moose in the western part, and bear predation in the eastern part (Butler 2004, Woolington 2004). Over the last 25 years, managed harvesting, predator control, an increase in caribou herds as an alternative predator food source, and consecutive mild winters have led to an increase and expansion of the moose population westward (Woolington 2004, Butler 2004).

The largest moose population in the study area is in the Nushagak drainage; the upper watershed (GMU 17B) has an estimated 2,800–3,500 moose, and the lower drainage (GMU 17C) has an estimated 2,900–3,600 moose (ADF&G 2011). These moose comprise about 73% of the total population in the Nushagak and Kvichak River watersheds. The Nushagak River drainage has large, healthy areas of riparian habitat, a major component of which is felt-leaf willow, a preferred browse species (Bartz 2005). The number of moose in the Kvichak River watershed was estimated at 2,000 in GMU 9B, and less than 400 moose in the portion of GMU 9C outside Katmai National Park (Butler 2008).

Fall trend counts have provided notoriously unreliable data on moose populations in GMU 17 (Woolington 2008). Suitable survey conditions, including complete snow coverage, light winds, and moose presence on winter range rarely occurs before antler drop. Regular population estimation surveys of portions of the unit during late winter provide the best population information; unfortunately they do not provide reliable information on sex and age composition (Woolington 2008).

Moose population estimates in northern Bristol Bay area are produced by a statistical model, which uses harvest ticket data from sport and subsistence hunters (Woolington 2004, Butler 2004). The ADF&G, Division of Subsistence suspects considerable subsistence harvest is unreported and that illegal harvest occurs (ADF&G 2011). Illegal harvest of moose in Unit 17 was probably more of a problem in the past than during recent years. Unit residents formerly pursued moose with snow machines during winter and spring, when both male and female moose were taken. Attitudes have changed following considerable efforts by resource management agencies, working with local communities to help hunters see the benefits of reducing illegal moose kills. It is now common to see moose near local villages throughout the winter (Woolington 2008).

## HUMAN USE (SUBSISTENCE, RECREATION)/ INTERACTION/MANAGEMENT

In Alaska 7,400 moose were harvested in 2007. Residents harvested 6,750 moose and 685 were taken by nonresident hunters (ADF&G 2011). The harvest of 7,400 moose yields approximately 3.5 million pounds (1,587,573 kg) of meat.

Harvest records from ADF&G for 1983 to 2002 indicate that GMUs 9 and 17 provided 7% of the total moose harvest in Alaska (Bureau of Land Management 2007). According to ADF&G, Division of Subsistence (http://www.adfg.alaska .gov/sb/CSIS/), local subsistence hunters from King Salmon, Naknek, and South Naknek harvested 19 moose in GMU 9B in 2007; total meat harvested was estimated at 10,206 pounds (4,629 kg). In unit 17B, local residents from Igiugig, Koliganek, and New Stuyahok harvested 88 moose in 2005 (the most recent year with available data); total meat harvested was estimated at 48,208 pounds (21,867 kg). Residents from Naknek and South Naknek harvested four moose from unit 17C with a total of 5,357 pounds (2,430 kg) of meat. In total, subsistence moose meat accounted for 63,771 pounds (28,987 kg) with an average of 128 pounds (158 kg) harvested per household (Table 8.4).

Moose are an important subsistence food species for residents in the area served by the Bristol Bay Area Health Corporation (Ballew 2004). In a survey on traditional food consumption conducted in 2002, 86% of respondents reported consumption of moose meat within the past year, at a median per capita consumption rate of five lb/yr (2.3 kg/yr) (Ballew 2004). Moose was the third greatest source of subsistence meat to residents of that region after salmon and caribou. Subsistence statistics (Table 8.4) suggest that, on average, 38% of individuals from villages in the area attempted to harvest a moose, with about 20% succeeding. Additionally, about 24% of individuals reported sharing their moose with others, while 44% received meat from others. In addition to being a source of subsistence meat, moose contributed to the local

**Table 8.4**  Subsistence statistics for moose harvest in GMUs 9B, 17B, and 17C

| GMU | Community name | Study year | Respondents using moose (%) | Respondents attempting harvest (%) | Moose successfully harvested (%) | Respondents sharing meat (%) | Respondents receiving meat (%) | Reported harvest (#) | Estimated harvest (#) | Reported harvest (lbs) | Estimated harvest (lbs) | Mean harvest/household (lbs) |
|---|---|---|---|---|---|---|---|---|---|---|---|---|
| 09B | King Salmon | 2007 | 33 | 31 | 10 | 10 | 24 | 5 | 9 | 2,700 | 4,849 | 55 |
| 09B | Naknek | 2007 | 48 | 23 | 5 | 5 | 47 | 4 | 10 | 2,160 | 5,357 | 29 |
| 09B | South Naknek | 2007 | 29 | 24 | 0 | 0 | 29 | 0 | 0 | 0 | 0 | 0 |
| 17B | Igiugig | 2005 | 100 | 50 | 42 | 75 | 67 | 6 | 6 | 3,240 | 3,510 | 270 |
| 17B | Koliganek | 2005 | 86 | 68 | 50 | 54 | 46 | 16 | 24 | 8,640 | 12,960 | 309 |
| 17B | New Stuyahok | 2005 | 94 | 65 | 51 | 43 | 65 | 30 | 58 | 16,200 | 31,738 | 331 |
| 17C | Naknek | 2007 | 48 | 23 | 5 | 5 | 47 | 4 | 10 | 2,160 | 5,357 | 29 |
| 17C | South Naknek | 2007 | 29 | 24 | 0 | 0 | 29 | 0 | 0 | 0 | 0 | 0 |
| Total | | | | | | | | | | 35100 | 63771 | |
| Mean | | | 58 | 38 | 20 | 24 | 44 | | | | | 128 |

(Data are from 2005 or 2007 and represent the most recent information available.)

(http://www.adfg.alaska.gov/sb/CSIS/)

economy, through jobs created as a result of nonresident hunters seeking remote fly-in or boat-in opportunities to take a trophy moose.

## RECOMMENDED RESEARCH

The decrease in consistent fall and spring snow conditions has created a pressing research need for improved methods of monitoring moose densities. Standard methods of estimating moose abundance in western Alaska rely on broad snow coverage so that the moose are in strong visual contrast and thus have very high detection rates for observers surveying the area from a small plane (e.g., Kellie and DeLong 2006). These conditions, which have recently been infrequent, are expected to become even more rare with the region's projected warmer and wetter winter climate (see Table 4 at https://westernalaskalcc.org/science/Science Workshop Materials/Background Materials/Climate Projections/Climate Projection Information and Maps.pdf). Recent collaborative efforts between staff at the ADF&G's Division of Wildlife Conservation, the USFWS, and the Western Alaska Landscape Conservation Cooperative have started to consider promising strategies to address this broad need. Similarly, the region's resource managers need a better understanding of how projected changes in the region's climate will impact habitats supporting moose and thus, moose distribution and abundance.

## SUMMARY

Moose are an important part of the Bristol Bay ecoregion, providing subsistence meat for local communities, influencing plant productivity and nutrient cycling, and impacting predator/prey dynamics of co-occurring species such as caribou, wolves, and brown and black bears. Moose were either rare or absent from the Bristol Bay area until the turn of the twentieth century, when they may have migrated into the area from middle Kuskokwim River drainages. Over the last 25 years, managed harvesting, predator control, an increase in caribou herds as an alternative predator food source, and consecutive mild winters have led to an increase and expansion of the moose population westward. In 2004 there were an estimated 8,100 to 9,500 moose in the Nushagak and Kvichak watersheds.

Habitat selection by moose is influenced by food availability, cover to avoid predators, and snow depth. Moose are ruminants classified as browsers; they consume large quantities of low-quality forage such as leaves, twigs, and bark of woody plants. Moose benefit from early successional stages that provide woody browse. Large delta floodplains typically have the highest density of moose; moose also inhabit shrub/scrub habitats, boreal forests, mixed forests, and stream valley/riparian zones. Home range sizes and migration movements of moose vary significantly among regions in Alaska, and also vary by gender. In Alaska, male and female moose select habitats differently, leading to their spatial segregation throughout most of the year. Adult males select for greatest forage abundance, while females select for cover availability, especially if they are with calves.

The dynamics of a moose population are determined by the number of individuals in the population, the population's rate of production of young, and its rate of loss through mortality. Moose breed only during a brief window in autumn, possibly triggered by photoperiod. This strategy ensures that calves are born in spring, when forage is high in nutrient quality and cows are most likely to produce sufficient milk to raise a calf. When cows are in prime health condition, and available habitat provides both sufficient forage and cover from predation, cows are more likely to successfully bear and raise twins. Moose die from a variety of causes including hunting, predation, starvation, accident, drowning, vehicle collision, parasites, and disease. Of all age classes, calves have the highest rates of mortality; old moose are also vulnerable to predation and starvation. Wildlife managers strive to create conditions under which healthy moose populations are available to support subsistence and other hunting opportunities.

## ACKNOWLEDGEMENTS

Marcus Geist of The Nature Conservancy provided GIS support including maps, watershed area calculations and land cover information and analysis. Numerous Federal and State agency wildlife biologists generously gave time and effort to review and improve sections of the report. Of special note are United States Geological Survey (USGS) biologist Layne Adams, National Park Service (NPS) biologists Buck Mangipane and Grant Hilderbrand, retired NPS biologist Page Spencer, and USFWS biologists Andy Aderman (Togiak NWR) and Dom Watts (Alaska Peninsula-Becharof NWR) who provided significant reviews, edits, information, and insight based on their considerable scientific expertise and knowledge. Heather Dean of the Environmental Protection Agency and Rich Harris of the University of Montana provided thorough and impressive editing of the initial and external review drafts of the report, respectively. Their attention to detail made this a far more consistent, precise, and readable report. Finally, this project would not have been possible without the support and encouragement from Ann Rappoport, long-time Anchorage Fish and Wildlife Field Office Field Supervisor, now retired.

The findings and conclusions in this chapter are those of the authors and do not necessarily represent the views of the USFWS.

# REFERENCES

ADF&G (Alaska Department of Fish and Game). 2011. Moose Species Profile, Alaska Department of Fish and Game, Juneau, AK.

Ballard, W.B., C.L. Gardner, and S.D. Miller. 1980. Influence of predators on summer movements of moose in southcentral Alaska. Proceedings of the North American Moose Conference Workshop 16:338–359.

Ballard, W.B., L.A. Ayres, P.R. Krausman, D.J. Reed, and S.G. Fancy. 1997. *Ecology of Wolves in Relation to a Migratory Caribou Herd in Northwest Alaska*. Wildlife Monographs 135:1–47.

Ballard, W.B., J.S. Whitman, and C.L. Garden. 1987. *Ecology of an Exploited Wolf Population in Southcentral Alaska*. Wildlife Monographs 98:1–54.

Ballard, W.B., J.S. Whitman, and D.J. Reed. 1991. *Population Dynamics of Moose in South-Central Alaska*. Wildlife Monographs 114:1–49.

Ballard, W.B. and V. Van Ballenberghe. 2007. Predator/prey relationships. Pages 247–273. In: A.W. Franzmann and C.C. Schwartz (Eds.), *Ecology and Management of the North American Moose*, University Press of Colorado, Boulder, Colorado.

Ballew, C., A. Ross, R.S. Wells, V. Hiratsuka, K.J. Hamrick, E.D. Nobmann, and S. Bartell. 2004. Final report on the Alaska Traditional Diet Survey, Alaska Native Health Board, Anchorage, Alaska.

Balsom, S., W.B. Ballard, and H.A. Whitlaw. 1996. Mature coniferous forest as critical moose habitat. Alces 32:131–140.

Bangs, E.E., T.N. Bailey, and M.F. Portner. 1989. Survival rates of adult female moose on the Kenai Peninsula, Alaska. Journal of Wildlife Management 53:557–563.

Bangs, E.E., S.A. Duff, and T.N. Bailey. 1985. Habitat differences and moose use of two large burns on the Kenai Peninsula, Alaska. Alces 21:17–35.

Barboza, P.S. and R.T. Bowyer. 2000. Sexual segregation in dimorphic deer: a new gastrocentric hypothesis. Journal of Mammalogy 81:473–489.

Bartz, K.K. and R.J. Naiman. 2005. Effects of salmon-borne nutrients on riparian soils and vegetation in Southwest Alaska. Ecosystems 8:529–545.

Bedard, J., M. Crete, and E. Audy. 1978. Short-term influence of moose upon woody plants of an early seral wintering site in Gaspe Peninsula, Quebec. Canadian Journal of Forest Research 8:407–415.

Bishop, R.H. and R.A. Rausch. 1974. Moose population fluctuations in Alaska. Naturaliste Canada (Quebec) 101:559–593.

Boer, A.W. 1992. Fecundity of North American moose (*Alces alces*): a review. Alces Supplement 1:1–10.

Boer, A.W. 2007. Interspecific relationships. Pages 337–349. In: A.W. Franzmann and C.C. Schwartz (Eds.), *Ecology and Management of the North American Moose*, University Press of Colorado, Boulder, Colorado.

Bowyer, R.T., V. Van Ballenberghe, and J.G. Kie. 1998. Timing and synchrony of parturition in Alaskan moose: long-term versus proximal effects of climate. Journal of Mammalogy 79:1332–1344.

Bowyer, R.T., V. Van Ballenberghe, J.G. Kie, and J.A.K. Maier. 1999. Birth-site selection by Alaskan moose: maternal strategies for coping with a risky environment. Journal Of Mammalogy 80:1070–1083.

Bowyer, R.T., B.M. Pierce, L.K. Duffy, and D.A. Haggstrom. 2001. Sexual segregation in moose: effects of habitat manipulation. Alces 37:109–122.

Bowyer, R.T., K.M. Stewart, S.A. Wolfe, G.M. Bundell, K.L. Lehmkuhl, P.J. Joy, T.J. McDonough, and J.G. Kei. 2002. Assessing sexual segregation in deer. Journal of Wildlife Management 66:536–544.

Bowyer, R.T. and J.A. Neville. 2003. Effects of browsing history by Alaskan moose on regrowth and quality of feltleaf willow. Alces 39:193–202.

Bubenik, A.B. 2007. Behavior. Pages 172–221. In: A.W. Franzmann and C.C. Schwartz (Eds.), *Ecology and Management of the North American Moose*, University Press of Colorado, Boulder, Colorado.

Butler, L.G. 2004. Unit 9 moose management report. Pages 113–120. In: C. Brown, (Ed.), Moose management report of survey and inventory activities: July 1, 2001–June 30, 2003, Alaska Department of Fish and Game, Juneau, Alaska.

Butler, L.G. 2008. Unit 9 moose management report. Pages 116–124. In: P. Harper (Ed.), Moose management report of survey and inventory activities July 1, 2005–June 30, 2007, Alaska Department of Fish and Game, Juneau, Alaska.

Cederlund, G. and H.K.G. Sand. 1991. Population dynamics and yield of a moose population without predators. Alces 27:31–40.

Child, K.N. 2007. Incidental mortality. Pages 275–302. In: A.W. Franzmann and C.C. Schwartz (Eds.), *Ecology and Management of the North American Moose*, University Press of Colorado, Boulder, Colorado.

Coady, J.W. 1980. History of moose in northern Alaska and adjacent regions. Canadian Field-Naturalist 94:61–68.

Coady, J.W. 1982. Moose. Pages 9092–922. In: J. A. Chapman and G. A. Feldhamer (Eds.), Wild mammals of North America, Johns Hopkins University Press, Baltimore, Maryland.

Crete, M., R.J. Taylor, and P.A. Jordan. 1981. Optimization of moose harvest in southwestern Quebec. Journal of Wildlife Management 45:598–611.

Crete, M. 1987. The impact of sport hunting on North American moose. Swedish Wildlife Research Supplement 1:553–563.

Crichton, V.F.J. 1992. Six year (1986/87–1991/92) summary of in utero productivity of moose in Manitoba, Canada. Alces 28:203–214.

Darby, W.R. and W.O. Pruitt. 1984. Habitat use, movements, and grouping behavior of woodland caribou, *Rangifer tarandus caribou*, in southeastern Manitoba. Canadian Field-Naturalist 97:184–190.

Davis, J.L. and A.W. Franzmann. 1979. Fire-moose-caribou interrelationships: a review and assessment. Proceedings of the North American Moose Conference Workshop 15:80–118.

Des Meules, P. 1964. The influence of snow on the behavior of moose. Rapport No. 3:51–73, Ministrere du tourisme, de la chasse et de la peche, Quebec.

Dodds, D.G. 1960. Food competition and range relationships of moose and snowshoe hare in Newfoundland. Journal of Wildlife Management 24:52–60.

Dodge, W.B., S.R. Winterstein, D.E. Beyer, and H. Campa. 2004. Survival, reproduction, and movements of moose in the western peninsula of Michigan. Alces 40:71–85.

Dussault, C., J.P. Ouellet, R. Courtois., J. Huot, L. Breton, and H. Jolicoeur. 2005. Linking moose habitat selection to limiting factors. Ecography 28:619–628.

Eastman, D.S. and R. Ritcey. 1987. Moose behavioral relationships and management in British Columbia. Swedish Wildlife Research Supplement 1:101–118.

Franzmann, A.W., C.C. Schwartz, and R.O. Peterson. 1980. Moose calf mortality in summer on the Kenai Peninsula, Alaska. Journal of Wildlife Management 44:764–768.

Franzmann, A.W. and C.C. Schwartz. 1985. Moose twinning rates: a possible condition assessment. Journal of Wildlife Management 49:394–396.

Franzmann, A.W. and C.C. Schwartz. 1986. Black bear predation on moose calves in highly productive versus marginal moose habitats on the Kenai Peninsula, Alaska. Alces 22:139–154.

Franzmann, A.W. and C.C. Schwartz. 1998. *Ecology and Management of the North American Moose*. Smithsonian Institution Press, Washington, D.C.

Franzmann, A.W. and C.C. Schwartz. 2007. *Ecology and Management of the North American Moose*. 2nd ed. University Press of Colorado, Boulder, Colorado.

Gasaway, W.C., R.O. Stephenson, J.L. Davis, P.E.K. Shepherd, and O.E. Burris (1983) *Interrelationships of Wolves, Prey, and Man in Interior Alaska*. Wildlife Monographs 84:1–50.

Gasaway, W.C., R.D. Boertje, D.V. Grangaard, D.G. Kelleyhouse, R.O. Stephenson, and D.G. Larsen. 1992. *The Role of Predation in Limiting Moose at Low Densities in Alaska and Yukon and Implications for Conservation*. Wildlife Monographs 120:1–59.

Geist, V. 1971. *Mountain Sheep: A Study in Behavior and Evolution*. University of Chicago Press, Chicago, Illinois.

Gillingham, M.P. and D.R. Klein. 1992. Late-winter activity patterns of moose (*Alces alces gigas*) in western Alaska. Canadian Journal of Zoology 70:293–299.

Grauvogel, C.A. 1984. Seward Peninsula moose population identity study. Federal Aid in Wildlife Restoration, Final Report, Alaska Department of Fish and Game, Juneau, Alaska. pp. 93.

Hauge, T.M. and L.B. Keith. 1981. Dynamics of moose populations in northeast Alberta. Journal of Wildlife Management 45:573–597.

Hofmann, R.R. 1973. The ruminant stomach. East African Monograph in Biology, East African Literature Bureau, Nairobi, Africa. pp. 54.

Hofmann, R.R. 1989. Evolutionary steps of ecophysiological adaptation and diversification of ruminants: a comparative view of their digestive system. Oecologia 78:443–457.

Houston, D.B. 1968. The Shiras moose in Jackson Hole, Wyoming. Technical Bulletin 1, Grand Teton National Historic Association. pp. 110.

Hundertmark, K.J., R.T. Bowyer, G.F. Shields, and C.C. Schwartz. 2003. Mitochondrial phylogeography of moose (*Alces alces*) in North America. Journal of Mammalogy 84:718–728.

Hundertmark, K.J. 2007. Home range, dispersal, and migration. Pages 303–336. In: A.W. Franzmann and C.C. Schwartz (Eds.), *Ecology and Management of the North American Moose*. University Press of Colorado, Boulder, Colorado.

Jordan, P.A. 1987. Aquatic forage and the sodium ecology of moose: a review. Swedish Wildlife Research Supplement 1:119–137.

Joyal, R. 1987. Moose habitat investigations in Quebec and management implications. Swedish Wildlife Research Supplement 1:139–152.

Karns, P. 2007. Population distribution, density, and trends. Pages 125–140. In: A.W. Franzmann and C.C. Schwartz (Eds.), *Ecology and Management of the North American Moose*. University Press of Colorado, Boulder, Colorado.

Keech, M.A., M.S. Lindberg, R.D. Boertje, P. Valkenburg, B.D. Taras, T.A. Boudreau, and K.B. Beckmen. 2011. Effects of predator treatments, individual traits, and environment on moose survival in Alaska. Journal of Wildlife Management 75:1361–1380.

Kellie, K.A. and R.A. DeLong. 2006. Geospatial survey operations manual. Alaska Department of Fish and Game. Fairbanks, AK, USA. https://winfonet.alaska.gov/sandi/moose/surveys/documents/GSPEOperationsManual.pdf.

Kelsall, J.P. 1969. Structural adaptations of moose and deer for snow. Journal of Mammalogy 50:302–310.

Kelsall, J.P. and W. Prescott. 1971. Moose and deer behavior in snow. Report Series 15, Canadian Wildlife Service, Ottawa, Canada. pp. 27.

Kelsall, J.P. and E.S. Telfer. 1974. Biogeography of moose with particular reference to western North America. Naturaliste Canada (Quebec) 101:117–130.

Kelsall, J.P., E.S. Telfer, and T.D. Wright . 1977, The effects of fire on the ecology of the boreal forest, with particular reference to the Canadian north: a review and selected bibliography. Occasional Paper 323, Canadian Wildlife Service, Ottawa, Canada.

Larsen, D.G., D.A. Gauthier, and R.L. Markel. 1989. Causes and rates of moose mortality in the southwest Yukon. Journal of Wildlife Management 53:548–557.

LeResche, R.E. 1974. Moose migration in North America. Naturaliste Canada (Quebec) 101:393–415.

LeResche, R.E., R.H. Bishop, and J.W. Coady. 1974. Distribution and habitats of moose in Alaska. Le Naturaliste Canadien 101: 143–178.

Lynch, G.M. and L.E. Morgantini. 1984. Sex and age differential in seasonal home range of moose in northwestern Alberta. Alces 20:61–78.

MacCracken, J.G., V. Van Ballenberghe, and J.M. Peek. 1993. Use of aquatic plants by moose—sodium hunger or foraging efficiency. Canadian Journal of Zoology 71:2345–2351.

MacCracken, J.G., V. Van Ballenberghe, and J.M. Peek. 1997. *Habitat Relationships of Moose on the Copper River Delta in Coastal South-Central Alaska*. Wildlife Monographs 136:1–52.

Maier, J.A.K., J.M. Ver Hoef, A.D. McGuire, R.T. Bowyer, L. Saperstein, and H.A. Maier. 2005. Distribution and density of moose in relation to landscape characteristics: effects of scale. Canadian Journal of Forest Research 35:2233–2243.

Markgren, G. 1969. Reproduction of moose in Sweden. Viltrevy 6:127–299.

McArt, S.H., D.E. Spalinger, W.B. Collins, E.R. Schoen, T. Stevenson, and M. Bucho. 2009. Summer dietary nitrogen availability as a potential bottom-up constraint on moose in south-central Alaska. Ecology 90:1400–1411.

Miller, G.S. 1899. A new moose from Alaska. Proceedings of Biological Society of Washington 13:57–59.

Miller, B.K. and J.A. Litvatitis. 1992. Habitat segregation by moose in a boreal forest ecotone. Acta Theriological 37:41–50.

Miquelle, D.G. and V. Van Ballenberghe. 1989. Impact of bark stripping by moose on aspen-spruce communities. Journal of Wildlife Management 53:577–586.

Miquelle, D.G., J.M. Peek, and V. Van Ballenberghe. 1992. *Sexual Segregation in Alaskan Moose*. Wildlife Monographs 122:1–57.

Modafferi, R.D. 1991. Train-moose kills in Alaska: characteristics and relationships with snowpack depth and moose distribution in Lower Susitna Valley. Alces 27:193–207.

Molvar, E.M., R.T. Bowyer, and V. Van Ballenberghe. 1993. Moose herbivory, browse quality, and nutrient cycling in an Alaskan treeline community. Oecologia 94:472–479.

Molvar, E.M. and R.T. Bowyer. 1994. Costs and benefits of group living in a recently social ungulate: the Alaskan moose. Journal Of Mammalogy 75:621–630.

Odum, E.P. 1983. Fundamentals of Ecology. 3rd ed. CBS College Publications, New York, New York.

Oehlers, S.A., R.T. Bowyer, F. Huettmann, D.K. Person, and W.B. Kessler. 2011. Sex and scale: implications for habitat selection by Alaskan moose *Alces alces gigas*. Wildlife Biology 17:67–84.

Oldemeyer, J.L., A.W. Franzmann, A.L. Brundage, P.D. Arneso, and A. Flynn. 1977. Browse quality and the Kenai moose population. Journal of Wildlife Management 41:533–542.

Oldemeyer, J.L. and W.L. Regelin. 1987. Forest succession, habitat management, and moose on the Kenai National Wildlife Refuge. Swedish Wildlife Research Supplement 1:163–179.

Osborne, T.O., T.F. Paragi, J.L. Bodkin, A.J. Loranger, and W.N. Johnson. 1991. Extent, causes, and timing of moose calf mortality in western interior Alaska. Alces 27:24–30.

Parker, G.R. and L.D. Morton. 1978. The estimation of winter forage and use by moose on clearcuts in northcentral Newfoundland. Journal of Range Management 31:300–304.

Parker, K.L., C.T. Robbins, T.A. Hanley. 1984. Energy expenditures for locomotion by mule deer and elk. Journal of Wildlife Management 48:474–488.

Peek, J.M. 1986. A review of wildlife management Prentice-Hall, Englewood Cliffs, New Jersey.

Peek, J.M. 2007. Habitat relationships. Pages 351–376. In: A.W. Franzmann and C.C. Schwartz (Eds.), *Ecology and Management of the North American Moose*, University Press of Colorado, Boulder, Colorado.

Peterson, R.L. 1952. A review of the living representatives of the genus *Alces*. Contributions to the Royal Ontario Museum of Zoology and Paleontology 34:1–30.

Peterson, R.L. 1955. North American moose. University of Toronto Press, Toronto, Ontario, Canada.

Peterson, R.O., J.D. Woolington, and T.N. Bailey. 1984. *Wolves of the Kenai Peninsula, Alaska*. Wildlife Monographs 88:1–52.

Phillips, R.L., W.E. Burg, and D.B. Sniff. 1973. Moose movement patterns and range use in northeastern Minnesota. Journal of Wildlife Management 37:266–278.

Pimlott, D.H. 1959. Reproduction and productivity of Newfoundland moose. Journal of Wildlife Management 23:381–401.

Pruitt, W.O. 1959. Snow as a factor in the winter ecology of the barren ground caribou. Arctic 12:159–179.

Quinn, T.P., S.M. Carlson, S.M. Gende, and H.B.J. Rich. 2009. Transportation of Pacific salmon carcasses from streams to riparian forests by bears. Canadian Journal of Zoology 87:195–203.

Regelin, W.L., C.C. Schwartz, and A.W. Franzmann. 1985. Seasonal energy metabolism of adult moose. Journal of Wildlife Management 49:394–396.

Renecker, L.A. and R.J. Hudson. 1986. Seasonal energy expenditures and thermoregulatory responses of moose. Canadian Journal of Zoology 64:322–327.

Renecker, L.A. and R.J. Hudson. 1989. Ecological metabolism of moose in aspen-dominated forests, central Alberta. Canadian Journal of Zoology 67:1923–1928.

Renecker, L.A. and C.C. Schwartz. 2007. Food habits and feeding behavior. Pages 403–440. In: A.W. Franzmann and C.C. Schwartz (Eds.), *Ecology and Management of the North American Moose*. University Press of Colorado, Boulder, Colorado.

Risenhoover, K.L. 1986. Winter activity patterns of moose in interior Alaska. Journal of Wildlife Management 50:727–734.

Risenhoover, K.L. 1989. Composition and quality of moose winter diets in interior Alaska. Journal of Wildlife Management 53:568–577.

Saether, B.E. 1987. Patterns and processes in the population dynamics of the Scandinavian moose (*Alces alces*): some suggestions. Swedish Wildlife Research Supplement 1:525–537.

Shelford, V.E. 1963. The ecology of North America University of Illinois Press, Urbana, Illinois.

Schwartz, C.C. 1992. Physiological and nutritional adaptations of moose to northern environments. Alces Supplement 1:139–155.

Schwartz, C.C. 2007. Reproduction, natality, and growth. Pages 141–171. In: A.W. Franzmann and C.C. Schwartz (Eds.), *Ecology and Management of the North American Moose*. University Press of Colorado, Boulder, Colorado.

Schwartz, C.C. and B. Bartley. 1991. Reducing incidental moose mortality: considerations for management. Alces 27:227–231.

Schwartz, C.C. and A.W. Franzmann. 1989. Bears, wolves, moose, and forest succession, some management considerations on the Kenai Peninsula, Alaska. Alces 25:1–10.

Schwartz, C.C. and A.W. Franzmann. 1991. *Interrelationship of Black Bears to Moose and Forest Succession in the Northern Coniferous Forest*. Wildlife Monographs 113:3–58.

Schwartz, C.C. and K.J. Hundertmark. 1993. Reproductive characteristics of Alaska moose. Journal of Wildlife Management 57:454–458.

Schwartz, C.C. and L.A. Renecker. 2007. Nutrition and energetics. Pages 440–478. In: A.W. Franzmann and C.C. Schwartz (Eds.), *Ecology and Management of the North American Moose*. University Press of Colorado, Boulder, Colorado. Schwartz, C.C., W.L. Regelin, and A.W. Franzmann. 1987. Protein digestion in moose. Journal of Wildlife Management 51:352–357.

Schwartz, C.C., M.E. Hubbert, and A.W. Franzmann. 1988. Energy requirements of adult moose for winter maintenance. Journal of Wildlife Management 52:26–33.

Servheen, G. and L.J. Lyon. 1989. Habitat use by woodland caribou in the Selkirk Mountains. Journal of Wildlife Management 53:230–237.

Singer, F.J. and J. Dalle-Molle. 1985. The Denali ungulate-predator system. Alces 21:339–358.

Spencer, D.L. and E.F. Chatelaine. 1953. Progress in the management of the moose in south central Alaska. Transactions of the North American Wildlife Conference 18:539–552.

Spencer, D.I. and J.B. Hakala. 1964. Moose and fire on the Kenai. Proceedings of Tall Timbers Fire Ecology Conference 3:11–33.

Stephenson, T.R., V. Van Ballenberghe, J.M. Peek, and J.G. MacCracken. 2006. Spatio-temporal constraints on moose habitat and carrying capacity in coastal Alaska: vegetation succession and climate. Rangeland Ecology & Management 59:359–372.

Telfer, E.S. and J.P. Kelsall. 1984. Adaptation of some large North American mammals for survival in snow. Ecology 65:1828–1834.

Testa, J.W. 2004. Population dynamics and life history trade-offs of moose (*Alces alces*) in south-central Alaska. Ecology 85:1439–1452.

Testa, J.W., E.F. Becker, and G.R. Lee. 2000. Movements of female moose in relation to birth and death of calves. Alces 36:155–162.

Timmerman, H.R. and M.E. Buss. 2007. Population and harvest management. Pages 559–616 in A. W. Franzmann and C. C. Schwartz (Eds.), *Ecology and Management of the North American Moose*. University Press of Colorado, Boulder, Colorado.

Thomson, R.N. 1991. An analysis of the influence of male age and sex ratio on reproduction in British Columbia moose (*Alces alces L.*) populations. Doctoral dissertation, University of British Columbia.

Thompson, I.D. and D.J. Euler. 1987. Moose habitat in Ontario: a decade of change in perception. Swedish Wildlife Research Supplement 1:181–193.

Van Ballenberghe, V. 1977. Migratory behavior of moose in southcentral Alaska. Pages 103–109. In: T. J. Peterle (Ed.), XIIIth International Congress of Game Biologists, Wildlife Management Institute, Washington, D.C.

Van Ballenberghe, V. and W.B. Ballard. 2007. Population dynamics. In: A.W. Franzmann and C.C. Schwartz (Eds.), *Ecology and Management of the North American Moose*. University Press of Colorado, Boulder, Colorado. pp. 223–246.

Van Ballenberghe, V., and D.G. Miquelle. 1990. Activity of moose during spring and summer in interior Alaska. Journal of Wildlife Management 54:391–396.

Van Ballenberghe, V. and D.G. Miquelle. 1993. Mating in moose: timing, behavior, and male access patterns. Canadian Journal of Zoology 71:1687–1690.

Van Ballenberghe, V., D.G. Miquelle, and J.G. MacCracken. 1989. Heavy utilization of woody plants by moose during summer at Denali National Park, Alaska. Alces 25:31–35.

Viereck, L.A. and C.T. Dyrness. 1980. A preliminary classification system for vegetation in Alaska. General Technical Report, PNW-106, U.S. Forest Service, Washington, D.C. pp. 38.

Weixelman, D.A., R.T. Bowyer, and V. Van Ballenberghe. 1998. Diet selection by Alaskan moose during winter: effects of fire and forest succession. Alces 34:213–238.

Woolington, J.D. 2004. Unit 17 moose management report, in: C. Brown (Ed.), Moose management report of survey and inventory activities: July 1, 2001–June 30, 2003, Alaska Department of Fish and Game, Juneau, Alaska. pp. 246–266.

Woolington, J.D. 2008. Unit 17 moose management report. Pages 246–268 in P. Harper (Ed.), Moose management report of survey and inventory activities July 1, 2005–June 30, 2007, Alaska Department of Fish and Game, Juneau, AK.

# 9

# BARREN GROUND CARIBOU

Kenneth Whitten, Alaska Department of Fish and Game (Retired)

## INTRODUCTION

Alaska is currently home to 31 herds of wild caribou (*Rangifer tarandus granti*), with a combined population of approximately 660,000 in 2014 (Harper and McCarthy 2015). Caribou herds are defined by their traditional and predictable use of calving areas, each of which are separate and distinct from the calving grounds of other herds (Skoog 1968). Use of other seasonal ranges is variable and less traditional. Caribou from different herds may overlap on seasonal ranges other than calving areas (Cameron et al. 1986). Historically, most caribou herds have fluctuated widely in numbers and in use of range (Skoog 1968).

Adult bull caribou in southwestern Alaska usually weigh between 159 to 182 kg (350 to 450 lbs); females weigh between 80 to 120 kg (175 and 225 lbs) [Alaska Department of Fish & Game (ADF&G) 1985]. Body weight can vary with environmental and nutritional conditions (Cameron 1994, Valkenburg et al. 2003). Caribou are the only members of the deer family in which both males and females grow antlers. Bulls begin to shed the velvet on their antlers between late August and early September, marking the start of breeding season. The largest bulls begin shedding (dropping) their antlers in late October, with smaller bulls losing their antlers later in the winter. Females shed velvet in September (Skoog 1968). Pregnant females usually keep their antlers until the calving season in the spring, whereas nonpregnant females lose their antlers about a month before calving begins (Figure 9.1). Some females never grow antlers (Whitten

**Figure 9.1.** Caribou bulls begin shedding their antlers in October, whereas pregnant females may keep theirs until spring calving season. *Photo credit: Carol Ann Woody.*

**Figure 9.2.** Male Caribou with a magnificent set of antlers. *Photo credit: Carol Ann Woody.*

1995). Caribou populations throughout the Bristol Bay region have declined recently and body weights and antler sizes are now relatively low. In the past, the area produced large-bodied animals with record-book antlers (Valkenburg et al. 2003) (see Figure 9.2).

## POPULATION HISTORY OF CARIBOU IN THE UPPER BRISTOL BAY REGION

Historical accounts from the early 1800s indicate that caribou were plentiful in the Bristol Bay region. There may have been a large herd that ranged from Bristol Bay across the Kuskokwim and Yukon River deltas all the way to Norton Sound. By the late 1800s caribou throughout this area had declined dramatically. Caribou numbers may have increased in the early 1930s, but were declining again by the late 1930s. Domestic reindeer were brought to the Bristol Bay region in the early 1900s, but by the 1940s, reindeer herds were widely neglected and either died out or assimilated into wild caribou populations (Skoog 1968, Woolington 2009, Colson et al. 2014). Caribou in the Nushagak drainage remained relatively scarce into the 1970s, at about 10,000–15,000 animals (Woolington 2009).

Over the past thirty years, caribou herds in southwest Alaska have continued to undergo significant changes in abundance. The Nushagak and Kvichak River watersheds are now used primarily by caribou from the Mulchatna herd. This herd grew rapidly during the 1980s and 1990s, from a population of about 18,600 animals in 1981 to a peak of approximately 200,000 in 1997. By 1999 the Mulchatna herd had declined to 175,000 and it continued to decline, to approximately 30,000 in 2008 (Valkenburg et al. 2003, Woolington 2009). Numbers have remained fairly stable since then. As the Mulchatna herd grew, it overlapped with and eventually assimilated the much smaller Kilbuck (or Qavilnguut) herd that formerly ranged infrequently into the western part of the Nushagak River watershed. By the late 1990s the Kilbuck herd had ceased to function as a distinct population (Woolington 2009).

The Northern Alaska Peninsula herd recovered from a population low of about 2,000 in the late 1940s to about 20,000 in 1984. The population remained at about 15,000 to 19,000 through 1993, but has since declined steadily to about 2,000 to 2,500 today (Butler 2009). For the most part, caribou of the Northern Alaska Peninsula herd remain well south of the Kvichak River watershed. However, from 1986 to 2000 many caribou from the Northern Alaska Peninsula herd wintered in the Kvichak River watershed, south of Lake Iliamna (Butler 2009). In the late 1980s and early 1990s, the Kvichak River watershed was also used by far more (up to 50,000) Mulchatna caribou (Woolington 2009). The two herds always returned to their traditional calving and summer ranges and remained distinct (Hinkes et al. 2005, Butler 2009, Woolington 2009).

The Nushagak Peninsula herd is a small population that was established in 1988 when caribou from the Northern Alaska Peninsula herd were translocated to the Nushagak Peninsula south of the Nushagak River delta, on the west side

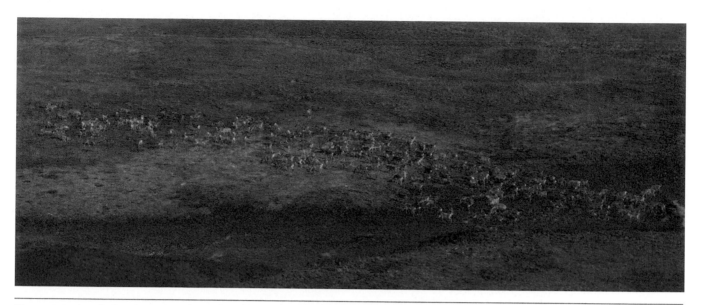

**Figure 9.3**  Herd of caribou on the move in summer. *Photo credit: Carol Ann Woody.*

of upper Bristol Bay. The Nushagak Peninsula had been unoccupied by caribou for approximately 100 years (Hotchkiss 1989, Paul 2009). The Nushagak Peninsula herd grew rapidly after its introduction, from 146 caribou to over 1,000 caribou in 1994. Growth continued at a slower rate to about 1,400 caribou in 1997. Population density peaked at approximately 1.2 caribou/km² (3.1/mi²). During the next decade, calf recruitment and adult female survival decreased and the population declined to 546 caribou in 2006 (Aderman 2009). The population remained at about 550 caribou until 2009 and then increased to 801 in 2011 (Aderman and Lowe 2011) (Figure 9.3).

## HABITAT

Spring calving grounds tend to be in open tundra areas or high and rugged mountains. Predator densities are often lower in such areas, but large caribou herds can also calve at high densities in sparsely forested terrain, where their sheer numbers and synchronized timing of births can swamp the effects of predators (Skoog 1968, ADF&G 1985).

During summer (mid-June to mid-August), caribou feed in open tundra, mountain, or sparsely forested areas. To avoid harassment from mosquitoes and other insects, caribou often gather on windswept ridges, glaciers, lingering snow drifts, gravel bars, elevated terrain, cinder patches, and beaches. Caribou near the coast may also avoid insects by standing head down and motionless on mudflats (Skoog 1968, ADF&G 1985).

Caribou often feed in forested areas in winter, especially where there are spruce-lichen associations. In addition to forested areas, caribou can also be found along ridge tops, on frozen lakes, and in bogs during winter (Skoog 1968, ADF&G 1985).

## FOOD HABITS

### Spring

From mid-April to mid-June, caribou usually eat catkins of willow (*Salix alaxensis, S. planifolia, S. pulchra,* and *S. glauca*), as well as grasses and sedges (*Carex bigelowii, C. membranacea, C. podocarpa,* and *Eriphorum vaginatum*). They also consume fruticose lichens, resin birch (*Betula glandulosa*), dwarf birch (*B. nana*), and horsetails (*Equisetum* spp.) (Skoog 1968, ADF&G 1985).

### Summer

From mid-June to mid-August, caribou typically consume willow leaves, resin birch, and dwarf birch, as well as sedges and grasses, especially grasses from the genera *Alopecurus, Arctagrostis, Dupontia, Festuca, Poa, Puccinellia, Calamagrostis,* and *Hierochloe*. They also eat horsetails, legumes (*Astragalus umbellatus*, *Lupinus arcticus*, *Hedysarum alpinum*, and *Oxytropis nigresens*), and forbs such as *Gentiana glauca*, *Swertia perennis*, *Sedum roseum*, *Antennaria monocephala*, *Artemisia arctica*, *Epilobium latifolia*, *Pedicularis* spp., *Petasites frigidus*, *Polygonum bistorta*, *Rumex arcticus*, and *Saxifraga* spp. (Skoog, 1968, ADF&G 1985).

## Fall

Caribou feed on grasses, sedges, and lichens throughout the fall. They also feed on willow and water sedge (*Carex aquatilis*) if they are available (ADF&G 1985). Caribou also feed on mushrooms when available (Skoog 1968).

## Winter

Caribou winter diets consist primarily of lichens (especially *Cladonia* spp. and *Cetraria* spp.), with smaller amounts of sedges and grasses, as well as horsetails, and the tips and buds of willows and dwarf shrubs (e.g., *Vaccinium uliginosum*). They may consume vegetation in muskrat (*Ondatra zibethicus*) pushups during winter, as well as aquatic vegetation in poorly drained coastal plains (ADF&G 1985, Skoog 1968).

# BEHAVIOR

## Seasonal Range Use and Migrations

Some caribou herds travel long distances between summer and winter ranges to find adequate sources of food and bear their calves in areas relatively free of predators (Skoog 1968, Whitten et al. 1992, Bergerud 1996, Griffith et al. 2002). Physical features on the landscape influence caribou migration routes. Caribou must make their way around open seawater, large lakes, swift rivers, rivers with floating ice, rocky regions in high mountains, volcanic cinder patches, glaciers, and burned areas. Features such as frozen lakes and rivers, as well as ridge tops, eskers, streambeds, and hard-surfaced snow drifts can aid caribou during winter migration (ADF&G 1985). Since the 1980s, calving areas, other seasonal ranges, and migration routes of the Mulchatna herd have varied widely. The Mulchatna herd has ranged extensively throughout most of the Nushagak and Kvichak River watersheds, but caribou from this herd also spend much of their time to the north in the Kuskokwim River drainage (Woolington 2009).

In contrast to most other migratory caribou herds, the Mulchatna herd does not use the same traditional calving ground each year, although its calving areas have remained distinct from those of any other herds. The Mulchatna herd calved in the Bonanza Hills area of the upper Mulchatna River watershed during the 1980s. In 1992, calving shifted west to the Mosquito River drainage in the upper Mulchatna watershed. From 1994 to 1999 calving generally occurred in the Upper Nushagak River watershed. From 2000 to 2002 calving was split between the Lower Nushagak River watershed and the South Fork of the Hoholitna River in the Kuskokwim River watershed. In 2003 and from 2005 to 2008, calving occurred near Kemuk Mountain in the Nushagak River watershed, as well as near the South Fork of the Hoholitna River in the Kuskokwim River watershed. In 2004, calving was widespread, from Dillingham in the Nushagak River watershed, north to the Holitna and Hoholitna Rivers in the Kuskokwim River watershed (Woolington 2009).

The Mulchatna herd often ranges widely across the Nushagak River drainage during summer and fall, but also frequently uses areas to the north and west in the Kuskokwim Mountains. During the 1980s much of the Mulchatna herd wintered north and west of Lake Iliamna in the Kvichak River watershed. In the 1990s most wintering shifted to the Kuskokwim Mountains. For the past decade part of the herd has wintered in the Nushagak and Mulchatna River watersheds while part of the herd has wintered in the Kuskokwim River watershed. In 2006–07 and 2007–08, up to 20,000 Mulchatna caribou wintered in the Lower Nushagak and Kvichak valleys with some going as far south as the Naknek valley in 2006–07 (Woolington 2009).

Mulchatna caribou are often widely dispersed during movements between seasonal ranges. In accordance with the highly variable locations of seasonal ranges, migration routes tend to vary from year to year (Woolington 2009).

Historically, the Northern Alaska Peninsula herd has spent most of its time in the Naknek River watershed and in areas south of it, far removed from the Kvichak and Nushagak River watersheds. From about 1986 through 2000, however, many caribou from the Northern Alaska Peninsula herd wintered in the Kvichak River watershed, south of Lake Iliamna. But since 2001, only a single radio-collared caribou from this herd has wintered north of the Naknek River (Butler 2009).

Like many small caribou herds, the Nushagak Peninsula herd is sedentary and spends the entire year on the Nushagak Peninsula, although a few caribou from this herd have made short forays (< 20 km and for < 1 month) off the Peninsula (Aderman and Woolington 2001). No overlap between the Nushagak Peninsula herd and the much larger migratory Mulchatna herd has yet been documented.

## Response to Disturbance

Industrial activities impact caribou by hindering or altering their movements or displacing them from preferred habitats. Barren-ground caribou on the North Slope of Alaska have avoided development such as exploration wells (Fancy 1983) and linear developments such as roads and pipelines (Dau and Cameron 1986) by distances of 2 to 5 km. Establishment of extensive, densely packed development with interconnecting road networks, high levels of traffic, aircraft activity, and ongoing construction or production activity around the Prudhoe Bay oilfields has resulted in general displacement of caribou from some areas (Griffith et al. 2002). Avoidance and displacement are most prevalent among females with young calves (Cameron et al. 1979, Cameron and Whitten 1980, Griffith et al. 2002). Similarly, woodland caribou in Canada typically avoided areas near mining sites by 1 to 5 km (Weir et al. 2007). Mining activities had the highest impact on caribou during calving season. Larger groups and females with young typically avoided mining sites more often than smaller groups and caribou without young (Weir et al. 2007). Weir et al. (2007) identified corridors such as roads and seismic lines as the greatest development impact on caribou because they increase the chance of encounters with humans and predators. The large Red Dog Mine in northwestern Alaska has had only limited and localized effects on caribou movements and distribution, in part because the mine occupies only a tiny fraction of the Western Arctic Caribou herd's otherwise pristine range. Mine operators and workers have implemented policies to minimize conflicts between traffic and caribou along the road from the mine to the port site (Dau 2009). In Norway movement patterns and range use by wild reindeer have been disrupted by combinations of highways and railroads, as well as by large hydroelectric developments (Nellemann et al. 2001, Nellemann et al. 2003). Impacts of development tend to be less when they occur on noncritical seasonal range in areas or at times when caribou are at low density relative to available range, or when similar habitats are available nearby (Griffith et al. 2002).

# INTERSPECIES INTERACTIONS

The interrelationships of wolves, caribou, and moose populations have been studied extensively in Alaska (Gasaway et al. 1983, National Research Council 1997, Mech et al. 1998). In large areas of interior Alaska, moose tend to persist for long time periods at low densities with population size regulated by high rates of predation by wolves and bears (Gasaway et al. 1992). In contrast, caribou are able to periodically escape regulation by predators and at least temporarily achieve high densities (Davis and Valkenburg 1991, Valkenburg 2001). Caribou population dynamics in the Nushagak and Kvichak River watersheds are consistent with this large pattern. Predation by wolves does not appear to be a major factor in regulating the Mulchatna herd, possibly due to rabies outbreaks that periodically reduce the wolf population (Valkenburg et al. 2003, Woolington 2009). Large, migratory caribou herds, such as the Mulchatna herd, may also avoid predation by moving seasonally to areas with few resident predators, or by using seasonal ranges erratically and unpredictably. Wolves in the Nushagak and Kvichak drainages are not known to follow migratory caribou (Woolington 2009).

In spite of consistent statements in management reports that predation was not a major limiting factor for the Mulchatna herd nor a likely cause of the recent decline in numbers, ADF&G and the Alaska Board of Game instituted an intensive management program for the Mulchatna herd in 2011, including active wolf control in part of the herd's range. Data on both wolf and caribou remain sketchy. Wolf harvest through normal hunting and trapping take was high in the initial year of the intensive management program, but very few wolves have been taken as part of the official control effort and few wolves were taken by any means during 2012 or 2013. Intensive management goals for increased calf:cow and bull:cow ratios are being met; however, the increases have occurred largely in the portions of the herd range outside of the wolf control area. For example, increased calf survival in 2014 may have been due to a shift in the calving area of the eastern portion of the herd (outside the control area) to a new area with presumably fewer predators (ADF&G 2015).

Intensive management, including wolf control was also implemented for the Northern Alaska Peninsula herd in 2010. Few wolves have been taken so far, and the program is too new to assess results (ADF&G 2015).

While bull and cow caribou are intermingled during fall when composition counts usually occur, the ratios of both bulls and calves to cows can vary widely from one geographic area to another. Counts should be distributed over the entire range of the herd in proportion to the local abundance of caribou.

Caribou range use and movement patterns in the Mulchatna and Kvichak River watersheds are inconsistent and irregular. The herd often divides into separate eastern and western segments, but there can be interchange between these segments within and/or among years. This complicates assessment of management actions such as wolf control applied only to part of the herd's range. It could also make potential effects of development activities difficult to measure if those effects are subsequently diluted across the entire population. More frequent monitoring of conventional radio-collars could lead to a better understanding of movements and distribution. Satellite collars and/or GPS collars would be even more useful by providing more complete data, especially when frequent inclement weather makes tracking conventional radio-collars by aircraft difficult or impossible.

More information is needed on range conditions and caribou nutritional status. Previous studies have indicated poor nutritional condition linked to lower productivity and possibly to increased incidence of disease and parasites. Current predation management assumes that habitat is not limiting for caribou. However, numerous management and research reports have suggested that habitat or nutrition is, in fact, limiting. If so, efforts to increase caribou numbers through predator control could exacerbate problems rather than alleviate them.

If large-scale industrial developments occur in the Mulchatna/Kvichak area, baseline data on caribou movements and use should be gathered before development actually occurs. It may be necessary to capture, mark, and monitor caribou annually in development and similar nondevelopment areas to assess potential differences in productivity, survival, or nutritional condition.

## SUMMARY

Caribou in the Nushagak and Kvichak drainages have historically fluctuated widely in numbers and distribution. Two separate herds, or populations, have traditionally used the area. The Mulchatna herd is the primary population in the area with the Northern Alaska Peninsula herd using only the Kvichak River watershed, and then only occasionally. A third population, the Nushagak Peninsula herd, was established by transplant of caribou from the Northern Alaska Peninsula herd in 1988. The Nushagak Peninsula herd remains small and confined to the Nushagak Peninsula. Natural factors such as climate and range/nutritional conditions are the driving forces in caribou population dynamics in this area. Predator numbers, especially for wolves, also vary widely in the area, possibly due to periodic disease outbreaks. In spite of lack of evidence directly linking predation to the recent declines of caribou, a predator control program has recently been authorized. Natural instability in caribou numbers and distribution are making the effectiveness of the wolf control program difficult to assess and will likely also make it difficult to assess effects of possible future industrial developments in the area, at least at the caribou population level.

## REFERENCES

Aderman, A.R. 2009. Population monitoring and status of the Nushagak Peninsula caribou herd, 1988–2008. Progress Report, Togiak National Wildlife Refuge, Dillingham, Alaska. pp. 29.

Aderman, A.R. and S.J. Lowe. 2011. Population monitoring and status of the Nushagak Peninsula caribou herd, 1988–2011. Draft progress report, Togiak National Wildlife Refuge, Dillingham, Alaska. pp. 29.

Aderman, A.R. and S.J. Lowe. 2012. Population monitoring and status of the Nushagak Peninsula caribou herd, 1988–2011. Progress report, Togiak National Wildlife Refuge, Dillingham, Alaska. pp. 29.

ADF&G (Alaska Department of Fish and Game). 1985. Alaska Habitat Management Guide, Southwest Region Volume 1: Fish and Wildlife Life Histories, Alaska Department of Fish and Game, Juneau, AK.

ADF&G (Alaska Department of Fish and Game). 2015. Annual report to the Alaska Board of Game on intensive management for caribou with wolf predation control in game management units 9B, 17B&C, and 19A&B, The Mulchatna caribou herd. Alaska Department of Fish and Game, Anchorage Alaska, Division of Wildlife Conservation, February 2015. Available at: http://www.adfg.alaska.gov/static/research/programs/intensivemanagement/pdfs/2015_gmu_9ce_intensive_management_annual_report.pdf.

Ballew, C., A. Ross, R.S. Wells, V. Hiratsuka, K.J. Hamrick, E.D. Nobmann, and S. Bartell. 2004. Final report on the Alaska Traditional Diet Survey, Alaska Native Health Board, Anchorage, AK.

Bergerud, A.T. 1996. Evolving perspectives on caribou population dynamics, have we got it right yet? Rangifer Special Issue 9:95–116.

Butler, L. 2009. Units 9C & 9E caribou management report. In: P. Harper (Ed.), Caribou management report survey and inventory activities July 1, 2006–June 30, 2008, Alaska Department of Fish and Game, Juneau, AK.

Cameron, R.D., K.R. Whitten, W.T. Smith, and D.D. Roby. 1979. Caribou distribution and group composition associated with construction of the Trans-Alaska Pipeline. Canadian Field-Naturalist 93:155–162.

Cameron, R.D. and K.R. Whitten. 1980. Influence of the Trans-Alaska Pipeline corridor on the local distribution of caribou. Pages 475–484. In: E. Reimers, et al. (Eds.), Proceedings of the 2nd International Reindeer/Caribou Symposium, Trondheim: Direktoratet for vilt og ferskvannsfisk, Roros, Norway, 1979.

Cameron, R.D., K.R. Whitten, and W.T. Smith. 1986. Summer range fidelity of radio-collared caribou in Alaska's Central Arctic Herd. Rangifer Special Issue 1:51–55.

Cameron, R.D. 1994. Reproductive pauses by female caribou. Journal of Mammalogy 75:10–13.

Colson, K.E., K.H. Mager, and K.J. Hundertmark. 2014. Reindeer introgression and the population genetics of caribou in Southwestern Alaska. Journal of Heredity 105(5): 585–596. Available: http://jhered.oxfordjournals.org/content/early/2014/05/19/jhered.esu030.full.pdf.

Dau, J.R. and R.D. Cameron. 1986. Effects of a road system on caribou distribution during calving. Rangifer Special Issue 1:95–101.

Dau, J.R. 2009. Units 21D, 22A, 22B, 22C, 22D, 22E, 23, 24, and 26A caribou management report. In: P. Harper (Ed.), Caribou management report of survey and inventory activities 1 July 2006 – 30 June 2008, Alaska Department of Fish and Game, Juneau, AK. pp. 176–239.

Davis, J.L. and P. Valkenburg. 1991. A review of caribou population dynamics in Alaska emphasizing limiting factors, theory, and management implications. Pages 184–209. In: C. E. Butler and S. P. Mahoney (Eds.), Proceedings of the Fourth North American Caribou Workshop, 1989, Newfoundland and Labrador Wildlife Division, St. John's, Newfoundland, Canada.

Fancy, S.G. 1983. Movements and activity budgets of caribou near oil drilling sites in the Sagavanirktok River Floodplain, Alaska. Arctic 36.

Gasaway, W.C., R.O. Stephenson, J.L. Davis, P.E.K. Shepherd, and O.E. Burris. 1983. Interrelationships of wolves, prey, and man in interior Alaska. Wildlife Monographs 84:1–50.

Griffith B., D.C. Douglas, N.E. Walsh, D.D. Young, T.R. McCabe, D.E. Russell, R.G. White, R.D. Cameron, and K.R. Whitten. 2002. The Porcupine caribou herd. Pages 8–37. In: D.C. Douglas, et al. (Eds.), Arctic Refuge coastal plain terrestrial wildlife research summaries, U.S. Geological Survey, Biological Resources Division, Biological Science Report USGS/BRD BSR-2002-0001.

Harper, P., and L McCarthy (Eds), 2015. Caribou management report of survey-inventory activities July 1, 2012–June 30, 2014. Alaska Department of Fish and Game, Species Management Report ADF&G/DWC/SMR-2013-3, Juneau.

Hinkes, M.T. and L.J. Van Daele. 1996. Population growth and status of the Nushagak Peninsula caribou herd in southwest Alaska following reintroduction, 1988–1993. Rangifer Special Issue 9:301–310.

Hinkes, M.T., G.H. Collins, L.J. Van Daele, S.D. Kovach, A.R. Aderman, J.D. Woolington, and R.J. Seavoy. 2005. Influence of population growth on caribou herd identity, calving ground fidelity, and behavior. Journal of Wildlife Management 69:1147–1162.

Hotchkiss, L.A. 1989. 1988–1989 annual progress report of the caribou reintroduction project on the Togiak National Wildlife Refuge, Dillingham, Alaska, Togiak National Wildlife Refuge, Dillingham, Alaska. pp. 8.

Mech, L.D., L.G. Adams, T.J. Meier, J.W. Burch, and B.W. Dale. 1998. The Wolves of Denali. University of Minnesota Press, Minneapolis, Minnesota.

National Research Council. 1997. Wolves, Bears, and Their Prey in Alaska. National Academy Press, Washington D.C.

Nellemann, C., I. Vistnes, P. Jordhoy, and O. Strand. 2001. Winter distribution of wild reindeer in relation to power lines, roads and resorts. Biological Conservation 101:351–360.

Nellemann, C., I. Vistnes, P. Jordhoy, O. Strand, and A. Newton. 2003. Progressive impact of piecemeal infrastructure development on wild reindeer. Biological Conservation 113:307–317.

Paul, T.W. 2009. Game transplants in Alaska. Technical Bulletin No. 4, 2nd Ed., Alaska Department of Fish and Game, Juneau, AK. pp. 150.

Skoog, R.O. 1968. Ecology of caribou (Rangifer tarandus granti) in Alaska. University of California, Berkeley, USA.

Valkenburg, P. 2001. Stumbling towards enlightenment: understanding caribou dynamics. Alces 37:457–474.

Valkenburg, P., R.A. Sellers, R.C. Squibb, J.D. Woolington, A.R. Aderman, and B.W. Dale. 2003. Population dynamics of caribou herds in southwestern Alaska. Rangifer Special Issue 14:131–142.

Weir, J.N., S.P. Mahoney, B. McLaren, and S.H. Ferguson. 2007. Effects of mine development on woodland caribou Rangifer tarandus distribution. Wildlife Biology 13:66–74.

Whitten, K.R. 1995. Antler loss and udder distension in relation to parturition in caribou. Journal of Wildlife Management 59:273–277.

Whitten, K.R., G.W. Garner, F.J. Mauer, and R.B. Harris. 1992. Productivity and early calf survival in the Porcupine Caribou Herd. Journal of Wildlife Management 56:201–212.

Woolington, J.D. 2009. Mulchatna caribou management report, Units 9B, 17, 18 south, 19A & 19B. Pages 11–31. In: P. Harper (Ed.), Caribou management report of survey and inventory activities 1 July 2006–1 July 2008, Alaska Department of Fish and Game, Juneau, AK.

Woolington, J. D. 2013. Mulchatna caribou management report, Units 9B, 17, 18 south, 19A & 19B. Pages 23–45. In: P. Harper (Ed.), Caribou management report of survey and inventory activities July 1, 2010–June 30, 2012. Alaska Department of Fish and Game, Species Management Report ADF&G/DWC/SMR-2013-3, Juneau.

# 10

# THE GRAY WOLF

Lori A. Verbrugge, U.S. Fish and Wildlife Service
Ashley E. Stanek, University of Alaska Anchorage
Buck Mangipane, Lake Clark National Park and Preserve

## INTRODUCTION

The gray wolf (Figure 10.1; *Canis lupus*) is the largest wild extant canid (Paquet and Carbyn 2003). The historic distribution of wolves once covered most of North America, but as the contiguous United States were settled during the past 250 years, wolves were widely persecuted due to their tendency to prey on livestock and pets (Mech 1995). By the 1970s, wolf populations in the contiguous U.S. were decimated, which led to their protection under the Endangered Species Act in 1978. The gray wolf is currently listed as *endangered* in most of the Lower 48 states, with the exceptions of: Minnesota, where they are considered threatened; the northern Rocky Mountain states of Idaho, Montana, the eastern

**Figure 10.1** A gray wolf along the coast of Amalik Bay, Katmai National Park and Preserve, Alaska. National Park Service (NPS). *Photo credit: D. Kopshever.*

third of Oregon and Washington, and north central Utah, where populations are considered *recovered*, and as a result are de-listed; in Wyoming, populations are considered *nonessential experimental* (United States Fish and Wildlife Service (USFWS) 2017). Wolves in Alaska are not considered threatened or endangered and the Alaska Department of Fish and Game (ADF&G) estimates from 7,000 to 11,000 wolves inhabit the state.

## HABITAT

Wolves are habitat generalists and their home ranges can encompass a variety of diverse habitats (Mech 1970, Mladenoff et al. 1995, Paquet and Carbyn 2003). Historically, gray wolves were distributed throughout the northern hemisphere in every habitat where large ungulates were found (Mech 1995). Prey abundance and availability strongly influence habitat use by wolves (Paquet and Carbyn 2003). Male and female wolves do not differ in habitat selection, and the pack maintains their territory throughout the year.

Wolf pups are born, protected, fed, and raised in natal and secondary den sites, a series of rendezvous sites, and surrounding areas (Paquet and Carbyn 2003). Dens provide shelter and are often located in a hole, rock crevice, hollow log, overturned stump, abandoned beaver lodge, or expanded mammal burrow (Paquet and Carbyn 2003). Rendezvous sites are areas where pups are left while pack members forage (Theberge 1969).

A misperception, common in years past, was that wolves needed wilderness to survive. More recent studies have shown that wolves do not need wilderness, but they do require adequate prey and a relatively low rate of human-caused mortality (Mech 1995, Mladenoff et al. 1999). The presence of roads has a complex impact on habitat selection by wolves. Roads benefit wolves by easing their travel and access to prey, but conversely roads are associated with human contact and increased wolf mortality through either intentional or accidental killing; cryptic behavior to avoid humans is documented (Mladenoff et al. 1999, Houle et al. 2010, Zimmermann et al. 2014). Near the Kenai National Wildlife Refuge (NWR) in Alaska, wolves regularly used a gated pipeline road presumably because it offered an easy travel corridor with little human use (Thurber et al. 1994). In that study, wolf absence from human-settled areas and heavily traveled roads seemed to be caused by wolf behavioral avoidance rather than direct human-caused mortality of wolves in those areas.

## FOOD HABITS

### Diet

Wolves are obligate carnivores whose use of prey depends largely on the availability and vulnerability of ungulates (Weaver 1994). Dietary habits such as preferred prey species and prey switching tactics vary substantially among wolf packs in different locations in response to local ecological relationships.

Wolves can exhibit flexible diets, and shift to nonungulate prey species when ungulate prey are scarce (Forbes and Theberge 1996) or to take advantage of seasonally abundant, nutritious alternate prey species such as salmon (Darimont et al. 2008). Some wolf packs require multiple prey species during the summer to meet the high energetic demands of reproduction (Paquet and Carbyn 2003) and because adult ungulates are less vulnerable to predation (and consequently less available as prey) during summer (Jędrzejewski et al. 2002, Metz et al. 2012). Availability of alternative prey species is also important for wolf packs in northwestern Alaska that rely on migratory caribou that move seasonally to calving grounds where they are inaccessible to wolves (Ballard et al. 1997).

In Algonquin Park, Ontario, beavers and snowshoe hares (*Lepus americanus*) are important to the winter diet of wolves as are scavenged moose carcasses (Forbes and Theberge 1992, Forbes and Theberge 1996). Other animals such as lemmings, voles, muskrat, and a variety of birds (especially waterfowl) and their eggs also supplement the wolf diet (Kuyt et al. 1981, Spaulding et al. 1998, Weibe et al. 2009), while fish and berries are consumed seasonally where available (Kohira and Rexstad 1997, Darimont and Paquet 2000).

Coastal wolves consume marine mammal carcasses, mussels, crabs, and even barnacles (Darimont and Paquet 2000). The Ilnik wolf pack on the Alaska Peninsula was found to preferentially use coastal habitat along Bristol Bay, where it was frequently observed consuming marine mammal carcasses that had washed ashore (Watts et al. 2010). In winter when Bristol Bay was frozen, the pack was documented using offshore sea ice, and wolves killed sea otters (*Enhydra lutris kenyoni*) near the coastline when the otters were trapped above the sea ice (Figure 10.2) (Watts et al. 2010).

Wolves on the Kenai Peninsula of Alaska were found to rely heavily on moose during the summer (Peterson et al. 1984). Moose comprised an estimated 97% of ingested prey biomass in summer, which was largely scavenged from old

**Figure 10.2**  Gray wolf with sea otter pup along the coast of Swikshak Bay, Katmai National Park and Preserve, Alaska. *Photo credit: NPS, Kaitlyn Kunce.*

kills; only 16% of moose carcasses found in association with wolves in summer were fresh kills. In contrast, 80% of moose consumed during the winter were fresh wolf-kills (Peterson et al. 1984). Kenai Peninsula wolves also ate snowshoe hares and beavers during the summer, and minor quantities of small rodents, birds, vegetation and other prey (Peterson et al. 1984). Scats from wolves in southcentral Alaska confirmed reliance on moose; but beavers and snowshoe hares were also commonly consumed (Ballard et al. 1987). Wolves in southcentral Alaska sometimes consumed caribou (Figure 10.3), muskrat, squirrels, voles, vegetation, and a variety of other dietary items (Ballard et al. 1987). In Lake Clark National Park and Preserve (LACL), and elsewhere in Alaska, Dall's sheep are an additional ungulate prey source. GPS collar locations

**Figure 10.3**  Gray wolf with scavenged caribou leg, Denali National Park. *Photo credit: NPS, Ken Conger.*

of wolves in the Lake Clark study showed some packs were regularly present in Dall's sheep habitat, and were observed consuming Dall's sheep (B. Mangipane, pers. comm.).

## Salmon as a Food Source

Foraging on salmon may have considerable adaptive value for wolves regardless of ungulate density. Foraging theory predicts an advantage to avoiding dangerous ungulate prey in favor of less dangerous alternatives such as salmon (Stephens and Krebs 1986). Salmon also offers superior nutritive value; in one study pink salmon contained more than four times as much energy per 100 g of meat than raw black-tailed deer (*Odocoileus hemionus columbianus*) (Darimont et al. 2008). Behavioral observations suggest that wolves may have a broad history of seasonal consumption of salmon in areas where the distributions of wolves and salmon overlap (Darimont et al. 2003).

Wolves have often been observed consuming only the head of salmon rather than the entire fish (Darimont et al. 2003). There are several possible explanations for this behavior. Wolves may be consuming only the most energetically valuable part of the prey item, or they may be targeting specific micronutrients such as omega-3 fatty acids (Gende et al. 2001). Wolves may also be selecting head tissue to minimize their exposure to parasites such as *Neorickettsia helminthoeca*, which can infect salmon and be fatal to canids (Darimont et al. 2003).

The use of salmon by wolves can occur throughout the year (Adams 2010, Stanek et al. 2017). During summer, salmon can be a reliable resource, and help to offset the reduced availability of ungulates (Darimont et al. 2008, Szepanski et al. 1999). Consumption of salmon in the fall may improve pup survivorship during weaning (Person 2001). Winter snow can preserve salmon carcasses buried underneath, enabling use by wolves and other scavengers for the rest of the winter (Carnes 2004). Carnes (2004) compared scats from different packs in the Copper and Bering River deltas, and noted increased consumption of salmon in winter; he hypothesized this might be related to unavailability of moose in those areas.

Several studies throughout Alaska have examined the importance of salmon to wolves. In the Copper and Bering River Deltas, late summer rendezvous sites for wolves were typically located alongside shallow spots in spawning areas or at bends where gravel bars extended out into streams (Carnes 2004). Researchers at these areas observed wolves, especially pups, waiting for spent salmon carcasses to float by (Carnes 2004). In southeast Alaska, marine protein comprised 18% of the lifetime total diet of Alexander Archipelago wolves; most of the marine contribution was likely salmon, although other marine organisms were probably also consumed (Szepanski et al. 1999). In Bristol Bay's Togiak NWR, wolves have been observed delivering intact salmon carcasses to their pups at rendezvous sites (Walsh 2011). Similar foraging behavior has been observed among wolves in other areas of Bristol Bay, including the Alaska Peninsula Game Management Units (GMUs) 9C and 9E, which extend from the Naknek River drainage through Port Moller, where wolves often transport captured salmon to den or rendezvous sites (D. Watts, pers. comm.) (For GMU map see Figure 7.4 in this volume, or refer to http://www.adfg.alaska.gov/index.cfm?adfg=huntingmaps.alaskamaps for a comprehensive listing of GMU maps in Alaska).

Salmon are not only a food resource for coastal wolves. Some Pacific salmon migrate long distances inland, returning to spawning grounds that may be hundreds or even thousands of miles from the ocean. A study in Denali National Park and Preserve (DNPP), documented substantial seasonal salmon consumption among these wolves who lived more than 1,200 river km (746 river mi) from the coast (Adams et al. 2010). DNPP wolves with territories where salmon were seasonally abundant and ungulates occurred at low densities ate the most salmon; salmon averaged 17% of their total long-term diet (Adams et al. 2010).

Between 2008 and 2012, LACL conducted a study of wolves in the region to acquire baseline information focusing on diet and general distribution. One component of this study examined the relative use of salmon compared to terrestrial prey by wolves (Stanek et al. 2017). In contrast to other studies, Stanek et al. (2017) estimated diet of individual wolves during multiple temporal periods by conducting stable isotope analysis of different tissues representing approximately summer, fall and early winter, and late winter, rather than over multiple years (Adams et al. 2010, Szepanski et al. 1999) or from a single tissue (Darimont et al. 2002). One of the key outcomes of this study was documenting the variability with which salmon was used by wolves in the Lake Clark region. Wolves consumed salmon in each season analyzed, though the variability between individuals, groups, and years, was substantial. In summer, the diets of some individuals regularly consisted of over 50% salmon while others consumed primarily terrestrial prey. In one year, two individuals were documented consuming salmon during the late fall and early winter, up to 89% by one individual; these individuals were consuming frozen salmon at a lakeshore at the time of capture in December 2008. Salmon consumption was generally consistent within social groups however, there were instances where within a single social group some individuals consumed salmon while others did not. Salmon are clearly an important resource for some wolves though the impact of this alternative resource is unclear.

## Dispersal of Marine-Derived Nutrients (MDNs) by Wolves

The influences of salmon on terrestrial systems are largely dependent on predators that remove salmon from streams, consume a portion, and leave the remains behind (Hilderbrand et al. 1999, Reimchen 2000). Abandoned salmon carcasses contribute to ecosystem processes when scavenging, decomposition, and fecal-urinary deposition provide marine derived nutrients (MDNs) to terrestrial systems that are typically nitrogen- and phosphorus-limited (Ben-David et al. 1998, Willson et al. 1998, Hilderbrand et al. 1999, Reimchen 2000).

Wolf behavior influences the distribution pattern of MDNs within terrestrial ecosystems because wolves often transport salmon some distance rather than consuming it in the stream or immediate vicinity. Wolves in British Columbia were observed consuming salmon on grass next to the river 70% of the time (Darimont et al. 2003). However, in LACL, preliminary data indicate that wolves move considerable distances over several days to feed on salmon. For example, in 2009, one wolf was documented traveling up to 64 km (40 mi) from a den site to feed on salmon. In 2010 and 2011, the same wolf traveled up to 24 km (15 mi) and 40 km (25 mi) to feed on salmon (B. Mangipane, pers. comm.). LACL wolves have been observed feeding on fish carcasses frozen into lake ice, and the backbones and heads left from human subsistence fishing (B. Mangipane, pers. comm., P. Spencer, pers. comm.).

## Interspecies Interactions: Response to Change in Salmon Populations/Distribution

Salmon are important seasonal prey for wolves in coastal regions and along major river systems (Kohira and Rexstad 1997, Adams et al. 2010), to the extent that in some coastal areas salmon may seasonally decouple the dependence of wolves on ungulate prey (Darimont et al. 2008). In DNPP, salmon were found to be a particularly important food item for wolves in areas with low ungulate density but high salmon abundance (Adams et al. 2010). The availability of salmon had a strong impact on abundance of wolves in the northwestern flats area of DNPP; wolves were only 17% less abundant in that area compared to the rest of the study area even though ungulate densities were 78% lower. The higher wolf population density facilitated by the availability of salmon was thought to result in increased overall predation pressure on ungulates in that system (Adams et al. 2010). Moose were the predominant ungulate in the northwestern flats but occurred at densities approaching the lowest in North America (Gasaway et al. 1992), and appeared to be limited by predation rather than nutritional constraints (Adams et al. 2010).

Under some circumstances, predation by wolves may limit or regulate ungulate populations, sometimes suppressing their numbers at very low densities (top-down forcing) (Gasaway et al. 1992). In other cases, ungulate populations are influenced mostly by carrying capacity of habitat regardless of wolf predation (bottom-up effects) (Ballard et al. 2001). The relative importance of top-down and bottom-up factors can change over time depending on habitat changes, weather conditions, and anthropomorphic disturbances (Bowyer et al. 2005).

The interrelationships of wolves, caribou, and moose populations have been characterized in several ecosystems. Alterations in moose densities can have a major influence on caribou populations, through their effect on wolf predation rates. In southeastern British Columbia, wolf population numbers can be suppressed due to a lack of available food in winter, when moose numbers are low, particularly when caribou over-winter in areas inaccessible to wolves (Seip 1992). Elevated moose densities may cause a concomitant rise in wolf numbers. If moose numbers later decline, wolves in the area will turn to caribou as an alternative food source, potentially affecting the caribou population. Industrial development can exacerbate this effect by increasing wolves' access to caribou via the creation of new linear corridors, such as roads and pipelines (James et al. 2004). These interrelationships may not be applicable to the southwest Alaska ecosystem, where caribou occupy tundra habitats in contrast to the low-density woodland caribou herds from the Canadian studies.

Wolves are coursing predators that actively pursue prey rather than passively ambushing them (Mech 1970, Mech et al. 1998). Consequently, in much of Alaska they typically select open or sparsely forested habitats that enable detection and pursuit of prey. Deep snow that hinders movement of ungulate prey or restricts them to small, forested patches often facilitates predation by wolves (Mech and Peterson 2003). For large ungulate prey such as moose, wolves often focus predation on calves, which tend to be the most vulnerable. In interior Alaska, wolves are most effective hunting in flat or rolling terrain covered with sparse boreal forest or tundra. In coastal rainforests, prey tend to be most vulnerable by wolves in open muskeg heaths at low elevations (Farmer et al. 2006).

# BEHAVIOR

## Wolf Packs

Gray wolves are territorial and social carnivores that typically live in packs of about six to eight animals although packs may include > 20 wolves (Mech and Boitani 2003) (Figure 10.4). Wolf packs typically consist of a single breeding pair, pups of the year, and their older siblings (Mech and Boitani 2003). Mating occurs during late January to March and gestation is approximately 63 days. Most commonly, each pack produces only a single litter of pups which are born in dens, usually during late April to May. Multiple litters, although rare, have been observed within some packs in Alaska (Ballard et al. 1987, Meier et al. 1995). Litter sizes range from 1 to 12 pups but more commonly number 4 to 6 (Fuller et al. 2003). Dens in coastal temperate rainforests are located within the root wads of living or dead trees (Person and Russell 2009). In boreal forest or tundra, dens are located in sandy areas or gravel eskers (Ballard and Dau 1983, McLoughlin et al. 2004). Wolves and their pups occupy dens from late April to early July, and then move to rendezvous sites where sequestered pups are fed by pack members until September or early October when they are sufficiently large to move with the pack (Mech et al. 1998, Packard 2003, Person and Russell 2009). Pup mortality during summer is affected strongly by availability of food (Fuller et al. 2003). Wolves usually remain within their natal packs until they reach sexual maturity at 22 to 24 months. At that age, some may disperse from their packs to find mates and establish their own packs. However, researchers reported dispersers ranging in age from 10 months to 5 years (Mech and Boitani 2003). Abundant prey may induce some wolves to defer dispersal until they are older, allowing packs to increase in size (Fuller et al. 2003, Mech and Boitani 2003). Dispersing wolves may travel hundreds of kilometers and traverse very difficult terrain before settling (Mech and Boitani 2003); they may even cross large bodies of water. For example, in southeastern Alaska, dispersing wolves were documented swimming 3 to 4 km (1.9–2.5 mi) in open ocean to move between islands (Person and Russell 2008).

## Range

Resident wolf packs occupy extensive territories that they attempt to defend from other wolves. Wolf territories tend to be smaller in summer, when packs remain closer to dens and home sites (Mech 1977), and larger in winter, when packs resume nomadic traveling as pups mature. In south-central Alaska, the average distance between dens of neighboring packs was 45 km (28 mi) (Ballard et al. 1987).

**Figure 10.4**    Gray wolf pack in Lake Clark National Park and Preserve, Alaska. *Photo credit: Buck Mangipane.*

Territories of wolf packs tend to be much larger in Alaska than in the continental U.S., due to the relatively low density of prey in Alaska. Territory and pack sizes are largely influenced by availability of prey (Mech and Boitani 2003). In general, wolf territory size is inversely related to the density of available prey (Fuller et al. 2003). As with wolves generally, average territory sizes of Alaskan wolf packs were consistently larger in winter than in summer (Table 10.1) (Peterson et al. 1984, Ballard et al. 1987, Ballard et al. 1997, Ballard et al. 1998, Burch et al. 2005, Adams et al. 2008). Wolves in north-central Minnesota had smaller winter territory size ($\overline{X} = 116$ km$^2$) and higher density (39/1,000 km$^2$ in mid-winter) than any of these Alaska packs, because of an abundant white-tailed deer (*O. virginianus*) population (Fuller 1989). Preliminary analysis of recent data from LACL indicates annual territory sizes from 524 km$^2$ to over 6,000 km$^2$ (202 mi$^2$ to over 2,316 mi$^2$). One pack in this study had territory sizes, of 1201 km$^2$ (464 mi$^2$), 1928 km$^2$ (744 mi$^2$), 1591 km$^2$ (614 mi$^2$), and 1667 km$^2$ (643 mi$^2$), in 2008–09, 2009–10, 2010–11, 2011–12, respectively (B. Mangipane, pers. comm.).

## Dispersal (Emigration)

Several studies in Alaska have documented emigration as a vital factor influencing the population dynamics of wolves (Peterson et al. 1984, Ballard et al. 1987, Ballard et al. 1997, Adams et al. 2008). Individuals may leave a pack and strike out on their own in response to low prey densities (Messier 1985). High rates of infectious disease (Ballard et al. 1997), social stress within the pack, or a lack of opportunity to achieve the high social status needed to successfully breed (Peterson et al. 1984) may also cause wolves to disperse to new territories.

Dispersal is a key mechanism that wolves use to colonize new habitats. Dispersing wolves experience a high rate of mortality, but when successful, they are able to establish a new pack (Peterson et al. 1984) or join an existing pack (B. Mangipane, pers. comm.). Successful colonization of new territory requires both a vacancy of suitable habitat and bonding with a mate (Rothman and Mech 1979).

## Seasonal Movements

Wolves in southcentral Alaska do not follow migratory movements of moose or caribou outside their territories, but do follow elevation movements of moose within their territories (Ballard et al. 1987). In most years, wolf packs in northwestern Alaska did not follow migratory caribou from the Western Arctic herd, but rather, switched to moose for prey during the winter and maintained year-round resident territories (Ballard et al. 1997). However, in years when moose densities were low, up to 17% of radio-collared wolf packs in northwest Alaska followed migratory caribou and then returned to their original territory for denning (Ballard et al. 1997).

In Bristol Bay, there is no evidence that wolf packs follow the Mulchatna caribou herd, although wolves are occasionally seen with the herd as it moves throughout the region (Woolington 2009). However, recent information in LACL shows that individual wolves followed caribou herds for a portion of the year (Figure 10.5). For example, GPS collar locations showed one adult male wolf followed caribou to calving grounds for approximately one week, and this was done in consecutive years. (B. Mangipane, pers. comm.). Packs are more likely to have established territories and take advantage of caribou when they move through those territories.

**Table 10.1.** Average territory sizes (km$^2$) of wolf packs in Alaska

| Region | Summer | Winter | Annual | Reference |
|---|---|---|---|---|
| Northwest | 621 | 1,372 | 1,868 | Ballard et al. 1997 |
| Denali National Park | — | — | 871 | Burch et al. 2005 |
| Central Brooks Range | — | — | 358–2,315[a] | Adams et al. 2008 |
| Kenai Peninsula | — | — | 466-864[b] | Peterson et al. 1984 |
| Southcentral | — | — | 1,644 | Ballard et al. 1987 |
| Lake Clark National Park and Preserve | — | — | 1967 | B. Mangipane, NPS, pers. comm. |

[a] Range of territory sizes estimated for wolf packs over the four-year study period
[b] Range of average annual territory sizes during study period
— not determined

Game Management Unit 9 includes the Kvichak River watershed and extends from LACL to False Pass. Wolf population estimates for the region are available, but they should be used with caution for several reasons. ADF&G groups GMUs 9 and 10 (the Aleutian Islands) for statistical purposes, so those wolf population estimates include lands outside the study area. Also, wolf population dynamics have been studied only informally in the region, and only limited descriptions of methods and results are available (Butler 2009). Methods included monitoring 10 wolf packs using radio-collar tracking, inferring trends through observations made during other fieldwork, reviewing reports from hunters and guides, and collecting responses to annual trapper questionnaires. Using these data, ADF&G estimated a total population of 350 to 550 wolves in GMUs 9 and 10 (Butler 2009). Wolf densities in GMU 9 and 10 are considered low to moderate, but wolf numbers in GMU 9 appear to have increased since the 1990s despite a decline in caribou populations. Possible explanations for this increase in wolves include an abundance of alternative prey such as marine mammal carcasses, salmon, or snowshoe hares; a population rebound following a high period of mortality from a rabies outbreak; or immigration of wolves from surrounding areas (Butler 2009). Data are not available to evaluate these hypotheses. Estimated wolf densities in GMU 9E (the Alaska Peninsula south to Port Moller) and the southwestern portion of GMU 9C (the Naknek watershed outside Katmai NPP) are 6–7 wolves/1,000 km$^2$ (D. Watts, pers. comm.).

## HUMAN USE/INTERACTION/MANAGEMENT

Reporting of wolves harvested in Alaska is mandatory, but reporting compliance is suspected to be weak in some areas (Ballard et al. 1997). The degree of reporting compliance within the Nushagak and Kvichak River watersheds, is unknown. The reported wolf harvests in GMUs 9, 10, and 17 for 2003–08 are summarized in Table 10.2.

Annual harvest of wolves in Alaska varies widely due to fur prices, hunter access to wolf habitat, predator control policies and practices, and population changes in response to prey populations. Hunter access is influenced by winter travel conditions (Woolington 2009), including snow depth and fuel prices. Wolves in the Bristol Bay area are typically hunted and trapped by local residents, but are also harvested opportunistically by nonresidents who are primarily hunting other species.

Trappers from southwestern Alaska indicated that wolves were the fourth most important species they targeted, behind river otters, beavers, and red foxes (in that order) (ADF&G 2010). State trapping regulations do not distinguish among different uses (e.g., *subsistence*, *recreational*, or *commercial*) (ADF&G 2011). Most rural Alaska communities are supported by a mixed subsistence-cash economy (Wolfe 1991). Trapping is one of many traditional subsistence activities that can provide a modest income for participants. Some harvested furs are sold to dealers, but others are used locally. Furs are often made into hand-crafted items, which are more valuable than the raw pelts (Wolfe 1991). Items commonly crafted with furs include mitts, coats, boots, fur ruffs, and slippers. In some rural areas, households use most of their harvested wolf pelts locally for ruffs, hats, and lining for winter gear, because imported materials are considered inferior (Wolfe 1991).

## RECOMMENDED RESEARCH

Wolf population numbers have not been well studied in Bristol Bay. Better regional wolf population estimates, gathered using scientifically rigorous methods, are needed to improve understanding of wolf populations in the region. Arising from the studies that have been conducted in the region, topics deserving future exploration include assessing whether salmon sustain wolves at higher numbers than the ungulate prey base alone could support, ultimately limiting

**Table 10.2** Wolf harvests in GMUs 9, 10, and 17 reported to the ADF&G

| Year | GMU 9 and 10[a] | GMU 17[b] |
|---|---|---|
| 2003–04 | 119 | 141 |
| 2004–05 | 64 | 60 |
| 2005–06 | 120 | 62 |
| 2006–07 | 85 | 79 |
| 2007–08 | 110 | 73 |

a = Butler 2009; b = Woolington 2009

the density-dependent feedback from a diminishing ungulate prey base, and determining the applicability of wolf studies from the continental United States, Canada, and even other regions in Alaska. Additionally, wolf populations can change quickly due to harvest, dispersal, and a high reproductive potential. Work in LACL-documented wolves dispersing throughout the entire region, for example from LACL to Togiak National Wildlife Refuge north-west of Bristol Bay, demonstrate the connectivity of the region. However, as these animals move far beyond agency boundaries, it is often challenging to continue studying particular individuals in a single-agency study. Long-term and region-wide studies would help to better capture the population dynamics, distribution, and behavior of wolves, which could potentially provide insights into how wolves affect ungulate populations. This type of project would be most informative if undertaken cooperatively among agencies, including the NPS, FWS, and the ADF&G.

## SUMMARY

There are currently estimated to be 7,000 to 11,000 gray wolves in Alaska. Gray wolves are territorial and social carnivores that typically live in packs of about six to eight animals. Wolves are habitat generalists and were historically distributed throughout North America wherever large ungulates were found. No wolf sub-populations are currently listed under the Endangered Species Act in Alaska, in contrast to the continental United States, but wolf territory sizes in Alaska are relatively large due to a lower density of prey in Alaska. Caribou and moose are important prey species for wolves, but a number of other dietary items are also consumed. Salmon provide a nutritious seasonal food resource for some wolves in the Bristol Bay region. Wolves can transfer marine-derived nutrients from salmon to the surrounding terrestrial ecosystem, both by fecal-urinary deposition and by transport of salmon prior to consumption.

In unexploited wolf populations, mortality rates are strongly influenced by prey availability. In Alaska, human causes of wolf mortality include trapping for commercial, recreational, and subsistence purposes, and targeted predator control programs in certain areas. In some rural communities, wolf fur is used in clothing and handicrafts, and provides a source of income to local trappers. Wolves have a prolific reproductive potential, so populations may rebound quickly following a discrete mortality event such as a lethal disease outbreak. Wolf population numbers have not been well studied in Bristol Bay. Better regional wolf population estimates, gathered using scientifically rigorous methods, are needed to improve understanding of wolf populations in the region.

Special Note: The findings and conclusions in this chapter are those of the authors and do not necessarily represent the views of the USFWS.

## ACKNOWLEDGEMENTS

Numerous Federal and State agency wildlife biologists generously gave time and effort to review and improve sections of this chapter. Of special note are United States Geological Survey (USGS) biologist Layne Adams, retired NPS biologist Page Spencer, and USFWS biologist Dom Watts with the Alaska Peninsula-Becharof National Wildlife Refuge; they all provided significant reviews, edits, information, and insight based on their considerable scientific expertise and knowledge. We also thank Cara Staab and Bruce Seppi from the Bureau of Land Management and retired ADF&G biologists Ken Whitten and Bob Tobey for their thoughtful reviews and contributions. Finally, this project would not have been possible without the support and encouragement of Ann Rappoport, long-time Anchorage Fish and Wildlife Field Office Supervisor, now retired

## REFERENCES

Adams, L.G., R.O. Stephenson, B.W. Dale, R.T. Ahgook, and D.J. Demma. 2008. Population dynamics and harvest characteristics of wolves in the Central Brooks Range, Alaska. Wildlife Monographs 170:1–25.

Adams, L.G., S.D. Farley, C.A. Stricker, D.J. Demma, G.H Roffler, D.C. Miller, and R.O. Rye. 2010. Are inland wolf-ungulate systems influenced by marine subsidies of Pacific salmon? Ecological Applications 20:251–262.

ADF&G (Alaska Department of Fish and Game). 2010. Trapper Questionnaire: Statewide Annual Report, July 1, 2008–June 30, 2009, Alaska Department of Fish and Game, Division of Wildlife Conservation, Juneau, AK.

ADF&G (Alaska Department of Fish and Game). 2011. 2011–2012 Alaska Trapping Regulations, Alaska Department of Fish and Game, Juneau, AK.

Ballard, W.B. and J.R. Dau. 1983. Characteristics of gray wolf den and rendezvous sites in south-central Alaska. Canadian Field-Naturalist 97:299–302.

Ballard, W.B., J.S. Whitman, and C.L. Garden. 1987. *Ecology of an exploited wolf population in southcentral Alaska.* Wildlife Monographs 98:1–54.

Ballard, W.B., L.A. Ayres, P.R. Krausman, D.J. Reed, and S.G. Fancy. 1997. *Ecology of wolves in relation to a migratory caribou herd in northwest Alaska.* Wildlife Monographs 135:1–47.

Ballard, W.B., M. Edwards, S.G. Fancy, S. Boe, and P.R. Krausman. 1998. Comparison of VHF and satellite telemetry for estimating sizes of wolf territories in northwest Alaska. Wildlife Society Bulletin 26:823–829.

Ballard, W.B., D. Lutz, T.W. Keegan, L.H. Carpenter, and J.C. deVos. 2001. Deer-predator relationships: a review of recent North American studies with emphasis on mule and black-tailed deer. Wildlife Society Bulletin 29:99–115.

Ben-David, M., T.A. Hanley, and D.M. Schell. 1998. Fertilization of terrestrial vegetation by spawning Pacific salmon: the role of flooding and predator activity. Oikos 83:47–55.

Boertje, R.D. and R.O. Stephenson. 1992. Effects of ungulate availability on wolf reproductive potential in Alaska. Canadian Journal of Zoology 70:2441–2443.

Bowyer, R.T., D.K. Person, and B.M. Pierce. 2005. Detecting top-down versus bottom-up regulation of ungulates by large carnivores: implications for conservation of biodiversity. Pages 342–361. In: J. C. Ray, et al. (Eds.), Large carnivores and the conservation of biodiversity, Island Press, Covelo, CA, USA.

Burch, J.W., L.G. Adams, E.H. Follmann, and E.A. Rexstad. 2005. Evaluation of wolf density estimation from radiotelemetry data. Wildlife Society Bulletin 33:1225–1236.

Burkholder, B.L. 1959. Movements and behavior of a wolf pack in Alaska. Journal of Wildlife Management 23:1–11.

Butler, L.B. 2009. Unit 9 and 10 wolf management report. In: P. Harper (Ed.), Wolf management report of survey and inventory activities: July 1, 2005–June 30, 2008, Alaska Department of Fish and Game, Juneau, AK.

Carnes, J.C. 2004. Wolf ecology on the Copper and Bering River Deltas, Alaska. Ph.D. dissertation. University of Idaho, Moscow, Idaho.

Darimont, C.T. and P.C. Paquet. 2000. The Gray Wolves (*Canis lupus*) of British Columbia's Coastal Rainforests: Findings from year 2000 pilot study and conservation assessment, Raincoast Conservation Society, Victoria, B.C.

Darimont, C.T., T.E. Reimchen, and P.C. Paquet. 2003. Foraging behavior by gray wolves on salmon streams in coastal British Columbia. Canadian Journal of Zoology 81:349–353.

Darimont, C.T., P.C. Paquet, and T.E. Reimchen. 2008. Spawning salmon disrupt trophic coupling between wolves and ungulate prey in coastal British Columbia. BioMed Central Ecology 8:1–12.

Farmer, C.D., D.K. Person, and R.T. Bowyer. 2006. Risk factors and mortality of black-tailed deer in a managed forest landscape. Journal of Wildlife Management 70:1403–1415.

Forbes, G.J. and J.B. Theberge. 1992. Importance of scavenging on moose by wolves in Algonquin Park, Ontario. Alces 28:235–241.

Forbes, G.J. and J.B. Theberge. 1996. Response by wolves to prey variation in central Ontario. Canadian Journal of Zoology 74:1511–1520.

Fuller, T.K. 1989. *Population dynamics of wolves in north-central Minnesota.* Wildlife Monographs 105:1–41.

Fuller, T.K., L.D. Mech, and J.F. Cochrane. 2003. Wolf population dynamics. Pages 161–191. In: L. D. Mech and L. Boitani (Eds.), *Wolves: behavior, ecology, and conservation.* University of Chicago Press, Chicago, IL.

Gasaway, W.C., R.D. Boertje, D.V. Grangaard, D.G. Kelleyhouse, R.O. Stephenson, and D.G. Larsen. 1992. *The role of predation in limiting moose at low densities in Alaska and Yukon and implications for conservation.* Wildlife Monographs 120:1–59.

Gende, S.M., T.P. Quinn, and M.F. Willson. 2001. Consumption choice by bears feeding on salmon. Oecologia 127:372–382.

Hilderbrand, G.V., T.A. Hanley, C.T. Robbins, and C.C. Schwartz. 1999. Role of brown bears (*Ursus arctos*) in the flow of marine nitrogen into a terrestrial ecosystem. Oecologia 121:546–550.

Houle, M., D. Fortin, C. Dussault, R. Courtois, and J.-P. Ouellet. 2010. Cumulative effects of forestry on habitat use by gray wolf (*Canis lupus*) in the boreal forest. Landscape Ecology 25:419–433.

James, A.R.C., S. Boutin, D.M. Hebert, and A.B. Rippin. 2004. Spatial separation of caribou from moose and its relation to predation by wolves. Journal of Wildlife Management 68:799–809.

Jędrzejewski, W., K. Schmidt, J. Theuerkauf, B. Jędrzejewska, N. Selva, K. Zub, and L. Szymura. 2002. Kill rates and predation by wolves on ungulate populations in Białowieża Primeval Forest (Poland). Ecology 83:1341–1356.

Kohira, M. and E.A. Rexstad. 1997. Diets of wolves, *Canis lupus*, in logged and unlogged forests of southeastern Alaska. Canadian Field-Naturalist 111:429–435.

Kuyt, E., B.E. Johnson, and R.C. Drewien. 1981. A wolf kills a juvenile whooping crane. Blue Jay 392:116–119.

McLoughlin, P.D., K.M. Walton, H.D. Cluff, P.C. Pacquet, and M.A. Ramsay. 2004. Hierarchal habitat selection by tundra wolves. Journal of Mammalogy 85:576–580.

Mech, L.D. 1970. The wolf: The ecology and behavior of an endangered species. American Museum of Natural History, Garden City, NY.

Mech, L.D. 1977. Productivity, mortality, and population trends of wolves in northeastern Minnesota. Journal of Mammalogy 58:559–574.

Mech, L.D. 1988. The arctic wolf: Living with the pack. Voyageur Press, Stillwater, Minnesota.

Mech, L.D. 1994. Regular and homeward travel speeds of arctic wolves. Journal of Mammalogy 75:741–742.

Mech, L.D. 1995. The challenge and opportunity of recovering wolf populations. Conservation Biology 9:270–278.

Mech, L.D., L.G. Adams, T.J. Meier, J.W. Burch, and B.W. Dale. 1998. The wolves of Denali. University of Minnesota Press, Minneapolis, Minnesota.

Mech, L.D. and L. Boitani. 2003. Wolf social ecology. Pages 1–34 in L. D. Mech and L. Boitani (Eds.), *Wolves: behavior, ecology, and conservation*. University of Chicago Press, Chicago, USA.

Mech, L.D. and R.O. Peterson. 2003. Wolf-prey relations. Pages 131–160. In: L.D. Mech and L.B. (Eds.), *Wolves: behavior, ecology, and conservation*. University of Chicago Press, Chicago, USA.

Meier, T.J., J.W. Burch, L.D. Mech, and L.G. Adams. 1995. Pack structure dynamics and genetic relatedness among wolf packs in a naturally regulated population. Pages 293–302. In: L.N. Carbyn, et al. (Eds.), Ecology and conservation of wolves in a changing world. Canadian Circumpolar Institute, Edmonton, Alberta.

Messier, F. 1985. Solitary living and extraterritorial movements of wolves in relation to social status and prey abundance. Canadian Journal of Zoology 63:239–245.

Metz, M.C., D.W. Smith, J.A. Vucetich, D.R. Stahler, and R.O. Peterson. 2012. Seasonal patterns of predation for gray wolves in the multi-prey system of Yellowstone National Park. Journal of Animal Ecology 81:553–563.

Mladenoff, D.J., T.A. Sickley, R.G. Haight, and A.P. Wydeven. 1995. A regional landscape analysis and prediction of favorable gray wolf habitat in the northern Great Lakes Region. Conservation Biology 9:279–294.

Mladenoff, D.J., T.A. Sickley, and A.P. Wydeven. 1999. Predicting gray wolf landscape recolonization: Logistic regression models vs. new field data. Ecological Applications 9:37–44.

Packard, J.M. 2003. Wolf behavior: reproductive, social and intelligent. Pages 35–65. In: L.D. Mech and L.Boitani (Eds.), *Wolves: behavior, ecology, and conservation*. University of Chicago Press, Chicago, USA.

Paquet, P.C. and L.N. Carbyn. 2003. Gray wolf. Pages 482–510. In: G.A. Feldhamer, B.C. Thompson and J.A. Chapman (Eds.), Wild Mammals of North America: Biology, Management, and Conservation, John Hopkins University Press, Baltimore, Maryland.

Person, D.K., M.D. Kirchhoff, V. Van Ballenberghe, G.C. Iverson, and E. Grossman. 1996. The Alexander Archipelago wolf: a conservation assessment. PNW-GTR-384., USDA Forest Service Gen. Tech. Rep. pp. 42.

Person, D.K. 2001. Wolves, deer and logging: Population vitality and predator-prey dynamics in a disturbed insular landscape. University of Alaska, Fairbanks.

Person, D.K. and A.L. Russell. 2008. Correlates of mortality in an exploited wolf population. Journal of Wildlife Management 72:-1540–1549.

Person, D.K. and A.L. Russell. 2009. Reproduction and den site selection by wolves in a disturbed landscape. Northwest Science 83.

Peterson, R.O., J.D. Woolington, and T.N. Bailey. 1984. *Wolves of the Kenai Peninsula*, Alaska. Wildlife Monographs 88:1–52.

Quinn, T.P. 2004. *The behavior and ecology of Pacific salmon and trout.* University of Washington Press, Seattle, WA.

Reimchen, T.E. 2000. Some ecological and evolutionary aspects of bear-salmon interactions in coastal British Columbia. Canadian Journal of Zoology 78:448–457.

Rothman, R.J. and Mech L.D. 1979. Scent-marking in lone wolves and newly formed pairs. animal behavior 27:750–760.

Seip, D.R. 1992. Factors limiting woodland caribou populations and their interrelationships with wolves and moose in southeastern British Columbia. Canadian Journal of Zoology 70:1494–1503.

Spaulding, R.L., P.R. Krausman, and W.B. Ballard. 1998. Summer diet of gray wolves, *Canis lupus*, in northwestern Alaska. Canadian Field-Naturalist 112:262–266.

Stanek, A.E., N. Wolf, G.V. Hilderbrand, B. Mangipane, D. Causey, and J.M. Welker. 2017. Seasonal foraging strategies of Alaskan gray wolves (*Canis lupus*) in an ecosystem subsidized by Pacific salmon (*Oncorhynchus* spp.). Canadian Journal of Zoology. 95: 555–563.

Stephens, D.W. and J.R. Krebs. 1986. Foraging Theory. Princeton University Press, Princeton, New Jersey.

Szepanski, M.M., M. Ben-David, and V. Van Ballenberghe. 1999. Assessment of anadromous salmon resources in the diet of the Alexander Archipelago wolf using stable isotope analysis. Oecologia 120:327–335.

Theberge, J.B. 1969. Observations of wolves at a rendezvous site in Algonquin Park. Canadian Field-Naturalist 83:122–128.

Thurber, J.M., R.O. Peterson, T.D. Drummer, and S.A. Thomasma. 1994. Gray wolf response to refuge boundaries and roads in Alaska. Wildlife Society Bulletin 22:61–68.

USFWS (U.S. Fish and Wildlife Service). 2017. Species Profile for Gray Wolf (*Canis lupus*). Refer to: https://ecos.fws.gov/ecp0/profile/speciesProfile?spcode=A00D, checked August 15, 2017.

Walsh, P. 2011. Personal communication from Supervisory Biologist, Togiak National Wildlife Refuge, to Ann Rappoport, U.S. Fish and Wildlife Service, August 17, 2011.

Watts, D.E., L.G. Butler, B.W. Dale, and R.D. Cox. 2010. The Ilnik wolf *Canis lupus* pack: use of marine mammals and offshore sea ice. Wildlife Biology 16:144–149.

Weckworth, B.V., S. Talbot, G.K. Sage, D.K. Person, and J. Cook. 2005. A signal for independent coastal and continental histories among North American wolves. Molecular Ecology 14:917–931.

Weaver, J.L. 1994. Ecology of wolf predation admidst high ungulate diversity in Jasper National Park, Alberta. University of Montana, Missoula.

Wiebe, N., G. Samelius, R.T. Alisauskas, J.L. Bantle, C. Bergman, R. de Carle, C.J Hendrickson, A. Lusignan, K.J. Phipps, and J. Pitt. 2009. Foraging behaviours and diets of wolves in the Queen Maud Gulf Bird Sanctuary, Nunavut, Canada. Arctic 62:399–404.

Willson, M.F. and R.H. Armstrong. 1998. Intertidal foraging for Pacific Sand-lance, *Ammodytes hexapterus*, by birds. Canadian Field-Naturalist 112.

Wolfe, R.J. 1991. Trapping in Alaska communities with mixed, subsistence-cash economies. Alaska Department of Fish and Game, Division of Subsistence, Juneau, AK.

Woolington, J.D. 2009. Unit 17 wolf management report. Pages 121–127. In: P. Harper (Ed.), Wolf management report of survey and inventory activities July 1, 2005–June 30, 2008, Alaska Department of Fish and Game, Juneau, AK.

Zimmermann, B., L. Nelson, P. Wabakken, H. Sand, and O. Liberg. 2014. Behavioral responses of wolves to roads: scale-dependent ambivalence. Behavioral Ecology (2014), 25(6), 1353–1364. doi:10.1093/beheco/aru134.

# 11

# SHOREBIRDS OF BRISTOL BAY

Susan Savage, U.S. Fish and Wildlife Service (retired)

## INTRODUCTION

Shorebirds, small to medium long-legged birds (called waders in Europe) with probing bills, are a diverse group, with species occurring on all continents and in all habitats ranging from sea level to the highest mountains. North American shorebirds fall into five families (Charadriidae, Haematopodidae, Recurvirostridae, Jacanidae, and Scolopacidae) within the order Charadriiformes; they are generally associated with water, particularly intertidal and estuarine environments especially during migration. Many species have broad geographic distributions with remarkable seasonal migrations, regularly spanning continents and frequently hemispheres. Several species engage in long, nonstop flight, but most rely on a series of sites where they stop to *refuel* for subsequent legs of their migrations. Many species flock, or come together, during these migrations (Figure 11.1) and in Alaska these large congregations may be viewed along coastlines in both the spring and the fall. Alaska intertidal areas, particularly Bristol Bay estuaries, are very important for shorebirds at these two different times of the year. During the late summer through autumn, the majority of the shorebird populations that nested in western Alaska (for the purpose of this discussion, the Yukon-Kuskoqim Delta, the Seward Peninsula and the northwest coast of the Alaska Peninsula) move to the benthic-rich intertidal communities of the Yukon-Kuskoqim Delta and Bristol Bay, where ample food supports them while they complete their molt and fatten for autumn migrations. Nonbreeding destinations include sites throughout North, Central, and South America (Cornell Laboratory of Ornithology: All About Birds 2016), the central Pacific islands, and Australasia (Bird Life International 2016). During the spring, hundreds of thousands of shorebirds migrate back to their western Alaska breeding grounds using staging grounds on the Copper River Delta and estuaries of Cook Inlet; from there they pass through a broad lowland corridor (the Lake Iliamna corridor) at the base of the Alaska Peninsula, linking Kamishak Bay in lower Cook Inlet to upper Bristol Bay. In most years, the migration through this corridor is direct, but in years with late spring arrival or adverse weather conditions, birds stop in large numbers at Bristol Bay estuaries until conditions improve farther west (Gibson 1967, Gill and Handel 1981, Bishop and Warnock 1998, Gill and Tibbitts 1999, Warnock et al. 2004). All of the major river outlets and lagoons of Bristol Bay (Izembek and Moffet Lagoons, Port Moller (Nelson Lagoon-Herendeen Bay), North Alaska Peninsula Coastal (Hook Lagoon, Meshik Bay, Cinder River Lagoon, Ugashik Bay, Egegik Bay), Nushagak and Kvichak Bays) have been identified as Important Bird Areas (Alaska Audubon 2016) (See Figure 15.1 in this volume); the State of Alaska also recognizes most of these areas as Alaska State Critical Habitat Areas (Port Moller, Port Heiden, Cinder River, Pilot Point, Egegik) (Alaska Department of Fish and Game (ADF&G) 2016). The Nushagak and Kvichak River bays have been recognized as Western Hemisphere Shorebird Reserve Network sites (Western Hemisphere Shorebird Reserve Network 2011). In addition to use of the intertidal areas, the uplands of Bristol Bay region[1] serve as feeding areas during high tide and important breeding grounds from May to early July for a broad diversity of shorebirds.

Thirty-two of 41 (~78%) of the shorebird species or subspecies that regularly occur in Alaska each year (Alaska Shorebird Group 2008) can be found in the Bristol Bay watershed; 22 of these 32 regularly nest there (Table 11.1). Shorebird populations worldwide are showing steady declines (Stroud et al. 2006) with causes most often attributed to loss or alteration of habitats and environmental contamination. Sixteen species that regularly occur in the Bristol Bay watershed have been ranked by the Alaska Shorebird Working Group (2008) as being of high conservation concern (Table 11.1).

---

[1] The coastal plain may extend inland from the shore to 100 km (60 miles) from the coast.

**Figure 11.2**   Whimbrels and Hudsonian godwits were marked with metal bands, color bands, and color flags on Chiloé Island, Chile in the mid-2000s. King Salmon and Naknek were the first locations north of Chiloé Island where marked birds were resighted, establishing an important hemispheric linkage (J. Johnson, pers. comm.) *Photo credit: Rod Cyr.*

**Figure 11.3**   **A**—This Pacific golden-plover (color marked as blue on right/hot pink over metal on left) was banded on the Alaska Peninsula at Port Heiden during the May 2009 breeding season in an effort by federal agencies to detect Avian Influenza in Pacific migrants. This bird was subsequently observed by plover biologists Dr. Phil and Mrs. Andrea Brunner (December 2009) and photographed by plover biologist Dr. O. Wally Johnson (March 2009) on wintering grounds in Hawaii (in **B**). Note the contrast between breeding and winter plumage. *Photo credit: A-USFWS, B-Dr. O. Wally Johnson.*

efficient way to survey migrating birds is by air, but covering the area from Togiak Bay to Izembek requires several days of flying and at least two refueling stops. Multiday flights are plagued by the often windy or low-ceiling conditions of the bay. Even studies of single bays or estuaries cannot be accomplished on foot or even in some cases by motorized vehicle within a single day. Aerial surveys require expert birders who recognize birds in flight from above; therefore, most aerial surveys categorize birds by size rather than species. Life history studies are also challenged by the size of the area and by the lack of roads; randomized sampling requires use of fixed-wing and rotary aircraft to reach remote areas with no roads.

The Bristol Bay/Alaska Peninsula lagoon system is one of the most important migratory shorebird stop-over areas in the state. Probably only the Copper River Delta and the Yukon-Kuskokwim Delta are more important (Isleib and Kessel

1973, Senner 1979, Gill and Handel 1990). The entire set of lagoons supports thousands to hundreds of thousands of individuals of multiple species, millions of shorebirds in total (Gill and Jorgenson 1979), which undertake post-breeding migrations to the Pacific coast of North America and across the Pacific Ocean to Australia, Southeastern Asia, and Oceania. For species that migrate directly across the ocean to Hawaii or other South Pacific islands [e.g., Pacific golden-plover, bar-tailed godwit (*Limosa lapponica*), ruddy turnstone (*Arenaria interpres*)], these lagoons provide the last stopover before their long overwater flights. Western sandpiper (*Calidris mauri*), dunlin (*C. alpina*), and long-billed dowitcher (*Limnodromus scolopaceus*) use the peninsula's lagoons to replenish energy reserves before departing nonstop for British Columbia and points south. The lagoons of Bristol Bay are also used by shorebirds as they migrate north in spring, providing an essential refueling location that enables species not only to succeed in reaching their breeding grounds, but also to begin breeding shortly thereafter. The relative importance of each lagoon/delta, including the deltas of the Bristol Bay region, is likely to vary annually and by species, and the loss of any one site may have an adverse effect on a species' ability to successfully migrate, and consequently add another factor to already declining populations.

## HABITAT

The geomorphology of Bristol Bay is shaped by the interaction between the shallow basin of the bay and the twice-daily tidal fluctuation of 2 m (6 ft.) in the southwestern part of the bay to an excess of 10 m (33 ft.) in Kvichak Bay. These features interact with the numerous river deltas to form an expansive intertidal zone dominated by sediment-dominated shorelines (61%), organic (e.g., estuarian, 23%), and eelgrass dominated shorelines (12% found primarily at Izembek Lagoon, Herendeen Bay/Nelson Lagoon, and Togiak Bay) (National Oceanographic Atmospheric Administration (NOAA)—National Marine Fisheries Service (NMFS) 2012). Only 1.2% of mapped shorelines are bedrock. A virtual tour of this coastline is available in part through the Alaska Shore Zone Flex Mapping website (NOAA—NMFS 2016). The estuarian areas, primarily found in the bays and lagoons, provide the primary shorebird and other avifaunal habitat, although shorebirds also find food along the exposed mudflats. Area measures of intertidal habitat are currently only available for more enclosed bays (see estimates in Table 11.2).

In winter, substantial shore ice forms along the coast and sea ice moves through the area with the tides. The supralittoral splash zone (Figure 11.4) varies from gradually sloping unvegetated or sparsely vegetated shore, to sandy bluffs (from glacial moraines or sand dunes) up to 30 m (100 ft) high. Beyond the shore zone, the region is characterized by a mosaic of wetland and tundra habitats, punctuated with low and tall shrub communities, located primarily along drainages. At higher latitudes and elevations, spruce, mixed spruce and deciduous, birch, or cottonwood forests give way to ericaceous dwarf shrub or sparsely vegetated substrates in the alpine zone.

Shorebirds inhabit the Bristol Bay watershed primarily during two phases of their annual life cycle: migration and breeding. During each phase, they make use of geographically distinct parts of the watershed. Shorebirds use the expansive intertidal and adjacent supralittoral areas during both spring and fall migrations. In spring, the mouths of major rivers (especially the Naknek River) are often the first areas to become ice-free, and provide critical feeding habitat in the

**Table 11.2** Estimated areas of bays and lagoons along the Bristol Bay coastline

| Bay or Lagoon | Estimated Area |
|---|---|
| Nushagak Bay[a] | 400 km$^2$, 154 mi$^2$ |
| Kvichak Bay[a] | 530 km$^2$, 205 mi$^2$ |
| Egegik Bay[b] | 128 km$^2$, 49 mi$^2$ |
| Ugashik Bay[b] | 282 km$^2$, 109 mi$^2$ |
| Cinder-Hook Lagoon[b] | 100 km$^2$, 39 mi$^2$ |
| Meshik[b] | 270 km$^2$, 104 mi$^2$ |
| Seal Islands[b] | 71 km$^2$, 27 mi$^2$ |
| Nelson Lagoon-Mud Bay[b] | 254 km$^2$, 98 mi$^2$ |
| Izembek-Moffett Lagoon[b] | 350 km$^2$, 135 mi$^2$ |
| TOTAL | 2,385 km$^2$, 920 mi$^2$ |

[a] Western Hemisphere Shorebird Reserve Network 2011,

[b] R. Gill (USGS) and K. Wohl (USFWS) pers. comm., 1996

**Figure 11.7** Whimbrels, black turnstones, and short-billed dowitchers returning from more northern or local breeding grounds share the beach with commercial set net salmon fishermen near Naknek, Alaska in July. *Photo credit: Rod Cyr.*

**Figure 11.8** A close-up of a Wilson's snipe nest among the dried sedges shows the typical olive green eggs—about 4 cm (1.5″) in length—wreathed with dark brown speckles. Four eggs are the normal clutch size for most small to medium-sized shorebirds. *Photo credit: USFWS.*

and short-billed dowitcher) prefer moist to wet herbaceous vegetation, whereas many of the plovers or montane breeders (e.g., American golden-plover (*Pluvialis dominica*), semipalmated plover, surfbird (*Aphriza virgata*), rock sandpiper), prefer dwarf shrub/lichen vegetation or even barren areas for nest sites. Several species are highly dependent on lake or river shores (e.g., spotted sandpiper (*Actitis macularius*), wandering tattler) (Petersen et al. 1991, Gill et al. 2015). A few species (marbled godwit and black turnstone) prefer the coastal fringe (Gill and Handel 1981, Gill et al. 2004). All of these shorebirds may feed in marine intertidal zones during breeding, depending on their proximity or their preference for feeding in these environments.

## FOOD HABITS

The shorebird group derives its name from the fact that many species spend migration, and often nonbreeding periods, associated with shore environments. In many cases, these are marine shores. Food is likely the most important factor controlling the movements of shorebirds throughout the Bristol Bay region. In spring, shorebirds need to acquire critical food resources, not only to fuel their migration, but also for some species, to assure that they arrive on the breeding grounds with sufficient reserves to initiate nesting and egg production (Klaassen et al. 2006, Yohannes et al. 2010). Beginning in mid-summer and continuing into early autumn, a few Alaska shorebird species must again find food-rich areas to support the process of partial or complete molt and, for all but a few species, to fuel extended migration. Indeed, some of the longest migrations known to birds involve shorebird species (e.g., bar-tailed godwit, Pacific golden-plover; Figures 11.9 and 11.10) that use Bristol Bay intertidal areas in autumn (Gill et al. 2009, Battley et al. 2012, Johnson et al. 2011,

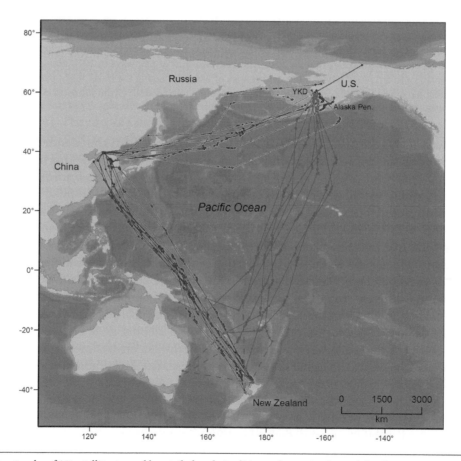

**Figure 11.9** Migration tracks of 17 satellite-tagged bar-tailed godwits (*Limosa lapponica baueri*); birds were tagged in either Alaska or New Zealand, 2006–2010, and tracked during one to three of the nonstop flight legs that comprise the annual journey. Red tracks show the leg from the nonbreeding grounds in New Zealand to the spring staging areas in the Yellow Sea region, gold tracks show Yellow Sea region to the breeding grounds in Alaska, and blue tracks show autumn staging grounds in western Alaska (including Bristol Bay estuaries) back to New Zealand. Small circles along track lines represent positions calculated from Argos data. Dashed lines denote movements interpolated from resightings subsequent to transmitters going off air. YKD = Yukon-Kuskokwim Delta. *Image credit: USGS Alaska Science Center adapted from Battley et al. 2012.*

the Nushagak and Kvichak River bay surveys, shorebird numbers peaked in mid-October (30,373 birds; Savage & Payne 2013) while for the Izembek-Moffet Lagoon system shorebird counts reached 36,000 birds in late July and remained around 30,000 birds until they peaked at 41,251 in mid-October. All of these studies, but especially the ones conducted in Nelson Lagoon, provide insight on the seasonality and duration of use by at least 25 shorebird species (Gill et al. 1977, Gill et al. 1978, Gill and Jorgenson 1979). Late summer and fall shorebird use is likely greater than spring use because of the addition of juveniles to the population, longer residence times, different pathways of migrants during different times of the year, or different use patterns of individual species.

The autumn migration is broken into phases based on species, age, sex, and individual breeding success. Species-specific use patterns have been reported for Nelson Lagoon on the central Alaska Peninsula (Gill and Jorgensen 1979) and patterns for other species common to the Bristol Bay watershed are reported from studies on the Yukon Delta (Gill and Handel 1981, Gill and Handel 1990). In general, the earliest species include: short-billed dowitchers (mid-June) (Figure 11.11); black-bellied plovers, whimbrel, dunlin and western sandpipers (late June); and turnstones, phalaropes and rock sandpipers (early to mid-July). Later species include: sanderlings, American golden-plovers (mid-August), sharp-tailed sandpipers (early September). Failed breeders move to the coastal zone sooner than successful breeders (Gill et al. 1983, Handel and Dau 1988). On the Y-K Delta, Gill and Handel (1990) observed three age-based patterns of intertidal use through the late summer. In the most common pattern (exemplified by western sandpipers) adults arrived first, followed by a period in which adults and juveniles occurred together, and finally juveniles appeared alone. In the second pattern (bar-tailed godwits, dunlin, and rock sandpipers) (Figure 11.12), adults appeared first, followed by a long period of use by both adults and juveniles. The third pattern was demonstrated by plovers, in which only juveniles used the intertidal zone in late summer. In addition, some species demonstrated a sex-specific pattern: female western sandpipers departed before males (Gill and Jorgensen 1979), but in pectoral sandpipers (*Calidris melanotos*) (Figure 11.13), males depart before females (Pitelka 1959). The specific migration patterns demonstrated by individual shorebirds with regard to micro- and macro-habitat use and timing will become clearer in the future as researchers continue to deploy satellite transmitters and geolocators.

Below freezing temperatures place extreme energetic demands, even on well-insulated birds. Compounded with food sources covered in shore ice, most shorebirds leave the area for warmer climes in winter. Mention must be made of two

**Figure 11.11**    Short-billed dowitchers and Hudsonian godwits on the Bristol Bay intertidal in August. *Photo credit: Rod Cyr.*

**Figure 11.12**   Bar-tailed godwits along the Bristol Bay shore near Naknek, Alaska. *Photo credit: Rod Cyr.*

**Figure 11.13**   The pectoral sandpiper is infrequently detected along the Bristol Bay Coast. *Photo credit: Rod Cyr.*

Although Alaska is known as a nursery ground for shorebirds, systematic inventory of breeding shorebird distribution or abundance over large areas is limited to studies from montane areas in Katmai and Lake Clark NPPs (Ruthrauff et al. 2007), ANIA (Ruthrauff and Tibbitts 2009), from lowlands of the northern Alaska Peninsula (Savage et al. in press), and the adjacent Kilbuck and Ahklun Mountains (Petersen et al. 1991). Species-specific studies include: marbled godwit (Mehall-Niswander 1997, Gill et al. 2005b) and Pacific golden-plover (Savage and Johnson 2005). This work provides some basis for understanding the distribution of breeding species; additional analysis of the point-transect data from Katmai and the northern Alaska Peninsula lowlands is ongoing to evaluate species-habitat relationships and perhaps some level of breeding density estimates (Handel et al. 2013). For montane areas in Katmai and Lake Clark, the most common species found during the breeding season (in May) were semipalmated plover, spotted sandpiper, wandering tattler, greater and lesser yellowlegs, surfbird, least sandpiper, and Wilson's snipe (Ruthrauff et al. 2007). For lowland areas of the northern Alaska Peninsula, the most common shorebird species found during May were greater yellowlegs, least sandpiper (Figure 11.16), dunlin, short-billed dowitcher, Wilson's snipe, and red-necked phalarope (Savage et al. in press). In the area north and west of the Bristol Bay watershed, Petersen et al. (1991) found black-bellied plover, semipalmated plover, spotted sandpiper, greater yellowlegs, western sandpiper, rock sandpiper, dunlin, Wilson's snipe, and red-necked phalarope to be the most common breeding shorebirds. Ruthrauff et al. (2007) extended the breeding range of several alpine shorebirds [wandering tattler, surfbird, and Baird's sandpiper (*C. bairdii*)] and confirmed these and another three species (black-bellied plover, American golden-plover, Pacific golden-plover), previously only known as migrants, to be breeders in Katmai NPP. Savage and Johnson (2005) further extended the breeding range of Pacific golden-plover and black-bellied plover to the Meshik River drainage. Other species recently confirmed to breed in the northern Bristol Bay watershed include whimbrel and Hudsonian godwit (Ruthrauff et al. 2007).

Mention should be made of the marbled godwits (Figure 11.17) of Bristol Bay. Three subspecies are found in disjunct breeding populations of North America (Gratto-Trevor 2000). Gibson and Kessel (1991) described the Alaska subspecies (*Limosa fedoa beringiae*); it is found only in the Ugashik and Cinder River drainages of Bristol Bay. Time-budgets and habitat use of breeding birds have been documented (Mehall-Niswander 1997). The population has been estimated to be 1,352 birds (859–2,204, 95% CI) (Gill et al. 2005b); a population this small is vulnerable to an acute large-scale pollution event, or to smaller scale chronic perturbations such as nesting disturbance or over-hunting.

Shorebird productivity, survivorship, and mortality are affected by many factors that vary by species, region, and annual conditions. These parameters are not known specifically for the Bristol Bay region, and may not be known at all for

**Figure 11.16** The least sandpiper, a common breeder in the Bristol Bay lowlands of the Alaska Peninsula, searches for invertebrate prey in the clear, shallow waters of Ugashik Lake outlet. *Photo credit: Robert Dreeszen.*

**Figure 11.17** The *L. f. beringiae* subspecies of marbled godwit is found breeding only in the Ugashik/Cinder River area of the Alaska Peninsula. *Photo credit: Rod Cyr.*

many shorebird species. Productivity may be affected by life history (e.g., age at first reproduction, annual participation in breeding), seasonal abundance of food resources, weather, flooding, predation, and other forms of disturbance. Productivity in birds is measured in various ways, including proportion of eggs hatched, proportion of successful nests, and proportion of young fledged. Shorebird productivity is low with pairs producing, at most, four chicks per season. Most small and medium-sized birds, including shorebirds, suffer from high mortality during their first weeks and months of life (for an extensive discussion of shorebird mortality and survival, see Colwell 2010). Some species may experience complete reproductive failure in a region during some years.

Survival may be affected by food availability, weather, predation, and human-caused mortality (e.g., collisions with buildings, domestic cat predation, and contaminated or otherwise degraded habitats). Human disturbance and habitat degradation are significantly greater along the migratory paths and nonbreeding grounds of most shorebirds than on the northern breeding grounds. The U.S. Geological Survey's Bird Banding Lab maintains longevity records for banded birds (USGS Bird Banding Lab 2011). However, these paint extremes of survival. Warnock (2001) summarizes annual adult survival rates of shorebirds: they *typically range from 60–70% in small species and 85–95% in larger species; survivorship of shorebirds in their first year is often less than 50%* (Evans and Pienkowski 1984, Evans 1991, Jackson 1994, Sandercock and Gratto-Trevor 1997, Warnock et al. 1997, Reed et al. 1998). These survival rates are reiterated in Sandercock (2003).

## POPULATIONS AND SUBPOPULATIONS

Shorebird populations throughout North America have been experiencing declines especially during the 1990s and early 2000s (Morrison et al. 2006; Alaska Shorebird Group 2008). Although accurate population data are lacking for most shorebirds, of the 32 regularly occurring shorebird species in the Bristol Bay watershed, 11 are suspected to be declining and 10 are thought to be stable; information is insufficient to make a determination for 10 species (Andres et al. 2012). Only one species (semipalmated plover) using the Bristol Bay watershed is thought to be increasing in abundance.

Of shorebird species found in Bristol Bay, ten include subspecies designation (Table 11.2). Most represent the only subspecies found in North America (e.g., bar-tailed godwit) or in Alaska (e.g., ruddy turnstone). Two shorebird species

# 12

# BALD EAGLES[1]

**Lowell H. Suring, Northern Ecologic L.L.C.**
**Maureen de Zeeuw, Bureau of Ocean Energy Management**

## INTRODUCTION

Bald eagles range across North America, but are most abundant in Alaska where approximately half of the world population occurs (Buehler 2000). During the first half of the 20th century in Alaska, the abundance of bald eagles and their attraction to human food sources made them vulnerable to bounty hunters (DeArmond 2010). Bounties were paid on bald eagles from 1917 to 1953 primarily to protect salmon resources, but also to protect Dall sheep (*Ovis dalli*); domestic sheep (*Ovis aries*); and farmed arctic foxes (*Alopex lagopus*) and red foxes (*Vulpes vulpes*) from predation by bald eagles (Crabb 1923, Bailey 1993, DeArmond 2010). However, even before the end of the bounty bald eagles were valued by Americans (King 2010). Bald eagles and their nests receive protections through the Bald and Golden Eagle Protection Act (16 U.S.C. 668-668c), enacted in 1940, and amended several times since then. This Act prohibits anyone, without a permit issued by the Secretary of the Interior, from taking bald eagles, including their parts, nests, or eggs. Additionally, this Act establishes federal responsibility for the protection of bald eagles, and requires consultation with the United States Department of the Interior (USDI) Fish and Wildlife Service (FWS) to ensure activities do not adversely affect bald eagle populations.

A large apex predator as well as opportunistic scavenger, bald eagles are a key species of most of the regional trophic webs across coastal Alaska. Success of breeding populations of bald eagles is an indicator of the condition of both freshwater and marine ecosystems (Stalmaster 1987, Bennett et al. 2006). Consequently, bald eagles were a proposed management indicator species (MIS) for National Forest lands in Alaska (Sidle and Suring 1986) and selected as a MIS for the Tongass National Forest (USDA Forest Service 2008:3-230–3-241). They were also included as a vital sign for long-term monitoring in the southwest Alaska national parks (Bennett et al. 2006). On the Alaska Peninsula from the Katmai coast south to Unimak Island, the bald eagle population has been considered healthy (Savage and Hodges 2006). Both the number of adult eagles and occupied nests increased from 1983 to 2000, and appear to have remained stable (Savage and Hodges 2000, 2006).

Although the Bristol Bay watershed is relatively pristine, there are potential impacts to bald eagles in this area. These impacts could result from resource extraction, oil spills, increasing human recreation (sport hunting, sport fishing, tourism), climate change, and effects on food sources due to harvest of salmon and other fish species (Gende et al. 1997, Buehler 2000, Hilderbrand et al. 2004, Martínez-Abraína et al. 2010, Harvey et al. 2012, Marcot et al. 2015). The purpose of this chapter is to provide a synthesis of information on bald eagle populations and their habitat in the Bristol Bay watershed that may serve as a basis for evaluating potential effects of perturbations or development on bald eagles.

---

[1]This chapter is based on the following unpublished report: de Zeeuw, M. and L.H. Suring. 2013. Bald eagles. Pages 92–105; *in* P.J. Brna and L.A. Verbrugge, editors. Wildlife resources of the Nushagak and Kvichak River watersheds, Alaska. USDI Fish and Wildlife Service, Anchorage Fish and Wildlife Field Office, Anchorage, Alaska, USA.

The proportion of bald eagles that overwinter in the Bristol Bay watersheds has not been quantified, and potential links between overwintering sites and open water or winter prey accessibility remain to be studied. Ritchie and Ambrose (1987) reported that records of bald eagles overwintering in northern boreal forests are rare and that one reason for this is that the waterbodies are frozen. They have observed bald eagles in winter along the Tanana River in Interior Alaska, where open water probably provides access to spawning salmon and waterfowl. Bald eagles often congregate along the ice/open water interface on the Naknek River where wintering common mergansers (*Mergus merganser*) and common goldeneyes (*Bucephala clangula*) are often found [i.e., during the 1986–2010 King Salmon-Naknek Christmas bird counts 0–48 adult bald eagles (average 18) were recorded (S. Savage, Biologist, Alaska Peninsula/Becharof NWRs, pers. comm., 2011)]. Some bald eagles overwinter near Dillingham and the surrounding area, but in much lower densities than found in summer (M. Swaim, Biologist, Togiak NWR, pers. comm., 2011). Some of those overwintering bald eagles obtain human garbage at the city dump. Ritchie and Ambrose (1987) also suggested that overwintering on the breeding grounds may provide a competitive edge in territory selection and with early initiation of nesting.

## FOOD HABITS

While bald eagles are primarily fish eaters, they have a variable diet that can include birds, mammals, and crustaceans, and even human garbage (Knight and Knight 1983, Anthony et al. 1999; summaries by Sidle et al. 1986, Stalmaster 1987, and Armstrong 2010). Specific food habits of bald eagles vary spatially according to specific prey availability and abundance at a particular site (Figure 12.3). Bald eagles nesting near and foraging at seabird colonies during the summer may take primarily bird prey (DeGange and Nelson 1982). Knight et al. (1990) determined that in the Pacific Northwest, diet varied among sites, with bird prey items found under nests generally outnumbering fish items. This is likely the case at some sites on Togiak NWR, including Cape Peirce and Cape Newenham, which support high densities of breeding seabirds (M. Swaim, Biologist, Togiak NWR, pers. comm., 2011).

Although the diet of bald eagles tends to vary temporally, based on prey availability and abundance, nesting bald eagles rely primarily on the availability of salmon resources (Hansen 1987). Also, inland bald eagles whose nests are close to

**Figure 12.3**    After a successful fishing trip, a bald eagle eats its catch (Arctic char [*Salvelinus alpinus*]) along the gravel banks of the Ugashik River on the Alaska Peninsula. Bald eagles will select fish over any other prey when a choice is available. *Photo credit: Robert Dreeszen.*

spawning streams have higher nesting success than those whose nests are more distant (Gerrard et al. 1975). However, when salmon resources are scarce during late winter and early spring, coastal populations of bald eagles often shift their diet to birds (Isleib 2010, Wright and Schempf 2010). Imler and Kalmbach (1955) found that birds averaged nearly 20% of stomach-contents by volume over the course of a year, but could range up to 86% and may be especially high during the colder months. In other areas, mammalian prey may be utilized because it is as available, or more available, than birds in the winter. For example, on the Kenai NWR bald eagles may seasonally shift from a diet of primarily fish to snowshoe hare (*Lepus americanus*) or mammalian carrion (Bangs et al. 1982a).

Specific information on food habits of bald eagles in the Bristol Bay watershed is generally not available. However, bald eagles in the Bristol Bay watershed eat the five species of Pacific salmon that occur in Bristol Bay [i.e., sockeye salmon (*Oncorhynchus nerka*), Chinook salmon (*Oncorhynchus tsawytscha*), coho salmon (*Oncorhynchus kisutch*), pink salmon (*Oncorhynchus gorbuscha*), and chum salmon (*Oncorhynchus keta*)] (S. Savage, Biologist, Alaska Peninsula/Becharof NWRs, pers. comm., 2011). Bald eagles in the winter along the Naknek River have been observed to take small fish, which may include eulachon (*Thaleichthys pacificus*) (S. Savage, Biologist, Alaska Peninsula/Becharof NWRs, pers. comm., 2011). Pacific herring (*Clupea pallasi*) may also be an important resource for bald eagles in the Togiak area. In early spring, bald eagles were observed catching large rainbow trout (*Oncorhynchus mykiss*) (S. Savage, Biologist, Alaska Peninsula/Becharof NWRs, pers. comm., 2011). Bald eagles in Bristol Bay may also scavenge dead marine mammals as well (S. Savage, Biologist, Alaska Peninsula/Becharof NWRs, pers. comm., 2011). Bald eagles in Alaska also congregate to feed on other species of anadromous and shallow-water spawning fish, particularly Dolly Varden (*Salvelinus malma malma*), Pacific herring, and eulachon (Armstrong 2010). Armstrong (2010) also summarized the importance cited in several studies of the marine fish Pacific sand lance (*Ammodytes hexapterus*) to bald eagles and other marine-associated birds and mammals.

Independent of prey availability, energetic requirements may influence prey selection at some level as well. During the breeding season, many birds choose large fish over small fish (Jenkins and Jackman 1994). Diets of nesting bald eagles are much more variable than those of nonbreeders (Hansen et al. 1984, Hansen et al. 1986). Nonbreeders are able to range further for preferred food items, (e.g., in late fall birds may leave the Chilkat Valley in southeast Alaska to go to British Columbia and Washington, where salmon may still be available). Feeding of young is, as Stalmaster (1987) says, an enormous chore; bald eaglets may consume > 50% of their weight in food each day (Stewart 1970). Subsequently, breeding birds may exploit a variety of food resources within their home range to meet these energy needs.

Fresh salmon and salmon carcasses provide an ideal food resource for bald eagles, because they are large fish which become available in great numbers when they enter shallow water to spawn. Shallow water also increases the likelihood that fish will be available to bald eagles because the limited depth of water brings fish closer to the surface (Livingston et al. 1990). After spawning salmon die their carcasses accumulate on stream banks, river bars, lake and marine shores, and tidal flats (Armstrong 2010). This provides a significant seasonal pulse of marine derived nutrients, including nitrogen and phosphorous, to the generally oligotrophic streams and lakes of northern Pacific watersheds (Willson et al. 1998, Naiman et al. 2002, Hilderbrand et al. 2004). Although spawned-out salmon are low in fat and considered a relatively low-energy food source (Christie and Reimchen 2005), their large size, availability, sheer numbers, and other factors, [including cold air temperatures which can increase efficiency of digestion of some prey (Stalmaster and Gessaman 1982)] contribute to their value to bald eagles. Salmon are approximately 79% edible flesh as compared to hares (*Lepus americanus*) (71%) and ducks (68%) (Stalmaster 1981). Although wet metabolizable energy was lowest for salmon compared to hares and ducks, based on prey size, a bald eagle would require only 57 salmon/year compared to 87 hares or 135 ducks (Stalmaster 1981, Stalmaster and Gessaman 1982). Besides size, numbers, and availability in shallow water, other unique aspects of salmon life history may contribute to their importance to bald eagles. For example, large numbers may be frozen into river ice in the winter, to become available again in spring as food sources for bald eagles and others (Hansen et al. 1984).

Armstrong (2010) reported that several studies have shown correlations between bald eagle abundance and the abundance of spawned-out salmon. Simply put, bald eagles in southeast Alaska, the Kenai NWR, and many other parts of their range likely depend on salmon (Armstrong 2010, Bangs et al. 1982b). However, the nature of the bald eagle-salmon relationship is complex (Hansen et al. 1986). Fluctuations in salmon abundance (and other bald eagle food sources) from year to year apparently cause bald eagles to be limited by food availability (Stalmaster and Gessaman 1984). Salmon abundance affects not only bald eagle abundance and distribution, but also their breeding and behavior. Bald eagles, in turn, affect the riparian ecosystem and other areas they inhabit by distributing marine-derived nutrients in their excretions (Gende et al. 2002).

Bald eagles are opportunistic foragers that exhibit complex social feeding behaviors. Bald eagles use a variety of methods to obtain food, including active hunting and killing, scavenging carcasses, and theft (pirating or kleptoparasitizing) (Stalmaster 1987). They are visual predators that locate their prey by sight (Restani et al. 2000). Foraging methods chosen by bald eagles vary according to complex relationships among conspecifics or other predators or competitors, as well as seasonal variability of food sources. Bald eagles may search for prey themselves, or follow other birds or even mammals to a concentrated food source (McClelland et al. 1982, Knight and Knight 1983, Harmata 1984). Also, as summarized in Armstrong (2010), bald eagles will not only steal fish or force other predators away from fish, but exploit fish that are injured or driven to the surface by others, and scavenge crippled or dead fish or fish parts left by other predators such as humans or bears (*Ursus* spp.). Bald eagles in the Chilkat Valley in southeast Alaska typically competed among themselves for salmon (Hansen et al. 1984). Dominance in bald eagles may be based on several conditions that often include age and size (Stalmaster 1987, Garcelon 1990).

Bald eagle foraging on salmon usually takes place when the fish arrive in streams to spawn because the fish are found in shallow waters, swimming or floating near the surface, or washed up or stranded on banks, and are relatively easy prey for bald eagles. Armstrong (2010) stressed what he believed to be a particularly important relationship with bears who pull salmon out of deep pools where they may otherwise be inaccessible to bald eagles. The bears then often transport and discard portions of carcasses so that bald eagles can then scavenge them (Armstrong 2010). Brown bears (*Ursus arctos*) often eat only the brains and eggs of salmon, leaving a significant portion of the flesh for other animals, including bald eagles (Gende et al. 2001).

According to Stalmaster (1987), bald eagles generally appear to prefer stealing food to scavenging and prefer scavenging to killing. Hansen et al. (1984) observed higher frequencies of stealing over scavenging salmon carcasses in the Chilkat Valley in southeast Alaska, even though the cost and benefits of both may be equal. However, others have found that when food is scarce, bald eagles will choose scavenging over stealing if both methods are available (Knight and Knight-Skagen 1988). During times of food scarcity, bald eagle feeding strategy may switch to one of more active hunting, particularly of large gulls and waterfowl, and some bald eagles may steal ducks from hunters or scavenge in garbage dumps (Wright and Schempf 2010).

# BEHAVIOR

Breeding bald eagles occupy and defend territories during the nesting period (Mahaffy and Frenzel 1987). A territory includes an active nest and may include one or more inactive nests that may be maintained but are not used for nesting in a given year (Hansen et al. 1984). They maintain the same territory year after year, using the same nest or an alternate nest within the same territory (Steidl et al. 1997, Watts 2015). The defended territory contains the nest trees, and favored perches and roost(s). Territories have been reported to range from 0.2–4.2 km$^2$ (Garrett et al. 1993), but size varies according to site and other parameters (Stalmaster et al. 1985). The territory is within a larger home range. Bald eagles, unlike many other birds, do not necessarily use a territory to monopolize food, but commonly range out of their territory to obtain food communally at a site of abundant prey (Stalmaster 1987).

In any given year, not all territories will be occupied, and not all occupants will attempt to reproduce (Stalmaster 1987). During the nonbreeding season, or if not breeding, bald eagles generally do not defend territories (Armstrong 2010), although a pair may remain close to their nest or return to their territory regularly over the winter (Gende 2010). Bald eagles in newly established territories are often highly sensitive to disturbance and prone to abandon nest sites during the courtship and nest building stage, although even well-established birds are often highly sensitive (Gende et al. 1998). Information is not currently available on the characteristics of bald eagle territories (e.g., size, use patterns, average number of nests, variability according to habitat type) in the Bristol Bay watershed.

When a food resource is concentrated, bald eagles will often forage in large flocks (Stalmaster 1987). This is true for scavenging and stealing, as can occur when carrion is present, and for hunting and killing when there are large aggregations of forage fish like eulachon or sand lance (Stalmaster and Gessaman 1984, Stalmaster 1987, Willson and Armstrong 1998). In the winter, when food availability is limited (e.g., by iced-over rivers, limited daylight), bald eagles aggregate in large flocks and become very aggressive, often pirating food from other birds. An available food source will initially draw bald eagles to a site, but then the presence of large numbers of bald eagles will attract additional birds.

At night, nonbreeding and wintering bald eagles may congregate in communal roost areas (Hansen et al. 1980). The same roost areas are used for several years. Roosts are often in locations that are protected from the wind by vegetation or terrain that will provide a favorable thermal environment. The use of these protected sites helps minimize the energy stress encountered by wintering birds. Communal roosting may also assist bald eagles in finding food. However, the use of communal roosts is poorly documented in Alaska (USDI FWS 2009).

Diurnal and tidal cycles affect the daily activity patterns of fish, as well as enhancing and inhibiting the hunting conditions for bald eagles (Hansen et al. 1986). Even though the population of bald eagles in southeast Alaska is nonmigratory, individuals will leave their territories to visit foraging areas for several days at a time (Kralovec 1994). Pairs also return to their breeding territory periodically over the course of the winter. Variations in these daily patterns lead to local movements of bald eagles.

The extent to which bald eagles are migratory or nonmigratory varies with breeding site and the severity of its climate, particularly in winter; whether the individual is adult or subadult; and year-round food availability (Buehler 2000). Bald eagles breeding in coastal Alaska typically remain in the vicinity of their nest sites throughout the year. For example, the southeast Alaska adult population is mostly nonmigratory, remaining in its rainforest habitat year-round (Sidle et al. 1986). Adults in Aleutian Island populations are generally resident as well (Sherrod et al. 1976). Wintering grounds for migratory Alaska bald eagles are not well understood, but it is suspected that interior bald eagles winter in the Intermountain West and Pacific Northwest (Ritchie and Ambrose 1996). Local movement patterns of bald eagles and the extent of overwintering and migration, and how those may vary with age, food availability, or other factors are not well understood for the Bristol Bay watershed. It is known that at least some adults and subadults overwinter in Bristol Bay (see Wright 2010 for summary).

# INTERSPECIES INTERACTIONS

Prey availability has a strong influence on bald eagle reproduction, habitat use, and territorial behavior in Alaska. The studies of Hansen et al. (1984) suggested that salmon availability in spring is tightly correlated with if and when adult bald eagles will lay eggs in a given year, although this has not been studied specifically in the Bristol Bay watershed. Bald eagles also preferentially select nest sites near stable food supplies, (e.g., salmon in the Chilkat Valley). These studies also indicate that food availability (i.e., salmon) during the nesting period regulates the survival rate of offspring. Hansen et al. (1984) also determined that while breeding adults commonly defend feeding territories, they did not do so when salmon become overabundant. Fall and winter habitat use is often correlated with salmon availability, as has been clearly demonstrated in the Chilkat Valley (Hansen et al. 1984)—and is likely true in Bristol Bay.

Bald eagles defend vulnerable young against predators (Stinson et al. 2001). However, bald eagles are less aggressive with other species than they are with other bald eagles, with which antagonistic interactions regularly occur during feeding and territory defense. One exception are ospreys (*Pandion haliaetus*), which bald eagles commonly prevent from nesting nearby (Stalmaster 1987), although this behavior has not been investigated in the Bristol Bay area. Great horned owls (*Bubo virginianus*), which nest earlier than bald eagles, and osprey, which nest later, each may occupy bald eagle nests in southwest Alaska (S. Savage, Biologist, Alaska Peninsula/Becharof NWRs, pers. comm., 2011).

# BREEDING, PRODUCTIVITY, SURVIVAL, AND MORTALITY

The timing of bald eagle nesting varies by latitude; in Alaska it begins with courtship and nest building in February and ends when the young fledge by late August into early September (Figure 12.4). However, in the Bristol Bay watershed, initiation may not occur until mid- to late March (S. Savage, Biologist, Alaska Peninsula/Becharof NWRs, pers. comm., 2011). The young are attended by the adults near the nest for several weeks after fledging (Buehler 2000). A pair's territory frequently contains more than one nest (Haines and Pollock 1998). Only one nest is used in a given breeding year, although bald eagles do not necessarily breed every year. The territory (and pair bond) is usually maintained for life (Jenkins and Jackman 1993).

Whether or not adult bald eagles breed in a given year (i.e., proportion of occupied nests) and how early they may initiate in a given year appear to be related to food availability (Hansen 1987), particularly in spring (Hansen et al. 1984). These studies suggest that there may be a natural long-term population cycle, at least in southeast Alaska's Chilkat Valley, resulting from a saturated breeding habitat and a surplus of nonbreeders, who then compete for food and cause productivity to decline. The decline may result in lower recruitment into the nonbreeding population, leading to lower competition, and ultimately increased productivity. Annual occupancy rates at known nest sites within Togiak NWR varied from 45–88% between 1986–2006 (M. Swaim, Biologist, Togiak NWR, pers. comm., 2011). The lowest occupancy mortality included loss of nests (eggs and nestlings) to spring storms, parental desertion or death, and predation by gulls, black bears (*Ursus americanus*), and other predators (see Stalmaster 1987 for a summary.). Although eggs tend to have a higher mortality rate than nestlings, nestlings also kill each other in fights, die from starvation when more aggressive nest mates receive the majority of feedings from the parent, and fall prematurely from nest trees.

Top-level predators, such as bald eagles, are believed to be especially vulnerable to many contaminants and can be used as sentinel species for evaluating contamination in their habitats (Welch 1994, Holl and Cairns 1995, Leith 2014). Eggs from bald eagles in the Aleutian Islands contained elevated levels of organochlorine pesticides, but concentrations of these contaminants and mercury were significantly higher in eggs from Kiska Island than in eggs from the other islands (Anthony et al. 1999). In contrast, polychlorinated biphenyl concentrations were higher in eggs from Adak, Amchitka, and Kiska islands than in those from Tanaga Island. The most likely source of these contaminants in bald eagles was from their diets, which were variable spatially and temporally. Similar findings reported in Anthony et al. (2007) indicated that contaminant concentrations in bald eagle eggs were influenced more by point sources of contaminants and geographic location than trophic status of bald eagles among the different islands.

Stout and Trust (2002) reported that mean cadmium, chromium, mercury, and selenium concentrations in tissues in bald eagles from Adak Island were consistent with levels observed in other avian studies and were below toxic thresholds. However, elevated concentrations of chromium and mercury in some individuals may warrant concern. Furthermore, although mean polychlorinated biphenyl and *pp′*-dichlorodiphenyldichloroethylene concentrations were below acute toxic thresholds, they were surprisingly high given Adak Island's remote location. Fish from selected national parks in Alaska, including LACL and Katmai NPP, had significantly higher concentrations of mirex, hexachlorobenzene, chlordanes, and polychlorinated biphenyls than fish from the western contiguous United States (Flanagan Pritz et al. 2014), indicating that bald eagles in Bristol Bay may be at risk to environmental contamination.

## POPULATION, DISTRIBUTION, AND ABUNDANCE

Bald eagles are one of the most abundant raptors in Alaska, with a population estimated at > 58,000 (Hodges 2011). Most Alaskan bald eagles occur in the vicinity of the southern coast (from Dixon Entrance to Bristol Bay), and secondarily along interior rivers and lakes (Schempf 1989). Savage and Hodges (2006) estimated that 2,775 adult bald eagles were present along the Alaska Peninsula Gulf Coast in 2005.

Surveys of nests and calculations of nest densities and occupancy rates are commonly conducted, although nesting rates display considerable temporal and spatial variability. Nesting density is considered to be generally correlated with food availability (Dzus and Gerrard 1993). Although Leighton et al. (1979) found density of breeding bald eagles in Saskatchewan to be correlated with mean April temperatures. The densities of nests in inland river areas of southeast Alaska were highly variable among sites and years. Hodges (1979) suggested this may have been correlated with food abundance and weather conditions. Nests in the Susitna River watershed, though, are thought to be more uniformly distributed (Ritchie and Ambrose 1996). For Interior Alaska populations, Ritchie and Ambrose (1996) surmised that densities were greatest in areas adjacent to coastal areas and where weather is somewhat milder, and prey more seasonally accessible and diverse.

Bald eagle nesting densities in southeast Alaska varied from 0.33–0.50 (or higher on Admiralty Island) active nests/km of coastline and 0.25–0.38 active nests/km of river (Robards and King 1966, Hodges 1979, Hansen et al. 1984). Population and nesting density was also reported to be high on Kodiak NWR, where almost 1,000 nests were located within an area of about 8,000 km$^2$ in 2002 (Zwiefelhofer 2007). Along the Gulkana River in interior Alaska, 0.01–0.08 active nests/km were documented (Byrne et al. 1983).

A comprehensive survey has not been completed for bald eagles or their nests in the Bristol Bay watershed. Available data indicate that nest density may be almost as high in portions of the region as elsewhere in Alaska, excluding the highest known densities in southeast Alaska and Kodiak Island. The USDI FWS Bald Eagle Nest Database has approximately 230 nest records for the study area (Table 12.2). However, 61 of those records are from the 1970s and 1980s, and those nests may not have persisted on the landscape. The remaining 169 records were collected between 2003 and 2006. A fixed-wing survey of adult bald eagles was conducted by the USDI FWS in 2006 along main-stem portions of some Alaskan rivers which documented 50 bald eagle nests along portions of the Nushagak, Mulchatna, and Kvichak rivers. Of those, 24 were identified as active. Database records for 2004 and 2005 are from a project contractor survey (not flown for the USDI FWS Bald Eagle Nest Database) that was conducted along the north side of Lake Iliamna. The 2004 and 2005 surveys recorded 75 total nests in this area (S. Lewis, Biologist, FWS, pers. comm. 2011). This appears to be a relatively

**Table 12.2** Summary of surveys for bald eagle nests in the Bristol Bay watershed

| Survey | Survey dates and results | |
|---|---|---|
| USDA Fish and Wildlife Service bald eagle nest surveys (recorded in bald eagle nest database) | 1970–1990<br><br>61 nest records | 2003–2006<br><br>169 nest records |
| Nushagak and Multchatna rivers survey by FWS | 2006<br><br>50 nest records (24 active) | |
| North side of Lake Illiamna; survey by contractor | 2004–2005<br><br>74 nest records | |

high nesting density, although it is not known which or how many of those nests were active, nor the density of active territories. In 2003, three nests were recorded (one active and one empty nest in the Lower Nushagak drainage, and one of unknown status on an islet off the north shore of Iliamna Lake).

Some site-specific surveys conducted in portions of southwest Alaska have reported numbers of individual bald eagles. For example, summer activity surveys for Katmai NPP identified between 50 and 87 individuals in the Naknek Lake drainage between 1991 and 1997 (Savage 1997). Although systematic efforts have not been made to identify fall bald eagle congregation sites in the Bristol Bay area (Wright 2010), congregation sites are known to exist in surrounding areas (e.g., Port Moller, Savonoski River), and are believed to be related to late-spawning sockeye salmon, fall runs of coho salmon, and fall-staging waterfowl. Although bald eagle densities are greatest overall in southeast Alaska, salmon also appear to be a major driving force for the Bristol Bay watershed population of bald eagles, so some comparisons may be inferred. In the Chilkat Valley, fall and winter bald eagle densities in habitats adjacent to foraging areas may be 10 times those of the same habitats (e.g., gravel bars, cottonwood stands) located far from food sources (Hansen et al. 1984).

# RECOMMENDED RESEARCH

All aspects of the ecological relationships of bald eagles in the Bristol Bay watershed lack adequate research and published results. Specific, local information currently available is limited to the results of sporadic aerial surveys of nest sites which provide locations of nests and data on nest occupancy and nest success (Dewhurst 2010, Wright 2010). The results of these surveys are available only in unpublished agency reports and databases. Other aspects of the natural history of bald eagles in this area are only described through anecdotal observations of biologists working in this area. More comprehensive information on habitat use patterns and primary threats to bald eagles in the Bristol Bay watershed is necessary to evaluate the effects of potential development in this area (e.g., Snyder 2014) on these birds.

Understanding the habitat relationships of bald eagles in Bristol Bay watersheds—particularly nesting habitat—is crucial to understanding where they may occur and why. Habitat for a specific organism includes those areas that have a combination of resources and environmental conditions that promotes occupancy by the species and allows individuals to survive and reproduce (Morrison et al. 2006). Selection of nest sites often involves a host of specific features ranging from local factors like the thermal environment to broader factors like proximity to foraging areas (Janes 1985). Habitat models should be developed to provide a quantitative means of determining the potential of a site as bald eagle nesting habitat using the information in the Alaska Bald Eagle Nest Atlas (http://www.fws.gov/alaska/mbsp/mbm/landbirds/alaskabaldeagles/default.htm) and available remotely sensed data that describe landscape characteristics (e.g., Livingston et al. 1990). Such models may be used to resolve conflicts regarding developments near active and historical nests by focusing management and mitigation efforts on habitat attributes deemed critical to breeding bald eagles in the Bristol Bay watershed.

The potential for mineral extraction in the Bristol Bay watershed brings with it the potential for environmental contamination (Hinck et al 2006). Top-level predators, such as bald eagles are especially vulnerable to many environmental contaminants and can be used as sentinel species for contaminated areas (Welch 1994, Holl and Cairns 1995, Leith 2014). Most contaminant studies of bald eagles have involved relatively noninvasive egg, blood, and feather collection techniques. Of those, only two studies reported exclusively on Alaska bald eagle populations, both of those were from the Aleutian Archipelago (Estes et al. 1997, Anthony et al. 1999). Tissue studies of contaminants in bald eagles are also available, but are of limited applicability to Alaska bald eagles (e.g., Anthony et al. 1993). It is therefore important to begin noninvasive collections of bald eagle tissues (e.g., feathers) from the Bristol Bay watershed and complete an evaluation of

Witter, L.A., and S.A. Anderson. 2011. Bald eagle nest survey Katmai National Park & Preserve, Alaska. USDI National Park Service Natural Resource Data Series NPS/KATM/NRDS—2011/311.

Witter, L.A., and B.M. Mangipane. 2011. Bald eagle nest survey Lake Clark National Park and Preserve, Alaska. USDI National Park Service Natural Resource Data Series NPS/LACL/NRDS—2011/313.

Wright, B.A., and P. Schempf. 2010. Introduction. Pages 8–18 *in* B.A. Wright and P. Schempf, editors. Bald eagles in Alaska. Hancock House Publishers, Surrey, British Columbia, Canada.

Wright, J.M. 2010. Bald eagles in western Alaska. Pages 251–257 *in* B.A. Wright and P. Schempf, editors. Bald eagles in Alaska. Hancock House Publishers, Surrey, British Columbia, Canada.

Zwiefelhofer, D. C. 2007. Comparison of bald eagle (*Haliaeetus leucocephalus*) nesting and productivity at Kodiak National Wildlife Refuge, Alaska, 1963-2002. Journal of Raptor Research 41:1–9.

# SECTION III

## WILDLIFE RESOURCES OF BRISTOL BAY: MARINE SYSTEMS

**Table 13.1**  Cetaceans found in Bristol Bay and peripheral rivers

| Common name | Scientific name (genus species) | Presence | NMFS stock designation | Status[1]/trend |
|---|---|---|---|---|
| **Mysticetes** | | | | |
| North Pacific right whale | *Eubalaena japonica* | Seasonal | Eastern North Pacific[1] | Endangered/unknown |
| Gray whale | *Eschrichtius robustus* | Seasonal | Eastern North Pacific | Recovered/increasing |
| Humpback whale | *Megaptera novaeangliae* | Seasonal | Central and Western North Pacific[1,2] | Endangered/increasing |
| Fin whale | *Balaenoptera physalus* | Seasonal | Northeast Pacific[1] | Endangered/unknown |
| Minke whale | *Balaenoptera acutorostrata* | Year-round | Alaska | unknown |
| **Odontocetes** | | | | |
| Sperm whale | *Physeter macrocephalus* | Rarely | North Pacific[1,3] | Endangered/unknown |
| Beluga | *Delphinapterus leucas* | Year-round | Bristol Bay | increasing |
| Killer whale | *Orcinus orca* | Seasonal | Gulf of Alaska, Aleutian Islands, and Bering Sea Transient and Eastern North Pacific Alaska Resident | unknown |
| Pacific white-sided dolphin | *Lagenorhynchus obliquidens* | Year-round | North Pacific | unknown |
| Dall's porpoise | *Phocoenoides dalli* | Seasonal | Alaska[3] | unknown |
| Harbor porpoise | *Phocoena phocoena* | Seasonal | Bering Sea | unknown |

[1] U.S. Endangered Species Act listed species.

[2] The Hawaii Distinct Population Segment within the Central North Pacific stock was delisted in September 2016.

[3] Stock Assessment range map (Muto et al. 2016) does not include Bristol Bay, but this species is known to occur there.

northward, following sperm whales across the North Pacific. By 1840, sperm whales were in low numbers. At the same time *whalebone* (baleen) prices began to rise dramatically. Baleen could be shaped similarly to today's plastics and maintain flexibility. Although this product had been used for centuries, it was the Parisian and English desire for hoop skirts that ratcheted up demand (Bockstoce and Burns 1993). The whalers began hunting right whales and, eventually bowhead whales. Shore-based whaling stations were established in the mid-1800s and continued operations into the early 1900s, primarily targeting gray and humpback whales (e.g., Reeves et al. 1985, Clapham et al. 1997). Innovation in ship design in the 1900s also allowed whalers to hunt the faster blue and fin whales (Clapham and Baker 2002). Although there had been earlier attempts to manage pelagic whaling, it was not until the 1930s that international protections were put in place for some species (Clapham and Baker 2002, Ivashchenko and Clapham 2014). Despite these protections, large-scale illegal Soviet and Japanese whaling continued to decimate populations in the North Pacific into the 1970s (Ivashchenko and Clapham 2014).

Enactment of the MMPA in 1972 prohibited the *take* and importation of marine mammals by U.S. citizens. The term *take* is defined as: *to harass, hunt, capture, or kill, or attempt to harass, hunt, capture, or kill any marine mammal*. In 1994, the MMPA was amended to allow specific exceptions to the take prohibitions that would allow some takes incidental to industrial and commercial fishing activities, for Alaska Native subsistence purposes, and scientific research. NMFS manages marine mammal stocks by limiting these activities when marine mammal resources become depleted. This may involve closing areas to human activity or reducing the number of allowable takes. When the MMPA was amended in 1994, one new requirement included preparation of assessments for each marine mammal stock under U.S. jurisdiction. These assessments are divided across regions of the U.S. and include information on survey effort, abundance, distribution, sources of mortality, and habitat concerns.

The Alaska marine mammal stock assessments (e.g., Muto et al. 2016) include stocks that inhabit the Gulf of Alaska, Bering Sea, Chukchi Sea, Beaufort Sea, and adjacent waters such as Bristol Bay (see Table 13.1 and Figure 13.1). Some stocks are present only seasonally in Alaska waters and their geographic ranges extend into the North Pacific. Bristol

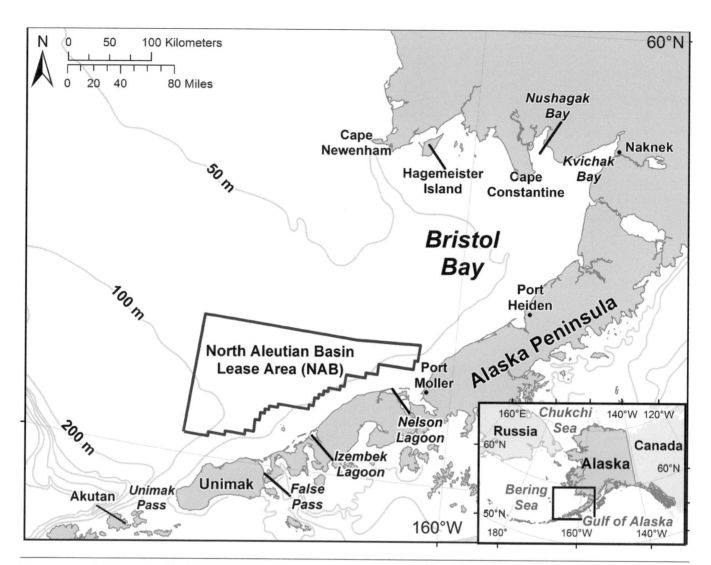

**Figure 13.1** Place names and bathymetric features mentioned in the text. *Map Credit: Kim Shelden.*

Bay provides a unique environment for these species that traverse the North Pacific and Gulf of Alaska, seeking passes through the Aleutian Island chain into the Bering Sea before arriving in Bristol Bay to feed (Figure 13.1). The bay also provides year-round habitat for at least three species (Table 13.1). However, data are limited because systematic aerial and vessel surveys have occurred infrequently, rarely include all of Bristol Bay, and typically are conducted only during the summer season (e.g., Leatherwood et al. 1983, Lowry et al. 1987, Frost et al. 1992, Moore et al. 2002, Hobbs and Waite 2010, Friday et al. 2012, 2013). The NMFS Platforms of Opportunity Program (POP) database also provides another source for Bristol Bay marine mammal sightings (e.g., Braham and Dahlheim 1982, Kajimura and Loughlin 1988).

The POP began in 1971 to solicit information about sightings of marine mammals made by personnel on various types of vessels (Boucher and Boaz 1989). The NMFS Marine Mammal Laboratory maintains this database of marine mammal observations which includes data gathered from: (1) the pelagic fur seal program (1958 to 1974); (2) the Dall's Porpoise Research Program (operated from NOAA and U.S. Coast Guard ships from 1975 to 1980); (3) an Outer Continental Shelf Environmental Assessment Program (OCSEAP) dedicated summer vessel cruise in 1980; and (4) POP observers on NOAA or other ships (including U.S. Coast Guard personnel, Federal fisheries observers, fisheries personnel, ferry operators, tourists, or other private boat operators). The sighting information includes date, time, location, sighting conditions, species, and number of animals. Participation in the program is completely voluntary and there is no quantification of effort; therefore, useful information includes only the date and location of animals. The following sections on mysticetes and odontocetes present available sighting data, historic and present-day status, and habitat use within Bristol Bay.

# MYSTICETES

## Exploitation and Current Status

Three of the five species of baleen whales that are currently known to occur in Bristol Bay waters are listed as Endangered under the U.S. Endangered Species Act (Table 13.1): North Pacific right whale, humpback whale, and fin whale. Only eastern North Pacific gray whales have been removed from the list after recovering to precommercial whaling numbers (Punt and Wade 2012). Minke whales, at under 8 m in length, were considered too small to chase by commercial whalers. Much of the commercial whaling during the period 1839–1904 occurred in lower latitudes of the North Pacific (Townsend 1935), where some of these species spend the winter months breeding and calving. One exception, the critically endangered eastern North Pacific right whale population, was also hunted in Bristol Bay (Townsend 1935) and was decimated by illegal Soviet whaling in the Bering Sea and Gulf of Alaska in the 1960s (Ivashchenko and Clapham 2012). In 2008, North Pacific right whales were recognized as a species separate from the North Atlantic (*Eubalaena glacialis*) and southern hemisphere right whales (*Eubalaena australis*), and now number in the low 30s in the Bering Sea (Wade et al. 2011). Humpback, fin, and occasionally minke whales were also hunted from the shore whaling station on Akutan Island during the period 1912–1939 (Reeves et al. 1985). Unimak Pass, located east of Akutan (Figure 13.1), provides the primary entry point for cetaceans migrating from the Gulf of Alaska into the Bering Sea and Bristol Bay. Some gray whales also migrate through False Pass on the east side of Unimak Island (Barrett-Lennard et al. 2011).

Aboriginal hunting of baleen whales by prehistoric and early contact peoples was documented in ethnological interviews and archaeological finds in the northern Gulf of Alaska (Yarborough 1995). Yarborough (1995) also mentioned the importance of baleen whales to the Yupik in the Bering Sea, and to subarctic Aleuts and Eskimos. Village excavation sites along the western Alaska Peninsula, particularly those less than 1,000 years old, typically included a few large whale bones at one or two house sites (Maschner 2004). A unique discovery of a whale bone house occurred near Izembek Lagoon (Figure 13.1). Originally thought to have been built around A.D. 900–1100 (McCartney 1974), the structure of the house included 34 whale mandibles. Reexamination of the site suggests this house was built during the Izembek Phase, A.D. 1250–1475 (Maschner 2004).

Reeves and Smith (2006) based their review of temperate aboriginal whaling operations in southern Alaska on McCartney's (1984) study. They noted gray and humpback whales were likely the primary targets of eastern Pacific whaling societies. When considering the length of the 34 whale mandibles (> 5 m) in Izembek, McCartney (1995) thought they were likely from blue whales (*Balaenoptera musculus*) and not humpback or gray whales. Clarke (1945) included blue whales among the rorquals [i.e., fin, minke, humpback, and sei whales, (*Balaenoptera borealis*)] observed in the Bering Sea, and they were among the catch reported at Akutan (Reeves et al. 1985), but there are few confirmed sightings north of the Aleutian Islands (Leatherwood et al. 1983). Sei whales are also observed infrequently in the Bering Sea (Leatherwood et al. 1983, Friday et al. 2012, 2013) and were reported only one time in the catch record from Akutan (Reeves et al. 1985). In 1975, one long-dead specimen was found on a beach in northeast Bristol Bay near Cape Constantine (Leatherwood et al. 1983). Present-day hunting of baleen whales does not occur in Bristol Bay.

## Habitat

Baleen whales are observed in greatest numbers in Bristol Bay during the summer feeding period. This is particularly true for species that give birth during the winter months in lower latitude regions and undertake long-distance, seasonal migrations, such as gray, humpback, fin, and possibly right whales. It is during this time that whales replenish their blubber stores after long periods of fasting during the migration and on the wintering grounds, where prey such as zooplankton and small, schooling fish tend to be sparse. For lactating and newly pregnant females who must support their calves, a poor feeding season may jeopardize calf survival.

During the spring, most gray whales (Figure 13.2a, b, c) migrate to Alaska from Mexico, entering Bristol Bay from the Gulf of Alaska via Unimak Pass or False Pass (Pike 1962, Barrett-Lennard et al. 2011). Feeding occurs in the region as the whales continue their northbound migration into the Chukchi Sea (Nerini 1984, Ferguson et al. 2015). Gray whales are opportunistic feeders, primarily consuming zooplankton (Figure 13.3) in the benthos, leaving large feeding pits in the soft substrate, but they also feed within the water column and at the surface (Nerini 1984). Prey documented in stomachs of whales killed in the Chukchi and northern Bering Sea consisted primarily of grammaridean amphipods, but also included bottom-dwelling isopods, mysids, mollusks, polychaetes, hydroids, decapod crustaceans, cumaceans, and gastropods (Rice and Wolman 1971). Euphausiids, a fast swimming krill, were also found in the baleen of a whale killed

**Figure 13.2** Gray whale (*Eschrichtius robustus*) (a) surfacing, (b) flukes, (c) head. *Photos taken under NOAA, NMFS, AFSC, MMPA/ESA research permit—photo credit: M. Gosho.*

**Figure 13.3** Examples of zooplankton prey consumed by baleen whales. Types of prey include meroplankton (animals that spend part of their life cycle as plankton) such as the larval forms of mollusks and crustaceans, and holoplankton (plankton for their entire life cycle) such as copepods and euphausiids (krill). *Photo credit: M. Wilson and J. Clark, NOAA, AFSC.*

in northern California waters (Rice and Wolman 1971). Moore et al. (2007) noted the prevalence of cumaceans (*lollipop shrimp*) near foraging gray whales off Kodiak Island, Alaska. The authors concluded that these flexible foragers may be taking advantage of atypical prey as their population continues to increase in number and traditional foraging habitats may be affected by climate change. Gray whales are observed throughout the summer in Bristol Bay. In the fall, migrants again pass through Bristol Bay as they head south to breeding and nursery lagoons in Mexico (Rugh 1984). Ferguson et al. (2015) delineated Biologically Important Areas (BIAs) for migrating (northbound) and feeding gray whales, primarily within waters less than 25 m deep along the shoreline, based on systematic surveys conducted within Bristol Bay (Figure 13.4).

Humpback whales (Figure 13.5a, b, c) migrate to Alaska to feed during the summer months. From 2004–2006, the Structure of Populations, Levels of Abundance, and Status of Humpbacks (SPLASH) project was conducted to determine migratory destinations from low-latitude winter nursery areas to high-latitude summer feeding areas. It was not possible to obtain genetic samples from all migratory destinations, including northern Bristol Bay; however, it is likely that humpbacks feeding within Bristol Bay belong to either the Central North Pacific stock or the Western North Pacific stock (Barlow et al. 2011, Muto et al. 2016). Humpbacks consume a range of prey from euphausiids and copepods (see Grant and Radenbaugh this volume), to fish and squid (Witteveen et al. 2011). Ferguson et al. (2015) defined the humpback feeding BIA within Bristol Bay based on results from systematic vessel surveys conducted along the shelf region (Figure 13.4).

Fin whales (Figure 13.6) found in this region are considered part of the Northeast Pacific stock, which ranges from the Chukchi Sea south to the Washington coast (Muto et al. 2016), although it appears likely that two migratory stocks mingle in the Bering Sea in July and August (Mizroch et al. 2009). Friday et al. (2013) noted that fin whales were more abundant in the Bering Sea in colder years, which may be due to shifts in the distribution of their prey. Fin whale

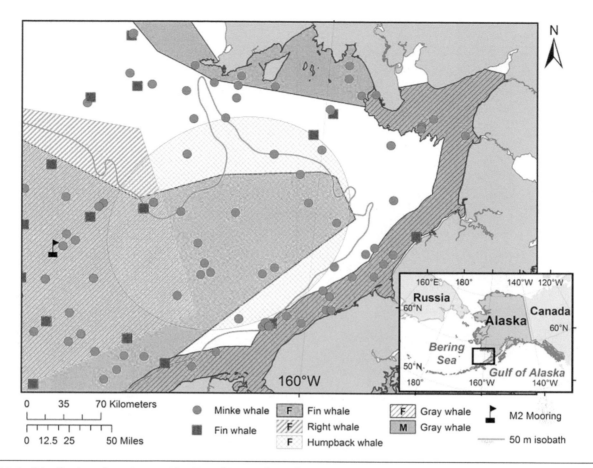

**Figure 13.4**  Distribution of mysticetes within Bristol Bay. Biologically important areas (BIAs) for feeding (F) and migration (M) are shown as shaded regions (Ferguson et al. 2015). Sighting locations (symbols) were included for species without BIAs (i.e., minke whales: circles), and to note locations of fin whales (squares) in waters < 50 m deep (Leatherwood et al. 1983, Moore et al. 2002, Friday et al. 2012, NMFS Platforms of Opportunity Program observations).

vocalizations (i.e., calls) were detected on a passive acoustic recorder attached to a biophysical mooring near the entrance to Bristol Bay (Figure 13.4); call rates increased following peaks in euphausiid biomass (Stafford et al. 2010). Ferguson et al. (2015) include Bristol Bay, where water depth is greater than 50 m, as part of the feeding BIA for fin whales. There have been few sightings within coastal waters of Bristol Bay (Berzin and Rovnin 1966). Leatherwood et al. (1983) reported a single sighting of two fin whales northeast of Port Heiden during their May-early June aerial surveys, and three sightings were recorded in the NMFS POP database; all occurred in the 1980s.

North Pacific right whales (Figure 13.7) were hunted in Bristol Bay during the period of commercial whaling from 1839–1904, with catches reported in July, August, and September along the Alaska Peninsula and northern shoreline (Townsend 1935, Shelden et al. 2005). It appears that some of these whales may migrate to Hawaii in the winter (Kennedy et al. 2011); however, calving grounds were never reported by commercial whalers (Scarff 1986) and have not been discovered to this day. It is possible that some whales may remain within the Bering Sea year-round (Shelden et al. 2005). Their primary prey are copepods, and a correlation was found between an increase in right whale vocalizations (at the M2 mooring; Figure 13.4) and peaks in their prey biomass (Stafford et al. 2010). Present-day distribution within the Bering Sea appears to be primarily within the area designated as Endangered Species Act (ESA) critical habitat. Ferguson et al. (2015) included this area as a feeding BIA for eastern North Pacific right whales.

Minke whales (Figure 13.8) in the Bristol Bay region are managed as a single stock within Alaska waters (Muto et al. 2016). At about eight meters in length, they are the smallest of the baleen whales found in the Bering Sea. These whales tend to be cryptic, presenting a low profile when surfacing and typically travel alone. They also have been reported to avoid ships (Palka and Hammond 2001). However, surface active whales have been observed feeding near fishing fleets in northern Bristol Bay (Leatherwood et al. 1983). The authors described individual whales swimming rapidly through

**Figure 13.5**    Humpback whale (*Megaptera novaeangliae*) (a) surfacing, (b) breaching, (c) flukes. *Photos taken under NOAA, NMFS, AFSC, MMPA/ESA research permit—photo credit: P. Clapham, B. Rone, and S. Mizroch.*

**Figure 13.6**   Fin whale (*Balaenoptera physalus*). *Photo taken under NOAA, NMFS, AFSC, MMPA/ESA research permit—photo credit: D. Ellifrit.*

**Figure 13.7a, b**   North Pacific right whale (*Eubalaena japonica*) (a) surfacing, (b) aerial view. *Photos taken under NOAA, NMFS, AFSC, MMPA/ESA research permit—photo credit: A. Kennedy and B. Rone.*

**Figure 13.8**    Minke whale (*Balaenoptera acutorostrata*). *Photo taken under NOAA, NMFS, AFSC, MMPA/ESA research permit—photo credit: B. Rone.*

schools of unidentified fish or in close proximity to working herring boats in May, June, and August. Minke whales, and all of the odontocetes (with the exception of belugas), were not included in the BIA analysis conducted by Ferguson et al. (2015). This was largely due to the paucity of data on where feeding, migration, and reproduction occur in the Bering Sea and Aleutian Islands for these species. Observations during systematic surveys show minke whales are present in all seasons within Bristol Bay (Leatherwood et al. 1983) and their distribution includes both coastal areas and offshore waters (Figure 13.4).

# ODONTOCETES

## Exploitation and Current Status

Toothed whales, dolphins, and porpoises that inhabit Bristol Bay range from occasional visitors to year-round residents. The smallest of these odontocetes are the harbor porpoises and the largest, the sperm whale.

Similar to the baleen whales, sperm whales were hunted by commercial whalers, primarily in low-latitude waters of the North Pacific (Townsend 1935, Mizroch and Rice 2006). Sperm whales were also hunted from the shore station on Akutan, with catches consisting almost exclusively of males (Reeves et al. 1985). Soviet factory fleets operating in the North Pacific in the 1960s routinely falsified catch records, thereby under-reporting total catch and the size/sex of sperm whales killed (Ivashchenko et al. 2013, Ivashchenko and Clapham 2014). Aboriginal hunting of this species primarily occurs in the tropical Pacific (Reeves and Smith 2006), although small whales appear to have been hunted or scavenged in the northern Gulf of Alaska (Yarborough 2000).

The Akutan whaling station also reported catches of belugas (Reeves et al. 1985), presumably from the Bristol Bay stock. These whales are an important subsistence resource in Bristol Bay, dating from prehistoric Yupik societies (Yarborough 1995) to present day Alaska Native hunters (Lowry et al. 2008). Currently, about 20 to 30 whales are taken annually in Bristol Bay by Alaska Native subsistence hunters (Muto et al. 2016). During the 1950s and 1960s, 127 Bristol Bay belugas were killed for research and eight were live-captured for aquaria (Lowry et al. 2008). Killer whales are known to prey on this beluga population as well (Frost et al. 1992). Their potential impact on commercial salmon fisheries led to predator control actions being taken from the mid-1950s to 1978. These actions included dynamiting near belugas, playing killer whale calls, and harassment using power boats (Lowry et al. 2008). Although the population is small compared to other western Alaska stocks, at about 1,000 whales, it appears to be healthy and increasing in number at 4.8% per year (Lowry et al. 2008). In 2008, NMFS, in collaboration with the Bristol Bay Native Association, aquaria, universities, and the State of Alaska, began a health assessment study of this population (e.g., Norman et al. 2012, Castellote et al. 2014, Citta et al. 2016), in part as a baseline study for comparison to the critically endangered beluga population that resides year-round in Cook Inlet, Alaska (Shelden et al. 2015). One important finding, based on results from satellite-tagging studies (Citta et al. 2016), suggests this population remains year-round in Bristol Bay.

Killer whales and harbor porpoise were occasionally reported in the catch record at Akutan (Reeves et al. 1985). Ethnological accounts, faunal remains, and pictographs suggest aboriginal hunting of killer whales, Dall's, and harbor porpoises occurred in the northern Gulf of Alaska (Yarborough 1995, Shelden et al. 2014), but no records were found to indicate these species were also hunted for subsistence north of the Aleutian Islands. Killer whales within Bristol Bay include two ecotypes (genetically distinct and adapted to unique environmental conditions): (1) mammal eaters, also known as *transients* (and recently renamed *Bigg's killer whale* in honor of the scientist who first described them), and (2) fish eaters or *residents*. For management purposes, these whales are included in the *Gulf of Alaska, Aleutian Islands, and Bering Sea Transient Stock* and the *Eastern North Pacific Alaska Resident Stock*, respectively (Muto et al. 2016).

The Bering Sea stock of harbor porpoise is managed based on abundance estimates obtained during aerial surveys in Bristol Bay in 1991 (Dahlheim et al. 2000) and 1999 (Hobbs and Waite 2010). Population estimates ranged from 10,946 (CV = 0.300) harbor porpoise in 1991 to 48,215 (CV = 0.223) in 1999. This dramatic increase in porpoise numbers in Bristol Bay may be the result of changes in survey design, areas surveyed, and analyses, although densities were also greater in 1999 (Muto et al. 2016).

Stock assessment reports do not include Bristol Bay within the range of the Alaska stock of Dall's porpoise (Muto et al. 2016), although Leatherwood et al. (1983) reported sightings within the bay in all seasons but spring. The stock assessments also note that sightings of Pacific white-sided dolphins are infrequent in the southern Bering Sea (Muto et al. 2016). However, these dolphins have been observed regularly near Port Moller (Figure 13.9) and in all seasons within the southern Bering Sea (Waite and Shelden, in review).

The ranges of the Alaska stocks of beaked whales (family Ziphiidae) do not include Bristol Bay (Muto et al. 2016). Cuvier's (*Ziphius cavirostris*), Stejneger's (*Mesoplodon stejnegeri*), and Baird's (*Berardius bairdii*) beaked whales are typically found in pelagic waters of the Bering Sea (Friday et al. 2012, 2013), but only a single stranded dead specimen of each species has been recorded in Bristol Bay (Leatherwood et al. 1983). Some temperate and tropical species, such as Risso's dolphins (*Grampus griseus*) and pilot whales (*Globicephala* spp.), have been reported in the Bering Sea and Bristol Bay (Kajimura and Loughlin 1988), although these sightings appear to be atypical (Leatherwood et al. 1983).

## Habitat

The largest of the toothed whales, the sperm whale (Figure 13.10), appears to be a rare visitor to Bristol Bay. Sightings in the Bering Sea typically occur seasonally along the shelf break and oceanic frontal zones (Mizroch and Rice 2013). A small number of encounters have been reported in the Bering Sea, usually in waters deeper than 100 meters (Friday et al. 2012, 2013). Leatherwood et al. (1983) did not find these whales during their surveys but attributed this to poor sighting conditions and the tendency for high-latitude sperm whales to be solitary, long-diving males. Sperm whales consume cephalopods, fishes, skates, and sharks (Rice 1989). One of two POP sightings in Bristol Bay (Figure 13.9) noted damaged fish in the haul, suggesting possible sperm whale depredation (when whales actively remove or partially consume fish hooked on the longline). Though rarely observed in the Bering Sea, this type of depredation is widespread in the Gulf of Alaska (Perez 2006, Sigler et al. 2008).

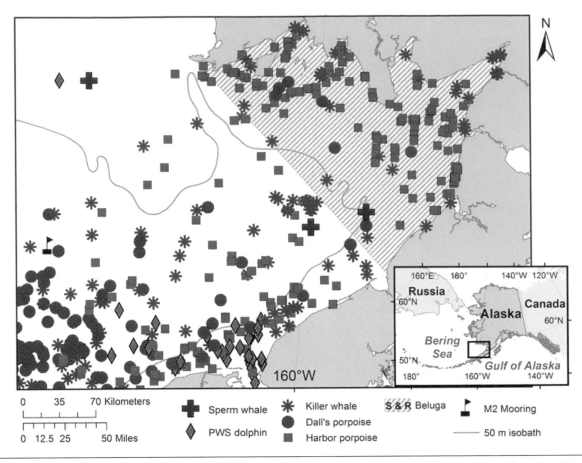

**Figure 13.9** Distribution of odontocetes within Bristol Bay. Biologically important areas (BIAs) for the small and resident (S & R) beluga population (Ferguson et al. 2015), and sighting locations (symbols) for all other species with unknown BIAs are shown (Braham and Dahlheim 1982, Leatherwood et al. 1983, Lowry et al. 1987, Kajimura and Loughlin 1988, Frost et al. 1992, Hobbs and Waite 2010, Friday et al. 2012, 2013, Waite and Shelden, in review, NMFS Platforms of Opportunity Program observations). PWS = Pacific white-sided.

**Figure 13.10**   Sperm whale (*Physeter macrocephalus*). *Photo taken under NOAA, NMFS, AFSC, MMPA/ESA research permit—photo credit: V. Burkanov.*

The Bristol Bay stock of belugas (Figure 13.11) remains in the region year-round, moving to deeper waters in the bay during the winter (Citta et al. 2016), and entering the Nushagak and Kvichak River estuaries and smaller rivers of these watersheds during the summer to feed on salmon runs and to give birth. Fish are their primary prey during the summer period. Of 101 stomachs examined that contained prey, 100 contained fish, primarily salmonids (Quakenbush et al. 2015). Belugas follow salmon runs far up the Nushagak, Kvichak, and Snake rivers and these areas may provide some protection from Bigg's/transient killer whales. Ferguson et al. (2015) included the eastern portion of Bristol Bay as a BIA for this small, resident population (Figure 13.9).

Killer whale presence within the bay (Figure 13.9) has ranged from few or no sightings reported during extensive observation efforts (Braham and Dahlheim 1982, Leatherwood et al. 1983, Lowry et al. 1987, Kajimura and Loughlin 1988, Friday et al. 2012, 2013, POP observations) to multiple sightings within a two-year period (Frost et al. 1992). Killer whale strandings have also been reported (Lowry et al. 1987, Barbieri et al. 2013). When killer whales do enter Bristol Bay, it is typically during the summer months, and is likely associated with concentrations of Pacific salmon (for resident killer whales: Figure 13.12a) and belugas (for Bigg's/transient killer whales: Figure 13.12b) in Nushagak and Kvichak bays. According to Lowry et al. (2008), Bristol Bay area residents have noted increased presence of killer whales in Nushagak Bay.

**Figure 13.11a, b** Beluga whale (*Delphinapterus leucas*) (a) aerial view of a beluga whale pod, (b) mother and calf. *Photos taken under NOAA, NMFS, AFSC, MMPA/ESA research permit by L. Morse, and MMPA/ESA permit 14210 by LGL Research Associates.*

**Figure 13.12a, b**   Killer whale (*Orcinus orca*) (a) breaching, (b) pod of killer whales. *Photos taken under NOAA, NMFS, AFSC, MMPA/ESA research permit—photo credit: H. Fearnbach and D. Ellifrit.*

Notably, in September 2011, three killer whales entered the Nushagak River and remained for three weeks. Two died, both females, one of which was pregnant (Pemberton 2011, Barbieri et al. 2013, S. Raverty, unpublished data). Genetic analysis confirmed that the animals were Bigg's killer whales (K. Parsons, NMFS, NWFSC, Seattle, WA, pers. comm.). Bigg's killer whales have also been observed preying on Steller sea lions (*Eumetopias jubatus*), harbor seals (*Phoca vitulina*), and walrus (*Odobenus rosmarus*) in Bristol Bay (Frost et al. 1992) (see also Chapter 14).

Harbor porpoise (Figure 13.13) tend to be cryptic, travel alone or in small groups, and avoid vessels. Harbor porpoise prey include small, schooling fishes. These porpoise are generally found in shallow water less than 100 meters deep

(Figure 13.9) (Leatherwood et al. 1983, Kajimura and Loughlin 1988, Dahlheim et al. 2000, Hobbs and Waite 2010, Friday et al. 2012, 2013, POP observations). Leatherwood et al. (1983) noted the near absence of harbor porpoise in the Bering Sea during winter. Numbers increased from spring through summer, then declined again, presumably as porpoise left the area with the formation of ice along the shoreline (i.e., *shore-fast ice*) and sea ice. Peaks in sightings in late spring and late summer likely coincide with anadromous fish runs. Calving also occurs during the summer months in Alaska waters (Leatherwood et al. 1983).

Dall's porpoise (Figure 13.14) are found only in the North Pacific and the Bering Sea and tend to prefer deeper waters. These porpoises are often highly visible when surfacing, creating a rooster-tail spray behind them as they race through the water, often approaching vessels to bow ride. They consume small, schooling fish and cephalopods (Kajimura et al. 1980). Sighting locations reported in Bristol Bay (Figure 13.9) included both coastal and offshore waters (Leatherwood et al. 1983, Kajimura and Loughlin 1988, Friday et al. 2012, 2013, POP observations). Dall's porpoise were found in all seasons in the Bering Sea, with peaks in fall and winter sightings in shelf waters of Bristol Bay (Leatherwood et al. 1983).

Pacific white-sided dolphins (Figure 13.15) consume prey similar to the porpoises (Kajimura et al. 1980). Their highly localized distribution near Port Moller (Kajimura and Loughlin 1988, Friday et al. 2012, 2013, Waite and Shelden, in review, POP observations) may suggest summer site-fidelity (i.e., that they return to this area year after year) possibly for calving, breeding, or foraging (Figure 13.9). Sightings have been reported during all seasons in the vicinity of Unimak Pass (Waite and Shelden, in review, POP observations).

**Figure 13.13**  Harbor porpoise (*Phocoena phocoena*). *Photo taken under NOAA, NMFS, AFSC, MMPA/ESA research permit: photo credit: C. Emmons.*

**Figure 13.14**  Dall's porpoise (*Phocoenoides dalli*). *Photo taken under NOAA, NMFS, AFSC, MMPA/ESA research permit—photo credit: K. Stafford.*

**Figure 13.15**    Pacific white-sided dolphin (*Lagenorhynchus obliquidens*). *Photo taken under NOAA, NMFS, NWFSC, MMPA/ESA research permit.*

# CONSERVATION CONCERNS

## Fisheries Interactions

The Bristol Bay region is an important area for whales, dolphins, and porpoises for feeding, migration, and reproduction. The diversity of prey available to these cetaceans and the importance of this region to commercial and subsistence fisheries has the potential for conflict, whether by entanglement, ship strike, prey depredation, or competition for limited resources. Numerous commercial fisheries operate in and near Bristol Bay. These include the Bering Sea and Aleutian Islands groundfish trawl, longline, jig, and pot fisheries, and salmon set gillnet and drift gillnet fisheries. Since 1972, observers have been deployed by NOAA/NMFS to collect catch and bycatch data from U.S. commercial fishing and processing vessels. *Bycatch* refers to any species that is not the target species for that particular fishery and also includes documenting interactions with marine mammals. In 2013, significant changes occurred within the NOAA fisheries observer program, which monitors halibut and groundfish operations. This included placing observers on a percentage of smaller vessels (between 40–57.5 ft. in length) that had previously been unmonitored (e.g., NMFS 2015). Observer coverage of these fisheries within Bristol Bay has been sparse and none of the salmon fisheries have been monitored (Muto et al. 2016) as nearly all fishing occurs within State of Alaska waters not Federal. In addition to commercial fisheries, subsistence gillnetters also target salmon in Bristol Bay.

Interactions between baleen whales and these fisheries have not been reported in Bristol Bay. However, humpback whales have been observed depredating on groundfish catch and entangled in trawl nets, longlines, and pot gear in the Bering Sea (Perez 2003, 2006, Breiwick 2013, Allen et al. 2014, Helker et al. 2015). Minke and gray whales have also become entangled in trawl gear in the Bering Sea, with minkes suspected of depredation (Perez 2003, 2006, Breiwick 2013). Fishing gear entanglement was suspected for one right whale photographed in the Bering Sea (line scars were visible in the image) and another near British Columbia (Muto et al. 2016), and fin whale entanglements in trawl gear have been reported in the Gulf of Alaska (Perez 2003, 2006, Allen et al. 2014, Helker et al. 2015). In most instances, the whales are freed alive from the gear, but serious injuries and deaths do occur. The number of serious injuries and deaths allowed per year are calculated within each species stock assessment report (e.g., Muto et al. 2016). This *Potential Biological Removal (PBR) Level* is defined by the MMPA as "the maximum number of animals, not including natural mortalities, that may be removed from a marine mammal stock while allowing that stock to reach or maintain its optimum sustainable population." The PBR level is the product of the following factors: the minimum population estimate of the stock, one-half the maximum theoretical or estimated net productivity rate of the stock at a small population size, and a recovery factor of between 0.1 and 1.0.

Of the odontocetes, sperm whales, killer whales, and Dall's porpoise approach fishing vessels and actively prey on the catch. Though rarely reported in the Bering Sea, sperm whale interactions with longline fisheries are a frequent occurrence in the Gulf of Alaska and injuries and entanglements have occurred (Perez 2003, 2006, Sigler et al. 2008, Breiwick 2013, Allen et al. 2014). Killer whales have been observed in the Bering Sea depredating catches on longline, trawl, and pot gear, consuming fish processing waste, becoming entangled in longlines and trawls, and getting struck by fishing vessel propellers (Perez 2003, 2006, Breiwick 2013, Peterson et al. 2013, Allen et al. 2014, Helker et al. 2015, Dahlheim and Breiwick, in review). Dall's porpoise have been killed in all gear types used in the Bering Sea groundfish fishery (Perez 2003, 2006, Breiwick 2013, Helker et al. 2015). While depredation has not been observed during encounters with Pacific white-sided dolphins and harbor porpoise in the Bering Sea, both species have been killed in trawl gear, and a Pacific white-sided dolphin was killed by longline gear (Perez 2003, Breiwick 2013). Leatherwood et al. (1983) noted that harbor porpoise are also vulnerable as fisheries bycatch, given their close association with coastal net fisheries for salmon, herring, and cod in Alaska waters. Although harbor porpoise entanglements in subsistence nets have been reported in the Bering Sea, none of these occurred in Bristol Bay (Allen et al. 2014, Helker et al. 2015).

Belugas are also susceptible to incidental mortality in fisheries operations. One beluga caught and killed in trawl gear in the Bering Sea was not sampled and, therefore, stock identity was never confirmed (Helker et al. 2015). According to Muto et al. (2016), commercial salmon set gillnet and drift gillnet fisheries in Bristol Bay, which combined had 2,845 active permits in 2010, have never been monitored by fisheries observers. Beluga entanglements in these fisheries have been reported in the past (Frost et al. 1984) and they also have been entangled and killed in Bristol Bay personal-use/subsistence salmon set-nets (Allen et al. 2014, Helker et al. 2015). Belugas killed incidentally in the personal-use or commercial salmon fisheries are usually butchered for subsistence use and included in the harvest data reported by the Bristol Bay Native Association.

## Offshore and Coastal Development

Disturbance by anthropogenic noise is an increasing concern for most cetacean species. Bristol Bay is adjacent to the North Aleutian Basin lease sale planning area (Figure 13.1), a region proposed for oil and gas exploration and development (see Chapter 24). Elevated levels of sound from anthropogenic sources include those associated with seismic testing, construction, and shipping, which may alter or disrupt migration, feeding, and reproductive behaviors and potentially mask the ability of these species to communicate (e.g., Clark et al. 2009, Moore et al. 2012). Currently, there are no active leases from past sales and this planning area was not included by the Bureau of Ocean Energy Management in the 2012–2017 lease schedule.

Increased shipping activity associated with development may also lead to ship strikes (e.g., Jensen et al. 2015). Collisions with vessels are already observed where these species interact with commercial fisheries in Alaska waters (Allen et al. 2014, Helker et al. 2015). It is anticipated that vessel traffic into the eastern Bering Sea will increase as more areas of the Arctic remain ice-free for longer periods (Ellis and Brigham 2009), with possible expansion of some routes traversing Bristol Bay. This may include not only commercial transport but tourism and personal use vessels as well.

With increased shipping there is the potential for coastal development. Nearshore species such as harbor porpoise and belugas are particularly vulnerable to physical habitat modifications that may include construction of docks and other over water structures, filling shallow areas, and dredging. Discharge from vessels and nonpoint source runoff from industrial development and waste management are also areas of concern, not only because of direct exposure, but also ingestion and bioaccumulation through the prey consumed by some cetacean species. Ingestion of marine debris, particularly plastics and microplastics that are more likely to accumulate within semi-enclosed basins (e.g., Fossi et al. 2012, 2016) such as Bristol Bay, are increasingly becoming an issue for concern given their potential toxicological effects. With the potential for new development in Bristol Bay, mitigation, enforcement, and risk assessment measures will need to be in place to minimize disturbance to the cetacean species in this area.

## Climate Change

With changes in the Arctic due to loss of multiyear sea ice, areas previously inaccessible are now available to a number of cetacean species found in Bristol Bay. Observations of humpback, fin, gray, and killer whales and harbor porpoise are occurring more often and for longer periods north of Bering Strait (e.g., Stafford et al. 2007, Higdon et al. 2009, MacLeod 2009). Ice-free waters and shifts in prey distribution are likely driving these range extensions. Temperate and tropical cetacean species may find their way to Bristol Bay, along with changes in phytoplankton, zooplankton, and fish assemblages (Grebmeier et al. 2006). Given the wide distribution and flexible feeding behaviors of many of these cetacean species, they may be less sensitive to climate change compared to the influx of anthropogenic activities that may accompany it (e.g., Laidre et al. 2008, Heide-Jorgensen et al. 2010, Moore et al. 2012).

## RECOMMENDED RESEARCH

Systematic visual surveys conducted from vessels and aircraft have infrequently focused solely on Bristol Bay (e.g., Hobbs and Waite 2010). The costs of such surveys have limited how often they are conducted. Year-round passive acoustic monitoring (such as that at the M2 mooring) has provided valuable information on the presence of endangered whales such as North Pacific right whales and fin whales (e.g., Stafford et al. 2010). Of course, this relies on the animals being vocal, and analysis of data collected during these year-long deployments is also costly. Satellite tagging and tracking of individual whales has provided new insights on movement patterns for Bristol Bay belugas (e.g., Citta et al. 2016). Similar projects have been proposed for other species such as North Pacific right whales (e.g., Shelden et al. 2005), but remain unfunded. It is critical to obtain baseline data for Bristol Bay given the potential for development (petroleum, shipping, fisheries, and coastal) within this region.

## SUMMARY

Bristol Bay provides a unique habitat for a diversity of cetaceans from small porpoises to large baleen whales. Many of these species have been subjected to commercial whaling and subsistence hunting, and now face potential threats from interactions with fisheries, offshore and coastal development, and climate change. Much of what we know about these species is limited to anecdotal accounts, spatially and temporally restricted systematic surveys, and a few species-specific studies. Several species appear to occur in Bristol Bay year-round: belugas, Pacific white-sided dolphins, and minke whales. Others increase in number during the ice-free months: harbor porpoise and Dall's porpoise; while others rely on the region for foraging during the summer months: gray whales, humpback whales, fin whales, sperm whales, killer whales, and North Pacific right whales. Baleen whales that give birth in warmer climes, such as humpback and gray whales, typically bring their calves to the same feeding areas that they were taken to as calves. Bristol Bay is also a nursery area for belugas born during the summer months in the estuaries of the Kvichak and Nushagak rivers. Still, there is much we need to learn in terms of habitat requirements for many of these species.

## REFERENCES

Allen, B. M., V. T. Helker, and L. A. Jemison. 2014. Human-caused injury and mortality of NMFS-managed Alaska marine mammal stocks, 2007-2011. U.S. Dept. of Commerce, NOAA Tech. Memo. NMFS-AFSC-274, 84 p. http://www.afsc.noaa.gov/Publications/AFSC-TM/NOAA-TM-AFSC-274.pdf.

Barbieri, M. M., S. Raverty, M. B. Hanson, S. Venn-Watson, J. K. B. Ford and J. K. Gaydos. 2013. Spatial and temporal analysis of killer whale (*Orcinus orca*) strandings in the North Pacific Ocean and the benefits of a coordinated stranding response protocol. Marine Mammal Science 29(4):E448-E462. DOI: 10.1111/mms.12044.

Barlow, J., J. Calambokidis, E. A. Falcone, C. S. Baker, A. M. Burdin, P. J. Clapham, J. K. Ford, C. M. Gabriele, R. LeDuc, D. K. Mattila and T. J. Quinn. 2011. Humpback whale abundance in the North Pacific estimated by photographic capture-recapture with bias correction from simulation studies. Marine Mammal Science 27(4):793–818. DOI: 10.1111/j.1748-7692.2010.00444.x.

Barrett-Lennard, L. G., C. O. Matkin, J. W. Durban, E. L Saulitis and D. Ellifrt. 2011. Predation on gray whales and prolonged feeding on submerged carcasses by transient killer whales at Unimak Island, Alaska. Marine Ecology Progress Series 421:229–241. http://dx.doi.org/10.3354/meps08906.

Berzin, A. A. and A. A. Rovnin. 1966. The distribution and migrations of whales in the northeastern part of the Pacific, Chukchi and Bering seas. Izvestia TINR0 58:179–207.

Braham, H. W. and M. E. Dahlheim. 1982. Killer whales in Alaska documented in the Platforms of Opportunity Program. Report of the International Whaling Commission 32:643–646.

Breiwick, J. M. 2013. North Pacific marine mammal bycatch estimation methodology and results, 2007–2011. U.S. Department of Commerce, NOAA Technical Memorandum NMFS-AFSC-260, 40 p.

Boucher, G. C. and C. J. Boaz. 1989. Documentation for the marine mammal sightings database of the National Marine Mammal Laboratory. U.S. Department of Commerce, NOAA Technical Memorandum NMFS F/NWC-159, 60 p.

Castellote, M., T. A. Mooney, L. Quakenbush, R. Hobbs, C. Goertz and E. Gaglione. 2014. Baseline hearing abilities and variability in wild beluga whales (*Delphinapterus leucas*). The Journal of Experimental Biology 217(10):1682–1691.

Citta, J. J., L. T. Quakenbush, K. J. Frost, L. F. Lowry, R. C. Hobbs and H. Aderman. 2016. Movements of beluga whales (*Delphinapterus leucas*) in Bristol Bay, Alaska. Marine Mammal Science 32(4):1272–1298. doi: 10.1111/mms.12337.

Clapham, P. J. and C. S. Baker. 2002. Modern whaling. Pages 1328–1332 *in* W. F. Perrin, B. Würsig and J. G. M. Thewissen, editors. Encyclopedia of Marine Mammals, Academic Press, New York, NY.

Clapham, P. J., S. Leatherwood, I. Szczepaniak and R. L. Brownell. 1997. Catches of humpback and other whales from shore stations at Moss Landing and Trinidad, California, 1919–1926. Marine Mammal Science 13(3):368–394.

Clark, A. H. 1945. Animal life of the Aleutian Islands. Pages 31–61 *in* H. B. Collins, A. H. Clark and E. H. Walker, editors. The Aleutian Islands: their people and natural history (with keys for the identification of the birds and plants). Smithsonian Institution War Background Studies, no. 21. Washington, D.C., 129 p.

Clark, C. W., W. T. Ellison, B. L. Southall, L. Hatch, S. M. Van Parijs, A. Frankel and D. Ponirakis. 2009. Acoustic masking in marine ecosystems: intuitions, analysis, and implication. Marine Ecology Progress Series 395:201–222.

Dahlheim, M. and J. Breiwick. In review. Killer whale (*Orcinus orca*) interactions, injuries, and mortality related to Alaska fishing operations. Submitted to Fishery Bulletin.

Dahlheim, M., A. York, R. Towell, J. Waite and J. Breiwick. 2000. Harbor porpoise (*Phocoena phocoena*) abundance in Alaska: Bristol Bay to southeast Alaska, 1991–1993. Marine Mammal Science 16(1):28–45.

Ellis B. and L. Brigham, editors. 2009. Arctic marine shipping assessment 2009 report. Arctic Council. 194 p. http://www.arctic.noaa.gov/detect/documents/AMSA_2009_Report_2nd_print.pdf (November 2015).

Ferguson, M. C., J. M. Waite, C. Curtice, J. T. Clarke and J. Harrison. 2015. Biologically important areas for cetaceans within U.S. Waters—Aleutian Islands and Bering Sea region. Aquatic Mammals 41(1):79–93. DOI 10.1578/AM.41.1.2015.79.

Fossi, M. C., C. Panti, C. Guerranti, D. Coppola, M. Giannetti, L. Marsili and R. Minutoli. 2012. Are baleen whales exposed to the threat of microplastics? A case study of the Mediterranean fin whale (*Balaenoptera physalus*). Marine Pollution Bulletin 64(11):2374–2379.

Fossi, M. C., L. Marsili, M. Baini, M. Giannetti, D. Coppola, C. Guerranti, I. Caliani, R. Minutoli, G. Lauriano, M. Grazia Finoia, F. Rubegni, S. Panigada, M. Bérubé, J. Urbán Ramírez and C. Panti. 2016. Fin whales and microplastics: The Mediterranean Sea and the Sea of Cortez scenarios. Environmental Pollution 209 (2016);68–78, ISSN 0269-7491. http://dx.doi.org/10.1016/j.envpol.2015.11.022.

Friday, N. A., J. M. Waite, A. N. Zerbini and S. E. Moore. 2012. Cetacean distribution and abundance in relation to oceanographic domains on the eastern Bering Sea shelf: 1999–2004. Deep Sea Research Part II: Topical Studies in Oceanography 65:260–272.

Friday, N. A., A. N. Zerbini, J. M. Waite, S. E. Moore and P. J. Clapham. 2013. Cetacean distribution and abundance in relation to oceanographic domains on the eastern Bering Sea shelf in June and July of 2002, 2008, and 2010. Deep-Sea Research Part II 94:244–256.

Frost, K. J., L. F. Lowry and R. R. Nelson. 1984. Belukha whale studies in Bristol Bay, Alaska. Pages 187–200 *in* Proceedings of the workshop on biological interactions among marine mammals and commercial fisheries in the southeastern Bering Sea. Oct. 18–21, 1983, Anchorage AK. Alaska Sea Grant Report 84-1.

Frost, K. J., R. B. Russell and L. F. Lowry. 1992. Killer whales, *Orcinus orca*, in the southeastern Bering Sea: recent sightings and predation on other marine mammals. Marine Mammal Science 8(2):110–119. http://www.researchgate.net/profile/Lloyd_Lowry/publication/241688276_KILLER_WHALES_ORCINUS_ORCA_IN_THE_SOUTHEASTERN_BERING_SEA_RECENT_SIGHTINGS_AND_PREDATION_ON_OTHER_MARINE_MAMMALS/links/0c96051ca0aefc8a40000000.pdf.

Grebmeier, J. M., J. E. Overland, S. E. Moore, E. V. Farley, E. C. Carmack, L. W. Cooper, K. E. Frey, J. H. Helle, F. A. McLaughlin and S. L. McNutt. 2006. A major ecosystem shift in the northern Bering Sea. Science 311(5766):1461–1464.

Heide-Jørgensen, M., K. Laidre, D. Borchers, T. Marques, H. Stern, and M. Simon. 2010. The effect of sea-ice loss on beluga whales (*Delphinapterus leucas*) in West Greenland. Polar Research 29:198–208. doi: 10.1111/j.17518369.2009.00142.x.

Helker, V. T., B. M. Allen and L. A. Jemison. 2015. Human-caused injury and mortality of NMFS-managed Alaska marine mammal stocks, 2009–2013. U.S. Deparment of Commerce, NOAA Technical Memorandum NMFS-AFSC-300, 94 p. doi:10.7289/V50G3H3M. http://www.afsc.noaa.gov/Publications/AFSC-TM/NOAA-TM-AFSC-300.pdf.

Higdon, J. W. and S. H. Ferguson. 2009. Loss of Arctic sea ice causing punctuated change in sightings of killer whales (*Orcinus orca*) over the past century. Ecological Applications 19(5):1365–1375.

Hobbs, R. C. and J. M. Waite. 2010. Abundance of harbor porpoise (*Phocoena phocoena*) in three Alaskan regions, corrected for observer errors due to perception bias and species misidentification, and corrected for animals submerged from view. Fishery Bulletin 108(3):251–267.

Ivashchenko, Y. V. and P. J. Clapham. 2012. Soviet catches of bowhead (*Balaena mysticetus*) and right whales (*Eubalaena japonica*) in the North Pacific and Okhotsk Sea. Endangered Species Research 18: 201–217.

Ivashchenko, Y. V., P. J. Clapham and R. L. Brownell, Jr. 2013. Soviet catches of whales in the North Pacific: revised totals. Journal of Cetacean Research and Management 13(1): 59–71.

Ivashchenko, Y. V. and P. J. Clapham. 2014. Too much is never enough: the cautionary tale of Soviet illegal whaling. Marine Fisheries Review 76(1-2):1–21.

Jensen, C. M., E. Hines, B. A. Holzman, T. J. Moore, J. Jahncke and J. V. Redfern. 2015. Spatial and temporal variability in shipping traffic off San Francisco, California. Coastal Management 43(6):575–588. http://dx.doi.org/10.1080/08920753.2015.1086947

Kajimura, H. and T.R. Loughlin. 1988. Marine mammals in the oceanic food web of the eastern subarctic Pacific. Bulletin of the Ocean Research Institute University of Tokyo 26(II):187–223.

Kajimura, H., C. H. Fiscus and R. K. Stroud. 1980. Food of the Pacific white-sided dolphin, *Lagenorhynchus obliquidens*, Dall's porpoise, *Phocoenoides dalli*, and northern fur seal, *Callorhinus ursinus*, off California and Washington; with appendices on size and food of Dall's porpoise from Alaskan waters. U.S. Department of Commerce, NOAA Technical Memorandum NMFS F/NWC-2, 30 p.

Kennedy, A. S., D. R. Salden and P. J. Clapham. 2011. First high- to low-latitude match of an eastern North Pacific right whale (*Eubalaena japonica*). Marine Mammal Science 28(4):E539–E544. doi: 10.1111/j.1748-7692.2011.00539.x.

Laidre, K. L., I. Stirling, L. Lowry, Ø. Wiig, M. P. Heide-Jørgensen and S. Ferguson. 2008. Quantifying the sensitivity of Arctic marine mammals to climate-induced habitat change. Ecological Applications 18(2):S97-S125.

Leatherwood, S., A. E. Bowles and R. R. Reeves. 1983. Aerial surveys of marine mammals in the southeastern Bering Sea. U.S. Deparment of Commerce, NOAA, OCSEAP Final Report 42(1986):147–490. http://www.arlis.org/docs/vol1/OCSEAP2/authorindex.html.

Lowry, L. F., R. R. Nelson and K. J. Frost. 1987. Observations of killer whales, *Orcinus orca*, in western Alaska: sightings, strandings, and predation on other marine mammals. Canadian Field Naturalist 101:6–12.

Lowry, L. F., K. J. Frost, A. Zerbini, D. DeMaster and R. R. Reeves. 2008. Trend in aerial counts of beluga whales (*Delphinapterus leucas*) in Bristol Bay, Alaska, 1993–2005. Journal of Cetacean Research and Management 10(3):201–207. http://www.north-slope.org/assets/images/uploads/Lowry_et_al_bbBelugaTrend_2008.pdf.

MacLeod, C. D. 2009. Global climate change, range changes and potential implications for the conservation of marine cetaceans: a review and synthesis. Endangered Species Research 7(2):125–136.

Maschner, H. D. 2004. Traditions past and present: Allen McCartney and the Izembek Phase of the western Alaska Peninsula. Arctic Anthropology 41(2): 98–111.

McCartney, A. P. 1974. Prehistoric cultural integration along the Alaska Peninsula. Anthropological Papers of the University of Alaska 16(1):59–84.

McCartney, A. P. 1995. Whale size selection by precontact hunters of the North American Western Arctic and subarctic. Pages 83–108 *in* A. P. McCartney, editor. Hunting the largest animals: Native whaling in the western Arctic and subarctic. Studies in Whaling no. 3, occasional publications no. 36, University of Alberta, Canada.

Mizroch S. A. and D. W. Rice. 2006. Have North Pacific killer whales switched prey species in response to depletion of the great whale populations? Marine Ecology Progress Series 310:235–246.

Mizroch S. A. and D. W. Rice. 2013. Ocean nomads: distribution and movements of sperm whales in the North Pacific shown by whaling data and discovery marks. Marine Mammal Science 29(2):E136-E165. DOI:10.1111/j.1748-7692.2012.00601.x.

Mizroch, S. A., D. Rice, D. Zwiefelhofer, J. Waite and W. Perryman. 2009. Distribution and movements of fin whales in the North Pacific Ocean. Mammal Review 39(3):193–227.

Moore, S. E., K. M. Wynne, J. C. Kinney, and J. M. Grebmeier. 2007. Gray whale occurrence and forage southeast of Kodiak, Island, Alaska. Marine Mammal Science 23(2):419–428.

Moore, S. E., J. M. Waite, N. A. Friday and T. Honkalehto. 2002. Cetacean distribution and relative abundance on the central–eastern and the southeastern Bering Sea shelf with reference to oceanographic domains. Progress in Oceanography 55(1):249–261.

Moore, S. E., R. R. Reeves, B. L. Southall, T. J. Ragen, R. S. Suydam and C. W. Clark. 2012. A new framework for assessing the effects of anthropogenic sound on marine mammals in a rapidly changing Arctic. BioScience, 62(3):289–295.

Muto, M. M., V. T. Helker, R. P. Angliss, B. A. Allen, P. L. Boveng, J. M. Breiwick, M. F. Cameron, P. J. Clapham, S. P. Dahle, M. E. Dahlheim, B. S. Fadely, M. C. Ferguson, L. W. Fritz, R. C. Hobbs, Y. V. Ivashchenko, A. S. Kennedy, J. M. London, S. A. Mizroch, R. R. Ream, E. L. Richmond, K. E. W. Shelden, R. G. Towell, P. R. Wade, J. M. Waite and A. N. Zerbini. 2016. Alaska marine mammal stock assessments, 2015. U.S. Department of Commerce, NOAA Technical Memorandum NMFS-AFSC-323, 300 p. doi:10.7289/V5/TM-AFSC-323.

NMFS (National Marine Fisheries Service). 2015. North Pacific groundfish and halibut observer program 2014 annual report. National Oceanic and Atmospheric Administration, 709 West 9th Street. Juneau, Alaska 99802. 101 p. https://alaskafisheries.noaa.gov/sites/default/files/annualrpt2014.pdf (November 2015).

Nerini, M. 1984. A review of gray whale feeding ecology. Pages 423–448 *in* M. L. Jones, S. Swartz and S. Leatherwood, editors. The gray whale: *Eschrichtius robustus*. Academic Press Inc., New York, NY.

Norman, S. A., C. E. C. Goertz, K. A. Burek, L. T. Quakenbush, L. A. Cornick, T. Romano, T. Spoon, W. Miller, L. A. Beckett and R. C. Hobbs. 2012. Seasonal hematology and serum chemistry of wild beluga whales (*Delphinapterus leucas*) in Bristol Bay, Alaska. Journal of Wildlife Diseases 48(1):21–32.

Palka, D. L. and P. S. Hammond. 2001. Accounting for responsive movement in line transect estimates of abundance. Canadian Journal of Fisheries and Aquatic Sciences 58(4):777–787.

Pemberton, M. 2011. Alaska killer whale in river was pregnant, vets say. Huffington Post. http://www.huffingtonpost.com/2011/10/12/alaska-river-killer-whale-pregnant_n_1007718.html (November 2015).

Perez, M. A. 2003. Compilation of marine mammal incidental take data from the domestic and joint venture groundfish fisheries in the U.S. EEZ of the North Pacific, 1989–2001. U.S. Department of Commerce, NOAA Technical Memorandum NMFS-AFSC-138, 145 p.

Perez, M. A. 2006. Analysis of marine mammal bycatch data from trawl longline and pot groundfish fisheries of Alaska 1998–2004, defined by geographic area, gear type, and target groundfish species. U.S. Department of Commerce, NOAA Technical Memorandum NMFS-AFSC-167, 194 p.

Peterson, M. J., F. Mueter, D. Hanselman, C. Lunsford, C. Matkin and H. Fearnbach. 2013. Killer whale (*Orcinus orca*) depredation effects on catch rates of six groundfish species: implications for commercial longline fisheries in Alaska. ICES Journal of Marine Science 70(6):1220–1232.

Pike, G. C. 1962. Migration and feeding of the gray whale (*Eschrichtius gibbosus*). Journal of the Fisheries Board of Canada 19(5): 815–838.

Punt, A. E. and P. R. Wade. 2012. Population status of the eastern North Pacific stock of gray whales in 2009. Journal of Cetacean Research and Management 12(1):15–28.

Reeves, R. R. and T. D. Smith. 2006. A taxonomy of world whaling. Pages 82–101 *in* J. A. Estes, D. P. DeMaster, D. F. Doak, T. M. Williams and R. L. Brownell, Jr., editors. Whales, whaling, and ocean ecosystems. University of California Press, Berkeley, CA.

Reeves, R. R., S. Leatherwood, S. A. Karl and E. R. Yohe. 1985. Whaling results at Akutan (1912–39) and Port Hobron (1926–37), Alaska. Report of the International Whaling Commission 35, 441–457.

Rice, D. W. 1989. Sperm whale: *Physeter macrocephalus*. Pages 177–233 *in* S. H. Ridgway and R. Harrison, editors. Handbook of marine mammals. Volume 4. River dolphins and the larger toothed whales. Academic Press, New York, NY.

Rice, D. W. and A. A. Wolman. 1971. The life history and ecology of the gray whale (*Eschrichtius robustus*). The American Society of Mammalogists, Special Publication No. 3, 141 p.

Rugh, D. 1984. Census of gray whales at Unimak Pass, Alaska, November–December 1977–1979. Pages 225–248 *in* M. L. Jones, S. Swartz and S. Leatherwood, editors. The gray whale: *Eschrichtius robustus*. Academic Press Inc., New York, NY.

Scarff, J. E. 1986. Historic and present distribution of the right whale (*Eubalaena glacialis*) in the eastern North Pacific south of 50°N and east of 180°W. Report of the International Whaling Commission (Special Issue 10):43–63.

Shelden, K. E. W., S. E. Moore, J. M. Waite, P. R. Wade and D. J. Rugh. 2005. Historic and current habitat use by North Pacific Right Whales *Eubalaena japonica* in the Bering Sea and Gulf of Alaska. Mammal Review 35:129–155.

Shelden, K. E. W., B. A. Agler, J. J. Brueggeman, L. A. Cornick, S. G. Speckman and A. Prevel-Ramos. 2014. Harbor porpoise, *Phocoena phocoena vomerina*, in Cook Inlet, Alaska. Marine Fisheries Review 76(1-2):22–51.

Sigler, M. F., C. R. Lunsford, J. M. Straley and J. B. Liddle. 2008. Sperm whale depredation of sablefish longline gear in the northeast Pacific Ocean. Marine Mammal Science 24(1):16–27.

Stafford, K. M., S. E. Moore, M. Spillane and S. Wiggins. 2007. Gray whale calls recorded near Barrow, Alaska, throughout the winter of 2003–04. Arctic 60(2):167–172.

Stafford, K. M., S. E. Moore, P. J. Stabeno, D. V. Holliday, J. M. Napp and D. K. Mellinger. 2010. Biophysical ocean observation in the southeastern Bering Sea. Geophysical Research Letters 37. DOI 10.1029/2009GL040724.

Townsend, C. H. 1935. The distribution of certain whales as shown by logbook records of American whaleships. Zoologica NY 19:1–50.

Wade, P. R., A. Kennedy, R. LeDuc, J. Barlow, J. Carretta, K. Shelden, W. Perryman, R. Pitman, K. Robertson, B. Rone, J. Carlos Salinas, A. Zerbini, R. L. Brownell, Jr. and P. Clapham. 2011. The world's smallest whale population. Biology Letters 7:83–85.

Waite, J. M. and K. E. W. Shelden. In review. The northern extent of Pacific White-Sided Dolphin, *Lagenorhynchus obliquidens*, distribution in the eastern North Pacific. Submitted to Northwestern Naturalist.

Witteveen, B. H., G. A. J. Worthy, R. J. Foy and K. M. Wynne. 2011. Modeling the diet of humpback whales: An approach using stable carbon and nitrogen isotopes in a Bayesian mixing model. Marine Mammal Science 28(3):E233–250. http://dx.doi.org/10.1111/j.1748-7692.2011.00508.x.

Yarborough, L. F. 1995. Prehistoric use of cetacean species in the northern Gulf of Alaska. Pages 63–82 *in* A. P. McCartney, editor. Hunting the largest animals: Native whaling in the western arctic and subarctic. Studies in Whaling No. 3, occasional publications no. 36, University of Alberta, Canada.

Yarborough, L. F. 2000. Prehistoric and early historic subsistence patterns along the north Gulf of Alaska Coast. Ph.D. dissertation, University of Wisconsin-Madison, 434 p.

# 14

# PINNIPEDS OF BRISTOL BAY

Anne Hoover-Miller, Pacific Rim Research

## INTRODUCTION

Pinnipeds, *feather-footed* marine mammals inhabiting Bristol Bay, include phocid (true or earless) seals, otariids (eared seals) and odobenids (walruses). The eight species present in Bristol Bay forage on abundant and diverse concentrations of fish and invertebrates and use ice and terrestrial haul-out sites as platforms for giving birth, caring for young, molting, and resting between foraging trips.

Five species of phocid seals associate with Bristol Bay. Harbor seals (*Phoca vitulina*), are the predominant pinniped in Bristol Bay and are year-around residents using shoreline haulouts throughout the bay. The other four species are pagophilic (ice-loving) species, and include the spotted (*P. largha*), ringed (*P. hispida*), bearded (*Erignathus barbatus*), and ribbon seals (*Histriophoca fasciata*), which use pack and shore-fast ice that advances into Bristol Bay in the winter and spring. Two species of otariids, Steller (northern) sea lions (*Eumetopias jubatus*) and northern fur seals (*Callorhinus ursinus*), also are present. Steller sea lions occupy a rookery and haulouts near Port Moller, but seasonally use several haulouts in northern Bristol Bay. Northern fur seals are not residents but forage and occasionally rest at shoreline haulouts, especially near Unimak Pass. Pacific walruses (*Odobenus rosmarus*) are iconically associated with ice and haul out on sea ice in winter and spring where they breed and give birth. During the summer, most move north with the retreating ice but thousands of males remain in Bristol Bay, occupying several terrestrial haulouts. The diversity of life history strategies exhibited by these marine mammals attests to the ecological breadth and dynamics of Bristol Bay on which they rely.

All pinnipeds undergo a common series of events each year that vary somewhat in timing and duration (Figure 14.1). In spring or early summer, adults begin to reproduce. Females give birth to offspring from March through July. Phocid seals have a relatively short nursing period, lasting roughly a month, after which the pup is weaned and parental care ceases. Females breed with males near the time of weaning or shortly thereafter. Walruses and Steller sea lions exhibit extended care of offspring that lasts one to two years or longer. Female Steller sea lions breed about 11 days after giving birth. Northern fur seals do not breed in Bristol Bay. Like Steller sea lions they breed about a week after giving birth but limit care of their pups to about four months. Breeding for female walruses is less frequent than other pinnipeds, occurring once every two to three years, or longer. Breeding occurs after weaning previous calves and, unlike other pinnipeds, months prior to the calving season. Like humans, embryonic development of young takes about eight to nine months, or in the case of walruses 11 months, but pinniped births occur at the same time each year. To accommodate the length of pregnancy beyond the period of gestational growth, females undergo embryonic diapause, where implantation of the newly fertilized egg in the uterus and subsequent embryonic development is suspended for about 4 months, which allows females to complete their molt and, in the case of phocid seals, recover from the energetic demands of pup rearing.

Once each year, pinnipeds molt, replacing their old coat for new pelage (fur). During the molt, pinnipeds spend additional time hauled out on land or ice to warm their skin which promotes hair growth. At this time foraging is reduced. For adults, the molt follows breeding and the combination of reproductive events and the molt is energetically demanding. By the end of molt, adults are often in their poorest condition of the year despite the abundance of prey available

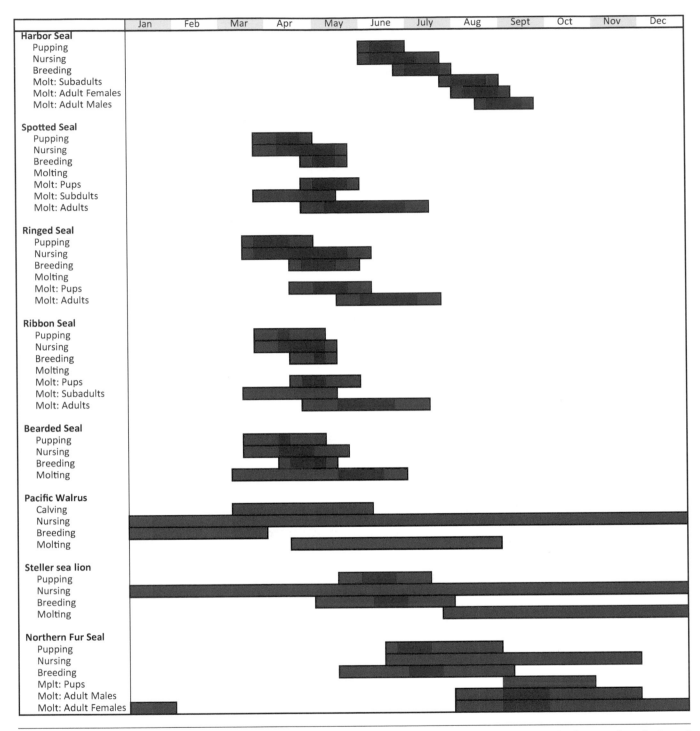

**Figure 14.1**    Timeline of annual life history events (parturition, care of dependent young, breeding, and molting) of pinnipeds in the Bristol Bay region. Yellow marks the time range in which events occur; orange marks periods of time when events are most frequent.

during the spring and summer. Subadult pinnipeds are less constrained than adults in movements and activities. They also undergo an annual molt during the summer, but are able to spend the remainder of the year foraging, some spending little time at haulouts.

The seasonal and interannual extent of ice strongly influences the distribution and abundance of pinnipeds in Bristol Bay. In a warming climate regime, extensive sea ice coverage that reaches Bristol Bay during the winter and spring is forecasted to become less frequent (Boveng et al. 2009, Ray et al. 2016). As the maximum extent of ice retreats northward, the composition of species present in Bristol Bay, their behaviors, and habitat use will change (Ray et al. 2016).

# Habitats

Bristol Bay includes diverse habitats that dramatically change between winter and summer. Marine ecological zones, or ecoregions, in Alaska have been described by Piatt and Springer (2007) based on distributions of birds and associated forage fish (Figure 14.2) which are influenced by (1) seafloor bathymetry and composition, (2) marine characteristics (salinity, temperature, and chemical composition) of local water masses and currents, and (3) the timing and magnitude of plankton blooms that fuel the food web. Ecological zones within Bristol Bay have been associated with the depth of the bay with an inner or coastal domain from shoreline to 50 m, a middle domain from 50–100 m, and outer domain beyond 100 m. Most of Bristol Bay is included in the coastal domain, characterized by shallow waters less than 50 m depth. The Alaska Coastal Current, which flows from the Gulf of Alaska, follows the 50 m depth contour across the mouth of Bristol Bay and northwestward toward the Bering Strait. Circulation resulting from the Alaska Coastal Current along the 50 m depth contour generates upwelling in central Bristol Bay that fuels mixing of nutrients and biological productivity. Within inner Bristol Bay, although tides are strong, currents are generally weak, estuarine, and strongly influenced by river discharge and coastal runoff (National Marine Fisheries Service (NMFS) 2013).

Near the head of Bristol Bay, the tidal range can exceed 9.1 m (30 feet), and quickly wash low-lying mudflats used by harbor seals for haulouts. Tidal ranges diminish toward the mouth of Bristol Bay. In Port Moller, the tidal range is about 4.6 m (15 feet) while in Platinum, about 45 km (28 miles) northeast of Cape Newenham, the maximum tidal range is about 3.4 m (11 feet) (https://tidesandcurrents.noaa.gov/tide_predictions.html). Bristol Bay's seafloor is relatively smooth and gently slopes on roughly a 2.4% grade. The seabed is covered by sediments that provide habitat for bottom-dwelling

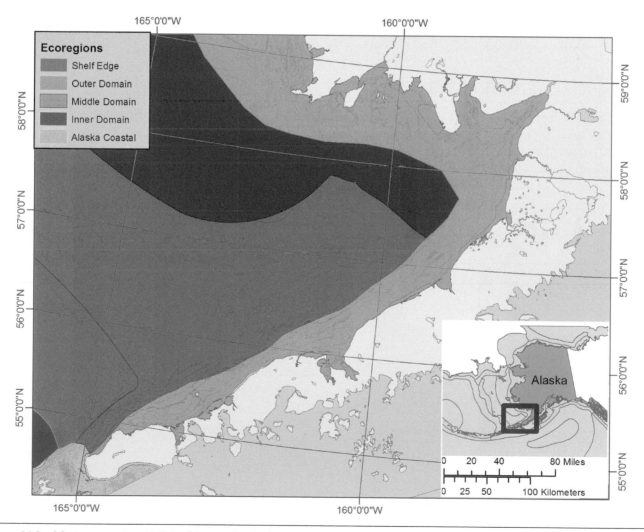

**Figure 14.2**  Marine ecoregions in Bristol Bay. Darker bathymetric contour under the coastal and inner domains represent the 50 m depth contour. *Source credit: Piatt and Springer 2007.*

(benthic) fish and invertebrates, including clams, worms, amphipods, shrimps, crabs, and other organisms consumed by bearded seals, walruses, and other pinnipeds. Nearshore, sediments consist of poorly sorted gravel sands which transition to well-sorted fine sands in the central bay where walruses feed (Sharma et al. 1972).

Rivers and estuaries provide biologically rich habitats that attract foraging marine mammals in the spring, summer, and fall. Smelts and salmon rely on rivers and associated estuarine habitats when reproducing. Pinnipeds often concentrate at rivers and associated deltas as fish move upstream to reproduce as well as when salmon smolts leave rivers for the sea. Herring concentrate and spawn on shorelines in northern Bristol Bay and along the north side of the Alaska Peninsula in the spring, attracting diverse aggregations of foraging marine birds and mammals.

Ice begins to form along the south facing shorelines during the early fall and can be found in the northern sections of Bristol Bay in October and November. As ice forms from shore, northerly winds push it southward, while new ice forms near the coast. The ice edge continues moving southward until it reaches its maximum extent in late March or April at which time it begins a rapid retreat. The position of the southernmost ice edge is set by a balance of wave action, winds, melting rate, and southern growth of new ice. Exposure to sea and swell causes ice near the open water to undergo rafting and ridging creating floes approximately 10–20 m in diameter and 2–5 m thick. Further into the pack, floes are subjected to less swell and compression from the sea. Protection from sea conditions allows dimensions of the floes to be larger but the thickness of the ice is reduced. Although ice-associated seals can be found on ice throughout Bristol Bay, greatest concentrations have been located near the mouth of Bristol Bay, where seals exhibit species-specific ice preferences that influence their distribution and produce areas of denser concentrations of each species (Braham et al. 1984).

Shoreline haulouts used by harbor seals, spotted seals, Steller sea lions and walruses are comprised of mud and sand bars, rocky beaches, and offshore rocks and islets. The availability of haul-out sites can be strongly influenced by tidal cycles and weather conditions. At locations with steep beaches or access to elevated haulouts—such as on large rocks or islets favored by Steller sea lions—tides and surf have less effect than on low-lying mudflats that are prevalent among many of the harbor seal haulouts along the north side of the Alaska Peninsula.

## PINNIPED IDENTIFICATION

The first step in identifying pinnipeds is determining whether they are otariids (sea lions and fur seals), phocids (true seals), or odobenids (walruses) (Table 14.1). Several features help distinguish them, including their shape, whether they can walk on all four flippers, whether they have external ear pinnae (outer ear covers), and how they propel themselves through the water. Once the family is identified, body characteristics and life history strategies help distinguish each species.

**Table 14.1**   Identifying characteristics of the three families of pinnipeds and representatives of each family present in Bristol Bay

| Pinniped Family: | Otariid | Phocid | Odobenid |
|---|---|---|---|
| | | | |
| **Identifying Characteristics:** | | | |
| External Ear Pinnae | Yes | No | No |
| Walk on four flippers | Yes | No | Yes |
| Propel through water | Front Flippers | Hind Flippers | Hind Flippers |
| **Species in Bristol Bay:** | | | |
| | Steller Sea Lion<br>Northern Fur Seal | Harbor Seal<br>Spotted Seal<br>Ringed Seal<br>Bearded Seal<br>Ribbon Seal | Pacific Walrus |

## Identifying Characteristics of Pinnipeds Associated with Bristol Bay

### Harbor Seal

> Pinniped family: Phocidae (true seal)
> Common name: Harbor seal
> Scientific name: *Phoca vitulina*
> Alaska Native names: *isuwiq* (Alutiiq), *qasigiaq* (Inupiaq)

Harbor seals (Figure 14.3) are medium-sized phocid seals that express some sexual dimorphism in size. In the eastern Aleutian Islands harbor seals are large. Adult males have a mean length of 176 cm (69 in) and weight of 95 kg (209 lbs) compared with adult females that average 162 cm (64 in) and weigh 66 kg (145 lbs). At birth, pups weigh about 11 kg (24 lbs) and are about 78 cm (31 in) long (Burns and Gol'tsev 1984).

The color of recently molted harbor seals ranges from shades of silver-white to dark grey in two basic patterns that are retained through life. The dark phase has a dark grey background with light rings and light and black blotches and spots. The light phase has light sides and belly with black blotches or spots. The back often is grey with light and dark blotches and light rings. Intermediate patterns also are expressed. After molting, the coloration gradually fades to shades of tan and brown. Unlike other phocids seals in the Bering sea, nearly all harbor seal pups molt their long, fluffy, off-white colored, lanugo pelage prior to birth and are born with an adult-like coat, similar in appearance to recently molted harbor seals (Burns and Gol'tsev 1984, Hoover-Miller 1994).

**Figure 14.3**    Resting harbor seals with examples of dark and light-phase coloration. *Photo credit: Alaska Department of Fish and Game.*

## Spotted Seal

Pinniped family: Phocidae (true seal)
Common name: Spotted seal, largha seal
Scientific name: *Phoca largha*
Alaska Native names: *issuriq* (central Yupik), *gazigyaq* (St Lawrence Island Yupik), and *qasigiaq* or *kasegaluk* (northern Inupiaq)

Spotted seals (Figure 14.4) are medium-sized phocid seals that look nearly identical to harbor seals. In the Bering sea, adult females typically weigh between 65–115 kg (143–253 lbs) and are 151–169 cm (59–67 in) long, while adult males weigh between 85–110 kg (187–243 lbs) and are 161–176 cm (63–69 in) long. At birth, pups weigh between 7–12 kg (15–26 lbs) and are 75–92 cm (30–36 in) long. By the end of nursing, pups may more than triple their birth weight to 30 kg (66 lbs) or more (Boveng et al. 2009).

Pelage markings are similar to light-phase harbor seals. Pelage usually consists of a light-colored background, often with a darker grey back. Dark grey and black spots typically are scattered on the body. The belly often is silvery-white with fewer spots, although some young seals have numerous spots on the belly. Older seals develop more vivid and contrasting spottiness. Many young seals have a broad, dark band that extends along the middle of the back from the head to tail. With age, the band becomes more mottled, with the appearance of distinct oval rings then fades in older animals. Male seals may have brighter and more contrasted spot patterns than females (Boveng et al. 2009).

Unlike most harbor seals, spotted seal pups are born with white- or cream-colored lanugo pelage (Figure 14.5) about an inch long, which provides insulation throughout the nursing period while pups develop an insulating layer of blubber. Lanugo is usually shed around the time of weaning and is replaced by a short, smooth adult-like coat (Boveng et al. 2009).

**Figure 14.4**    Spotted seal on pack ice. *Photo credit: Michael Cameron, NOAA.*

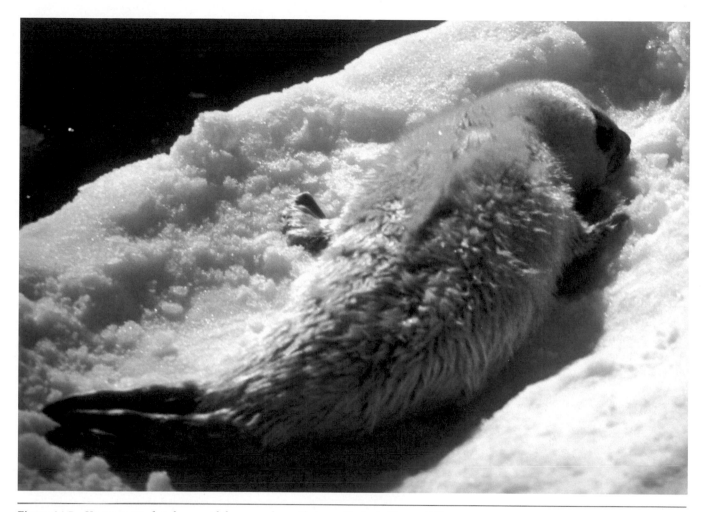

**Figure 14.5** Young spotted seal pup with lanugo pelage. *Photo credit: Anne Hoover-Miller.*

## Distinguishing Spotted Seals from Harbor Seals

Harbor seals and spotted seals are the most abundant seals in Bristol Bay and are closely related, probably diverging from a common ancestor about one million years ago. Spotted seals are difficult to distinguish from light-phase harbor seals except for their genetics, subtle differences in skull characteristics and life history attributes associated with the spotted seal's affiliation with ice. Shaughnessy and Fay (1977) distinguish light-phase harbor seals from spotted seals by noting four characteristics that often, but not always, differ between each species:

1. The grey saddle of harbor seals are often made up of closely packed, sometimes overlapping, blackish spots and blotches, of which a few are often set off by contrasting whitish rings;
2. Spots on the sides and belly of harbor seals are often very widely spaced and tend to be palest on the belly;
3. Spots of harbor seals are often larger than on spotted seals and tend to coalesce, forming irregular blotches; and
4. There is a tendency, at all ages, for the areas around the eyes and snout of harbor seals to be pale.

They considered the first and last characteristics to be most indicative of light-phase harbor seals.

## Ribbon Seal

Pinniped family: Phocidae (true seal)
Common name: Ribbon Seal (striped seals, banded seals)
Scientific name: *Histriophoca fasciata*
Alaska Native names: *qasruliq* (central Yupik), *kukupak* (St. Lawrence Island Yupik), *iglagayax* (Unangan), *qaigulik* (northern Inupiaq)

Ribbon seals (Figure 14.6) are medium-sized phocids, sexually dimorphic in coloration but not in size. Adults typically attain lengths of approximately 150–175 cm (4.9–5.7 ft) and weigh about 70–90 kg (154–198 lbs). Newborn pups are approximately 86 cm (34 in) long and weigh about 9.5 kg (21 lbs). The head of ribbon seals appears rounded with a short snout and large eyes (Boveng 2008).

Adults show a striking banded pattern with four light-colored ribbons on a background of darker pelage. One ribbon encircles the neck and nape, another encircles the trunk (near the hips), and two lateral ovals encircle each fore flipper, from the lower neck to the midsection. Adult males have bright white ribbons on a dark brown to black pelage. Ribbon patterns vary in shape and width and sometimes become fused. Adult females are brown to silvery grey with ribbons similar to males but exhibiting less contrast. Pups are born with a thick, wooly, white lanugo coat that is molted after 3–5 weeks. Once molted, their coat is counter-shaded dark grey on the back and light grey toward the belly. Banding is faint or absent in young seals and gradually develops over three years with each successive molt (Boving 2008).

Unlike the other Bering sea phocids seals, that move across the ice by bouncing on their belly, with or without the aid of their fore flippers, ribbon seals move across the ice very rapidly in a serpentine motion with alternating front claws digging into the ice while the head and hips move in a side to side motion (Boveng et al. 2008).

**Figure 14.6**    Ribbon Seal mother with pup. *Photo credit: Anne Hoover-Miller.*

# Ringed Seal

Pinniped family: Phocidae (true seal)
Common name: Ringed seal
Scientific name: (*Phoca hispida* or *Pusa hispida*)
Alaska Native names: *natchiq* (Inupiat), *niknik* (Yupik), *nayir* (Cup'ig, Nunivak Island)

Ringed seals (Figure 14.7) are the smallest of all pinnipeds and are not sexually dimorphic in size or coloration. Adults are typically about 1.5 m (5 ft) in length and weigh about 70 kg (154 lbs). At birth, pups are about 60–65 cm (25 in) long and weigh about 4.5 kg (10 lbs). Ringed seals look similar to a small harbor seal but have relatively shorter snouts and larger eyes (Kelly et al. 2010, Garlich-Miller et al. 2011, Mac Donald and Winfred 2008).

Ringed seals vary greatly in coloration but, like harbor seals, have two basic variations, a dark and light phase. The light phase consists of a dark grey saddle with light superimposed rings. The side and belly surfaces are light colored with or without darker spots. The dark phase has a dark background with light rings overall. Ringed seals have strong claws that are used for digging and maintaining breathing holes in ice more than 7 feet thick and for excavating subnivean lairs for protection from weather and predators (Kelly et al. 2010).

**Figure 14.7**  Adult ringed seal. *Photo credit: John Jansen, AFSC, NOAA Fisheries.*

## Bearded Seal

Pinniped family: Phocidae (true seal)
Common name: Bearded seal (square flipper)
Scientific name: *Erignathus barbatus nauticus*
Alaska Native names: *mukluk* (Yupik), *oogruk* (Inupiat)

Bearded seals (Figure 14.8) are not sexually dimorphic. Adults, the largest phocids seal resident in the Bering Sea, average 2.1–2.4 m (7 ft) long and weigh up to 360 kg (793 lbs). Pups weigh an average 33.6 kg (74 lbs) and are 131 cm (4 ft) long. Pups are born with a subcutaneous layer of fat and grow fast at an estimated 2.8–3.6 kg/d (6-8 lbs/d). Pups are weaned 12–18 days and weigh about 85 kg (187 lbs).

Mature bearded seals have a distinguishing shape with a wide girth and small head in proportion to body size. Both sexes and ages have relatively long, bushy, vibrissae, square-shaped fore flippers and four teats, rather than two that are present in the other phocids seals (Cameron et al. 2010).

The coat of bearded seals is gray to brown without distinct markings, although the back may be slightly darker. The coat is comprised of short straight hairs. Some individuals may have yellow or rust colors on the head and fore flippers, which is thought to be the result of foraging on the seafloor and trapping oxidized iron compounds in the hair (Cameron et al. 2010).

Unlike other Arctic phocids, which initially develop a nearly white lanugo, bearded seal pups have a grey-brown wavy lanugo. Some pups completely shed their lanugo coat prior to birth while others are completely covered with lanugo. At the time of weaning, all lanugo is shed and the pup's pelage more closely resembles adult pelage. Molted pups are countershaded with blue-gray to gray-brown above and a lighter ventral surface. Lighter marking may be present on the back crown and face that can form a T shape on the forehead of darker fur (Cameron et al. 2010).

**Figure 14.8**    Adult bearded seal. *Photo credit: Gavin Brady, NOAA.*

## Walrus

Pinniped family: Odobenidae (walruses)
Common name: Pacific walrus
Scientific name: *Odobenus rosmarus divergens*
Alaska Native names: *aivik* (Inupiaq), *asveq* or *kaugpak* (Yupik), *ayveq* (St. Lawrence Island Yupik), amak or *amaghak* (Aleut), *amgaada* (Unangan)

Pacific walruses (Figure 14.9) are among the largest pinnipeds and differ in size and appearance among sexes. Adult females average 270 cm (8.8 ft) and weigh 830 kg (1,830 lbs), but can reach lengths of up to 305 cm (10 ft) and weigh 1,134 kg (2,500 lbs). Males grow for a longer period of time than females and average 320 cm (10.5 ft) in length and weigh 1,210 kg (2,668 lbs) but can reach 370 cm (12 ft) long and weigh as much as 1,814 kg (4,000 lbs) pounds. At birth, calves are about 65 kg (143 lbs) and 113 cm (44 in) long. Calves shed their soft white lanugo coat before birth and are born with a short coat of dark grey fur. Their grey coat is replaced two or three months later with a juvenile coat of brown fur. Adults have short sparse tawny pelage which they molt annually during the summer. Walrus flippers are bare of hair. The skin of adult males often has large nodules that are absent in females and subadults (Garlich-Miller et al. 2011).

Walrus lack outer ear covers (pinnae). Their nostrils point upward on their squarish snouts that appear to protrude from their robust neck. Male walruses have more massive, block-shaped heads than females. Both males and females have tusks which are enlarged canine teeth. Tusks become externally visible in two-year-old walrus and grow throughout their life. Tusks of males are more massive and straighter than tusks of females which are slimmer and more curved. Males use tusks in threat displays and fights in order to establish dominance. Tusks are used by both sexes to establish and defend positions when hauled out and for attachment to ice flows when resting in the water or hauling out (Garlich-Miller et al. 2011).

**Figure 14.9** Adult female walrus. *Photo credit: Anne Hoover-Miller.*

## Steller Sea Lion

Pinniped family: Otariidae (eared seals)
Common name: Steller sea lion (also northern sea lion)
Scientific name: *Eumetopias juatus*
Alaska Native names: *qawax* (Aleut), *wiinaq* (Alutiiq), *uginaq*, sometimes *apakcuk* (central Yup'ik), *qawax* (Unangan)

Steller sea lions (Figure 14.10) are the largest otariid and differ in size and shape between sexes. Adult males average 282 cm (9.3 ft) (maximum 325 cm (10.7 ft)) in length while females are 228 cm (7.5 ft) (maximum 290 cm (9.5 ft)). Males weigh an average 566 kg (1,248 lbs) (maximum 1,120 kg (2,469 lbs)) compared to females which weigh 263 kg (580 lbs) (maximum 350 kg (772 lbs)). The chest and neck of adult males are massive and muscular, shielded by long, coarse hair on their chest and shoulders, especially during the breeding season. Females and juveniles are slimmer and look similar to each other. Skin without fur, present on the flippers and belly, is black. Newborn pups are about 1 m (3.3 ft) long and weigh 16–23 kg (35-51 lbs). Pups are born with a wavy dark brown, thick coat that molts to lighter brown, adult-like pelage after three months. By the end of their second year, pups have taken on the same pelage color as adults (NMFS 2008, Hoover 1988).

**Figure 14.10**    Adult males, females, and pups Steller sea lions on rookery. *Photo credit: Ocean Alaska Science and Learning Center.*

## Northern Fur Seal

Pinniped family: Otariidae (eared seals)
Common name: Northern fur seal
Scientific name: *Callorhinus ursinus*
Alaska Native name: *laaqudax* (Unangan)

Northern fur seals (Figure 14.11) are strongly sexually dimorphic. Adult Males weigh 200–275 kg (450–600 lbs) and are 2 m (6.5 ft) long, while females weigh 36–50 kg (65–110 lbs) and are 1.3 m (4.2 ft). The head of northern fur seals is rounded with large eyes and a short, pointed snout; their outer ear pinnae is located low on the neck. The hind flippers of northern fur seals are the longest among otariids and can measure up to one-fourth of their total body length. Adults have long white vibrissae, while young have black vibrissae (Gentry 1998, Wynne and Folkens 2012).

The bodies of northern fur seals are covered in thick fur. The fur is made up of guard hairs covering permanent dense brown underfur. The underfur typically is not visible in dry animals but appears as brown streaks in wet animals. Due to its density (46,500 hairs per square centimeter), the underfur always remains dry and provides highly efficient insulation from the cold water. Long guard hairs form the outer coat are molted each year. Northern fur seal flippers are black, hairless, and assist in regulating the animal's body temperature by dispersing heat into the air or water, or gaining heat through absorbing sunlight while on land or when held above the water while at sea (Gentry 1998, Wynne and Folkens 2012).

Males have a stocky body that contrasts with its small head and short snout. The males vary from black to reddish in color and generally develop a lighter mane over their shoulders as they age. Adult males develop a sagittal crest (a bony ridge along the top of the skull) and thick fur on the top of the head that elevates the crown of the head relative to the snout. Females are brown to grey with a lighter grey, silver, or cream on their throats, chest, and stomach, and a buff-colored belly and neck with light markings on the face. Pups are born black and molt into juvenile pelage when about three months old. When observed at sea, northern fur seals appear very dark and frequently are seen porpoising (swimming with a series of arcing jumps out of the water) or resting on the surface with one or more of their long flippers curved above the water (Gentry 1998, Wynne and Folkens 2012).

**Figure 14.11**   Adult male northern fur seal with females and pups on rookery. *Photo credit: Rolf Ream, NOAA.*

with respect to the condition of females (Temte 1994, Bowen et al 2003). In Bristol Bay, births primarily begin the second week of June and continue through the end of June.

Pups typically swim within an hour of birth, a trait reinforced by tidal encroachment of haul-out sites and disturbances caused by birds preparing to feed on the expelled placenta. The first hours after birth are critical for the development of the mother-pup bond. Olfactory cues are important for mother-pup recognition and individual differences in pup vo-calizations allow mothers to identify their pups above and below the water's surface. Mothers are vigilant of their pups, but disturbances or separations before the mother-pup bond is formed may result in permanent abandonment and sub-sequent death of the pup.

The nursing period lasts three to six weeks—during which time the pup increases weight from about 8–10 kg (18-24 lbs) at birth to roughly 25 kg (55 lbs) at weaning. Weaning is typically abrupt with no overt association between the mother and pup afterward. Shortly after weaning, pups often disperse from their natal area to explore (Small et al. 2005), while the newly independent mothers begin foraging to restore blubber that was depleted by nursing (Cordes and Thompson 2013). Females typically enter estrus within two weeks of weaning their pups and then breed (Bigg 1969).

During the breeding period, males conduct aquatic displays which involve slapping their flippers on the water's surface and blowing streams of bubbles underwater near an established display site. Dominance relationships are established between males though females ultimately choose which males to mate with (Hayes et al. 2004). Neck wounds are fre-quently seen on adult males, likely resulting from aggressive encounters between males. After breeding, female harbor seals undergo embryonic diapause in which the embryo stops developing and remains dormant in the uterus for one and a half to three months while the female completes her annual molt and resumes foraging. The timing of implantation varies and appears to be related to the condition of females, where implantation occurs earlier among females in better body condition (Bowen et al. 2003).

In mid- to late summer, harbor seals undergo their annual molt, where they lose their old pelage and grow new hairs. During the shedding portion of the molt, seals spend additional time at haulouts to warm their skin and promote hair growth, while foraging is reduced. Yearlings typically molt first, followed by subadults, adult females, and then adult males (Daniel et al. 2003).

Harbor seals can see and hear well above and below the water. Although they are color blind, they are sensitive to the blue-green spectrum and are adept at brightness discrimination (Scholtyssek et al. 2014). Seals use broadband click trains, mostly at night or in the absence of light, which may assist with capturing small fish in the dark. Vibrissae are used as tactile receptors when touching objects (Grant et al. 2013) and when making nose to nose contact with other seals. Facial whiskers also serve as receptors that help detect low-frequency vibrations, such as those made by nearby swimming fish (Murphy et al. 2015).

## Pagophilic Seals

Pagophilic seals (spotted, ringed, ribbon, bearded) use sea ice and shore-fast ice for pupping, breeding, molting, and plat-forms for resting between foraging bouts. Ice in the Bering Sea is highly dynamic—forming each winter and retreating each summer. The maximum extent of ice usually extends into Bristol Bay for at least part of the winter. When the ice retreats in spring, most ringed and bearded seals move north with the receding ice. Spotted seals exhibit a mixed strat-egy, some moving north while others shift to terrestrial haulouts during ice-free periods. Ribbon seals adopt a pelagic lifestyle—swimming extensively and rarely hauling out on shore during the ice-free period (Boveng et al. 2008, 2009, Kelly et al. 2010, Cameron et al. 2010). Ice-associated species have adopted different life history attributes and ice prefer-ences that allow them to benefit from enhanced spring productivity in the Bering Sea while reducing direct competition through ice and foraging preferences (Burns et al. 1981, Cooper et al. 2009).

## Spotted Seals [2]

Spotted seals occur in the Bering, Okhotsk, and Chukchi seas and occupy the ice front during winter and spring. Highest concentrations occur near Bristol Bay (Figure 14.13) and in Kavaginski Bay on the Siberian coast, within 25 km of the southern edge of the pack ice. During April, adults associate as triads with an adult male, adult female, and a pup. During

---

[2]Primary source Boveng et al. (2009)

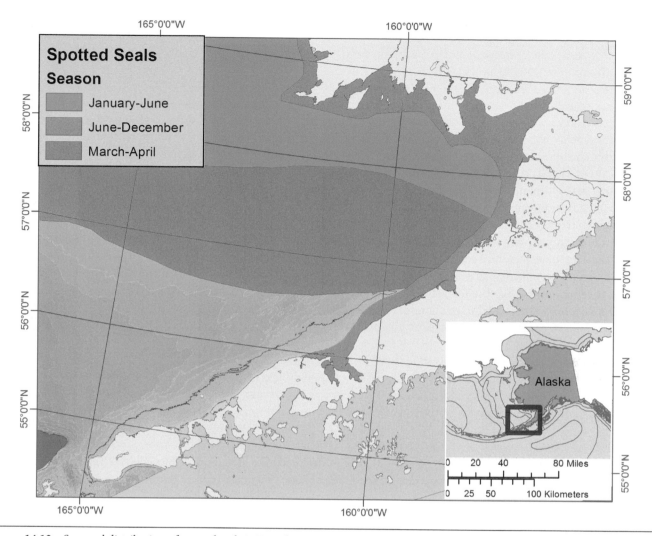

**Figure 14.13**   Seasonal distribution of spotted seals in Bristol Bay. 50 m depth contour is shown in dark blue. *Data source: Smith 2010.*

the molt, spotted seals congregate in small herds and move northward to coastal and estuarine habitats in the northern Bering and Chukchi seas (Braham et al. 1984). During the summer, some spotted seals remain at Bristol Bay, often in association with harbor seals (Frost et al. 1982, MacDonald and Winfree 2008).

Spotted seals and harbor seals are the most closely related species of phocid seals, sharing a common ancestor approximately 1.1 million years ago (Higdon et al. 2007). Spotted seals look very similar to light-phased harbor seals but developed different life history attributes for using sea ice as a substrate for pupping, breeding, molting, and resting. Even so, spotted seals have retained more affiliation for terrestrial haulouts than other pagophilic species and in Russia and Japan commonly pup and molt on shore (Nesterenko and Katin 2010). To accommodate the seasonal presence of ice, the life-cycle calendar of spotted seals is earlier than for harbor seals by about one month (Figure 14.1). Unlike the polygynous, hierarchical social system of harbor seals, spotted seals adopted an annually monogamous reproductive strategy where a male, female, and pup associate as a triad on the ice throughout the nursing period.

Spotted seal pups are born with a white wooly coat called lanugo and retain it for two to four weeks after birth. Pups rarely enter the water prior to weaning, although females will encourage them to leave the ice when disturbed. Harbor seal pups, on the other hand, typically shed their lanugo in utero, thus are born with an adult-like coat, enabling them to swim within an hour of birth. Harbor seal pups spend many hours each day, exploring the marine habitat with their mother in contrast to spotted seal pups that postpone learning about the marine environment until they have been weaned and are independent.

During pupping, female spotted seals are primarily attentive to their pups while males are attentive to females. Females typically do not associate with pups other than their own and appear to show less ability than harbor seals in distinguishing their pup from others. Instead, spotted seals appear to orient to the triad's ice floe and care for the pup present on the floe, even if not her own (Burns et al. 1972).

Spotted seals forage over the continental shelf, in waters < 200 m deep. The diets of spotted seals strongly overlap with harbor seals and show seasonal and regional differences. In spring and summer, when spotted seals are associated with sea ice, primary prey includes many schooling fishes, including walleye pollock, Pacific herring, Arctic cod, Pacific sand lance, capelin, and saffron cod, as well as greenlings, eelpouts, sculpins, flatfishes, squid, octopus, and crustaceans. In the summer, some seals gather near rivers where they frequently prey on runs of spawning salmon. In fall and winter, herring, capelin, smelt, saffron cod, Arctic cod, and non-fish prey items, such as octopuses, small crabs, and shrimp comprise a large part of the diet. In winter, dominant species encompass walleye pollock, capelin, Pacific sand lance, Arctic cod, and shrimp. Younger animals predominantly consume crustaceans, and older seals primarily eat fish. Amphipods, euphausiids, and other crustaceans are the dominant and often the only prey in newly weaned spotted seal pups, although a few fish, including Pacific sand lance, are occasionally consumed by older pups. As seals age, fish and larger shrimp become more important in the diet, with fish the dominant prey of adult seals.

Outside the pupping/breeding season, spotted seals and harbor seals overlap in habitat use. Pelage differences associated with the molt can help distinguish each species in May and June. In spring and early summer, newly molted spotted seals will have less brown and brighter contrast in their pelage relative to harbor seals, whose pelage will be browner with fading mottling and spots as they prepare for the molt. Except for dark-phase harbor seals, once molted, the two species are difficult to distinguish.

As the ice retreats, some spotted seals move north with the ice, while others shift to using nearshore areas and coastal haulouts, often adjacent to harbor seal haulouts in Bristol Bay, including in Nanvak Bay and Port Heiden, and the eastern Aleutians (Quakenbush 1988, Allen and Angliss 2015). Spotted seals also have been observed hauled out on a barrier island near Egegik in summer and at several locations on the north and western side of Hagemeister Island (Quakenbush et al. 2011a)

In winter, when sea ice provides a haul-out option that is away from land and closer to the winter distribution of prey, spotted seals move offshore and onto the ice. Spotted seals prefer shallow ice-covered waters over the continental shelf and avoid feeding in deep, open water beyond the shelf break (Lowry et al. 2000).

## Ribbon Seal[3]

Worldwide, ribbon seals are only located in the north Pacific region and concentrate in the Bering Sea and Sea of Okhotsk. They are not commonly observed in Bristol Bay (Figure 14.14). Based on surveys conducted during the mid-1970s the entire population has been estimated at 240,000 seals with 90,000–100,000 ribbon seals in the Bering Sea. Current population estimates and trends are unknown (Allen and Angliss 2015). Ribbon seals are not well adapted for maintaining breathing holes in winter sea ice and associate with ice front or edge zone of the seasonal pack ice. They generally prefer new, stable, moderately thick, white ice floes with an even surface, which are typically located in the inner zone of the ice front and rarely occur near shore. In spring, highest numbers of ribbon seals are located in central and western Bering Sea with lower densities found near the mouth of Bristol Bay (Braham et al. 1984).

Ribbon seals are deep divers and have several associated anatomical characteristics, including larger organ size and a large volume of blood with high concentrations of hemoglobin, which support oxygen storage during diving. They have larger eyes and a more streamlined body shape compared to the other phocids seals in the Bering Sea. Their skull is comparatively short and wide, with a short snout. Unlike other seals, some ribbon seals, particularly males, have an air sack extending from the lower end of the trachea. When fully developed it extends down the right side of the body to the level of the fore flipper and potentially assists with buoyancy at sea and sound production.

The diet of ribbon seals is not well known as a low proportion of stomachs examined contained prey. More than 50 prey species have been identified in ribbon seal stomachs that include fishes, cephalopods, and crustaceans. Pollock and

---

[3]Primary source Boveng et al. (2008)

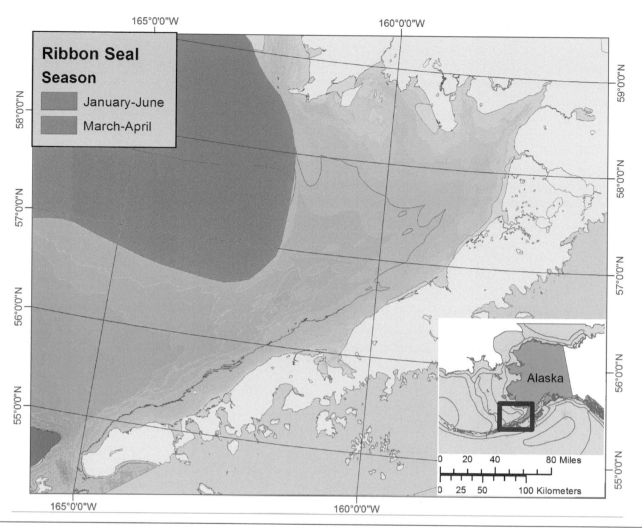

**Figure 14.14** Seasonal distribution of ribbon seals in Bristol Bay. 50 m depth contour is shown in dark blue. *Data source: Smith 2010.*

magistrate armhook squid are dominant prey; other commonly consumed prey include arctic cod, Pacific cod, saffron cod, Pacific sand lance, smooth lumpsucker, eelpouts, capelin and flatfish. Other cephalopods are consumed including several species of squid and octopus (Quakenbush and Citta 2008, Lowry et al. 1982).

The normal lifespan of a ribbon seals is estimated at around 20 years, with a maximum age of about 30 years. In the Bering Sea, female ribbon seals first ovulate at a young age relative to other seals (average 2.2 years; range 1–5 years) and pregnancy rates have been high (95%) during the past several decades, indicating sufficient prey resources are available to them. Ribbon seals also are less wary than other seals, suggesting they infrequently face predation in ice environments.

The social system of ribbon seals is not well understood. Females give birth from mid-March through early May and nurse their pups for three to four weeks. Mothers may leave their pups for extended periods, presumably to forage. As with other phocid seals, the pups are abruptly weaned and then abandoned—left to fend for themselves. Adult male ribbon seals do not accompany females during the early part of the nursing period and little is known about their breeding structure. Two types of vocalizations have been described. One is an intense downward frequency sweep and the other resembles a broadband puffing sound. Based on the seasonal timing and analogy with sounds made by other seals, the sounds are thought to be related to reproduction and/or territorial behavior. From mid-May through June, ribbon seals may rest on the ice much of the day as weaned pups develop self-sufficiency and older seals complete their molt, which lasts two to three weeks. Upon completing the molt, seals enter a pelagic period when they disperse widely, traveling and foraging in coastal areas as well as the interior of the Bering Sea. They rarely haul out during this time.

## Ringed Seal[4]

Ringed seals are the smallest and most ice-adapted of the arctic phocids and are distributed throughout the Holarctic. The total abundance of ringed seals has been difficult to estimate due to their widespread distribution in remote locations. Minimum estimates indicate more than one million ringed seals reside in Alaska waters and several million reside throughout the Arctic.

In the Bering Sea and North Pacific, ringed seals range as far south as the Sea of Okhotsk and Sea of Japan on the western side and the southern extent of winter sea ice in the Bering Sea and Bristol Bay (Figure 14.15). In Bristol Bay, they frequent the northern coast and in spring, feed on concentrations of smelts at stream mouths (Fall et al. 2013). Ringed seals are capable of diving for at least 39 minutes and to depths of over 500 m, though most dives last less than 10 minutes.

The diet of ringed seals includes higher concentrations of crustaceans and smaller fish (5–10 cm long) than typically consumed by adult harbor, spotted, or ribbon seals. The diet of ringed seals differs by region and age, and has changed across decades (Quakenbush et al. 2011b). Invertebrate prey items predominantly include mysids, amphipods, and decapods (crabs and shrimp). Decapods were the most dominant class and include three families of shrimp (Hippolytidae, Pandalidae, and Crangonidae). Predominant amphipods consumed include Ampeliscid, Gammarid, and Hyperiid. Mysids are commonly consumed by ringed seals, but more commonly eaten in summer than winter. Mollusks are infrequently consumed, but gastropods and bivalves are eaten. Dominant species of fish eaten by ringed seals include Arctic

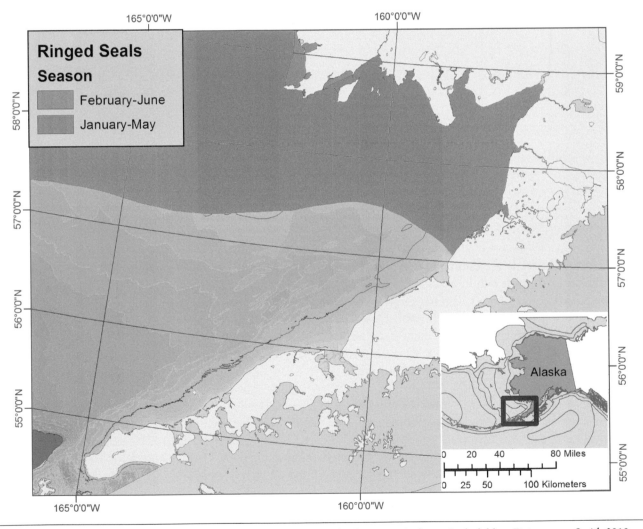

**Figure 14.15**    Seasonal distribution of ringed seals in Bristol Bay. 50 m depth contour is shown in dark blue. *Data source: Smith 2010.*

---

[4]Primary source Kelly et al. (2010)

cod, saffron cod, sculpins, rainbow smelt and walleye pollock. Ringed seals also consume Pacific herring, capelin, Pacific sand lance, and pricklebacks. In general, fish have been found more commonly in the stomachs of adults and subadults, and invertebrates dominate the diet of young seals.

Ringed seals typically are solitary, aggregating only during the molt in May and June. During the winter, highest concentrations of breeding adults occur 5–40 km offshore, in fast ice, stable pack ice, or among pressure ridges, while juveniles and subadults tend to be found further offshore (Braham et al. 1984). Subadults that wintered in the Bering Sea migrate northward in April, while adults and pups migrate later.

Ringed seals are the only Arctic seal that maintain breathing holes in shore-fast ice. Using their claws, females dig birth lairs in the snow above breathing holes; there they give birth and raise pups in subnivean (snow covered) lairs. Males also make and use multiple resting lairs. During the molt, lairs are not used and ringed seals preferentially haul out on top of the ice where they are exposed to sunlight. The range of ringed seals expands southward as sea ice advances in the winter and north as sea ice retreats in the spring; adult seals return to the same breeding areas each year.

Female ringed seals give birth to a single pup within a subnivean lair which provides protection from predators and severe weather. Pups are born covered a white wooly hair (lanugo) that provides insulation until they build a thick blubber layer four to six weeks after birth. During the nursing period, pups can quadruple their birth weight. Some females give birth on drifting pack ice, and are often more vulnerable to polar bear predation. Females nurse their pups for up to two months on stable shore-fast ice, but as few as three to six weeks on moving pack ice due to the earlier breakup of the ice. Lairs provide an essential shelter for newborn pups but warm temperatures and rains, which have become more frequent in recent years, can collapse the roofs of birth lairs, exposing pups to predators and adverse weather before pups are sufficiently fat to stay warm.

Ringed seal pups are more aquatic than other ice-associated seal pups, spending nearly half of their time in the water during the nursing period, alternating roughly six hour periods on the ice with spans of eight hours in the water. While swimming, pups are able to dive for 12 minutes and as deep as 89 m, although most last about 1 minute. Like older ringed seals, pups use multiple breathing holes and lairs. To avoid predation by polar bears and Arctic foxes, mothers may physically move their pups from the birth lair to alternate lairs.

Males rut from late March to mid-May and exhibit territorial-like behavior, marking breathing holes and lairs with a strong gasoline-like scent generated from glands on their faces. Males typically breed in April or May. Females typically mate with males within a month of giving birth, prior to weaning her pups. Overall, males appear to be territorial and polygynous, but they may associate with an individual female until she is ready to breed, suggesting monogamous or mixed breeding systems involving guarding rather than territoriality. In May and June, ringed seals molt, spending about half of their time hauled out on the ice to warm their skin and promote hair regeneration.

Currently, ringed seals are growing faster, and are maturing at younger ages than previous years, indicating that the females are in a positive nutritional state (Quakenbush et al. 2011b). The average age of sexual maturity for adult females ranged between 5.0–6.4 years from 1965–1984. Since 1999, however, the average age of maturity diminished to 3.2 years. The average life span of ringed seals is 15–28 years, although some have exceeded 40 years.

## Bearded Seal[5]

Bearded seals are circumpolar in distribution. In the Pacific, they range from the Arctic Ocean south to Sakhalin Island on the west and the winter ice edge on the east. Bearded seals are broadly distributed on pack ice in the Bering Sea during winter and spring and occur primarily in waters less than 200 m deep near polynyas (an area of open water surrounded by ice), leads (large fractures within a large span of sea ice), and regions of thin ice, where they can feed on the sea floor. In the eastern Bering Sea, the highest densities have been observed near St. Lawrence Island, southeast of St. Matthew Island and south of Nunivak Island. Bearded seals move northward into the Chukchi Sea as the ice retreats (Braham et al. 1984)

Bearded seals prefer sea ice with natural access to water, but are able to make breathing holes in thinner ice and break holes in ice up to 10 cm thick with their wide, stout head. In heavier ice, they use their strong foreflipper claws to maintain breathing holes. They are found in a broad range of ice types but avoid areas of continuous thick shore-fast ice. Bearded seals usually rest near the edges of floes where they can easily escape into the water. With the exception of females with pups, bearded seals typically are solitary in April, although, on occasion, several seals may rest on the same floe. Bearded seals may use land haulouts in late summer and early autumn until ice floes appear near the coast. Use of

---

[5]Primary source Cameron et al. (2010)

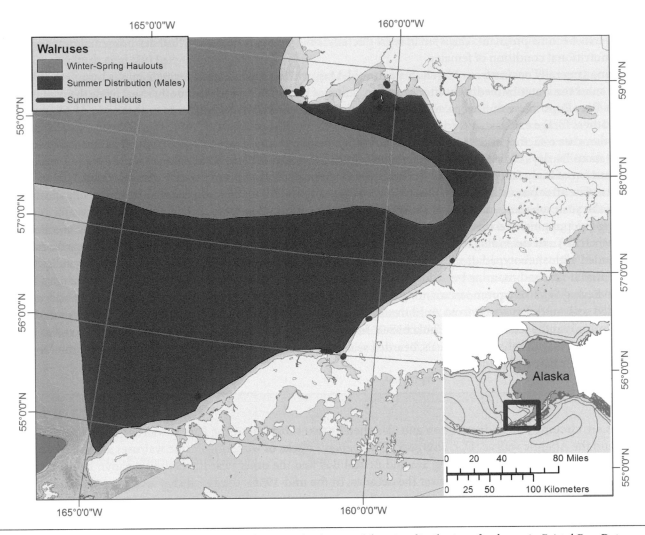

**Figure 14.17**    Winter-spring haul-out distribution and summer haul-out and foraging distribution of walruses in Bristol Bay. *Data source: Smith 2010.*

Beach, and Maggy Beach) and Cape Newenham in the Togiak NWR, and Cape Seniavin (on the Alaska Peninsula). Less consistently used haulouts include Hagemeister Island, Crooked Island, Twin Islands, and Cape Constantine, in the Togiak NWR and Walrus Islands State Game Sanctuary in northern Bristol Bay, and Amak Island on the Alaska Peninsula.

Walruses are social and gregarious. They tend to travel and haul out on ice or land in densely packed groups of up to several thousand animals. Walruses lay in close physical contact, with younger animals often resting on top of others. Stampedes from a haulout can cause injuries and deaths, especially to calves and younger walrus.

Calves are carefully cared for and typically have higher survival rates than other marine mammals. Calves can swim at birth but require constant attention, especially to maintain body temperature. Mothers normally remain with their calves for one to two years and do not breed until after their calf is weaned. Calves depend primarily on their mother's milk for the first year and are gradually weaned in their second year. After separation from their mother, female juveniles tend to remain with groups of adult females, while male juveniles gradually separate and associate with other males.

Walruses consume large amounts of food and strongly influence the seafloor structure and ecology. Although capable of diving to more than 250 m (820 ft), walruses usually forage in waters 80 m (262 ft) or less in areas of soft sand and mud. Foraging is widespread in Bristol Bay, but is most-concentrated south of Togiak Bay and westward to the 50 m depth contour (Jay and Hills 2005). Foraging trips can last from a few hours to several days.

Walruses root in the bottom sediment with their muzzles and rely on hundreds of sensitive, stiff vibrissae for detecting food and other objects as small as three millimeters in the seafloor. Their tongue works like a piston in their cylindrical shaped mouth to generate jets of water to disperse sediment and expose prey, as well as for creating a suction, to extract the soft parts of the prey from their shells (Garlich-Miller et al. 2011). They can use their foreflippers, nose, and jets of water to extract prey buried up to 32 cm (12.6 in) deep and may leave large troughs in the sediment.

The diet of walruses is diverse and includes more than 100 taxa of benthic invertebrates including clams, snails, worms, and many other invertebrates. Occasionally they consume fish and even marine mammals. Although bivalve mollusks (clams) comprise the bulk of their diet, the diversity of their diet appears reflective of the local benthic invertebrate community. Walruses swallow invertebrates without their shells, removing the soft parts by suction. Atlantic walruses have been found to consume an average of 53 bivalves per dive and may consume 29–74 kg (64–174 lbs) of food per day. Fay and Lowry (1981) estimated that at the population levels present in the early 1980s, walrus could have consumed two to four times the estimated annual sustainable yield of surf clams in a local Bristol Bay clam fishery zone.

Walruses using land haulouts are susceptible to injury due to disturbances. In order to reduce incidences of disturbance and reduce competition for foods near haulouts, protective measures have been undertaken to restrict seasonal vessel traffic near Round Island and other islands in the Walrus Islands State Game Sanctuary and near Cape Peirce.

State regulation prohibits vessel transit within three miles of Round Island and Cape Peirce in the Walrus Islands State Game Sanctuary; fishing is also restricted in the Walrus Islands State Game Sanctuary. Federal regulations further affect vessel transit and fishing activities further offshore in those areas.

## Steller Sea Lions[7]

Steller sea lions are distributed along the North Pacific Ocean rim from northern Japan, through the Aleutian Islands and Bering Sea, along Alaska's southern coast, and south to California. Steller sea lions in Alaska are partitioned into two populations segments (Eastern and Western) which transitions at Cape Suckling (144° W). Sea lions in the western stock have experienced a population decline of nearly 90% since the 1950s, with the steepest phase of the decline occurring from the mid-1980s to 1994 when the number of pups diminished by 70%. Steller sea lions in the western stock were listed as *endangered* under the Endangered Species Protection Act in 1997. Critical habitats within 20 nautical miles of rookeries and important haulouts have been established and protective measures have been undertaken to reduce disturbance and competition with fisheries. Recent population trends of sea lions for the Aleutian Islands (2000–2012) indicate increasing numbers of nonpups of (2.39%/yr ) and pups (3.3%/yr) (Allen and Angliss 2015).

In the eastern Bering Sea, Steller sea lions routinely use haulouts throughout the Aleutian and Pribilof islands. Sea Lion Rock, located near Amak Island, is the only rookery in Bristol Bay and is associated with the Eastern Aleutian Island group of rookeries. Sea lions also regularly haul out on rocks near Amak Island and several sites on Unimak Island (Figure 14.12). They seasonally occupy several haulouts in northern Bristol Bay including Cape Newenham, Cape Peirce, Hagemeister Island, Twin Islands, and Round Island (MacDonald and Winfree 2008).

In the 1980s, roughly 5,000 sea lions were estimated to be in Bristol Bay during the spring and summer (Frost et al. 1982). Recent counts indicate the presence of roughly 2,000 sea lions at Bristol Bay haulouts, but do not include those absent from haulouts during surveys. Surveys conducted in June 2014 identified 592 sea lions plus four pups on Amak Island and nearby rocks, 504 sea lions and 185 pups at the Sea Lion Rock Rookery near Amak Island, and 594 sea lions and one pup on Oksenof Point, Unimak Island (Fritz et al. 2015a). Surveys conducted in June 2015 in northern Bristol Bay counted 194 sea lions at Cape Newenham (Fritz et al. 2015b).

Sea lions have a polygynous reproductive system, where males establish territories in early May and maintain them for about 40 days without feeding. More favorable territories, those where males breed most frequently, tend to have access to water. Females choose which males to breed with and do not necessarily breed with the male on whose territory she cares for her pup. Mating occurs from May through July—peaking in mid- to late June. Female sea lions first breed between three and eight years of age. Males may attain physiological maturity before seven years of age, but are seldom able to establish and defend a territory until eight years or older. Females normally breed annually but reproductive failures are common. Birth rates are estimated at 55%–65%.

Females arrive on rookeries in late May through early July, a few days before giving birth. A few females may be accompanied by offspring from the previous year. Females give birth to a single pup and breed about 11 days after parturition. If offspring from previous years are present, the mother may wean the offspring, nurse both offspring, or abandon the pup and nurse the older offspring. Pups normally stay on land for about two weeks, and then spend increasing amounts of time in intertidal areas and swimming near shore. Mothers go on repeated foraging expeditions once their pup is about

---

[7]Primary sources National Marine Fisheries Service (2008) and Hoover (1988)

10 days old. Initially, females spend about 23 hours ashore followed by about 16 hours at sea (Maniscalco et al. 2006). As pups grow older, females spend less time on shore. In summer, while on breeding rookeries, adult females that are attending pups tend to remain within 20 nautical miles of the rookery. Females with pups begin dispersing from rookeries to haulouts when pups are about 2.5 months old. Most pups are weaned at about one-year-old while others are nursed an additional year or longer.

During the mid-1980s, evidence of reproductive failure and reduced rates of body growth indicated that nutritional stress was adversely affecting populations. The percentage of females that carried their pregnancy to late gestation fell to 67% during the 1970s, and to 55% in the 1980s. Lactating females were less likely to become pregnant that nonlactating females, with only 30% of lactating females carrying a pup to full term in the 1980s, as opposed to 63% in the 1970s.

In water, Steller sea lions forage in habitats extending from near shore to beyond the continental shelf break, although most remain on the continental shelf. Sea lions are social feeders—often traveling together in small groups. Although large numbers of sea lions (hundreds) may depart haulouts to forage, typically, sea lions leave haulouts in smaller groups of 2–12 individuals. When feeding on schooling species of prey such as herring, sea lions tend to aggregate in larger groups than when feeding on more dispersed prey. The prey of Steller sea lions is diverse and includes pollock, cods, herring, smelt, salmon, sculpins, sablefish, Atka mackerel, flatfish, rays, sandfish, sand lance, poachers, and eelpout, as well as octopus, squid, crabs, shrimp, snails, and marine mammals. Sea lions show regional and seasonal shifts in diets and consume prey of a wide size range extending from a few inches to more than up to 60 cm (2 ft) long.

The population decline of Steller sea lions has been associated with insufficient prey resources that have affected reproductive success and juvenile survival. Factors identified as threats to the recovery of the western stock of Steller sea lions include environmental variability affecting prey, competition with commercial fisheries affecting prey availability, and predation by killer whales. Toxic substances potentially can affect survival and reproduction of sea lions and their prey. Other sources of mortality—including incidental take by fisheries, Alaska Native subsistence harvest, illegal shooting, entanglement in marine debris, disease, parasitism, and disturbance from vessel traffic and tourism—have the potential to affect sea lions adversely, but are currently considered to have a relatively low impact on the recovery of sea lions.

Critical habitats have been designated for the western stock of Steller sea lions that include both terrestrial haulouts and rookeries and marine habitats that include a 20-nautical-mile buffer around all major haulouts and rookeries; associated terrestrial, air, and aquatic zones; and three large offshore foraging areas. No-entry zones have been established around Steller sea lion rookeries, including Sea Lion Rocks (near Akun Island), that prohibits vessel approach within three nautical miles of the rookery. A complex suite of fishery management measures has been implemented to minimize competition between fishing and the endangered population of Steller sea lions in critical habitat areas.

# Northern Fur Seal[8]

Northern fur seals breed at only six locations in the North Pacific and Bering Sea. Three of those locations (Pribilof Islands, Bogoslof Island, and San Miguel Island) are in the United States; the other three (Commander Islands, Kuril Island, and Robben Islands) are located in Russia. Of the worldwide population, more than 50% breed in dense aggregations on the Pribilof Islands. In the 1940s and early 1950s, about 2.1 million fur seals resided on the Pribilof Islands, but that number had declined to an estimated 721,935 seals in 2006. Bogoslof Island, an active volcano, is a relatively new rookery where small numbers were observed in the late 1970s and the first known pups were born in 1980. By 2006, the Bogoslof population exceeded 12,000 seals.

Male fur seals become sexually mature at five to seven years of age and begin competing for a territory after seven to nine years of age. Territorial males will fast while defending territories until early August. Despite not holding territories, immature males also fast at haul-out sites, losing an estimated 20–30% of their body weight. Most females become sexually mature between the ages of four and seven, averaging at about five years old; and then they will pup annually.

Males arrive at the rookeries first and establish territories. They arrive in descending order by age, with the youngest males not returning until mid-August or later. Females arrive at the rookery between mid-June through early August, peaking in early July. Females give birth to a single pup within two days of arrival. They will nurse the pup for about four months, and breed with males only three to eight days after parturition. As with Steller sea lions, females alternate onshore attendance with at sea foraging trips; however, foraging trips for fur seals are longer, lasting 3–10 days, and alternate with one to two day visits to the rookery to feed pups. Mothers and pups recognize each other through their vocalizations,

[8]Primary sources National Marine Fisheries Service (2007a, b)

a recognition that persists for at least four years. In early August, males leave the islands. Females leave the island as soon they wean their pups, and most adult females and juveniles migrate south from the Pribilof Islands in October. Pups remain longer, leaving the island in early November, after the older animals have departed, and typically do not return for another 22 months.

Fur seals are pelagic and highly migratory, rarely coming ashore except during the breeding and pupping season. They migrate during early winter through the eastern Aleutian Islands into the North Pacific Ocean, then southward to waters off the coast of British Columbia through California. Most adult males winter south of the Aleutian Islands and east into the Gulf of Alaska. Females are more prevalent in the southern parts of their range, in waters extending from southern California north to southeast Alaska.

Most important feeding habitats of fur seals are located in upwelling areas and are rarely found close to shore. Sightings of fur seals in Bristol Bay are concentrated mostly in the western portions of the bay and the Unimak Pass area as well as in the Togiak NWR. Haul out is considered rare compared to other pinnipeds (USFWS 1983).

In the Bering Sea, important summer and fall foods of northern fur seals include walleye pollock, capelin, herring, northern smoothtongue (a deep-sea smelt), and squid. Pollock is particularly important around the Pribilof Islands and other inshore areas from July to September. Capelin is the main prey consumed near Unimak Pass during June to October. A large number of other types of prey are eaten in smaller quantities. In the Eastern Bering Sea, over the continental shelf, 96% of walleye pollock consumed are from the 0 and 1 age classes (< 20 cm; 7.9 in); only 4% were from the age of two and older. Large, adult pollock were most frequently consumed by fur seals that were foraging over the outer domain of the continental shelf. North of Unimak Pass, in the outer shelf domain, fur seals also eat Atka mackerel while northern smoothtongue and gonatid squid are more frequently eaten by fur seals over continental slope and oceanic waters.

Because of their thick, high-quality pelage, northern fur seals have been subject to high levels of commercial harvests. Initial commercial harvests at the Pribilof Islands began in 1786 with the Russians who took 70,000 or more annually through 1839. In 1847 harvest of females was stopped and annual harvests progressively declined to about 30,000–35,000 seals. During the first two years following the purchase of Alaska by the United States, harvests were unregulated and high. Approximately 240,000 fur seals were taken in 1868 alone; pelagic sealing resulted in additional harvests. In 1911, Great Britain (for Canada), Japan, Russia, and the United States ratified the Treaty for the Preservation and Protection of Fur Seals and Sea Otters, which prohibited pelagic sealing and required a reduction in the harvest of seals on land. In 1973, a harvest moratorium was initiated on St. George Island, and in 1984 it was determined that no commercial harvest could be conducted under existing domestic law (Marine Mammal Protection Act) and the commercial harvest on St. Paul Island was terminated. Subsistence harvests are allowed and still continue. In 1988, the Pribilof Island population was designated as *depleted* under the Marine Mammal Protection Act because numbers had declined by more than 50% since the 1950s. Since 2009, trends show that the Pribilof Islands population continues to decline.

The Eastern Pacific northern fur seal stock has been adversely affected by both natural and human-related factors. Natural threats include trauma, starvation, disease, predation (killer whales, Steller sea lions, and foxes), and environmental change. Of human activities, subsistence harvests, poaching on land or at sea, entanglement and indirect effects of fishing, human presence, construction, vehicles and vessel noise and development, and environmental contamination affect the seals. Pups have experienced high mortality on the rookery. Hookworm disease was responsible for 45% of pup mortality from 1974–1977, although incidences of hookworm have declined on St. Paul Island in recent years. Pup mortality on the rookery does not explain the current population decline. Ongoing research is examining foraging success and other factors affecting the health, condition, and survival of fur seals.

## THREATS TO PINNIPED POPULATIONS AND MANAGEMENT STRATEGIES

Pinniped populations are being challenged by alterations of the marine environment associated with climate change. In a warming climate, winter ice becomes less abundant and persistent, resulting in less habitat for ice-associated seals during critical phases of their life cycle. Ocean acidification and warming temperatures will likely alter the food base for the prey that pinnipeds depend on. Toxic compounds (e.g., demoic acid, paralytic shellfish poisoning) associated with certain algal blooms are more frequently found in warmer marine conditions. Those compounds are biomagnified up the food chain, and are known to affect the behavior and survival of pinnipeds adversely. As the distribution and movements of pinnipeds and other organisms shift in response to climate change, new diseases can be introduced to populations that lack specific antibodies to protect them. Large, widespread die-offs of pinnipeds have been observed in seals of the North

Atlantic Ocean. Although, pinnipeds of the North Pacific have been less affected, a recent unusual mortality event from an unidentified disease caused widespread illness and mortality of ice-associated seals and walruses in the Bering and Chukchi seas.

Human activities also threaten pinnipeds through disturbances at haulouts, mineral resource developments (e.g., oil, gas, and mining), coastal development near haulouts, fishery interactions, and other activities. Human activities, including approaches by land, aircraft, and vessels adversely affect resting pinnipeds and can disrupt care of young offspring. In the case of walruses and harbor seals, disturbances can cause hundreds to thousands to flee into the water at one time, resulting in increased energy expenditures and the potential for injury.

Fisheries affect pinnipeds through direct interactions with fishing gear and fishermen and indirect interactions involving competition for fish and alterations of marine habitats. Pollution from coastal development, military sources, and other activities from both local and distant sources are being incorporated in the marine food web and concentrated in pinnipeds. Although oil development in the North Aleutian Basin is currently withdrawn from future Federal oil and gas leases, vessel traffic through Unimak Pass and other coastal areas has caused spills of oil and other toxic substances that adversely affect pinnipeds and the resources on which they depend.

Bristol Bay is managed by a mosaic of federal, state, and tribal entities that govern marine resource allocation and protection (Figure 14.18). Pinnipeds that inhabit Bristol Bay are important subsistence resources for Alaska Natives. Comanagement agreements between federal agencies and Alaska Native organizations have strengthened the voice of Alaska Natives in resource management while sustaining traditional foods, practices, and values. All pinniped populations in the U.S. receive federal protection through the Marine Mammal Protection Act. The Endangered Species Act

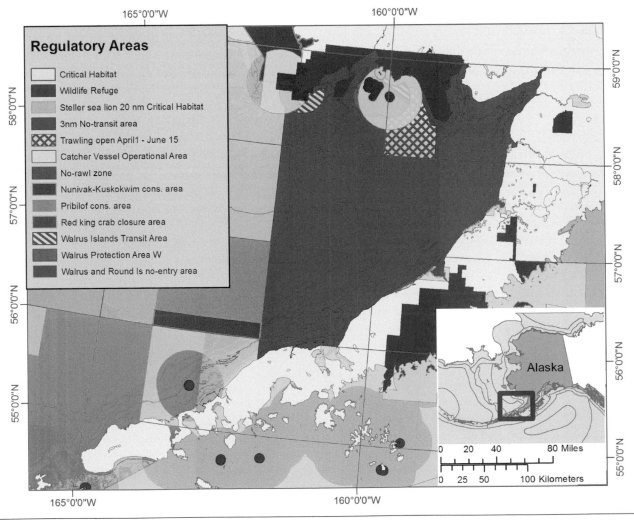

**Figure 14.18**　Geographic extents of regulatory measures that influence fish prey abundance, ecosystem structure, and disturbances of pinnipeds—particularly Steller sea lions and Pacific walruses.

further enhances the regulatory strength and breadth of measures needed to support the recovery of the western stock of Steller sea lions. Specific pinniped protection measures that have been implemented in Bristol Bay include:

1. Fishery restrictions and no-transit zones around major Steller sea lion rookeries and haulouts to reduce competition for resources and provide protection from disturbances,
2. No-transit zones around Round Island, Walrus Islands, and Cape Peirce to reduce disturbances, and
3. Prohibition of bottom trawling (except for one section southeast of Round Island, which remains seasonally open to a yellowfin sole trawl fishery) in order to safeguard benthic resources that walruses, bearded seals, and other organisms rely on.

## RECOMMENDED RESEARCH

In order to maintain the health and diversity of pinnipeds and other organisms in Bristol Bay, integrated research and management are necessary for enhancing the understanding of and responses to changing marine conditions. In an atmosphere of diminishing funds, monitoring, research, and enforcement are being scaled back. Strategically designed, long-term research is needed to provide an integrated understanding of marine ecosystem change throughout Bristol Bay that crosses geographic and jurisdictional boundaries as well as encompasses ecological parameters beyond those associated with the dominant commercial fisheries. Coordination of research and management efforts that involve all stakeholders facilitates the inclusion of local knowledge and the timely exchange of information among stakeholders. As importantly, coordination of management goals and regulations between federal and state waters and across species is needed to promote adaptive management practices while ensuring that commercially harvested species are sustainably managed with consideration of broader ecosystem dynamics.

## SUMMARY

Bristol Bay provides land, ice, and foraging habitats for five species of true seals (harbor, spotted, ringed, ribbon, and bearded), walruses, Steller sea lions, and northern fur seals. In late winter and spring, ice-associated pinnipeds concentrate on sea ice where they give birth and care for their young, breed, molt, and rest between foraging excursions. The reproductive and molting cycle of harbor seals, Steller sea lions, and northern fur seals, which rely on terrestrial haulouts in Bristol Bay and the Pribilof Islands, occurs a month or two later. Spotted seals and adult male walruses, which are associated with sea ice during the spring reproductive period can be found at terrestrial haulouts in northern Bristol Bay and at Cape Seniavin on the Alaska Peninsula during the summer. Northern fur seals occasionally haul out, but primarily forage when in Bristol Bay.

More than 18,000 harbor seals are distributed throughout Bristol Bay, with the highest numbers being along the north side of the Alaska Peninsula. Seasonally, harbor seals move northward into the Togiak NWR as the winter ice retreats. The only substantial breeding location for Steller sea lions is near Amak Island, in southwest Bristol Bay, after the winter ice retreats. Steller sea lions also use haulouts in the Togiak NWR.

The diet of pinnipeds involves three general categories of prey:

1. Pelagic fish—including pollock and a variety of forage fish such as herring, smelts, and cods;
2. Invertebrates—including squid, octopus, shrimp, and euphausiids; and
3. Bottom dwelling species—including sculpin and other benthic fishes, shrimp, crab, and clams.

Most pinnipeds consume a mixture of fish and invertebrates, but walruses and bearded seals concentrate on bottom-dwelling species. Young animals, especially those less than one year old, tend to eat smaller and slower moving prey than adults. Estuarine habitats, especially during fish spawning events, and upwelling zones associated with the 50 meter bathymetric contour are highly productive regions that provide rich foraging opportunities for pinnipeds.

Bristol Bay also is the site of major fisheries, including salmon, Pacific herring, and yellowfin sole. Complex regulations (Figure 14.18) have been established to reduce the adverse impacts of fisheries on pinnipeds, especially walruses and endangered Steller sea lions. The likelihood of oil and gas development has been reduced but pinnipeds are susceptible to disturbances from vessel, aircraft and foot traffic, coastal development, and other human activities. Climate change will strongly affect pinniped populations in Bristol Bay. Winter ice is forecasted to be thinner and spend less time in Bristol

Bay, providing less habitat for ice-associated seals. The composition and concentration of fish species will likely change. Toxic algal blooms and impacts from ocean acidification are becoming stronger and more frequent. Resource managers, involving multiple federal, state, and tribal agencies and organizations, supported by a well-designed and coordinated ecosystem-based marine research program, need to remain vigilant in detecting and responding to those changes in a coordinated manner.

# REFERENCES

Allen, B.M., and R.P. Angliss. 2015. Alaska Marine Mammal Stock Assessments, 2014. U.S. Dep. Commer., NOAA Tech. Memo. NMFS AFSC-301, 304 p. doi:10.7289/V5NS0RTS.

Bigg. M.A. 1969. The Harbour Seal in British Columbia. Fish. Res. Board Can. Bull. 172. 33 p.

Boveng, P.L., J.L. Bengtson, T.W. Buckley, M.F. Cameron, S.P. Dahle, B.A. Megrey, J E. Overland, and N.J. Williamson. 2008. Status review of the ribbon seal (*Histriophoca fasciata*). U.S. Dep. Commer., NOAA Tech. Memo. NMFS-AFSC-191, 115 p.

Boveng, P.L., J.L. Bengtson, T.W. Buckley, M.F. Cameron, S.P. Dahle, B.P. Kelly, B.A. Megrey, J.E. Overland, and N.J. Williamson. 2009. Status Review of the Spotted Seal (*Phoca largha*). U.S. Dept. Commerce, NOAA Tech. Memo. NMFS-AFSC-200, 153 p.

Bowen, D.W., S.L. Ellis, S.J. Iverson and D.J. Boness. 2003. Maternal and Newborn Life-History Traits during Periods of Contrasting Population Trends: Implications for Explaining the Decline of Harbour Seals (*Phoca vitulina*), on Sable Island. Journal of Zoology, 261, pp 155–163. doi:10.1017/S0952836903004047.

Braham, H.W., J.J. Burns, G.A. Fedoseev, B.D. Krogman. 1984. Habitat Partitioning by Ice-Associated Pinnipeds: Distribution and Density of Seals and Walruses in the Bering Sea, April 1976. Pp 25–47. In: F.H. Fay and G.A. Fedoseev, (Eds). Soviet-American Cooperative Research on Marine Mammals. Volume 1—Pinnipeds. National Oceanic and Atmospheric Administration Technical Report 12. National Marine Fisheries Service.

Burns, J.M., J. Van Lanen, D. Withrow, D. Holen, T. Askoak, H. Aderman, G. O'Corey-Crowe, G. Zimpelman, and B. Jones. 2013. Integrating Local Traditional Knowledge and Subsistence use Patterns with Aerial Surveys to Improve Scientific and Local Understanding of the Iliamna Lake seals. Report to the North Pacific Research Board Final Report 1116. 189 p.

Burns J.J. and V. N. Gol'tsev. 1984. Comparative Biology of Harbor Seals, Phoca vitulina Linnaeus, 1758, of the Commander, Aleutian, and Pribilof Islands. Pp 17–24. In: F.H. Fay and G.A. Fedoseev, (Eds), Soviet-American Cooperative Research on Marine Mammals. Volume 1—Pinnipeds. National Oceanic and Atmospheric Administration Technical Report 12. National Marine Fisheries Service.

Burns, J.J., L.H. Shapiro, and F.H. Fay. 1981. Ice as Marine Mammal Habitat in the Bering Sea. The Eastern Bering Sea Shelf: Oceanography and Resources, 2, 781–797.

Burns, J.J., G.C. Ray, F.H. Fay, and P.D. Shaughnessy. 1972. Adoption of a Strange Pup by the Ice-Inhabiting Harbor Seal, *Phoca vitulina largha*. Journal of Mammalogy, 53(3), 594–598. http://doi.org/10.2307/1379048.

Cameron, M.F., J.L. Bengtson, P.L. Boveng, J.K. Jansen, B.P. Kelly, S.P. Dahle, E.A. Logerwell, J.E. Overland, C.L. Sabine, G.T. Waring, and J.M. Wilder. 2010. Status Review of the Bearded Seal (*Erignathus barbatus*). U.S. Dept. Commerce, NOAA Tech. Memo. NMFS-AFSC-211, 246 p.

Cooper, M.H., S.M. Budge, A.M. Springer, G. Sheffield. 2009. Resource Partitioning by Sympatric Pagophilic Seals in Alaska: Monitoring Effects of Climate Variation with Fatty Acids. Polar Biol. 32:1137–1145.

Cordes, L.S., P.M. Thompson. 2013. Variation in Breeding Phenology Provides Insights into Drivers of Long-Term Population Change in Harbour Seals. Proceedings of the Royal Society B 280: 20130847. http://dx.doi.org/10.1098/rspb.2013.0847.

Daniel, R.G., L.A. Jemison, G.W. Pendleton, and S.M. Crowley. 2003. Molting Phenology of Harbor Seals on Tugidak Island, Alaska. Marine Mammal Science, 19: 128–140.

Fall, J.A., N.S. Braem, C.L. Brown, L.B. Hutchinson-Scarbrough, D.S. Koster, and T.M. Krieg. 2013. Continuity and Change in Subsistence Harvests in Five Bering Sea Communities: Akutan, Emmonak, Savoonga, St. Paul, and Togiak. *Deep Sea Research Part II: Topical Studies in Oceanography*, 94, 274–291.

Fay, F.H., and L.F. Lowry. 1981. Seasonal Use and Feeding Habits of Walruses in the Proposed Bristol Bay Clam Fishery Area. Final Report to the North Pacific Fishery Management Council. Contract No. 80. Pages 3–61.

Fritz, L, K. Sweeney, R. Towell and T. Gelatt. 2015a. Results of Steller Sea Lion Surveys in Alaska, June–July 2014. Memorandum to: D. DeMaster, J. Bengtson, J. Balsiger, J. Kurland, and L. Rotterman dated 28 January 2015. Available AFSC, National Marine Mammal Laboratory, NOSS, NMFS 7600 Sand Point Way, NE, Seattle WA 98115. 14 p.

Fritz, L, K. Sweeney, R. Towell and T. Gelatt. 2015b. Results of Steller Sea Lion Surveys in Alaska, June–July 2015. Memorandum to: D. DeMaster, J. Bengtson, J. Balsiger, J. Kurland, and L. Rotterman dated 28 December 2015. Available AFSC, National Marine Mammal Laboratory, NOSS, NMFS 7600 Sand Point Way, NE, Seattle WA 98115. 16 p.

Fritz, L, K. Sweeney, R. Towell, and T. Gelatt, Tom. 2015c. Steller sea Lion Haulout and Rookery Locations in the United States for 2015-05-14 (NCEI Accession 0129877). Version 1.2. NOAA National Centers for Environmental Information. Dataset. doi:10.7289/V58C9T7V [Accessed 19 April 2016].

Frost, K.J. and L.F. Lowry. 1986. Sizes of Walleye Pollock *Theragra chalcogramma*, Consumed by Marine Mammals in the Bering Sea. Fish. Bull. 84: 192–197.

Frost, K.J., L.F. Lowry, and J.J. Burns. 1982. Distribution of Marine Mammals in the Coastal Zone of the Bering Sea during Summer and Autumn. U.S. Dept. Commerce, NOAA. OCSEAP Final Report 20:365–561.

Garlich-Miller, J., J.G, MacCracken, J. Snyder, R, Meehan, M. Myers, J.M. Wilder, E. Lance, and A, Matz. 2011. Status Review of the Pacific Walrus (*Odobenus rosmarus divergens*). U.S. Fish and Wildlife Service. Anchorage, Alaska. 155 p.

Gentry, R.L. 1998. *Behavior and Ecology of the Northern Fur Seal*. Princeton University Press.

Grant, R. S. Wieskotten, N. Wengst, T. Prescott, G. Dehnhardt. 2013. Vibrissal Touch Sensing in the Harbor Seal (*Phoca vitulina*): How Do Seals Judge Size? Journal of Comparative Physiology A 199(6):521–533.

Hauser, D.D., C.S. Allen, H.B. Rich Jr., and T.P. Quinn. 2008. Resident Harbor Seals (*Phoca vitulina*) in Iliamna Lake, Alaska: Summer Diet and Partial Consumption of Adult Sockeye Salmon (*Oncorhynchus nerka*). Aquatic Mammals, 34(3), 303.

Hayes, S.A., D.P. Costa, J.T. Harvey, and B.J. Boeuf. 2004. Aquatic Mating Strategies of the Male Pacific Harbor Seal (*Phoca vitulina richardii*): Are Males Defending the Hotspot? Marine Mammal Science, 20(3), 639–656.

Higdon, J.W., O.R. Bininda-Emonds, R.M. Beck, S.H. Ferguson. 2007. Phylogeny and Divergence of the Pinnipeds (Carnivora: Mammalia) Assessed Using a Multigene Dataset. BMC Evolutionary Biology, 7(1):1.

Hiruki-Raring, L., P. Goddard, T. Kushin, and H. Huber. 2013. Laaqudax the Northern Fur Seal. K–6 Grade Curriculum. Alaska Fisheries Science Center. National Marine Fisheries Service, Seattle, WA.

Hoover-Miller, A.A. 1988. Steller Sea Lions (*Eumetopias jubatus*). In: *Selected Marine Mammals of Alaska: Species Accounts with Management Recommendations*. Jack Lentfer (Ed.). Marine Mammal Commission. Washington D.C.

Hoover-Miller, A.A. 1994. Harbor Seal (*Phoca vitulina*) Biology and Management in Alaska. Contract number T75134749. Marine Mammal Commission, Washington D.C. 45 p.

Jay, C.V., and S. Hills. 2005. Movements of Walruses Radio-Tagged in Bristol Bay, Alaska. Arctic 58(2):192–202.

Jemison, L.A., G.W. Pendleton, C.A. Wilson, and R.J. Small. 2006. Long-Term Trends in Harbor Seal Numbers at Tugidak Island and Nanvak Bay, Alaska. Marine Mammal Science, 22(2), 339–360.

Kelly, B.P., J.L. Bengtson, P.L. Boveng, M.F. Cameron, S.P. Dahle, J.K. Jansen, E.A. Logerwell, J.E. Overland, C.L. Sabine, G.T. Waring, and J.M. Wilder. 2010. Status Review of the Ringed Seal (*Phoca hispida*). U.S. Dept. Commerce, NOAA Tech. Memo. NMFS-AFSC-212, 250 p.

London, J.M., K.M. Yano, E.L. Richmond, D.E. Withrow, S.P. Dahle, J.K. Jansen, H.L. Ziel, G.M. Brady, and P.L. Boveng. 2015. Observed Haul-Out Locations for Harbor Seals in Coastal Alaska. Alaska Fisheries Science Center, National Oceanic and Atmospheric Administration.

Lowry, L.F., K.J. Frost, D.G. Calkins, G.L. Swartzman, and S. Hills. 1982. Feeding Habits, Food Requirements and Status of Bering Sea Marine Mammals. North Pacific Fisheries Management Council. Anchorage, Alaska, Doc 19 and 19a. 574 p.

Lowry, L.F., VN. Burkanov, K.J. Frost, M.A. Simpkins, R. Davis, D.P. DeMaster, R. Suydam, and A, Springer. 2000. Habitat Use and Habitat Selection by Spotted Seals (Phoca largha) in the Bering Sea. Can. J. Zool. 78:1959–1971.

MacDonald, R. and M. Winfree. 2008. Marine Mammal Haulout Use in Bristol Bay and Southern Kuskokwim Bay, Alaska, 2006. Status Report of the 2006 Marine Mammal Monitoring Effort at Togiak National Wildlife Refuge. U.S. Fish and Wildlife Service. Togiak, AK. 50 p.

Maniscalco, J.M, P. Parker, S. Atkinson. 2006. Interseasonal and Interannual Measures of Maternal Care Among Individual Steller Sea Lions (*Eumetopias jubatus*). Journal of Mammalogy 87 (2) 304–311; DOI: 10.1644/05-MAMM-A-163R2.1.

Murphy, C.T., Coleen Reichmuth, and D. Mann. 2015. Vibrissal Sensitivity in a Harbor Seal (*Phoca vitulina*). Journal of Experimental Biology. 218, 2463–2471.

National Marine Fisheries Service. 2007a. Steller Sea Lion and Northern Fur Seal Research Final Programmatic Environmental Impact Statement. Volume 1. Office of Protected Resources. National Marine Fisheries Service. Silver Springs Md. 1112 p.

National Marine Fisheries Service. 2007. Conservation Plan for the Eastern Pacific Stock of northern Fur Seal (*Callorhinus ursinus*). National Marine Fisheries Service, Juneau, Alaska.

National Marine Fisheries Service. 2008. Recovery Plan for the Steller Sea Lion (*Eumetopias jubatus*). Revision. National Marine Fisheries Service, Silver Spring, MD. 325 p.

National Marine Fisheries Service. 2013. Biological Characterization: an Overview of Bristol, Nushagak, and Kvichak Bays; Essential Fish Habitat, Processes and Species Assemblages. National Marine Fisheries Service Alaska Region Report. Juneau, AK. 59 pages.

Nesterenko, V.A. and I.O. Katin. 2010. Cycle of Transformation of the Spotted Seal (*Phoca largha*, Pallas, 1811) Onshore Associations in Peter the Great Bay of the Sea of Japan. Zoology of Vertebrates. Russian Journal of Marine Biology 36(1):47–55.

O'Corry-Crowe, G.M., K.K. Martein, and B.L. Taylor 2003. The Analysis of Population Genetic Structure in Alaskan Harbor seals, Phoca vitulina, as a Framework for the Identification of Management Stocks. Administrative Report LJ-03-08. National Marine Fisheries Service Southwest Fisheries Science Center La Jolla, CA 92037.

Piatt, J.F., and A.M. Springer. 2007. Marine Ecoregions of Alaska. Pp. 522–526. In: Robert Spies (Ed.), Long-term Ecological Change in the Northern Gulf of Alaska. Elsevier, Amsterdam. Data layer access: http://alaska.usgs.gov/science/biology/nppsd/marine_ecoregions.php.

Quakenbush, L. 1988. Spotted Seal. Pp 107–124. In: J. W. Lentfer (Ed.), *Selected Marine Mammals of Alaska: Species Accounts with Research and Management Recommendations*. Marine Mammal Commission, Washington D.C.

Quakenbush, L., and J. Citta. 2008. Biology of the Ribbon Seal in Alaska. Final Report to: National Marine Fisheries Service. Report to National Marine Fisheries Service. Alaska Department of Fish and Game. Fairbanks. AK. 46 p.

Quakenbush, L., J. Citta, and J. Crawford. 2009. Biology of the Spotted Seal (*Phoca largha*) in Alaska from 1962–2008. Report to National Marine Fisheries Service. Alaska Department of Fish and Game. Fairbanks. AK. 66 p.

Quakenbush, L., J. Citta, and J. Crawford. 2011a. Biology of the Bearded Seal (*Erignathus barbatus*) in Alaska, 1961–2009. Final Report to National Marine Fisheries Service. Alaska Department of Fish and Game. Fairbanks. AK. 71 p.

Quakenbush, L., J. Citta, and J. Crawford. 2011b. Biology of the Ringed Seal (*Phoca hispida*) in Alaska, 1960–2010. Final Report to National Marine Fisheries Service. Alaska Department of Fish and Game. Fairbanks, AK. 72 p.

Ray, G.C., G.L. Hufford, J.E. Overland, I. Krupnik, J. McCormick-Ray, K. Frey, and E. Labunski. 2016. Decadal Bering Sea Seascape Change: Consequences for Pacific Walruses and Indigenous Hunters. *Ecological Applications*. 26(1) 24–41.

Rugh, D.J., and K.E. Shelden. 1997. Spotted Seals, Phoca largha, in Alaska. *Marine Fisheries Review*, 59(1), 1.

Scholtyssek, C., A. Kelber, and G. Dehnhardt. 2015. Why Do Seals Have Cones? Behavioural Evidence for Colour-Blindness in Harbour Seals. Animal Cognition, 18, 551–560. http://doi.org/10.1007/s10071-014-0823-3.

Sharma, G.D., A.S. Naidu, D.W. Hood. 1972. Bristol Bay: Model Contemporary Graded Shelf. AAPG Bulletin. 56(10) 2000–2012.

Small, R.J., L.F. Lowry, J.M. ver Hoef, K.J. Frost, R.A. Delong, and M.J. Rehberg. 2005. Differential Movements by Harbor Seal Pups in Contrasting Alaska Environments. Marine Mammal Science, 21: 671–694. doi: 10.1111/j.1748-7692.2005.tb01259.x.

Smith, M. 2010. Arctic Marine Synthesis: Atlas of the Chukchi and Beaufort Seas. Audubon, Alaska and Oceana. http://ak.audubon.org/arctic-marine-synthesis-atlas-chukchi-and-beaufort-seas (Accessed May 22, 2016). Associated GIS data sets accessed through AOOS Arctic Data Integration Portal: Arctic Marine Synthesis (http://portal.aoos.org).

Temte J.L. 1994. Photoperiod Control of Birth Timing in the Harbour Seal (Phoca vitulina). Journal of Zoology 233:369–384.

U.S. Fish and Wildlife Service. 1983. Migratory Birds and Marine Mammals of the Bristol Bay region. Wildlife Narratives for the Bristol Bay Cooperative Management Plan. U.S. Fish and Wildlife Service Refuge Planning. Anchorage, AK 94 p.

Vania, J.S., E. Klinkhart, and K. Schneider. 1969. Harbor seals. Alaska Department of Fish and Game, Anchorage. Fed. Aid Wildlife Restoration. Annual Project Segment Report W-14-R33 and W-17-1, Workplan G, 1 Jan–31 Dec. 1968.

Wynne, K., and P. Folkens, 2012. Guide to Marine Mammals of Alaska. Alaska Sea Grant College Program. University of Alaska, Fairbanks, AK.

Withrow, D.E., J.C. Cesarone, J.K. Jansen, and J.L. Bengtson. 2001. Abundance and Distribution of Harbor Seals (*Phoca vitulina richardsi*) in Bristol Bay and along the North Side of the Alaska Peninsula during 2000. Pages 69–82. In: Lopez, A. L. and R. P. Angliss (Eds.), Marine mammal Protection Act and Endangered Species Act Implementation Program 2000. Annual Report to the Office of Protected Resources. National Marine Mammal Laboratory. Seattle, WA.

Yurk, H. and A.W. Trites. 2000. Experimental Attempts to Reduce Predation by Harbor Seals on Out-Migrating Juvenile Salmonids. Transactions of the American Fisheries Society, 129(6), 1360–1366.

# 15

# THE IMPORTANCE OF BRISTOL BAY TO MARINE BIRDS OF THE WORLD

**Nils Warnock and Melanie Smith**
Audubon, Alaska

## INTRODUCTION

Fueled by the richly productive waters of Bristol Bay, a diverse (over 100 species, see Appendix 15.1 at the end of this chapter) and staggering abundance of marine birds (shorebirds, waterfowl, and seabirds) numbering into the millions use the region throughout the year (Schneider and Shuntov 1993, Hunt et al. 2002, Schamber et al. 2010), making Bristol Bay one of the most productive areas in the world for marine birds. For thousands of years, humans living along Bristol Bay, the Alaska Peninsula, and the Aleutians have relied on its rich bird resources (Maschner 1999, Causey et al. 2005). However, written descriptions of the region's bird life only began during the last 250 years. The Russian explorer and naturalist Georg Wilhelm Steller (who named the Steller's eider (*Polysticta stelleri*) and Steller's Jay (*Cyanocitta stelleri*) among other animals) produced some of the first detailed notes on Alaskan birds during his ill-fated expedition in the 1740s (Gabrielson and Lincoln 1959). Additionally, naturalists on largely Russian expeditions (in addition to a few key British ones), fueled by the fur trade during 1741 to 1867, began laying the ornithological groundwork for the Bristol Bay region (Gabrielson and Lincoln 1959). Early observations in the Aleutians and Bristol Bay area, albeit opportunistic, began in the late 1800s, thanks in part to naturalists of the U.S. Army Signal Corps and the U.S. Coast Survey, (e.g. Dall 1873, Turner 1886, Nelson 1887). One of the earliest U.S. Signal Corps stations was just below the confluence of the Wood and Nushagak rivers at Nushagak (also named Fort Alexander). Station pioneer at Nushagak, Charles McKay [for whom the McKay's Bunting (*Plectrophenax hyperboreus*) is named (Osgood 1904)], has been described as Bristol Bay's first resident scientist (Branson 2012). Beginning in 1881 until his untimely death in 1883 in a boating accident, McKay made various bird collections through the area that are referenced in Osgood's 1902 biological reconnaissance (Osgood 1904).

McKay's opportunistic surveys noted the plentiful and diverse avifauna of the region. However, systematic scientific surveys of terrestrial and marine birds did not begin until the 1950s (King and McKnight 1969, Bartonek and Gibson 1972, Arneson 1980, Gill et al. 1981, Hodges et al. 1996), driven in part by programs such as the Outer Continental Shelf Environment Assessment Program as well as the U.S. Fish and Wildlife Service's (USFWS) focus on breeding waterfowl. The first aerial offshore bird surveys in Bristol Bay were likely flown by biologists with the USFWS (King and McKnight 1969) and the Alaska Department of Fish and Game (ADF&G) in October 1969 (ADF&G 1977). Some of the first work documenting and monitoring seabird colonies in north Bristol Bay, including at Cape Peirce, occurred in the 1970s (Dick and Dick 1971, Petersen and Sigman 1977). Monitoring of seabird colonies in north Bristol Bay continues today, especially through the efforts of the Alaska Maritime National Wildlife Refuge (NWR) and Togiak NWR (USFWS 2009).

During the 1970s, federal biologists also began surveying the Alaska Peninsula with a special focus on shorebirds (e.g., Gill and Jorgensen 1979; for more details see Savage, this volume). Annual aerial surveys, especially for waterfowl of conservation concern [e.g., Steller's eider, emperor goose (*Chen canagica*)], began in the early 1980s (e.g., Dau and Mallek 2006, Larned 2008), adding greatly to the knowledge about the abundance and phenology of waterbirds in the Bristol

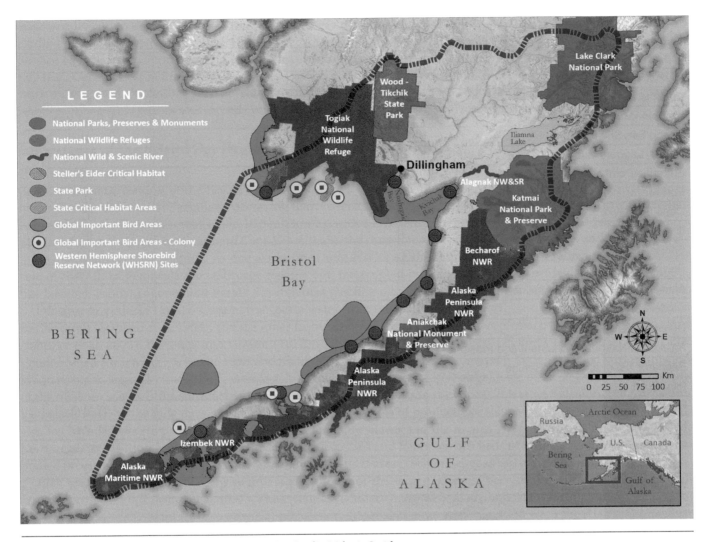

**Figure 15.1** Map of Bristol Bay Important Areas. *Map Credit: Melanie Smith.*

of Izembek (Smith et al. 2014) (Figure 15.1). At the main drainage of the Bristol Bay watershed, the areas around Kvichak and Nushagak bays support globally significant IBAs (Table 15.2).

There have been a number of shorebird studies in different parts of the Bristol Bay region (particularly along the Alaska Peninsula); however relatively few comprehensive surveys of Bristol Bay have been dedicated to shorebirds (for further details see Savage, this volume). Important sites include Nelson Lagoon/Mud Bay, Egegik Bay, Nushagak Bay, Kvichak Bay, Cinder-Hook Lagoon, and Port Heiden (Gill and Handel 1981, Gill and Sarvis 1999, Gill et al. 2001, Alaska Shorebird Group 2008, Savage and Payne 2013).

## Key Marine Birds of the Bristol Bay Area

Depending on the time of year, the Bristol Bay region may support low millions to tens of millions of marine birds. During summer months, an estimated 8 to 13 million nonbreeding seabirds come to feed in Bristol Bay (Dept. of Commerce 1977). The most abundant pelagic, nonbreeding seabird—numbering up to the tens of millions—is the short-tailed shearwater (Schneider and Shuntov 1993, Carey et al. 2014). The Bristol Bay region is also important to breeding seabirds and waterbirds. Over one million seabirds breed in the region, the most abundant being common murres followed by black-legged kittiwakes and tufted puffins (*Fratercula cirrhata*) (Table 15.2). Low millions of shorebirds use the Bristol Bay region (e.g., Gill and Handel 1981), the majority staging (sensu Warnock 2010) at the area's rich coastal estuaries during fall, and to a lesser degree, spring, making it one of the most important shorebird areas in North America (Brna and

**Table 15.2**  Numbers of breeding seabirds and colonies in the Bristol Bay region[1]

| Species | Sowls et al. 1978 | NPSCD 2015 |
|---|---|---|
| Double-crested cormorant | 1696 | 1070 |
| Pelagic cormorant | 15,262 | 11,387 |
| Red-faced cormorant | 3311 | 3797 |
| Unidentified cormorant spp. | 4327 | 2748 |
| Common eider | | 540 |
| Black oystercatcher | | 8 |
| Glaucous gull | | 40 |
| Glaucous-winged gull | 30,442 | 23,597 |
| Mew gull | 300 | 210 |
| Black-legged kittiwake | 493,604 | 260,070 |
| Arctic tern | 1430 | 2062 |
| Aleutian tern | 1130 | 955 |
| Common murre | 1,037,300 | 622,128 |
| Thick-billed murre | 2 | 135 |
| Unidentified murre spp.[2] | 318,044 | 353,316 |
| Pigeon guillemot | 1631 | 1702 |
| Parakeet auklet | 2485 | 1914 |
| Crested auklet | 100 | 50 |
| Least auklet | | 8 |
| Horned puffin | 4205 | 2922 |
| Tufted puffin | 89,140 | 89,123 |
| **Total** | **2,004,409** | **1,377,782** |

[1] Data from Sowls et al. (1978) and the North Pacific Seabird Colony Database (accessed in June 2015). Note that the Seabird Colony database includes data from Sowls et al. (1978), but colony numbers are updated through the mid-1990s.

[2] Based on survey results from the region, the majority of unidentified murres are probably common murres.

Verbrugge 2013, Savage this volume). The terrestrial areas around Bristol Bay also support lesser numbers of breeding shorebirds of at least 24 species (Appendix 15.1).

Adjacent to the waters of Bristol Bay, the Bristol Bay Lowlands is one of the most important waterfowl breeding areas in Alaska, averaging more than 350,000 ducks [mostly greater scaup (*Aythya marila*), northern pintail (*Anas acuta*), and black scoters (*Melanitta americana*)]; 15,000 tundra swans (*Cygnus columbianus*); 3000 loons; 5000 sandhill cranes (*Grus canadensis*); and more than 4000 geese in the Lowlands, particularly to the east and southeast of Bristol Bay (Platte and Butler 1995, Hodges et al. 1996, Mallek and Groves 2011). Even larger numbers of waterfowl use the area during the nonbreeding season, exploiting the relatively ice-free (winter) and food rich resources of Bristol Bay. North Bristol Bay, around Cape Peirce, is also an important passage site for the molt migration of white-winged (*Melanitta fusca*), and to a lesser degree, surf scoters (*M. perspicillata*) and black scoters, with over 50,000 birds passing by in late June to early July (Herter et al. 1989).

# KEY BRISTOL BAY SPECIES

Marine bird species for which a large percentage of their population (either globally or continentally for North America) concentrate within the coastal and marine areas of the Bristol Bay region include:

- Emperor goose
- Brant

- Cackling goose (*Branta hutchinsii*)
- Tundra swan
- Steller's eider
- King eider (*Somateria spectabilis*)
- Black scoter
- Long-tailed duck (*Clangula hyemalis*)
- Short-tailed shearwater
- Bar-tailed godwit
- Marbled godwit (*Limosa fedoa*)
- Dunlin (*Calidris alpina*)
- Common murre
- Black-legged kittiwake

Here we provide population estimates, information on their conservation status, and what is known of their concentrations and distributions in Bristol Bay.

## Emperor Goose

### Population and Status

The emperor goose (Figure 15.2) primarily breeds in the Bering Sea region of western Alaska with some breeding in northeastern Russia (Schmutz et al. 2011). The global population of emperor goose is now estimated at over 85,000 birds—most of which breed on the Yukon-Kuskokwim Delta and stage along the Alaska Peninsula (Schmutz et al. 2011, Stehn and Wilson 2014). In southwest Alaska, the mean spring count (1992 to 2012) is 36,368 birds (Larned 2012). The emperor goose is an Audubon WatchList species (Warnock 2017) and an International Union for the Conservation of Nature (IUCN) Red List *Near Threatened* species. It is a species of conservation concern due to an approximately 50% global population reduction during the last 50 years; in Alaska, the population declined from 139,000 in 1964 to 42,000 in 1986 (Schmutz et al. 2011). Currently, fall counts on the Alaska Peninsula show a 0.2% per year increase (Mallek and Dau 2011). Due to a rapidly declining population, all hunting of the emperor goose was closed in 1987 (Wilson and Dau 2014b), and only recently reopened (Warnock 2017).

### Bristol Bay

In some years, the entire population of the emperor goose migrates through Izembek NWR in spring and fall (Petersen and Gill 1982, Hupp et al. 2008, Stehn and Wilson 2014). In late April, over 70% of emperor geese are found between Nelson Lagoon and Port Heiden; and about 4% from Egegik Bay to Nanvak Bay (Dau and Mallek 2011). Key late-September sites include Cinder River Estuary—12,844 (21.4%), Seal Islands—20,834 (34.8%), and Nelson Lagoon and adjacent estuaries—14,256 (23.8%) (Mallek and Dau 2011). Small numbers (low thousands) of emperor geese overwinter in Bristol Bay; Izembek winter counts (1980 to 2014) range from 542 to 5,139 birds (Wilson and Dau 2014b), and winter numbers vary depending on ice conditions (Gill et al. 1981, Wilson and Dau 2014a). Numbers of emperor geese are

**Figure 15.2**    Emperor goose in flight. *Photo credit: Milo Burcham.*

generally higher on northcentral Alaska Peninsula in the fall than in the spring (Gill et al. 1981). The emperor goose is listed as an uncommon breeder in the Togiak NWR bird checklist (USFWS 2016).

The emperor goose stands out in the herbivorous goose world in that they mainly feed on benthic invertebrates (Schmutz et al. 2011). Alaska Peninsula staging areas are especially important to the emperor goose because over 86% of the population is concentrated in this area for four to five months of the year.

## Brant

### Population and Status

Brant (Figure 15.3) are Holarctic breeders with four subpopulations breeding in North America (Lewis et al. 2013). The global population of Brant is estimated to be about 560,000 birds (BirdLife International 2015). The Pacific Flyway portion of the brant population (Pacific black brant) is made up of 132,000 to 162,000 birds breeding in Alaska, the western Canadian Arctic, and the eastern Russian Arctic, and wintering primarily from Alaska to Mexico (Pacific Flyway Council 2002). The Pacific black brant is an Audubon WatchList species and a BirdLife species of concern due to an approximately 40% reduction in the global population; however, the last 18 years of mid-winter population counts indicate a stable population in Alaska (Stehn et al. 2011).

### Bristol Bay

Almost the entire population of Pacific black brant uses the Izembek Lagoon area for up to four weeks in spring and up to six weeks in fall (Pacific Flyway Council 2002). Currently, about 40% of the Pacific black brant population winters along the Alaska Peninsula and that proportion is growing at over 7% per year, presumably related to more favorable conditions resulting from milder winters on the Alaska Peninsula (Ward et al. 2009, Wilson and Dau 2014). Other important sites include Chagvan and Nanvak bays (Pacific Flyway Council 2002). The seagrass beds around Izembek are some of the largest in the world, and eelgrass composes as much as 99% of the brant's diet during this period (Derksen and Ward 1993, Ward et al. 2005, Shaughnessy et al. 2012).

## Cackling Goose

### Population and Status

In 2004, the American Ornithological Union split four populations of the Canada goose (*Branta canadensis*)—Taverner's, Richardson's, Aleutian, and cackling—and placed them into a new species, the cackling goose (Banks et al. 2004). Cackling geese breed across the Arctic of North America and the global population ranges from 920,000 to 1,400,000 individuals (Birdlife International 2015). The cackling goose has an IUCN Red List category of *Least Concern* and populations appear to be increasing dramatically (BirdLife International 2015). Recent population estimates for the *Branta hutchinsii minima* subspecies of cackling goose (hereafter referred to as the cackling goose) range from 281,000 to 312,000 birds (USFWS 2014). The cackling goose nests on the Yukon–Kuskokwim Delta of western Alaska, mainly winters in the

**Figure 15.3** Flying Pacific black brant. *Photo credit: Milo Burcham.*

Willamette and Lower Columbia River valleys of Oregon and Washington, and migrates through the Bristol Bay region (Pacific Flyway Council 1999).

## Bristol Bay

Few cackling geese appear to use the Bristol Bay area in the spring. At one time, cackling geese may have bred down to the Alaska Peninsula, but that is not currently the case (Pacific Flyway Council 1999). Bristol Bay has long been noted as a place of importance for cackling geese during fall months (Nelson and Hansen 1959, from Pacific Flyway Council 1999). Subsequent surveys have documented almost the entire population staging at Bristol Bay, mainly at Cinder Lagoon and Ugashik Bay, from September through October (Gill et al. 1997). With an increasing population since the 1990s, the distribution of cackling geese in the Bristol Bay region has changed and numbers have increased at Izembek Lagoon (Pacific Flyway Council 1999). In fall, many cackling geese appear to migrate nonstop from the Alaska Peninsula to their wintering grounds (Gill et al. 1997), so they forage and fatten intensively on the intertidal marshes and coastal wetlands of Bristol Bay, eating vegetation such as *Puccinellia, Triglochin*, and *Hippuris* (Pacific Flyway Council 1999).

## Tundra Swan

### Population and Status

Tundra swans (Figure 15.4) breed across Arctic North America (Limpert and Earnst 1994) with a global population estimate of over 300,000 individuals (Birdlife International 2015). In Alaska, the North Slope population of tundra swans winters in the eastern United States and the western Alaska breeding population winters in the western United States (Pacific Flyway Council 2001). The western population averages 133,300 birds (1995 to 2004) (Conant and Groves 2005). Globally, the tundra swan has an IUCN Red List category of *Least Concern* (BirdLife International 2015).

### Bristol Bay

In spring, swans are among the earliest waterfowl to migrate to the Bristol Bay region—generally arriving mid- to late March (Wilk 1988). Approximately 15,000 tundra swans breed on the Bristol Bay lowlands (Conant and Groves 2005), accounting for over 10% of the western breeding population. Uniquely, less than 1,000 tundra swans breed at the southern end of the Alaska Peninsula and winter on Unimak Island and near Izembek Lagoon (Pacific Flyway Council 2001); this is apparently the only nonmigratory breeding population of tundra swans in North America (USFWS 1998). Tundra

**Figure 15.4**    Standing tundra swan. *Photo credit: John and Karen Hollingsworth, USFWS.*

swans have been noted to be sensitive to human disturbance, especially along roads (Sowl and Poetter 2004). Swans are herbivores and on the north shoreline side of the Alaska Peninsula have been observed feeding on pond weed (*Potamogeton filiformis* and *P. praelongus*) (Wilk 1988).

## Steller's Eider

### Population and Status

Steller's eider (Figure 15.5) breed across Arctic Russia and Alaska (Fredrickson 2001) with a global population estimate of 110,000 to 125,000 individuals (Birdlife 2015). In southwest Alaska, the mean spring count (1992 to 2012) was 81,925 (Larned 2012). Steller's eiders are on various special status lists including on: Audubon Alaska's WatchList, IUCN's Red List (as *Vulnerable*), the State of Alaska's species of special concern list, and the U.S.' s federally endangered species list (as *Threatened*). Based on population estimates of birds staging in southwest Alaska, the Alaskan population has declined by 2.3% per year from 1992 to 2011, or 46% over 20 years (Larned 2012).

### Bristol Bay

In the spring, Steller's eiders migrate along the Bristol Bay coast of the Alaska Peninsula, cross Bristol Bay toward Cape Peirce, then continue northward along the Bering Sea (from Larned 2012). In late summer or early fall, Steller's eiders begin arriving to Bristol Bay from Alaskan and Russian breeding grounds (Dau et al. 2000). Numbers of Steller's eiders appear to be particularly high during mild winters (Gill and Petersen 1981). Petersen (1981) found that in the fall along the north side of the Alaska Peninsula, adult female Steller's eiders are primarily molting at Izembek Lagoon, adult males at Nelson and Izembek lagoons, and subadult birds primarily at Nelson Lagoon. The majority of the Pacific population of Steller's eiders comes to the region to molt (and are particularly vulnerable since they are flightless during much of the molt stage) and a significant proportion of these birds winter in the Bristol Bay area (Larned 2008, 2012). During the fall and winter, Steller's eiders at Nelson Lagoon feed on blue mussels (*Mytilus edulis*), other bivalves—mostly clams (*Macoma balthica*), and gammarid amphipods in near-shore marine waters (Petersen 1981).

**Figure 15.5**    Male Steller's eider on breeding grounds in northern Alaska. *Photo credit: Milo Burcham.*

## King Eider

### Population and Status

The king eider (Figure 15.6) is a circumpolar Arctic breeder with two populations in North America, one in the western Arctic and one in the eastern Arctic (Powell and Suydam 2012, Birdlife International 2015). The global population of king eiders is estimated to be about 790,000 to 930,000 individuals (Birdlife International 2015) with a western population (Alaska and western Canada) of about 350,000 individuals (Powell and Suydam 2012). King eiders concentrate close to shore, in shallow water with low salinity (Phillips et al. 2006). During the winter, king eiders are found in areas with low ice concentrations. The inability of eiders to fly away from disturbance during wing molt may make them vulnerable to catastrophic events (Phillips et al. 2006). In southwest Alaska, the mean spring count of king eiders (1992 to 2010) was 166,841 birds (Larned 2012) with a great deal of variation (e.g., see Larned 2008), probably because this species is not targeted during surveys. King eiders are an Audubon WatchList species. Migration counts of the western population of king eiders from Point Barrow show a decline of 55% between 1976 and 1996 (Powell and Suydam 2012).

### Bristol Bay

King eiders are known to use Bristol Bay throughout the year, with spring and fall migration peaks and breeding season use by subadult birds (Schamber et al. 2010). Large spring concentrations of 200,000+ have been recorded in the shallow waters from Cape Constantine to Kvichak Bay with lesser numbers off of Port Heiden and Port Moller (Larned 2008, 2012, Schamber et al. 2010). Tracks of satellite tagged birds marked on breeding grounds in northern Alaska and the western Canadian Arctic reveal three main wintering regions for king eiders of which the Bristol Bay region, especially the inner bay, is one (Phillips et al. 2006, Oppel et al. 2008), although winter use is variable (Schamber et al. 2010). King eiders generally feed on benthic invertebrate, but their specific diet in the Bristol Bay region is unknown (Schamber et al. 2010).

## Black Scoter

### Population and Status

The black scoter (Figure 15.7) breeds mainly in North America (small numbers in far eastern Russia) with an eastern Canada Arctic and a western Alaska Arctic breeding population (Bordage and Savard 2011). The global black scoter population is estimated at 350,000 to 560,000 birds (BirdLife International 2015). Currently, there are believed to be about 200,000 black scoters in the western North American population, of which about 75% live in Alaska (USFWS 2005). In southwest Alaska, the mean spring count (1992 to 2012) was 38,055 birds with an April 2000 high count of 55,538 birds (Larned 2012). The black scoter is listed as an Audubon WatchList species and an IUCN Red List *Near Threatened* species. In Alaska, the species appears to be declining (Stehn et al. 2006, Platte and Stehn 2013).

### Bristol Bay

Black scoters begin arriving in the Bristol Bay area in April and most depart by November (Schambler et al. 2010). In spring (mid-April), the highest counts of migrating black scoters come from Izembek Lagoon to Port Heiden. In inner Bristol Bay, Kvichak Bay (> 5,000 birds) was particularly important (Larned 2008, 2012). The Bristol Bay Lowlands are second only to the Yukon-Kuskokwim Delta in importance as a breeding area for black scoters with an estimated 46,104 breeding birds (from May 1993 to 1994 surveys, Platte and Butler 1995). Overall, a significant proportion of the Pacific black scoter population (up to 75%) appears to use the northeastern area of Bristol Bay at some point in their life cycle (Schambler et al. 2010). Black scoters tend to concentrate closer to shore than king eiders, feeding on benthic invertebrates in shallow water and are found in areas with low ice concentrations (Schambler et al. 2010).

## Long-Tailed Duck

### Population and Status

Long-tailed ducks (Figure 15.8) are a circumpolar Arctic and subarctic breeder with a global population of 6,200,000 to 6,800,000 birds (BirdLife International 2015). A mean of 89,500 birds (2001 to 2010) breed in Alaska (Mallek and Groves 2011). Within Alaska, breeding long-tailed ducks have declined 40% from 1957 to 2010 (Mallek and Groves 2011). The

**Figure 15.6**    Pair of king eiders on the breeding grounds in northern Alaska. *Photo credit: Milo Burcham.*

**Figure 15.7**    Pair of black scoters. *Photo credit: Milo Burcham.*

**Figure 15.8**    Long-tailed duck in winter. *Photo credit: Milo Burcham.*

North American Breeding Waterfowl Survey indicates a steady long-term downward trend between the 1970s and early 1990s, and a stable population since (Birdlife International 2015).

## Bristol Bay

The mean spring count of long-tailed ducks (1992 to 2010) in southwest Alaska was 20,860 birds, mainly dispersed between Kvichak Bay and the Izembek area (Larned 2012). During this period, over 20% of the Alaska population may pass though Bristol Bay as they head north to breeding grounds. Small numbers (< 5,000) winter in the Izembek NWR complex (Wilson and Dau 2014), and at least some of the long-tailed ducks that winter along the Alaska Peninsula breed on the Yukon-Kuskokwim Delta (Petersen et al. 2013). Long-tailed ducks are a deep-diving sea duck (> 60 m), feeding on a variety of invertebrates and other small animals (Robertson and Savard 2002).

## Short-Tailed Shearwater

### Population and Status

Short-tailed shearwaters breed in Australia and Tasmania with a global population estimated to be greater than 23 million individuals (Birdlife International 2015). As many as 16 million short-tailed shearwaters migrate from southern breeding grounds in Australia northward to feed in the rich waters of the Bering Sea during summer months (Schneider and Shuntov 1993, Carey et al. 2014), making it the most abundant marine bird species to use the Bristol Bay region. In the Bering Sea, the highest numbers of shearwaters occur between May and September, concentrated particularly over the continental shelf in the southern Bering Sea (Denlinger 2006), and often associated with the 50-m isobaths (Hunt et al. 2002a). Globally, the short-tailed shearwater has an IUCN Red List category of *Least Concern* (BirdLife International 2015) and a low conservation status among Alaskan seabirds (USFWS 2009), although the population is suspected to be declining (Schumann et al. 2014).

### Bristol Bay

Short-tailed shearwaters occur in the Bristol Bay region between May and September before heading south to their Australasian breeding grounds (Denlinger 2006). They are the most abundant pelagic bird species within Bristol Bay waters during July and August (Bartonek and Gibson 1972). These shearwaters feed on crustaceans, squid, and fish (Denlinger 2006) and are prone to large die-offs in the Bering Sea. Past die-offs appeared to be the result of starvation in warm-water years when populations of euphausiids (*Thysanoessa* spp.), a favored prey, became unavailable (Baduini et al. 2001).

## Bar-Tailed Godwit

### Population and Status

Bar-tailed godwits breed in Arctic regions of Europe and Asia with one population (*Limosa l. baueri*) breeding in western and northern Alaska (McCaffery and Gill 2001). The global population of bar-tailed godwits is estimated at 1,100,000 to 1,200,000 individuals (BirdLife International 2015), of which approximately 90,000 (range 80,000 to 120,000) breed in Alaska (Andres et al. 2012). Bar-tailed godwits are an Audubon WatchList species and have a conservation category of *High Concern* in the Alaska Shorebird Conservation Plan (Alaska Shorebird Group 2008). The Alaska population is estimated to be declining at close to 3% per year (Piersma et al. 2016) due to tidal habitat loss on the Yellow Sea coast (Ma et al. 2015).

### Bristol Bay

The Bristol Bay region is particularly important to fall migrating bar-tailed godwits, with over one-third of the Alaska breeding population staging at Egegik Bay (over 30,000 birds) prior to migrating to wintering areas in New Zealand and eastern Australia (Gill and McCaffery 1999, Gill and Sarvis 1999). Peak counts occur in September to early October (Gill and Handel 1981). Bar-tailed godwits often move back and forth between the Kuskokwim Shoals on the Yukon-Kuskokwim Delta and the Alaska Peninsula before migrating south; some individuals do this multiple times within a matter of days (Gill et al. 2009, Battley et al. 2011). Between 3,000 and 10,000 bar-tailed godwits have been observed at Cinder Lagoon, Port Heiden, and Nelson Lagoon/Mud Bay. A small number of them may breed in the Bristol Bay region, on the Nushagak Peninsula, the southern extent of *baueri's* breeding range (McCaffery and Gill 2001). Feeding on clams and polychaete worms on Egegik's mudflats (Warnock, pers. obs.), these birds will put on fat loads that are 55% of their body mass in a matter of weeks (Piersma and Gill 1998) before embarking on nonstop migratory flights lasting over eight days (Gill et al. 2009, see Savage this volume).

## Marbled Godwit

### Population and Status

The majority of marbled godwits breed in the grasslands of the Prairie Pothole region of the U.S. and Canada, but a small population (*Limosa f. beringiae*) breeds in Alaska (Gratto-Trevor 2000). The global population of marbled godwits is estimated at 173,500 individuals with 1,000 to 3,000 *L. f. beringiae* breeding on the Alaska Peninsula (Andres et al. 2012). Marbled godwits are an Audubon WatchList species and are a species of *High Concern* in the Alaska Shorebird Conservation Plan (Alaska Shorebird Group 2008). The Alaska breeding population is thought to be stable (Andres et al. 2012), but it is poorly studied. While immature marbled godwits have been noted along the Alaska Peninsula since at least the early 1880s (collected near Ugashik, July 1881; Osgood 1904, noted in Murie 1959), the first nests of this species on the Alaska Peninsula were found in the 1990s (North et al. 1996).

## Bristol Bay

The Bristol Bay region is of critical importance to the *beringiae* subspecies of marbled godwits. All *beringiae* godwits breed on the central Alaska Peninsula, restricted to low-lying regions between Port Heiden and Ugashik (Alaska Shorebird Group 2008, Ruthrauff and Tibbitts 2009), with major fall staging sites at Ugashik Bay and Cinder-Hook Lagoon (Gill and Sarvis 1999). Marbled godwits mainly feed on benthic invertebrates (Gratto-Trevor 2000).

# Dunlin

## Population and Status

Dunlin are a Holarctic breeder with three populations breeding in North America (Warnock and Gill 1996). The global population of dunlin is estimated to number from 4,600,000 to 6,500,000 individuals (BirdLife International 2015). Two populations of dunlin breed in Alaska, *Calidris a. pacifica*, western Alaska breeders and *C. a. arcticola*, northern Alaska breeders (Warnock and Gill 1996). Most, if not all, of the dunlin on the Alaska Peninsula appear to be *pacifica* (Gill et al. 2013). The *pacifica* population is estimated to number 550,000 birds (Andres et al. 2012). The dunlin is an Audubon Watchlist species and has a conservation category of *High Concern* (both races) in the Alaska Shorebird Conservation Plan (Alaska Shorebird Group 2008). Reliable trend data are lacking for *pacifica* dunlin, although there are indications that this is a population in decline (Fernandez et al. 2008).

## Bristol Bay

Spring time surveys between May 3 and May 11 at Port Heiden, have recorded > 45,000 dunlin in some years (Gill et al. 2001). Some dunlin breed along the coastal fringe of the Bristol Bay region (Murie 1959, Gill et al. 1981, Brna and Verbrugge 2013). The region is particularly important to post-breeding dunlin during the fall migration period. In some years, over 30% of *pacifica* dunlin may stop and use the Nelson Lagoon-/Mud Bay wetland complex (184,000 birds were counted by plane on September 22, 1993), but most of the estuaries between Nelson Lagoon to inner Bristol Bay support numbers of migrating dunlin and other small shorebirds (Gill et al. 2001, Fernandez et al. 2010, Brna and Verbrugge 2013). Dunlin feed on benthic invertebrates during migration (Warnock and Gill 1996).

# Common Murre

## Population and Status

Common murres (Figure 15.9) have a circumpolar breeding distribution with a global population estimate of 13 to 21 million birds (USFWS 2009, BirdLife International 2015). In Alaska, the breeding population is estimated to be 2.8 million birds (USFWS 2009). Globally, the common murre has an IUCN Red List category of *Least Concern* (BirdLife International 2015) and a low conservation priority among Alaskan seabirds (USFWS 2009). This is one of the most abundant seabirds in Alaska and Bristol Bay. For recent decades, the overall population trend for common murres in Alaska appears to be stable (Dragoo 2014). The Cape Peirce murre colony has declined 1.5% per year from 1989 to 2013, although most of the decline is due to a drop in the population between 1997 and 1998 (Dragoo et al. 2014). Overall, the Round Island population has increased 0.6% per year, but it has been declining since 2008 (Dragoo et al. 2014).

## Bristol Bay

Common murres occur in the waters of Bristol Bay year round (Denlinger 2006). Common murres breed in dense colonies and close to a million may breed in the Bristol Bay region (including unidentified murres, Table 15.2), with 75% of the birds breeding in five main colonies in northeastern Bristol Bay: Round Island (145,724 birds), Bird Rock off of Cape Newenham (102,632 birds), Shaiak Island (102,368 birds), North Twin Island (98,231 birds), and Jagged Mountain (42,646 birds) (USFWS 2015). A main food item for murres is juvenile pollock (*Theragra chalcogramma*), but there is considerable variation among regions (Dragoo et al. 2010). Murres are voracious feeders, daily consuming 10–30% of their body mass (Denlinger 2006). Large die-offs of up to 100,000 birds—attributed to spring storms and starvation—have been recorded for common murres on the north side of the Alaska Peninsula (Bailey and Davenport 1972).

**Figure 15.9**    Common murre on nesting island. *Photo credit: Milo Burcham.*

## Black-Legged Kittiwake

### Population and Status

Black-legged kittiwakes (Figure 15.10) breed across much of the Holarctic and number 17 to 18 million birds globally (BirdLife International 2015). The Pacific population of black-legged kittiwake is about 2.6 million birds, 50% of which are estimated to live in Alaska (USFWS 2009). These kittiwakes breed in enormous cliff-side colonies in Alaskan waters, and after the breeding season they move into deeper waters along outer ocean shelves (USFWS 2009). Black-legged kittiwakes are of moderate conservation concern in Alaska (USFWS 2009) and are an Audubon Watchlist species. Population trends for black-legged kittiwake colonies in Alaska are mixed (Dragoo et al. 2014), with significant declines in the Gulf of Alaska (Wernock 2017).

### Bristol Bay

Black-legged kittiwakes are common in the Bristol Bay region from spring through fall and uncommon during winter months (Denlinger 2006). In the 1970s, close to 500,000 black-legged kittiwakes bred in Bristol Bay; however, more recent estimates are only half of that (260,070 birds) (Table 15.2). The largest colony, 80,000 birds, is at Round Island (USFWS 2015). At Cape Peirce, from 1990 to 2013, the colony has decreased annually by 2.6%, although since around 2005 the colony appears to have stabilized (Dragoo et al. 2014). The black-legged kittiwake colony at Round Island has declined about 1.9% per year since 1999 (Dragoo et al. 2014). Black-legged kittiwakes feed on fish and invertebrates with considerable variation in diet among breeding localities (Dragoo et al. 2010). Kittiwakes tend to be surface feeders, and their breeding productivity is closely tied with availability of their prey (Schneider and Shuntov 1993, Denlinger 2006).

**Figure 15.10**    Black-legged kittiwake on ice. *Photo credit: Milo Burcham.*

## Other Species

While Bristol Bay is particularly important to populations of birds listed above, other species still deserve mention. Nearly 90,000 tufted puffins breed in the region, the majority on Shaiak Island (USFWS 2015). The *couesi* race of rock sandpiper (*Calidris ptilocnemis*) breeds in the Aleutians and along the Alaska Peninsula to as far north as the Ugashik Bay area; most of these birds remain throughout the year except for severe winters (Gill et al. 2002). Izembek Lagoon is a key breeding and post-breeding site for these birds as well as the barrier islands of the Seal Islands where September counts have ranged from 15,000 to 30,000 birds (Gill et al. 2001). The breeding area for the *tschuktschorum* race of rock sandpiper is from the northern part of Bristol Bay northward (Gill et al. 2002). The most common breeding ducks in the Bristol Bay Lowlands and other areas of the region are greater scaup and lesser scaup (*Aythya affinis*), which are grouped together due to the difficulty of distinguishing species differences during aerial waterfowl surveys—although most are thought to be greater scaup (Stehn et al. 2006). The 20 year average of scaup in the Bristol Bay Lowlands is 93,340 birds (Stehn et al. 2006). A significant proportion of Aleutian terns (*Onychoprion aleuticus*), a species thought to be in significant decline in Alaska, breed in colonies along the Alaska Peninsula and into the northeastern Bristol Bay region (USFWS 2009).

Not only do the inter-tidal flats and waters of Bristol Bay support nationally and internationally significant numbers of birds, but the birds that use the Bristol Bay area provide important subsistence value to residents of Alaska. A yearly average of 37,500 birds (from 2001–2005) were harvested for subsistence use in Bristol Bay, mainly ducks and geese, including 1,000–5,000 brant/year (Wentworth 2005). Converted to usable weights, the subsistence harvest of birds provided an average of 75,000 pounds of meat annually to Bristol Bay residents from 2001 to 2005—about 30 pounds of meat per household.

## Conservation Issues

The Bristol Bay region, including the Bristol Bay Borough, the Lake and Peninsula Borough, and the Dillingham Census area, is home to fewer than 8,000 people (Alaska Population Projections 2007), thus the region does not exhibit many of the environmental problems that impact marine birds in more densely populated areas of the world—such as large-scale, human caused habitat loss and alteration (e.g., Ma et al. 2015) or pollution. Still, significant threats pose a risk to marine birds, especially those related to climate change, potential resource extraction (oil and gas development, mining), bycatch from fishing, disruption of sensitive bottleneck sites for birds (both in Bristol Bay and beyond), and the potential for disruption of the rich food chain that supports the millions of marine birds in the Bristol Bay area.

Climate change resulting from both anthropogenic and natural impacts [e.g., cyclical climate patterns like the Pacific Decadal Oscillation (PDO)] has and will continue to impact the biota of the Bering Sea and Bristol Bay by changing sea and air temperatures, water chemistry, impacting the extent and timing of sea ice, and by affecting the food web, among other things (Brodeur et al. 1999, Hunt et al. 2002b, Schumacher et al. 2003). Mirroring Arctic Alaska (Wendler et al. 2014), mean temperatures have been increasing (particularly in the winter season) and winter ice has been decreasing in Bristol Bay (Ward et al. 2009, see also http://climate.gi.alaska.edu/). Additionally, the Bristol Bay region has a high ocean acidification risk index (Mathis et al. 2015). With these changing conditions, in the recent decades (1990s into the mid-2000s) much of the Bering Sea witnessed declining benthic prey populations, increasing pelagic fish, retreating sea ice, and increasing air and ocean temperatures (Grebmeier et al. 2006); and conditions continue to change (Coyle et al. 2011). These changes impact waterbirds in the Bristol Bay region in different ways.

The marine prey base for seabirds and waterbirds in the Bering Sea is especially impacted by water temperature and ice cover (Hunt et al. 2002b, Eisner et al. 2014). Irons et al. (2008) showed that common murre populations in the Arctic increased the most with moderate cooling of sea surface temperatures, but decreased when sea surface temperature shifts were large (in either directions). The winter distribution of ice-associated sea ducks, such as king eiders, is influenced by the extent of sea ice in the Bering Sea and Bristol Bay, and decreasing amounts of sea ice may shift their winter distribution northward in the Bering Sea (Oppel et al. 2008). Ocean acidification and its direct negative effects on calcifying species like mussels, crabs, and shrimp (e.g. Bechmann et al. 2011) could negatively impact many marine birds of Bristol Bay, including scoters and eiders, but also birds foraging on calcified invertebrates in the intertidal, like shorebirds (e.g., bar-tailed godwit, rock sandpiper) and some of the waterfowl (e.g., emperor goose, Steller's eider, sandhill crane).

For brant, with higher average winter temperatures and lower amounts of shore-fast ice cover along the Alaska Peninsula, the wintering population has grown significantly (7% per year between 1986 and 2004), presumably because of better access to food supplies (mainly eelgrass beds) and reduced thermoregulatory demands (Ward et al. 2009). Ward et al. (2009) also note that with an increasing proportion of Pacific black brant wintering along the Alaska Peninsula, severe winter events with extreme ice events have the potential to negatively impact greater numbers of brant. This pattern of increasing use of Izembek by brant in the winter because of increasing temperatures could be reversed by another climate issue, rising sea levels. Shaughnessey et al. (2012) modeled eelgrass response to globally predicted sea-level rise at seven key wetlands used by brant along the Pacific Coast between Alaska and Mexcio, including Izembek Lagoon, and found that accessible eelgrass habitat would decrease in winter months (particularly December) as sea levels rise, forcing brant to winter at more mid-latitude sites.

Brant also highlight the potential impact that climate change has on global wind patterns. Ward et al. (2009) found that the numbers of brant migrating south to Mexico from Izembek Lagoon was positively correlated with the number of days of strong northwesterly winds in November—winds favorable for southward migration. Years with warm phases of the PDO had fewer days of strong northwest winds than years with cold phases. Other fall migrating geese and shorebirds (including the cackling goose, bar-tailed godwit, whimbrel (*Numenius phaeopus*), and dunlin) use similar weather patterns to depart on long, nonstop migrations, and the impact of climate variability on these wind-selected avian migrants is still being studied (Gill et al. 2005, 2009, 2014).

Marine bird mortality in Alaskan waters due to bycatch entanglement and capture by fishing equipment, especially gill nets and long lines, has been and continues to be significant, although bycatch of seabirds is improving because of better management (Melvin and Parrish 2001, Uhlmann, 2003, Manville 2005, USFWS 2009, Dietrich and Fitzgerald 2010). During the 1970s and 80s, large-scale, high seas gill nets spanning hundreds of kilometers killed hundreds of thousands of seabirds per year in the North Pacific (Manville 2005). Short-tailed shearwaters were particularly impacted by the huge Japanese mothership salmon driftnet fishery from 1952 to 1988 (Uhlmann 2003), but these fisheries have largely stopped (Denlinger 2006). In Alaska, trawl fisheries account for a significant proportion of the total shearwater bycatch, although bycatch is thought to be relatively small (Denlinger 2006).

Currently, the longline fisheries have replaced the gill net fisheries as the most significant contributor to seabird bycatch in Alaskan waters. From 1993 to 1997, an average of over 11,000 birds was documented as bycatch in the Bering Sea/Aleutian Island longline fisheries (Stehn et al. 2001). The species most impacted by the groundfish fisheries (at least numerically) was the northern fulmar (*Fulmarus glacialis*). Of total bycatch in the longline fisheries in the Bering Sea/Aleutian Islands, northern fulmars comprised 59% (7431 individuals per year) of the catch, while in the trawling fisheries, northern fulmars comprised > 53% of the total bycatch between 1998 and 2003 (Denlinger 2006). Other species impacted by the longline fisheries include gulls, albatross [Laysan (*Phoebastria immutabilis*), black-footed (*P. nigripes*), and a few short-tailed (*P. albatrus*)], and sooty (*Puffinus griseus*) and short-tailed shearwaters (Manville 2005, Dietrich and Fitzgerald 2010).

A serious potential threat to waterbirds of the Bristol Bay area comes from proposed open pit mining within the watersheds in the upper reaches of the Bristol Bay basin that flow to Bristol Bay. Tailings left behind by these mines typically have high concentrations of copper, which can be toxic to fish, wildlife, and invertebrates in freshwater and marine systems (Eisler 1997). As seen in recent history, tailings at large mines can quickly spill and travel hundreds to thousands of kilometers (Bacsujlaky 2004). The ensuing pollution could potentially disrupt the salmon-based food web flowing from the Bristol Bay watersheds that Bristol Bay's marine birds rely on (Wipfli and Baxter 2010, Chambers et al. 2012, EPA 2014). King eiders and black scoters have been identified as a species particularly vulnerable to disruptions to their food chain through mining accidents in the Bristol Bay region (Schamber et al. 2010).

While the Bristol Bay area has a low human density, real threats remain with regard to human disturbance of birds at sites where birds congregate. Izembek Lagoon is a site where this threat crystalizes. In 1985, a road project was identified in the Bristol Bay Management Plan to connect the small towns of King Cove and Cold Bay (U.S. Department of Commerce 1985, USFWS 1998). The approximately 50 km road would cross ~16 km of Izembek NWR lands, including lands designated as wilderness, bisecting a 5-km wide isthmus separating Izembek and Kinzarof lagoons—areas where globally significant numbers of brant, emperor geese, and Steller's eiders molt, fatten for migration, and increasingly, overwinter. USFWS biologists concluded that "Increased access to previously remote staging areas, particularly through use of off road vehicles, could interrupt foraging activities and displace birds from feeding areas, compromising migration readiness and survival . . ." (Sowl and Poetter 2004, p. 11). In December 2013, Secretary of Interior, Sally Jewell, signed a Record of Decision (USFWS 2013) on the proposed road corridor through the Izembek NWR, reaffirming the Department of Interior's opposition to the road.

Oil spills from oil and gas development or shipping accidents can directly and indirectly cause mortality to marine birds (Jenssen 1984, Wiens 1995) as well as impair their productivity by disrupting and/or destroying the food web they rely on (Schamber et al. 2010). There is a long history of oil exploration in Bristol Bay (Sherwood et al. 2006, Sherwood this volume). However, the lack of any major oil and gas finds coupled with the 2014 order by President Obama to disallow oil and gas drilling in Bristol Bay reduced any near-term potential threat that oil and gas development presents to marine birds in the region. Regardless, oil spills due to shipping accidents remain a potential threat (USFWS 2009), especially since shipping is increasing in the Bering Sea due to decreasing Arctic sea ice (Arctic Marine Shipping Assessment 2009).

## RECOMMENDED RESEARCH

Given the global importance of the Bristol Bay region to marine birds of various species, there are still significant and basic gaps in our knowledge about the ecology of marine birds in the area. Overall, better monitoring of most marine bird species is warranted in all seasons. Currently, the main sources of information come from a handful of annual or biannual surveys that focus on a few key species, including the biannual emperor goose surveys, the annual Steller's eider surveys, and the annual USFWS breeding pairs surveys (mainly for waterfowl). While these surveys form the foundation for what we know about marine bird use of the region, they are limited in their ability to collect data for other species (i.e., the emperor goose and the Steller's eider surveys are strictly coastal). At-sea surveys have been sporadic at best. Certain special status species for which the region is particularly important would benefit from more detailed species-specific survey efforts around Bristol Bay, including Aleutian terns and bar-tailed godwits. For marbled (*Brachyramphus marmoratus*) and Kittlitz's murrelets (*Br. brevirostris*), both species of concern, it is still unknown how important the region is to them. A synthesis of species distribution and phenology from existing datasets like the North Pacific Pelagic Seabird Database would be useful.

Given the high biomass of marine birds that use the region, a better understanding of the food base that supports these birds is of great importance. Except for commercially important species (e.g., crabs), the invertebrate fauna of the region has not been well surveyed. Little is known about what the shorebirds prey on when using coastal estuaries of Bristol Bay or when they are on their breeding grounds.

Coupled with a better understanding of the food base for marine birds in the Bristol Bay region is the need to understand how climate change is impacting this region, its food web, and the waterbirds that rely on it. Certain seabirds have already been shown to be impacted by sea temperature and changes in prey composition. These types of studies need to be expanded to other groups of waterbirds in the region. How will ocean acidification affect bivalve populations in Bristol Bay and what impacts will this have on birds—such as eiders, emperor geese, and shorebirds—that rely on these bivalves? As temperatures continue to warm, especially in winter months, how will the distribution of bird populations change in the region? What will the effect of decreasing sea ice have on marine productivity and waterbird populations?

Likewise, given the sheer scale of some of the proposed large-scale mining activities, especially those in the upper reaches of the Bristol Bay watershed, potential impacts of mining on the food web and bird populations of the region need to be better understood. Finally, demographic studies of the waterbirds of Bristol Bay are largely lacking with the exception of the seabird breeding studies carried out by the USFWS's Alaska Maritime National Wildlife Refuge staff. Demographic data are key to understanding the health of waterfowl populations in the area. Colony studies of birds like Aleutian terns might elucidate why these birds are declining in the state of Alaska.

## SUMMARY

Bristol Bay is an ecologically rich marine area with a productive food-web that nourishes abundant, globally significant, and highly diverse marine bird populations. In the past 40 years, as ornithologists have begun to systematically survey birds in the region, a better understanding of this salmon-centered ecosystem has emerged and begun to elucidate the attraction for millions of shorebirds, seabirds, and waterfowl to Bristol Bay each year. This region boasts vast eelgrass beds in coastal lagoons, high densities of zooplankton such as euphausids in the pelagic zone, rich densities of invertebrates such as crabs and mussels in the benthos, and the tens of millions of salmon occupying the whole of the marine, estuarine, and freshwater ecosystem.

The single most abundant seabird in the region, numbering in the tens of millions, is the short-tailed shearwater, which breeds in Australia during the northern winter then comes to Alaska to forage on euphausids and small fish in the Bering Sea during the northern summer. At least 75 marine bird species breed in the Bristol Bay region, many of them in globally significant numbers. Perhaps the entire global populations of the emperor goose and the Pacific black brant

**Appendix 15.1**    Regularly occurring marine birds in the Bristol Bay region[1]

| Species | | Breeding status |
|---|---|---|
| Greater white-fronted goose | *Anser albifrons* | B[2] |
| Emperor goose | *Chen canagica* | B |
| Snow goose | *Chen caerulescens* | |
| Brant | *Branta bernicla* | B |
| Cackling goose | *Branta hutchinsii* | |
| Canada goose | *Branta canadensis* | B |
| Trumpeter swan | *Cygnus buccinator* | B |
| Tundra swan | *Cygnus columbianus* | B |
| Gadwall | *Anas strepera* | B |
| Eurasian wigeon | *Anas penelope* | |
| American wigeon | *Anas americana* | B |
| Mallard | *Anas platyrhynchos* | B |
| Northern shoveler | *Anas clypeata* | B |
| Northern pintail | *Anas acuta* | B |
| Green-winged teal | *Anas crecca* | B |
| Canvasback | *Aythya valisineria* | |
| Redhead | *Aythya americana* | |
| Ring-necked duck | *Aythya collaris* | B |
| Greater scaup | *Aythya marila* | B |
| Lesser scaup | *Aythya affinis* | |
| Steller's eider | *Polysticta stelleri* | |
| Spectacled eider | *Somateria fischeri* | |
| King eider | *Somateria spectabilis* | |
| Common eider | *Somateria mollissima* | B |
| Harlequin duck | *Histrionicus histrionicus* | B |
| Surf scoter | *Melanitta perspicillata* | B |
| White-winged scoter | *Melanitta fusca* | B |
| Black scoter | *Melanitta americana* | B |
| Long-tailed duck | *Clangula hyemalis* | B |
| Bufflehead | *Bucephala albeola* | B |
| Common goldeneye | *Bucephala clangula* | B |
| Barrow's goldeneye | *Bucephala islandica* | |
| Common merganser | *Mergus merganser* | B |
| Red-breasted merganser | *Mergus serrator* | B |
| Red-throated loon | *Gavia stellata* | B |
| Pacific loon | *Gavia pacifica* | B |
| Common loon | *Gavia immer* | B |
| Horned grebe | *Podiceps auritus* | B |
| Red-necked grebe | *Podiceps grisegena* | B |
| Northern fulmar | *Fulmarus glacialis* | |
| Sooty shearwater | *Puffinus griseus* | |
| Short-tailed shearwater | *Puffinus tenuirostris* | |
| Fork-tailed storm-petrel | *Oceanodroma furcata* | |
| Leach's storm-petrel | *Oceanodroma leucorhoa* | |

| Species | | Breeding status |
|---|---|---|
| Double-crested cormorant | *Phalacrocorax auritus* | B |
| Red-faced cormorant | *Phalacrocorax urile* | B |
| Pelagic cormorant | *Phalacrocorax pelagicus* | B |
| Sandhill crane | *Grus canadensis* | B |
| Black oystercatcher | *Haematopus bachmani* | B |
| Black-bellied plover | *Pluvialis squatarola* | B |
| American golden-plover | *Pluvialis dominica* | B |
| Pacific golden-plover | *Pluvialis fulva* | B |
| Semipalmated plover | *Charadrius semipalmatus* | B |
| Spotted sandpiper | *Actitis macularius* | B |
| Solitary sandpiper | *Tringa solitaria* | |
| Wandering tattler | *Tringa incana* | B |
| Greater yellowlegs | *Tringa melanoleuca* | B |
| Lesser yellowlegs | *Tringa flavipes* | B |
| Whimbrel | *Numenius phaeopus* | B |
| Bristle-thighed curlew | *Numenius tahitiensis* | |
| Hudsonian godwit | *Limosa haemastica* | B |
| Bar-tailed godwit | *Limosa lapponica* | B |
| Marbled godwit | *Limosa fedoa* | B |
| Ruddy turnstone | *Arenaria interpres* | |
| Black turnstone | *Arenaria melanocephala* | B |
| Red knot | *Calidris canutus* | |
| Surfbird | *Calidris virgata* | B |
| Sharp-tailed sandpiper | *Calidris acuminata* | |
| Sanderling | *Calidris alba* | |
| Dunlin | *Calidris alpina* | B |
| Rock sandpiper | *Calidris ptilocnemis* | B |
| Baird's sandpiper | *Calidris bairdii* | B |
| Least sandpiper | *Calidris minutilla* | B |
| Pectoral sandpiper | *Calidris melanotos* | B |
| Semipalmated sandpiper | *Calidris pusilla* | |
| Western sandpiper | *Calidris mauri* | B |
| Short-billed dowitcher | *Limnodromus griseus* | B |
| Long-billed dowitcher | *Limnodromus scolopaceus* | |
| Wilson's snipe | *Gallinago delicata* | B |
| Red-necked phalarope | *Phalaropus lobatus* | B |
| Red phalarope | *Phalaropus fulicarius* | |
| Pomarine jaeger | *Stercorarius pomarinus* | |
| Parasitic jaeger | *Stercorarius parasiticus* | B |
| Long-tailed jaeger | *Stercorarius longicaudus* | B |
| Common murre | *Uria aalge* | B |
| Thick-billed murre | *Uria lomvia* | B |
| Pigeon guillemot | *Cepphus columba* | B |
| Marbled murrelet | *Brachyramphus marmoratus* | B |
| Kittlitz's murrelet | *Brachyramphus brevirostris* | B |

## Offshore

Fisheries surveys of the offshore waters of Bristol Bay have been conducted since the 1930s. The Alaska Fisheries Science Center (AFSC) has conducted annual bottom trawl surveys in the eastern Bering Sea offshore and outer Bristol Bay waters since 1982 using standardized gear and repeatable methods. These surveys identify numerous groundfish species inhabiting the eastern Bering Sea and Bristol Bay, generally deeper than the 15–20 m contour (Lauth 2010).[1] The more common species represented in the surveys are cod and pollock (*Gadidae*); fifteen species of flatfish (*Pleuronectiformes*); forage fish species such as herring, eulachon, capelin, smelts, sand lance, and sandfish; and dozens of other species well represented, such as skate (*Rajidae*), poachers (*Psychrolutidae*), greenling (*Hexagrammos*), rockfish (*Scorpaenidae*), sculpin (*Cottidae*), crab (*Cancer*), and salmon. In Table 16.1 we identify all fish species known to inhabit these waters.

The hundreds of fish and invertebrate species that inhabit Bristol Bay waters contribute to trophic levels at various life stages; tides and currents transport and distribute larval marine fish and invertebrate species from offshore to nearshore nursery areas (Norcross et al. 1984, Lanksbury et al. 2007). The relationship between marine and nearshore processes and species presence in Bristol Bay has been well documented in the life histories of species such as walleye pollock, red king crab (*Paralithodes camtschaticus*), and yellowfin and rock sole. Larval forms of each species are transported and concentrated in nutrient-rich nearshore habitat. These four species illustrate relevant examples of recognized marine species with population segments that in a larval or juvenile phase rely on nearshore marine habitat (depths less than 20 m) for refuge and nutrition (Nichol 1998).

Walleye pollock are generally recognized as a pelagic species spawning in open marine waters (Bailey et al. 1999). As Coyle (2002) notes, pollock in their larval and juvenile forms are known to be transported into nearshore nursery zones: the current carries the eggs and larvae along the north shore of the Alaska Peninsula and into the nearshore nursery zones of Bristol Bay (Napp et al. 2000). A recent investigation of trophic interactions shows that juvenile pollock feed on euphasiid and mysiid populations nearshore, especially mysiids, which have been shown to be more abundant in the diets of pollock found in the northern nearshore zones than those found in deep water (Aydin 2010).

Bristol Bay is also home to the second-largest population of red king crab (Dew and McConnaughey 2005, Chilton et al. 2011). Although red king crab of both genders and several stages of maturity occur throughout central Bristol Bay, immature larvae and juveniles are often concentrated along nearshore areas. The Aleutian North Shore and Bering Coastal currents transport larval king crab from the eastern Bering Sea to inner Bristol Bay (Dew and McConnaughey 2005). Larval red king crab (smaller than 2 mm) settle in cobble and gravel substrates of Kvichak Bay[2] (Armstrong et al. 1981, McMurray 1984, Loher et al. 1998); juveniles are present along the nearshore zone in the Togiak district (Armstrong et al. 1993, Ormseth 2009). These juvenile phases inhabit nearshore rocks, shell hash, or a variety of biological cover in shallow depths (from 5 to 70 meters).

Yellowfin and rock sole are among several species of flatfish that inhabit the eastern Bering Sea and for which nearshore substrates (depths less than 30 meters) in Bristol Bay are optimal habitat (McConnaughey and Smith 2000, Lauth 2010; Table 16.1). Life histories of these species and other flatfish take advantage of the same currents that transport larvae into nearshore nursery areas (Nichol 1998, Wilderbuer et al. 2002, Norcross and Holladay 2005, Lanksbury et al. 2007, Cooper et al. 2011). Larval and juvenile yellowfin sole are abundant in shallow nearshore areas along the northern shore and Togiak Bay (Nichol 1998, Wilderbuer et al. 2002, Ormseth 2010). These findings for Pollock, red king crab, yellowfin, and rocksole substantiate our understanding of nearshore and estuary zones as nutrient-rich fish nurseries, providing juvenile fish species with greater forage opportunity in the form of abundant invertebrate populations. For further reading see Chapters 17 and 18.

## Bristol Bay Salmon

The ecological role of Bristol Bay salmon is complex. Salmon facilitate energy and nutrient exchange across multiple trophic levels from terrestrial headwaters through estuarine and marine ecosystems. Each species migrates through these waters at slightly different times depending on life history and watershed of origin. Because of their abundance,

---

[1] All species were found east of the 162° west longitude line and in waters deeper than 15 meters. Because the surveys represent a snapshot of species present at a particular time, they may not represent complete species diversity. Also, because standardized trawl gear mesh is size selective, juvenile and larval specimens of a species may not be well represented. It is important to note that salmon species at any life stage may not be well represented due to seasonality of surveys, species migration, and pelagic marine life stage.

[2] Larval red king crab were present on substrates less than 70 to 80 feet (approximately 21 to 24 meters) at mean low water in Kvichak Bay.

**Table 16.1** Fish species list: species listed have been identified in the NOAA-AFSC Bering Sea Bottom Trawl Surveys between 1982–2010 (Lauth 2010)

| Fish species common name | Family | Scientific name |
|---|---|---|
| Chinook salmon | *Salmonidae* | *Oncorhynchus tshawytscha* |
| Chum salmon | | *Oncorhynchus keta* |
| Steelhead | | *Oncorhynchus mykiss* |
| Pacific cod | *Gadidae* | *Gadus macrocephalus* |
| Walleye pollock | | *Gadus chalcogrammus* |
| Arctic cod | | *Boreogadus saida* |
| Saffron cod | | *Eleginus gracilis* |
| Sablefish | *Anoplopomatidae* | *Anoplopoma fimbria* |
| Eulachon | *Osmeridae* | *Thaleichthys pacificus* |
| Capelin | | *Mallotus villosus* |
| Rainbow smelt | | *Osmerus mordax* |
| Smelt unident. | | *Osmeridae* |
| Pacific herring | *Clupeidae* | *Clupea pallasi* |
| Pacific sand lance | *Ammodytidae* | *Ammodytes hexapterus* |
| Pacific sandfish | *Trichodontidae* | *Trichodon trichodon* |
| Pacific halibut | *Pleuronectidae* | *Hippoglossus stenolepis* |
| Yellowfin sole | | *Limanda aspera* |
| Northern rock sole | | *Lepidopsetta polyxystra* |
| Rock sole unident. | | *Lepidopsetta sp.* |
| Flathead sole | | *Hippoglossoides elassodon* |
| Dover sole | | *Microstomus pacificus* |
| Rex sole | | *Glyptocephalus zachirus* |
| Butter sole | | *Isopsetta isolepis* |
| Sand sole | | *Psettichthys melanostictus* |
| Starry flounder | | *Platichthys stellatus* |
| Alaska plaice | | *Pleuronectes quadrituberculatus* |
| Arrowtooth flounder | | *Atheresthes stomias* |
| Kamchatka flounder | | *Atheresthes evermanni* |
| Longhead dab | | *Limanda proboscidea* |
| Sanddab unident. | | *Citharichthys sp.* |
| Northern rockfish | *Scorpaenidae* | *Sebastes polyspinis* |
| Big skate | *Rajidae* | *Raja binoculata* |
| Bering skate | | *Bathyraja interrupta* |
| Starry skate | | *Raja stellulata* |
| Alaska skate | | *Bathyraja parmifera* |
| Aleutian skate | | *Bathyraja aleutica* |
| Whitespotted greenling | *Hexagrammos* | *Hexagrammos stelleri* |
| Rock greenling | | *Hexagrammos lagocephalus* |
| Kelp greenling | | *Hexagrammos decagrammus* |
| Smooth lumpsucker | | *Aptocyclus ventricosus* |
| Greenling unident. | | *Hexagrammidae* |

| Fish species common name | Family | Scientific name |
|---|---|---|
| Sawback poacher | *Psychrolutidae* | *Leptagonus frenatus* |
| Gray starsnout | | *Bathyagonus alascanus* |
| Sturgeon poacher | | *Podothecus accipenserinus* |
| Aleutian alligatorfish | | *Aspidophoroides bartoni* |
| Arctic alligatorfish | | *Ulcina olrikii* |
| Warty poacher | | *Chesnonia verrucosa* |
| Bering poacher | | *Occella dodecaedron* |
| Wolf-eel | *Anarhichadidae* | *Anarrhichthys ocellatus* |
| Bering wolffish | | *Anarhichas orientalis* |
| Threaded sculpin | *Gymnocanthus sp.* | *Gymnocanthus pistilliger* |
| Arctic staghorn sculpin | | *Gymnocanthus tricuspis* |
| Armorhead sculpin | | *Gymnocanthus galeatus* |
| Northern sculpin | | *Icelinus borealis* |
| Sculpin unident. | | *Cottidae* |
| Hookhorn sculpin | *Artediellus sp.* | *Artediellus pacificus* |
| Irish lord | | *Hemilepidotus sp.* |
| Red Irish lord | | *Hemilepidotus hemilepidotus* |
| Yellow Irish lord | | *Hemilepidotus jordani* |
| Ribbed sculpin | *Triglops sp.* | *Triglops pingeli* |
| Brightbelly sculpin | | *Microcottus sellaris* |
| Warty sculpin | | *Myoxocephalus verrucosus* |
| Great sculpin | | *Myoxocephalus polyacanthocephalus* |
| Plain sculpin | | *Myoxocephalus jaok* |
| Pacific staghorn sculpin | *Myoxocephalus sp.* | *Leptocottus armatus* |
| Antlered sculpin | | *Enophrys diceraus* |
| Spinyhead sculpin | | *Dasycottus setiger* |
| Crested sculpin | | *Blepsias bilobus* |
| Eyeshade sculpin | | *Nautichthys pribilovius* |
| Sailfin sculpin | | *Nautichthys oculofasciatus* |
| Bigmouth sculpin | | *Hemitripterus bolini* |
| Thorny sculpin | | *Icelus spiniger* |
| Spatulate sculpin | | *Icelus spatula* |
| Variegated snailfish | *Liparis sp.* | *Liparis gibbus* |
| Snailfish unident. | | *Liparidinae* |
| Daubed shanny | *Stichaeidae* | *Lumpenus maculatus* |
| Snake prickleback | | *Lumpenus sagitta* |
| Decorated warbonnet | | *Chirolophis decoratus* |
| Bearded warbonnet | | *Chirolophis snyderi* |
| Polar eelpout | | *Lycodes turneri* |
| Giant wrymouth | *Cryptacanthodidae* | *Cryptacanthodes giganteus* |

distribution, and overall economic importance, Bristol Bay sockeye salmon have been more extensively studied than other salmonids in the region. Generally, once in marine waters juvenile salmon spend their first summer in relatively shallow waters on the southeastern Bering Sea shelf, feeding, growing and eventually moving offshore into the Bering Sea basin and North Pacific Ocean (Myers et al. 2007, Farley et al. 2011, Farley 2012, pers. comm.).

## Range and Distribution

The Magnuson-Stevens Act defines essential fish habitat (EFH) as *waters and substrate necessary to fish for spawning, breeding, feeding, or growth to maturity*. For salmon, EFH consists of those fresh and marine waters needed to support healthy stocks in order to provide long-term sustainable salmon fisheries. Because of the broad range and distribution of salmon in Alaskan waters, all marine waters over the continental shelf in the Bering Sea extending north to the Chukchi Sea and over the continental shelf throughout the Gulf of Alaska and in the inside waters of the Alexander Archipelago are defined as EFH for all juvenile salmon (Echave et al. 2011). EFH for immature and mature Pacific salmon (*Oncorhynchus spp*) includes nearshore and oceanic waters, often extending well beyond the shelf break (Echave et al. 2011).

In their emigration phase, anadromous juvenile salmon occupy shallows of estuaries and nearshore zones, although timing, duration, and abundance vary throughout the year depending on species, stock, and life history stage (Groot and Margolis 1991, Quinn 2005). Nearshore and estuarine habitats act as transition zones supporting osmoregulatory changes (the physiological changes by which smolt adapt between fresh and salt water) (Hoar 1976 and 1988, Clarke and Hirano 1995, Dickhoff et al. 1997). Studies have shown that subyearling salmon in the Pacific Northwest move repeatedly between zones of low and high salinity, and although no studies have yet shown Bristol Bay salmon to behave similarly, the Pacific Northwest studies suggest that such behavior may be integral to the survival and growth of young salmon (Healey 1982, Simenstad et al. 1982, Simenstad 1983, Thom 1987, Levings 1994, Levings and Jamieson 2001).

The eastern Bering Sea shelf is an important nursery ground for juvenile and subadult Bristol Bay sockeye salmon (Farley et al. 2009). Early models of eastern Bering Sea and North Pacific salmon stocks describe migrations and broad distributions to the south and east in winter and spring and to the north and west in summer and fall (French et al. 1975, French et al. 1976, Rogers 1987, Burgner 1991, Shuntov et al. 1993). These studies were the first to suggest that population migrations crossed the Aleutian Island chain into the North Pacific (Myers et al. 1996, Myers 2011, pers. comm.). Recent investigations incorporating genetic (DNA) and scale pattern analysis validate these observations (Bugaev 2005, Farley et al. 2005, Habicht et al. 2005, Habicht et al. 2007, Myers et al. 2007). Investigations conducted in autumn 2008 and winter 2009 substantiate the migration of juvenile Bristol Bay sockeye salmon from the southeastern Bering Sea shelf to the North Pacific, south of the Aleutian Island chain (Habicht et al. 2010, Farley et al. 2011, Seeb et al. 2011).

In their first oceanic summer and fall, juveniles are distributed on the southeastern Bering Sea shelf, and by the following spring immature salmon are distributed across a broad region of the central and eastern North Pacific. In their second summer and fall, immature fish migrate to the west in a band along the south side of the Aleutian chain and northward through the Aleutian passes into the Bering Sea. In subsequent years, immature fish migrate between their summer/fall feeding grounds in the Aleutians and Bering Sea and their winter habitat in the North Pacific. In their last spring, maturing fish migrate across a broad, east-west front from their winter/spring feeding grounds in the North Pacific, northward through the Aleutian passes into the Bering Sea, and eastward to Bristol Bay (Farley et al. 2011).

More than 55% of 1-year-old ocean sockeye salmon sampled during the 2009 winter survey in the North Pacific were from Bristol Bay stocks. These broad seasonal shifts in distribution likely reflect both genetic adaptations and behavioral responses to environmental cues (e.g., prey availability and water temperature) that are mediated by bioenergetic constraints (Farley et al. 2011). This extensive range and distribution suggest that Bristol Bay sockeye salmon contribute to the trophic dynamics in the Bering Sea as well as the North Pacific.

## Salmon Contribution to Trophic Levels

A recent evaluation was conducted by the AFSC Ecosystem Modeling Team to assess the contribution of Nushagak and Kvichak River sockeye salmon to trophic dynamics of the southeastern Bering Sea shelf and North Pacific ecosystems (Gaichas and Aydin 2010). Using estimates of outbound salmon smolt survival and adult returns, researchers calculate that these two rivers account for nearly 70% (56,000 of 81,100 tons) of adult salmon biomass in the southeastern Bering Sea. In the open ocean, sockeye salmon represent 47% of total estimated salmon biomass present in the eastern subarctic gyre (Aydin et al. 2003). Bristol Bay sockeye salmon from the Nushagak and Kvichak Rivers compose 26% of total sockeye salmon biomass and 12% of total salmonid biomass in the entire eastern subarctic gyre. The Nushagak and Kvichak Rivers produce a significant portion of all salmon in offshore marine ecosystems and the majority of salmon in the southeastern Bering Sea shelf, thus producing the majority of juveniles and returning adults in the salmon biomass (Gaichas and Aydin 2010). The AFSC's evaluation indicates sockeye salmon from these river systems rank among the top ten forage groups, comparable to Pacific herring or eulachon as a nutritional source for other marine species in the Bering Sea and North Pacific. One study supports this rational, indicating that outbound salmon smolt export substantial levels of nitrogen and phosphorus seaward (Moore and Schindler 2004).

Napp, J.M., Kendall, A.W., and Schumacher, J.D., 2000. A synthesis of biological and physical processes affecting the feeding environment of larval walleye pollock (Theragra chalcogramma) in the eastern Bering Sea. *Fisheries Oceanography*, 9(2), p. 147–162.

Neher, T. D. H., Rosenberger, A. E., Zimmerman, C. E., Walker, C. M., and Baird, S. J. 2014. Use of glacier river-fed estuary channels by juvenile Coho Salmon: transitional or rearing habitats? Environmental Biology of Fishes, 97(7), 839–850.

Nichol, D. R. 1998. Annual and between sex variability of yellowfin sole, *Pleuronectes asper*, spring-summer distributions in the eastern Bering Sea. Fish. Bull., U.S. 96:547–561.

NOAA. 1987. Bering, Chukchi, and Beaufort Seas: Coastal and Ocean Zones. Strategic Assessment: Data Atlas. United States Department of Commerce.

NOAA. 1998. Biogeorgraphic Regions of the NERRS. Silver Spring, MD: NOAA.

Norcross, B. L. and R. F. Shaw. 1984. Oceanographic and estuarine transport of fish eggs and larvae: a review. Transactions of the American Fisheries Society 113, 153–165.

Norcross, B. L., B. A. Holladay, and F. J. Muter. 1995. Nursery area characteristics of pleuronectids in coastal Alaska, USA. Neth. J. Sea Res. 34 (1–3), 161–175.

Norcross, B. L. and B. A. Holladay. 2005. Feasibility to design and implement a nearshore juvenile flatfish survey—Eastern Bering Sea. Final Technical Report to the Cooperative Institute for Arctic Research. Award # NA17RJ1224. 42 p.

North Pacific Fisheries Management Council. 2013. Website last accessed on March 26, 2013, http://alaskafisheries.noaa.gov/npfmc/conservation-issues/habitat-protections.html.

O'Keefe, T. C. and R. T. Edwards. 2002. Evidence for hyporheic transfer and removal of marine derived nutrients in a sockeye stream in Southwest Alaska. Am. Fish. Soc. Symp. 33: 99–107.

Ormseth, O. 2009. Utilization of nearshore habitat by fishes in Nushagak and Togiak Bays. NOAA- AFSC/REFM, EFH Status Report for Project 2009–12.

Orsi, J. A., M. V. Sturdevant, J. M. Murphy, D. G. Mortensen, and B. L. Wing. 2000. Seasonal habitat use and early marine ecology of juvenile Pacific salmon in southeastern Alaska. N. Pac. Anad. Fish Comm. Bull. 2:111–122.

Parker, R. R. 1968. Marine mortality schedules of pink salmon of the Bella Coola River, Central British Columbia. Journal of the Fisheries Research Board of Canada 25:757–794.

Pauly, D., A. W. Trites, E. Capuli, and V. Christensen. 1998a. Diet composition and trophic levels of marine mammals. ICES (International Council for the Exploration of the Sea) Journal of Marine Science 55:467–481.

Powers, S. P., M. A. Bishop, and G. H. Reeves. 2006. Estuaries as essential fish habitat for salmonids: Assessing residence time and habitat use of coho and sockeye salmon in Alaska estuaries. North Pacific Research Board Project Final Report 310. 65 p.

Quinn, T. P. 2005. Behavior and Ecology of Pacific Salmon and Trout. University of Washington Press and the American Fisheries Society.

Radenbaugh, T. 2010. Personal Communication. Assistant Professor Environmental Science. University of Alaska Fairbanks, Bristol Bay Campus, Bristol Bay Environmental Science Lab. Discussion regarding recent surveys and data collection in Nushagak and Kvichak Bays.

Radenbaugh, T. 2011. Personal Communication. Assistant Professor Environmental Science. University of Alaska Fairbanks, Bristol Bay Campus, Bristol Bay Environmental Science Lab. Discussion regarding recent surveys and data collection in Nushagak and Kvichak Bays.

Radenbaugh, T. 2012. Benthic Faunal Zones of Nushagak Bay, *In Press*.

Reed, R. K. and P. J. Stabeno. 1994. Flow along and across the Aleutian Ridge. J. Mar. Res. 52:639–648.

Reimchen, T. E. 1992. Mammal and bird utilization of adult salmon in stream and estuarine habitats at Bag Harbour, Moresby Island. Canadian Parks Service.

Reimchen, T. E. 1994. Further studies of predator and scavenger use of chum salmon in stream and estuarine habitats at Bag Harbour, Gwaii Haanas. Technical report prepared for Canadian Parks Service. Queen Charlotte City, British Columbia, Canada.

Reimchen, T. E., D. Mathewson, M. D. Hocking, and J. Moran. 2002. Isotopic evidence for enrichment of salmon-derived nutrients in vegetation, soil, and insects in riparian zones in coastal British Columbia. American Fisheries Society Symposium. XX: 1–12.

Reimchen, T. E., D. Mathewson, M. D. Hocking, J. Moran, and D. Harris. 2003. Isotopic evidence for enrichment of salmon-derived nutrients in vegetation, soil, and insects in riparian zones in coastal British Columbia. In: Nutrients in Salmonid Ecosystems: Sustaining Production and Biodiversity (ed. Stockner J), p. 59–69. American Fisheries Society Symposium 34, Bethesda.

Reimers, P. E. 1971. The length of residence of juvenile fall chinook salmon in Sixes River, Oregon. Dissertation. Oregon State University, Corvallis, OR.

Rice, T. R. and R. L. Ferguson. 1975. Response of estuarine phytoplankton to of estuarine phytoplankton to environmental conditions. In: Physiological ecology of estuarine organisms. Edited by F.J. Vernberg. University of South Carolina Press, Columbia, South Carolina. p. 1–43.

Rich, W. H. 1920. Early history and seaward migration of Chinook salmon in the Columbia and Sacramento rivers. Fish. Bull. 37:1–74.

Richey, J. E., M. A. Perkins, and C. R. Goldman. 1975. Effects of Kokanee salmon (*Oncorhynchus nerka*) decomposition on the ecology of a subalpine stream. Journal of the Fisheries Research Board of Canada 32:8 17–820.

Rogers, D. E. 1987a. Pacific Salmon. In: The Gulf of Alaska. D.W. Hood and S.T. Zimmerman (eds) Washington DC: NOAA Dept. Commerce, p. 461–475.

Rodin, V. E. 1989. Population biology of the king crab, Paralithodes camtschatica Tilesius, in the north Pacific Ocean. Pages 133–144 in B. R. Melteff, Coordinator. Proceedings of the international symposium on king and Tanner crabs. Report AK-SG-90-04. University of Alaska Sea Grant Program, Anchorage, AK.

Schindler, D. A., M. D. Scheuerell, J. W. Moore, S. M. Gende, O. B. Francis, and W. J. Palen. 2003. Pacific salmon and the ecology of coastal ecosystems. Frontiers in Ecology and the Environment 1:31–37.

Schumacher, J. D., T. H. Kinder, D. J. Pashinski, and R. L. Charnell. 1979. A structural frontover the continental shelf of the eastern Bering Sea. Journal of Physical Oceanography 9:79–87.

Schumacher, J. D. and P. J. Stabeno. 1998. The continental shelf of the Bering Sea. In: The Sea: the Global Coastal Ocean Regional Studies and Synthesis, Volume XI. A.R. Robinson and K. H. Brink (eds). New York: John Wiley and Sons, p. 869–909.

Seeb, L. W., J. E. Seeb, C. Habicht, E. V. Farley Jr., and F. M. Utter. 2011. Single-nucleotide polymorphism genotypes reveal patterns of early juvenile migration of sockeye salmon in the eastern Bering Sea. Transactions of the American Fisheries Society 140:734–748.

Sharma, G. D., A. S. Naidu, and D. W. Hood. 1972. A model contemporary graded shelf. American Association of Petroleum Geologists Bulletin, 56: 2000–2012.

Sheaves, M., Baker, R., Nagelkerken, I., and Connolly, R. M., 2015. True value of estuarine and coastal nurseries for fish: incorporating complexity and dynamics. *Estuaries and Coasts*, 38(2), p. 401–414.

Sheaves, M., R. Baker, and R. Johnston. 2006. Marine nurseries and effective juvenile habitats: an alternative view. Marine Ecology-Progress Series 318: 303–306.

Shuntov, V. P., V. I. Radchenko, V. V. Lapko, and Yu. N. Poltev. 1993. Distribution of salmon in the western Bering Sea and neighboring Pacific waters. J. Ichthyol. 33(7): 48–62.

Simenstad, C. A., K. L. Fresh, and E. O. Salo. 1982. The role of Puget Sound and Washington coastal estuaries in the life history of Pacific Salmon: an unappreciated function. Pages 343–364 in V. S. Kennedy, editor. Estuarine Comparisons. Academic Press, New York.

Simenstad, C. A. 1983. The ecology of estuarine channels of the Pacific Northwest coast: A community profile. FWS/OBS-83/05. U.S. Fish and Wildlife Service, Olympia, Washington. 181 p.

Smith, K. R. and R. A. McConnaughey. 1999. Surficial sediments of the eastern Bering Sea continental shelf: EBSSED database documentation. U.S. Department of Commerce, NOAA Technical Memorandum. NMFS-AFSC-104. 41 p.

Stabeno, P. J., N. A. Bond, N. B. Kachel, S. A. Salo, and J. D. Schumacher. 2001. On the temporal variability of the physical environment over the south-eastern Bering Sea, Fisheries Oceanography, 10, 81–98.

Stabeno, P. J. and G. L. Hunt Jr. 2002. Overview of the inner front and southeast Bering Sea carrying capacity programs. Deep-Sea Research II, this issue (PII: S09670645(02)00339-9).

Stabeno, P. J., R. K. Reed, and J. M. Napp. 2002. Transport through Unimak Pass, Alaska. Deep Sea Res. II 49:5919–5930.

Stabeno, P. J., N. B. Kachel, and M. E. Sullivan. 2005. Observations from moorings in the Aleutian Passes: temperature, salinity and transport. Fish. Oceanogr. 14(Suppl. 1):39–54.

Stockner, J. G. 1987. Lake fertilization: The enrichment cycle and lake sockeye salmon (*Oncorhynchus nerka*) production. Pages 198–215 In: H. D. Smith, L. Margolis, and C. C Wood, editors. Sockeye salmon (*Oncorhynchus nerka*) population biology and future management. Canadian Special Publications Fisheries and Aquatic Sciences.

Stockner J. G. and MacIsaac E. A. 1996. British Columbia lake enrichment programme: two decades of habitat enhancement for sockeye salmon. Regul Rivers Res Manag 12:547–561.

Stockner, J. G., E. Rydin and P. Hyenstrand. 2000b. Cultural oligotrophication: causes and consequences for fisheries resources. Fisheries, 25: 7–14.

Stockwell, D. A., T. E. Whitledge, S. I. Zeeman, K. O. Coyle, J. M. Napp, R. D. Brodeur, A. I. Pinchuk, and G. L. Hunt Jr. 2001. Anomalous conditions in the southeastern Bering Sea, 1997: nutrients, phytoplankton, and zooplankton. Fisheries Oceanography 10, 99–116.

Straty, R. R. 1977. Current Patterns and Distribution of River Waters in Inner Bristol Bay, Alaska. NOAA Technical Report, NMFS SSRF-713. U.S. Dept. of Commerce.

Straty, R. R. and I. W. Jaenicke. 1980. Estuarine influence of salinity, temperature and food on the behavior, growth and dynamics of Bristol Bay sockeye salmon, p. 247–265. In W. J. McNeil and D. C. Himsworth (eds.), Salmonid Ecosystems of the North Pacific. Oregon State University Press, Corvallis, OR.

Sugai, S. F. and D. C. Burrell. 1984. Transport of dissolved organic-carbon, nutrients, and trace metals from the Wilson and Blossom Rivers to Smeaton Bay, Southeast Alaska. Can. J. Fish. Aquat. Sci. 41(1):180–190.

Thedinga, J. F., S. W. Johnson, K. V. Koski, J. M. Lorenz, and M. L. Murphy. 1993. Potential effects of flooding from Russell Fiord on salmonids and habitat in the Situk River, Alaska. National Marine Fisheries Service, Alaska Fisheries Science Center Processed Report 93-01, Auke Bay Laboratory, Juneau, AK.

Thedinga, J. F., S. W. Johnson, and K V. Koski. 1998. Age and marine survival of ocean-type chinook salmon (*Oncorhynchus tshawytscha*) from the Situk River, Alaska. Alaska Fishery Bulletin 5 (2):143–148.

Thedinga J. F., S. W. Johnson. A. D. Neff and M. R. Lindeberg. 2008. Fish assemblages in shallow nearshore habitats of the Bering Sea. Trans Am Fish Soc 137:1157–1164.

Thorpe, J. E. 1994. Salmonid fishes and the estuarine environment. Estuaries, 17: 73–93. Thom, R. M. 1987. The biological importance of Pacific Northwest estuaries. Northwest Environmental Journal 3(1):21–42.

U.S. CFR. 2016. United States Code of Federal Regulations. 50 CFR Part 679—Fisheries of the Exclusive Economic Zone off Alaska. Bristol Bay Trawl Closure Area (679.22(a)(9)) Accessed 10/10/17–https://alaskafisheries.noaa.gov/sites/default/files/679b22.pdf

United States Geological Survey (USGS). 2011. USGS- GIS Topography Data Sets.URL: http://nhd.usgs.gov/wbd_data_citation.html. Last accessed on Tuesday, September 6, 2011 at 4:05 PM.

U.S. Fish and Wildlife Service. Conservation Planning Assistance. 2009. Studies of Anadromous Fish in Knik Arm. A Literature Review. Prepared by Prevel-Ramos, A., Brady, J. A., Houghton, J., Dec. 2009.

Von Biela, V. R., C. E. Zimmerman, B.R. Cohn, and J.M., Welker. 2013. Terrestrial and marine trophic pathways support young-of-year growth in a nearshore arctic fish. Polar Biology 36:137–146.

Warner, I. M., and P. Shafford. 1981. Forage fish spawning surveys: southern Bering Sea. Pages 1–64 in Environmental assessment of the Alaskan continental shelf. National Oceanic and Atmospheric Administration, Final Report 10, Boulder, CO.

Weitkamp, L. A. and M. V. Sturdevant. 2008. Food habits and marine survival of juvenile Chinook and coho salmon from marine waters of Southeast Alaska. Fisheries Oceanography 17:380–395.

Wiedmer, M. 2013. Personal Communication. Doctoral Candidate Fisheries Research, University of Washington, School of Fisheries. Discussion regarding the movement of coho salmon and use of marine estuaries and overwinter rearing in fresh water tributaries in the Bristol Bay region.

Wilderbuer, T. K., A. B. Hollowed, W. J. Ingraham Jr, P. D. Spencer, M. E. Conners, N. A. Bond, and G. E. Walters. 2002. Flatfish recruitment response to decadal climatic variability and ocean conditions in the eastern Bering Sea. Prog. Oceanogr. 55, 235–247.

Willette, T. M., R. T. Cooney and K. Hyer. 1999. Predator foraging mode shifts affecting mortality of juvenile fishes during the subartic spring bloom. Can. J. Fish. Aquat. Sci. 56:364–376.

Wilkinson T., E. Wiken, J. Bezaury-Creel, T. Hourigan, T. Agardy, H. Herrmann, L. Janishevski, C. Madden, L. Morgan, and M. Padilla. 2009. Marine Ecoregions of North America. Commission for Environmental Cooperation. Montreal, Canada. 200 p.

Willson, M. F. and K. C. Halupka. 1995. Anadromous fish as keystone species in vertebrate communities. Conservation Biology 9:489–497.

Willson, M. F., S. M. Gende, and A. H. Marston. 1998. Fishes and the forest. Bioscience 48:455–462.

Wilson, M. F., S. M. Gende, and P. A. Bisson. 2004. Anadromous fishes as ecological links between ocean, fresh water, and land. In: Food Webs at the Landscape Level (eds Polis, G.A., Power, M. E. & Huxel, G. R.). The University of Chicago Press, Chicago, p. 284–300.

Wipfli, M. S., J. Hudson, and J. Caouette. 1998. Influence of salmon carcasses on stream productivity: response of biofilm and benthic macroinvertebrates in southeastern Alasksa, U.S.A. Canadian Journal of Fisheries and Aquatic Sciences 55:1503–1511.

Yamashita, Y., T. Otake, and H. Yamada. 2000. Relative contributions from exposed inshore and estuarine nursery grounds to the recruitment of stone flounder, (*Platichthys bicoloratus*), estimated using otolith Sr:Ca ratios. Fish Oceanogr 9:316–327.

# 17

# MARINE INVERTEBRATES
# OF BRISTOL BAY

W. Stewart Grant and Aaron Baldwin, Alaska Department of Fish and Game
Todd A. Radenbaugh, University of Alaska Fairbanks

## INTRODUCTION

Bristol Bay (*Iilgayaq* in Yup'ik) is a large, shallow semi-enclosed estuary that is an extension of the larger Bering Sea, which covers about 2,000,000 square km. Bristol Bay itself is about 400 km long about 290 km wide and is bounded by the Yukon-Kusikuwim river deltas to the north, by the Alaska Peninsula to the south, and by the open eastern Bering Sea to the north and west (Figure 17.1). The adjoining continental shelf beyond the bay is one of the largest continental shelves globally with oceanic and faunal dynamics that influence the functioning of ecosystems in Bristol Bay (Hunt et al. 2010). In addition to the vast oceanic domain of Bristol Bay, the numerous rivers and estuaries along the southern and eastern shores of Bristol Bay increase productivity by providing nutrients and creating a diversity of habitats.

The invertebrates of Bristol Bay range in size from microscopic copepods to large crustaceans and cephalopods, and can be found as drifting plankton in the water (pelagic), living on the bottom (epibenthic) or in the mud and sand of the bay (infauna). Together, these species support one of the largest biomasses of marine mammals and birds in the southeastern Bering Sea (Hunt et al. 2008, and Shelden and Waite, Hoover-Miller and Savage this volume). Forage fish and invertebrates in these estuaries support the early ocean phases of five species of Pacific salmon, including some of the world's largest remaining populations of wild Chinook and sockeye salmon; in 2015 more than 58 million sockeye salmon returned to Bristol Bay (Alaska Department of Fish and Game (ADF&G) 2015). Large populations of crangonid shrimps and gammarid amphipods in the estuaries and coastal zones are especially important prey species for salmon (Figure 17.2). The diversity of species in estuaries additionally helps to support numerous top predators, including a vibrant population of beluga whales (*Delphinapterus leucas*) (Quackenbush et al. 2015). Bristol Bay also borders the waters of the Aleutian Island Archipelago, which has one of the world's largest diversities of cold water cnidarians (e.g., jellyfish (Figure 17.3), anenomes, and corals) and echinoderms (e.g., sea stars, sea urchins) (Heifetz 2002, Heifetz et al. 2005, Jewett et al. 2012). However, the intricacies of the food webs in the estuaries, coastal areas, and Bristol Bay are not fully understood—in particular, how organisms living on the bottom and in the water column are integrated into a remarkable ecosystem that supports a large biomass and diversity of species.

The diversity of lower trophic level invertebrates in Bristol Bay supports large fisheries of red king and Tanner crabs, pollock, and five species of Pacific salmon in Bristol Bay. In the broader Bering Sea, these fisheries represent more than 40% of the finfish and shellfish harvests in the United States (Stebeno et al. 2012). Commercial harvests of Pacific salmon alone from Bristol Bay recently had a value of $1.5 billion (Knapp et al. 2013, Rinella et al. this volume), and shellfish harvests have reached as much as $120 million during some years in the past few decades (Fitch et al. 2012). The invertebrate shellfish harvests in Bristol Bay have included red king crabs, Tanner crabs, Dungeness crabs, and whelks. In addition to commercial fisheries, high levels of productivity in Bristol Bay are of considerable economic and cultural importance to local residents. Traditionally, three Alaskan Native groups have depended on Bristol Bay biodiversity for subsistence fishing and hunting (Van Stone 1972, Radenbaugh and Wingert-Pederson 2011, Borras and Knott this volume). These groups use a variety of species from shellfish at the bottom of the food web to fish, birds, and vertebrates (such as with fishermen, to survey marine areas using a statistically sound survey design to be able to address questions more rigorously than can be done with catch data from fishermen. The most common survey methods use nets (trawls) towed behind a vessel that are designed to capture organisms in particular parts of the marine ecosystem. For example, bongo nets are used at the surface to capture small organisms, such as fish larvae and zooplankton. Midwater trawls are designed to capture schooling fish in the water, such as Alaska pollock, and bottom trawls capture fishes and invertebrates near or on the bottom. Trawl surveys in Bristol Bay indicate that invertebrates fall into two spatial groups, generally separated by the isobath at a depth of 50 m. These groups are defined by occurrences of key species in the bottom-trawl surveys, which have revealed large-scale patterns of community organization (Walters and McPhail 1982, Yeung and McConnaughey 2006). However, these patterns can change, as they did in response to the ocean-climate shifts from 1982 to 2002 that altered water temperatures in Bristol Bay and reorganized suitable habitats. For example, changes in the distributions and abundances of red king crabs occurred because of the appearance of a bottom cold pool in Bristol Bay driven by climate changes (Loher and Armstrong 2005).

Planktonic invertebrates are prey for coastal foraging fishes, especially salmon smolts that are produced by large spawning populations in the rivers and streams emptying into the bay (e.g., Straty and Jaenicke 1980). Large marine mammals—for example, walruses—forage on bivalves such as clams buried in shallow-water areas (Jay et al. 2001, Bornhold et al. 2005), and resident beluga whales forage on shrimp and amphipods (Figure 17.8) (Quakenbush et al. 2015). Our understanding of the ecologies of the various invertebrates in Bristol Bay is skewed toward commercially important species. However, the importance of other species is underscored by their roles in energy cascades through food webs and often by the physical structure they provide for the juvenile stages of fishes and larger invertebrates. The diversity of invertebrates is here outlined by major taxonomic groups.

**Figure 17.8**    These crustaceans called amphipods are a mainstay in the diets of salmon. *Photo credit: T. Radenbaugh.*

## PORIFERA: SPONGES

Sponges have structurally simple bodies composed of two major cell types, one of which includes flagellated individuals that draw food-laden water through pores in the body of the sponge formed by a scaffold of structural individuals (Figure 17.9). Hence, these organisms are efficient suspension feeders. Research trawl surveys show that sponges occur only in the deeper waters of Bristol Bay, but are found in large numbers in the Bering Sea and around the Aleutian Islands (Lauth and Acuna 2007, Stone et al. 2011). One species that can sometimes occur in Bristol Bay has a stalked, trumpet-shaped body that extends into the water for more efficient suspension feeding (Figure 17.10). Sponges provide physical relief on the flat bottom and are refuges from predation for small fishes and juvenile red king crabs (NOAA 2015a).

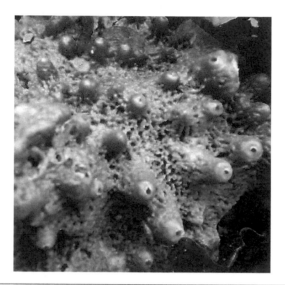

**Figure 17.9**   A close-up photograph of the sponge *Halichondria panicea* show incurrent and excurrent pores. *Photo credit: A. Baldwin.*

**Figure 17.10**   The trumpet-shaped sponge *Semisuberites* is a stalked sponge with a body that protrudes into the water to better feed on minute organic particles in the water. *Photo credit: Jeremy Chevalier.*

# TUNICATES (SEA SQUIRTS, ASCIDIANS)

Tunicates occur occasionally in Bristol Bay, but a diversity of species inhabits the Bering Sea. These attached filter-feeding organisms occur as solitary individuals (Figure 17.20) or cluster into colonies. Solitary tunicates are attached to a hard surface directly or by a stalk, such as the sea onion *Boltenia ovifera*. Colonial tunicates create jelly-like masses encrusted on rocks and submerged wood. Both solitary and colonial tunicates have two siphons for filter feeding: one channels water into the body and the other expels water after food particles have been filtered out. These animals can reproduce asexually by budding new adult-looking individuals, or sexually by producing tadpole-like larvae with several morphological features of chordates. Larvae swim in the plankton for a short period before settling onto a hard surface to metamorphose into the sack-like, bottom-dwelling adult form. One of the most common tunicates in the Bering Sea is the solitary sea potato *Styela rustica*, which in some years makes up at least 5% of the biomass of benthic invertebrates (Jewett and Feder 1981). Tunicates are generally not preyed upon, but are efficient suspension feeders on micro-plankton. Tunicates provide structure by making thick patches on the bottom that protect juvenile stages of crabs and fishes (McMurray et al. 1984, Stevens and Kittaka 1998). Tunicates are more abundant in habitats with salinities greater than 20 practical salinity units (psu), so they occur more frequently in coastal areas than in estuaries.

**Figure 17.20**    This solitary tunicate *Halocynthia aurantium* has two siphons, one to draw water into its body, where small organic particles are removed, and another siphon to expel water. *Photo credit: A. Baldwin.*

# ECHINODERMS

Echinoderms are exclusively marine animals and are ecologically and biologically diverse. They are generally character-ized by five-faceted radial symmetry and include five taxonomic classes: sea stars, sea urchins and sand dollars, sea cu-cumbers, basket stars, and sea lilies and feather stars. Representatives of all groups inhabit Bristol Bay. The coast of Alaska and the Aleutian Island Archipelago have the greatest diversity of echinoderms, especially sea stars, in the world. While most of this diversity is located in the deeper waters of the Bering Sea Basin and around the Aleutian Islands, Bristol Bay is home to a large number of echinoderms that are adapted to shallow water mud and gravel bottoms. Sea stars and basket stars form the largest biomass of epibenthic species in Bristol Bay (Yeung and McConnaughey 2006).

## Sea Cucumbers

These soft-bodied echinoderms have long bodies stretched along the oral-anal axis giving them the appearance of a large worm. This form is unlike those of sea stars and brittle stars, which have a flat compact body with five or more arms. Sea cucumbers are deposit feeders of organic matter and are important for recycling nutrients and ventilating muddy bottoms. In Bristol Bay, they are found in mostly marine waters, since their range is limited by freshwater runoff. Sea cucumbers are harvested by scuba divers in southeast Alaska and several hundred pounds annually are sold to food mar-kets largely in Asia. Sea cucumbers have not been harvested in Bristol Bay.

## Sea Urchins and Sand Dollars

Sea urchins graze on algae or sea grasses growing in wave-protected waters. Sea urchins in the genus *Strongylocentrotus* are important grazers and can severely limit the growth of seaweeds on hard substrates (Figure 17.21). When predation by sea otters is reduced by harvests of sea otter pelts, sea urchin populations explode and overgraze bottom algae (Figure 17.22).

**Figure 17.21** This sea urchin *Strongylocentrotus pallidus* lives on algae and organic matter on soft bottoms. When urchins are abundant they may overgraze algae and reduce the productivity of an area. *Photo credit: A. Baldwin.*

**Figure 17.22**   Sea otters (*Enhydra lutris*) are a major predator of sea urchins. The decimation of the sea otter population in the North Pacific by fur hunters led to an explosion of sea urchins, which shifted near-shore ecosystems in many areas. *Photo credit: NOAA.*

Overgrazing changes the physical structure of a shallow-water ecosystem and can lead to major food-web shifts that ripple through the various trophic levels, sometimes affecting the abundances of marine mammals (Estes et al. 2005, Stewart and Konar 2012). Sand dollars such as *Echinarachnius* are bottom feeders, bulldozing a through the organic accumulations that have settled out of the water column.

## Sea Stars

At least 120 sea star species in at least 45 genera are recognized in Alaskan marine waters, with at least an additional 25 species in the process of being formally described (www.jaxshells.org/starfish.htm). Sea stars are voracious predators of whelks and buried clams. Sea stars themselves are generally not preyed upon, except by some other sea stars. The sea star *Asterias amurensis* (Figure 17.23) dominated the epibenthic biomass in near-shore areas of Bristol Bay in trawl surveys

**Figure 17.23**   The sea star *Asterias* is a voracious predator of clams, whelks, and polychaete worms. The sea star on the right is showing its bottom side with rows of sucker appendages that help it pry open clams and mussels. Also seen in the photo are prickly green sea urchins and blue mussels. The fronds of a brown seaweed *Fucus* can also be seen. *Photo credit: T. Radenbaugh.*

from 1982 to 2002 (Yeung and McConnaughey 2006). Many of these sea stars occur in the Nushagak and Kvichak River estuaries in waters less than five meters and with salinities as low as 15 psu, where they scavenge on spawned-out salmon carcasses during summer and autumn runs. Hence, sea stars can drive ecosystem dynamics in both nearshore and offshore habitats.

### Brittle Stars and Basket Stars

Brittle stars are bottom feeders and basket stars are mobile suspension feeders, capturing floating bits of organic material from the water with extensible rays that uncoil to trap particles and coil to bring the food to the mouth on the central disc (Figure 17.24). These species make up a small, but significant biomass of the invertebrate epibenthic community of central Bristol Bay and are most abundant in fully marine waters below 10 m (Yeung and McConnaughey 2006).

## MOLLUSCS

Molluscs are largely represented in Bristol Bay by bivalves (clams), gastropods (whelks and nudibranchs), and cephalopods (squids and octopuses) (Baxter 1983). Bivalves are suspension feeders that inhabit soft bottoms where they burrow into the substrate for protection from predators such as sea stars. Whelks move on the surface or plow through muddy bottoms, preying on small invertebrates and bivalves. Some groups of large whelks in the genera *Neptunea* and *Buccinum* can be locally abundant (MacIntosh and Sommerton 1981, Yeung and McConnaughey 2006).

### Bivalves, Mussels and Clams

Bristol Bay has a diverse group of bivalves (McDonald et al. 1981). Most bivalves are nourished by suspension feeding using a pumping gill and two muscular siphons: one to draw in water containing food particles and another to expel water. Both epibenthic (mussels) and endobenthic (clams) bivalves are found in Bristol Bay. Mussels generally live in marine

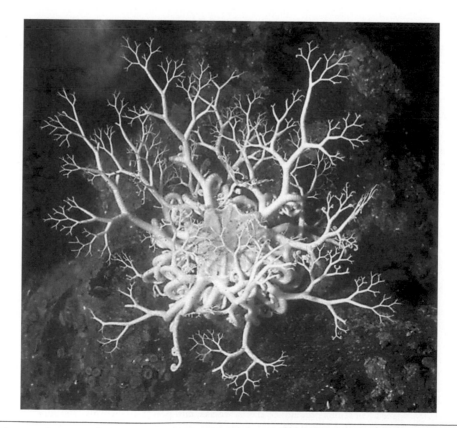

**Figure 17.24**  The basket star *Gorgonocephalus eucnemis* uses its agile arms with finely articulated tentacles to capture organic particles from the water. *Photo credit: NOAA.*

and estuarine waters just below the low tide mark and easily attach to most hard surfaces, such as glacial dropstones, or other mussels. Clams bury themselves into soft sediments with a muscular foot and feed using long siphons that reach the water-sediment interface. The soft sediments of mud, sand, and gravel in Bristol Bay are ideal habitats for burrowing bivalves. The abundant bivalves are preyed on by walrus, thus walrus foraging marks are found over large areas of Bristol Bay (Bornhold et al. 2005, Sheffield and Grebmeier 2009). Large pits and long furrows in the sediment in waters less than about 60 m deep are tell-tale signs of walrus foraging on bottom invertebrates (Bornhold et al. 2005). In the shallower and fresher waters of estuaries, bivalves generally include only blue mussels (*Mytilus*) and *Macoma* clams.

The diversity and abundance of clams in Bristol Bay has attracted the interest of industry to explore the possibility of commercially viable harvests. The supply of clam meat from Atlantic Ocean suppliers for the production of chowders has varied in the past few decades, and a benthic survey was mounted to assess the distributions and abundances of Arctic surf clams, *Mactromeris* (= *Spisula*) *polynyma*, throughout Bristol Bay (Hughes et al. 1971). These clams were the dominant bivalve in 230 trawls with their greatest abundances in shallow waters north of the Alaska Peninsula from Port Moller to Ugashik Bay. These clams yielded a high meat mass to total whole animal mass ratio that would provide an excellent commercial product. In addition to Arctic surf clams, the clams *Tellina lutea*, *Serripes gronlandicus*, and *S. laperousii* were relatively common, and several others were less common, including razor clams *Siliqua alta*, softshell clams *Mya arenaria*, and macoma clams *Macoma calcarea* and *M. middendorfii*. These species generally showed several size modes, indicating vibrant populations with regular recruitments. The results of this survey indicated that the clam resource in Bristol Bay had been underestimated and would potentially support commercial harvests in some areas. However, only subsistence harvests of clams now occur in Bristol Bay.

## Gastropods, Whelks

The Bering Sea and Bristol Bay have high diversities of gastropod species. Large hyperbenthic predatory gastropods in the genera *Buccinum* and *Neptunea* can be locally abundant. Species of *Neptunea* have consistently been found in large numbers and abundances in bottom trawl surveys in the Bering Sea (Smith et al. 2011). The most abundant neptune whelk is *N. pribiloffensis*, but *N. lyrata*, *N. heros*, and *N. ventricosa* are also common in the outer reaches of Bristol Bay (Shimek and Gardner 1979, Shimek 1984, Smith et al. 2011). Additional species of *Neptunea* are found in deeper waters of the Bering Sea but may occasionally be found in more shallow waters. These include the whelks *N. borealis*, *N. amianta*, and *N. insularis*. Whelks develop tall columns of egg capsules (Figure 17.25), or single-layer patches on hard substrates, with juveniles emerging as miniature adults without progressing through a planktonic larval phase. Although little is known about the feeding ecologies of these whelks in Bristol Bay, they are likely important predators and scavengers on benthic animals. The diets of the various species of *Neptunea* overlap somewhat, but with some resource partitioning to avoid competition in areas where they co-occur (Shimek 1984). Experiments show that some species of *Neptunea* can live without food for several months (Tamburri and Barry 1999). Whelks can be preyed upon by sea stars and octopuses and by fishes and marine mammals (Smith et al. 2011). Whelks are not currently being targeted in commercial fisheries, even though substantial numbers appear in bottom trawls.

## Cephalopods, Squids and Octopuses

Only a few species of pelagic cephalopods, such as octopuses and squids, have been observed in the near-shore domain of the southeastern Bering Sea, and none is common in the waters of Bristol Bay. Most octopuses and squids (Figure 17.26) are found in the deeper waters along the edge of the continental shelf, in waters between 200 and 750 meters and in even deeper abyssal areas (Kubodera 1991, Jorgensen 2009). The large, bottom-dwelling octopus, *Enteroctopus dofleini* (Figure 17.27), is often caught as by-catch in Pacific cod and red king crab pots, and in some years this by-catch has closed the cod pot fishery because the fishery might endanger the octopus population. Mid-water squids occur in some abundance in Bristol Bay in some years, so that juvenile salmon often switch from foraging on zooplankton to squid before migrating into the Bering Sea or North Pacific Ocean (Ruggerone et al. 2005).

**Figure 17.25**  Whelks can be abundant in the deeper waters of Bristol Bay. These whelk *Neptunea lyratra* egg capsules are in a column that allow the eggs to be ventilated by passing water currents. *Photo credit: A. Baldwin.*

**Figure 17.26**  Various species of squids inhabit the pelagic waters of Bristol Bay. This squid *Berryteuthis magister* is a stealthy predator of small fishes and invertebrates. *Photo credit: A. Baldwin.*

**Figure 17.27**   Octopuses are highly developed mollusks that inhabit epibenthic habitats. *Enteroctopus dofleini* is the most common octopus species in Bristol Bay and the northeastern Pacific. *Photo credit: A. Baldwin.*

# CRUSTACEANS

## Barnacles (*Cirripedia*)

Barnacles are sessile, encrusting crustaceans common on hard substrates in shallow waters, such as rocks, whelk and clam shells, and the bodies of large crabs. The mobile planktonic larvae stages of barnacles are typical of many other crustaceans until settlement. Unlike most other crustacea, barnacles metamorphose and settle head down on hard surfaces, then begin to feed with modified legs by raking food particles from the water (Figure 17.28). Because barnacles are immobile after they settle, they must settle close enough to other barnacles so they can reproduce. Barnacles are not abundant in coastal and estuarine areas because of the lack of suitable settling surfaces, or because strong tidal currents

**Figure 17.28**   The abundance of barnacles in Bristol Bay is generally limited by the lack of hard rocky substrates. However, barnacles can be found on dead shells of clams and scallops and the carapaces of Tanner and king crabs. This photograph shows an individual of *Chirona evermanni* feeding with articulated tentacles (cirri) that originate early in development from leg appendages. *Photo credit: A. Baldwin.*

scour rocks that would otherwise provide a home. Barnacles in the genus *Balanus* can be locally abundant on beds of blue mussel shells and on large rocks in channels that are free of sediments.

## Miscellaneous Crabs

Bristol Bay harbors a large number of crabs that are ecologically and taxonomically diverse. Numerous species of small hermit crabs (Figure 17.29) prey and scavenge on other invertebrates and fish, and each species often specializes on a particular food type. Larger crabs, such as the hair crab *Erimacrus isenbeckii* (Figure 17.30) and Dungeness crabs *Metacarcinus magister*, are sporadically abundant and have been harvested to a small extent. Hair crabs are generally harvested in deeper waters offshore from Bristol Bay, but can be found in small numbers in shallower waters. Fishery openings for hair crabs have been contentious, as the fishery can conflict with the red king crab fishery in Bristol Bay (Fitch et al. 2012). However, recent stock assessments of hair crabs indicate that the stock in Bristol Bay is depressed and is unable to support a sustained fishery. A small sporadic fishery for Dungeness crabs occurs along the northern shore of the Alaska Peninsula (Fitch et al. 2012).

## Shrimps

Shrimps are hyperbenthic species that occasionally burrow. They are abundant in the eastern Bering Sea, but population sizes tend to be small in Bristol Bay compared to populations in the Gulf of Alaska and southeast Alaska. After hatching

**Figure 17.29**    Hermit crabs inhabit empty shells of whelks and move along the bottom scavenging and preying on small mollusks and other crustaceans. Here, a colorful individual of *Elassochirus tenuimanus* awaits an unsuspecting animal to prey on. *Photo credit: A. Baldwin.*

**Figure 17.30**   The hairy crab *Erimacrus isenbeckii* has sometimes been commercially harvested. *Photo credit: A. Baldwin.*

in March and April from eggs carried by females over the winter, larvae spend several weeks in the plankton before settling to the bottom, where they inhabit muddy and sandy areas, or areas covered with shell hash. Shrimps are generally bottom scavengers, but may prey on small worms, crustaceans and mollusks. These shrimps generally develop as males that mature in about 1.5 years, on average, then change (metamorphose) into females that become reproductively mature, on average, at 2.5 to 4.5 years of age, depending on species (Charnov and Anderson 1989, Orensanz et al. 1998). They reproduce annually or biennially and are highly fecund, so when environmental conditions are favorable and food is available, populations can exponentially increase (Figure 17.31).

**Figure 17.31**   This shrimp *Lebbeus groenlandicus* is often found in association with the sea anenome *Cribinopsis fernaldi*. *Photo credit: A. Baldwin.*

Several species of pandalid shrimps are abundant in the Bering Sea and have been the targets of trawl fisheries and, to a small extent, of pot fisheries. A fishery by Japanese boats in the Bering Sea peaked in the 1950s, but collapsed in the 1960s (Orensanz et al. 1998). After several years of low population numbers, biomass surveys in the southeastern Bering Sea showed that pandalid shrimps, as a group, rebounded from 1991 to 1993, but have subsequently dropped to low quantities (Aydin and Mueter 2007). These population fluctuations were likely due to climatic shifts and the Pacific Decadal Oscillation (PDO) (Aydin and Mueter 2007). The population numbers of shrimps in the Bering Sea and Bristol Bay are presently low and are not targeted for harvest (Orensanz et al. 1998, Yeung and McConnaughey 2006). Cragon shrimps are abundant in coastal areas and in estuaries that they use as spawning habitats in early summer. In the Nushagak and Kvichak estuaries and coastal zones, these shrimps represent the largest biomass during late summer (Radenbaugh and Wingert 2008, Ormseth, et al. 2009, Harwell et al. 2015).

## Tanner Crabs

Bristol Bay supports large populations of Tanner crabs *Chionoecetes bairdi* (Figure 17.32), which are broadly distributed from the Kuril Islands to the Bering Sea and beyond to Oregon. These crabs inhabit warmer and deeper waters than snow crabs and breed in large pods (Figure 17.33). Year-class strength in Bristol Bay is influenced by water temperature, upwelling, and predation intensity (Rosenkranz et al. 2001). Unusually cold bottom temperatures adversely influence the reproductive cycles of Tanner crabs by delaying the developments of gametes and embryos, which produce a mismatch between the emergence of larvae and the spring plankton bloom. In some years, northeastern winds push surface waters offshore that are replaced with upwelled bottom water carrying nutrients. Episodes of upwelling stimulate primary production and lead to higher abundances of copepods, which are preferred by small Tanner crabs. Currents across the bay, driven by northeastern winds, carry drifting larvae to suitable settlement areas of fine sediments in the outer portions of Bristol Bay. Year-class strength in Bristol Bay is additionally influenced by the intensity of predation on early life-history stages from juvenile and adult sockeye salmon transiting the basin, from resident Pacific cod, and from carnivorous invertebrates. Tanner-crab abundances are inversely related to the abundances of these predators (Rosenkranz et al. 2001).

A fishery for Tanner crabs began in the 1970s and has continued sporadically to the present at moderate levels (Fitch et al. 2012). The fishery for Tanner crabs was closed from 1999 to 2005 because of low stock levels, but has since reopened, either as a limited directed fishery or as by-catch in the red king crab fishery (Fitch et al. 2012). Biomass estimates of Tanner crabs in the eastern Bering Sea since 1973 indicate an overall decline in Tanner crabs from a peak of about 290,000 tons in 1975 to lows of 8–10 tons in 1985 and 1977 (NOAA 2015c). In 1995, the ex-vessel value of Tanner crab in the eastern Bering Sea was $11.7 million, but only $1.41 million in 2007 (NOAA 2015c).

**Figure 17.32**   Tanner crabs *Chionoecetes bairdi* are abundant in Bristol Bay and support a large commercial harvest. *Photo credit: ADF&G.*

**Figure 17.33** Tanner crabs congregate into large pods, most likely to protect themselves from predators and to mate. *Photo credit: NOAA—B. Stevens.*

## Snow Crabs

Snow crabs, *Chionoecetes opilio*, range from the Bering Sea, sporadically across the Arctic, and into the Northwestern Atlantic Ocean. In Alaskan waters, these crabs are most abundant in the colder waters of the central southeastern Bering Sea shelf, where they support a large fishery (Aydin et al. 2007, Fitch et al. 2012). As a result of cycles in larval recruitment and shifts in ocean conditions, these crabs are found in small numbers in some years in central Bristol Bay (Zheng and Kruse 2001, Orensanz et al. 2007). Over the last few decades, the biomass of snow crabs in the eastern Bering Sea peaked at about 470,000 tons in 1990, but declined to about 100,000 tons in 2001 (NOAA 2015b). The ex-vessel price for snow crab peaked in 1995 at $6.90 per pound, but was about $4.00 per pound in 2009 (NOAA 2015b).

Molecular markers show that Tanner and snow crabs hybridize in areas where their ranges overlap in the Bering Sea (Merkouris et al. 1998, Smith et al. 2005). While first generation hybrids are most common, genetic introgression in subsequent generations is also apparent. The hybridization between these two commercially important crabs is problematic for the management of harvests in some areas of the Bering Sea. Additionally, molecular markers also show that the Bristol Bay population of Tanner crabs differs genetically to a small degree from the population around the Pribilof Islands, indicating that these populations are demographically distinct and should be managed as separate populations (Merkouris et al. 1998). Harvest quotas based on a single population unit can potentially lead to the loss of the weaker population.

## Red King Crabs

Populations of red king crab (*Paralithodes camtschaticus*) (Figure 17.34) occur across the North Pacific from the northern Sea of Japan, throughout the Bering Sea to British Columbia, and into the Chukchi Sea (Figure 17.8). They can be abundant in deeper parts of central Bristol Bay (Feder et al. 2005, Dew 2010). Males and females become sexually mature at about seven years of age when they start an annual summer migration into deep offshore waters. In late winter and spring, crabs move into shallow water where they mate, before the females molt. Anywhere from 40,000–500,000 eggs are fertilized and extruded onto the female's abdomen. Eggs are carried for about 11 months before hatching, which is usually in April during the spring plankton bloom. The planktonic crab larvae progress through several developmental stages before settling to the bottom two to three months after hatching. Juveniles remain in shallow waters with a depth of less than 25 meters for their first year and seek out course substrates, such as boulders, cobble, shell hash, bryozoans, and ascidians, where they lead a solitary existence (McMurray et al. 1984, Stevens and Machintosh 1991, Stevens 2003). Juveniles molt several times a year until the age of three, then annually as adults. For a given age class, females tend to be smaller than males. At two to four years of age, juveniles congregate into dense pods, often consisting of thousands of

**Figure 17.34**  Red king crabs *Paralithodes camtschaticus* support a lucrative commercial fishery in Bristol Bay. *Photo credit: S. Grant.*

crabs (Dew 1990). Juveniles continue to pod until about age four, when they move into deeper waters of the Bering Sea and join the adults.

Populations of red king crabs in Bristol Bay have historically supported lucrative harvests, which take place in late fall after adults have mated and moved into deeper water. Harvest guidelines are based on sex (males only), minimal carapace size (6.5 inches), timing, and harvest area in an effort to allow crabs to reproduce. The population in Bristol Bay has been managed separately from other stocks since a major population crash in the 1980s. Genetic markers confirm this approach to management and indicate that populations in Bristol Bay are genetically unique, differing from populations of red king crab in Norton Sound, the Aleutians, and the Gulf of Alaska (Grant and Cheng 2012, Grant et al. 2014). These genetic differences indicate demographic differences among populations. For example, Norton Sound red kings are smaller than Bristol Bay crabs and show different patterns of recruitment from Bristol Bay crabs that may reflect adaptive differences between populations (Zheng and Kruse 2000).

The harvest management of red king crabs in Bristol Bay is overseen by NOAA, who regularly survey the population to track its location and population size. The population in Bristol Bay has declined in abundance over the past several decades. About 60 million kilograms (kg) of legal males were harvested in 1980 and were worth about $115 million ex-vessel (Kruse et al. 2010). This was followed by a drop to about 1 million kg in 1985 (Dew 2008). The Bristol Bay population has since recovered somewhat, but harvests are limited to a fraction of what they were at peak harvests (Kruse et al. 2010). The standing stock of red king crab in the Bristol Bay population is now estimated at two to five million kg (FMP 2011) and is much lower than at the height of the harvest in the 1970s when nearly three hundred boats participated in the red king crab fishery, generating an ex-vessel value of $40–$100 million. The reason for the long-term downturn in this population is the subject of some debate. On the one hand, the precipitous decline in the Bristol Bay population in

Grant, W.S., D. Zelinina, and N. Mugue. 2014. Population Genetics and Phylogeography of Red King Crab: Implications for Management and Stock Enhancement. Pages 47–72. In: B. Stevens (Ed.), *The King Crabs*. CRC Press, Boca Raton, FL.

Haflinger, K. 1981. A Survey of Benthic Infaunal Communities of the Southeastern Bering Sea shelf. Pages 1091–1104. In: D.W. Hood and J.A. Calder (Eds.), *The Eastern Bering Sea shelf: Oceanography and Resources*. Vol. 2. Office of Marine Pollution Assessment, NOAA, University of Washington Press, Seattle, WA.

Hare, J.A., W.E. Morrison, M.W. Nelson, M.M. Stachura, E.J. Teeters, R.B. Griffis, M.A. Alexander, J.D. Scott, L. Alade, R.J. Bell, A.S. Chute, K.L. Curti, T.H. Curtis, D. Kircheis, J.F. Kocik, S.M. Lucey, C.T. McCandless, L.M. Milke, D.E. Richardson, E. Robillard, H.J. Walsh, M.C. McManus, K.E. Marancik, and C.A. Griswold. 2016. A Vulnerability Assessment of Fish and Invertebrates to Climate Change on the Northeast U.S. Continental Shelf. PLoS One 11: e0146756.

Hare, S.R. and N.J. Mantua. 2000. Empirical Evidence for North Pacific Regime Shifts in 1977 and 1989. *Progress in Oceanography*. 47:103–146.

Hartwell, I, D. Apeti, T. Pait, T. Radenbaugh, and R. Britton. 2015. Bioeffects Assessment in Bristol Bay, Alaska: Characterization of Benthic Habitats, Animal Condition, and Contaminant Baseline Assessment. Poster. Alaska Marine Science Symposium, Anchorage, AK.

Heifetz, J. 2002. Coral in Alaska: Distribution, Abundance, and Species Associations. Hydrobiologia. 471:19–28.

Heifetz, J., B.L. Wing, R.P. Stone, P.W. Malecha and D.L. Courtney. 2005. Corals of the Aleutian Islands. Fisheries Oceanography. 14 (Suppl. 1):131–138.

Hilborn, R., T.P. Quinn, D.E. Schindler, and D.E. Rogers. 2003. Biocomplexity and Fisheries Sustainability. Proceedings of the National Academy of Sciences. 100: 6564–6568.

Hughes, S.E., R.W. Nelson, and R. Nelson. 1977. Initial Assessments of the Distribution, Abundance, and Quality of Subtidal Clams in the S.E. Bering Sea. Report of A Cooperative Industry-Federal-State of Alaska study. US Department of Commerce NOAA, Seattle, WA.

Hunt Jr., G.L., C. Baduini, and J. Jahncke. 2002b. Diets of Short-Tailed Shearwaters in the Southeastern Bering Sea. Deep-Sea Research II. 49:6147–6156.

Hunt, G.L., Jr., P.J. Stabeno, S. Strom, and J. M. Napp. 2008. Patterns of Spatial and Temporal Variation in the Marine Ecosystem of the Southeastern Bering Sea, with Special Reference to the Pribilof Comanin. Deep-Sea Research II 55:1919–1944.

Hunt, G., N. Bond, P. Stabeno, C. Ladd, C. Mordy, T. Whitledge, R. Feely et al. 2010. Status and Trends of the Bering Sea Region, 2003–2008. Pages 196–267. In: S. M. McKinnell and M. Dagg, editors. *The Marine Ecosystems of the North Pacific Ocean; Status and Trends*. PICES Special Publication 4, 393 p.

Hunt, G.L. Jr., K.O. Coyle, L.B. Eisner, E.V. Farley, R.A. Heintz, F. Muetr, J.M. Napp, J.E. Overland, P.H. Ressler, S. Salo, and P.H. Stabeno. 2011. Climate Impacts on Eastern Bering Sea Foodwebs: a Synthesis of New Data and an Assessment of the Oscillating Control Hypothesis. ICES Journal of Marine Science. 68:1230–1243.

Hunt, G.L., P. Stabeno, G. Waltrs, E. Sinclair, R.D. Brodeur, J.M. Napp, and N.A. Bond. 2002a. Climate Change and Control of the Southeastern Bering Sea Pelagic Ecosystem. Deep-Sea Research II. 49:5821–5853.

Jay, C.V., S.D. Farley, and G.W. Garner. 2001. Summer Diving Behavior of Male Walruses in Bristol Bay, Alaska. Marine Mammal Science. 17:617–631.

Jewett, S.C. and H.M. Feder. 1981. Epifaunal Invertebrates of the Continental Shelf of the Eastern Bering and Chukchi Seas. Pages 1131–1153. In: D.W. Hood and J.A. Calder Eds.), The Eastern Bering Sea shelf: Oceanography and Resources. Vol. 2. Office of Marine Pollution Assessment, NOAA, University of Washington Press, Seattle, WA.

Jewett, S.C., R.N. Clark, H. Chenelot, S. Harper, and M.K. Hoberg. 2012. Sea Stars of the Nearshore Aleutian Archipelago. Pages 144–172. In: Steller D, Lobel L (Eds.), *Diving for Science*. 2012. Proceedings of the American Academy of Underwater Sciences 31st Symposium. AAUS, Dauphin Island, AL.

Jorgensen, E.M. 2009. Field Guide to Squids and Octopods of the Eastern North Pacific and Bering Sea. Alaska Sea Grant College Program, University of Alaska, Fairbanks. 93 p.

Jouzel, J., V. Masson-Delmotte, O. Cattani, G. Dreyfus, S. Falourd, G. Hoffmann, B. Minster et al. 2007. Orbital and Millennial Antarctic Climate Variability over the past 800,000 years. Science 5839:793–796.

Kachel, N.B., L.L. Hunt, Jr., S.A. Salo, J.D. Schumache, P.J. Stabeno, and T.E. Whitledge. 2002. Characteristics and Variability of the Inner Front of the Southeastern Bering Sea. Deep-Sea Research II. 49:5889–5909.

Kaufman, D.S., T.A. Ager, N.J. Anderson, P.M. Anderson, J.T. Andrews, P.J. Bartlein, L.B. Brubaker, L.L. Coats, L.C. Cwynar, M.L. Duvall, A.S. Dyke, M.E. Edwards, W.R. Eisner, K. Gajewski, A. Geirsdottir, F.S. Hu, A.E. Jennings, M.R. Kaplan, M.W. Kerwin, A.V. Lozhkin, G.M. MacDonald, G.H. Miller, C.J. Mock, W.W. Oswald, B.L. Otto-Bliesner, D.F. Porinchu K. Rühland, J.P. Smol, E.J. Steig, and B.B. Wolfe. 2004. Holocene Thermal Maximum in the Western Arctic (0–180°W). Quaternary Science Reviews 23:529–560.

Keigwin, L.D., J.P. Donnelly, M.S. Cook, N.W. Driscoll, and J. Brigham-Grette. 2006. Rapid Sea-Level Rise and Holocene Climate in the Chukchi Sea. Geology. 34:861e864.

Knapp, G., M. Guetttabi, and S. Goldsmith. 2013. The Economic Importance of the Bristol Bay Salmon Industry. Bristol Bay Regional Seafood Development Association. http://www.iser.uaa.alaska.edu/Publications/2013_04-TheEconomicImportanceOfTheBristolBaySalmonIndustry.pdf.

Kober, K.M. and G. Bernardi. 2013. Phylogenomics of Strongylocentrotid Sea Urchins. BMC Evolutionary Biology 13:88.

Kruse, G.H., J. Zheng, and D.L. Stram. 2010. Recovery of the Bristol Bay Stock of Red King Crabs under a Rebuilding Plan. ICES Journal of Marine Science 67:1866–1874.

Kubodera, T. 1991. Distribution and Abundance of the Early Life Stages of Octopus, *Octopus dofleini*. Wulker, 1910 in the North Pacific. Bulletin of Marine Science 49:235–243.

Lambert, P. 2000. Sea Stars of British Columbia, Southeast Alaska, and Puget Sound. Royal British Columbia Museum.

Lauth, R.R. and E. Acuna. 2007. Results of the 2006 Eastern Bering Sea Continential Shelf Bottom Trawl Survey of Groundfish and Invertebrate Resources. U.S. Department of Commerce, NOAA Technical Memorandum NMFS-AFSC-176, 175 p.

Lauth (2011) Results of the 2010 eastern and northern Bering Sea continental shelf bottom trawl survey of groundfish and invertebrate fauna. U.S. Dept. Commerce, NOAA Tech Memo. NMFS-AFSC-227, 256 p.

Lees, D. C. 2006. Guide to Intertidal Bivalves in Southwest Alaska National Parks. National Park Service, Inventory and Monitoring Program, Rep. NPS/AKRSWAN/NRTR-2006/02, 65 p.

Limpinsel, D. 2013. Biological Characterization: An Overview of Bristol, Nushagak, and Kvichak Bays; Essential Fish Habitat, Processes, and Species Assemblages. Report, National Marine Fisheries Service, Alaska Region. 53 p.

Lisiecki, L.E. and M.E. Raymo. 2005. A Pliocene-Pleistocene stack of 57 globally distributed benthic $\delta^{18}$O records. Paleoceanography 20:PA1003.

Loher, T. and D.A. Armstrong. 2005. Historical Changes in the Abundance and Distribution of Ovigerous Red King Crabs (*Paralithodes camtschaticus*) in Bristol Bay (Alaska), and Potential Relationship with Bottom Temperature. Fisheries Oceanography. 14:292–306.

MacIntosh, R.A. and D.A. Somerton. 1981. Large Marine Gastropods of the Eastern Bering Sea. Pages 1215–1228. In: D.W. Hood and J.A. Calder (Eds.), The Eastern Bering Sea shelf: Oceanography and Resources. Vol. 2. NOAA, Office of Marine Pollution Assessment, University of Washington Press, Seattle, WA.

Mantua, N.J. and S.R. Hare. 2002. The Pacific Decadal Oscillation. Journal of Oceanography. 581:35–44.

McDonald, J., H.M. Feder, and M. Hoberg. 1981. Bivalve Mollusks of the Southeast Bering Sea. Pages 1155–1204. In: D.W. Hood and J.A. Calder (Eds.), *The Eastern Bering Sea Shelf: Oceanography and Resources*. Vol. 2. NOAA, Office of Marine Pollution Assessment, University of Washington Press, Seattle, WA.

McMurray, G., A.H. Vogel, P.A. Fishman, D.A. Armstrong, and S.C. Jewett. 1984. Distribution of Larval and Juvenile Red King Crabs (*Paralithodes camtschaticus*) in Bristol Bay. Pages 267–47. In: Outer Continental Shelf Environmental Assessment Program. Report No. 53. NOAA Office of Marine Pollution Assessment, Alaska Office, Anchorage, AK.

Merkouris, S.E., L.W. Seeb, and M.C. Murphy. 1998. Low Levels of Genetic Diversity in Highly Exploited Populations of Alaskan Tanner crabs, *Chionoecetes bairdi*, and Alaskan and Atlantic snow crabs, *C. opilio*. Fishery Bulletin, U. S. 96:525–537.

Miller, K.G, M.A. Kominz, J.V. Browning, J.D. Wright, G.S. Mountain, M.E. Katz, P.J. Sugarman, B.S. Cramer, N. Christie-Blick, and S.F. Pekar. 2005. The Phanerozoic Record of Global Sea-Level Change. Science 310:1293–1298.

Miller, R.J., J. Hocevar, R.P. Stone, and D.V. Fedorov. 2012. Structure-Forming Corals and Sponges and Their Use as Fish Habitat in Bering Sea Submarine Canyons. PLoS One 7:e33885.

National Research Council. 1996. The Bering Sea Ecosystem. National Academy Press, Washington, D.C.

NOAA. 2012. Coastal Habitat Mapping Program: Bristol Bay data summary report December 2012. Alaska Fisheries Science Center, Auk Bay. https://alaskafisheries.noaa.gov/shorezone/logs/bb12_summaryrpt1212.pdf (accessed October 29, 2015).

NOAA. 2015a. Habitat Areas of Particular Concern Eastern Bering Sea Invertebrates Species Synopses with Density-Distribution Maps 1984–88 and 1994–98. RACE Division, Alaska Fisheries Science Center, Auke Bay. http://www.afsc.noaa.gov/groundfish/HAPC/EBScontents.htm (accessed August 13, 2015).

NOAA 2015b. Snow Crab *Chionoecetes opilio*. NOAA Fisheries Service, Alaska Fisheries Science Center http://www.afsc.noaa.gov/Eduction/factsheets/10_opilio_Fs.pdf (accessed November 1, 2015).

NOAA 2015c. Southern Tanner crab *Chionoecetes bairdi*. NOAA Fisheries Service, Alaska Fisheries Science Center http://www.afsc.noaa.gov/Eduction/factsheets/10_bairdi_Fs.pdf (accessed November 1, 2015).

Ohashi, R., A. Yamaguchi, K. Matsuno, R. Saito, N. Yamada, A. Iijima, N. Shiga, and I. Imai. 2013. Interannual Changes in the Zooplankton Community Structure on the Southeastern Bring Sea shelf during summers of 1994–2009. Deep-Sea Research II. 94:44–56.

Orensanz, J.M., J. Armstrong, D. Armstrong, and R. Hilborn. 1998. Crustacean Resources are Vulnerable to Serial Depletion–the Multi-Faceted Decline of Crab and Shrimp Fisheries in the Greater Gulf of Alaska. Reviews in Fish Biology and Fisheries. 8:117–176.

Orensanz, J.M., B. Ernst and D.A. Armstrong. 2007. Variation of Female Size and State at Maturity in Snow Crab (*Chionoecetes opilio*) (Brachyura: Majidae) from the Eastern Bering Sea. Journal of Crustacean Biology. 27:576–591.

Ormsethm O.A., B.L. Norcross, and B. Holladay. 2009. Utilization of Nearshore Habitat by Fishes in Nushagak and Togiak Bays (Bristol Bay). NOAA Essential Fish Habitat project status report number: 2009–12, Seattle, WA.

Overland, J.E., N.A. Bond, and J.M. Adams. 2002. The Relation of Surface Forcing of the Bering Sea to Large-Scale Climate Patterns. Deep-Sea Research II. 49:5855–5868.

Overland, J.E., M. Wang, K.R. Wood, D.B. Percival, and N.A. Bond. 2012. Recent Bering Sea Warm and Cold Events in a 95-Year Context. Deep-Sea Research II. 65–70:6–13.

Quackenbush, L.T., R.S. Suydam, A.L. Bryan, L.F. Lowry, K.J. Frost, and B.A. Mahoney. 2015. Diet of Beluga Whales, *Delphinapterus leucas*, in Alaska from Stomach Contents, March–November. Marine Fisheries Review 77: 70–84.

Radenbaugh, T.A. and S Wingert. 2008. Nushagak Bay: Monitoring for Ecosystem Health, Nushagak Estuary Biodiversity Project Poster 36. American Fisheries Society, Alaska Chapter, Fairbanks, AK.

Radenbaugh, T.A. and S. Wingert-Pederson. 2011. Values of Nushagak Bay: Past, Present, and Future. In: North by 2020: Alaskan Perspectives on Changing Circumpolar Systems, Lovecraft, A. and H. Eicken (eds). Pages 95–110. University of Alaska Press, Fairbanks, AK.

Radenbaugh, T.A. 2013. Temperature Tolerance of the Isopod Saduria entomon. In: Responses of Arctic Marine Ecosystems to Climate Change F.J. Mueter, Danielle M.S. Dickson, Henry P. Huntington, James R. Irvine, Elizabeth A. Logerwell, Stephen A. MacLean, Lori T. Quakenbush, and Cheryl Rosa (Eds.), 28th Lowell Wakefield Fisheries Symposium, Alaska Sea Grant, Fairbanks, AK.

Regier J.C., J.W. Shultz, A. Zwick, A. Hussey, B. Ball, R. Wetzer, J.W. Martin, and C.W. Cunningham, (2010) Arthropod Relationships Revealed by Phylogenomic Analysis of Nuclear Protein-Coding Sequences. Nature 463: 1079–1083.

Rosenkranz, G.E., A.V. Tyler, and G.H. Kruse. 2001. Effects of Water Temperature and Wind on Year-Class Success of Tanner crab in Bristol Bay, Alaska. Fisheries Oceanography. 10:1–12.

Ruggerone, G.T., J.L. Nielsen, and J. Bumgarner. 2007. Linkages between Alaskan Sockeye Salmon Abundance, Growth at Sea, and Climate, 1955–2002. Deep-Sea Research II. 54:2776–2793.

Ruggerone, G.T., E. Farley, J. Nielsen, and P. Hagen. 2005. Seasonal Marine Growth of Bristol Bay sockeye Salmon (*Oncorhynchus nerka*) in Relation to Competition with Asian Pink Salmon (*O. gorbuscha*) and the 1977 Ocean Regime Shift. Fishery Bulletin 103:355–370.

Schwab, W.C. and B.F. Molnia. 1987. Unusual Bed Forms on the North Aleutian Shelf, Bristol Bay, Alaska. Geo-Mar Letters 7:207–215.

Sharma, D.S. 1979. The Alaska Shelf: Hydrographic, Sedimentary, and Geochemical Environment. Springer-Verlag, New York.

Sheffield, G. and J.M. Grebmeier. 2009. Pacific Walrus (*Odobenus rosmarus divergens*): Differential Prey Digestion and Diet. Marine Mammal Science. 25:761–777.

Shimek, R.L. 1984. The diets of Alaskan *Neptunea*. Veliger 26:274–281.

Shimek, R.L. and L.A. Gardner. 1979. Natural History of Inter-Tidal Alaskan *Neptunea pribiloffensis* and *N. lyrata*. American Zoologist. 19:865.

Sigler, M.F., P.J. Stabeno, L.B. Eisner, J.M. Napp, and F.J. Mueter. 2014. Spring and Fall Phytoplankton Blooms in a Productive Subarctic Ecosystem, the Eastern Bering Sea, during 1995–2011. Deep-Sea Research II. 109:71–83.

Smith, C.T., W.S. Grant, and L.W. Seeb. 2005. A Rapid, High-Throughput Technique for Detecting Tanner crabs *Chionoecetes bairdi* Illegally Taken in Alaska's Snow Crab Fishery. Transactions of the American Fisheries Society. 134:620–623.

Smith, K.R., R.A. McConnaughey, and C.E. Armistead. 2011. Benthic Invertebrates of the Eastern Bering Sea: a Synopsis of the Life History and Ecology of Snails in the Genus *Neptunea*. NOAA Technical Memorandum NMFS-AFSC-231, Auke Bay, Alaska. 59 p.

Stabeno, P.J., N.A. Bood, N.B. Kachel, S.A. Salo, and J.D. Schumacher. 2001. On the Temporal Variability of the Physical Environment over the Southeastern Bering Sea. Fisheries Oceanography. 10:81–98.

Stabeno, P.J., J. Napp, C. Mordy, and T. Whitledge. 2010. Factors Influencing Physical Structure and Lower Trophic Levels of the Eastern Bering Sea shelf in 2005: Ice, Tides and Winds. Progress in Oceanography. 85:180–196.

Stebeno, P.J., N.B. Kachel, S.E.Moore, J.M.Napp, M. Sigler, A. Yamaguchi, and A.N. Zerbini. 2012. Comparison of Warm and Cold Years on the Southeastern Bering Sea Shelf and Some Implications for the Ecosystem. Deep-Sea Research II, 65–70:31–45.

Stern, G. 2015. The Great Arctic Experiment. Science. 350: 520–521.

Stevens, B.G. 2003. Settlement, Substratum Preference, and Survival of Red King Crab *Paralithodes camtschaticus* (Tilesius, 1815) Glaucothoe on Natural Substrata in the Laboratory. Journal of Experimental Marine Biology and Ecology. 283:63–78.

Stevens, B.G. and J. Kittaka. 1998. Postlarval Settling Behavior, Substrate Preference, and Time to Metamorphosis for Red King Crab *Paralithodes camtschaticus*. Marine Ecology Progress Series. 167:197–206.

Stevens, B.G. and R.A. MacIntosh. 1991. Cruise results Supplement, Cruise 91-1 Ocean Hope 3: 1991 Eastern Bering Sea Juvenile Red King Crab Survey, May 24–June 1, 1991. National Marine Fisheries Service, Alaska Fisheries Science Center, Kodiak Fisheries Research Center, Kodiak, AK.

Stewart, N.L. and B. Konar. 2012. Kelp Forests Versus Urchin Barrens: Alternate Stable States and Their Effect on Sea Otter Prey Quality in the Aleutian Islands. Journal of Marine Biology. 2012:492308, 12 p.

Stoecker, D.K., A. Weigel, and J. Goes. 2014. Microzooplankton Grazing in the Eastern Bering Sea in summer. Deep-Sea Research II. 109:145–156.

Stone, R.P. and S.K. Shotwell 2007. State of Deep Coral Ecosystems in the Alaska Region: Gulf of Alaska, Bering Sea and the Aleutian Islands. Pages 65–108. In: The State of Deep Coral Ecosystems of the United States. NOAA Technical Memorandum CRCP-3, Silver Spring, MD.

Stone, R.P., H. Lehnert and H. Reiswig. 2011. A Guide to the Deep-Water Sponges of the Aleutian Island Archipelago. NOAA Professional Paper NMFS 12, Seattle, WA.

Strasburger, W.S., N. Hillgruber, A.I. Pinchuk, and F.J. Mueter. 2014. Feeding Ecology of Age-0 Walleye Pollock (*Gadus chalcogrammus*) and Pacific Cod (*Gadus macrocephalus*) in the Southeastern Bering Sea. Deep-Sea Research II. 109:172–180.

Straty, R.R. 1977. Current Patterns and Distribution of River Waters in Inner Bristol Bay, Alaska. NOAA Technical Report, NMFS SSRF-713. U.S. Dept. of Commerce.

Straty, R.R. and H.W. Jaenicke. 1980. Estuarine Influence of Salinity, Temperature, and Food on the Behavior, Growth and Dynamics of Bristol Bay Sockeye Salmon. Pages 247–265. In: W.J. McNeil and D.C. Himsworth (Eds.), *Salmonid Ecosystems of the North Pacific*. Oregon State University Press, Corvallis, OR.

Struck, T.H., C. Paul, N. Hill, S. Hartmann, C. Hösel, M. Kube, B. Lieb, A. Meyer, R. Tiedemann, G.N. Purschke, and C. Bleidorn. 2011. Phylogenomic Analyses Unravel Annelid Evolution. Nature. 471:95–98.

Tamburri, M.N. and J.P. Barry. 1999. Adaptations for Scavenging by Three Diverse Bathyal Species, *Eptatretus stouti*, *Neptunea amianta* and *Orchomene obtusus*. Deep-Sea Research I. 46:2079–2093.

Tremblay, J.E., C. Michel, K.A. Hobson, M. Gosselin, and N.M. Price. 2006. Bloom Dynamics in Early Opening Waters of the Arctic Ocean. Limnology and Oceanography 51:900–912.

Tynan, C.T., D.P. DeMaster, and W.T. Peterson. 2001. Endangered Right Whales on the Southeastern Bering Sea shelf. Science. 294:1894.

Van Stone, J.W. 1972. Nushagak: an Historic Trading Center in Southwestern Alaska. Fieldiana Anthropology, Anthropological Series Volume 62. Field Museum of Natural History, Chicago, IL.

Walters, G.E. and J.J. McPhail. 1982. An Atlas of Demersal Fish and Invertebrate Community Structure in the Eastern Bering Sea: Part 1, 1978–81. Report. RACE NOAA, Seattle, WA.

Wing, B.L. and D.R. Barnard. 2004. A Field Guide to Alaskan Corals. U.S. Department of Commerce, NOAA Technical Memorandum NMFS-AFSC-146, 67 p.

Yeung, C. and R.A. McConnaughey. 2006. Community Structure of Eastern Bering Sea Epibenthic Invertebrates from Summer Bottom-Trawl Surveys 1982–2002. Marine Ecology Progress Series. 318:47–62.

Zheng, J. and G.H. Kruse. 2000. Recruitment Patterns of Alaskan Crabs in Relation to Decadal Shifts in Climate and Physical Oceanography. ICES Journal of Marine Science. 57:438–451.

Zheng, J. and G.H. Kruse. 2001. Spatial Distribution and Recruitment Patterns of Snow Crabs in the Eastern Bering Sea. Pages 233–253. In: *Spatial Processes and Management of Marine Populations*. Alaska Sea Grant College Program AK-SG-01-02, Fairbanks, AK.

**Appendix 17.1**   List of common invertebrates in Bristol Bay, Alaska.[1] Numerous other species occur in Bristol Bay, but have not been sampled with appropriate equipment or have not been correctly identified. Many species are small and are not easily captured by commonly used sampling devices. Taxonomy followed the World Register of Marine Species (WoRMS) and Regier et al. (2011). We have also followed standard taxonomic convention for listing the names of researchers who originally described a species. These are referred to as taxonomic authorities and are placed after the genus and species names (in Latin). When a genus name has been changed from the original description, the authority is placed in parentheses.

| Major groups of species (Taxonomic level and name) | Common name | Distribution |
|---|---|---|
| **Sponges** (Phylum Porifera) | | |
| *Aphrocallistes vastus* Schulze, 1886 | Cloud sponge | North Pacific, Mexico to Japan |
| *Halichondria panicea* (Pallas, 1766) | Breadcrumb sponge | Bering Sea, Arctic Ocean, North Atlantic, Mediterranean |
| *Semisuberites cribrosa* (Miklucho-Maclay, 1870) | | Bering Sea, Arctic Ocean, North Pacific, North Atlantic |
| *Stelletta clarellade* Laubenfels, 1930 | | Alaska to Baja California |
| *Suberites ficus* Johnston, 1842 | Sea orange or stone sponge | North Atlantic, North Pacific |
| **Sea Anemones, hydras, jellyfish, and corals** (Phylum Cnidaria) | | |
| **Corals, sea anemones** (Class Anthozoa) | | |
| **Tube dwelling anemones** (Order Ceriantharia) | | |
| Unidentified species | | |
| | | |
| **Stony corals, polyps** (Subclass Hexacorallia) | | |
| **Sea anemones** (Order Actinaria) | | |
| *Anthopleura artemisia* (Pickering in Dana, 1848) | Burrowing anemone | Bering Sea to California |
| *Cribrinopsis fernaldi* Siebert and Spaulding, 1976 | Chevron tentacled anemone | Bering Sea, Northeast Pacific |
| *Liponema brevicornis* (McMurrich, 1893) | Tentacle shedding anemone | Bering Sea, Northeast Pacific, Generally at depth |
| *Metridium farcimen* (*giganteum*) (Brandt, 1835) | Giant anemone | Bering Sea, Northeast Pacific |
| *Metridium dianthus* (*senile*) (Ellis, 1768) | Clonal plumose anemone | Bering Sea, North Pacific, North Atlantic |
| *Stomphia coccinea* Müller, 1776 | Sea anemone, escape swimming | North Pacific, North Atlantic |

| Major groups of species (Taxonomic level and name) | Common name | Distribution |
|---|---|---|
| *Gorgonocephalus eucnemis* Müller and Troschel, 1842 | Basket star | Bering Sea, Northeast Pacific, North Atlantic; young associated with soft coral *Gersemia rubiformis* |
| *Ophiopholis aculeata* (Linnaeus, 1767) | Daisy brittle star | Bering Sea, Northeast Pacific |
| *Ophiura sarsi* Lütken, 1855 | Notched brittle star | Arctic Ocean, Bering Sea, North Pacific, North Atlantic |
| **Molluscs** (Phylum Mollusca) | | |
| **Squids, Octopus** (Class Cephalopoda) | | |
| *Benthoctopus leioderma* (Berry, 1911) | Smoothskin octopus | Bering Sea, occasionally in shallow water, North Pacific |
| *Berryteuthis magister* (Berry, 1911) | Magister armhook squid | Bering Sea, occasionally in shallow water, North Pacific |
| *Enteroctopus dofleini* (Wülker, 1910) | Giant Pacific octopus | Bering Sea, outer shelf, North Pacific |
| *Rossia pacifica* Berry, 1911 | Stubby squid | Bering Sea, occasionally in shallow water, North Pacific |
| | | |
| **Bivalves, Clams** (Class Bivalvia) | | |
| *Astarte borealis* (Schumacher, 1817) | Northern astarte | Bering Sea, Arctic Ocean, Northwest Atlantic |
| *Chlamys rubida* (Hinds, 1845) | Smooth pink scallop | Bering Sea, Northeast Pacific to Puget Sound |
| *Ciliatocardium* (*Clinocardium*) *ciliatum* (Fabricius, 1780) | Hairy cockle | Bering Sea, North Atlantic |
| *Cyclocardia crebricostata* (Krause, 1885) | Many-ribbed clam | Bering Sea, Beaufort Sea, North Pacific |
| *Hiatella arctica* (Linnaeus, 1767) | Arctic clam | Arctic, North Pacific, North Atlantic |
| *Keenocardium* (*Clinocardium*) *californiense* (Deshayes, 1839) | California cockle | Bering Sea, North Pacific |
| *Macoma balthica* (Linnaeus, 1767) | Baltic macoma clam | Bering Sea (?), Arctic, North Atlantic |
| *Macoma nasuta* (Conrad, 1837) | Bent-nose clam | Bering Sea, Northeast Pacific to Mexico |
| *Mactromeris polynyma* (Stimpson, 1860) | Arctic surf clam | Bering Sea, North Atlantic |
| *Modiolus modiolus* Linnaeus, 1758 | Horse mussel | North Atlantic, North Pacific |
| *Musculus discors* (Linnaeus, 1767) | Discordant mussel | Bering Sea, Arctic Ocean, North Atlantic |
| *Mya arenaria* Linnaeus, 1758 | Softshell clam | Bering Sea (invasive), Northeast Pacific, Northwest Atlantic |
| *Mytilus trossulus* Gould, 1850 | Blue bay mussel | North Atlantic, North Pacific |
| *Panomya norvegica* (Spengler, 1793) | Arctic rough mya clam | Bering Sea, North Atlantic |
| *Patinopecten caurinus* (Gould, 1850) | Weathervane scallop | Northeastern Pacific, Bering Sea |
| *Pododesmus macrochisma* (Deshayes, 1839) | Green false jingle (soft oyster) | Chukchi Sea, Bering Sea, Northeast Pacific to Mexico |
| *Saxidomus gigantea* (Deshayes, 1839) | Butterclam | Bering Sea, Northeast Pacific to California |

| Major groups of species (Taxonomic level and name) | Common name | Distribution |
|---|---|---|
| *Serripes groenlandicus* (Mohr, 1786) | Greenland cockle | Bering Sea, Arctic Ocean, North Atlantic |
| *Serripes laperousii* (Deshayes, 1839) | Broad cockle | Bering Sea, North Pacific |
| *Siliqua patula* (Dixon, 1788) | Pacific razor clam | Bering Sea, Northeast Pacific to N. California |
| *Silqua alta* (Broderip and Sowerby, 1829) | Alaska razor clam | Chukchi Sea, Bering Sea, North Pacific |
| *Tellina lutea* W. Wood, 1828 | Alaska great-tellin clam | Bering Sea, Arctic, North Pacific |
| *Yoldia hyperborea* (Gould, 1841) | Northern yoldia clam | Bering Sea, Arctic Ocean, North Atlantic |
| *Yoldia aeolica* (*seminuda*) (Valenciennes, 1846) | Crisscorssed yoldia clam | North Pacific, Bering Sea |
| | | |
| **Whelks** (Class Gastropoda) | | |
| *Ademete solida* (*regina*) (Aurivillius, 1885) | Noble admete, nutmeg whelk | Bering Sea, Arctic Ocean |
| *Aforia circinata* (Dall, 1873) | Keeled aforia | Bering Sea, Northeast Pacific |
| *Arctomelon stearnsii* (Dall, 1872) | Alaska volute | Bering Sea, North Pacific |
| *Arctomelon tamikoae* (Kosuge, 1970) | Volute | Bering Sea, Northwest Pacific |
| *Aulacofusus brevicauda* ( = *Colus spitzbergensis*) (Deshayes, 1832) | | Bering Sea, North Atlantic |
| *Beringius beringii* (Middendorff, 1848) | | Bering Sea, Arctic Ocean |
| *Beringius kennicottii* (Dall, 1871) | | Bering Sea |
| *Beringius stimpsoni* (Gould, 1860) | | Bering Sea, Chukchi Sea |
| *Boreotrophon alaskanus* (Dall, 1902) | Alaskan trophon | Bering Sea, Northwest Pacific |
| *Buccinum angulosum* J.E. Gray, 1839 | Angular whelk | Bering Sea, Beaufort Sea, North Atlantic |
| *Buccinum oedematum* Dall, 1907 | Swollen whelk | Bering Sea, North Pacific |
| *Buccinum plectrum* Stimpson, 1865 | Sinuous whelk | Bering Sea, North Atlantic |
| *Buccinum polare* J.E. Gray, 1839 | Thin whelk | Bering Sea, Beaufort Sea, North Atlantic |
| *Buccinum scalariforme* Møller, 1842 | Ladder whelk | Bering Sea, Beaufort Sea |
| *Buccinum tricarinatum* Dall, 1877 | Ribbed whelk | Bering Sea, Aleutians |
| *Clinopegma magnum* (W.H. Dall, 1895) | Helmet whelk | Bering Sea, Northwest Pacific |
| *Cryptonatica affinis* (Gmelin, 1791) | Arctic moonsnail | Bering Sea, Circum-Arctic, Northwest Pacific, North Atlantic |
| *Cryptonatica aleutica* (Dall, 1919) | Aleutian moonsnail | Bering Sea, Northwest Pacific |
| *Cryptonatica russa* (A.A. Gould, 1859) | Rusty moonsnail | Bering Sea, Northwest Pacific |
| *Dendronotus* sp. | Nudibranch | Bering Sea, North Pacific |
| *Euspira pallida* (Broderip and Sowerby, 1829) | Pale moonsnail | Bering Sea, Beaufort Sea, North Atlantic |
| *Fusitriton oregonensis* J.H. Redfield, 1846 | Oregon triton, hairy triton | Bering Sea, North Pacific |
| *Grandicrepidula* (*Crepidula*) *grandis* (Middendorff, 1849) | Great slipper limpet | Bering Sea, Aleutians, Northwest Pacific |
| *Latisipho* (*Colus*) *hypolispus* (Dall, 1891) | Oblique whelk | Bering Sea, Chukchi Sea, Gulf of Alaska |
| *Limacina helicina* (Phipps, 1774) | Pteropod | Bering Sea, Arctic Ocean, North Pacific, North Atlantic |

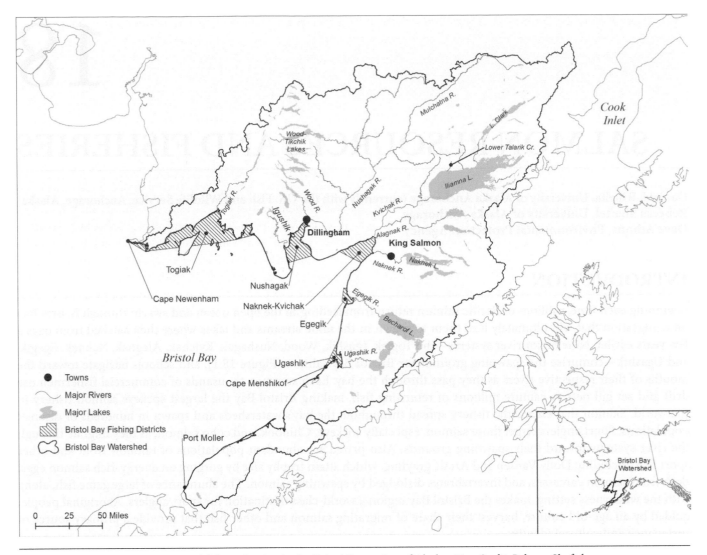

**Figure 18.1** Major river systems and fishing districts in the Bristol Bay region of Alaska. *Map Credit: Rebecca Shaftel.*

dug into the streambed (or lake bottom) gravel by a female trout or salmon where eggs are deposited and buried for incubation. Finally, *fry* and *smolt* both refer to juvenile salmon; the former to those that are rearing in fresh water and the latter for those that are undergoing physiological transformation in preparation for migration to the ocean.

## ECOLOGY AND LIFE HISTORY OF BRISTOL BAY FISHES

### General Salmon Life History

Five species of Pacific salmon are native to North American waters—pink (*Oncorhynchus gorbuscha*), chum (*O. keta*), sockeye (*O. nerka*), coho (*O. kisutch*), and Chinook (*O. tshawytscha*)—and all have spawning populations in the Bristol Bay region. These species share a rare combination of life history traits that contribute to their biological success, as well as their status as cultural icons around the North Pacific rim. These traits—anadromy, homing, and semelparity—are described briefly in the following paragraphs.

All Pacific salmon hatch in fresh water, migrate to sea for a period of relatively rapid growth, and return to fresh water to spawn. This strategy, termed *anadromy*, allows salmon to capitalize on the resource-rich marine environment, where growth rates are much faster than in fresh water. Thus, anadromy allows salmon to attain larger body size, mature more quickly, and maintain larger spawning populations than would be possible with a nonmigratory life history (McDowall 2001). The evolution of anadromy appears to be driven by a disparity in productivity between adjacent freshwater and marine environments (Gross et al. 1988). Productivity of fresh waters generally decline with latitude while that of oceans increases, and the disparity between the two peaks within the range of Pacific salmon (Gross et al. 1988).

When salmon enter fresh water to spawn, the vast majority return to the location where they were spawned. By this means, termed *homing*, salmon increase juvenile survival by returning to spawn in an environment with proven suitability (Cury 1994). Another adaptive advantage of homing is that it fosters reproductive isolation that enables populations to adapt to their particular environment (Blair et al. 1993, Dittman and Quinn 1996, Eliason et al. 2011). For instance, populations that travel long distances to reach inland spawning sites develop large lipid reserves to fuel the migration (Quinn 2005), since adult salmon generally do not feed after entering fresh water. As another example, sockeye salmon fry from populations that spawn downstream of nursery lakes are genetically programmed to migrate upstream after emergence, while fry from populations that spawn upstream of nursery lakes are programmed to migrate downstream (Burgner 1991). Examples of adaptations are many, and include heritable anatomical, physiological, and behavioral traits. Without homing, gene flow would occur throughout the species, making adaptation to specific freshwater conditions impossible; in this sense, homing counteracts the dispersal effects of anadromy (McDowall 2001). Homing is not absolute, however, and a small amount of straying ensures that amenable habitats are colonized by salmon (e.g., Milner and Bailey 1989).

Pacific salmon, quite famously, die after spawning only once. This trait, termed *semelparity*, serves to maximize the investment in one reproductive effort at the expense of any future reproductive effort. In salmon, it may have evolved as a response to the high cost of migration to spawning streams and the associated reduction in adult survival (Roff 1988). The evolution of semelparity in Pacific salmon was accompanied by increased egg size, thus, while long migrations may have been a prerequisite, the driving force behind the evolution of semelparity was likely the increase in egg mass and associated increase in juvenile survival (Crespi and Teo 2002).

As salmon approach sexual maturity, the countershading and silvery sheen that hide them at sea give way to characteristic spawning colors, often with hues of red (Figure 18.2 ). Males develop hooked snouts (the generic name *Oncorhynchus*

**Figure 18.2**  While in the ocean, sockeye salmon are bright silver and their flesh is red due to the carotenoids in their diet. Once they enter freshwater, the carotenoids in their flesh begin migrating to their skin, ultimately giving them their characteristic red spawning color, which helps attract mates. *Photo credit: Michael Melford.*

refers to this trait) and protruding teeth, and their previously bullet-shaped bodies become laterally flattened. These spawning colors and secondary sexual characteristics, which develop to varying degrees among species and even among populations, probably serve multiple purposes on the spawning grounds, including species recognition, sex recognition, and territorial displays.

With few exceptions, preferred spawning habitat consists of gravel-bedded stream reaches with moderate depth and current (30–60 cm deep and 30–100 cm per second, respectively; Quinn 2005). Females excavate a spawning redd in the gravel to receive the eggs, which are fertilized by one or more competing males as they are released and subsequently buried by the female. The seasonality of spawning and incubation is roughly the same for all species of Pacific salmon, although the timing can vary somewhat by species, population, and region. In general, salmon spawn during summer or fall and the fry emerge from the spawning gravel the following spring. While in the gravel, the embryos develop within their eggs and then hatch into fry that continue to subsist on yolk sacs. After emerging from the gravel, basic life history patterns of the five species differ in notable ways.

## SPECIES-SPECIFIC LIFE HISTORY AND ECOLOGY

### Sockeye Salmon

Sockeye salmon originate from river systems along the North American and Asian shores of the North Pacific and Bering Sea, roughly from the latitude of the Sacramento River to that of Kotzebue Sound. The largest North American populations occur between the Columbia and Kuskokwim rivers (Burgner 1991). Spawning sockeye salmon are readily identified by their striking red bodies with green heads and tails; males additionally develop a large hump in front of the dorsal fin (Figure 18.3).

**Figure 18.3**    Male sockeye spawners in fresh water are deep bodied, have large heads with elongated jaws, hooked snouts, and exposed teeth. They look quite different than their form in the marine environment. *Photo credit: Michael Melford.*

Sockeye are unique among salmon in that most populations rely on lakes as the primary freshwater rearing habitat. Some sockeye salmon spawn within the nursery lake where their young will rear. Others spawn in nearby stream reaches, and their fry migrate to the nursery lake after emerging from spawning redds. Sockeye salmon are by far the most abundant salmon species in the Bristol Bay region (Salomone et al. 2011), undoubtedly due to the abundance of accessible lakes in this landscape (Figure 18.1). Tributaries to Iliamna Lake, Lake Clark, and the Wood Tikchik Lakes are major spawning areas, and juveniles rear in each of these lakes. On average, Iliamna Lake (at 2600 km², the largest lake in Alaska) produces more sockeye salmon than any other lake in the Bristol Bay region. Juveniles in Bristol Bay systems rear for one or two years in their nursery lakes (West et al. 2009), feeding primarily on zooplankton in the limnetic zone (Burgner 1991). Some populations do not use lakes, and such riverine sockeye salmon spawn and rear throughout the Nushagak River watershed.

Bristol Bay sockeye salmon typically spend two or three years at sea (West et al. 2009), feeding on a range of invertebrates, small fish, and squid (Burgner 1991). They return at an average weight of 2.7 kg (based on recent commercial catches; Salomone et al. 2011).

## Chinook Salmon

Chinook salmon spawn in streams on both shores of the North Pacific and Bering Sea, roughly from the latitude of central California to that of Point Hope. There are more than a thousand North American spawning populations and a much smaller number in Asia. These populations tend to be relatively small, however, making Chinook salmon the rarest of North America's Pacific salmon species (Healey 1991). They are also the largest of the Pacific salmon; at least one specimen over 60 kg has been reported, but most weigh less than 23 kg (Mecklenburg et al. 2002).

Chinook salmon exhibit two distinct behavioral forms. The *stream type* form is predominant in Bristol Bay, as well as other areas of northern North America, Asia, and the headwaters of Pacific Northwest rivers (Healey 1991). These fish spend one or more years as juveniles in fresh water, range widely at sea, and return to spawning streams during spring or summer. *Ocean type* Chinook salmon, by contrast, migrate to sea soon after hatching, forage primarily in coastal marine waters, and return to spawning streams in the fall (Healey 1991). In fresh water, juvenile Chinook salmon tend to occupy flowing water and feed on aquatic insects. At sea, Chinook salmon are generally piscivorous (Brodeur 1990) and feed higher on the food chain than other salmon species (Satterfield and Finney 2002).

Chinook salmon spawn and rear throughout the Nushagak River watershed and in many tributaries of the Kvichak River. Some life history data are available from adults returning to the Nushagak River, Bristol Bay's largest Chinook salmon run. Essentially all Nushagak River Chinook salmon spend one year rearing in fresh water, and the vast majority (typically > 90% of a given brood year) spend two to four years at sea (Gregory Buck, ADF&G, unpublished data). Fish that spend four years at sea are the dominant age class and comprise approximately 43% of the average return, followed by those that spend three years (35%) and two years (17%) at sea. Chinook salmon individuals in recent Bristol Bay commercial catches have averaged 7.5 kg (Salomone et al. 2011).

## Rainbow Trout

Rainbow trout (*Oncorhynchus mykiss*) are native to western North America and the eastern coast of Asia, although their popularity as a sport fish has led to introduced populations around the world. Bristol Bay's rainbow trout are of the coastal variety (sensu Behnke 1992), which ranges from the Kuskokwim River to southern California. While classified in the same genus as the Pacific salmon, there are some key differences. Foremost, rainbow trout are not genetically programmed to die after spawning, making iteroparity (i.e., repeat spawning) a key life history trait of most populations. Also, most coastal watersheds support resident populations that rear entirely in fresh water in addition to anadromous populations (i.e., steelhead), although only the resident form occurs near the northern and southern limits of their distribution (Behnke 1992), including Bristol Bay rivers. Finally, rainbow trout spawn in the spring, as opposed to summer or early fall, although their spawning habitat and behavior is similar to that of salmon.

Bristol Bay rainbow trout tend to mature slowly and grow to a relatively large size. For example, 90% of spawners in Lower Talarik Creek, a tributary to Iliamna Lake, were more than seven years old; the majority of these were longer than 500 mm and a few exceeded 800 mm (years 1971–1976; Russell 1977). Growth was fastest for fish between four and six years of age and winter growth appeared to be minimal (Russell 1977).

Bristol Bay trout utilize complex and varying migratory patterns that allow them to capitalize on different stream and lake habitats for feeding, spawning, and wintering. For example, fish from Lower Talarik Creek migrate downstream to Iliamna Lake after spawning. From there, they appear to utilize a variety of habitats, as some tagged individuals have been recovered in other Iliamna Lake tributaries and in the Newhalen and Kvichak rivers (Russell 1977). In the Alagnak River watershed, a number of rainbow trout life history types have been identified, each with their own habitat use and seasonal migratory patterns (Meka et al. 2003). These consist of lake, lake-river, and river residents, the latter of which range from nonmigratory to highly migratory (Meka et al. 2003). Individuals comprising each of these life history types migrate in order to spend the summer in areas with abundant spawning salmon (Meka et al. 2003).

Eggs from spawning salmon are a major food item for Bristol Bay trout and are likely responsible for much of the growth attained by these fish. Upon arrival of spawning salmon in the Wood River watershed, rainbow trout shifted from consuming aquatic insects to primarily salmon eggs for a fivefold increase in ration and energy intake (Scheuerell et al. 2007). With this rate of intake, a bioenergetics model predicts a 100-g trout to gain 83 g in 76 days; without the salmon-derived subsidy, the same fish was predicted to lose five g (Scheuerell et al. 2007). Rainbow trout in Lower Talarik Creek were significantly fatter (i.e., higher condition factor) in years with high salmon spawner abundance than in years with low abundance (Russell 1977).

## Coho Salmon

Coho salmon are native to coastal watersheds in western North America and eastern Asia, approximately from the latitude of the Sacramento River to that of Point Hope (Sandercock 1991). Coho salmon occur in relatively small populations, and are second only to Chinook salmon in rarity.

Most Alaskan coho salmon populations tend to spend two years in fresh water and one year at sea (Sandercock 1991). Few age data exist for Bristol Bay, but samples from two years on the Nushagak River indicated that approximately 90% of spawning coho salmon shared this age structure, while the remaining fish had spent either one year or three years in fresh water (West et al. 2009). Coho salmon individuals in recent Bristol Bay commercial catches have averaged 3.0 kg (Salomone et al. 2011).

At sea, coho salmon consume a mix of fish and invertebrates (Brodeur 1990). Their trophic position is intermediate for Pacific salmon; Chinook salmon consume more fish while sockeye, pink, and chum salmon eat more zooplankton and squid (Satterfield and Finney 2002).

As juveniles in fresh water, coho salmon feed primarily on aquatic insects, although salmon eggs and flesh can be important nutritional subsidies (Heintz et al. 2010, Rinella et al. 2012, Armstrong et al. 2013). They utilize a wide range of lotic and lentic freshwater habitats, including stream channels, off-channel sloughs and alcoves, beaver ponds, and lakes. Coho salmon distribute widely into headwater streams, where they are often the only salmon species present (Woody and O'Neal 2010, King et al. 2012, ADF&G Anadromous Waters Catalog). Production of juvenile coho is often limited by the extent and quality of available wintering habitats (Nickelson et al. 1992, Solazzi et al. 2000).

## Pink Salmon

Pink salmon spawning populations occur on both sides of the North Pacific and Bering Sea, as far south as the Sacramento River and northern Japan. At the northern end of their range, small populations spawn along the North American and Asian shores of the Arctic Ocean. The most abundant salmon overall (Irvine et al. 2009), pink salmon have a simplified life history that relies little on freshwater rearing habitat. They typically spawn in shallow, rocky stream reaches relatively low in the watershed and their young migrate to sea soon after emerging (Heard 1991).

Essentially all pink salmon breed at two years of age, and this strict two-year life cycle results in genetic isolation of odd- and even-year spawning runs, even within the same river system. For reasons not entirely clear, large disparities between odd- and even-year run sizes occur across geographic regions and extend over many generations. An extreme example is the Fraser River, in southern British Columbia, where millions of pink salmon return during odd-numbered years, yet no fish return during even-numbered years (Riddell and Beamish 2003). In Bristol Bay rivers, even-year runs dominate the returns (Salomone et al. 2011).

Pink salmon are the smallest of the Pacific salmon species, with individuals in recent Bristol Bay commercial catches averaging 1.6 kg (Salomone et al. 2011). Sexually mature males become highly laterally compressed and develop a massive dorsal hump, hence the commonly used nickname *humpy*.

## Chum Salmon

Chum salmon spawn on both shores of the Bering Sea and North Pacific, extending south to the latitude of Japan and California (Salo 1991). Scattered spawning populations also occur on the Asian and North American shores of the Arctic Ocean. Populations tend to be relatively large, and chum salmon are the third most abundant species, behind pink and sockeye salmon.

Chum salmon, like pink salmon, migrate to sea soon after emerging from spawning gravel. Across their range, the vast majority spend two to four years at sea (Salo 1991), and at least one year's run in the Nushagak River showed similar age structure (West et al. 2009). At sea, chum salmon consume a range of invertebrates and fishes, and gelatinous material is commonly found in stomachs leading to speculation that jellyfish may be a common prey item (Brodeur 1990, Azuma 1992). Individuals in recent Bristol Bay commercial catches have averaged 3.1 kg (Salomone et al. 2011).

# BRISTOL BAY FISHERIES AND FISHERIES MANAGEMENT

## Historical Perspective on Commercial Salmon Fisheries

Salmon have long been an important driver in Alaska's economy and have played an important role in the state's history. Commercial fishing interests were among the original supporters of the purchase of Alaska from Russia in 1867 (King 2009). The first canneries were established eleven years later and by the 1920s salmon surpassed mining as Alaska's major industry as Alaska became the world's principal salmon producer (Ringsmuth 2005).

In the early years, fish packing companies essentially had a monopoly on the harvest of salmon. Packers in Bristol Bay and elsewhere built industrial fish traps, constructed of wood pilings and wire fencing with long arms that guided schools of migrating salmon into holding pens (King 2009). In Bristol Bay, packing interests also upheld a federal ban on fishing with motorized boats until 1951. Ostensibly a conservation measure, this law served to protect obsolete cannery-owned sailboat fleets by excluding independent Alaska-based fishermen who largely used motorized boats by this time (Troll 2011).

Salmon harvest peaked in 1936 then declined steadily for many years, leading to a federal disaster declaration in the 1950s (King 2009). A lack of scientific management, poor federal oversight, excessive harvest during World War II, and natural changes in ocean conditions contributed to the decline (King 2009).

Declining salmon runs, along with Alaskans' desire for more control over their fisheries, was a significant factor in the drive toward statehood (Augerot 2005, King 2009). In 1955, Alaskans began to develop a state constitution that included provisions intended to preserve Alaska's fisheries and, unique among state constitutions, to guarantee equal access to fish and game for all residents. Alaska became a state in 1959, the year that marked the lowest salmon harvest since 1900 (King 2009). Statehood was a turning point for Alaska's salmon fisheries, with the end of federal management, fish traps, and undue control of the resource by the canning industry. With the mandate for equal access came decentralization of the fishing industry, and thousands of independent fishermen began harvesting salmon for market to the canneries (Ringsmuth 2005) (Figure 18.4).

When the Alaska Department of Fish and Game (ADF&G) assumed management of the fisheries in 1960, restoring salmon runs to their former abundance became a primary objective. Inventorying fish stocks, understanding basic ecology, and improving run strength forecasting were central research goals. Of particular importance was the development and application of methods for counting salmon runs in spawning streams, which allowed the establishment of escapement goals and management based on scientific principles of sustained yield. Bristol Bay salmon research has been conducted primarily by ADF&G staff and researchers at the University of Washington's Alaska Salmon Program (see http://fish.washington.edu/research/alaska/). The latter, funded largely by the salmon processing industry, began researching factors controlling sockeye salmon production in 1947. While the scope of their investigations has expanded over the years, sockeye salmon monitoring is still a focus and represents the world's longest-running program for monitoring salmon and their habitats.

Over time, a number of state and federal policy changes have affected Bristol Bay salmon fisheries. A 1972 constitutional amendment set the stage for a bill that limited participation in Alaska commercial salmon fisheries. For fisheries around Alaska, this legislation set an optimum number of permits, which were then issued by the state based on an individual's fishing history. Permits are owned by the individual fisherman and are transferable, making them a limited and valuable asset (King 2009). The Fishery Conservation and Management Act of 1976, commonly known as

**Figure 18.4** Sockeye salmon on the *slime line* about to be processed into fillets or prepared for canning. *Photo credit: Michael Melford.*

the Magnuson-Stevens Act, was introduced to Congress by the late senator Ted Stevens as a means to curtail high-seas salmon fishing. In response to intensive Japanese gill netting in the western Aleutians and Bering Sea since 1952, this legislation extended America's jurisdiction from 12–200 miles (19–322 km) offshore. This ensured that salmon produced in Alaskan rivers would be harvested and processed locally and gave Alaska's fishery managers much more control in deciding when and where salmon are harvested. Both the Policy for the Management of Sustainable Salmon Fisheries (5 AAC 39.222) and the Policy for Statewide Salmon Escapement Goals (5 AAC 39.223) were adopted in the winter of 2000–2001 (Baker et al. 2009). The former established a comprehensive policy for the regulation and management of sustainable fisheries and the latter defined procedures for establishing and updating salmon escapement goals, including a process for public review of associated allocation disputes.

The ADF&G is responsible for managing fisheries under the sustained yield principle. Fishing regulations, policies, and management plans are enacted by the Board of Fisheries, which it does in consultation with the ADF&G, advisory committees, the public, and other state agencies. The Board of Fisheries consists of seven citizens, appointed by the governor and confirmed by the legislature, that serve three-year terms. Eighty-one advisory committees, whose members are elected in local communities around the state, provide local input. While regulations and management plans provide the framework for fisheries regulation, local fisheries managers are ultimately responsible for their execution. They are delegated with the authority to make *emergency orders*—in-season changes to fishing regulations—that allow rapid adjustments to changing conditions, often with very short notice. Managers use them to provide additional protection to fish stocks when conservation concerns arise and to liberalize harvest when surplus fish are available. Management plans directed at specific fish stocks are often based on anticipated scenarios and give specific directions to managers, making the in-season management process predictable to ADF&G, commercial fishermen, and the public. Alaska's management of its salmon fishery has proven successful; it was the second fishery in the world to be certified as well-managed by the Marine Stewardship Council (Hilborn 2006) and is regarded as a model of sustainability (Hilborn et al. 2003a, King 2009).

## Current Management of Commercial Salmon Fisheries

While all five species of Pacific salmon are harvested in Bristol Bay, sockeye salmon dominate the runs and harvest by a large margin (Table 18.1). Salmon return predominately to nine major river systems, located on the eastern and northern sides of the bay, and are harvested in five fishing districts near the river mouths, which allows managers to regulate harvest individually for the various river systems (Figure 18.1). The Naknek-Kvichak district includes those two rivers as well as the Alagnak River. The Nushagak district includes the Nushagak, Wood, and Igushik rivers. The Egegik, Ugashik, and Togiak districts include only the rivers for which they are named.

Fishing is conducted with drift or set gill nets. Drift gill nets have a maximum length of 150 fathoms (274 m) and are fished from boats no longer than of 32 ft. (9.8 m) in length (Figure 18.5). Set gill nets are fished from beaches, often with the aid of an open skiff, and have a maximum length of 50 fathoms (91 m) (Figure 18.6). There are approximately 1900 drift gill net permits and 1000 set gill net permits in the Bristol Bay salmon fishery, of which around 90% are fished on a given year (1990–2010 average; Salomone et al. 2011).

**Table 18.1** Mean harvest by species and fishing district, 1990–2009. Unpub. data, Paul Salomone, ADF&G Area Management Biologist

|  | Naknek-Kvichak | Egegik | Ugashik | Nushagak | Togiak | Total |
|---|---|---|---|---|---|---|
| Sockeye | 8,238,895 | 8,835,094 | 2,664,738 | 5,478,820 | 514,970 | 25,732,517 |
| Chinook | 2,816 | 849 | 1,402 | 52,624 | 8,803 | 66,494 |
| Chum | 184,399 | 78,183 | 70,240 | 493,574 | 158,879 | 985,275 |
| Pink* | 73,661 | 1,489 | 138 | 50,448 | 43,446 | 169,182 |
| Coho | 4,436 | 27,433 | 10,425 | 27,754 | 14,234 | 84,282 |

*Pink salmon data are from even-numbered years only since harvest is negligible during the smaller odd-year runs.

**Figure 18.5** Commercial fishermen pick sockeye salmon from their drift gill net during a Bristol Bay opener. To improve the quality of the fish, many fishers now bleed and ice fish prior to delivery to a tender or processor. Some entrepreneurs even process and market the salmon they catch in new creative ways, while processors are developing value-added products instead of just canning fish, as was done in the past. *Photo credit: Michael Melford.*

**Figure 18.6**   Commercial setnetters are often family groups that work together during fishing season. The 50-fathom gill nets are anchored to shore and fish more or less perpendicular to the beach. Crews sometimes get more than a thousand fish when the run is peaking, as illustrated here. *Photo credit: Michael Melford.*

The management of the Bristol Bay sockeye salmon fishery is focused on allowing an adequate number of spawners to reach each river system while maximizing harvest in the commercial fishery (Salomone et al. 2011). This balancing act is achieved through the establishment of escapement goals which represent the optimum range of spawners for a given river system. Escapement goals are established using a time series of spawner counts spanning many years, where a spawning run of a given size (i.e., stock) can be linked to the number of its offspring returning in subsequent years (i.e., recruits). Established stock-recruit models (Ricker 1954, Beverton and Holt 1957) are then used to estimate the number of spawners expected to result in the largest salmon catch in subsequent years, or the maximum sustained yield (Baker et al. 2009). In theory, spawning runs that are too small or large can result in reduced recruitment. With the former, too few eggs are deposited. With the latter, superimposition of spawning redds can diminish egg viability and competition in nursery lakes can reduce growth and survival (Schindler et al. 2005a, Ruggerone and Link 2006). Once escapement goals are set, the timing and duration of commercial fishery openings are then adjusted during the fishing season (i.e., in-season management) to ensure that escapement goals are met and any additional fish are harvested. Escapement goals are periodically reviewed and updated based on regulatory policies, specifically, the Policy for the Management of Sustainable Salmon Fisheries (5 AAC 39.222) and the Policy for Statewide Salmon Escapement Goals (5 AAC 39.223).

Each of Bristol Bay's nine major river systems has an escapement goal for sockeye salmon (Table 18.2), and in-season management of the commercial fishery is used to keep escapement in line with the goals. Management responsibility is divided among three managers: one for the Naknek, Kvichak, and Alagnak rivers; one for the Nushagak, Wood, Igushik, and Togiak rivers; and one for the Ugashik and Egegik rivers. Fishery openings are based on information from a number of sources, including preseason forecasts, the test fishery at Port Moller, early performance of the commercial fishery, and in-river escapement monitoring.

Preseason forecasts are the expected returns of the dominant age classes in a given river system, based on the number of spawning adults that produced each age class. In the Port Moller test fishery, gill netting at standardized locations provides a daily index of the overall number of fish entering Bristol Bay (Flynn and Hilborn 2004), with approximately seven days' lead before they enter the commercial fishing districts. Genetic samples from the test fishery are analyzed within four days (Dann et al. 2009) to give managers an advance estimate of run strength for each of the nine major river systems. As salmon move upstream, escapement is monitored with counting towers on each of the major rivers, except the Nushagak where a sonar system is used. Counting towers are elevated platforms along small to medium-sized (10–130 m wide), clear rivers from which migrating salmon are visually counted (Woody 2007). The Nushagak River's DIDSON sonar uses sound waves to detect and enumerate migrating salmon. Since tower and sonar monitoring occurs well upstream of the commercial fishery, all information regarding the performance of the fishery must be analyzed on a continual basis to ensure escapement levels will be met (Clark 2005, Salomone et al. 2011).

The fishery is typically opened on a schedule in the early part of the season, during which time the frequency and duration of openings are primarily based on preseason forecasts and management is conservative. As the fishing season progresses and more information becomes available, managers make constant adjustments to fishing time and area. If the escapement goal is expected to be exceeded at a given monitoring station, the fishery is opened longer and more frequently. If the escapement goal is not expected to be reached, the fishery is closed. If the escapement goal is not reached, the fishery is closed. Fishing time is opened and closed using emergency orders, and fishermen often learn of changes only a few hours before they go into effect. Since the bulk of the sockeye salmon harvest occurs during a short timeframe—from the last week of June until the middle of July—this short warning system is needed to maximize fishing time while ensuring that escapement levels are met. Migrating fish move quickly through the fishing districts, and delaying an opener by one day during the peak of the migration can forego the harvest of a million salmon. This is a significant loss of revenue to individual fishers and is compounded by the missed revenue of workers, processors, and marketers (Clark 2005). The fishery will periodically close *de facto* during the peak of the season when catch rates exceed processing capacity and processors stop buying fish. This lack of buyers can also curtail salmon harvest early and late in the season when numbers of fish do not warrant keeping processing facilities operational.

In-season management is also used to help meet an escapement goal for Chinook salmon on the Nushagak River (Table 18.3), where escapement is monitored by sonar. There are also Chinook salmon goals for the Togiak, Alagnak,

**Table 18.2**  Sockeye salmon escapement goals for Bristol Bay's nine major river systems

| River | Escapement Range (thousands) |
|---|---|
| Kvichak | 2,000–10,000 |
| Alagnak | 320 minimum |
| Naknek | 800–1,400 |
| Egegik | 800–1,400 |
| Ugashik | 500–1,200 |
| Wood River | 700–1,500 |
| Igushik | 150–300 |
| Nushagak-Mulchatna | 340–760 |
| Togiak | 120–170 |

**Table 18.3**  Chinook and chum salmon escapement goals for Bristol Bay river systems

| River | Species | Escapement Goal |
|---|---|---|
| Togiak | Chinook | 9,300 minimum |
| Nushagak | Chinook | 40,000-80,000 |
| Nushagak | Chum | 190,000 minimum |
| Alagnak | Chinook | 2,700 minimum |
| Naknek | Chinook | 5,000 minimum |
| Egegik | Chinook | 450 minimum |

Naknek, and Egegik Rivers and a chum salmon goal for the Nushagak River (Table 18.3), but in-season management is not used to help attain these goals (Baker et al. 2009).

Bristol Bay salmon fisheries are regarded as a management success (Hilborn et al. 2003a, Hilborn 2006). Hilborn (2006) lists four contributing factors: "(1) a clear objective of maximum sustainable yield; (2) the escapement-goal system, which assures maintenance of the biological productive capacity; (3) management by a single agency with clear objectives and direct line responsibility; and (4) good luck in the form of lack of habitat loss and good ocean conditions since the late 1970s (Figure 18.7)."

## Description of Sport Fisheries

The sport fisheries in Bristol Bay's river systems are regarded as world class. In contrast to commercial fisheries in which harvested fish can be sold, sport fishing is done by hook and line and any fish caught are solely for personal use. ADF&G notes that "The BBMA [Bristol Bay Management Area] contains some of the most productive Pacific salmon, rainbow trout, Arctic grayling, Arctic char, and Dolly Varden waters in the world. The area has been acclaimed for its sport fisheries since the 1930s" (Dye and Schwanke 2009). Similar views prevail in the popular sport fishing literature, where articles praising Bristol Bay as a destination are common. For example, *Fly Rod and Reel* (Williams 2006) says, "No place on earth is wilder or more beautiful or offers finer salmonid fishing." Over the years, many other articles in *Field and Stream, Fly Fisherman, Fish Alaska Magazine, Fly Rod and Reel, Salmon Trout Steelheader, World Angler*, and other magazines have touted the high quality fishing and wilderness ambiance.

Large numbers of salmon and trout are caught in Bristol Bay's sport fisheries each year (see below), but the area is best known for its rainbow trout fishing. The ADF&G (1990) notes, "Wild rainbow trout populations of the region are world

**Figure 18.7** The Bristol Bay commercial fishery is carefully managed by the ADF&G. The commercial fishery is opened and closed by emergency order to ensure sufficient fish escape fisheries to both conserve biodiversity and to sustain future runs of salmon. Here commercial fishers set drift gill nets during an opener. *Photo credit: Michael Melford.*

famous and are the cornerstone to a multimillion dollar sport fishing industry." Articles in the sport fishing press laud the trout fisheries, especially those of the Kvichak River watershed. *Fish Alaska Magazine* calls the Iliamna Lake watershed "One of the greatest trophy trout fisheries in the world . . . the crown of Alaska's sport fishing" (Weiner 2006) and names seven Bristol Bay river systems, five of which are in the Nushagak or Kvichak watersheds, in a rundown of Alaska's top ten spots for trophy rainbow trout (Letherman 2003). Rainbow trout up to 76 cm long can be caught in many areas of the Kvichak River and other watersheds (Randolph 2006) and 43% of clients at remote Bristol Bay sport fishing lodges reported catching a rainbow trout longer than 66 cm on their most recent trip (Duffield et al. 2006).

Unlike commercial fisheries, whose salient features tend to be readily quantifiable (e.g., economics, sustainability), the quality of a sport fishery can hinge on personal and subjective attributes. Despite the potential to catch high numbers of sizeable fish, Bristol Bay anglers rate aesthetic qualities as most important in selecting fishing locations. Of 11 attributes that capture different motivations and aesthetic preferences, including *catching and releasing large numbers of fish* and *chances to catch large or trophy-sized fish*, Alaska resident and nonresident anglers picked the same top five (from Duffield et al. 2006):

- Natural beauty of the area,
- Being in an area with few other anglers,
- Being in a wilderness setting,
- Chance to catch wild fish, and
- Opportunities to view wildlife

The Bristol Bay region is not linked to the state's highway system, and roads associated with the major communities provide very limited access. Small aircraft with floats are the primary source of fishing access followed by boats based out of communities and remote lodges (Dye and Schwanke 2009). A range of services are available for recreational anglers. Anglers willing to pay $7,500–$9,500 a week can stay in a plush remote lodge and fly to different streams each day with a fishing guide (Purnell 2011). Modest river camps, with cabins or wall tents, are a lower-budget option. Many self-guided expeditions center on multiday raft trips that use chartered aircraft for transport to and from access points along a river.

Site-specific data regarding participation, effort, and harvest have been collected from sport fishing guides and businesses since 2005 (Sigurdsson and Powers 2011). In 2010, the most recent year for which data are available, 72 businesses and 319 guides operated in the Kvichak and Nushagak watersheds (Table 18.4; Dora Sigurdsson, ADF&G, unpub. data). In addition, Table 18.4 shows figures for 2005, the first year of data collection, and 2008, a peak year.

## Management of Sport Fisheries

The ADF&G's Division of Sport Fish manages recreational fisheries in the BBMA, which includes all fresh waters flowing into Bristol Bay between Cape Menshikof, on the bay's southeast shore, and Cape Newenham in the northwest. Four local management plans guide sport fishing regulations in the Bristol Bay region (in addition to several statewide plans). The Nushagak-Mulchatna King Salmon Management Plan, the Nushagak-Mulchatna Coho Salmon Management Plan, and the Kvichak River Drainage Sockeye Salmon Management Plan call for sport fishing bag limit reductions or closures by emergency order during poor runs. The Southwest Alaska Rainbow Trout Management Plan instated conservative trout management uniformly throughout the region, replacing the fragmentary restrictions that had been established over the previous decades. Sport fishing regulations are updated annually and can be accessed on the ADF&G website: http://www.adfg.alaska.gov/index.cfm?adfg=fishregulations.sport.

**Table 18.4** The number of businesses and guides operating in the Nushagak and Kvichak watersheds in 2005, 2008, and 2010

| Watershed | 2005 | | 2008 | | 2010 | |
|---|---|---|---|---|---|---|
| | Businesses | Guides | Businesses | Guides | Businesses | Guides |
| Kvichak River (including Alagnak River) | 53 | 204 | 59 | 274 | 46 | 211 |
| Nushagak River (including Wood River) | 67 | 199 | 60 | 245 | 47 | 162 |
| Kvichak and Nushagak combined[1] | 91 | 336 | 92 | 426 | 72 | 319 |

[1]Business and guide totals are not additive because a business and/or guide can operate in multiple watersheds.

The Division of Sport Fish uses the annual Statewide Harvest Survey, mailed to randomly-selected licensed anglers, to monitor effort, catch, and harvest. Between 1997 and 2008, days of angler effort within the BBMA ranged from 83,994–111,838 (Dye and Schwanke 2009). Total annual sport harvest for the same period ranged from 39,362–71,539 fish, of which sockeye, Chinook, and coho salmon comprised the majority (Dye and Schwanke 2009). Resident fish species, including rainbow trout, Dolly Varden, Arctic char, Arctic grayling, northern pike, and whitefish, are also harvested in the BBMA (Dye and Schwanke 2009), but harvest rates are lower for these species than for salmon, likely due to restrictive bag limits, the popularity of catch-and-release fishing (Dye and Schwanke 2009), and smaller population sizes.

## Chinook Salmon

In the Nushagak River watershed, the general season runs from May 1 to July 31 for Chinook salmon, although some areas close on July 24 in order to protect spawners. The daily limit is two per day, only one of which can be over 28 inches (71 cm). The annual limit is four fish. The Nushagak-Mulchatna King Salmon Management Plan calls for an in-river return of 75,000 fish with a spawning escapement of 65,000 fish. The guideline harvest for the sport fishery is 5,000 fish, although restrictions are triggered if the in-river return falls below 55,000 fish. In other Bristol Bay watersheds, the daily limit for Chinook salmon is three and the annual limit is five, although there are additional restrictions in the Wood and Naknek watersheds.

The major Chinook salmon sport fisheries in the BBMA include the Nushagak, Naknek, Togiak, and Alagnak rivers and the average annual harvest is 11,100 fish for the period from 1997–2008. The largest individual fishery takes place in the Nushagak River, where harvest from 2003–2007 averaged 7,281, approximately 58% of the total Bristol Bay sport harvest for that period (Dye and Schwanke 2009).

## Sockeye Salmon

Sockeye salmon fishing is open year-round with a daily limit of five fish. Runs enter rivers starting in late June, peak in early July, and continue into late July or early August. The Kvichak River Drainage Sockeye Salmon Management Plan places restrictions on the sport fishery to avoid conflicts with subsistence users when the escapement falls below the minimum sustainable escapement goal of two million (M) fish. Restrictions include reductions in the daily bag limit and closure of areas to sport fishing that are also used for subsistence harvest.

Sockeye salmon are the most abundant salmon species in the BBMA. Recent annual sport harvest ranged from 8,444–23,002 fish (Dye and Schwanke 2009). The two locations that support the largest sport harvest are the Kvichak River, near the outlet of Iliamna Lake, and the Newhalen River, just above Iliamna Lake (Dye and Schwanke 2009). Other watersheds that support moderate harvests of sockeye salmon include the Naknek and Alagnak Rivers and the Wood River lake system (Dye and Schwanke 2009).

## Rainbow Trout

Due to their relatively small spawning populations and their popularity as a game fish, fishing regulations for rainbow trout are more restrictive than those for any other species. The Southwest Alaska Rainbow Trout Management Plan (ADF&G 1990) calls for conservative management, allows limited harvest in specific areas, and bans stocking of hatchery trout (although stocking had not been practiced previously). Special management areas were created to preserve a diversity of sport fishing opportunities: eight catch-and-release areas, six fly-fishing catch-and-release areas, and eleven areas where only single-hook artificial lures can be used (Dye and Schwanke 2009).

Only single-hook artificial lures can be used in the Kvichak River watershed, and all sport fishing is closed from April 10 through June 7 to protect spawning trout. From June 8 through October 31 anglers are allowed to keep one trout per day, except on selected streams where no harvest is allowed. From November 1 through April 9, when anglers are few, the daily limit increases to five fish although only one may be longer than 20 inches (51 cm). Rainbow trout fishing regulations are similarly restrictive in other watersheds across the BBMA.

The most popular rainbow trout fisheries are found in the Kvichak and Naknek watersheds, portions of the Nushagak and Mulchatna watersheds, and streams of the Wood River lakes system (Dye and Schwanke 2009). Field surveys and the Statewide Harvest Survey show that harvest has decreased over the past decade but that total catch and effort have remained stable or increased (Dye and Schwanke 2009). The annual BBMA-wide harvest between 1997 and 2008 averaged 1,900 fish, but the catch estimate over this period was nearly 100 times greater (183,000 fish; Dye and Schwanke 2009).

Although the fishery is widespread, approximately eighty percent of the total catch (144,400 fish) was from the eastern portion of the BBMA, where the Naknek and Kvichak river systems are located. Eastern BBMA streams with estimated sport catches greater than 10,000 fish in 2008 included the Naknek, Brooks, Kvichak, Copper, and Alagnak rivers (Dye and Schwanke 2009).

# PACIFIC SALMON ABUNDANCE TRENDS, WITH REFERENCE TO BRISTOL BAY

Pacific salmon species, from most to least abundant overall, are pink, sockeye, chum, coho, and Chinook salmon (Ruggerone et al. 2010). The relative abundance of Pacific salmon species relates to their life histories, as species that rear for one or more years in stream habitats (i.e., coho and Chinook salmon) can be constrained by habitat availability more so than species that rear in lakes (i.e., sockeye salmon) or go to sea soon after emerging from spawning redds (i.e., pink and chum salmon) (Quinn 2005). The highest Pacific-wide salmon harvest occurred in 2007 and totaled 513 million fish, over 300 million of which were pink salmon (Irvine et al. 2009). Approximately five billion juvenile salmon are released annually from hatcheries around the North Pacific (Irvine et al. 2009), although none are reared or released in the Bristol Bay region.

## Sockeye Salmon

### Size of Bristol Bay, Kvichak River, and Nushagak River Sockeye Salmon Runs

Escapement monitoring within the Bristol Bay watersheds has been conducted since the 1950s, when the ADF&G established counting towers on the nine major river systems. When combined with commercial, subsistence, and sport harvest, data from escapement monitoring allows estimates of total run sizes. A synthesis of salmon runs for 12 regions around the North Pacific also extends back to the 1950s, allowing comparisons of wild sockeye salmon runs between Bristol Bay and other regions for the period 1956–2005 (Ruggerone et al. 2010). The average global abundance of wild sockeye salmon over that period was 65.3 M fish, and Bristol Bay constituted the largest regional proportion of that total at 46% (Figure 18.8). Total runs to Bristol Bay ranged from a low of 3.5 M in 1973 to a high of 67.3 M in 1980 (Figure 18.9), with an annual average of 29.8 M. The region with the second largest wild runs is southern British Columbia/

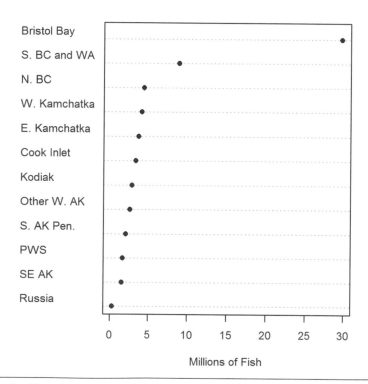

**Figure 18.8**   Average run sizes of wild sockeye salmon for regions around the North Pacific, 1956–2005.

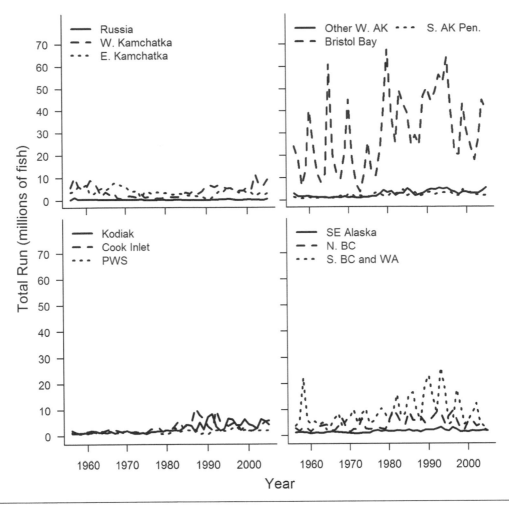

**Figure 18.9**   Run sizes of wild sockeye salmon for regions around the North Pacific, 1956–2005. Each graph shows three regions organized from west to east across the North Pacific.

Washington, which averaged 14% of the total (Figures 18.8 and 18.9), or 8.9 M salmon. Other regions that produce high abundances of wild sockeye salmon include the Kamchatka Peninsula, northern British Columbia, Cook Inlet, and Kodiak Island (Figures 18.8 and 18.9; Ruggerone et al. 2010).

The overwhelming majority of sockeye salmon originate from wild fish, although some are produced in hatcheries, released as juveniles to forage at sea, and captured when they return to their site of release as adults. Hatchery production of sockeye salmon started in 1977 and accounted for an annual average of 3 M fish, or 4% of the world total, during the 10-year period from 1995–2005 (Ruggerone et al. 2010). There are currently no fish hatcheries in the Bristol Bay region, but limited hatchery production has occurred in the past (Rowse and Kaill 1983). Alaskan regions with major hatchery production of sockeye salmon include Prince William Sound, Cook Inlet, and the Kodiak Archipelago, which produced a respective 1.0, 0.9 and 0.6 M hatchery fish, on average, from 1995–2005 (Ruggerone et al. 2010).

Although the Alagnak River is part of the Kvichak watershed and the Wood River is part of the Nushagak watershed, we report sockeye salmon data separately for these systems (unless noted otherwise) because the ADF&G monitors runs on each. On average, the Kvichak River has the largest sockeye salmon run in Bristol Bay, averaging 10.4 M fish annually between 1956 and 2010 (Figure 18.10). Iliamna Lake provides the majority of the rearing habitat for sockeye salmon in the Kvichak watershed, followed by Lake Clark where the estimated proportion of the escapement ranges from 7–30% (Young 2005). Runs exceeding 30 M fish have occurred three times in the Kvichak River: 47.7 M, 34.6 M and 37.7 M fish returned in 1965, 1970 and 1980, respectively (Tim Baker, ADF&G, unpub. data). Those runs accounted for 57%, 49%, and 40% of world production of sockeye salmon during those years (Ruggerone et al. 2010). The Egegik River supports

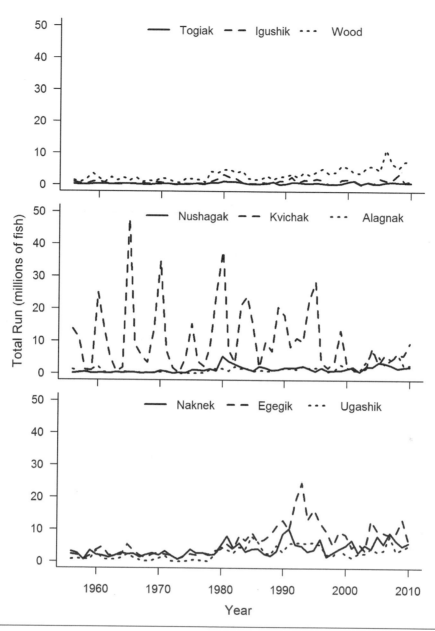

**Figure 18.10** Run sizes of wild sockeye salmon for Bristol Bay's major river systems, 1956–2010. Each graph shows three river systems listed from west to east across Bristol Bay.

Bristol Bay's second largest run, averaging 6.3 M fish from 1956–2010 (Figure 18.10). The Nushagak and Wood rivers host smaller runs that averaged 1.3 and 3.3 M fish, respectively, from 1956–2010.

The Kvichak River sockeye salmon runs are not only the largest in Bristol Bay, but also the largest in the world (Figures 18.10 and 18.11). As noted above, runs to the Kvichak River have averaged 10.4 M fish, and this number climbs to 11.9 M fish when runs to the Alagnak River are included (Tim Baker, ADF&G, unpub. data). The Fraser River system supports the world's second largest run, with an average of 8.1 M fish for the same period (Catherine Michielsens, Pacific Salmon Commission, unpub. data). Other major producers outside of Bristol Bay include the Copper, Kenai, Karluk, and Chignik rivers in Alaska and the Skeena River in British Columbia (Figure 18.11). The Kamchatka Peninsula in Russia also has rivers with large sockeye salmon runs, but abundances for individual rivers were not readily available. The combined runs for the western and eastern Kamchatka Peninsula averaged less than five M sockeye salmon during the period from 1952–2005 (Ruggerone et al. 2010).

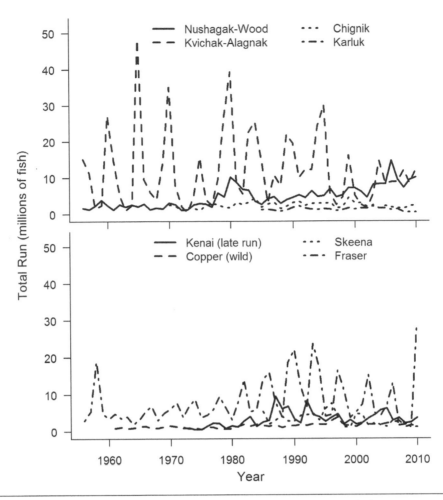

**Figure 18.11** Run sizes of wild sockeye salmon for major spawning rivers around the North Pacific from 1956–2010. The top graph includes time series for the Nushagak-Wood and Kvichak-Alagnak systems from 1956–2010, the Chignik River from 1970–2010, and the Karluk River from 1985–2010. The bottom graph shows the Kenai River late run from 1972–2010, the Copper River wild run from 1961–2010, the Skeena River from 1985–2010, and the Fraser River from 1956–2010. Rivers are listed in the graphs as they occur from west to east across the North Pacific.

## Factors Affecting Bristol Bay Sockeye Salmon Abundance

Changes in the ocean and freshwater environments that affect sockeye salmon abundances and trends across the North Pacific are many. A major driver is climatic variability such as the Pacific Decadal Oscillation (PDO), a recurring pattern of atmospheric and ocean conditions over the North Pacific (Mantua et al. 1997). The PDO alternates between warm and cold phases, each of which can least from years to decades. The warm phase of the PDO is characterized by warmer than average winter sea surface temperatures along the western coastline of North America and increased stream flows around the Gulf of Alaska, both of which are linked to increased salmon survival (Mantua et al. 1997, Ruggerone et al. 2007). There are three regime shifts documented in the recent climate record that correlate with salmon productivity: 1947, 1977, and 1989. From 1947–1977, the PDO was in a cool phase marked by low productivity for Alaskan and British Columbia sockeye salmon. The PDO shifted to a warm phase in 1977, after which most North American populations increased (Figure 18.9). For Bristol Bay populations, this warm phase corresponded with increased marine growth and, in turn, increased abundances and numbers of recruits (returning adults) generated per spawner (Ruggerone et al. 2007). Bristol Bay populations more than doubled during this warm phase and remained high until the mid-90s, when declines in the Kvichak and other rivers reduced the overall abundance (Figures 18.9 and 18.10) (Ruggerone et al. 2010). Biological indicators suggest that decreased productivity associated with a cool phase began in 1989, while climate indices point to a short-lived reversal from 1989–1991, followed by a return to a warm phase (Hare and Mantua 2000). Late marine growth and adult length-at-age of Bristol Bay sockeye salmon decreased after the 1989 regime shift, potentially reducing productivity (Ruggerone et al. 2007).

Another factor affecting sockeye salmon productivity is competition with increasing numbers of hatchery-produced fish released into the North Pacific. Alaska produces the most hatchery pink salmon in the world, averaging 42 M fish for the period 1995–2005, followed next by Russia, with 12.6 M for the same period (Ruggerone et al. 2010). Approximately 75% of the pink salmon hatchery production in Alaska occurs in Prince William Sound, with other facilities located in the Kodiak Archipelago, Cook Inlet, and Southeast Alaska. Japan dominates the production of hatchery chum salmon, with 67.3 M fish returning on average for 1995–2005 (Ruggerone et al. 2010). Coming in a distant second behind Japan, Southeast Alaska averaged 9.7 M hatchery chum salmon for the same period (Ruggerone et al. 2010). Bristol Bay sockeye salmon smolts that migrated to sea during even-numbered years and interacted with dominant odd-year Asian pink salmon experienced decreased growth, survival, and adult abundance compared to the smolts that migrated during odd-numbered years (Ruggerone et al. 2003). Additionally, Kvichak River sockeye salmon productivity was negatively correlated with a running three-year mean of Kamchatka pink salmon abundances (Ruggerone and Link 2006).

In the freshwater environment, spawning and rearing habitats can limit sockeye salmon populations through negative density dependence. The amount of suitable spawning habitat is fixed within a given system, so when spawning densities are high and suitable spawning sites are occupied, females will dig nests on top of existing nests, dislodging many of the previously laid eggs, or die without spawning (Semenchenko 1988, Essington et al. 2000). As such, the amount of available spawning habitat can impose an upper limit on potential fry production (Schindler et al. 2005b). In nursery lakes, juvenile growth rates decrease with rearing densities (Kyle et al. 1988, Schindler et al. 2005a), leading to decreased survival for small individuals in the subsequent marine stage (Koenings et al. 1992). Together, these processes limit the number of recruits potentially produced by a large spawning run.

Sockeye salmon run size in the Kvichak River watershed follows a five-year cycle that is unique among Bristol Bay's river systems (Figure 18.10). Previous hypotheses for the cycle included natural depensatory mechanisms, such as predation and fishing. Since the first escapement goal was established for the Kvichak River in 1962 until the most recent change in 2010, the escapement goals were managed to match the cycle year. Most recently, off-cycle years had an escapement goal range of 2–10 M spawners, while pre-peak and peak cycle years were managed for escapement of 6–10 M spawners (Baker et al. 2009). In 2010, the escapement goal was changed to one goal for all years of 2–10 M spawners. Ruggerone and Link (2006) recently analyzed the population characteristics of Kvichak River sockeye salmon and found that the cycle is likely perpetuated by three factors: the dominance of a 5-year life history, density dependence during pre-peak and peak cycle years that reduces productivity in off-cycle years, and higher interception rates during off-cycle years. In regard to the latter, Kvichak salmon were shown to have high interception rates in the Egegik and Ugashik fisheries in years when these runs were more than double the Kvichak River run, which depressed the number of returning recruits during off-cycle years.

In recent years, the ADF&G has developed genetic stock identification methods, which are being used to reanalyze past interceptions of Kvichak salmon from the mixed stock fisheries on the east side of the bay (Dann et al. 2009). It is anticipated that current brood tables from which total runs by system are reconstructed will change as this analysis progresses (Tim Baker, ADF&G, pers. comm.), giving researchers a more accurate understanding of the dynamics of Bristol Bay stock composition and return dynamics.

## The Decline in Kvichak River Sockeye Salmon Runs

From 1977 through 1995, during the warm PDO phase, Bristol Bay runs averaged almost 41 M fish annually, while runs to the Kvichak River averaged nearly 15 M, comprising about 36% of the entire Bristol Bay run (Table 18.5). Beginning in 1996, with the return of the 1991 brood year, Kvichak River runs dropped to an average of 4.7 M fish, comprising less than 14% of the total Bristol Bay run (Table 18.5). This decline was accompanied by a decline in productivity, as expressed by the number of recruits generated per spawner (R/S). Bristol Bay systems averaged approximately two R/S prior to the 1977 regime shift, and R/S increased substantially for many systems, such as the Egegik and Ugashik Rivers, during the subsequent warm phase (Hilborn 2006). R/S for the Kvichak River averaged 3.2 for the 1972–1990 broods, but five of the nine broods from 1991 onward failed to replace themselves (i.e., R/S < 1). Productivity also decreased during this time in two other systems on the east side of Bristol Bay, the Egegik and Ugashik rivers (Ruggerone and Link 2006). The decline in the Kvichak River run led the ADF&G to classify it as a stock of yield concern in 2001 (Morstad and Baker 2009), indicating an inability to maintain a harvestable surplus. The Kvichak River run was further downgraded to a stock of management concern in 2003, based on failure to meet escapement goals.

Ruggerone and Link (2006) analyzed the decline in the Kvichak River run starting with the 1991 brood year and identified a number of potential factors. The number of smolts per spawner declined by 48% and smolt-to-adult survival

**Table 18.5** Mean annual sockeye salmon run size for Bristol Bay's nine major river systems, 1956–2010, and percent of total by river system. Rivers are listed from east to west across Bristol Bay (unpub. data, Tim Baker, Area Management Biologist, ADF&G)

| Rivers | 1956–1976 | % | 1977–1995 | % | 1996–2010 | % | 1956–2010 | % |
|--------|-----------|-----|-----------|------|-----------|------|-----------|------|
| Ugashik | 882,458 | 4.6 | 4,123,115 | 10.1 | 3,522,697 | 10.1 | 2,722,023 | 8.8 |
| Egegik | 2,320,059 | 12.0 | 9,100,953 | 22.2 | 8,402,365 | 24.1 | 6,321,361 | 20.4 |
| Naknek | 2,200,534 | 11.4 | 4,454,164 | 10.9 | 5,251,810 | 15.1 | 3,811,227 | 12.3 |
| Alagnak | 514,544 | 2.7 | 1,360,651 | 3.3 | 3,008,922 | 8.6 | 1,487,121 | 4.8 |
| Kvichak | 10,482,754 | 54.3 | 14,784,340 | 36.1 | 4,757,008 | 13.7 | 10,407,190 | 33.6 |
| Nushagak | 392,574 | 2.0 | 1,919,420 | 4.7 | 1,933,461 | 5.6 | 1,340,272 | 4.3 |
| Igushik | 516,021 | 2.7 | 1,349,775 | 3.3 | 1,341,581 | 3.9 | 1,029,198 | 3.3 |
| Wood | 1,707,120 | 8.8 | 3,150,620 | 7.7 | 5,834,787 | 16.8 | 3,331,511 | 10.7 |
| Togiak | 305,069 | 1.6 | 661,011 | 1.6 | 742,696 | 2.1 | 547,384 | 1.8 |
| Total | 19,321,134 | | 40,904,050 | | 34,795,327 | | 30,997,285 | |

declined by 46%, suggesting that factors in both freshwater and marine habitats were involved. The average number of smolts out-migrating from the Kvichak River during the years 1982–1993 was approximately 150 M, which declined to an approximate average of 50 M from 1994–2001. The declines were accompanied by a shift in the dominant age structure of Kvichak spawners from 2.2 (i.e., two years in fresh water followed by two years at sea) to 1.3, indicating that salmon were spending less time in fresh water and more time at sea. Across the nine monitored Bristol Bay watersheds, the decrease in the proportion of 2.2 salmon explained a large amount of the decreased R/S and run size. A decrease in spawner length-at-age starting in 1991 may have contributed to lower reproductive potential, since smaller females produce fewer eggs. Competition with Asian pink salmon also may have played a role. Abundances of Asian pink salmon have been linked to decreased size at age of returning Bristol Bay sockeye salmon in addition to decreased abundance during even-year migrations when interactions are highest (Ruggerone et al. 2003). Kamchatka pink salmon were especially abundant from 1994–2000 and would have interacted at sea with age-1 sockeye from the 1991 and subsequent brood years. The three eastern Bristol Bay stocks that experienced the largest declines during the 1990s (Kvichak, Egegik, and Ugashik rivers) have greater overlap with Asian pink salmon in their marine distribution than other stocks that did not decline significantly (Ruggerone and Link 2006).

In 2009, following several years of improvement, the ADF&G upgraded the Kvichak's classification to a stock of yield concern (Morstad and Baker 2009). Since 2004, Bristol Bay runs have again totaled more than 40 million fish annually and in 2010 the Kvichak run increased to over 9.5 million fish, equating to 23% of the total for the Bay.

## Chinook Salmon

The total commercial harvest of Chinook salmon in the North Pacific ranged between three and four million fish until the early 90s, while recent total catches have decreased to one to two million fish (Eggers et al. 2005). Lacking escapement data for many runs, commercial harvest is a good surrogate for salmon abundance and suggests a decline in Chinook salmon abundance in recent decades. The U.S. makes over half of the total commercial catch, followed by Canada, Russia, and Japan (Heard et al. 2007). Recreational and subsistence harvest is significant for this species and totaled approximately one million annually in 2003–2004 (Heard et al. 2007). Washington dominates hatchery production of Chinook salmon, with over one billion juveniles released annually from 1993–2001 (Heard et al. 2007).

The Columbia River historically produced the largest Chinook salmon run in the world, with peak runs (spring, summer, and fall combined) estimated at 3.2 M fish during the late 1800s (Chapman 1986). Peak catches for the Columbia River summer-run Chinook salmon occurred at this time, until overfishing decimated the run. Fishing effort then shifted to the fall run, which suffered a similar demise in the early 1900s. There are currently five stocks of Chinook salmon in the Columbia River watershed listed under the Endangered Species Act (ESA) and the majority of the current runs are hatchery fish (70%, 80%, and 50% of the spring, summer, and fall runs, respectively) (Heard et al. 2007).

Currently, the largest runs of Chinook salmon in the world originate from three of the largest watersheds that drain to the North Pacific: the Yukon, Kuskokwim, and Fraser rivers (Table 18.6). Total Chinook salmon escapements to the

**Table 18.6** Nushagak River Chinook average run sizes for 2000–2009, in comparison to other rivers across the North Pacific. Other rivers are sorted in order of decreasing run size

| Watershed | Region | Average run size (2000–2009) | Area[15] (km²) |
|---|---|---|---|
| Nushagak R. | Bristol Bay, Western Alaska | 151,348[1] | 31,383 |
| Fraser R., total run | Bristish Columbia, Canada | 287,475[2] | 233,156 |
| Kuskokwim R., total run | Western Alaska | 284,000[3] | 118,019 |
| Yukon R., total run | Western Alaska | 217,405[4] | 857,996 |
| Harrison R. (trib. of Fraser R.) | Bristish Columbia, Canada | 98,257[5] | 7,870 |
| Taku R. | Southeast Alaska | 78,081[6] | 17,639 |
| Copper R. | Southcentral Alaska | 75,081[7] | 64,529 |
| Kenai R. (early and late runs) | Southcentral Alaska | 70,976[8] | 5,537 |
| Skeena R. | Bristish Columbia, Canada | 63,356[9] | 51,383 |
| Yukon R., Canadian mainstem | Yukon Territory, Canada | 59,346[10] | 323,800 |
| Nass R. | Bristish Columbia, Canada | 31,738[11] | 20,669 |
| Grays Harbor (Chehalis R. + 5 others) | Washington | 23,964[12] | 6,993 |
| Skagit R. | Washington | 18,286[13] | 8,234 |
| Nehalem R. | Oregon | 12,267[14] | 2,193 |

[1]Unpublished data, Gregory Buck, ADF&G

[2]Pacific Salmon Commission 2011, pg. 88

[3]Unpublished data, Kevin Schaberg, ADF&G

[4]Average from 2000-2004, Spencer et al. 2009, pg. 28

[5]Pacific Salmon Commission 2011, pg. 88

[6]McPherson et al. 2010, pg. 14

[7]Unpublished data, Steve Moffitt, ADF&G

[8]Begich and Pawluk 2010, pg. 69

[9]Pacific Salmon Commission 2011, pg. 87

[10]Howard et al. 2009, pg. 35

[11]Pacific Salmon Commission 2011, pg. 87

[12]Pacific Salmon Commission 2011, pg. 90

[13]Pacific Salmon Commission 2011, pg. 89

[14]Pacific Salmon Commission 2011, pg. 93

[15]Watershed area from the Riverscape Analysis Project 2010 (http://rap.ntsg.umt.edu).

Kuskokwim and Yukon rivers have not been quantified directly due to their large watershed area, but recent total run estimates based on mark-recapture studies put them at 217,000 and 265,000 fish, respectively (Molyneaux and Brannian 2006, Spencer et al. 2009). On the Fraser River, the average size of the spring, summer, and fall Chinook salmon runs combined (including the Harrison River) for the most recent ten-year period (2000–2009) was 287,000 fish (PSC 2011).

Chinook salmon sport and commercial harvests in the Nushagak River are larger than all of the other systems in Bristol Bay combined (Dye and Schwanke 2009, Salomone et al. 2011). The Nushagak produces runs that are periodically at or near the world's largest (Figure 18.12), which is remarkable considering its relatively small watershed area (Table 18.6). Runs consistently number over 100,000 fish, while runs greater than 200,000 fish have occurred eleven times between 1966 and 2010 (Figure 18.12). An especially productive six-year period from 1978–1983 produced three runs greater than 300,000 fish (Figure 18.12). Other rivers that produce large runs of Chinook salmon include the Copper, Kenai, and Taku rivers in Alaska and the Skeena and Harrison rivers in British Columbia (Table 18.6). The Harrison River is the dominant fall run for the Fraser River watershed.

A sustainable escapement goal (SEG) was implemented for Nushagak Chinook salmon in 2007 with a target of 40,000–80,000 fish. Sonar counts used to estimate escapement were initiated in 1989 and, since that time, the Nushagak run has consistently met the minimum escapement for the current SEG and was over the SEG 12 times (Gregory Buck, ADF&G, unpub. data). The Nushagak Chinook salmon stock is considered stable (Heard et al. 2007, Dye and Schwanke 2009) in contrast to Chinook stocks on the Kuskokwim and Yukon rivers, which experienced declines starting in the late 1990s.

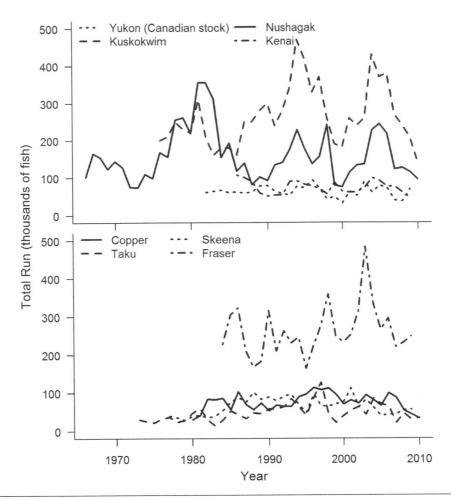

**Figure 18.12**   Run sizes of wild Chinook salmon for major spawning rivers around the North Pacific from 1966–2010. The top graph shows total runs for the Yukon River (Canadian stock) from 1982–2009, the Kuskokwim River from 1976–2010, the Nushagak River from 1966–2010, and the Kenai River from 1986–2010. The bottom graph shows total runs for the Copper River from 1980–2010, the Taku River from 1973–2010, the Skeena River from 1977–2009, and the Fraser River from 1984–2009. Rivers are organized from west to east across the North Pacific.

Chinook salmon in the Yukon and Kuskokwim rivers were listed as stocks of yield concern in 2000 (Estensen et al. 2009, Howard et al. 2009). The Yukon River stock is still listed but the Kuskokwim River Chinook salmon stock was delisted as a stock of concern in 2007, based on higher than normal runs starting in 2004 (Estensen et al. 2009). There are eight additional Chinook salmon stocks listed as management or yield concerns starting in 2010 or later, mostly from the Cook Inlet Region: Chuitna River, Theodore River, Lewis River, Alexander Creek, Willow Creek, Goose Creek, Sheep Creek, and Karluk River (Kodiak Island).

The decline in Yukon and Kuskokwim Chinook salmon that began in the late 1990s may have resulted from the 1997–1998 El Niño (Kruse 1998, Myers et al. 2010). That event was characterized by sea surface temperatures at least 2°C higher than normal in the Bering Sea, along with weak winds and high solar radiation that led to two anomalous phytoplankton blooms (Kruse 1998). The decline in Chinook salmon that persisted after the 1997–1998 El Niño indicate that multiple ocean age classes were affected by this event (Ruggerone et al. 2009).

Chinook salmon hatchery production contributes to harvests in both southeast and southcentral Alaska. The average number of returning hatchery Chinook salmon in Alaska for 2000–2009 was 118,000 fish annually and, in 2009, hatchery Chinook salmon contributed 16% of the total commercial harvest (White 2010). There are no salmon hatcheries located in western Alaska and none of the total runs for the Alaskan rivers listed in Figure 18.12 or Table 18.6 include contributions from hatcheries (Yukon, Kuskokwim, Nushagak, Kenai, Copper, and Taku rivers). Salmon enhancement programs for Chinook salmon in British Columbia are significant; for the period of 1990–2000, hatchery releases averaged approximately 50 million fish annually and hatcheries contributed approximately 30% to the total Canadian catch (MacKinlay et al. Undated). The Chehalis River hatchery in the Harrison River watershed and the Chilliwack River, Inch Creek, and Spius Creek hatcheries in the Fraser River watershed all contribute to the Chinook salmon runs on those systems (FOC 2011).

# THREATENED AND ENDANGERED PACIFIC SALMON AND CONSERVATION PRIORITIES

Although it is difficult to quantify the true number of extinct salmon populations around the North Pacific, estimates for the western U.S. (California, Oregon, Washington, and Idaho) have ranged from 106 to 406 populations (Nehlsen et al. 1991, Augerot 2005, Gustafson et al. 2007). Chinook salmon had the largest number of extinctions followed by coho salmon and then either chum or sockeye salmon (Nehlsen et al. 1991, Augerot 2005). Many of the patterns of population extinction are related to time spent in fresh water. Interior populations have been lost at a higher rate than coastal populations, stream-maturing Chinook salmon and steelhead trout (which may spend up to nine months in fresh water before spawning) had higher losses than their ocean-maturing counterparts, and species that relied on fresh water for rearing (Chinook, coho, and sockeye salmon) had higher rates of extinction than pink or chum salmon, which go to sea soon after emergence (Gustafson et al. 2007). Salmon populations in the southern extent of their range have suffered higher extinction rates and are considered at higher risk than populations further to the north (Brown et al. 1994, Kope and Wainwright 1998, Rand 2008). No populations from Alaska are known to have gone extinct.

In addition to the large number of populations that are now extinct, there are many that are considered at risk due to declining populations. The Columbia River watershed dominated the list of at-risk stocks identified by Nehlson et al. (1991), contributing 76 stocks to the total of 214 for the Pacific Northwest (California, Oregon, Washington, and Idaho). Approximately half of the 214 stocks evaluated were listed as high risk because they failed to replace themselves (fewer than one R/S) or had recent escapements below 200 individuals. More recent analyses continue to highlight the declines in the Pacific Northwest. A detailed assessment of salmon populations in the Columbia River watershed from 1980–2000 showed that many are declining and this trend is heightened when hatchery fish are excluded (McClure et al. 2003). A comparison between time periods reflecting both high and low ocean productivity for Columbia River salmon populations further indicates that the declining trends are not due to the regime shift of 1977 (McClure et al. 2003). An analysis of over 7,000 populations across the North Pacific found that over 30% of sockeye, Chinook, and coho salmon populations were at moderate or high risk of extinction and that the Pacific Northwest had the highest concentrations of high-risk populations (Augerot 2005).

A detailed assessment of sockeye salmon populations across the North Pacific highlights threats for this species in British Columbia (Rand 2008). At the global population level, sockeye salmon are considered a species of least concern. Eighty subpopulations were identified for assessment, five of which are extinct and 26 did not have the necessary data with which to conduct a status assessment. Of the remaining 49 subpopulations, 17 were identified as threatened and two as nearly threatened. British Columbia has 12 threatened (vulnerable, endangered, or critically endangered) subpopulations, 70% of the worldwide total. Three key threats to sockeye salmon were identified: mixed stock fisheries that lead to high harvests of small, less productive populations; poor marine survival rates and high rates of disease in adults due to changing climatic conditions; and negative effects of enhancement activities such as hatcheries and engineered spawning channels (Rand 2008). Twenty-five subpopulations were assessed for Alaska: 10 were data deficient, 12 were of least concern (including the one subpopulation identified for Bristol Bay), one subpopulation in the eastern Gulf of Alaska was listed as vulnerable (four of six streams had declining trends—Bering, East Alsek, Italio, and Situk rivers), and two subpopulations in Southeast Alaska (McDonald and Hugh Smith Lakes) were listed as endangered. Both the Hugh Smith and McDonald Lake populations were listed as stocks of management concern by the ADF&G in 2003 and 2009, respectively (Piston 2008, Eggers et al. 2009). Both were delisted within four years after runs met escapement goals for several consecutive years following implementation of successful fishing restrictions (Piston 2008, Regnart and Swanton 2011).

Government agencies in the U.S. and Canada are tasked with identifying and protecting salmon populations that are at risk. In the U.S., the National Marine Fisheries Service (NMFS) manages listings of salmon species under the ESA. Salmon populations considered for listing under the ESA must meet the definition of an evolutionarily significant unit (ESU); it must be substantially reproductively isolated from other nonspecific population units and it must represent an important component of the evolutionary legacy of the species (Federal Register 58612, November 20, 1991). Current determinations for the U.S. include one endangered and one threatened ESU for sockeye salmon; two threatened ESUs for chum salmon; one endangered, three threatened, and one ESU of concern for coho salmon; two endangered, seven threatened, and one ESU of concern for Chinook salmon; and one endangered, ten threatened, and one ESU of concern for steelhead trout (NMFS 2010). All listed ESUs occur in the Pacific Northwest. In Canada, the Committee on the Status of Endangered Wildlife in Canada (COSEWIC) conducts status assessments to determine if a species is at risk nationally. The Minister of the Environment and the federal cabinet then decide whether to list the species under the Species at Risk Act (SARA). Currently, COSEWIC status assessments have recommended listing two endangered sockeye salmon populations, one endangered coho salmon population, and one threatened Chinook salmon population, but none of these assessments have resulted in legal listings under SARA (COSEWIC 2009). On the Asian side of the Pacific, no information was found regarding listings of threatened or endangered salmon populations under a legal framework. Other assessments of Asian salmon distribution and status have relied on interviews with fishery biologists due to the scarcity of data and the dominance of hatcheries in Japanese fisheries (Augerot 2005, Rand 2008).

The causes leading to extinction and continued population declines are numerous and the effects of interacting factors within watersheds confound analyses. In California, the building of dams that eliminated access to upstream spawning and rearing areas and the destruction of coastal habitat from extensive logging were major contributors to the decline of coho salmon populations (Brown et al. 1994). Heavy fishing pressure at the end of the nineteenth century followed by extensive impacts to river habitats from agriculture, logging, mining, irrigation, and hydroelectric dams led to the extensive decline of Columbia River salmon by the mid-twentieth century (Chapman 1986, McConnaha et al. 2006).

Restoration activities to help restore salmon habitat and populations in the Pacific Northwest require huge expenditures with results that are often difficult to measure due to annual variation, the time lapse between restoration action and effect on the population, and changing climate and ocean conditions (GAO 2002). Approximately $1.5 billion was spent on Columbia River salmon and steelhead trout restoration efforts for the period of 1997 through 2001 (GAO 2002), yet predicted outcomes rarely take climate change into account. For example, models predicting the outcome of restoration on Snohomish River Chinook salmon habitat showed that increased temperature is causing winter precipitation to fall as rain instead of snow in high elevation watersheds, resulting in salmon declines through higher flows during egg incubation, lower flows during spawning, and increased temperatures during prespawning (Battin et al. 2007). Often used as mitigation for lost habitat, salmon hatcheries have resulted in decreased survival of the wild populations they are intended to support (NRC 1996, Naish et al. 2008). Impacts of hatchery fish include overfishing of wild populations in mixed-stock fisheries (Hilborn and Eggers 2000), competition with wild salmon in both fresh water and the ocean (Ruggerone and Nielsen 2009), and a reduction in life history diversity making populations more susceptible to climate variability (Moore et al. 2010).

Due to the high costs of restoration and the difficulty in predicting or measuring outcomes, some have argued that the best way to protect salmon for future generations is to create salmon sanctuaries that maintain intact and connected habitats throughout the watershed from headwaters to the ocean (Rahr et al. 1998, Lichatowich et al. 2000, Rahr and Augerot 2006). Protecting entire watersheds is especially important to sockeye, Chinook, and coho salmon, which spend one or more years rearing in fresh water prior to entering the ocean. These sanctuaries would provide habitat for salmon populations with heightened resilience to factors outside of management control, such as climate change and changes in the ocean environment. The salmon populations in Bristol Bay meet all the criteria for selecting sanctuaries across the North Pacific by having intact habitats, abundant populations, and a high diversity of life history patterns (Schindler et al. 2010). In addition Pinsky et al. (2009) characterized salmon watersheds with high conservation value across the North Pacific as the top 20% (out of 1046 total) based on abundance and run timing diversity. Bristol Bay, the Kamchatka Peninsula, and coastal British Columbia all had clusters of watersheds with high conservation value yet fewer than 9% of these watersheds had greater than half of their area under protected status.

## Key Habitat Elements of Bristol Bay River Systems

No published materials specifically address the question: *Why do Bristol Bay watersheds support so many salmon?* While this isn't particularly surprising given the complexity and scope of the question, it does require us to draw on experts and a diverse body of literature to posit an answer. Obviously, the simplest answer is *Habitat*. But what is it about the habitat in Bristol Bay watersheds that allows them to sustain such prolific fisheries? Our inquiry led us to the conclusion that interplay between the quantity, quality, and diversity of habitats in these river systems accounts for their productivity. The major habitat attributes discussed here were identified in personal communications with Dr. Tom Quinn (University of Washington) and Dr. Jack Stanford (University of Montana).

## Habitat Quantity

An obvious feature of the Bristol Bay watershed is the abundance of large, low-elevation lakes that are accessible to anadromous salmon (Figure 18.13). Lakes cover 7.9% of the Bristol Bay region, which is substantially higher than other major salmon-producing regions (Table 18.7). Lakes cover 13.7% of the Kvichak River watershed and 2.7% of the Nushagak River watershed (Table 18.7). Lakes in the Nushagak River watershed are concentrated in the Wood River subwatershed, where lakes cover 11.3% of the landscape. With the exception of one lake (Chikuminuk Lake), all major lakes in the Bristol Bay region are accessible to anadromous salmon.

Since watershed elevations in the Bristol Bay region are relatively low (Table 18.7), barriers to fish migration are few and a large proportion of the watershed can be accessed by salmon. The Nushagak and Kvichak watersheds have over 58,000 km of streams (National Hydrography Dataset), of which 7,671 km (13%) have been documented as anadromous fish streams (ADF&G 2011 Anadromous Waters Catalog) (Figure 18.13). Since fish use must be documented firsthand by field biologists, a large proportion of anadromous fish habitat undoubtedly remains undocumented. For example, a recent survey targeted 135 undocumented headwater (i.e., 1st- and 2nd-order) stream reaches with low to moderate gradient (i.e., < 10% channel slope) north of Iliamna Lake (Woody and O'Neal 2010). Of these stream reaches, 16% were dry or nonexistent, 53% had juvenile salmon, 66% had resident fish, and 3% contained no fish at the time of sampling (Woody and O'Neal 2010).

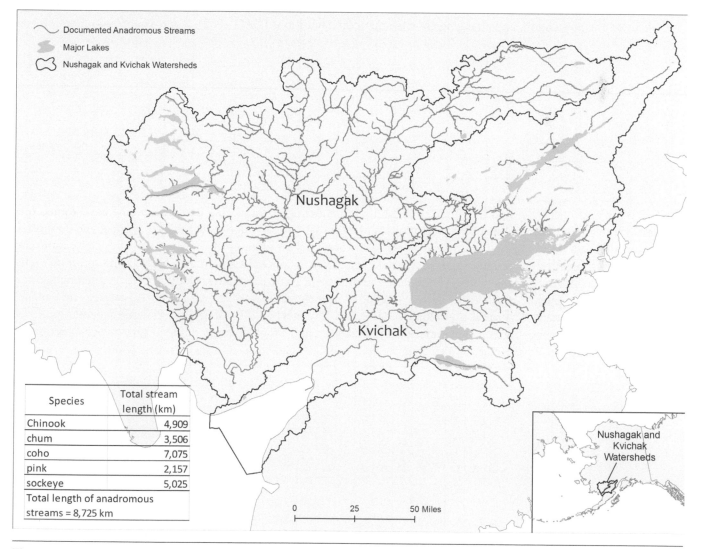

| Species | Total stream length (km) |
|---------|--------------------------|
| Chinook | 4,909 |
| chum | 3,506 |
| coho | 7,075 |
| pink | 2,157 |
| sockeye | 5,025 |
| Total length of anadromous streams = 8,725 km | |

**Figure 18.13**  Documented anadromous fish streams in the Nushagak and Kvichak watersheds. Data are from the ADF&G's Anadromous Waters Catalog for 2015.

**Table 18.7** Comparison of landscape features potentially important to sockeye salmon production for watersheds across the North Pacific (top portion of table) and across the Bristol Bay watershed (bottom portion of table). All landscape data are from the Riverscape Analysis Project (Luck et al. 2010)

| Watershed | Location | Watershed area (km²) | Mean watershed elevation (m) | Number of lakes > 1 km² | Average elevation of lakes (m) | % Lake coverage in watershed | Mean annual sockeye run millions of fish, 1990–2005[+] |
|---|---|---|---|---|---|---|---|
| Kamchatka | Russia | 53,598 | 549 | 82 | 15 | 0.4 | 3.2 |
| Kenai | Central Alaska | 5,537 | 522 | 2 | 97 | 2.9 | 5.2 |
| Copper | Prince William Sound | 64,529 | 1,194 | 9 | 448 | 0.5 | 3.0 |
| Fraser | British Columbia | 233,156 | 1,188 | 119 | 763 | 1.6 | 10.7 |
| Columbia | Washington | 669,608 | 1,328 | 68 | 1,212 | 0.2 | 0.2 |
| Bristol Bay | Western Alaska | 88,233 | 269* | 69 | 219* | 7.9* | 42.8 |
| | | | | | | | |
| Togiak | Bristol Bay | 4,600 | 322 | 6 | 160 | 1.4 | 0.7 |
| Igushik | Bristol Bay | 2,126 | 74 | 2 | 15 | 3.3 | 1.3 |
| Nushagak (inc. Wood) | Bristol Bay | 35,237 | 250 | 20 | 325 | 2.7 | 6.0 |
| Kvichak (inc. Alagnak) | Bristol Bay | 25,328 | 340 | 29 | 193 | 13.7 | 10.9 |
| Naknek | Bristol Bay | 9,624 | 312 | 8 | 230 | 8.3 | 5.2 |
| Egegik | Bristol Bay | 7,117 | 168 | 1 | 4 | 16.5 | 11.0 |
| Ugashik | Bristol Bay | 4,201 | 104 | 3 | 4 | 9.9 | 3.8 |

*Some figures for Bristol Bay represent the weighted average of individual Bristol Bay watersheds.

[+]Salmon abundance sources: Kamchatka, Fraser, and Columbia are from Ruggerone et al. 2010 (Fraser and Columbia rivers were combined into one region "Southern B.C. and Washington"); Kenai is from sockeye brood tables for Kenai River (pers. comm. Pat Shields, 2011); Copper is from sockeye brood tables for Copper River (pers. comm. Jeremy Botz, 2011); Bristol Bay and individual watersheds within Bristol Bay are from sockeye brood tables for Bristol Bay (pers. comm. Tim Baker, 2011).

## Habitat Quality

In addition to the overall abundance of salmon habitat, there are a number of habitat attributes that likely contribute to the productivity of Bristol Bay's river systems. First of all, Bristol Bay streambeds tend to have abundant gravel, which is essential substrate for salmon spawning and egg incubation (Bjornn and Reiser 1991, Quinn 2005). Several Pleistocene glacial advances have left behind a complex landscape of gravel-rich moraines, melt-water deposits, and outwash plains (Stilwell and Kaufman 1996, Hamilton and Kleiforth 2010). As stream channels meander and cut through these deposits, gravel and other sediments are captured and formed into riffles, bars, and other habitat features. In a survey of 76 wade-able stream reaches across the Kvichak and Nushagak watersheds, gravel (2–64 mm) was the dominant substrate, covering 56% (±15%) of each streambed (D.J. Rinella, unpub. data).

Groundwater inputs to streams and lakes are also an important feature of salmon habitat in the Kvichak and Nush-agak watersheds. Rainwater and melting snow infiltrate the extensive glacial deposits and saturate pore spaces below the water table, thus recharging the groundwater aquifer (Power et al. 1999). Ponds are common on the Bristol Bay land-scape and contribute disproportionately to groundwater recharge (Rains 2011). Once in the aquifer, groundwater flows slowly downhill and eventually surfaces in areas of relatively low elevation, like stream channels or lake basins. Areas of groundwater upwelling are heavily used by spawning sockeye salmon because they provide circulation, stable flows, and stable temperatures (Burgner 1991). These habitats include lake beaches and spring-fed ponds, creeks, and side channels (Burgner 1991). Studies in the Wood River system of Bristol Bay demonstrate the importance of groundwater upwell-ing to spawning sockeye salmon. In lakes, densities of beach spawners were highest at sites with the strongest upwelling, while spawners were absent at beach sites with no upwelling (Burgner 1991). Beach spawners comprised substantial portions of the spawning populations in three of the four main Wood River lakes from 1955–1962: 47% in Nerka, 87% in Beverly, 59% in Kulik, but only 3% in Aleknagik (Burgner et al. 1969). In a spring-fed tributary to Lake Nerka, the distribution of sockeye salmon spawners also corresponded with areas of groundwater upwelling (Mathisen 1962). Large numbers of sockeye salmon in the Kvichak River system also spawn in lake beaches, spring-fed ponds, and other ground-water-associated habitats (Morstad 2003). In addition to spawning sockeye salmon, groundwater is an important habitat feature for other salmon species and life history stages. Chum salmon have been shown to preferentially spawn in areas of groundwater upwelling (Salo 1991, Leman 1993). Groundwater also maintains ice-free habitats used extensively by wintering fishes, helps to maintain streamflow during dry weather, and provides thermal refuge during periods of warm water (Reynolds 1997, Power et al. 1999).

Salmon themselves are another important habitat feature of Bristol Bay watersheds. Each year, the region's spawn-ing salmon populations convey massive amounts of energy and nutrients from the North Pacific to fresh waters. These marine-derived nutrients (MDN), released as excreta, carcasses, and energy-rich eggs, enhance the productivity of fresh-water ecosystems, making Pacific salmon classic examples of keystone species that strongly effect the ecosystems where they spawn (Willson and Halupka 1995, Power et al. 1996).

Salmon contain limiting nutrients (i.e., nitrogen and phosphorus) and energy (i.e., carbon-based molecules) in the same relative proportions as needed for growth by rearing fishes, making MDN an ideal fertilizer for salmon ecosystems (Wipfli et al. 2004). Given the high densities of spawning salmon in some streams, MDN subsidies can be large. On aver-age, spawning sockeye salmon import 50,200 kg of phosphorus and 397,000 kg of nitrogen to the Kvichak River system and 12,700 kg of phosphorus and 101,000 kg of nitrogen to the Wood River system each year (Moore and Schindler 2004). In high latitudes, the importance of MDN is magnified by the low ambient nutrient levels in freshwater systems (Gross et al. 1988, Perrin and Richardson 1997). In Iliamna Lake, for example, nitrogen inputs from spawning salmon greatly exceed inputs from the watershed (Kline et al. 1993).

Resident fishes (e.g., rainbow trout, Dolly Varden, Arctic grayling) and juvenile salmon of species that rear for ex-tended periods in streams (i.e., coho and Chinook salmon) derive clear and substantial nutritional benefits through the consumption of salmon eggs and flesh and other food sources related to spawning salmon. In streams in the Nushagak River watershed, for example, ration size and energy consumption among rainbow trout and Arctic graying increased by 480–620% after the arrival of spawning salmon (Scheuerell et al. 2007). The increase in rainbow trout diet was attribut-able to salmon eggs, salmon flesh, and maggots that colonized salmon carcasses, while the increase in Arctic grayling diet was attributable to consumption of benthic invertebrates dislodged by spawning salmon (Scheuerell et al. 2007). A bioenergetics model suggested that these subsidies were responsible for a large majority of the annual growth of these fish populations (Scheuerell et al. 2007). In a stream in the Kvichak River watershed, Dolly Varden moved into ponds where sockeye salmon spawned and fed almost entirely on salmon eggs (Denton et al. 2009). The growth rate of these Dolly Varden increased threefold while salmon eggs were available (Denton et al. 2009). Work elsewhere in Alaska has found that growth rates and energy storage among coho salmon and Dolly Varden are greatest in streams with high densities of spawning salmon (Rinella et al. 2012). These and other studies indicate that the availability of MDN enhances growth rates (Bilby et al. 1996, Wipfli et al. 2003, Giannico and Hinch 2007), body condition (Bilby et al. 1998), and energy storage (Heintz et al. 2004) of stream-dwelling fishes, likely leading to increased chances of survival to adulthood.

MDN is also linked with bottom-up effects on aquatic food webs. In streams, increased standing stocks of biofilm (Wipfli et al. 1998, Wipfli et al. 1999, Johnston et al. 2004, Mitchell and Lamberti 2005) and macroinvertebrates (Claeson et al. 2006, Lessard and Merritt 2006, Walter et al. 2006) have been associated with MDN inputs, which may translate to increased food resources for fishes. In Bristol Bay lakes, MDN can comprise a major proportion of the annual nutrient budget (Mathisen 1972, Koenings and Burkett 1987, Schmidt et al. 1998). Salmon-derived nitrogen is ultimately taken up by juvenile sockeye salmon (Kline et al. 1993); however, it is not clear if these inputs have measurable effects on sockeye salmon populations (Schindler et al. 2005b, Uchiyama et al. 2008).

The importance of MDN to fish populations is perhaps most clearly demonstrated in cases where MDN supplies are disrupted by the depletion of salmon populations. The prolonged depression of salmon populations in the Columbia River watershed is a prime example, where a chronic nutrient deficiency hinders the recovery of endangered and threatened Pacific salmon (Gresh et al. 2000, Petrosky et al. 2001, Achord et al. 2003, Peery et al. 2003, Scheuerell et al. 2005, Zabel et al. 2006) and diminishes the potential of expensive habitat improvement projects (Gresh et al. 2000). Density-dependent mortality has been documented among juvenile Chinook salmon, despite the fact that populations have been reduced to a fraction of historic levels, suggesting that nutrient deficits have reduced the carrying capacity of spawning streams in the Columbia River watershed (Achord et al. 2003, Scheuerell et al. 2005). A population viability analysis has indicated that declines in MDN have likely contributed to low productivity of juvenile salmon and that increasing the productivity could lead to large increases in the salmon population (Zabel et al. 2006). Diminished salmon runs, thus, present a negative feedback loop where the decline in spawner abundance reduces the capacity of streams to produce new spawners (Levy 1997). Fisheries managers recognize the importance of MDN in sustaining the productivity of salmon systems and are now attempting to supplement nutrient stores by planting hatchery salmon carcasses and analogous fertilizers in waters throughout the Pacific Northwest (Stockner 2003, Shaff and Compton 2009).

In addition to their inherent natural productivity, Bristol Bay watersheds have not been subjected to anthropogenic disturbances that have contributed to declining salmon populations elsewhere. For example, Nehlsen et al. (1991) reviewed the status of native salmon and steelhead trout stocks in California, Oregon, Washington, and Idaho. They found that 214 stocks appeared to face a risk of extinction; of these, habitat loss or modification was a contributing factor for 194. These cases were in addition to at least 106 stocks that had already gone extinct (Nehlsen et al. 1991). A National Research Council committee (NRC 1996), convened to review the population status of Pacific Northwest salmon, summarized that:

> The ecological fabric that once sustained enormous salmon populations has been dramatically modified through heavy human exploitation—trapping, fishing, grazing, logging, mining, damming of rivers, channelization of streams, ditching and draining of wetlands, withdrawals of water for irrigation, conversions of estuaries, modification of riparian systems and instream habitats, alterations to water quality and flow regimes, urbanization, and other effects.

Thus, it is generally agreed that a complex and poorly understood combination of factors—with direct and indirect effects of habitat degradation at the fore—are responsible for declining Pacific Northwest salmon stocks (NRC 1996, Gregory and Bisson 1997, Lackey 2003).

In watersheds of the Bristol Bay region, including the Nushagak and Kvichak rivers, human habitation is confined to a few small towns and villages, roads are few, and large-scale habitat modifications are absent. The Riverscape Analysis Project, using spatial data from the Socioeconomic Data and Applications Center (Sanderson et al. 2002), ranked 1574 salmon-producing watersheds around the North Pacific based on an index of human footprint (http://rap.ntsg.umt.edu/humanfootprintrank; accessed 9/1/11). Of these, the Kvichak River ranked 197, the Nushagak (exclusive of the Wood River) ranked 131, and the Wood River ranked 332. Additionally, invasive fishes and riparian plants, which can negatively impact native fish populations, have not been introduced to Bristol Bay's watersheds.

## Habitat Complexity

Schindler, Seeb, and Seeb detail the remarkable habitat complexity in Bristol Bay's watersheds, the genetic and life history diversity it promotes in salmon populations, and how genetic and life history diversity, in turn, contribute to the reliability of salmon runs and the resilience of salmon populations to perturbations such as climate change (See Chapter 22 of this volume).

Here we will briefly describe the role of habitat complexity in sustaining Bristol Bay's fisheries. Since salmon adapt to conditions within their specific environments, a high level of habitat diversity fosters a correspondingly high level of population and life history diversity by optimizing behavioral and physiological traits like timing of spawning, egg size, and the size and shape of spawning adults (Hilborn et al. 2003b). The result is a stock complex comprised of hundreds of distinct spawning populations, each adapted to its particular spawning and rearing environment. This complexity is compounded by variation within each spawning population, likely in ways that are not yet fully understood. One clear example is variation in the amount of time spent rearing in fresh water and at sea (Hilborn et al. 2003b). Within a given cohort, most individuals rear for either one or two years in fresh water, although a small number may spend three years or go to sea shortly after hatching (i.e., zero years in fresh water). Once at sea, most fish will rear for an additional two or three years, although a few will rear for as little as one year or as many as five years. This life history complexity superimposed on localized adaptations results in a high degree of biological complexity within the stock complex.

These layers of biocomplexity result in a situation where different stocks within the complex show asynchronous patterns of productivity (Rogers and Schindler 2008). This is because differences in habitat and life history lead to different population responses, despite exposure to the same prevailing environmental conditions. For example, a year with low stream flows might negatively impact populations that spawn in small streams but not those that spawn in lakes (Hilborn et al. 2003b). Asynchrony in population dynamics of Bristol Bay sockeye salmon has been demonstrated at both the local scale (i.e., individual tributaries) and the regional scale (i.e., major river systems) (Rogers and Schindler 2008). The latter is demonstrated nicely by the relative productivity of Bristol Bay's major rivers during different climatic regimes (Hilborn et al. 2003b), where small runs in the Egegik River were offset by large runs in the Kvichak prior to 1977, but declining runs in the Kvichak River in the 2000s were in turn offset by large runs in the Egegik River (see Figure 18.10).

Population and life history diversity within Bristol Bay sockeye salmon populations can be equated to spreading risk with a diversified portfolio of financial investments (Schindler et al. 2010). Under any given set of conditions, some assets perform well, while others perform poorly, but maintenance of a diversified portfolio stabilizes runs over time. Within the sockeye salmon stock complex, the portfolio of population and life history diversity greatly reduces year-to-year variability in run size, making the commercial salmon fishery much more reliable than it would be otherwise. In addition, portfolio effects stabilize and extend the availability of salmon to aquatic and riparian food webs. Poor runs in some habitats will be offset by large runs in others, allowing mobile predators and scavengers (e.g., bears, eagles, rainbow trout) to access areas of relatively high spawner density each year (Schindler et al. 2010). Different populations vary in the timing of spawning, which substantially extends the period when salmon are occupying spawning habitats (Schindler et al. 2010).

Since a diversified salmon stock complex is contingent upon a complex suite of habitats, an important question becomes: How does habitat diversity in Bristol Bay watersheds compare to that in other salmon-producing regions? The Riverscape Analysis Project calculated remotely sensed indices of physical habitat complexity, allowing comparisons among salmon producing watersheds at the North Pacific Rim scale (Luck et al. 2010, Whited et al. 2012). Rankings of overall physical complexity were based on several attributes:

- Variation in elevation;
- Floodplain elevation;
- Density of floodplains and stream junctions;
- Human footprint;
- The proportion of watershed covered by glaciers, floodplains, and lakes; and
- The elevation and density of lakes.

While the characterization of habitat complexity at this broad spatial scale is necessarily imprecise and certainly fails to detect nuanced habitat features, it does seem to quantify attributes that are important to salmon as it explained general patterns in salmon abundance in validation watersheds (Luck et al. 2010). Overall physical complexity was relatively high for the watersheds considered in this assessment. Of the 1574 Pacific Rim watersheds characterized, the Kvichak River ranked the 3rd highest, the Wood River ranked 4th, and the Nushagak River (exclusive of the Wood River) ranked 44th in an analysis of physical complexity (http://rap.ntsg.umt.edu/overallrank; accessed 9/1/11).

The studies reviewed here demonstrate how biocomplexity in salmon populations provides resilience to environmental change. This resilience can break down when habitats are degraded or when the genetic diversity that allows salmon to utilize the full complement of available habitats is diminished. The loss of habitat diversity and associated loss of population diversity has contributed to declines of salmon fisheries that were prolific at one time, including those in the Sacramento (Lindley et al. 2009) and Columbia rivers (Bottom et al. 2005, Moore et al. 2010). Lindley et al. (2009), while summarizing causes for the crash in Sacramento River fall Chinook salmon, highlighted the importance of life history diversity:

> In conclusion, the development of the Sacramento-San Joaquin watershed has greatly simplified and truncated the once-diverse habitats that historically supported a highly diverse assemblage of populations. The life history diversity of this historical assemblage would have buffered the overall abundance of Chinook salmon in the Central Valley under varying climate conditions.

The demise of wild salmon in the Columbia, Sacramento, and many other rivers in the Pacific Northwest highlights the dangers associated with incremental habitat degradation and reinforces the notion that maintaining the productivity of Bristol Bay's salmon-based ecosystems can only be accomplished through conserving the region's natural range biodiversity and habitat complexity.

## RECOMMENDED RESEARCH

Climate change is altering aquatic ecosystems, particularly in high latitude regions like Alaska where air temperatures are increasing at relatively high rates. Climate-related impacts are driven by complex and interrelated changes in thermal and hydrologic regimes that include increasing water temperature (Mauger et al. 2017), earlier depletion of snowmelt-derived streamflow (Stewart et al. 2005), earlier ice breakup (Schindler et al. 2005a), and loss of glacial ice mass (Milner et al. 2009). Effects to fish are context-specific and difficult to predict, but may include:

- Changes to the abundance, taxonomic composition, and timing of prey availability;
- Altered growth rates (depending on food supply relative to changing metabolic rates);
- Accelerated development of embryos and juveniles, potentially leading to mismatches between life stage transitions and key food resources;
- Changes in migration patterns and habitat use due to thermal barriers, shifting climate envelopes, and altered streamflow; and
- Increased susceptibility to contaminants, disease, and parasites.

Understanding the nature, magnitude, and scope of these changes will allow appropriate fishery management decisions and will depend on rigorous monitoring of habitat conditions and fish populations, in addition to modeling efforts to project changes to habitats and consequences for fish. In regard to fish monitoring, basic information on abundance, distribution, and growth is scarce for resident fish species and salmon other than sockeye. The potential for the development of one or more large-scale copper sulfide mines in Bristol Bay's Nushagak and Kvichak watersheds reinforces the need for long-term monitoring in this region. Pebble Limited Partnership collected baseline biological and hydrologic data around the Pebble deposit to support the eventual permitting process, but climate change is causing baselines to shift and interactions with mining impacts could compound risks to fish populations (Wobus et al. 2015).

## SUMMARY

In this chapter we reviewed the biology, ecology, and management of the fishes of the Bristol Bay watersheds, with emphasis on sockeye salmon, Chinook salmon, and rainbow trout. Bristol Bay's nine major river systems produce nearly half of the world's wild sockeye salmon plus abundant populations of Chinook salmon, rainbow trout, and other fishes that support the largest sockeye salmon fishery in the world, culturally essential subsistence fisheries, and world-class sport fisheries in a wilderness setting. Bristol Bay's productivity as a salmon-producing region can be attributed to: (1) abundant habitat in the form of salmon-accessible lakes (the primary rearing habitat for sockeye salmon) and streams; (2) high-quality habitat supported by groundwater-rich gravel deposits, substantial inputs of salmon-derived nutrients and energy, and minimal human disturbance; and (3) a diverse assemblage of spawning and rearing habitats that stabilize salmon runs and prolong the availability of salmon to mobile consumers.

# REFERENCES

Achord, S., P.S. Levin, and R.W. Zabel. 2003. Density-dependent mortality in Pacific salmon: the ghost of impacts past? Ecology Letters 6:335–342.

ADF&G (Alaska Department of Fish and Game). 1990. Southwest Alaska rainbow trout management plan. Anchorage, AK.

Armstrong, J.B., D.E. Schindler, C.P. Ruff, G.T. Brooks, K.E. Bentley, and C.E. Torgersen. 2013. Diel horizontal migration in streams: Juvenile fish exploit spatial heterogeniety in thermal and trophic resources. Ecology 94:2066–2075.

Augerot, X. 2005. Atlas of Pacific salmon : the first map-based status assessment of salmon in the North Pacific. University of California Press. Los Angeles, CA.

Azuma, T. 1992. Diel feeding habits of sockeye and chum salmon in the Bering Sea during the summer. Nippon Suisan Gakkaishi 58:2019–2025.

Baker, T.T., L.F. Fair, F.W. West, G.B. Buck, X. Zhang, S. Fleischmann, and J. Erickson. 2009. Review of salmon escapement goals in Bristol Bay, Alaska, 2009. Alaska Department of Fish and Game, Anchorage, AK.

Battin, J., M.W. Wiley, M.H. Ruckelshaus, R.N. Palmer, E. Korb, K.K. Bartz, and H. Imaki. 2007. Projected impacts of climate change on salmon habitat restoration. Proceedings of the National Academy of Sciences 104:6720–6725.

Behnke, R.J. 1992. Native trout of western North America. American Fisheries Society Monograph 6, Bethesda, MD.

Beverton, R.J.H. and S.J. Holt. 1957. On the dynamics of exploited fish populations. The Blackburn Press.

Bilby, R.E., B.R. Fransen, and P.A. Bisson. 1996. Incorporation of nitrogen and carbon from spawning coho salmon into the trophic system of small streams: Evidence from stable isotopes. Canadian Journal of Fisheries and Aquatic Sciences 53:164–173.

Bilby, R.E., B.R. Fransen, P.A. Bisson, and J.K. Walter. 1998. Response of juvenile coho salmon (Oncorhynchus kisutch) and steelhead (Oncorhynchus mykiss) to the addition of salmon carcasses to two streams in southwestern Washington, USA. Canadian Journal of Fisheries and Aquatic Sciences 55:1909–1918.

Bjornn, T.C. and D.W. Reiser. 1991. Habitat requirements of salmonids in streams. Pages 83–138 in W. R. Meehan, editor. Influences of forest and rangeland management on salmonid fishes and their habitats. American Fisheries Society, Bethesda, MD.

Blair, G.R., D.E. Rogers, and T.P. Quinn. 1993. Variation in life history characteristics and morphology of sockeye salmon in the Kvichak River system, Bristol Bay, Alaska. Transactions of the American Fisheries Society 122:550–559.

Bottom, D.L., C.A. Simenstad, J. Burke, A.M. Baptista, D.A. Jay, K.K. Jones, E. Casillas, and M.H. Schiewe. 2005. Salmon at river's end: The role of the estuary in the decline and recovery of Columbia River salmon. National Marine Fisheries Service, Seattle, WA.

Brodeur, R.D. 1990. A synthesis of the food habits and feeding ecology of salmonids in marine waters of the North Pacific. Fisheries Research Institute, University of Washington, Seattle, WA.

Brown, L.R., P.B. Moyle, and R.M. Yoshiyama. 1994. Historical decline and current status of coho salmon in California. North American Journal of Fisheries Management 14:237–261.

Burgner, R.L. 1991. Life history of sockeye salmon (Oncorhynchus nerka). Pages 1–118 in C. Groot and L. Margolis, editors. Pacific Salmon Life Histories. University of Washington Press, Seattle.

Burgner, R.L., C.J. DiCostanzo, R.J. Ellis, G.Y. Harry, Jr., W.L. Hartman, O.E. Kerns, Jr., O.A. Mathisen, and W.F. Royce. 1969. Biological studies and estimates of optimum escapements of sockeye salmon in the major river systems in southwestern Alaska. Fishery Bulletin 67:405–459.

Chapman, D.W. 1986. Salmon and steelhead abundance in the Columbia River in the 19th century. Transactions of the American Fisheries Society 115:662–670.

Claeson, S.M., J.L. Li, J.E. Compton, and P.A. Bisson. 2006. Response of nutrients, biofilm, and benthic insects to salmon carcass addition. Canadian Journal of Fisheries and Aquatic Sciences 63:1230–1241.

Clark, J.H. 2005. Bristol Bay salmon, a program review. Alaska Department of Fish and Game, Anchorage, AK.

COSEWIC. 2009. Wildlife Species Search. Page Searchable database of wildlife species with listing status. Committee on the Status of Endangered Wildlife in Canada.

Crespi, B.J. and R. Teo. 2002. Comparative phylogenetic analysis of the evolution of semelparity and life history in salmonid fishes. Evolution 56:1008–1020.

Cury, P. 1994. Obstinate nature: an ecology of individuals. Thoughts on reproductive behavior and biodiversity. Canadian Journal of Fisheries and Aquatic Sciences 51:1664–1673.

Dann, T.H., C. Habicht, J.R. Jasper, H.A. Hoyt, A.W. Barclay, W.D. Templin, T.T. Baker, F.W. West, and L.F. Fair. 2009. Genetic stock composition of the commercial harvest of sockeye salmon in Bristol Bay, Alaska, 2006–2008. Alaska Department of Fish and Game, Anchorage, AK.

Denton, K., H. Rich, and T. Quinn. 2009. Diet, movement, and growth of Dolly Varden in response to sockeye salmon subsidies. Transactions of the American Fisheries Society 138:1207–1219.

Dittman, A.H. and T.P. Quinn. 1996. Homing in Pacific salmon: Mechanisms and ecological basis. Journal of Experimental Biology 199:83–91.

Duffield, J., D. Patterson, C. Neher, and O.S. Goldsmith. 2006. Economics of Wild Salmon Watersheds: Bristol Bay, Alaska. University of Montana, Missoula, MT.

Dye, J.E. and C.J. Schwanke. 2009. Report to the Alaska Board of Fisheries for the recreational fisheries of Bristol Bay, 2007, 2008, and 2009. Alaska Department of Fish and Game, Anchorage, AK.

Eggers, D.M., S.C. Heinl, and A.W. Piston. 2009. McDonald Lake sockeye salmon stock status and escapement goal recommendations, 2008. Alaska Department of Fish and Game, Anchorage, AK.

Eggers, D.M., J.R. Irvine, M. Fukuwaka, and V.I. Karpenko. 2005. Catch trends and status for north Pacific salmon. North Pacific Anadromous Fish Commission.

Eliason, E., T. Clark, M. Hague, L. Hanson, Z. Gallagher, K. Jeffries, M. Gale, D. Patterson, S. Hinch, and A. Farrell. 2011. Differences in Thermal Tolerance Among Sockeye Salmon Populations. Science 332:109–112.

Essington, T., T. Quinn, and V. Ewert. 2000. Intra- and inter-specific competition and the reproductive success of sympatric Pacific salmon. Canadian Journal of Fisheries and Aquatic Sciences 57:205–213.

Estensen, J.L., D.B. Molyneaux, and D.J. Bergstrom. 2009. Kuskokwim River salmon stock status and Kuskokwim area fisheries, 2009; a report to the Alaska Board of Fisheries. Alaska Department of Fish and Game, Anchorage, AK.

FLBS. 2011. Riverscape Analysis Project: Physical complexity ranking results. Flathead Lake Biological Station, University of Montana, Missoula, MT.

Flynn, L. and R. Hilborn. 2004. Test fishery indices for sockeye salmon (Oncorhynchus nerka) as affected by age composition and environmental variables. Canadian Journal of Fisheries and Aquatic Sciences 61:80–92.

FOC. 2011. Salmonid Enhancement Program. Fisheries and Oceans Canada.

GAO. 2002. Columbia River basin salmon and steelhead federal agencies' recovery responsibilities, expenditures, and actions. General Accounting Office, Washington, D.C.

Gardiner, W.R. and P. Geddes. 1980. The influence of body composition on the survival of juvenile salmon. Hydrobiologia 69:67–72.

Giannico, G.R. and S.G. Hinch. 2007. Juvenile coho salmon (Oncorhynchus kisutch) responses to salmon carcasses and in-stream wood manipulations during winter and spring. Canadian Journal of Fisheries and Aquatic Sciences 64:324–335.

Gregory, S.V. and P.A. Bisson. 1997. Degradation and loss of anadromous salmon habitat in the Pacific Northwest. Pages 277–314 in D. J. Stouder, P. A. Bisson, and R. J. Naiman, editors. Pacific salmon and their ecosystems. Chapman and Hall, New York, NY.

Gresh, T., J. Lichatowich, and P. Schoonmaker. 2000. An estimation of historic and current levels of salmon production in the Northeast Pacific ecosystem: Evidence of a nutrient deficit in the freshwater systems of the Pacific Northwest. Fisheries 25:15–21.

Gross, M.R., R.M. Coleman, and R.M. McDowall. 1988. Aquatic productivity and the evolution of diadromous fish migration. Science 239:1291–1293.

Gustafson, R.G., R.S. Waples, J.M. Myers, L.A. Weitkamp, G.J. Bryant, O.W. Johnson, and J.J. Hard. 2007. Pacific salmon extinctions: Quantifying lost and remaining diversity. Conservation Biology 21:1009–1020.

Hamilton, T.D. and R.F. Kleiforth. 2010. Surficial geology map of parts of the Iliamna D-6 and D-7 Quadrangles, Pebble Project Area, southwestern Alaska. Department of Natural Resources, Fairbanks, AK.

Hare, S. and N. Mantua. 2000. Empirical evidence for North Pacific regime shifts in 1977 and 1989. Progress in Oceanography 47:103–145.

Healey, M.C. 1991. Life history of Chinook salmon (Oncorhynchus tshawytscha). Pages 311–394 in C. Groot and L. Margolis, editors. Pacific Salmon Life Histories. UBC Press, Vancouver, BC.

Heard, W.R. 1991. Life history of pink salmon (Oncorhynchus gorbuscha). Pages 119–230 in C. Groot and L. Margolis, editors. Pacific Salmon Life Histories. UBC Press, Vancouver, BC.

Heard, W.R., E. Shevlyakov, O.V. Zikunova, and R.E. McNicol. 2007. Chinook salmon—trends in abundance and biological characteristics. North Pacific Anadromous Fish Commission Bulletin 4:77–91.

Heintz, R.A., B.D. Nelson, J. Hudson, M. Larsen, L. Holland, and M. Wipfli. 2004. Marine subsidies in freshwater: Effects of salmon carcasses on lipid class and fatty acid composition of juvenile coho salmon. Transactions of the American Fisheries Society 133:559–567.

Heintz, R.A., M.S. Wipfli, and J.P. Hudson. 2010. Identification of Marine-Derived Lipids in Juvenile Coho Salmon and Aquatic Insects through Fatty Acid Analysis. Transactions of the American Fisheries Society 139:840–854.

Hilborn, R. 2006. Fisheries success and failure: The case of the Bristol Bay salmon fishery. Bulletin of Marine Science 78:487–498.

Hilborn, R., T.A. Branch, B. Ernst, A. Magnusson, C.V. Minte-Vera, M.D. Scheuerell, and J.L. Valero. 2003a. State of the world's fisheries. Annual Review of Environment and Resources 28:359–399.

Hilborn, R. and D. Eggers. 2000. A review of the hatchery programs for pink salmon in Prince William Sound and Kodiak Island, Alaska. Transactions of the American Fisheries Society 129:333–350.

Hilborn, R., T. Quinn, D. Schindler, and D. Rogers. 2003b. Biocomplexity and fisheries sustainability. Proceedings of the National Academy of Sciences of the United States of America 100:6564–6568.

Howard, K.G., S.J. Hayes, and D.F. Evenson. 2009. Yukon River Chinook salmon stock status and action plan 2010; a report to the Alaska Board of Fisheries. Alaska Department of Fish and Game, Anchorage, AK.

Irvine, J.R., M. Fukuwaka, T. Kaga, J.H. Park, K.B. Seong, S. Kang, V. Karpenko, N. Klovach, H. Bartlett, and E. Volk. 2009. Pacific salmon status and abundance trends. North Pacific Anadromous Fish Commission.

Johnston, N.T., E.A. MacIsaac, P.J. Tschaplinski, and K.J. Hall. 2004. Effects of the abundance of spawning sockeye salmon (Oncorhynchus nerka) on nutrients and algal biomass in forested streams. Canadian Journal of Fisheries and Aquatic Sciences 61:384–403.

King, B. 2009. Sustaining Alaska's fisheries: Fifty years of statehood. Alaska Department of Fish and Game, Anchorage, AK.

King, R.S., C.M. Walker, D.F. Whigham, S. Baird, and J.A. Back. 2012. Catchment topography and wetland geomorphology drive macroinvertebrate community structure and juvenile salmonid distributions in southcentral Alaska headwater streams. Freshwater Science.

Kline, T.C., J.J. Goering, O.A. Mathisen, P.H. Poe, P.L. Parker, and R.S. Scalan. 1993. Recycling of elements transported upstream by runs of Pacific salmon: II. δ15N and δ13C evidence in the Kvichak River watershed, Bristol Bay, Southwestern Alaska. Canadian Journal of Fisheries and Aquatic Sciences 50:2350–2365.

Koenings, J.P. and R.D. Burkett. 1987. Population characteristics of sockeye salmon (Oncorhynchus nerka) smolts relative to temperature regimes, euphotic volume, fry density, and forage base within Alaskan lakes. Sockeye salmon (Oncorhynchus nerka) population biology and future management, Canadian Special Publications of Fisheries and Aquatic Sciences 96:216–234.

Koenings, J.P., H.J. Geiger, and J.J. Hasbrouck. 1992. Smolt-to-adult survival patterns of sockeye salmon (Oncorhynchus nerka): effects of smolt length and geographic latitude when entering the sea. Canadian Journal of Fisheries and Aquatic Sciences 50:600–611.

Kope, R. and T. Wainwright. 1998. Trends in the status of Pacific salmon populations in Washington, Oregon, California, and Idaho North Pacific Anadromous Fish Commission Bulletin 1:1–12.

Kruse, G.H. 1998. Salmon run failures in 1997–1998: A link to anomalous ocean conditions?

Kyle, G.B., J.P. Koenings, and B.M. Barrett. 1988. Density-dependent, trophic level responses to an introduced run of sockeye salmon (Oncorhynchus nerka) at Frazer Lake, Kodiak Island, Alaska. Canadian Journal of Fisheries and Aquatic Sciences 45:856–867.

Lackey, R. 2003. Pacific Northwest salmon: Forecasting their status in 2100. Reviews in Fisheries Science 11:35–88.

Leman, V.N. 1993. Spawning sites of chum salmon, Oncorhynchus keta, microhydrological regime and viability of progeny in redds (Kamchatka River basin). Journal of Ichthyology 33:104–117.

Lessard, J.L. and R.W. Merritt. 2006. Influence of marine-derived nutrients from spawning salmon on aquatic insect communities in southeast Alaskan streams. Oikos 113:334–343.

Letherman, T. 2003. Lair of the leviathon: top ten spots for trophy rainbows. Fish Alaska Magazine.

Levy, S. 1997. Pacific salmon bring it all back home. Bioscience 47:657–660.

Lichatowich, J.A., G.R. Rahr, S.M. Whidden, and C.R. Steward. 2000. Sanctuaries for Pacific Salmon. Pages 675–686 in E. E. Knudsen, C. R. Steward, D. D. MacDonald, J. E. Williams, and D. W. Reiser, editors. Sustainable Fisheries Management: Pacific Salmon. Lewis Publishers, Boca Raton, LA.

Lindley, S.T., C.B. Grimes, M.S. Mohr, W. Peterson, J. Stein, J.T. Anderson, L.W. Botsford, D.L. Bottom, C.A. Busack, T.K. Collier, J. Ferguson, J.C. Garza, A.M. Grover, D.G. Hankin, R.G. Kope, P.W. Lawson, A. Low, R.B. MacFarlane, K. Moore, M. Palmer-Zwahlen, F.B. Schwing, J. Smith, C. Tracy, R. Webb, B.K. Wells, and T.H. Williams. 2009. What caused the Sacramento River fall Chinook stock collapse? , National Oceanic and Atmospheric Administration, Santa Cruz, CA.

Luck, M., N. Maumenee, D. Whited, J. Lucotch, S. Chilcote, M. Lorang, D. Goodman, K. McDonald, J. Kimball, and J. Stanford. 2010. Remote sensing analysis of physical complexity of North Pacific Rim rivers to assist wild salmon conservation. Earth Surface Processes and Landforms 35:1330–1343.

MacKinlay, D., S. Lehmann, J. Bateman, and R. Cook. Undated. Pacific salmon hatcheries in British Columbia. Fisheries and Oceans Canada, Vancouver BC.

Mantua, N.J., S.R. Hare, Y. Zhang, J.M. Wallace, and R.C. Francis. 1997. A Pacific interdecadal climate oscillation with impacts on salmon production. Bulletin of the American Meteorological Society 78:1069–1079.

Mathisen, O.A. 1962. The effect of altered sex ratios on the spawning of red salmon. Pages 137–248 in T. S. Y. Koo, editor. Studies of Alaska red salmon. University of Washington, Seattle, WA.

Mathisen, O.A. 1972. Biogenic enrichment of sockeye salmon lakes and stock productivity. Verh. Internat. Verein. Limnol. 18:1089–1095.

Mauger, S., R. Shaftel, J.C. Leppi, and D.J. Rinella. 2017. Summer temperature regimes in southcentral Alaska streams: watershed drivers of variation and potential implications for Pacific salmon. Canadian Journal of Fisheries and Aquatic Sciences 74:702–715.

McClure, M., E. Holmes, B. Sanderson, and C. Jordan. 2003. A large-scale, multispecies status, assessment: Anadromous salmonids in the Columbia River Basin. Ecological Applications 13:964–989.

McConnaha, W.E., R.N. Williams, and J.A. Lichatowich. 2006. Introduction and background of the Columbia River salmon problem. Pages 1–28 in R. N. Williams, editor. Return to the River: Restoring Salmon to the Columbia River. Elsevier Academic Press, San Francisco.

McDowall, R.M. 2001. Anadromy and homing: Two life-history traits with adaptive synergies in salmonid fishes? Fish and Fisheries 2:78–85.

Mecklenburg, C.W., T.A. Mecklenburg, and L.K. Thorsteinson. 2002. Fishes of Alaska. American Fisheries Society, Bethesda, Maryland.

Meka, J.M., E.K. Knudsen, D.C. Douglas, and R.B. Benter. 2003. Variable migratory patterns of different adult rainbow trout life history types in a Southwest Alaska watershed. Transactions of the American Fisheries Society 132:717–732.

Milner, A.M. and R.G. Bailey. 1989. Salmonid colonization of new streams in Glacier Bay National Park, Alaska, USA. Aquaculture and Fisheries Management 20:179–192.

Milner, A.M., L.E. Brown, and D.M. Hannah. 2009. Hydroecological response of river systems to shrinking glaciers. Hydrological Processes 23:62–77.

Mitchell, N.L. and G.A. Lamberti. 2005. Responses in dissolved nutrients and epilithon abundance to spawning salmon in southeast Alaska streams. Limnology and Oceanography 50:217–227.

Molyneaux, D.B. and L.K. Brannian. 2006. Review of escapement and abundance information for Kuskokwim Area salmon stocks. Alaska Department of Fish and Game, Anchorage, AK.

Moore, J.W., M. McClure, L.A. Rogers, and D.E. Schindler. 2010. Synchronization and portfolio performance of threatened salmon. Conservation Letters 3:340–348.

Moore, J.W. and D.E. Schindler. 2004. Nutrient export from freshwater ecosystems by anadromous sockeye salmon (Oncorhynchus nerka). Canadian Journal of Fisheries and Aquatic Sciences 61:1582–1589.

Morstad, S. 2003. Kvichak River sockeye salmon spawning ground surveys, 1955–2002. Alaska Department of Fish and Game, Anchorage, AK.

Morstad, S. and T.T. Baker. 2009. Kvichak River sockeye salmon stock status and action plan, 2009: a report to the Alaska Board of Fisheries. Alaska Department of Fish and Game, Anchorage, AK.

Myers, K.W., R.V. Walker, N.D. Davis, J.A. Armstrong, W.J. Fournier, N.J. Mantua, and J. Raymond-Yakoubian. 2010. Climate-ocean effects on Chinook salmon. School of Aquatic and Fishery Sciences, University of Washington, Seattle, WA.

Naish, K.A., J.E. Taylor, P.S. Levin, T.P. Quinn, J.R. Winton, D. Huppert, and R. Hilborn. 2008. An evaluation of the effects of conservation and fishery enhancement hatcheries on wild populations of salmon. Advances in Marine Biology 53:61–194.

Nehlsen, W., J. Williams, and J. Lichatowich. 1991. Pacific salmon at the crossroads—stocks at risk from California, Oregon, Idaho, and Washington. Fisheries 16:4–21.

Nickelson, T.E., J.D. Rodgers, S.L. Johnson, and M.F. Solazzi. 1992. Seasonal changes in habitat use by juvenile coho salmon (Oncorhynchus kisutch) in Oregon coastal streams. Canadian Journal of Fisheries and Aquatic Sciences 49:783–789.

NMFS. 2010. ESA Salmon Listings: Listing determinations (endangered, threatened, or species of concern) for anadromous fish, including Pacific salmon and steelhead. National Marine Fisheries Service, Seattle, WA.

NRC. 1996. Upstream: Salmon and Society in the Pacific Northwest. National Research Council, Washington, D.C.

Peery, C.A., K.L. Kavanagh, and J.M. Scott. 2003. Pacific salmon: Setting ecologically defensible recovery goals. Bioscience 53:622–623.

Perrin, C.J. and J.S. Richardson. 1997. N and P limitation of benthos abundance in the Nechako River, British Columbia. Canadian Journal of Fisheries and Aquatic Sciences 54:2574–2583.

Petrosky, C.E., H.A. Schaller, and P. Budy. 2001. Productivity and survival rate trends in the freshwater spawning and rearing stage of Snake River chinook salmon (Oncorhynchus tshawytscha). Canadian Journal of Fisheries and Aquatic Sciences 58:1196–1207.

Pinsky, M.L., D.B. Springmeyer, M.N. Goslin, and X. Augerot. 2009. Range-Wide Selection of Catchments for Pacific Salmon Conservation. Conservation Biology 23:680–691.

Piston, A.W. 2008. Hugh Smith Lake sockeye salmon adult and juvenile studies, 2007. Alaska Department of Fish and Game, Anchorage, AK.

Power, G., R. Brown, and J. Imhof. 1999. Groundwater and fish—insights from northern North America. Hydrological Processes 13:401–422.

Power, M.E., D. Tilman, J.A. Estes, B.A. Menge, W.J. Bond, L.S. Mills, G. Daily, J.C. Castilla, J. Lubchenco, and R.T. Paine. 1996. Challenges in the quest for keystones. Bioscience 46:609–620.

PSC. 2011. 2010 Annual Report of Catches and Escapements. Pacific Salmon Commission.

Purnell, R. 2011. Abode of the blessed. Fly Fisherman.

Quinn, T.P. 2005. The Behavior and Ecology of Pacific Salmon and Trout. University of Washington Press, Seattle, WA.

Rahr, G. and X. Augerot. 2006. A proactive sanctuary strategy to anchor and restore high-priority wild salmon ecosystems.in R. T. Lackey, D. H. Lach, and S. L. Duncan, editors. Salmon 2100: The Future of Wild Pacific Salmon.

Rahr, G., J. Lichatowich, R. Hubley, and S. Whidden. 1998. Sanctuaries for native salmon: A conservation strategy for the 21st-century. Fisheries 23:6–36.

Rains, M. 2011. Water sources and hydrodynamics of closed-basin depressions, Cook Inlet Region, Alaska. Wetlands 31:377–387.

Ramstad, K.M., C.A. Woody, and F.W. Allendorf. 2010. Recent local adaptation of sockeye salmon to glacial spawning habitats. Evolutionary Ecology 24:391–411.

Rand, P.S. 2008. Oncorhynchus nerka. IUCN Red List of Threatened Species.

Randolph, J. 2006. Fabulous Bristol Bay. Fly Fisherman.

Regnart, J. and C.O. Swanton. 2011. Southeast Alaska stock of concern recommendations. Alaska Department of Fish and Game, Juneau, AK.

Reynolds, J.B. 1997. Ecology of overwintering fishes in Alaskan waters. Pages 281–302 in A.M. Milner and M.W. Oswood, editors. Freshwaters of Alaska: ecological syntheses. Springer-Verlag, New York, NY.

Ricker, W. 1954. Stock and recruitment. Journal of the Fisheries Research Board of Canada 11:559–623.

Riddell, B.E. and R.J. Beamish. 2003. Distribution and monitoring of pink salmon (Oncorhynchus gorbuscha) in British Columbia, Canada. Department of Fisheries and Oceans Canada, Nanaimo, BC, Canada.

Rinella, D.J., M.S. Wipfli, C.A. Stricker, R.A. Heintz, and M.J. Rinella. 2012. Pacific salmon (Oncorhynchus spp.) runs and consumer fitness: growth and energy storage in stream-dwelling salmonids increase with salmon spawner density. Canadian Journal of Fisheries and Aquatic Sciences 69:73–84.

Ringsmuth, K.J. 2005. Snug Harbor Cannery: a beacon on the forgotten shore 1919–1980. National Park Service, Lake Clark National Park.

Roff, D.A. 1988. The evolution of migration and some life history parameters in marine fishes. Environmental Biology of Fishes 22:133–146.

Rogers, L.A. and D.E. Schindler. 2008. Asynchrony in population dynamics of sockeye salmon in southwest Alaska. Oikos 117:1578–1586.

Rowse, M.L. and W.M. Kaill. 1983. Fisheries Rehabilitation and Enhancement in Bristol Bay a Completion Report. Alaska Department of Fish and Game, Juneau, AK.

Ruggerone, G.T. and M.R. Link. 2006. Collapse of Kvichak River sockeye salmon production brood years 1991–1999: population characteristics, possible factors, and management implications. North Pacific Research Board, Anchorage, AK.

Ruggerone, G.T. and J.L. Nielsen. 2009. A review of growth and survival of salmon at sea in response to competition and climate change. American Fisheries Society Symposium 70:241–265.

Ruggerone, G.T., J.L. Nielsen, and B.A. Agler. 2009. Climate, growth and populations dynamics of Yukon River Chinook salmon. North Pacific Anadromous Fish Commission Bulletin 5:279–285.

Ruggerone, G.T., J.L. Nielsen, and J. Bumgarner. 2007. Linkages between Alaskan sockeye salmon abundance, growth at sea, and climate, 1955–2002. Deep-Sea Research II 54:2776–2793.

Ruggerone, G.T., R.M. Peterman, and B. Dorner. 2010. Magnitude and trends in abundance of hatchery and wild pink salmon, chum salmon, and sockeye salmon in the North Pacific Ocean. Marine and Coastal Fisheries: Dynamics, Management, and Ecosystem Science 2:306–328.

Ruggerone, G.T., M. Zimmermann, K.W. Myers, J.L. Nielsen, and D.E. Rogers. 2003. Competition between Asian pink salmon (Oncorhynchus gorbuscha) and Alaskan sockeye salmon (Oncorhynchus nerka) in the North Pacific Ocean. Fisheries Oceanography 122:209–219.

Russell, R. 1977. Rainbow trout life history studies in Lower Talarik Creek–Kvichak drainage. Alaska Department of Fish and Game, Anchorage.

Salo, E.O. 1991. Life history of chum salmon (Oncorhynchus keta). Pages 231–310 in C. Groot and L. Margolis, editors. Pacific Salmon Life Histories. UBC Press, Vancouver, BC.

Salomone, P., S. Morstad, T. Sands, M. Jones, T. Baker, G. Buck, F. West, and T. Kreig. 2011. 2010 Bristol Bay Area Management Report. Alaska Department of Fish and Game, Anchorage, AK.

Sandercock, F.K. 1991. Life history of coho salmon (Oncorhynchus kisutch). Pages 395–446 in C. Groot and L. Margolis, editors. Pacific Salmon Life Histories. UBC Press, Vancouver, BC.

Sanderson, E., M. Jaiteh, M. Levy, K. Redford, A. Wannebo, and G. Woolmer. 2002. The human footprint and the last of the wild. Bioscience 52:891–904.

Satterfield, F. and B. Finney. 2002. Stable isotope analysis of Pacific salmon: insight into trophic status and oceanographic conditions over the last 30 years. Progress in Oceanography 53:231–246.

Scheuerell, M.D., P.S. Levin, R.W. Zabel, J.G. Williams, and B.L. Sanderson. 2005. A new perspective on the importance of marine-derived nutrients to threatened stocks of Pacific salmon (Oncorhynchus spp.). Canadian Journal of Fisheries and Aquatic Sciences 62:961–964.

Scheuerell, M., J. Moore, D. Schindler, and C. Harvey. 2007. Varying effects of anadromous sockeye salmon on the trophic ecology of two species of resident salmonids in southwest Alaska. Freshwater Biology 52:1944–1956.

Schindler, D.E., R. Hilborn, B. Chasco, C.P. Boatright, T.P. Quinn, L.A. Rogers, and M.S. Webster. 2010. Population diversity and the portfolio effect in an exploited species. Nature 465:609–U102.

Schindler, D.E., P.R. Leavitt, C.S. Brock, S.P. Johnson, and P.D. Quay. 2005b. Marine-derived nutrients, commercial fisheries, and production of salmon and lake algae in Alaska. Ecology 86:3225–3231.

Schindler, D.E., D.E. Rogers, M.D. Scheuerell, and C.A. Abrey. 2005a. Effects of changing climate on zooplankton and juvenile sockeye salmon growth in southwestern Alaska. Ecology:198–209.

Schmidt, D.C., S.R. Carlson, G.B. Kyle, and B.P. Finney. 1998. Influence of carcass-derived nutrients on sockeye salmon productivity of Karluk Lake, Alaska: Importance in the assessment of an escapement goal. North American Journal of Fisheries Management 18:743–763.

Semenchenko, N.N. 1988. Mechanisms of innate population control in sockeye salmon, Oncorhynchus nerka. Journal of Ichthyology 28:149–157.

Shaff, C. and J. Compton. 2009. Differential incorporation of natural spawners vs. artificially planted salmon carcasses in a stream food web: Evidence from delta N-15 of juvenile coho salmon. Fisheries 34:62–72.

Sigurdsson, D. and B. Powers. 2011. Participation, effort, and harvest in the sport fish business/guide licensing and logbook programs, 2010. Alaska Department of Fish and Game, Anchorage, AK.

Solazzi, M., T. Nickelson, S. Johnson, and J. Rodgers. 2000. Effects of increasing winter rearing habitat on abundance of salmonids in two coastal Oregon streams. Canadian Journal of Fisheries and Aquatic Sciences 57:906–914.

Spencer, T.R., J.H. Eiler, and T. Hamazaki. 2009. Mark-recapture abundance estimates for Yukon River Chinook salmon 2000–2004. Alaska Department of Fish and Game, Anchorage, AK.

Stilwell, K.B. and D.S. Kaufman. 1996. Late-Wisconsin glacial history of the northern Alaska Peninsula, southwestern Alaska, U.S.A. Arctic and Alpine Research 28:475–487.

Stewart, I.T., D.R. Cayan, and M.D. Dettinger. 2005. Changes toward earlier streamflow timing across western North America. Journal of Climate 18:1136–1155.

Stockner, J.G. 2003. Nutrients in salmonid ecosystems: Sustaining production and biodiversity. American Fisheries Society.

Troll, T. 2011. Sailing for salmon: the early years of commercial fishing in Alaska's Bristol Bay, 1884–1951. Nushagak-Mulchatna/ Wood Tikchik Land Trust, Dillingham AK.

Uchiyama, T., B.P. Finney, and M.D. Adkison. 2008. Effects of marine-derived nutrients on population dynamics of sockeye salmon (Oncorhynchus nerka). Canadian Journal of Fisheries and Aquatic Sciences 65:1635–1648.

Walter, J.K., R.E. Bilby, and B.R. Fransen. 2006. Effects of Pacific salmon spawning and carcass availability on the caddisfly Ecclisomyia conspersa (Trichoptera:Limnephilidae). Freshwater Biology 51:1211–1218.

Weiner, M. 2006. Crown jewel: Iliamna. Fish Alaska Magazine.

West, F., L. Fair, T. Baker, S. Morstad, K. Weiland, T. Sands, and C. Westing. 2009. Abundance, age, sex, and size statistics for pacific salmon in Bristol Bay, 2004. Alaska Department of Fish and Game, Anchorage, AK.

White, B. 2010. Alaska salmon enhancement program 2009 annual report. Alaska Department of Fish and Game, Anchorage, AK.

Whited, D.C., J.A. Lucotch, N.K. Maumenee, J.S. Kimball, and J.A. Stanford. 2012. A riverscape analysis tool developed to assist wild salmon conservation. Fisheries 37:305–314.

Williams, T. 2006. Pits in crown jewels. Fly Rod and Reel.

Willson, M.F. and K.C. Halupka. 1995. Anadromous fish as keystone species in vertebrate communities Conservation Biology 9:489–497.

Wipfli, M.S., J. Hudson, and J. Caouette. 1998. Influence of salmon carcasses on stream productivity: response of biofilm and benthic macroinvertebrates in southeastern Alaska, USA. Canadian Journal of Fisheries and Aquatic Sciences 55:1503–1511.

Wipfli, M.S., J.P. Hudson, and J.P. Caouette. 2004. Restoring productivity of salmon-based food webs: Contrasting effects of salmon carcass and salmon carcass analog additions on stream-resident salmonids. Transactions of the American Fisheries Society 133:1440–1454.

Wipfli, M.S., J.P. Hudson, J.P. Caouette, and D.T. Chaloner. 2003. Marine subsidies in freshwater ecosystems: Salmon carcasses increase the growth rates of stream-resident salmonids. Transactions of the American Fisheries Society 132:371–381.

Wipfli, M.S., J.P. Hudson, D.T. Chaloner, and J.R. Caouette. 1999. Influence of salmon spawner densities on stream productivity in Southeast Alaska. Canadian Journal of Fisheries and Aquatic Sciences 56:1600–1611.

Wobus C., R. Prucha, D. Albert, C. Woll, M. Loinaz, and R. Jones. 2015. Hydrologic alterations from climate change inform assessment of ecological risk to Pacific salmon in Bristol Bay, Alaska. PLoS ONE 10(12): e0143905. doi:10.1371/journal.pone.0143905.

Woody, C.A. 2007. Tower Counts. Pages 363–384 in D. H. Johnson, B. M. Shrier, J. S. O'Neal, J. A. Knutzen, X. Augerot, T. A. O'Neil, and T. N. Pearsons, editors. Salmonid Field Protocols Handbook. American Fisheries Society, Bethesda, MD.

Woody, C.A. and S.L. O'Neal. 2010. Fish surveys in headwater streams of the Nushagak and Kvichak river drainages, Bristol Bay, Alaska, 2008–2010. Fisheries Research and Consulting, Anchorage, AK.

Young, D.B. 2005. Distribution and characteristics of sockeye salmon spawning habitat in the Lake Clark watershed, Alaska. National Park Service, Port Alsworth, AK.

Zabel, R.W., M.D. Scheuerell, M.M. McClure, and J.G. Williams. 2006. The interplay between climate variability and density dependence in the population viability of Chinook salmon. Conservation Biology 20:190–200.

# SECTION V

# FRESHWATER ECOLOGY AND FISHERIES OF BRISTOL BAY

# 19

# FRESHWATER ENVIRONMENTS: WATER QUALITY OF THE NUSHAGAK AND KVICHAK WATERSHEDS

Kendra Zamzow, Center for Science in Public Participation

## INTRODUCTION

Nushagak Bay and Kvichak Bay represent two of the largest bodies of water within Bristol Bay. Fresh water flows through a breathtaking diversity of landscapes en route to the bays. The sheer scale is difficult to grasp. Lakes, critical to the life cycles of sockeye salmon and other wildlife, are prominent figures. There are deep, clear, narrow lakes that, interconnected, weave down from bare mountains to forested lowlands. Mountain valleys cup lakes that rise and fall as glacial melt enters and then dries up. Remnant spring snow blooms red as hiking boots crush cranberries that have lain throughout the winter snugged low to high dry tundra where headwater streams begin. Low-lying wetland tundra collects snowmelt in pools that come and go, evaporating into the summer air. Rivers run brown with organic material or sport the chalky gray of glacial sediment or volcanic ash. Most of the lotic waters are clear and incredibly pure, reflecting only the color of the sky.

There are eleven watersheds that feed Bristol Bay (Table 19.1). The focus in this chapter will be on seven of those watersheds: the Upper Nushagak, Mulchatna, Lower Nushagak, Wood River, Naknek, Lake Clark, and Lake Iliamna sub-basin watersheds. The remaining ones are the Port Heiden, Egegik Bay, Ugashik Bay, and Togiak sub-basin watersheds, all of which are at the extreme edges of the area and contribute less directly to water quality to upper Bristol Bay (Nushagak Bay and Kvichak Bay) (Figure 19.1).

**Table 19.1**   The United States Geological Survey (USGS) hydrologic unit codes (HUC) for the Kvichak Bay and Nushagak Bay freshwater systems. All are within the Alaska region (HUC code 19) and Southwest Alaska subregion (HUC code 1903). From http://water.usgs.gov/GIS/wbd_huc8.pdf and http://water.usgs.gov/wsc/sub/1903.html

| Basin | Basin name | Sub-basin | Sub-basin name | Sub-basin area (km²) |
|-------|-----------|-----------|----------------|----------------------|
| 190302 | Kvichak-Port Heiden | 19030201 | Port Heiden | 8,228 |
| | | 19030202 | Ugashik Bay | 11,155 |
| | | 19030203 | Egegik Bay | 7,184 |
| | | 19030204 | Naknek | 9,707 |
| | | 19030205 | Lake Clark | 9,277 |
| | | 19030206 | Lake Iliamna | 18,524 |
| 190303 | Nushagak River | 19030301 | Upper Nushagak | 12,984 |
| | | 19030302 | Mulchatna | 11,119 |
| | | 19030303 | Lower Nushagak | 7,809 |
| | | 19030304 | Wood River | 3,541 |
| | | 19030305 | Togiak | 12,365 |

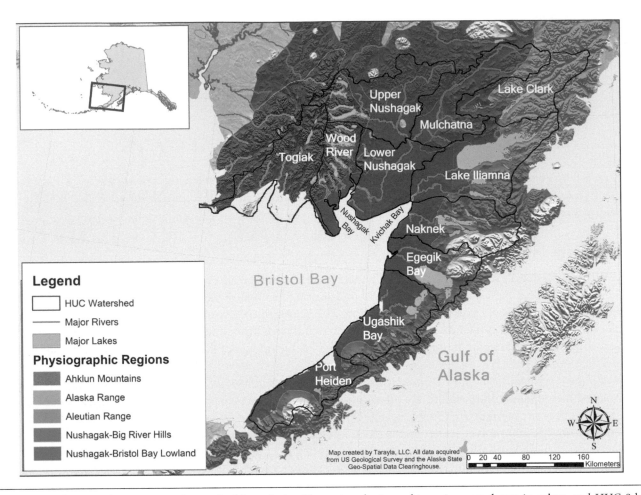

**Figure 19.1** Physiographic regions and watershed boundaries. The major physiographic regions are shown in colors, and HUC-8 level watershed boundaries are shown as black borders. The Nushagak drainage basin (HUC 190303) is made up of the Upper Nushagak, Mulchatna, Lower Nushagak, Wood River, and Togiak watersheds. All of these except the Togiak watershed enter Nushagak Bay. The Kvichak-Port Heiden drainage basin (HUC 190302) consists of the Lake Iliamna, Lake Clark, Naknek, Egegik Bay, Ugashik Bay, and Port Heiden watersheds. All but Ugashik and Port Heiden enter Kvichak Bay. *Map credit: Tarayla LLC. and Wahrhaftig 1965.*

Both Nushagak Bay and Kvichak Bay are each fed primarily by two rivers. The two rivers that enter Nushagak Bay are the Wood River and the Nushagak River. The two rivers that enter Kvichak Bay are the Naknek River and the Kvichak River. Fresh water in these rivers originates in the mountains and high hills that form an arc of watersheds, from the Ahklun Mountains in the west to the Nushagak Hills, the Alaska Range, and finally to the Aleutian Range that runs down the spine of the Alaska Peninsula, separating the Gulf of Alaska from Bristol Bay (Figure 19.1).

The Nushagak Bay drainage consists of four distinct smaller watersheds: the Wood River, the Upper Nushagak, the Mulchatna, and the Lower Nushagak. Although technically part of the Nushagak River drainage basin, the Togiak watershed does not drain into Nushagak Bay.

The Wood River sub-basin watershed consists of the Wood River and the Wood River Lakes, which are deep, interconnected, U-shaped lakes that weave through the remnants of glaciated valleys in the 1.6 million acre Wood-Tikchik state park. The other three sub-basin watersheds supply the Nushagak River, which is 450 km long, dropping down the landscape like a spinal cord fed by the capillary system of rivers and tributaries.

The Kvichak-Port Heiden drainage basin is also a hydrologic drainage basin. The smaller Lake Clark, Lake Iliamna, and Naknek Lake sub-basin watersheds feed into the upper part of Kvichak Bay through the Kvichak and Naknek Rivers. The Port Heiden and Ugashik Bay sub-basin watersheds, which are not covered in this chapter, enter the lower (southeast) area of Bristol Bay along the Alaska Peninsula, well below Kvichak Bay.

The Lake Clark and Lake Iliamna sub-basin watersheds collect water and send it to Kvichak Bay via the Kvichak River. Water originates in mountain glaciers, which make up nearly one-third of the watershed of Lake Clark. Icy water from glacial melt tumbles into the upper lake, travels down-lake to the Newhalen River, and out into Iliamna Lake, a vast island-studded lake that itself spills out through the Kvichak River.

Kvichak Bay is also fed by the Naknek River, south of Iliamna Lake on the Alaska Peninsula. Volcanoes collared with glaciers provide glacial melt. Glacial meltwater mixes with rivers that run through a valley of volcanic ash en route to Naknek Lake, and the lake outlet is through the Naknek River.

The focus of this chapter is to convey information on the geographical placement, water quality, and complex movement of freshwater flow into the Nushagak and Kvichak Bays, which make up the upper part of Bristol Bay. This freshwater primarily flows through seven sub-basin watersheds into four rivers, then into the Nushagak and Kvichak Bays. Physical, chemical, and biological aspects of the water quality of lakes, rivers, streams, and pools are discussed, along with the influence of glaciers, volcanoes, and salmon themselves in driving water quality. Most of the freshwater resources are cold, clear, and well oxygenated with neutral pH, reflecting the vast undeveloped and uninterrupted landscape. These freshwater resources are essential for the health and vitality of the ecosystems within the Bristol Bay region.

## Available Data

Despite the importance of Bristol Bay, water quality data is very limited. Data sets from the National Park Service (NPS), for Lake Clark National Park and Preserve (LACL) and Katmai National Park and Preserve (KATM), have a range of water quality information but cover a short time period. The University of Washington research stations on Iliamna Lake and Lake Aleknagik cover 50-plus years of salmon research, but with limited water quality information. In-depth water chemistry is available for streams at the site of the proposed Pebble mine, but the geographical scope is limited. The vast majority of waters have no data at all. This is surprising given the extraordinarily watery landscape, which can be observed on maps (http://maps.waterdata.usgs.gov/mapper/index.html; http://water.usgs.gov/streamer) or satellite imagery (https://www.google.com/maps/place/Iliamna+Lake); the greater the zoom magnification, the more water bodies that can be seen.

# PHYSICAL GEOGRAPHY AND WATERSHEDS

A watershed consists of waters originating from multiple sources moving to a common outlet. It can be a large geographic area that represents an entire state, a drainage basin, or a combination of drainage basins, or it can be a small distinct hydrologic feature. Physical geography (physiography) is the landscape through which waters travel. Watersheds and geography do not overlap neatly, but both are important to understanding water movement and water quality. Watersheds originate in and cross physical geographical areas, and aspects of the physical landscape frequently drive water quality.

## Physiography

Physiographic regions—defined by topography, geology, climate, and vegetation—are primary drivers of water temperature, oxygen, sediment/clarity, nutrients, and water chemistry. Five physiographic regions describe the landscape within the Bristol Bay region. Lakes, which are critical in the life cycles of sockeye salmon and other wildlife, are prominent features.

An arc, from west to east, contains hill and mountain regions: the Ahklun Mountains, the Nushagak-Big River Hills (referred to in this chapter as *Nushagak Hills*), the Alaska Range, and the Aleutian Range (Nowacki et al. 2001, Wahrhaftig 1965) come together in a partial bowl formation with these mountain ranges as the rim of a bowl, and the Nushagak-Bristol Bay Lowlands (referred to in this chapter as *Nushagak Lowlands* or simply *Lowlands*) as the bottom the bowl (Figure 19.1).

The steep Ahklun Mountains reach a height of 1,500 meters with glacial remnants, broad U-shaped valleys of glacial origin, and alpine tundra vegetation that transition to willow, birch, and alder at lower elevations. A striking series of east-west trending lakes, the Wood-Tikchik lakes, splay like thin fingers down the eastern slopes.

The Nushagak-Big River Hills are rounded ridges 450–750 meters in height with gentle slopes and broad valleys. There are numerous ponds but no glaciers. The hills do not enclose but do border one significant lake, Iliamna Lake.

In contrast, the Alaska Range consists of multiple glaciers and ice fields; the mountains are generally bare of vegetation on metamorphic rock and scree. Precipitation varies with elevation, from 200 cm at higher elevations to 89 cm at lower elevations (Brabets 2002). Lake Clark, a pristine lake in a glacially-carved deep valley, is encircled by the Alaska Range.

The Alaska Range tapers into the Aleutian Range and they may be categorized together as the Montane Region. This volcanic mountain chain is crumpled between the Gulf of Alaska and Bristol Bay. The glaciated, rounded ridges rise to

1,200 meters high with volcanoes up to 2,600 meters high; precipitation drops to around 43 cm at some low elevations (e.g., Port Alsworth). The area from Iliamna Lake down the Alaska Peninsula is also referred to as *Large Lake Country* with an extensive system of large lakes, dammed behind glacial moraines (Nature Conservancy 2004).

As fresh water from these high regions descends, it enters the Nushagak-Bristol Bay Lowlands. These low elevation (< 150 meters) rolling glacial moraines with broad basins contain numerous thaw lakes and wetlands which cover over half of the area. Precipitation is 46–81 cm and average annual temperatures are −1 to 3°C (McNab and Avers 1994). The Lowlands line the shores of most of Bristol Bay, including Nushagak Bay and Kvichak Bay.

## Watersheds

The United States Geological Survey (USGS) uses *hydrologic unit codes* (HUC) to define drainage areas of different sizes (http://water.usgs.gov/GIS/huc.html). There are 21 national major geographic regions that are further subdivided into sub-regions, basins, and sub-basins. Each is given an HUC label. Regions are represented by a two-digit HUC code, while sub-regions have a four-digit code, basins have a six-digit code, and sub-basins have an eight-digit code (http://water .usgs.gov/GIS/wbd_huc8.pdf). The Kvichak (HUC 190302) and Nushagak (HUC 190303) hydrologic drainage basins are nested within the Southwest Alaska sub-region (HUC 1903) in the Alaska region (HUC 19) (Table 19.1). Although any HUC level can be considered a watershed, for the purposes of this document we refer to basin-level watersheds as *drainage basins* and refer to sub-basin level watersheds simply as *sub-basin watersheds*.

### Watersheds of the Nushagak Drainage Basin

The Upper Nushagak sub-basin watershed contains three physiographic regions: the Ahklun Mountains, the Nushagak Hills, and the Lowlands. Along the Ahklun Mountains is the Wood-Tikchik State Park and associated lakes systems. There are at least 40 lakes in the Wood-Tikchik State Park. The largest of these form two chains—the Tikchik Lake system and the Wood River lake system (Figure 19.2). Lake water begins in the Ahklun Mountains and eventually ends up in Nushagak Bay.

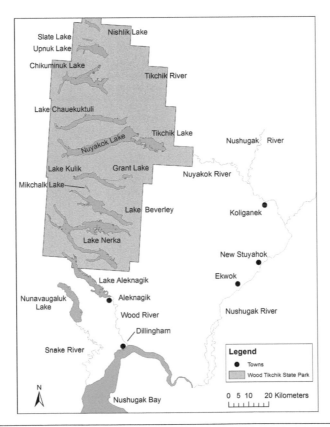

**Figure 19.2**  Wood River watershed. The figure shows the largest water bodies of the Wood River watershed. The northern string of lakes is the Tikchik Lake system. The southern string of lakes is the Wood River lake system. Waters from both systems enter the Nushagak River, although at different locations. *Map credit: Dan Young, NPS, Alaska.*

The Wood River sub-basin watershed has seven interconnected lakes: Grant Lake, Lake Kulik, Lake Mikchalk, Lake Beverley, Lake Nerka, Little Togiak Lake (an extension of Lake Nerka), and Lake Aleknagik. The Wood River is the outlet of Lake Aleknagik; it runs through forested woodlands of the Lower Nushagak sub-basin watershed to reach Nushagak Bay.

The Tikchik Lake system is made up of Nishlik Lake, Slate Lake, Upnuk Lake, Chikuminuk Lake, Lake Chauekuktuli, and Nuyakok Lake, all connected or separated by a short portage (Figure 19.2). Tikchik Lake is an extension of Nuyakok Lake, separated by a narrow channel. The Nuyakok River is the outlet of Tikchik Lake; it runs 58 km through the Lower Nushagak sub-basin watershed to join the main Nushagak River, which enters from the Nushagak Hills of the Upper Nushagak sub-basin watershed (Figure 19.3).

The Nuyakok River is one of two large tributaries to join the main trunk of the Nushagak River; the 260 km long Mulchatna River is the other (Figure 19.3). The Mulchatna River is entirely within the Mulchatna sub-basin watershed (Figure 19.1), and for most of its length, it is in the Nushagak Hills or the Lowlands, although it originates in a small mountain lake in the Alaska Range near Lake Clark. The Mulchatna River itself collects major tributaries, including the Chilikadrotna, originating near the Mulchatna headwater mountain lakes, and the Koktuli and Stuyahok rivers, which sweep in from the rolling hills north of Iliamna Lake (Figure 19.3). Parts of the Chilikadrotna and Mulchatna are two of the four federally designated *Wild and Scenic Rivers* in the Kvichak and Nushagak drainage basins (Table 19.2).

## Watersheds of the Kvichak Drainage Basin

Furthest to the north, the Lake Clark sub-basin watershed consists of Lake Clark and the tributaries that enter it (Figure 19.4). Lake Clark is almost entirely encircled by the steep glacial mountains of the Alaska Range, yet flattens out into a low-lying landscape at the southwest shore, just above Sixmile Lake. Sixmile Lake is a terminal extension of Lake Clark, separated from the main lake by a narrow channel. The outlet of Sixmile Lake is the Newhalen River, which connects to Iliamna Lake. The outlet of Iliamna Lake is the Kvichak River.

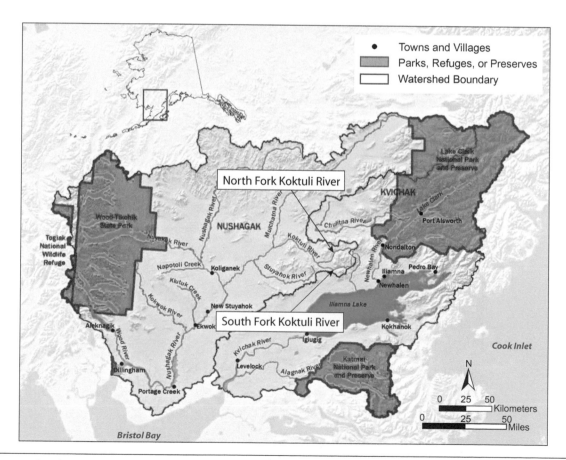

**Figure 19.3** Nushagak and Kvichak drainage basins. The main river and lake systems within the Nushagak and Kvichak drainage basins are shown. The figure does not show the boundaries of smaller watersheds within the drainage basins. *Figure credit: United States Environmental Protection Agency (USEPA 2014).*

**Table 19.2** Nationally designated Wild and Scenic Rivers. Rivers and river sections in federally managed lands of the Nushagak and Kvichak basins. The LACL boundaries include all of Lake Clark and the headwaters of the Mulchatna and Chilikadrotna Wild and Scenic River designation was given under Title VI, Part A of the Alaska National Interest Lands Conservation Act (ANILCA) of 1980. KATM = Katmai National Park and Preserve. LACL = Lake Clark National Park and Preserve. From http://www.rivers.gov/act.php

| Basin | Watershed | River | Length designated as wild and scenic | General location | Description of designated reach | Managing agency |
|---|---|---|---|---|---|---|
| Nushagak | Mulchatna | Chilikadrotna | 11 miles | Clear river on the western slopes of the Alaska Range, west of Lake Clark | The entire segment of the river within LACL | National Park Service, LACL |
| Nushagak | Mulchatna | Mulchatna | 24 miles | Clear river on the western slopes of the Alaska Range, west of Lake Clark | The entire segment of the river within LACL | National Park Service, LACL |
| Kvichak | Lake Clark | Tlikakila | 51 miles | Glacial outwash river that enters the north end of Lake Clark from the Alaska Range | The entire segment of the river within LACL | National Park Service, LACL |
| Kvichak | Lake Iliamna | Alagnak | 67 miles (all) | On the upper Alaska Peninsula. Begins at Kukaklek Lake and joins the Kvichak River shortly before Kvichak Bay. | From Kukaklek Lake to the west boundary of T13S, R43W and the entire Nonvianuk River | National Park Service, KATM |

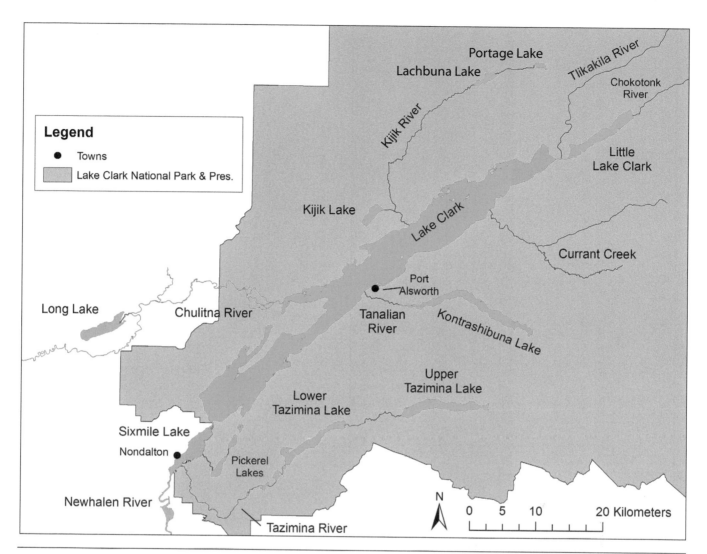

**Figure 19.4**   Lake Clark. The major rivers and lakes that drain into Lake Clark are shown. Sixmile Lake is a small lake separated from Lake Clark by a narrow reach. There are only two towns in the area—Port Alsworth on Lake Clark and Nondalton on Sixmile Lake. *Map credit: Dan Young, NPS, Alaska.*

Iliamna Lake is nearly ten times as large as Lake Clark (Table 19.3). The largest lake in Alaska and the largest sockeye salmon nursery lake in the world, Iliamna Lake is a centerpiece of the landscape, bordered by the Alaska Range, the Aleutian Range, the Nushagak Hills, and the Nushagak Lowlands (Figure 19.5). Although the watershed is referred to as *Lake Iliamna watershed*, the lake itself is referred to as *Iliamna Lake*. Other large lakes, such as Nonvianuk Lake, are within the Lake Iliamna sub-basin watershed. Iliamna Lake and Kukaklek Lake begin in mountains, but their outlet rivers are entirely in the Nushagak Lowlands physiographic region. The 80 km Kvichak River out of Iliamna Lake and the 103 km Alagnak River out of Kukaklek Lake run nearly parallel until they join shortly before entering Kvichak Bay. The Alagnak is another Wild and Scenic River (Table 19.2).

The Naknek sub-basin watershed consists of Naknek Lake, surrounding lakes, and the Naknek River (Figure 19.5). The lakes are in the Aleutian Range and the river is in the Nushagak Lowlands. Naknek Lake is unique in receiving a river of volcanic ash. One of the largest volcanic eruptions in the world, the Novarupta eruption of 1912, sent ash and hot gases 32 km into the air; within hours ash was falling on Seattle, and by the next day the ash cloud was moving over Virginia (Adleman 2002), (http://www.nps.gov/katm/planyourvisit/exploring-the-valley-of-ten-thousand-smokes.htm). The event turned the Ukak River valley into 64 square km of smoking ash up to 200 m thick (http://pubs.usgs.gov/gip// volcus/ustext.html). The Ukak River cuts a gorge through ash and underlying sandstone (Whipple et al. 2000), moving volcanic ash into one arm of Naknek Lake.

**Table 19.3** Characteristics of select lakes in the Nushagak and Kvichak basin freshwater systems. Watershed area for Lake Clark, Iliamna Lake, and Naknek Lake are from http://water.usgs.gov/GIS/wbd_huc8.pdf. na = not available, NPS = National Park Service

| Watershed | Lake | Watershed area (km²) | Lake area (km²) | Length (km) | Maximum depth (m) | Elevation (m) | Notes |
|---|---|---|---|---|---|---|---|
| Wood River[a] | Aleknagik | 2,850[b] | 83 | na | 99 | 26 | Non-glacial |
| | Nerka | na | 201 | na | 164 | 21 | Non-glacial |
| Upper Nushagak[a] | Tikchik | na | 58 | na | 45 | 95 | Non-glacial |
| | Nuyakok | na | 144 | na | 283 | 95 | Non-glacial |
| Mulchatna | Turquoise | 188[c] | 13[c] | 8[d] | 109[e] | 763[c] | Glacial |
| | Upper Twin | 320, Upper and Lower[c] | 15[c] | 18[d] | 99[e] | 604[c] | Glacial |
| | Lower Twin | | 8[c] | 8[d] | 44[e] | 604[c] | Glacial |
| Lake Iliamna | Iliamna Lake | 18,524[f] | 3,175[d] | 125[g] | 301[g] | 15[h] | Largest lake in Alaska |
| | Nonvianuk | na | 132[i] | na | na | 192[a] | Non-glacial |
| | Battle | na | 13 | na | 62 | 254 | Acidic and metalliferous |
| | Lake Clark | 7,620[i] | 370[g] | 66[i] | 322[i] | 77[k] | Glacial, 3rd largest lake within a National Park |
| Lake Clark | Kijik | 110[c] | 4[c] | 3[d] | 99[d] | 107[c] | Non-glacial |
| | Upper Tazimina | 725, Upper and Lower[c] | 15[c] | 14[d] | 102[d] | 235[c] | Non-glacial |
| | Lower Tazimina | | 12[c] | 12[d] | 62[d] | 194[c] | Non-glacial |
| Naknek | Naknek[a] | 9,707[f] | 610 | na | 173 | 10 | Largest lake in a National Park |
| | Brooks[a] | 775[f] | 75 | na | 79 | 19 | Non-glacial |
| | Idavain[i] | na | 11 | na | 21 | 223 | Non-glacial |

[a] Burgner et al. 1969; [b] USGS gage 15303000; [c] Chamberlain 1989; [d] Weeks 2003; [e] NPS, personal communication; [f] NPS estimate of watershed area from USGS: http://water.usgs.gov/GIS/wbd_huc8.pdf; [g] USGS gage 15305000; [h] Donaldson 1969; [i] LaPerriere 1997; [j] Young and Woody 2007; [k] Brabets and Ourso 2006

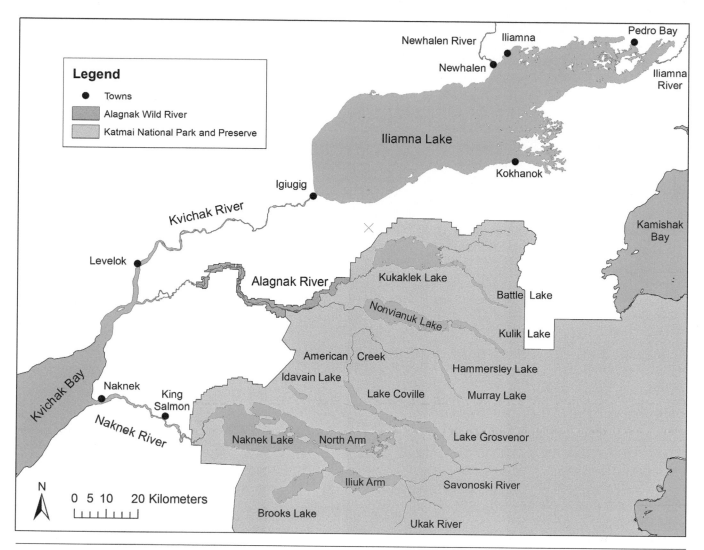

**Figure 19.5**  Lake Iliamna and Naknek watersheds. Both watersheds have prominent lakes and both feed into Kvichak Bay. *Map credit: Dan Young, NPS, Alaska.*

## WATER QUALITY DRIVERS

Physiography describes the broad regional landscape: mountains, lakes, wetlands, and so forth. Watersheds provide information on the specific water bodies. Due to the sheer number of water bodies in the region and lack of data, this chapter approaches water quality by discussing those that are important to the region or represent the diversity of landscapes and natural processes. For each water body, water quality is grouped into the categories of *temperature, oxygen, clarity; chemical water quality;* and *nutrients.*

Climate; seasonal patterns; surficial and subsurface geology; biologic inputs; and episodic flood, fire, volcanic, and seismic events provide nutrients, flush trace elements into the water column, affect water temperature, and/or change hydrologic flow and sediment loads. Broadly, we can consider these to be ways in which the atmosphere and landscape affect water bodies.

### Atmospheric Influences

A warming trend with increased precipitation is currently occurring across the Bristol Bay region, influenced by the Pacific Decadal Oscillation (PDO)—a long-term ocean temperature fluctuation that recurs every 20 to 30 years—and global warming. In northern latitudes, warming began around 1850 (Cohn 2009) and locally, this trend has been measured over 50 years, although a short-term local cooling event occurred in 2006–2010 (Alaska Salmon Program 2010, 2011).

Warming has impacted lake surface temperatures and productivity (Alaska Salmon Program 2010, 2011) and lake volumes (Brubaker et al. 2014). River temperatures also likely have been affected (Grazia and Mauger 2011). Furthermore, the extent and thickness of glaciers has been reduced (Brabets et al. 2004, Griffiths et al. 2011). Although warmer air has increased surface temperatures at some lakes (Alaska Salmon Program 2011, Moore and Shearer 2011) ironically, wetter conditions and increased glacial melt, attributed to the PDO and climate change, may be the reason for cool lake temperatures at other lakes (Moore and Shearer 2011, Brubaker et al. 2014).

Warmer air has also led to increased biological primary productivity in some oligotrophic lakes (Cohn 2009). Receding glaciers have led to an increase in vegetation, with particular expansion of alder (Cohn 2009, Brubaker et al. 2014). Alder (*Alnus* sp.) may be an important source of nitrogen to oligotrophic systems, boosting productivity (Cohn 2009, Harding and Reynolds 2014). Additionally, warm air melts the ice cover on lakes sooner, extending the ice-free period. The point at which most of the ice is gone is referred to as *ice-out*. With the ice gone, sunlight is able to penetrate into lake water, and surface waters are in contact with air again (Weeks 2003)—there is an increase in heat, light, and oxygen leading to higher density of zooplankton (Alaska Salmon Program 2011) (Figure 19.6).

Direct negative impacts from warming occur when vegetation is drought stressed (Cohn 2009), as when shallow water bodies shrink due to evaporation or lose hydrologic connections (Griffiths et al. 2011), and if streams warm above 17–20°C (USEPA 2001, Grazia and Mauger 2011). Warmer temperatures can have a negative impact on fish—particularly salmonids—affecting growth rates, susceptibility to disease, inability to remain in their habitat, and so forth (Table 19.4).

Wind and clouds also affect lakes. Winds cause deep mixing and prevent or disrupt the thermal stratification that warm surface temperatures would otherwise cause on large lakes (Shearer and Moore 2010). Increased cloud cover can reduce lake surface temperatures (Moore and Shearer 2011).

## Landscape Processes

Geology, vegetation, and disturbances such as volcanic eruptions and floods influence the delivery of nutrients, micronutrients, and minerals (Cohn 2009, Munk et al. 2010). These processes affect biota, including salmon. In a warming Earth, cold water prevents aquatic life from suffering thermal stress, clear water prevents silt from smothering eggs, groundwater bathes salmon eggs in oxygen and prevents streams from freezing to the bottom in winter. Well-mixed, oligotrophic lakes provide a deep cold euphotic zone in which young salmon feed. Landscapes of wetlands and riparian alder stands slow runoff and provide nutrients, spurring phytoplankton growth in low nutrient stream and lake systems. The landscape and the biology upon it are integrated, each affecting the other.

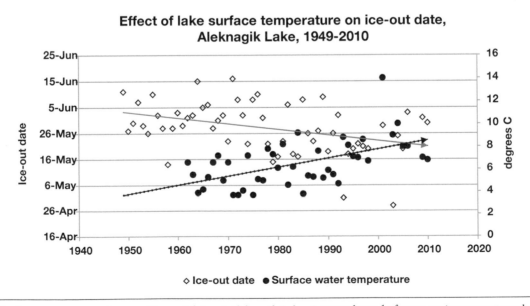

**Figure 19.6** Surface water temperatures and ice-out dates at Aleknagik Lake. A general trend of warmer air temperatures has caused ice to leave rivers and lakes earlier, which in turn has caused surface water in spring to be warmer than in the past—as mean temperatures measured June 21–26 at 0–20 meters depth. A cooling period occurred in 2006–2010. (From data published in Rogers and Rogers 1998 and Alaska Salmon Program 2006, 2007, 2011.)

**Table 19.4** Estimates of thermal conditions known to support life stages and biological functions of anadromous salmon. These numbers do not represent rigid thresholds, but rather represent temperatures above which adverse effects are more likely to occur. Important differences between various species and in how temperatures are expressed (e.g., instantaneous maximums, daily averages, etc.) are not reflected in this table. Note that Alaska water quality criteria (WQC) require less than 20°C in waters supporting freshwater aquatic life, except no more than 15°C on migration routes and rearing areas and no more than 13°C in spawning areas and for egg and fry incubation (from Grazia and Mauger 2011).

| Consideration | Condition | Temperature (°C) |
|---|---|---|
| Temperature of common summer habitat use | | 10–17 |
| Lethal temperatures (one week exposure) | Adults: | > 21–22 |
| | Juveniles: | > 23–24 |
| Adult migration | Blocked: | > 21–22 |
| Swimming speed | Reduced: | > 20 |
| | Optimal: | 15–19 |
| Gamete viability during holding | Reduced: | > 13–16 |
| Disease rates | Severe: | > 18–20 |
| | Elevated: | 14–17 |
| | Minimized: | < 12–13 |
| Spawning | Initiated: | 7–14 |
| Egg incubation | Optimal: | 6–10 |
| Optimal growth | Unlimited food: | 13–19 |
| | Limited food: | 10–16 |
| Smoltification | Suppressed: | > 11–15 |

## Geology

The geology of the surface and subsurface defines the material that can be eroded, and eroded material affects water chemistry. Surface geology drives surface water character and subsurface geology drives groundwater chemistry.

Erosion of subsurface bedrock such as calcite and dolomite (carbonate based), pyrites (sulfide based), feldspars and silicates (aluminum-silica based), or apatite (phosphate based) (Chamberlain 1989) contributes major ions such as calcium ($Ca^{2+}$), magnesium ($Mg^{2+}$), potassium ($K^+$), carbonate ($CO_3^{2-}$), bicarbonate ($HCO_3^-$), chloride ($Cl^-$), phosphate ($PO_4^-$), and sulfate ($SO_4^{2-}$), materials soluble enough that those on the surface eroded long ago. Underground they weather much more slowly through contact with groundwater. Groundwater transports these major ions, which control alkalinity and hardness, to surface water.

Commonly, surface water within the watersheds that are the focus of this chapter are very dilute and alkalinity is low (< 25 mg/L); however, groundwater upwelling can raise alkalinity by contributing major ions, and providing important buffering. *Buffering* refers to the ability of water to maintain pH without big swings up or down. In these waters, buffering is provided by $CO_3^{2-}$ accepting acid ($H^+$) to become $HCO_3^-$; in this way the concentration of free $H^+$ ions (pH) is not drastically changed. However, too much acidity can overwhelm the available carbonate, increase the concentration of $H^+$ ions, and cause the pH to drop. When waters are naturally very low in alkalinity, there is less carbonate available, and they tend to be susceptible to pH declines when acid enters. Sulfide ore, outcropped at the land surface or exposed during mining, classically produces acid that reduces stream pH and increases trace metal concentrations in water (Jambor et al. 2003); this is a potential risk in the Nushagak drainage (USEPA 2014).

Hardness is an important and related component. Cations [$Ca^{2+}$, $Mg^{2+}$, iron ($Fe^{3+}$)] contribute to water hardness; *harder* water moderates the toxicity of trace elements (copper, zinc, and so forth), in that they block sites where metals would enter the fish (competition for biological receptors).

Where the geology is dominated by limestone (calcium carbonate), limestone contributes calcium to increase hardness and increase carbonate to increase alkalinity. However, Bristol Bay waters are low in hardness and alkalinity. In most of these freshwater systems, $Ca^{2+} > Mg^{2+} \sim Na^+ > K^+$ with calcium less than 10 mg/L, and $HCO_3^- > SO_4^{2-} > Cl^- >$ nitrate ($NO_3^{2-}$)

with bicarbonate less than 25 mg/L (Burgner et al. 1969, Chamberlain 1989). For comparison, most aquariums will adjust hardness to 75–200 mg/L and adjust the alkalinity to 50–150 mg/L and generally waters are considered suited to aquatic life if they are over 20 mg/L (Wurts 2002, ADEC 2008). This does not mean the Bristol Bay waters are disadvantageous to aquatic life, but rather that biota has adapted to *dilute waters* and the water chemistry can be easily disrupted.

While groundwater is primarily responsible for contributing major elements such as calcium that substantially affect water composition, surficial rock weathering contributes minor elements such as iron (Fe), aluminum (Al), and manganese (Mn) (measured in mg/L), along with trace elements such as copper (Cu) (measured in µg/L) from harder minerals that persist above ground under wind and water erosion. These elements may be observed in a pulse of increased concentrations, although still quite small, in streams during snowmelt and rains, frequently carried integrated in eroded particulates.

The ability of water to carry an electrical current correlates with the sum of dissolved major ions and may be measured as electrical conductivity (EC) or specific conductance (SC) (EC corrected to 25°C). Dilute waters have low concentrations of these components and therefore have very low SC.

## Glaciers

Warm air increases glacial melt. Most of the melt occurs in the summer after the snow is gone. The effect of this is that glacially influenced lakes and streams receive cold water in spring from snowmelt and cold water in summer from glacial melt, extending the period of cold water temperatures and reducing the risk of thermal stress on aquatic life. Glaciers cover about 36% of the Lake Clark sub-basin watershed and store a substantial amount of the water that enters the lake (Brabets 2002); glaciers cover about 6% of the KATM, which contains Naknek Lake (Weeks 1999).

Glaciers are also the source of very fine silt-like material referred to as *glacial flour*, which creates turbidity in lakes and rivers (Brabets et al. 2004). Fish that spawn in glacial, silt-laden rivers may time reproduction around periods of low turbidity (Young and Woody 2007). As glaciers melt, more glacial flour is exposed and can be carried by winds and runoff (Brubaker et al. 2014).

Throughout most of the region, Pleistocene glacial retreat left a landscape of till and moraines mantled by soil, allowing shallow aquifers to develop. Surface water-groundwater exchange in these shallow aquifers are important drivers of surface and subsurface water temperature and oxygen (Woody and Higman 2011), and may play a role in nutrient storage (O'Keefe and Edwards 2002). They have also allowed development of extensive hyporheic zones, which is groundwater beneath the streambed that can extend up to 10 meters below the streambed (Marmonier et al. 1992, Stanford and Ward 1988) and up to three km laterally (Stanford and Gaufin 1974, Stanford and Ward 1988, Ward et al. 1998). The hyporheic zone provides small fish and juvenile stages of large fish, like salmon fry, protection from predators but also presents potential contaminant pathways (Woody and Higman 2011).

## Volcanoes

The largest single volcanic event in the area was the 1912 Novarupta eruption (Moore and Shearer 2011). The Ukak and Savonoski rivers still carry ash from this event. Although this was the largest eruption, eruptions still occur and affect water quality in the region. Ash from volcanoes can settle on lakes, reducing lake clarity and providing nutrients like phosphorus or trace elements like Cu (Munk et al. 2010).

Landscape processes and volcanic activity have resulted in an accumulation of some trace elements such as Cu, zinc, and mercury in Naknek Lake and Lake Clark since pre-1800 A.D., about 50% more than in other lakes found in Alaskan national parks (Munk et al. 2010). Nevertheless, throughout the Nushagak and Kvichak drainage basins, trace elements are in very low concentrations in streams and rivers, often below the limit of detection.

Ash that settles on land can continue to reduce clarity for years if it is carried into lakes by surface runoff. Clarity in Lake Clark was reduced after the 2009 Mt. Redoubt eruption and took more than 16 months to recover due to contributions from surface runoff (Moore and Shearer 2011) (Figure 19.7). The 1976 Mt. Augustine eruption did not cause any loss of clarity on Iliamna Lake, which is nearly ten times larger than Lake Clark, but did spur a burst of phytoplankton growth (Poe et al. 1977).

## Seasonal Hydrologic Patterns

The chemistry of lakes and streams in summer depends on the contribution of glacial melt, rains, and evaporation. Generally in the fall, an onset of rains and runoff is accompanied by a rise in water levels, increased sediment, and dilution

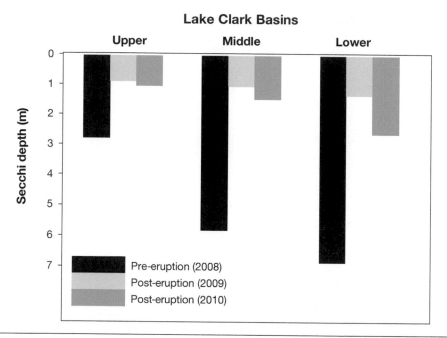

**Figure 19.7**  Effects of ashfall on Lake Clark clarity. Ash from the 2009 Mt. Redoubt volcanic explosion reduced clarity to near one meter in all Lake Clark basins. Sixteen months later, clarity was still reduced (from Moore and Shearer 2011).

of groundwater. Therefore the hydrograph (graphs showing the changing level of water over time) is relevant to water chemistry as well as the amount of water available in streams, lakes, ponds, and rivers.

Snowmelt, glacial melt, and rains control the hydrographs for the region. Lakes tend to follow one of two patterns: a single peak or a double peak in annual lake water level. Nonglacial lakes tend to have two peaks: one as the snow melts in May or June and another when rains arrive, usually in August or September. In glacial lakes, the rise that occurs with snowmelt persists due to summer glacial melt and through the fall with the onset of rain, not declining until winter. These patterns can shift, for instance if a summer is cool and rainy. The hydrograph is also relevant to tundra kettle ponds (shallow ponds), which may remain dry or as isolated water bodies until there is an increase in water levels, allowing the creation of a seasonal network of ponds, tributaries, and wetlands.

Seasons also control the relative contributions of surface water and groundwater to water bodies, which impacts water chemistry. Lakes and streams are traditionally frozen from October to May, and any water that flows in streams under the ice is supplied by groundwater. Groundwater is the source of major ions, which control the alkalinity and hardness of water. As ice and snow become surface water, the relative amount of groundwater in streams decreases due to dilution while surface runoff carries surficial material into streams. The result can be a decline in alkalinity and hardness accompanied by a brief spike in total suspended sediments (TSS) and minor elements (Al, Fe, Mn).

## Biological Inputs

As Pacific salmon feed in the ocean, they take in nutrients, increasing their weight a hundred to a thousand times or more. When they return to natal streams to spawn and then die, their carcasses provide a boost in nitrogen (N), phosphorus (P), and carbon (C) to oligotrophic systems. These are referred to as *marine-derived nutrients* (MDN) (Naiman et al. 2002, Gende et al. 2002, Janetski et al. 2009, Rinella et al. 2013, Harding and Reynolds 2014). Similarly, they may absorb mercury (Hg) in the ocean; although Alaska salmon are consistently low in Hg (Baker et al. 2009), anadromous lakes on the Alaska Peninsula contain higher concentrations of Hg (up to 250 ug/L in Brooks Lake) than nonanadromous lakes (mean ~25 ug/L) (Munk et al. 2010).

Terrestrial sources also provide nutrients. In addition to geologic sources of C and P mentioned earlier, wetlands provide C (Kline et al. 1990), and some riparian vegetation, especially alder, provides N (Naiman et al. 2002, O'Keefe and Edwards 2002, Cohn 2009). Kettle ponds and nonanadromous lakes that drain into anadromous waters can also be a source of N (Cohn 2009). In anadromous lakes with large watershed areas, terrestrially-sourced N exported from streams to the lake may be higher than MDN (Naiman et al. 2002). Importantly, there has been a trend of increasing primary productivity, C, and N in nonanadromous lakes and decreasing productivity, C, and N in anadromous systems in this region over the twentieth century (Cohn 2009).

The N and P concentrations are the sum balance of inputs, incorporation into biological vectors like zooplankton, out-migration of biological vectors like salmon smolt, and chemical oxidation, adsorption, and precipitation processes which can change seasonally and annually. Nutrient inputs during summer can be stored over winter in hyporheic zones, available for hatching fish eggs in spring.

Because systems are dynamic, it can be difficult to determine whether a system is N-limited or P-limited with small or incomplete data sets. Additionally, they may be seasonally limited or *co-limited* in N and/or P. In general, when the ratio of N:P is greater than 10, a system is likely P-limited (Brabets and Ourso 2006). Nonpolluted waters generally have a total P of less than 0.1 mg/L (Lind 1985).

While it is well documented that salmon provide MDN to invertebrates, resident fish, and some terrestrial vegetation, the full influence of a pulse of nutrients and its persistence in a system is complicated by many variables.

- MDN is incorporated into biota that directly consume carcasses or eggs, or indirectly through the food web (Kline et al. 1990, Gende et al. 2002).
- Stream discharge and sediment size affects MDN accumulation and cycling (Janetski et al. 2009).
- Natural runs of salmon provide more ammonia and P (through excretion) than artificial carcasses, but MDN from these carcasses may be more beneficial to benthic dwellers as they are not exposed to salmon digging up the stream bottom while building redds or nests (Janetski et al. 2009, Harding and Reynolds 2014).
- Ammonia and P may not persist in the water column over winter (Rinella et al. 2013), but they may still be available to aquatic life in spring if the hyporheic zone harbors benthics that store nutrients over winter (O'Keefe and Edwards 2002).
- Out-migrating smolt transport some MDN back to sea, but the relative amount removed is disputed and is likely site-specific (Donaldson 1967, Rogers 1977a, Naiman et al. 2002).

All these complications make the influence of MDN very site-specific. Tracing the source of N is also not clear-cut. Terrestrial and marine N sources are determined by stable isotopes (MDN is enriched in the $\delta^{15}N$ isotope), but various processes can mask the signature (Bartz 2002, Pinay et al. 2003, Cohn 2009). Many Southwest Alaska streams are lined with alder, which provide N. Where alder coverage is high, the contribution of N brought by salmon may be relatively low; where alder coverage is low, salmon may be important vectors (Naiman et al. 2002, O'Keefe and Edwards 2002). Additionally, the geomorphic differences in landforms (e.g., fine-sediment depositional areas, braided reaches, peat or conifer dominant) alter microbial denitrification and the signature of MDN N (Pinay et al. 2003), making it difficult to trace the N source.

In addition to influences on nutrients, biotic influences alter pH. Slight decreases in pH are observed with depth in Lake Clark, potentially due to photosynthesis which depletes available carbon dioxide causing a relatively greater percent of carbonic acid (Chamberlain 1989, LaPerrier 1997, Shearer and Moore 2010).

# LENTIC SYSTEMS (STILL WATERS)

Mountain lakes, wetland lakes, and some of the largest lakes in the nation are part of the Nushagak and Kvichak drainage basins (Table 19.3). The *splayed fingers* of interconnected lakes in the Ahklun Mountains belong to two systems: the Wood River Lakes and the Tikchik Lakes. Although they drain into two different sub-basin watersheds, they are all part of the Wood-Tikchik State Park, the largest state park in the nation at 0.7 million hectares. The lakes vary from 24 km to 72 km long, with their western edges in the mountains and their eastern shores on the gravel beaches and tundra of the Nushagak Lowlands (http://dnr.alaska.gov/parks/units/woodtik.htm).

Lake Clark, Iliamna Lake, and Naknek Lake are defining features of their watersheds. Iliamna Lake is the largest (3,175 km²) in Alaska (Shearer and Moore 2010) and the eighth largest in the United States. Naknek Lake is the largest (610 km²) (Burgner et al. 1969) within a national park and the third largest in Alaska, and Lake Clark is the third largest (341 km²) within a national park (D. Young, NPS, pers. comm.).

## Wood River Lakes

The Wood-Tikchik lake system sits astride a watershed divide (Figure 19.2). The Wood River Lakes drain through the short Wood River into Nushagak Bay; the Tikchik Lakes drain into the Nuyakok River in the Upper Nushagak sub-basin watershed and eventually into Nushagak Bay through a much longer route.

The Wood River sub-basin watershed is 3,510 km² (http://water.usgs.gov/GIS/wbd_huc8.pdf) and is covered in lakes, with at least 40 that are more than 3 km long. Foxhole-like trenches send lakes in a serpentine weave through the Ahklun Mountains, through 47 km long Lake Nerka, and finally to Lake Aleknagik (Rogers and Rogers 1998). The Wood River flows out of Lake Aleknagik to Nushagak Bay.

### Temperature, Oxygen, Clarity

Lake Aleknagik ice traditionally leaves between early May and mid-June (Rogers and Rogers 1998, Rogers et al. 1999). Water temperatures rise rapidly, snowmelt causes lake levels to rise, and approximately 20 days after ice-out, lake height will peak; by 30 days after ice-out, phytoplankton peak. If snowpack is low, a sharp drop in lake levels follows peak snowmelt until fall rains. Annual precipitation averages 51–89 cm (Hartman and Johnson 1984). Lake Nerka annual precipitation was higher; at 150 cm in 1952–1958. High water levels, followed by insulating snow pack on beaches as lake level dropped in winter, were determined to be important to salmon egg survival (Church and Burgner 2009).

Since 1949 a warming trend has affected lake temperatures and ice-out (Alaska Salmon Program 2011). Most recently, from 1993–2005, spring water temperatures were elevated, but in 2006–2010 temperatures entered a cooling period and conditions shifted back toward average (Figure 19.6). Based on air temperatures at Dillingham, warm conditions have returned since 2013 (www.wunderground.com).

Ice cover reduces the amount of light reaching water below the frozen layer, and prevents gas exchange, which can deplete dissolved oxygen and allow gases such as carbon dioxide and methane to build up in pockets (Weeks 2003 and pers. obs.). At the Wood River Lakes, earlier ice retreat has led to warmer lakes. This, in turn, has led to chlorophyll blooms and increased phytoplankton biomass, but reduced lake clarity; clarity has declined in all Wood River Lakes since 1968. Increased phytoplankton has, in turn, led to increases in zooplankton density—which are food for juvenile salmon—and juvenile salmon growth rates have increased (Rogers and Rogers 1998, Alaska Salmon Program 2007). Although growth rates have increased, actual juvenile salmon survival may not improve in part due to increased competition of fry for food resources (Rich et al. 2009).

### Chemical Water Quality

Lake Aleknagik, the most studied of the Wood River Lakes, is cold (generally < 14°C) and pH neutral (6.4–7.6). It is considered a calcium-bicarbonate type water (reflecting the major cation, calcium, and major anion, bicarbonate) (Tables 19.5–19.7). While bicarbonate can provide buffering, the low alkalinity (< 15 mg/L) and dilute waters indicate there is very little buffering capacity and the lake would respond quickly to changes in acid inputs, natural (e.g., from humic acids in wetlands) or anthropogenic (e.g., from acid rain or sulfide rock mining operations).

Trace elements [dissolved concentrations of arsenic (As), barium (Ba), cadmium (Cd), chromium (Cr), copper (Cu), lead (Pb), zinc (Zn), selenium (Se)] were below detection limits with the exception of Zn at 40 µg/L (http://nwis.waterdata.usgs.gov/nwis/, site #5921001584930000; Burgner et al. 1969) which is slightly above Alaska WQC for freshwater aquatic life (Table 19.7; note As, Ba, and Se are not included in the table). However, the data was collected in 1969. Modern analytical instruments have better sensitivity and might detect trace elements.

### Nutrients

Lake Aleknagik is oligotrophic, and where it empties into the Wood River, nitrate as N is 0.1 to 0.3 mg/L and phosphate 0.01 to 0.04 mg/L (USGS gage 15303000) (total N and total P were not measured). In 1975–1976, the University of Washington Fisheries Research Institute (FRI), which has conducted research on the Wood River Lakes since 1946, determined that declining salmon returns were related to limited food supply, competition, and predation (Rogers 1977b). The hypothesis was that the commercial fishery may have removed salmon to the extent that there were not enough carcasses to act as a storehouse of MDN, causing emerging fry to be nutrient deficient, particularly in P, resulting in smaller outmigrating smolt and a gradual decrease in lake rearing capacity (Rogers 1977a, b).

**Table 19.5** Basic water quality parameters of lake surface water. Data sets were frequently very limited, and water quality can change seasonally; therefore, the season in which they were sampled and the number of samples are provided. Ranges are given, except for temperature where the maximum temperature is listed. DO = dissolved oxygen, SC = specific conductance, n = number of samples. WQC = water quality criteria for freshwater aquatic life based on Alaska Department of Environmental Conservation (ADEC) regulations

| Watershed | Lake | Max temp (°C) | DO (mg/L) | pH | SC (µS/cm) | Secchi depth (m) | Season of data collection | n |
|---|---|---|---|---|---|---|---|---|
| Wood River[a] | Aleknagik | 19 | na | 6.4–7.6 | 31–52 | 5–12 | April–Oct | a |
| | Nerka | 13 | na | 7.1 | na | 7–14 | June–Oct | 14 |
| Upper Nushagak[b] | Tikchik | 12 | na | 6.6–7.6 | 49–73 | na | March–Oct | 12–20 |
| | Nuyakok | na | na | 7.3 | na | na | Aug | 2 |
| Mulchatna[c] | Turquoise | 14 | 11 | 7.3–8.0 | 59, 44 | <1 | July, Aug | 22 |
| | Upper Twin | 13 | 11 | 7.4–8.0 | 92 | 1–7 | July, Aug | 23 |
| | Lower Twin | 13 | 11 | 7.4–8.0 | 97–102 | 2–12 | July, Aug | 23 |
| Lake Iliamna | Iliamna[d, e] | 19 | 11[e] | 6–8.3[e] | 31–52 | 5–17 | June–Oct | d |
| | Nonvianuk[f] | 14 | 10 | 7.0–7.5 | 30–45 | na | July, August | 3 |
| | Battle[f] | 11 | 11 | 6.6–7.7 | 25–80 | 17 | July, August | 3 |
| Lake Clark | Clark[g] | 16 | 12 | 7.5–7.7 | 62 | 1–4 | May–Sept | g |
| | Kijik[g] | 10 | 11 | 7.2–7.7 | 71, 47 | 9–18 | May–Sept | g |
| | Upper Tazimina[i] | 9 | 11 | 6.6–7.5 | 22 | 5–13 | Aug, Sept | 4 |
| | Lower Tazimina[i] | 11 | 11 | 6.6–6.9 | 22–25 | na | August | 23 |
| Naknek | Naknek, W Basin | 16 | 11 | 8, 7.4[j] | 139 | 5 | May–Sept | l |
| | Naknek, Iliuk Arm | 9 | 12 | 7.8 | 144 | 0.5 | May–Sept | l |
| | Brooks | 11 | 12 | 7.8, 7.3[j] | 60–78 | 10 | May–Sept | l |
| | Idavain[f] | 14 | 11 | >8 | >80 | na | Aug | 1 |
| Alaska WQC | | <20[m] | 7–17 | 6.5–8.5 | na | na | | |

a A 60-year data set was used for temperature, SC, and Secchi depth; pH is based on 8 measurements. Data from Alaska Salmon Program 2011; USGS gages 15303000 and 592100158493000

b USGS gage 15302000, n = 12–20 depending on parameter

c Personal communication, K. Bartz, NPS, n = 2–3. Additional SC (44 µS/cm, n = 19) from Chamberlain 1989. Additional temperature, DO, pH, and Secchi depth from Russell 1980 (n = 1, August 1978)

d Alaska Salmon Program 2011, 60-year data set for temperature, SC, depth; pH = 7.3 from Burgner et al. 1969; Pebble Limited Partnership (PLP) 2011 (n = 12) for all

e PLP found SC to range from 0 to 1,448 µS/cm, with medians of 37–50 µS/cm at five sites, and DO to range from 8–14 with median of 11

f Data from LaPerriere 1996 and LaPerriere 1997

g Data from Moore and Shearer 2011, number of samples not provided and from Russell 1980 (n = 1). Lake Clark temperature collected daily Sept 2006–Aug 2010. Additional pH and SC for Kijik and Lachbuna lakes from Chamberlain 1989 (Kijik n = 46, Lachbuna n = 18). Additional temperature, DO, pH, and Secchi depth from Russell 1980 (n = 1, August 1978)

h USGS gages 15298010 and 15298010

i Data from Baldrige and Trihey 1982, n = 3 and Russell 1980, n = 1

j Data from Baldrige and Trihey 1982, n = 3. Additional pH and SC from Chamberlain 1989 (n = 20)

k USGS gage 15299000

l Data from Moore and Shearer 2011. Temperature collected daily Sept 2008–July 2010 in the West Basin; all other data is from 2010 only. Additional pH data at the West Basin and Brooks Lake from Burgner et al. 1969

m Alaska WQC require < 20°C in waters supporting freshwater aquatic life, except no more than 15°C on migration routes and rearing areas and no more than 13°C in spawning areas and for egg and fry incubation. http://dec.alaska.gov/commish/regulations/pdfs/18%20AAC%2070.pdf

**Table 19.6** Mean alkalinity and major ions in lakes of Bristol Bay watersheds. Ca = calcium, Mg = magnesium, Na = sodium, $HCO_3$ = bicarbonate, $SO_4$ = sulfate, $NO_3$ = nitrate. Chloride and potassium, not shown, were usually < 0.5 mg/L, except in Upper Tazimina and Iliamna Lakes where concentrations to 1 mg/L were documented. n = number of samples

| Watershed | Lake | Alkalinity (mg/L as $CaCO_3$) | $Ca^{2+}$ (mg/L) | $Mg^{2+}$ (mg/L) | $Na^+$ (mg/L) | $HCO_3^-$ (mg/L) | $SO_4^{2-}$ (mg/L) | $NO_3^-$ (mg/L) | n | Data source |
|---|---|---|---|---|---|---|---|---|---|---|
| Wood River | Aleknagik | 14[a] | 5[a] | 1[a] | 1[a] | 16 | 3 | < 0.5 | 1, 52[a], 10[a] | USGS gage 592100158493000 (anions, n = 1); Burgner et al. 1969 (alkalinity, n = 52 and cations, n = 10) |
|  | Nerka | 12 | 5 | 1 | 0.5 | na | na | 0.2 | 14 | Burgner et al. 1969 |
| Upper Nushagak | Tikchik | 22 | 9 | 1 | 1 | 27 | 6 | 0.4 | 12 | USGS gage 15302000 |
|  | Nuyakok | na | 8 | 2 | 0.4 | na | na | 0.1 | 2 | Burgner et al. 1969 |
| Mulchatna | Turquoise | na | 7 | 0.5 | 1 | 20 | 3 | < 0.1 | 19 | Chamberlain 1989 |
|  | Lower Twin Lakes | na | 11 | 0.5 | 1 | 29 | 7 | 0.2 | 21 | Chamberlain 1989 |
| Lake Iliamna | Iliamna | 15 | 5–6 | 1 | 1, 2[d] | 19[e] | 4[d], 6[a,e] | < 0.2[a,d] | 1[e], 45[a], 27[a], 12[d] | USGS gage 59455154070900 (n = 1); Burgner et al. 1969 (alkalinity, n = 45 and cations, n = 27); PLP 2011 (n = 12) |
|  | Nonvianuk | 9[a] | 5 | 1 | 2, 3[a] | 12 | 1 | 0.02[a] | 3, 1[a] | Burgner et al. 1969; LaPerriere 1996, 1997 |
|  | Battle | 2 | 4 | 1 | 2 | 2 | 16 | na | 3 | LaPerriere 1996, 1997 |
|  | Clark | 20 | 7[a] | 1[a] | 1[a] | 25 | na | 0.8 | 19 | USGS gage 15299000; Burgner et al. 1969 (n = 10) |
|  | Kijik | 34[b] | 8 | 1 | 1 | 18 | 5 | 2 | 11 | Chamberlain 1989 |
| Lake Clark | Upper Tazimina | 11[c] | 3 | 0.4 | 1,2[c] | 7 | 5,6[c] | 0.8 | 19 | Chamberlain 1989 |
|  | Lower Tazimina | 12 | 3 | 0.4 | 4, 6[c] | 8 | 4, 8[c] | 0.7 | 2, 20 | Baldrige and Trihey 1982 (alkalinity, n = 2); Chamberlain 1989 (n = 20) |
| Naknek | Naknek, excluding Iliuk arm | 29 | 17[a,f] | 3, 4[a] | 7[f], 10[a] | 37[f] | 29[f] | < 0.1 | 254, 19, 3[f] | Burgner et al. 1969 (alkalinity, n = 254 and major ions, n = 19); LaPerriere 1996, 1997 |
|  | Brooks | 27 | 9[a,f] | 2[a,f] | 4[a,f] | 35[f] | nd[f] | 0.01[a] | 248, 20, 3[f] | Burgner et al. 1969 (alkalinity, n = 248 and cations, n = 20); LaPerriere 1996 |
|  | Idavain | na | 6 | 1 | 3 | 26 | 2 | na | 3 | LaPerrier 1996 |

[a] Data and number of samples from Burgner et al. 1969

[b] Data from Russell 1980, n = 1

[c] Data from Baldrige and Trihey 1982, n = 3. Note that potassium for Upper Tazimina is 6 mg/L and for Lower Tazimina is 3 mg/L, n = 3.

[d] Data from PLP 2011 Appendix B, n = 12

[e] Data from USGS gage site

[f] Data from LaPerriere 1996; also note Cl is 19 mg/L in Naknek Lake, 9 mg/L in Brooks Lake, 4 mg/L in Battle Creek, and 4–9 mg/L in other Alaska Peninsula lakes, compared to 0.5–1 mg/L in most Bristol Bay watershed lakes

(D. Young, NPS, pers. comm.). In summer, the surface can reach 16°C with temperatures at depth near 4°C year round (Chamberlain 1989, Shearer and Moore 2010, Moore and Shearer 2011). Thermal stratification occurs, and turnover events have been observed in June and October, with mixing to between 20–50 m (Weeks 2003).

Late freeze-ups, warm winter water, and cool summer water temperatures suggest the weather is trending toward warmer and wetter. Warm and wet conditions on montane glaciers drive cold water into the northern basin of the lake, and peak lake level occurs in July. The summer of 2010 was cool and wet, providing both glacial melt and precipitation that resulted in cooler lake surface temperatures (9–11°C) (Moore and Shearer 2011). The lake is fully oxygenated (Weeks 2003) with dissolved oxygen increasing with depth and colder water, and can be supersaturated (Moore and Shearer 2011). The depth of oxygen and the euphotic zone is due to wind events that frequently prevent stratification.

Glacial flour severely reduces clarity in the upper lake. The Tlikakila River contributes about one million tons of sediment annually, mostly as glacial flour. It is this sediment that forms the delta, separating Lake Clark from Little Lake Clark (Figure 19.4). Currant Creek, also in the upper lake, contributes sediment to a lesser degree (Chamberlain 1989). A turbidity gradient extends from where the glacial rivers enter the upper lake, gradually clearing toward the lower lake. At the peak, turbidity is pushed furthest down-lake and the euphotic zone is most reduced. In 2008, the upper lake had clarity of about one meter and the lower lake had about six meters (Moore and Shearer 2011).

In addition to glacial input, volcanic eruptions can deposit ash and cause reduced clarity in the lake. After Mt. Redoubt, 30 miles due east of Lake Clark, erupted in 2009, clarity across Lake Clark dropped to about three meters, obscuring any gradient from glacial flour (Figure 19.7). The ash fell directly on the lake, but also on the surrounding hills and was carried into the lake as runoff the following spring. Sixteen months after the original eruption, the upper lake still remained at one meter of clarity and the lower lake had only improved to about three meters (Moore and Shearer 2011).

Prior eruptions of Mt. Redoubt in 1989–1990 also severely limited light penetration. This had the effect of decreasing the water column available for phytoplankton photosynthesis by more than 70% (Weeks 2003).

## Chemical Water Quality

Lake Clark is calcium-bicarbonate type ($Ca^{2+}$ as the dominant cation, $HCO_3^-$ as the dominant anion) (Table 19.6), however, alkalinity is low (20 mg/L); the pH is neutral (~7.6), decreasing with depth (Table 19.5) (Burgner et al. 1969, Stottlemeyer and Chamberlain 1987, Chamberlain 1989, Moore and Shearer 2011).

Differences in geology provide Lake Clark with generally higher concentrations of major ions than the Wood River Lakes but lower concentrations than Alaska Peninsula lakes. Geology also drives changes between the upper and lower sections of the lake. At the upper lake, tributaries flow from the Alaska Range, which has calcite, dolomite, and calc-silicate bedrock and provides $Ca^{2+}$, $Mg^+$, and bicarbonate. This material provides some buffering and hardness to the water. The southeast shores of Lake Clark and Sixmile Lake are in the Aleutian Range. Bedrock here is primarily K/Na/Al silicates with less calcite and dolomite so there is less material to provide alkalinity. Specific conductivity, reflecting the ions, was generally around 50 µS/cm at Lake Clark and surrounding smaller lakes with extremes at Tazimina Lakes (20–25 µS/cm) (Table 19.5) and at small, deep, clear Portage Lake (354 µS/cm) (Cohn 2009).

## Nutrients

Lake Clark is somewhat higher in nitrate (0.8 mg/L) than other regional lakes (Burgner et al. 1969, Chamberlain 1989). Flooding of alluvial plains—extensively covered with alder as glaciers have receded—could contribute N to upper Lake Clark (Cohn 2009), while observed seasonal decreases in N may be due to biological uptake. Phosphorous was below the limit of detection; phosphate was not measured (USGS gage 15299000).

Studies at Lake Clark again highlight the dynamic nature of nutrient cycling. Lake Clark (anadromous) may be limited by either N or P depending on season and depth (Wilkens 2002). Despite the higher nitrate (measured as the anion, not as N) and low phosphate, Lake Clark was observed to be N limited, although P stimulated production at some lake locations (Chamberlain 1989). This could change if Lake Clark responds over time in the same manner of Kontrashibuna Lake, with N and C increasing over time while P remains stable (Cohn 2009) (see the *Glacial Rivers with Lakes* section later on in this chapter).

## Iliamna Lake

Iliamna Lake is the largest lake in Alaska at 3,175 km² and 124 km long, and is the largest sockeye salmon rearing lake in the world. The eastern end of the lake in the Aleutian Range is deep (over 275 meters) and contains numerous islands. The lake becomes shallower in the western end where it borders the flat tundra of the Nushagak Lowlands.

## Temperature, Oxygen, and Clarity

Iliamna is a clear lake with 5–17 meters of clarity (Alaska Salmon Program 2011, PLP 2011). Dissolved oxygen is near saturation at the surface (8–14 mg/L) and frequently saturated at depth (PLP 2011).

From data collected at the lake since 1962, a long-term warming trend is evident (Rich et al. 2009, Kendall et al. 2010, Alaska Salmon Program 2011). There was a recent cooling period (2006–2010): the summers of 2006, 2009, and 2010 had the 11th, 12th and 16th coldest air temperatures on record, respectively, and July 2010 was the wettest on record, with August 2006 being the second wettest (Alaska Salmon Program 2010, 2011). However, the annual mean air temperature between 1962 and 2006 has increased by about 1.8°C (Kendall et al. 2010), spring air temperatures have warmed by 3.3°C (Rich et al. 2009), water temperatures are about 3°C warmer (Alaska Salmon Program 2007), and ice-out occurs about 10 days earlier than in the 1960s (Alaska Salmon Program 2010, 2011). In 2015 and 2016, residents reported that the lake froze for 1–4 weeks and ice did not get thick enough to travel on; in 2016 residents boated throughout the winter (K. Jensen, pers. comm.).

## Chemical Water Quality

Consistent with an oligotrophic lake, Iliamna Lake waters are dilute with low buffering capacity. Samples collected in 1961–1962, 1976, and 2005–2007 show the lake has low alkalinity (< 18 mg/L), low SC (< 80 µS/cm), and low total dissolved solids (TDS) (< 80 mg/L) with neutral pH (~7.0) (USGS site 594555154070900, Burgner et al. 1969, PLP 2011). These same studies showed that major cations and nitrate were similar to Lake Aleknagik—that is, calcium-bicarbonate dominant with most major ions in low concentrations—with the exception of slightly higher sulfate in Iliamna (Table 19.6).

Minor elements were in low concentration (Burgner et al. 1969, PLP 2011) (Table 19.7). In sampling across five sites, all samples remained well below any WQC. For example, total Al had a maximum 46 µg/L with medians < 11 µg/L for nearly all sites (PLP 2011), while WQC is 87 µg/L.

Trace elements were also in low concentrations (PLP 2011) (Table 19.7). Copper exceeded WQC, presuming a hardness-based WQC at 2.1 µg/L (ADEC 2008), and only at one site. Copper at Pile Bay was well below criteria (< 1.5 µg/L) for nine samples and over criteria for three samples (2.2 µg/L, 3.0 µg/L, 4.4 µg/L) all in May or June. Lead (Pb) was below detection limit (< 0.02 µg/L) with the exception of a single sample at Northeast Bay at 0.65 µg/L (PLP 2011); the elevated concentration is not related to runoff and may be a bad data point.

In dilute streams, it is common to see brief spikes in minor and trace element concentrations at snowmelt and with rainy periods as runoff washes erosional material into surface waters. This pattern was not evident in Iliamna Lake; minor elements such as Fe and Al were not elevated with runoff events in May; lake turnover would be expected to make any runoff signal difficult to detect.

## Nutrients

Sockeye salmon may provide ~25% of the N to nursery lakes in general in Alaska, and are important sources of phosphorous (Naiman et al. 2002).

P in salmon carcasses contributed 2%–60% of the P in Iliamna Lake (Rogers 1977a). However, total P was less than 0.1 mg/L in all surface water samples (Donaldson 1967, PLP 2011) and at depth, P was not detected above 0.1 mg/L in 1964–1967, but was reported to be 0.2–0.7 mg/L in fall (August 2005) at the Upper Talarik Creek outlet and at 0.2 mg/L (September 2007) at Pile Bay (PLP 2011). The reason for these differences is not known.

Hypothetically, when salmon carcasses rot, P is released into the water where it is taken up by periphyton and phytoplankton which act as a reservoir of nutrients that move into the food web the following year (Donaldson 1967, Rogers 1977a, Gende et al. 2002, Rinella et al. 2013). Visible algal growth on littoral substrate was observed only after peak escapement years, the high return years for salmon (Donaldson 1967).

P leaves the system through detrital settling and out-migration of salmon smolts. Studies conflict on the degree to which nutrients are exported by salmon. Although Iliamna Lake salmon runs in 1963–1965 were estimated to lose some salmon-derived P via out-migrating smolt (Donaldson 1967, Rogers 1977a), later studies estimated that nearly all of the MDN remained (Naiman et al. 2002). Adult sockeye from the years 1957–1998 were estimated to bring 70,000,000 kg of N and 3,300,000 kg of P into Iliamna Lake, and about 15 times more N and P deposited than were removed by out-migrating smolt (Naiman et al. 2002).

## Alagnak River Lakes

The drainages above flow from Lake Clark through the Newhalen River into Iliamna Lake, out through the Kvichak River to Kvichak Bay. A parallel drainage is present south of Iliamna Lake at the top of the Alaska Peninsula. Alaska Peninsula lake systems straddle the Aleutian Range and the Nushagak Lowland physiographic regions and flow into Kvichak Bay (Figures 19.1, 19.5).

The Alagnak River lakes are high-altitude lakes (192–254 meters elevation) in the Aleutian Range, located north of Naknek Lake and west of McNeil River (LaPerriere 1997). The two large lakes are Kukaklek Lake (176 km$^2$), which is the source of the Alagnak River, and Nonvianuk Lake (132 km$^2$) where the Nonvianuk River begins, entering the Alagnak River within about 30 km (Figure 19.5).

At the head of Kukaklek is Battle Lake (13 km$^2$) and at the head of Nonvianuk, a nonglacial lake, is glacially influenced Kulik Lake (28 km$^2$) (LaPerriere 1997). The Alagnak River joins the Kvichak River near its mouth (Weeks 1999) (Figure 19.5). The most unusual of the four lakes is Battle Lake, a calcium-sulfate lake fed by a mineralized tributary, Iron Springs Creek.

Basic water quality (DO, alkalinity, pH, major cations; n = 7–10) and temperature data (n = 17–20) was collected at Kukaklek and Nonvianuk Lakes in the summer of 1961 (Burgner et al. 1969). More extensive baseline data was collected in 1990–1992 at four lakes in the Alagnak River drainage (Nonvianuk, Battle, Kukaklek, and Kulik) and at six in the Naknek River drainage (discussed in the next section) (LaPerriere 1996).

### Temperature, Oxygen, and Clarity

Despite difference in sizes, DO, SC and pH were uniform throughout the water column of all four of the lakes studied over three summers. Temperature was generally uniform with depth, indicative of turnover, but thermoclines did develop inconsistently in each of the lakes except Battle Lake (LaPerriere 1996). Generally, the lakes maintained oxygenated but warm bottom conditions in summer, subject to annual variations (e.g., 10°C at 55 meters at Battle Lake in 1990 and 5°C in 1991) (LaPerriere 1996). Most of the lakes, including Battle Lake, are extremely clear (Table 19.5).

### Chemical Water Quality

The influence of geology is striking. The Alagnak lakes had very low alkalinity (~10 mg/L) and low Ca$^{2+}$, Mg$^+$, Na$^+$, and bicarbonate relative to Naknek and Brooks Lakes (alkalinity ~27 mg/L) (Burgner et al. 1969), possibly due to granitic bedrock material (Weeks 2003). Despite low alkalinity, pH is neutral even at Battle Lake where alkalinity is only two mg/L (LaPerriere 1997) (Table 19.6).

All the lakes were calcium dominant, but not all were bicarbonate dominant (Table 19.6). Nonvianuk Lake is calcium-bicarbonate, but with high chloride (4 mg/L), Kukaklek is calcium-bicarbonate-chloride (Cl$^-$ at 6 mg/L), Battle Lake is calcium-sulfate dominant, and Kulik Lake is calcium dominant but with roughly equal proportions of bicarbonate (8 mg/L), sulfate (5 mg/L), and chloride (4 mg/L) (LaPerriere 1997). This is a distinctly different pattern than the calcium-bicarbonate lakes of the Naknek River system and of all the lakes previously discussed in this chapter.

Kulik Lake is at the head of Nonvianuk Lake. Streams entering Kulik Lake exhibited low bicarbonate (0–17 mg/L) and high sulfate (0–14 mg/L) relative to streams entering Nonvianuk Lake (bicarbonate 17–25 mg/L, sulfate < 0.3 mg/L), although pH remained neutral (6.8–7.5) and SC was low (15–58 µS/cm) (LaPerriere 1996).

Battle Lake, reflecting incoming tributaries, had proportionally higher chloride and sulfate (sulfate at 16 mg/L represents 71% of total anions) due to very low bicarbonate (Table 19.6) (LaPerriere 1996, 1997). One tributary, Iron Springs Creek, had a pH of four with high aluminum and no signs of aquatic life; Iron Springs Creek itself issued from Iron Springs Lake, an acidic lake (pH 3.6) with elevated sulfate, aluminum, and zinc (LaPerriere 1996, 1997). The sulfate that enters Battle and Kulik lakes is probably due to weathering of sulfide rock material, although a geothermal spring also appeared to enter Kulik Lake (temperature 16°C, SC 416 µS/cm, bicarbonate 16 mg/L, sulfate 195 mg/L).

### Nutrients

At the Alagnak River lakes, primary productivity was related to P, which was generally lower than P in Naknek River lakes and in much lower concentration than N. Total P was < 0.005 mg/L and total N was frequently 0.2–0.4 mg/L; the higher N may be due more to low primary productivity rather than to high N inputs (Weeks 2003). In Battle Lake, a depth profile showed nutrients were highest at the surface and at 5–20 m depth (LaPerriere 1996).

## Naknek River Lakes

The Naknek River watershed is 9,707 km$^2$ and primarily fed by Naknek Lake (610 km$^2$), the largest lake in the NPS and the third largest lake in Alaska (Table 19.3). High elevation (~500 meters) small lakes flow into low elevation (~32 meters) Coville and Grosvenor Lakes, which exit to the Savonoski River and into the Iliuk Arm of Naknek Lake (10 meters elevation) (LaPerriere 1997). Brooks Lake (anadromous) and Idavain Lake (nonanadromous) flow over waterfalls into Naknek Lake (Figure 19.5).

Naknek Lake has multiple basins with depths that range from 170 meters in Iliuk Arm and the North Arm to 75 meters in the East and West Basins (Moore and Shearer 2011).

Proximity to the coast influences water chemistry, as do glaciers and volcanoes. Along the Aleutian Range are six volcanoes that have been active in the past century and six more that are considered active today (Weeks 1999). The massive 1912 Novarupta eruption, the largest of the twentieth century, created the ash-filled Valley of Ten Thousand Smokes and profoundly influenced the water chemistry of Naknek Lake. Volcanoes can increase soil fertility, and contribute nutrients, toxic elements, ash, and sediment.

### Temperature, Oxygen, and Clarity

Due to frequent coastal winds, water mixes deeply and temperature and oxygen are uniform with depth (LaPerriere 1997, Weeks 1999), although some stratification may occur in deeper basins, and in Naknek Lake the North Arm warms more slowly in the spring and cools more slowly in the fall than the West Basin (Moore and Shearer 2011).

Temperatures were taken at seven Naknek River system lakes in 1990–1992, and Naknek Lake has had consistent temperature monitoring only since 2008. Lakes usually freeze from December to early May, and summer surface water temperatures reach 12–16°C, with depths near 7°C (LaPerriere 1997, Moore and Shearer 2011). Wind events can cool surface temperatures, as in 2010 when winds dropped temperatures from 10°C to 6°C at the North Arm of Naknek Lake (Moore and Shearer 2011). As with Lake Clark, the summer of 2010 was cool (Moore and Shearer 2011), although temperatures were higher again in 2013–2015 (www.wunderground.com for King Salmon, AK).

The Iliuk Arm of Naknek Lake, which receives glacial sediment and ash flows, has only one meter of clarity compared to the North Arm at about 9 meters and 9–17 meters in Brooks Lake (Weeks 1999, Moore and Shearer 2011). Similar to the turbidity gradient at Lake Clark, Naknek Lake has a turbidity gradient from the Iliuk Arm, decreasing toward the main body of the lake.

### Chemical Water Quality

Alaska Peninsula lakes are higher in cations than other lakes in the Kvichak and Nushagak basins (Burgner et al. 1969). In addition to geology, in this area major ions are driven by marine influences (Na$^+$, Cl$^-$, sulfate) and geothermal sources (sulfate). As a testament to the influence of the sea, Na$^+$ is in higher concentration than Mg$^{2+}$; major ions follow Ca$^{2+}$ > Na$^+$ > Mg$^{2+}$ > K$^+$ (Table 19.6) and decrease with lake elevation (LaPerriere 1996, 1997). Despite the higher sodium, most Naknek River lakes are calcium-bicarbonate types (bicarbonate of 25–35 mg/L, other anions < 7 mg/L). Naknek Lake had neutral pH (Table 19.5) but was the *saltiest* of the lakes. It is a calcium-bicarbonate sulfate type with high concentrations of major ions relative to other lakes (means as bicarbonate 37 mg/L, sulfate 29 mg/L, chloride 19 mg/L) (LaPerriere 1997). SC (mean 139 µS/cm) was constant with depth, and higher than other lake systems throughout the Kvichak and Nushagak basins.

Alkalinity is 29 mg/L (mean), indicating good buffering ability. Elevated alkalinity may be a reflection of microbial communities as well as geology. Sulfate- and nitrate-reducing bacteria may thrive in Naknek Lake (Schindler et al. 1980), raising alkalinity to the highest in the region.

Elements such as silica (Si), Al, Pb, Cu, molybdenum (Mo), Cr, As, and fluoride (F) may pulse into lakes after initial volcanic eruptions (LaPerriere 1996, Weeks 1999, Munk et al. 2010); sediment cores show that lakes may take 75–100 years to return to a pre-eruption state (Munk et al. 2010). In the long term, weathering may continue to release elements. Naknek Lake shows a steady increase in minor and trace element accumulation in sediment; with about a 50% increase in trace element accumulation since pre-1800 A.D. Brooks Lake had a steady accumulation until around 1600 A.D. followed by a decrease. However, this accumulation has not resulted in elevated metals in the water column. Analysis for a suite of trace elements observed they were below detection limits, although the most sensitive analytical methods were not applied (LaPerriere 1996). Other elements [Al, Fe, Mn, barium (Ba), and Si, strontium (Sr)] were detected, but below

WQC for freshwater aquatic life; boron (B) was only found at Naknek Lake, and may come from volcanic seeps (LaPerriere 1996).

## Nutrients

Alaska Peninsula lakes are historically N limited (Brooks, Naknek, Becharof, Coville) or P limited (Grosvenor, Kulik) (Goldman 1960, LaPerriere 1996). Nitrate was observed to be lower in Alaska Peninsula lakes than in Wood River lakes or lakes from the upper Kvichak basin (Burgner et al. 1969). In one study, Naknek and Brooks Lake were limited by $Mg^{2+}$ and nitrate until midsummer, when P became the limiting nutrient (Goldman 1960). In later studies of Naknek and Brooks Lakes, total N was 0.24 mg/L and 0.14 mg/L respectively, and total P was < 0.005 mg/L (LaPerriere 1996, from three samples 1990–1992). Primary production was lower in Naknek Lake than Brooks Lake (Weeks 1999), but high sodium might limit phytoplankton in Brooks Lake (Burgner et al. 1969). Given the evidence of increased terrestrial N at Kontrashibuna Lake with glacial retreat, the ratio of N and P could be changing at Naknek Lake as well.

A study of Alaska Peninsula lakes found the mass of zooplankton was larger than the mass of algae in several lakes (LaPerriere 1996). This is not uncommon in oligotrophic lakes, where microbes and extremely small plankton that should be counted as phytoplankton are small enough to pass unmeasured through filtering equipment. This *inverted pyramid* of phytoplankton was only observed in sockeye salmon nursery lakes (Brooks, Kukaklek, Nonvianuk) and two small lakes where sockeye do not reproduce (Idavain and Murray Lakes). Idavain Lake, which was only sampled for a single summer, had an order of magnitude more zooplankton (dry weight as $mg/m^3$) and at least twice as much phytoplankton ($mg/m^3$ as chlorophyll a) than any of the other 10 lakes (LaPerriere 1996). Productivity in nonanadromous lakes may be boosted by nitrogen entering from shallow kettle ponds; in Idavain Lake and other nonanadromous lakes with high organic content the amount of $\delta^{15}N$ in sediments ($\sim 4\ ^0/_{00}$) was similar to that found in many anadromous systems, and zooplankton values were high ($\sim 9\ ^0/_{00}$) (Cohn 2009).

Ash from volcanic eruptions did not affect long-term water quality at Brooks Lake, but did cause spruce trees around the lake to go through accelerated growth for a few years, possibly related to Novarupta ash that supplied P, magnesium, calcium, and other elements to the soil (Weeks 1999).

## Nonanadromous Lakes

There are some lakes with high water quality that do not host salmon or other anadromous species due to physical barriers such as waterfalls or rapids.

### Tazimina Lakes

Upper and Lower Tazimina Lakes are an example of nonglacial mountain lakes. Set in glacial outwash, a shallow aquifer ensures groundwater and surface runoff, rather than glacial melt, provides much of the water in the drainage. The Tazimina Lakes also serve as sediment settling basins; while clarity is only 5 m in the Upper Tazimina, it is 12 meters in the Lower Tazimina (Russell 1980). Lower Tazimina Lake is the source of the clear-running Tazimina River, which enters Sixmile Lake at the terminus of Lake Clark (see *Clear Rivers* section in this chapter).

The lakes are clear and cold with high DO and neutral pH and seasonal fluxes in turbidity (Table 19.5), but have the most dilute waters (SC 20–25 µS/cm) of all the lakes reviewed (Table 19.5) and have a diluting effect on Lake Clark (Stottlemeyer and Chamberlain 1987, Chamberlain 1989). Although the Tazimina Lakes, Tazimina River, and Sixmile Lake are higher in $K^+$ and $Na^+$ than the rest of Lake Clark, they have the lowest $Ca^{2+}$ and $Mg^{2+}$, lowest minerals, and low bicarbonate and alkalinity—second only to Battle Lake (Baldrige and Trihey 1982, Chamberlain 1989) (Table 19.6). With little buffering capacity, they are susceptible to changes in water quality with small changes in the environment. Of three samples collected, all trace elements were below the limit of detection, with the exception of copper (3 µg/L) and zinc (5–6 µg/L) (Baldrige and Trihey 1982) (Table 19.7).

The nutrient profiles of glacial and nonglacial lakes differ. A study of three nonanadromous water bodies—Kontrashibuna Lake (deep, glacially fed mountain lake) and Portage and Lower Tazimina Lakes (deep, nonglacial mountain lakes)—found lower $\delta^{15}N$ in the glacially fed lake, although the total carbon to nitrogen ratio was about the same for all three (Cohn 2009).

## Kontrashibuna Lake

Kontrashibuna Lake, in the Chigmit Mountains, is a 21 km long, 107 meters deep mountain lake with only 2 meters of clarity. As with the Tazimina Lakes, Kontrashibuna water quality is clear throughout the summer and has neutral pH, but is dilute with low SC (40 uS/cm); it has elevated $Cl^-$ (1.2 mg/L), but low $Ca^{2+}$ and $Mg^{2+}$ (Chamberlain 1989).

Kontrashibuna Lake, like Lake Clark, may be limited by either N or P depending on season and depth (Wilkens 2002). Portage Lake and Kontrashibuna Lake in the mountains around Lake Clark were studied in 2004 (Cohn 2009). The watershed for Portage Lake, the deepest clear lake around Lake Clark, is much smaller than that feeding Kontrashibuna Lake (10 km$^2$ versus 507 km$^2$), yet N as nitrate + nitrite is higher (1.6 mg/L at Portage; 0.5 mg/L at Kontrashibuna). Observed over time, total N and total C have been increasing at Kontrashibuna Lake since about 1975, while remaining unchanged at Portage Lake since the 1860s. Total P has remained the same at both lakes (~0.07%) consistently. The C:N ratio suggests the N in Kontrashibuna Lake may be from diatoms in lake surface waters (Cohn 2009).

## Small Lakes and Ponds

Innumerable small lakes are part of the Bristol Bay freshwater environment, providing water storage, groundwater recharge, acting as sediment-settling basins, and harboring aquatic life. However, there is not much information on these systems. Headwater lakes and tundra ponds are discussed here; sediment-settling lakes are discussed in the *Glacial Rivers with Lakes* section of this chapter.

## Glacial Mountain Lakes

Between 2012 and 2014, the NPS collected basic water quality once per year on five small lakes of LACL that are within the Mulchatna watershed (K. Bartz, unpub. data). Turquoise, Upper Twin Lake, and Lower Twin Lake are high elevation (600–760 m), mountain glacially fed lakes and the source of the Mulchatna and Chilikadrotna; Turquoise and Lower Twin also have data from 1985–1987 (Chamberlain 1989). Fishtrap and Snipe Lakes are small, shallow lakes in the Chilikadrotna wetland complex below Twin Lakes (discussed in the next section) with no prior data.

The mountain lakes are cold, dilute, calcium-bicarbonate lakes with neutral pH throughout the water column (Tables 19.5 and 19.6). Turquoise Lake was more turbid than the other lakes, although it had lower TDS (38 mg/L compared to about 62 mg/L). Turbidity varied between years for the Twin Lakes. Upper Twin Lake acts as a settling pool for glacial sediment, and Lower Twin Lake was consistently clearer.

## Wetland Lakes

The two wetland lakes, at 28 meters (Fishtrap) and 20 meters (Snipe) deep, were warmer (15°C at the surface in 2013–2014) than the mountain lakes and consistently had Secchi readings near 6 meters. All five lakes sampled by NPS were saturated in dissolved oxygen throughout the water column with the exception of Fishtrap Lake, which had 60–70% saturation at 15–20 meters depth. At Fishtrap Lake, pH was observed to decline with depth, from pH 7.6 at the surface to pH 6.9 at the bottom in 2013, and from pH 8.1 at the surface to pH 7.3 at depth in 2014. Snipe Lake was the most alkaline of the five NPS lakes (pH > 8), and had the highest TDS (88 mg/L) and SC (138 µS/cm). The drivers behind water quality differences such as the higher alkalinity in Snipe Lake and the change in pH with depth at Fishtrap Lake have not been analyzed yet.

## Tundra Ponds

Innumerable small ponds cross the tundra landscape with no stream outlet or only temporally connecting to streams and rivers during spring and fall floods when formerly disconnected areas become a large-scale wetland. These ponds play a role in storing water and recharging groundwater aquifers. A recent EPA guidance clarified when these types of ephemeral and isolated pools could come under the Clean Water Act (CWA), determining that if they, or the wetlands they were in, could be traced to navigable waters, then the CWA applied (Federal Register 2014).

In a study by Rains (2011), the water chemistry of perched and *flow-through* ponds was reported. Perched ponds sit above the water table, collect runoff, and lose water through evaporation and groundwater recharge. Flow-through ponds receive groundwater in addition to surface water, tend to remain full through most of the year, and have vegetated banks. Perched ponds recharge groundwater at much higher rates than flow-through ponds or the broader landscape. Perched ponds had extremely dilute water (SC 3 µS/cm), reflecting contributions from precipitation, relative to water in

**Table 19.9**  Mean alkalinity and major ions in rivers of Bristol Bay watersheds. Ca = calcium, Mg = magnesium, Na = sodium, $HCO_3$ = bicarbonate, $SO_4$ = sulfate, $NO_3$ = nitrate chloride and potassium, not shown, commonly < 0.5 mg/L, except Upper Tazimina and Iliamna Lake to 1 mg/L. n = number of samples

| Watershed | River/Creek | Alkalinity (mg/L as $CaCO_3$) | $Ca^{2+}$ (mg/L) | $Mg^{2+}$ (mg/L) | $Na^+$ (mg/L) | $HCO_3^-$ (mg/L) | $SO_4^{2-}$ (mg/L) | $NO_3^-$ (mg/L) | DOC (mg/L) | n | Data source |
|---|---|---|---|---|---|---|---|---|---|---|---|
| Wood River | Silver Salmon Creek | 16 | 6 | 1 | 2 | 19 | 3 | < 0.5 | na | 1 | USGS gage 15303010 |
| Mulchatna | Swan River | na | 8 | 2 | 2 | na | 2 | < 0.01 | na | 4-17 | BBNA unpublished |
| | Koktuli River | 19 | 5 | 1 | 2 | 22 | 2 | 0.3 | na | 24 | BBNA unpublished |
| Lower Nushagak | Nushagak River | 20 | 7 | 1 | 2 | 24 | 4 | 0.3 | 2.5 | 18[a] | USGS gage 15302500 |
| | Tlikakila River | 21 | 7-8 | 1 | 1 | 27 | 4-5 | 1 | 0.4 | 11, 9 | USGS gage 15297970, Chamberlain 1989 |
| | Currant Creek | na | 7 | 1 | 1 | 14 | 7 | 1 | 1[b] | 11 | Chamberlain 1989 |
| | Tanalian River | 12 | 6-7 | 1 | 1-2 | 12-15 | 7 | 0.6 | na | 5, 3 | USGS gages 15298000 and 15298010, Chamberlain 1989 |
| | Kijik River | 26 | 11 | 1 | 2 | 25 | 7 | 0.7-1 | 0.6 | 1-9[c] | USGS gage 15297990, Chamberlain 1989 |
| Lake Clark | Chulitna River | 32 | 10 | 2 | 2 | 38 | 6 | 0.2 | 5 | 22 | USGS gage 15298040 |
| | Nikabuna tributary | 27 | 6 | 2 | 2 | 24 | 1 | 1 | 4 | 3 | Zamzow 2011 |
| | Tazimina River, upper | 13 | 2 | < 0.5 | 2 | na | 6 | na | na | 1 | Baldrige and Trihey 1982, river mile 11.6 |
| | Tazimina River, lower | 13 | 2 | < 0.5 | 2 | na | 7 | na | na | 1 | Baldrige and Trihey 1982, river mile 1.7 |
| | Kvichak River | 20[d] | 6 | 1 | 2 | 25[d] | na | 1[d] | na | 19 | Levelock Village Council Environmental Program |
| Lake Iliamna | Yellow Creek | na | 5 | 2 | 4 | na | na | na | na | 19 | Levelock Village Council Environmental Program |
| | Alagnak River | na | 5 | 1 | 3 | na | na | na | na | 18 | Levelock Village Council Environmental Program |
| Naknek | Battle Creek | na | 4 | 1 | 2 | 5 | 7 | na | na | 3 | LaPerriere 1996 |
| | Ukak River | na | 33 | 8 | 24 | 51 | 70 | na | na | 1 | LaPerriere 1996 |
| | Savonoski River | na | 21 | 5 | 4 | 32 | 25 | na | na | 1 | LaPerriere 1996 |

[a] For dissolved organic carbon (DOC), n = 1 on the Nushagak

[b] From USGS gage 15297980, n = 1; all major ion information is from Chamberlain 1989

[c] n = 9 for alkalinity, n = 5 for nitrate, n = 1 for DOC, n = 8 for Ca, Mg, Na, $HCO_3$, $SO_4$

[d] From USGS gage 15300500, n = 1

**Table 19.10** Select minor and trace element concentrations in select rivers of Bristol Bay. Averages or ranges are shown. If WQC is exceeded, the range is shown. The Tanalian, and Chulitna, shown in Tables 19.6 and 19.7, did not have minor or trace element data and is not included here. Additional trace element data is available for some sites, for example at the Kvichak River. T = total (unfiltered) metal concentration, D = dissolved (filtered), Al = aluminum, Fe = iron, Mn = manganese, Cd = cadmium, Cu = copper, Pb = lead, Zn = zinc, n = number of samples, BBNA = Bristol Bay Native Association, WQC = water quality criteria for freshwater aquatic life

| Watershed | River/Creek | Al (µg/L) | Fe (µg/L) | Mn (µg/L) | Cd (µg/L) | Cu (µg/L) | Pb (µg/L) | Zn (µg/L) | n | Data Source |
|---|---|---|---|---|---|---|---|---|---|---|
| Wood River | Silver Salmon Creek | na | 100 (T) | 2 (T) | na | na | na | na | 1 | USGS 59210015849300 |
| Mulchatna | Swan River | 50 (D) | 466 (D) | 34 (D) | < 0.02 (D) | 1.1 (D) 0.4–3.0 | 0.04 (D) | 2 (D) | 4 | BBNA, unpublished |
| Mulchatna | Koktuli River | 50 (D) | 145 (D) | 16 (D) | < 0.02 (D) | 0.57 (D) | 0.03 (D) | 2 (D) | 14 | USGS gage 15302000 |
| Mulchatna | Koktuli headwaters | 7–1100 (T), 3–45 (D) | 99–1540 (T), 32–673 (D) | 10–110 (T), 5–97 (D) | < 0.1 (T), < 0.06 (D) | 0.7–20.4 (T), 0.5–15.9 (D) | < 0.01–0.8 (T), 0.01–0.3 (D) | 0.5–19 (T), < 0.5–20 (D) | 39 | PLP 2011 (site SK-134A, SK-136B), Zamzow 2011 (sites SK-31 and SK-51) |
| Lower Nushagak | Nushagak River | < 30 (T) | 466 (D) | 27 (D), < 10–40 | < 0.02 (D) | < 2–4 (T) | < 1–6 (T) | < 3–30 (T) | 2–10 | USGS gage 15302500 |
| Lower Nushagak | Tlikakila River | na | 38 (D) | 6 (D) | na | na | na | na | 11 | USGS gage 15299000 |
| Lower Nushagak | Chulitna River | 4–45 (D) | 118–368 (D) | 7–62 (D) | < 0.03 (D) | 0.5–1.3 (D) | 0.02–0.13 (D) | < 2 (D) | 22 | USGS gage 15298040 |
| Lower Nushagak | Nikabuna tributary | 382 (T) 70–1028 (T) 14 (D) | 791 (T) 334–1510 (T) 190 (D) | 66 (T) 33–103 (T) 48 (D) | 0.009 (T) 0.005 (D) | 0.8 (T) 0.3 (D) | 0.26 (T) 0.05–0.6 (T) 0.008 (D) | 3.1 (T) 1.8 (D) | 3 | Zamzow 2011 |
| Lake Clark | Tazimina River, upper | na | 16 | < 0.2 | < 2 | < 2 | 0.2 | < 1 | 1 | Baldrige and Trihey 1982, river mile 11.6 |
| Lake Clark | Tazimina River, lower | na | 11 | < 6 | < 2 | 7 | < 0.1 | < 1 | 1 | Baldrige and Trihey 1982, river mile 1.7 |
| Lake Clark | Kvichak River | 64 (T) (18–225) | 74 (T) | 6 (T) | < 0.5 (T) | 0.6 (T) < 1–2.2 | 0.08 (T) < 0.2–0.56 | 3 (T) | 19 | Levelock Village Council Environmental Program |
| Lake Clark | Yellow Creek | 156 (T) (56–533) | 2,163 (T) 1,600–3,550 | 72 (T) 47–184 | < 0.5 (T) | 0.4 (T) | 0.08 (T) | 4 (T) | 19 | Levelock Village Council Environmental Program |
| Lake Iliamna | Alagnak River | 94 (T) (20–251) | 362 (T) | 20 (T) | < 0.5 (T) | 1.7 (T) < 1–9.9 | 0.06 (T) | 4 (T) | 18 | Levelock Village Council Environmental Program |
| Lake Iliamna | Battle Creek | 70–90 | 30–83 | 21 | < 4 | < 3 | < 40 | 14–55 | 3 | LaPerriere 1996 |
| Naknek | Ukak | 6270–9070 | 6250–8040 | 120 | < 4 | 9–13 | < 40 | 17–27 | 3 | LaPerriere 1996 |
| Naknek | Savonoski | 6950–7930 | 8390–9320 | 150 | < 4 | 7–12 | < 40 | 20–45 | 3 | LaPerriere 1996 |
| Alaska WQC | | 87 | 1,000 | 50 | 0.09[a] | 2.74[a] | 0.54[a] | 36[a] | | http://dec.alaska.gov/commish/regulations/pdfs/18%20AAC%2070.pdf |

[a] Hardness—dependent criteria; value given is for 25 mg/L hardness

Minor and trace element concentrations were low. Cu (unfiltered) was nondetect in 15 of 19 samples and exceeded WQC in two samples: in June 2007, concentrations were 8.5 μg/L, associated with snowmelt and high turbidity, and in September 2012 concentrations were 4 μg/L, possibly associated with rain runoff, although turbidity was low at 1 NTU.

## Lower Nushagak River

Four villages lie along the Nushagak River. Koliganek sits just above the confluence with the Mulchatna; downstream of the confluence are New Stuyahok, Ekwok, and Portage Creek. No water quality information is available for the upper part of the Nushagak River.

A USGS gage (15302500) was located at Ekwok on the Lower Nushagak. The gage covered a drainage area of 25,510 km$^2$ and discharge ranges from 164 m$^3$/s (5,800 cfs) to 2,095 m$^3$/s (74,000 cfs). Data were collected inconsistently from March to November over seven years (1956–1986) (n = 18). Despite high discharge, turbidity and TSS were generally low (< 10 mg/L), although there were exceptions, particularly during snowmelt or rainy periods. Dissolved oxygen was high and temperatures were generally low (< 12°C), except in July when they could reach 16°C (Table 19.8).

The pH varied from 6.0–6.9, and alkalinity (14–26 mg/L) and SC were low (< 65 μS/cm). These conditions are similar to those on the Nuyakok River, a main tributary of the Nushagak River well upstream of Ekwok (USGS gage 15302000).

Waters are calcium-bicarbonate and dilute with low concentrations of major ions (Table 19.9). Sulfate concentrations varied from 1–9 mg/L without any seasonal pattern. Most trace elements (Be, Cd, Co, Cu, Pb, Mo, V) were below the limit of detection for most of the samples, but two unfiltered Cu samples were elevated (10 μg/L, 20 μg/L) as were two Pb samples (5 and 6 μg/L); more frequent analysis with more sensitive methods is needed to determine if these are outliers. Also found, but in low concentrations, were Al, Fe, Mn, As, Ba, Cr, Ni, Sr, Zn (see Table 19.10 for Al, Fe, Mn, Zn).

Nutrients were low, with mean total N concentration of 0.6 mg/L (maximum 1.3 mg/L), mean organic N at 0.4 mg/L (maximum 1.1 mg/L), and total N as nitrate at 2 mg/L (maximum 5 mg/L). Total P averaged 0.04 mg/L (maximum 0.16 mg/L) as P, with ortho-phosphate nondetect and total P as phosphate at 0.1 mg/L (maximum 0.5 mg/L). Dissolved organic C was 1–5 mg/L with the highest concentrations in September.

The conditions in the Kvichak and Nushagak—low turbidity, neutral pH, dilute waters with some DOC—provide ideal water quality for aquatic life; more information should be collected to determine if elevated Cu and Pb concentrations are common.

## Glacial Rivers with Lakes

Lakes in glacial systems play a role in changing turbid, flashy waters to clear-running streams that are better suited for aquatic life by settling sediment and moderating flow and temperature. As glacial lakes accept and settle turbidity, they lose biological productivity; they may have only 5% the productivity of clear lakes (Brabets and Ourso 2006). A demonstration of this is on the Kijik River system; although similar in size and physical characteristics, Kijik Lake had nearly five times the zooplankton biomass as Lachbuna Lake, potentially due to differences in turbidity, as well as differences in MDN since salmon spawn in Kijik Lake but cannot access Lachbuna for spawning due to a barrier.

Both the Tanalian River and Kijik River are born in glaciers, but eventually run clear and cold into Lake Clark (Table 19.8).

### Tanalian River and Kontrashibuna Lake

Kontrashibuna Lake has only two meters of clarity, but the Tanalian River issuing from it is clear (Weeks 2003) (Figure 19.4). A USGS gage (15298000) collecting data on the Tanalian from May to October (1954–1956, 1999–2000) found that TSS for flows up to 168 m$^3$/s (5,920 cfs) suspended solids did not exceed 5 mg/L, indicating the effectiveness of the lake in trapping sediment.

### Kijik River with Portage, Lachbuna, and Kijik Lakes

In the Neacola Mountains, Portage Lake feeds Lachbuna Lake via the Kijik River (Figure 19.4). Portage Lake, no longer fed by glaciers or snowfields, is clear but the Kijik River is turbid as it enters Lachbuna Lake. Neither lake hosts anadromous fish.

Lachbuna, a 30 meter deep glacial lake, acts to settle the sediment, and the Kijik River emerges clear with less than 4 mg/L TSS for flows of ≤ 5 meters³/s (183 cfs), except during the June freshets (snowmelt runoff) when TSS was 21 mg/L (discharge of 6.4 meters³/s or 226 cfs) (Brabets and Ourso 2006). The highest TSS, measured at the Kijik River outlet at Lake Clark, was in the spring of 1999 with 123 mg/L (discharge of 50 meters³/s or 1,750 cfs). The Portage-Lachbuna lake system had the highest alkalinity of the lakes in this review (Chamberlain 1989).

Kijik Lake (~4 km²) is anadromous, 100 m deep, and fed by nonglacial runoff and groundwater springs; Kijik Lake drains to the Kijik River and eventually Lake Clark. It is clear and cold with neutral pH; no information on minor or trace elements was available. Nitrate in Kijik Lake was highest early in summer and declined throughout the summer, similar to Lake Clark. Potassium and nitrate were higher at the inlet site than at the outlet, suggesting they were retained in the lake, while Ca was higher at the outlet than the inlet, suggesting possible groundwater inputs (Stottlemeyer and Chamberlain 1987).

Despite no anadromous fish to provide MDN, the $\delta^{15}N$ signature in sediment at Portage Lake is becoming more enriched and $\delta^{13}C$ more depleted, in contrast to the signature at Kontrashibuna Lake. This may reflect lake turnover or the rate of N and C cycling, but the reasons are not entirely clear. Portage Lake has a small watershed (10 km²) with low alder mass, yet has higher inorganic N than Kontrashibuna Lake, which appears to be receiving increasing terrestrial N over time with glacial retreat that allows alder expansion (Cohn 2009).

## Glacial Rivers without Lakes

Mountain rivers without lake systems are colder and more turbid than rivers that pass through sediment-settling lakes. The nearly 150 glaciers in the Tlikakila drainage are thinning, releasing entrapped silt and, along with contributions from Currant Creek and Chokotonk River, create highly turbid water in upper Lake Clark, with TSS ranging from 5–710 mg/L on the Tlikakila alone (USGS gage site 15297970) (Table 19.8). The adjacent Chokotonk and Currant systems (USGS gages 15297930 and 15297980) are smaller drainages than Tlikakila (430 km² each versus 1,610 km²) but TSS loads can be significant at around 100 mg/L. These systems also contribute calcium and sulfate to Lake Clark (Table 19.9) due to a difference in the underlying bedrock or possibly related to the small drainage size, which limits the amount of time water is in contact with surface material (Chamberlain 1989).

High precipitation, averaging 180–300 cm, is also a hydrologic driver. The Tlikakila River is dominated by snowmelt in June and by glacial melt and storm runoff in July and August. Annual flow averages 75% snowmelt, 11% glacial melt, 14% rainwater, and 1% groundwater (Brabets et al. 2004). Despite icy, turbid conditions, a 2007 study found 21% of radio-tagged fish in Lake Clark spawned in the Tlikakila; importantly, spawning coinciding with a decline in turbidity (Young and Woody 2007).

These systems are low in DOC (< 1 mg/L), high in DO (11–15 mg/L) with neutral pH (7.1–8.2) (Table 19.8). The Tlikakila has slightly higher alkalinity (13–37 mg/L versus 11–24 mg/L) and SC (31–117 µS/cm versus < 75 µS/cm) than the Chokotonk and Currant; major ions in the Tlikakila and Currant are similar (Table 19.9). Total N is higher in the Tlikakila (0.5 mg/L) than in the two smaller glacial streams (0.2 mg/L) but total P is similar (mean 0.1–0.2 mg/L). Whether the higher N reflects drainage vegetation or contributions from salmon carcasses was not reported.

## Organic-Rich Rivers

Streams and rivers that flow through wetlands contribute C to the ecosystem. The extensive wetland complexes of the Bristol Bay landscape are important for nutrient cycling and maintaining low TSS loads, helping to keep dissolved metal concentrations low and C concentrations high.

### Chulitna River

The Chulitna River basin (3,000 km²) is twice as large as the Tlikakila drainage (Figure 19.4.). The 145 km long slow-moving Chulitna River originates with two arms meeting in lowlands to provide warm, low-sediment water to the lower end of Lake Clark, in contrast to the cold, high-sediment glacial rivers entering the north end of the lake. Where the Chulitna empties into Lake Clark, temperatures are warmer (to 16°C) than Lake Clark, and diel temperatures can flux by 7°C. It is the only river from the Nushagak-Big River Hills region and the only nonglacial, organic river to enter Lake Clark. It supports important fish species such as humpback whitefish and northern pike, while anadromous salmon use appears limited to one clear water tributary—the Koksetna River (Russell 1980, Young and Woody 2007). A summary of data from five USGS gages within the Chulitna drainage is provided in Brabets (2013).

The first arm of the Chulitna headwaters is Caribou Lake, which feeds the Koksetna River. Caribou Lake is a small lake that is clear nearly to the bottom (3 meters) and is fed by snowmelt and springs (Weeks 2003). The Koksetna River has cold (< 10°C), neutral pH (7.3–7.9), dilute (< 100 µS/cm) water (Levelock Village Council Environmental Program, unpub. data) (USGS gage 601049154540600). The Koksetna is a flashy stream that runs through forested wetlands, but even at flows of 37 meters$^3$/s (1,300 cfs) suspended sediment is low (< 3 mg/L).

The second arm of the Chulitna originates in the Nikabuna Lakes wetland system. Within and surrounding this watershed, mining companies have leased over 1,000 km$^2$ for mining development (Brabets 2013). Just above Nikabuna Lakes, a USGS gage (600524155254200) recorded neutral pH (6.8–7.4) with maximum temperatures of 11°C. DO was lower in winter (7–9 mg/L) than in summer (10 mg/L) and SC varied seasonally: 62 µS/cm in summer and ~130 µS/cm in winter, indicating groundwater inputs that increase dissolved solids (Brabets 2013).

A tributary entering Nikabuna Lake had a full suite of metals analyzed (Zamzow 2011). Unfiltered concentrations of Al and Mn were elevated above WQC during snowmelt, associated with high TSS (10–39 mg/L) and unfiltered concentrations of Fe and Pb were elevated when TSS was 39 mg/L (Table 19.10). However, concentrations dropped quickly. For example, unfiltered Al at 1,028 µg/L on May 3, 2009 dropped to 70 µg/L by June 5, 2009. Other elements [Cu, Zn, Hg, As, Cr, Mo, Ni, uranium (U)] were well below WQC, and Sb, Cd, and Se were nondetect. Relatively high alkalinity (22–34 mg/L) and DOC (4 mg/L) (Table 19.9) provide some ability to ameliorate potentially negative effects of briefly elevated metals.

The lower Chulitna River, below the confluence of the Nikabuna system with the Koksetna River, is broad and less flashy. Wetlands contributed C and wetland organic acids did not reduce pH below neutral; the river has good alkalinity and neutral pH (Tables 19.8, 19.9), although these are influenced seasonally by groundwater inputs. In March, with a greater percent of groundwater, pH and DOC are lower (6.7–7.0; 1–2 mg/L) and alkalinity high (40–52 mg/L); in summer, with greater surface runoff, pH and DOC increase (pH 7.2–7.8, DOC 4–5 mg/L) and alkalinity drops (21–30 mg/L) (USGS gage 15298040). With a reduction in groundwater, calcium drops from 15 mg/L in March to 9 mg/L in summer.

A full suite of metals was analyzed at the lower Chulitna from 2010–2013, and dissolved metals were all in low concentrations, with the exception of Mn in March 2012 at 62 µg/L, slightly above the WQC of 50 µg/L (USGS gage 15298040) (Table 19.10). Low concentrations of minor and trace elements may be related to the low concentrations of suspended solids. In the lower Chulitna, TSS was < 10 mg/L even at discharge of 100 meters$^3$/s (3,500 cfs). High flows (> 77 meters$^3$/s or 2,700 cfs) could produce TSS of 17–44 mg/L (USGS gage 15298040).

## Yellow Creek

The Levelock community conducted baseline water quality sampling on Yellow Creek, a *stained water* tributary of the Kvichak River from May-September 2007–2014 (n = 19) (Levelock Village Council Environmental Program, unpub. data). Yellow Creek is a low-lying, meandering, small creek. Data on Yellow Creek were collected on the same days as data on the Kvichak River and Alagnak River.

The stream was cool at 2–13°C with one reading of 18°C when the stream was low. The pH was neutral with low SC (Table 19.8). Turbidity was elevated (4–8 NTUs) relative to the Kvichak and Alagnak Rivers, and DO was occasionally measured at 7–8 mg/L, slightly lower than the rivers and providing less than ideal conditions for aquatic life.

Although average Ca concentrations were slightly lower than the Kvichak and Alagnak Rivers, Mg, K, and Na were higher, with mean Na of 4 mg/L. This creek appears to have more in common with the similarly small, meandering, unnamed tributary of Nikabuna Lakes at the headwaters of the Chulitna River. Major ions and low trace element concentrations were similar to those at the Nikabuna tributary, and like it, Al, Fe, and Mn reached concentrations above WQC (Table 19.10): Fe averaged over 2,000 µg/L (WQC is 1,000 µg/L) and Mn averaged 72 µg/L (WQC is 50 µg/L). Elevated Al, Fe, and Mn at Yellow Creek were associated with periods of higher turbidity in August and September 2007.

## Mineralized Tributaries

As of 2011, about 2,000 km$^2$ of land in the Nushagak and Kvichak drainage basins were leased for mineral exploration (Woody and Higman 2011), representing about 22% of the total area in the combined watersheds of the Upper Nushagak, Mulchatna, Lake Clark, and Lake Iliamna watersheds (Bogan et al. 2012).

The proposed Pebble copper-gold mine straddles a watershed divide; mine facilities would cover the headwaters of the Koktuli River in the Mulchatna watershed, and the headwaters of Upper Talarik Creek in the Lake Iliamna watershed; tributaries of Nikabuna Lakes in the Chulitna drainage are nearby. Although outside Pebble, additional leases to other mining companies cover 1,000 km$^2$ within the Chulitna drainage of the Lake Clark watershed and just outside the LACL (Brabets 2013).

## Koktuli River Headwaters

The two largest tributaries of the Mulchatna River are the Chilikadrotna River and the Koktuli River. The headwaters of the upper Koktuli River split into two arms, the North Fork Koktuli River and South Fork Koktuli River (Figure 19.3).

In sampling conducted May 2009, June 2009, and June 2010, in wetlands at the headwaters of the North Fork, conditions were similar to those at Chulitna wetlands: with high DOC (4–6 mg/L), low TSS, and low metals, although pH was lower at snowmelt (~6.2), possibly due to an influx of wetland organic acids (Zamzow 2011).

Two headwater tributaries of the South Fork run across a mineralized zone, including the outcrop of a Cu ore body. Tributaries on the outcrop were cold (8°C) with neutral pH (6.8–7.3), low SC (54–78 µS/cm), and high dissolved oxygen (10–14 mg/L), although alkalinity varied between sampling in June 2009 and June 2010 (16 mg/L and 42 mg/L). The 2010 alkalinity was the highest in the Koktuli River system, only matched by tributaries less than 2 km away and on the same ore body of Upper Talarik Creek in the Lake Iliamna watershed. The lower 2009 alkalinity was probably related to the high amount of runoff (dilution) in June 2009, as deep winter snow melted quickly and the increased alkalinity was matched by an increase in concentration of major ions ($Ca^{2+}$, $Mg^{2+}$, $K^+$, $Na^+$, $SO_4^{2-}$, $Cl^-$) (Zamzow 2011).

Cu was elevated in one South Fork tributary (5 µg/L) but not the other (1 µg/L) and in both the sulfate was higher than most waters (10–20 mg/L), and much higher than sulfate in the nearby Upper Talarik Creek tributaries. All other metals (Al, Sb, As, Cd, Cr, Fe, Pb, Mn, Mo, Ni, Se, U, Zn) in the unfiltered form were below WQC. Sediment concentrations of Cu collected at a single outcrop tributary were significantly higher (101 mg/kg) than in sediment collected at other wetland sites on the North Fork Koktuli and Upper Talarik Creek (Zamzow, unpub. data).

Downstream of the ore outcrop is Kaskanak Mountain. A narrow stream on the mountain slope with a cobble bed and banks of mud and grass is within the mineralized zone but not near the outcrop. Temperatures, DO, SC, and sulfate were similar to the wetland streams, while DOC was lower (1–3 mg/L) (Zamzow 2011). The pH was low (6.2) in May 2009 with snowmelt, higher in June 2009 and June 2010 (~7.0), and alkalinity was low (11–20 mg/L). Several metals (Al, Mn, Cd, Cu, Pb, Zn) in the unfiltered form were above WQC and associated with TSS (9–16 mg/L) during snowmelt, but were generally low in the dissolved form. Other metals were low to nondetect (Sb, As, Cr, Hg, Mo, Ni, Se, U).

Compared to downstream South Fork Koktuli tributaries, these upper tributaries were higher in metals. Below the confluence of the South Fork Koktuli and North Fork Koktuli, the main stem of the Koktuli River was low in all minor and trace elements (Table 19.10).

## Ukak and Savonoski River

On the Alaska Peninsula in the Naknek watershed, the Ukak and Savonoski Rivers are still affected by the 1912 Novarupta eruption (Figure 19.5). In addition to the ash, tributaries to the Ukak and Savonoski rivers are influenced by glaciers (LaPerriere 1997). Both rivers have high $Ca^{2+}$, $Mg^{2+}$, bicarbonate, and sulfate; the Ukak has approximately equal amounts of $Ca^{2+}$ (33 mg/L) and Na+ (24 mg/L), and sulfate is not only very high (70 mg/L), it is higher than bicarbonate (51 mg/L) (Table 19.9). Recall that Naknek Lake, into which the Ukak and Savonoski rivers flow, was highly sodic (7–10 mg/L) (Table 19.6). Both flow into Iliuk Arm and transport ash and sediment, reducing clarity in Iliuk arm.

The very high levels of Al (over 6 mg/L), Fe, Mn, and Cu in these rivers (Table 19.10) caused a caution that they should not be used for drinking water (LaPerriere 1996). Other elements could be at toxic concentrations (e.g., Pb); more sensitive analytical methods need to be applied.

## Clear Rivers and Streams

The vast majority of streams are likely to be clear-running rivers. Here we describe some of the clear streams around Lake Clark and Iliamna Lake.

## Koktuli and Swan Rivers

The Koktuli River and the Swan River, a tributary of the Koktuli, are within the Mulchatna sub-basin watershed with several years of data (2005–2011) (BBNA, unpub. data). The Swan River is a 23 meter wide stream (< 14 meters$^3$/s, < 500 cfs) with a sand bottom flowing through forested wetlands; temperatures reach 19°C. The waters are diluted with pH and alkalinity well-suited for aquatic life (Tables 19.8, 19.9). N (as nitrate + nitrite) and total P were low (each averaging 0.013 mg/L). Dissolved Mn and dissolved Cu were each slightly elevated above WQC a single time on the Swan River, but all other metals were low or nondetect (Al, As, Cr, Fe, Pb, Ni, Se, Sr, Zn).

Koktuli River samples were collected 1.5 km downstream of the confluence with the Swan River, where the main stem of the Koktuli River is about 60 m wide in forested lowlands. At this point, the river drains 1,110 km$^2$ and flows reach 110 meters$^3$/s (3,900 cfs) (BBNA, unpub. data). Water quality was similar to the Swan River, except Ca$^{2+}$ was lower, nitrate higher, and all dissolved metals were in low concentrations (Tables 19.8–19.10). Despite low mineral concentrations along most of the river, the Koktuli headwaters had some of the highest trace element concentrations of the rivers outside the Alaska Peninsula.

## Tributaries of Iliamna Lake and Lake Clark

In the Kvichak drainage, single samples were collected from 19 tributaries (2–64 m wide) surrounding Iliamna Lake (D. Schindler, unpub. data). All were cold in July or August (8–17°C) with neutral pH (6.6–8.2) and low to very low SC (13–68 µS/cm); an SC of 13 µS/cm is not much more concentrated than rainwater.

Similarly, samples were collected at 15 wadeable streams in the Lake Clark watershed, with results showing all were cold (2–8°C) with neutral pH (7.1–7.9), high oxygen content (11–13 mg/L), and low SC (17–84 µS/cm) with the exception of two, Priest Rock Creek (112 µS/cm) and Tommy Creek (176 µS/cm) (Levelock Village Council Environmental Program, unpub. data).

An expanded study that included 17 sites in the Lake Iliamna watershed, 34 sites in the Mulchatna watershed, and 8 sites in the Upper Nushagak watershed showed remarkably similar results (Bogan et al. 2012) (Table 19.11).

## Tazimina River

The Tazimina River, the only river to enter Sixmile Lake, originates in Upper and Lower Tazimina Lakes in the Chigmit Mountains (Figure 19.4). The Tazimina River is unique in that it is the only water body within the Bristol Bay drainage to host a run-of-river hydroelectric power system that provides electricity to the communities of Nondalton (on Sixmile Lake), Newhalen, and Iliamna (on Iliamna Lake). Although the data is limited, the Tazimina River water chemistry is similar to lake chemistry (Tables 19.5–19.10), with some of the lowest concentrations of minor and trace elements (Chamberlain 1989). It is the largest spawning area in the extended Lake Clark area, utilized by rainbow trout and salmon, but fish are unable to reach the lakes due to a 250-foot waterfall.

# GROUNDWATER INFLUENCE

Groundwater springs are important water temperature regulators. At the inlet to Kijik Lake (Figure 19.4), where there are groundwater seeps, Little Kijik River water is about 7°C through the summer; at the outlet, Little Kijik River temperatures fluctuate with the warming lake surface (6°C in June and 16–19°C in July–August 2004) (Brabets and Ourso 2006).

Lachbuna Lake is hypothesized to have groundwater inputs near the lake floor (Moore and Shearer 2011), keeping temperatures cooler and more even throughout the water column. Lake temperatures drop gradually from 14°C at the surface to 6°C at 35 m in summer with no thermocline, compared to the sharper drop in Kijik Lake from near 18°C to 30 meters, below which waters were 4°C (Brabets and Ourso 2006). In both lakes, waters were near freezing at the surface in March and warmer (near 2.5°C) at 25 m (Brabets and Ourso 2006). The theory of groundwater seeps into the Lachbuna Lake bottom is supported by data that shows specific conductance increasing with depth in summer.

Groundwater also provides alkalinity, by carrying bicarbonate from subsurface material to surface waters, which may be important for buffering very dilute waters.

Groundwater springs keep river reaches open in winter, and most spawning areas are believed to be located near upwelling sites. At the Tlikakila River, the lower reach receives groundwater, as evidenced by open leads with temperatures above 0°C in winter when most of the river is frozen, offering critical overwintering habitat for fish. During late fall,

**Table 19.11** Water quality at wadeable streams. Samples were collected from 2008–2011 at 77 sites with five sites sampled every year. All streams were less than 457 m in elevation, wadeable, with hard-bottom stream substrates and shallow slopes (mean slope < 3%, maximum 8%). Most, although not all, were inaccessible by road. Sites with high specific conductance may be locations with a high percentage of groundwater input. In addition to water quality data, collected with a Hydrolab MiniSonde multiprobe, data was collected on physical habitat, riparian vegetation, stream discharge, macroinvertebrates, and diatoms. From Bogan et al. 2012

| Basin | Watershed | No. of sites | Water Temperature (°C) | Dissolved oxygen (mg/L) | pH | Specific conductance (μS/cm) | Discharge (m³/s) | Discharge (cfs) |
|---|---|---|---|---|---|---|---|---|
| Nushagak | Mulchatna | 34 | 8 (1–14) | 11 (9–13) | 7.2 (5.4–8.1) | 48 (5–140) | 0.8 (0.06–7) | 29 (2–248) |
| Nushagak | Upper Nushagak | 8 | 7 (4–10) | 12 (11–13) | 7.2 (6.6–7.6) | 47 (20–87) | 1.2 (0.06–3.4) | 44 (2–121) |
| Kvichak | Lake Clark | 15 | 6 (2–8) | 12 (11–13) | 7.4 (7.1–7.9) | 59 (17–176) | 1.1 (0.11–3.6) | 40 (4–127) |
| Kvichak | Lake Iliamna | 17 | 8 (4–17) | 11 (8–13) | 7.4 (6.9–7.9) | 44 (22–67) | 0.8 (0.1–2.8) | 29 (5–98) |

sockeye salmon spawn throughout the Tlikakila Rivers groundwater influenced reaches, which would offer incubating eggs protection from freezing temperatures. Open leads have also been observed at Upper Talarik Creek and nearby tributaries and rivers that supply Iliamna Lake and the Mulchatna River, likely providing similar benefits (Woody and Higman 2011).

Although data are limited, groundwater inputs may be frequent and significant drivers of aquatic habitat throughout the Bristol Bay watersheds.

## RECOMMENDED RESEARCH

There is a paucity of water quality data for freshwaters of Bristol Bay watersheds. With the exception of water temperature and clarity at Lake Clark collected by the NPS and water temperature, clarity, solar radiation, and chlorophyll at Iliamna Lake and Lake Aleknagik collected by the University of Washington, there are no long-term water quality data sets. The data available for major ions and minor and trace elements are often based on a handful of samples. Interesting research has been conducted on the impact of ashfall on lake clarity over several years, the historical record of trace elements, and other unique aspects of water quality in the Bristol Bay region, but long-term data sets of basic water quality need to be collected.

Future research should determine methods for expanding the data set, through institutions such as the NPS and the University of Washington, as well as through community-based sampling as is occurring on the Kvichak River and Alagnak River through the Levelock community and on the Koktuli River through the BBNA.

For most water bodies, data on minor and trace elements is either entirely lacking or analysis was performed with insensitive methods. More data on inorganic elements and their seasonal fluxes would provide information on the concentrations to which aquatic life is commonly exposed. Reports of unusually elevated Cu and Pb concentrations in the Kvichak and Lower Nushagak Rivers need to be verified or dismissed through further sampling. In combination with information on alkalinity and dissolved organic matter, an assessment could be made on the vulnerability of fresh water to changes. This baseline would be useful to determine how or if future development—roads, mines, housing—could impact water quality.

Trace element data can be expensive to collect, but temperature and major ion data is inexpensive and can locate groundwater seeps. Groundwater inputs may be a frequent and significant driver of habitat use and provide fish cool thermal refuge when stream temperatures rise. More data is needed on the extent and location of groundwater inputs and connections to spawning sites.

Last, the impacts of the regional (and global) warming trend need to be tracked. While cold water bodies may benefit from longer periods of light and warmer water, tundra ponds may shrink and decrease in number; hyporheic and ephemeral pathways may be negatively impacted; and shallow, warm streams may reach temperatures and oxygen levels that negatively impact salmon.

## SUMMARY

Freshwater systems in the Bristol Bay watersheds are generally cold, clear, oxygenated, and very dilute; as a generalization, lakes are oligotrophic and rivers have low concentrations of major ions.

No waters, including surveys of over 100 streams, were found to exceed 20°C and nearly all were fully oxygenated. Many lakes were oxygenated to depth due to wind-generated mixing. Glacial rivers poured cold water into lakes in summer; groundwater kept some lakes and streams cool in summer and prevented stream water and eggs laid on gravel lake beaches from freezing in winter. Innumerable tundra ponds capture snowmelt and provide groundwater recharge.

Due to low primary production, most fresh waters are clear. However, inputs of glacial sediment or volcanic ash can reduce clarity as observed at Lake Clark and the Iliuk Arm of Naknek Lake. Lakes present along glacial river systems act to trap sediment, changing turbid streams to clear ones, increasing productivity and usable aquatic habitat.

Lakes, rivers, and streams have a neutral pH within the range for aquatic life (6.5–8), even at mineralized tributaries, with the exception of Iron Springs Creek (pH 4.4) on the Alaska Peninsula.

With the exception of the Naknek River lakes, alkalinity is almost universally low (10–20 mg/L), and would be unable to buffer additional inputs of acid beyond the organic acids regularly infused through wetlands. Naknek River lakes are more alkaline and higher in major ions than other Bristol Bay lakes due in part to proximity to the coast. Most Bristol Bay watershed lakes, although dilute, have calcium-bicarbonate type water. However, Alagnak River lakes, also near the coast, have low concentrations of bicarbonate, which keeps alkalinity low. Naknek Lake is calcium-sulfate-bicarbonate due to high sulfate inputs.

Minor elements may increase briefly above WQC as erosional particulates enter water bodies during runoff. Trace elements like Cu and Zn can be toxic to aquatic life, but are low across the region, except at localized mineralized sites (Kaskanak Mountain, Iron Springs Creek, Ukak and Savonoski Rivers).

Regionally, water bodies are warming with global increases in temperature. Ice leaves earlier and in recent years, Lake Clark and Iliamna Lake froze only briefly, for a few weeks in February 2015. Less ice results in longer periods of light, improving conditions for zooplankton growth and potentially for juvenile salmon. Some increase in water temperature could be beneficial; however, shallow streams and small streams in late summer may approach temperatures that cause negative impacts on aquatic life, including salmon, and ponds and ephemeral pathways may dry up.

Together, the ponds, tributaries, streams, rivers, lakes, and groundwater systems provide a stunning example of undisturbed wilderness waters, with innumerable streams and tributaries providing a vast network of waterways, nearly all of which are extremely pure and suited for aquatic life.

# REFERENCES

Adleman, J. 2002. The great eruption of 1912. In Alaska Park Science: connections to natural and cultural resource studies in Alaska's National Parks. National Park Service. Anchorage, AK.

Alaska Department of Environmental Conservation (ADEC). 2008. Alaska water quality criteria manual for toxic and other deleterious organic and inorganic substances, draft. http://dec.alaska.gov/water/wqsar/wqs/index.htm.

Alaska Salmon Program. 2006. Alaska salmon research 2005. University of Washington School of Aquatic and Fishery Sciences, SAFS-UW-0602. Seattle, WA.

Alaska Salmon Program. 2007. Alaska salmon research 2006. University of Washington School of Aquatic and Fishery Sciences, SAFS-UW-0701. Seattle, WA.

Alaska Salmon Program. 2010. Alaska salmon research 2009. University of Washington School of Aquatic and Fishery Sciences, SAFS-UW-1001. Seattle, WA.

Alaska Salmon Program. 2011. Alaska salmon research 2010. University of Washington School of Aquatic and Fishery Sciences, SAFS-UW-1001. Seattle, WA.

Baker, M.R., D. Schindler, G. Holtgrieve, and V.L. St. Louise. 2009. Bioaccumulation and transport of contaminants: migrating sockeye salmon as vectors for mercury. Environmental Science and Technology (43): 8840–8846.

Baldrige, J.E. and E.W. Trihey. 1982. Potential effects of two alternative hydroelectric developments on the fishery resource of the lower Tazimina River, Alaska: a preliminary instream flow assessment, draft final report. Arctic Environmental and Information Data Center, Anchorage, AK.

Bogan, D., D. Rinella, and R. Shaftel. 2012. Baseline Macroinvertebrate and Diatom Surveys in Wadeable Streams of the Kvichak and Nushagak Watersheds, Bristol Bay, Alaska. Alaska Natural Heritage Program, University of Alaska Anchorage for The Nature Conservancy. Anchorage, AK.

Brabets, T. 2002. Water quality of the Tlikakila River and five major tributaries to Lake Clark, Lake Clark National Park and Preserve, Alaska, 1999–2001. U.S. Geological Survey Water Resources Investigation Report 02-4127.

Brabets, T.P., R.S. March, and D.C. Trabant. 2004. Glacial history and runoff components of the Tlikakila River basin, Lake Clark National Park and Preserve, Alaska. U.S. Geological Survey Scientific Investigations Report 2004-5057.

Brabets, T.P. and R.T. Ourso. 2006. Water quality, physical habitat, and biology of the Kijik River basin, Lake Clark National Park and Preserve, Alaska, 2004–2005. U.S. Geological Survey Scientific Investigations Report 2006-5153.

Brabets, T.P. 2013. Water quality and flow data, Chulitna River basin, Southwest Alaska, October 2009–June 2012. U.S. Geological Survey Open File Report 2013-1009.

Brubaker, M., C. Balluta, S. Flensburg, J. Skarada, and R. Drake. 2014. Climate change in Nondalton, Alaska: strategies for community health. Alaska Native Tribal Health Consortium Center for Climate and Health. Anchorage, AK.

Burgner, R.L., C.J. DiCostanzo, R.J. Ellis, G.Y. Harry, Jr., W.L. Hartman, O.E. Kerns, Jr., O.A. Mathisen, and W.F. Royce. 1969. Biological studies and estimates of optimum escapements of sockeye salmon in the major river systems in Southwestern Alaska. Fishery Bulletin 67 (2): 405–459.

Chamberlain, D.M. 1989. Physical and biological characteristics and nutrient limiting primary productivity, Lake Clark, Alaska. Master thesis, Michigan Technical University, Houghton, MI.

Church, W.A. and R.L. Burgner. 2009. Studies on the effect of winter climate on survival of sockeye salmon embryos in the Wood River Lakes, Alaska, 1952–1959. Fisheries Research Institute, University of Washington School of Fisheries SAFS-UW-0901. Seattle, WA.

Cohn, B.R. 2009. Recent paleoenvironmental change recorded in three non-anadromous lakes in Southwest Alaska: effects of climatic and vegetation dynamics on lake productivity. Thesis presented to the University of Alaska Fairbanks, Fairbanks, AK.

Donaldson, J.R. 1967. The phosphorus budget of Iliamna Lake, Alaska, as related to the cyclic abundance of sockeye salmon. Thesis submitted for Ph.D. at University of Washington.

Federal Register. 2014. Definition of the "Waters of the United States" under the Clean Water Act. 333 CRF Part 328. Volume 79: 22188–22274.

Gende, S.M., R.T. Edwards, M.F. Willson, and M.S. Wipfli. 2002. Pacific salmon in aquatic and terrestrial ecosystems. Bioscience 52 (10): 917–928.

Goldman, C.R. 1960. Primary productivity and limiting factors in three lakes of the Alaska Peninsula. Ecological Monographs 30:207–230.

Grazia, E. and S. Mauger. 2011. Overview of water temperature criteria developed by EPA, Oregon, Washington, Idaho, and Alaska and considerations for improving protection of Alaska's wild salmon from the impacts of thermal change. Cook InletKeeper, Homer, AK.

Griffiths, J.E., D.E. Schindler, L.S. Balistrieri, and G.T. Ruggerone. 2011. Effects of simultaneous climate change and geomorphic evolution on thermal characteristics of a shallow Alaska lake. Limnology and Oceanography 56 (1): 193–205.

Janetski, D.J., D.T. Chaloner, S.D. Tiegs, and G.A. Lamberti. 2009. Pacific salmon effects on stream ecosystems: a quantitative synthesis. Oecologia 159: 583–595.

Harding, J.N. and J.D. Reynolds. 2014. Opposing forces: evaluating multiple ecological roles of Pacific salmon in coastal stream ecosystems. Ecosphere 5 (12): 157.

Hartman, C.W. and P.R. Johnson. 1984. Environmental atlas of Alaska. University of Alaska, Fairbanks.

Hildrith, D. 2008. A Pilot Study to Conduct a Freshwater Fish Inventory of Tundra Ponds on the Bristol Bay Coastal Plain, King Salmon, Alaska, 2006. Alaska Fisheries Data Series Number 10, July 2008.

Jambor, J.L., D.W. Blowes, and I.M. Ritchie. 2003. Environmental aspects of mine wastes. Short course handbook, vol. 31, Mineralogical Association of Canada, Ottawa, 430 p.

Kendall, N.W., H.B. Rich, L.R. Jensen, and T.P. Quinn. 2010. Climate effects on inter-annual variation in growth of the freshwater mussel (*Anodonta beringiana*) in an Alaskan lake. Freshwater Biology 55:2339–2346.

Klein, E., E.E. Berg, and R. Dial. 2005. Wetland drying and succession across the Kenai Peninsula lowlands, south-central Alaska. Canadian Journal of Forest Research 35 (8): 1931–1941.

Kline, T.C., Jr., J.J. Goering, O.A. Mathisen, P.H. Poe, and P.L. Parker. 1990. Recycling of elements transported upstream by runs of Pacific salmon: I. $\delta15N$ and $\delta13C$ evidence in Sashin Creek, southern Alaska. Canadian Journal of Fisheries and Aquatic Sciences 47:136–144.

LaPerriere, J.D. 1996. Water quality inventory and monitoring, Katmai National Park and Preserve. Alaska Cooperative Fish and Wildlife Research Unit, University of Alaska, Fairbanks, AK.

LaPerriere, J.D. 1997. Limnology of two lake systems of Katmai National Park and Preserve Alaska: physical and chemical profiles, major ions, and trace elements. Hydrobiologia 354: 89–99.

Lind, O. 1985. Handbook of common methods in limnology, 2nd edition. Kendall Hunt Publishing. Dubuque, Iowa.

Marmonier, P., M.J. Dole-Olivier, M.É. Des Creuz Châtelliers. 1992. Spatial distribution of interstitial assemblages in the floodplain of the Rhŏne River. Regulated Rivers: Research and Management 7(1): 75–82.

McNab, W.H. and P.E. Avers. 1994. Ecological subregions of the United States: Chapter 3, Section 126 –Bristol Bay Lowlands. Prepared in cooperation with regional compiles and the ECOMAP team of the U.S. Forest Service. U.S. Forest Service WO-WSA-5. Accessed April 12, 2015 at http://www.fs.fed.us/land/pubs/ecoregions/toc.html.

Moore, J. and C. Shearer. 2011. Water quality and surface hydrology of freshwater flow systems in southwest Alaska: 2010 annual summary report. National Park Service Technical Report NPS/SWAN/NRTR-2011/428.

Munk, L., B. Cohn, and B. Finney. 2010. Historical trace element trends recorded in lake sediment cores from the Southwest Alaska Network of Parks. National Park Service (NPS) Natural Resource Technical Report NPS/SWAN/NRTR-2010/395.

Naiman, R.J., R.E. Bilby, D.E. Schindler, and J.M. Helfield. 2002. Pacific salmon, nutrients and the dynamics of freshwater and riparian ecosystems. Ecosystems (5): 399–417.

Nature Conservancy. 2004. Alaska Peninsula and Bristol Bay ecoregional assessment. Anchorage, AK.

Nowacki, G., P. Spencer, T. Brock, M. Fleming, and T. Jorgenson. 2001. Ecoregions of Alaska and neighboring Territories. U.S. Geological Survey Open File Report 02-297 (map). http://agdc.usgs.gov/data/usgs/erosafo/ecoreg/index.html.

O'Keefe, T.C. and R.T. Edwards. 2002. Evidence for hyporheic transfer and removal of marine-derived nutrients in a sockeye stream in Southwest Alaska. American Fisheries Society Symposium 33: 99–107.

Pebble Limited Partnership (PLP). 2011. Environmental baseline document 2004 through 2008, Appendix B, Iliamna Lake Study. Prepared by HDR Alaska, Inc. for Pebble Limited Partnership.

Pinay, G., T. O'Keefe, R. Edwards, and R.J. Naiman. 2003. Potential denitrification activity of the landscape of a Western Alaska drainage basin. Ecosystems 6:336–343.

Poe, P.H., B. Stables, C.A. Kreger, and O.A. Mathisen. 1977. The production of sockeye salmon in the Kvichak River system during peak cycle years. Fisheries Research Institute, University of Washington.

Rains, M.C. 2011. Water sources and hydrodynamics of closed-basin depressions, Cook Inlet region, Alaska. Wetlands 31:377–387.

Rich, H.B., T.P. Quinn, M.D. Scheuerell, and D.E. Schindler. 2009. Climate and intraspecific competition control the growth and life history of juvenile sockeye salmon (*Onchorynchus Nerka*) in Iliamna Lake, Alaska. Canadian Journal of Fisheries and Aquatic Sciences 66: 238–246.

Rinella, D.J., M.S. Wipfli, C.M. Walker, C.A. Stricker, and R.A. Heintz. 2013. Seasonal persistence of marine-derived nutrients in south-central Alaskan salmon streams. Ecosphere 4 (10): 122.

Rogers, D.E. 1977a. Will fertilization increase growth and survival of juvenile sockeye salmon in the Wood River Lakes? Part C of the final report "Collection and analysis of biological data from the Wood River Lake system, Nushagak district, Bristol Bay, Alaska." Fisheries Research Institute, University of Washington College of Fisheries, FRI-UW-7617-C. Seattle, WA.

Rogers, D.E. 1977b. How to increase catches of sockeye salmon in the Nushagak district. Part F of the final report "Collection and analysis of biological data from the Wood River Lake system, Nushagak district, Bristol Bay, Alaska." Fisheries Research Institute, University of Washington College of Fisheries, FRI-UW-7708. Seattle, WA.

Rogers, D.E. and B.J. Rogers. 1998. Limnology in the Wood River Lakes. Fisheries Research Institute, University of Washington School of Fisheries, FRI-UW-9807. Seattle, WA.

Rogers, D., T. Quinn, and B. Rogers. 1999. Alaska salmon research, annual report 1998 to Bristol Bay processors. Fisheries Research Institute, University of Washington School of Fisheries, FRI-UW-9904. Seattle, WA.

Russell, R. 1980. A fisheries inventory of waters in the Lake Clark National Monument area. Alaska Department of Fish and Game with National Park Service.

Schindler, D.W., R. Wagemann, R.B. Cook, T. Ruszcyzynski, and J. Propkopwich. 1980. Experimental acidification of Lake 223, Experimental Lakes Area: Background data and the first three years of acidification. Canadian Journal of Fisheries and Aquatic Sciences 37:342–354.

Shearer, J. and C. Moore. 2010. Water quality and surface hydrology of freshwater flow systems in southwest Alaska: 2009 annual summary report. National Park Service Technical Report NPS/SWAN/NRTR-2010/304.

Stanford, J.A. and A.R. Gaufin. 1974. Hyporheic communities of two Montana rivers. Science 185: 700–702.

Stanford, J.A. and J.V. Ward. 1988. The hyporheic habitat of river ecosystems. Nature 335: 64–66.

Stottlemeyer, R. and D. Chamberlain. 1987. Chemical, physical, and biological characteristics of Lake Clark and selected surface waters of Lake Clark National Park and Preserve, Alaska: 1986 progress report. Michigan Technical University, Houghton, MI.

USEPA (U.S. Environmental Protection Agency). 2001. Issue paper 5: summary of technical literature examining the physiological effects of temperature on salmonids, prepared as part of Region 10 temperature water quality criteria guidance development project. EPA-910-D-01-005.

USEPA (U.S. Environmental Protection Agency). 2014. An assessment of the potential mining impacts on salmon ecosystems of Bristol Bay, Alaska. EPA-910-R-14-001.

Wahrhaftig, C. 1965. Physiographic divisions of Alaska. U.S. Geological Survey, Professional Paper 482.

Ward, J.V., G. Bretschko, M. Brunke, D. Danielopol, J. Gibert, T. Gonser, and A.G. Hildrew. 1998. The boundaries of river systems: the metazoan perspective. Freshwater Biology 40: 531–569.

Weeks, D.P. 1999. Katmai National Park and Preserve, Alaska. Water resources scoping report. National Park Service Technical Report NPS/NRWRD/NRTR-99-226.

Weeks, D.P. 2003. Lake Clark National Park and Preserve, Alaska, Water resources scoping report. National Park Service Technical Report NPS/NRWRD/NRTR-2001/292.

Whipple, K.X., N.P. Snyder, and K. Dollenmayer. 2000. Rates and processes of bedrock incision by the Upper Ukak River since the 1912 Novarupta ash flow in the Valley of Ten Thousand Smokes, Alaska. Geology 28: 835–838.

Wilkens, A.X. 2002. The Limnology of Lake Clark. Masters Thesis. University of Alaska Fairbanks, Fairbanks, AK.

Woody, C.A. and B. Higman. 2011. Groundwater as essential salmon habitat in Nushagak and Kvichak River headwaters: issues related to mining. *For* Center for Science in Public Participation, Bozeman, MT.

Woody, C.A. and S.L. O'Neal. 2010. Fish surveys in headwater streams of the Nushagak and Kvichak River drainages of Bristol Bay, Alaska, 2008–2010. The Nature Conservancy, Anchorage, AK.

Wurts, W.A. 2002. Alkalinity and hardness in production ponds. World Aquaculture, 33(1): 16–17.

Young, D.B. and C.A. Woody. 2007. Spawning distribution of sockeye salmon in a glacially influenced watershed: the importance of glacial habitats. Transactions of the American Fisheries Society 136: 452–459.

Zamzow, K.L. 2011. Surface water quality near the Pebble prospect, Southwest Alaska, 2009–2010: Nushagak, Kvichak, and Chulitna drainage headwaters. Prepared for The Nature Conservancy, Anchorage, AK.

# 20

# MACROINVERTEBRATE AND DIATOM COMMUNITIES IN HEADWATER STREAMS OF THE NUSHAGAK AND KVICHAK RIVER WATERSHEDS, BRISTOL BAY, ALASKA

Daniel Bogan, Rebecca Shaftel, and Dustin Merrigan, University of Alaska Anchorage
Daniel Rinella, U.S. Fish and Wildlife Service

## INTRODUCTION

The structure of biological communities shifts in response to stressors in their physical and chemical environments, such as nutrient enrichment, toxic chemicals, increased temperature, and sedimentation. For this reason, aquatic macroinvertebrates (insects, crustaceans, etc.) and diatoms (single-celled algae with silica cell walls) are often monitored to assess and track the quality of aquatic environments (Barbour et al. 1999, Hill et al. 2000, Rosenberg et al. 2008). For example, a 2000–2004 survey of biological condition in wadeable streams throughout the 48 contiguous United States found that 28% of the streams were in good condition, 25% were in fair condition, 42% were in poor condition, and that nutrient enrichment and sedimentation were the leading causes of biological degradation (Paulsen et al. 2008). Nutrient enrichment can cause increased algal growth in otherwise healthy streams that are naturally low in nutrients, which can reduce dissolved oxygen required by invertebrates and fish and alter stream habitat. Excessive sedimentation buries stream substrates that are the primary habitat of benthic (i.e., associated with the stream bottom) macroinvertebrates and diatoms.

Because different communities operate on different spatial and temporal scales and are sensitive to different types of impacts (Hughes et al. 2000), monitoring multiple biological communities can enhance the ability to detect and diagnose ecological impairment (Karr and Chu 1999). Macroinvertebrates are the most commonly used community in aquatic monitoring (U.S. EPA 2002). They are mobile, have relatively long life cycles (1–2 years), are taxonomically and functionally diverse, and show a wide range of tolerance to physical and chemical stressors (Resh and Jackson 1993, Barbour et al. 1999). Diatoms are relatively sedentary primary producers with short life cycles (multiple generations/summer) that respond quickly to physical and chemical impacts. A considerable body of research has established diatom species optima for nutrients and trophic status (Van Dam et al. 1994) as well as diatom tolerance to acidification (Van Dam et al. 1994), organic pollution (Palmer 1969, Lange-Bertalot 1979), and sedimentation (Stevenson and Bahls 1999). These attributes can be quantified to detect general environmental condition and to diagnose specific causes of environmental impairment (Karr 1993).

The importance of diatoms and macroinvertebrates in monitoring biological condition is eclipsed by their importance in aquatic food webs. Diatoms are photosynthetic and nutritious and vast numbers of them live in the slippery biofilm layer that covers streambeds. They are an important energy source in streams, especially in settings where an abundance of sunlight reaches the stream channel (Mulholland et al. 2001), as in the treeless uplands of the Bristol Bay region. Along with leaves and other organic matter that falls into streams, diatoms in the biofilm layer are major food sources for the macroinvertebrate community (Wallace and Webster 1996), which, in turn, feed juvenile salmonids rearing in streams (Nielsen 1992).

The Aquatic Ecology Program at the University of Alaska Anchorage's Alaska Natural Heritage Program has been involved in biological monitoring of aquatic resources in Alaska since 1998. Standard operating procedures (SOPs) based on the U.S. Environmental Protection Agency's (USEPA) Rapid Bioassessment Protocols (Barbour et al. 1999) were developed for use in Alaska's wadeable streams (Major and Barbour 2001). The SOPs include macroinvertebrate and diatom sampling and measurements of water quality and instream and riparian habitat. These SOPs have been used to guide the calibration of a macroinvertebrate biological assessment index for the Alexander Archipelago ecoregion (i.e., southeast Alaska) (Rinella et al. 2005) and a macroinvertebrate and diatom index for the Cook Inlet Basin ecoregion (Rinella and Bogan 2007).

Very little publicly available data on these important aquatic communities exist for the Nushagak and Kvichak River watersheds in Bristol Bay, which is troubling in light of the intensive mineral exploration occurring in the area. Work collecting baseline biological data in wadeable streams of the Nushagak and Kvichak River watersheds began in 2008. Benthic macroinvertebrates and diatoms were sampled from a total of 77 streams over three years (2008, 2009, and 2010) (Figure 20.1) during May and June. The main objective was to characterize baseline habitat and biological conditions so that any future impacts from resource development or climate change can be recognized. In this chapter, we have summarized the physical and chemical characteristics of wadeable streams in the Nushagak and Kvichak River watersheds, in addition to describing their macroinvertebrate and diatom communities.

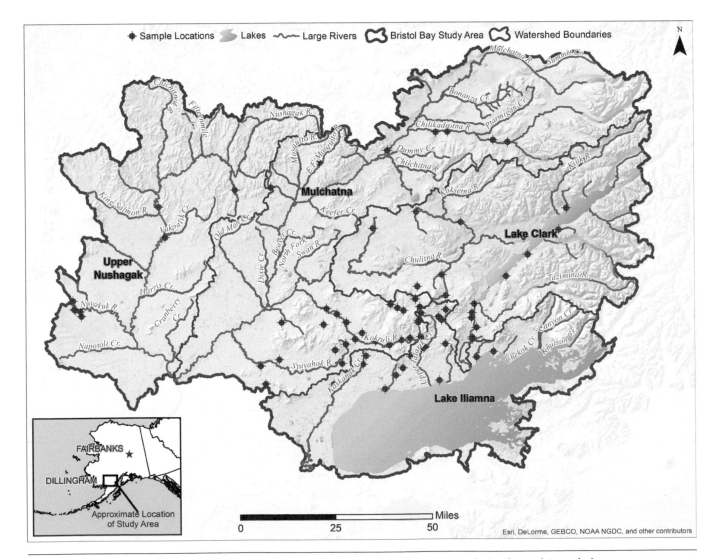

**Figure 20.1**    Location of 77 stream sampling sites in the Nushagak and Kvichak River watersheds of Bristol Bay, Alaska.

# METHODS

## Study Area and Site Selection

Seventy-seven stream sites were sampled for diatoms and macroinvertebrates, located within four major subwatersheds, delineated by the United States Geological Survey (USGS) 8-digit hydrologic unit codes (HUC) (Upper Nushagak, Mulchatna, Lake Clark, and Lake Iliamna; Figure 20.1). In total, these subwatersheds drain an area of approximately 50,000 km² and include the proposed Pebble Mine area, a massive porphyry deposit which has undergone extensive exploration for copper, gold, and molybdenum. The deposit is centered about twenty miles west of the village of Nondalton, and is surrounded by additional mining claims on Alaska state land, which comprises approximately 22% of the total area in the four subwatersheds (Alaska Department of Natural Resources data).

Prior to each field season sampling sites were selected on wadeable streams in the Nushagak and Kvichak River watersheds opportunistically based on the target sampling population and feasibility of access. Streams sampled met the following criteria: 460 m or less in elevation, hard-bottomed substrates, and wadeable along at least half of the stream reach length. The stream reach was the length of stream channel where biological samples were collected and physical habitat measurements were made. These streams were accessed by road, helicopter, float plane, power boat, raft, and hiking.

## Physical Habitat and Water Quality

Physical habitat and water quality measurements followed the USEPA's Wadeable Streams Assessment (2004). Each sampling site consisted of a stream reach with a length of 150 meters or 40 times the average stream wetted width, whichever was greater. Each reach was subdivided with 11 evenly spaced transects and the following was measured at each: channel wetted width, bankfull width (width of the stream at normal high flow), the width of any mid channel bars, bank height, bank angle, and bank undercut distance (Figure 20.2). Riparian canopy coverage was measured with six densiometer readings along each transect. Water depth, substrate size class, and embeddedness (the extent to which rocky substrate is encased in fine sediment) at five points along each transect was recorded. Additionally, the substrate size (USEPA 2004) at

**Figure 20.2**  Collecting stream habitat measurements at one of 11 evenly spaced transects along the stream reach.

midpoints between each of the 11 transects, for a total of 21 transects and 105 measurements, was recorded. Fish cover was estimated based on the extent of filamentous algae, macrophytes, big and small woody debris, live trees or roots, overhanging vegetation, undercut banks, and boulders at each transect using five areal cover classes (0%, < 10%, 10–40%, 40–75%, or > 75%). Between each of the 11 transects, pieces of large woody debris were counted within and above the bankfull channel according to several size classes. Riparian vegetation cover was characterized separately for canopy, understory, and ground cover along the entire stream reach using five areal cover classes (0%, < 10%, 10–40%, 40–75%, or > 75%) and several types (deciduous, coniferous, mixed, or none). To characterize channel slope and sinuosity, the compass azimuth (aspect) and channel slope were recorded between each pair of transects. At the center transect, stream discharge was measured using the velocity-area method and a Marsh-McBirney flow meter and temperature, dissolved oxygen, specific conductance, and pH were measured in-situ with a Hydrolab MiniSonde 5 multiprobe that was calibrated daily.

## Macroinvertebrates

Macroinvertebrate sampling and processing followed Major and Barbour (2001): a modification of the USEPA Rapid Bioassessment Protocols (Barbour et al. 1999) for use in Alaska. Macroinvertebrate samples were collected throughout a 100-m reach at each site with a 350-µm-mesh D-frame net (Figure 20.3). Each sample was a composite of 20 subsamples collected from various instream habitats in proportion to each habitat's abundance. Cobbles and gravel in riffles were the predominant substrate sampled, with submerged streambanks and large woody debris comprising smaller portions. To collect riffle samples an area of streambed approximately 0.14 m² (1.5 ft.²) to a depth of 10 cm (4 in.) was disturbed and each cobble and boulder rubbed by hand to ensure all macroinvertebrates were dislodged and swept into the net by the stream's current. Streambanks were sampled by making three successive sweeps of the net across a 0.14 m² (1.5 ft.²) area while rapidly jabbing the net into the substrate. Woody debris was sampled by manually scouring a 0.14 m² (1.5 ft.²) area of wood

**Figure 20.3** Sampling macroinvertebrates with a D-frame net.

immediately upstream of the net. All field samples were preserved with ethanol and returned to the Aquatic Ecology lab at the University of Alaska Anchorage for processing. In the lab, each macroinvertebrate sample was subsampled to obtain a fixed count of 300 ± 20% organisms to standardize the taxonomic effort across all sites. In addition, a five-minute search through the remaining sample was conducted to select any large or rare taxa that may have been missed during subsampling. Insects were identified to genus or lowest practical taxonomic level, including Chironomidae, and non-insects to a higher taxonomic level (usually family or order) using standard taxonomic keys (Weiderholm 1983, Pennak 1989, Wiggins 1996, Thorpe and Covich 2001, Stewart and Oswood 2006, Merritt et al. 2008).

## Diatoms

Diatom sampling and processing followed the USEPA's Rapid Bioassessment Protocols (Stevenson and Bahls 1999). Each diatom sampling reach consisted of four consecutive riffles; from each riffle four stones (cobble or large gravel) were selected, ensuring that algal coverage on the stones was visually representative of the riffle at large. A standardized surface area on each stone (4.5 cm diameter circle) was scrubbed with a small brush and the loosened algal layer was rinsed into a washtub (Figure 20.4). For each stream, algae from all stones (4 stones × 4 riffles) was composited into a single sample and preserved with Lugol's solution.

In the lab, each sample was homogenized and then 20 ml was transferred into a clean beaker. Nitric acid and heat were added to digest the diatom protoplasm and other organic material, thereby clearing the diatoms for easier identification. The acid digested diatom frustules were then neutralized through a series of rinses with deionized water and slide mounted using NAPHRAX mounting medium. For each sample site, a fixed count of 600 diatom valves was identified to species or lowest practical taxonomic level. The slide was then scanned and any rare taxa not discovered in the fixed count were recorded. The primary taxonomic references were Krammer and Lange-Bertalot (1986–1991) and Patrick and Reimer (1975).

**Figure 20.4** Collecting diatoms by scrubbing four rocks from four successive riffles.

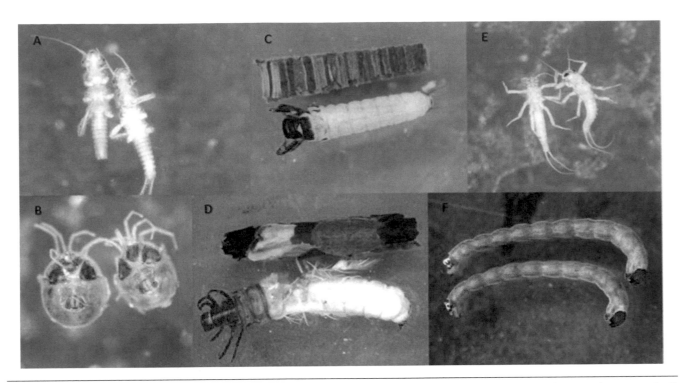

**Figure 20.5** Some common macroinvertebrates found in Bristol Bay area streams. A. *Zapada* sp.; B. *Hygrobates* sp. (water mite); C. *Brachycentrus americanus*; D. *Onocosmoecus* sp.; *Baetis bicaudatus*; F. *Orthocladius (Euorthocladius)* sp. (Chironomidae). *Photo credit: Dustin Merrigan.*

Mean diatom richness from the fixed count across the four subwatersheds ranged from 26 to 35 (Table 20.5). Individual site richness ranged from a low of 14 at a tributary of Sixmile Lake (just upstream of its outlet to the Newhalen River) to a high of 51 at Rock Creek (just upstream of its confluence with Groundhog Creek), both of which are in the Lake Clark subwatershed.

A total of 19 diatom taxa occurred at more than 50% of the sampled sites (Table 20.6). Only one taxon *Achnanthidium minutissimum*, a common widely distributed species, had a mean abundance greater than 10%. Taxa from the family Fragilariaceae were the most commonly encountered, with *Fragilaria capucina, Fragilaria vaucheriae, Staurosirella pinnata, Meridion ciruculare, Synedra ulna, Hannaea arcus*, and *Diatoma mesodon* all occurring in more than 75% of the sites. In addition, *Tabellaria flocculosa* (recently reclassified from the Fragilariaceae) was one of the most pervasive diatoms, encountered in 95% of sampled sites. The closely related diatoms *Encyonema silesiacum* and *E. minutum* were found in over 80% of sampled sites, while two monoraphid diatoms, *Psammothidium subatomoides* and *Cocconeis placentula*, were present in over 75% of sampled sites. Some common diatoms encountered are pictured in Figure 20.6.

**Table 20.5**   Taxonomic richness of diatoms by subwatershed

| Subwatershed | Total sites[1] | Richness[2] (no. of taxa) |
|---|---|---|
| Lake Clark | 18 | 27 (14–51) |
| Lake Iliamna | 17 | 35 (25–46) |
| Mulchatna R. | 33 | 30 (16–45) |
| Upper Nushagak R. | 8 | 26 (17–36) |

[1] 76 sites because one site was not sampled for diatoms.

[2] Richness is based on the 600-valve (300 organism) fixed count and does not include taxa identified in the secondary scan.

**Table 20.6**  Frequencies and abundances of common diatom taxa. Only taxa that have frequency greater than 50% are shown. Total frequency and frequency within each of the subwatersheds is provided along with the mean percent abundance across all of the sites

| Family | Species | Frequency (%) | Mean abundance (%) | Frequency in each subwatershed (%) | | | |
|---|---|---|---|---|---|---|---|
| | | | | Lake Clark | Lake Iliamna | Mulchatna River | Upper Nushagak R. |
| Fragilariaceae | *Fragilaria capucina* | 95 | 4 | 89 | 94 | 97 | 100 |
| Tabellareaceae | *Tabellaria flocculosa* | 95 | 6 | 94 | 94 | 94 | 100 |
| Achnanthidiaceae | *Achnanthidium minutissimum* | 92 | 17 | 100 | 100 | 91 | 63 |
| Fragilariaceae | *Fragilaria vaucheriae* | 88 | 2 | 83 | 88 | 91 | 88 |
| Cymbellaceae | *Encyonema silesiacum* | 87 | 2 | 94 | 82 | 82 | 100 |
| Fragilariaceae | *Staurosirella pinnata* | 86 | 6 | 83 | 94 | 82 | 88 |
| Cymbellaceae | *Encyonema minutum* | 80 | 2 | 67 | 94 | 88 | 50 |
| Fragilariaceae | *Meridion circulare* | 79 | 6 | 67 | 71 | 88 | 88 |
| Fragilariaceae | *Synedra ulna* | 79 | 2 | 72 | 82 | 79 | 88 |
| Achnanthaceae | *Psammothidium subatomoides* | 78 | 2 | 89 | 76 | 76 | 63 |
| Fragilariaceae | *Hannaea arcus* | 78 | 8 | 78 | 76 | 76 | 88 |
| Fragilariaceae | *Diatoma mesodon* | 76 | 7 | 56 | 71 | 91 | 75 |
| Achnanthidiaceae | *Eucocconeis laevis* | 75 | 2 | 100 | 71 | 73 | 38 |
| Aulacoseiraceae | *Aulacoseira alpigena* | 74 | 2 | 61 | 65 | 85 | 75 |
| Cymbellaceae | *Reimeria sinuata* | 71 | 3 | 83 | 88 | 58 | 63 |
| Achnanthidiaceae | *Planothidium haynaldii* | 67 | 3 | 39 | 76 | 76 | 75 |
| Fragllariaceae | *Pseudostaurosira brevistriata* | 63 | 4 | 50 | 76 | 61 | 75 |
| Gomphonemataceae | *Gomphonema micropus* | 57 | 1 | 61 | 53 | 55 | 63 |
| Cocconeidaceae | *Cocconeis placentula* | 53 | 2 | 83 | 65 | 33 | 38 |

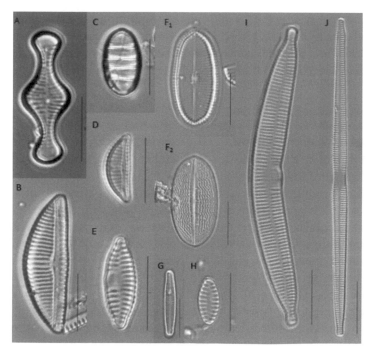

**Figure 20.6**  Some common diatoms found in Bristol Bay area streams. A. *Tabellaria flocculosa*; B. *Encyonema silesiacum*; C. *Diatoma mesodon*; D. *Encyonema minutum*; E. *Fragilaria vaucheriae*; F. *Cocconeis placentula* (1 = raphe bearing valve, 2 = rapheless valve); G. *Achnanthidium minutissimum*; H. *staurosirella pinnata*; I. *Hannaea arcus*; and J. *Synedra ulna*. Photo credit: Daniel Bogan.

Luís, A.T., P. Teixeira, S.F.P. Almeida, L. Ector, J.X. Matos, and E.A. Ferreira de Silva. 2009. Impact of acid mine drainage (AMD) on water quality, stream sediments and periphytic diatom communities in the surrounding streams of Aljustrel mining area (Portugal). Water, Air, and Soil Pollution 200:147–167.

Magurran, A. E., S. R. Baillie, S. T. Buckland, J. M. Dick, D. A. Elston, E. M. Scott, R. I. Smith, P. J. Somerfield, and A. D. Watt. 2010. Long-term datasets in biodiversity research and monitoring: assessing change in ecological communities through time. Trends in Ecology and Evolution 25:574–582.

Major, E.B. and M.T. Barbour. 2001. Standard operating procedures for the Alaska Stream Condition Index: a modification of the U.S. EPA Rapid Bioassessment Protocols. Fifth edition. Environment and Natural Resources Institute, University of Alaska Anchorage, Anchorage, AK.

Mauger, S. and T. Troll. 2014. Implementation plan: Bristol Bay regional water temperature monitoring network. Prepared for the Western Alaska LCC and the Southwest Alaska Salmon Habitat Partnership. Cook Inletkeeper, Homer, AK. 21 p.

Merritt, R.W., K.W. Cummins, and M.B. Berg (editors). 2008. An Introduction to the Aquatic Insects of North America. Fourth edition. Kendall/Hunt, Dubuque, IA.

Milner, A.M. and R.J. Piorkowski. 2004. Macroinvertebrate assemblages in streams of interior Alaska following alluvial gold mining. River Research and Applications 20:719–731.

Milner, A.M., A. Woodward, J.E. Freilich, R.W. Black, and V.H. Resh. 2016. Detecting significant change in stream benthic macroinvertebrate communities in wilderness areas. Ecological Indicators 60:524–537.

Mulholland, P.J., C.S. Fellows, J.L. Tank, N.B. Grimm, J.R. Webster, S.K. Hamilton, E. Marti, L. Ashkenas, W.B. Bowden, W.K. Dodds, W.H. McDowell, M.J. Paul, and B.J. Peterson. 2001. Inter-biome comparison of factors controlling stream metabolism. Freshwater Biology 46:1503–1517.

Nielsen, J.L. 1992. Microhabitat-specific foraging behavior, diet, and growth of juvenile Coho salmon. Transactions of the American Fisheries Society 121: 617–634.

Oswood, M.W. 1989. Community structure of benthic invertebrates in interior Alaskan (USA) streams and rivers. Hydrobiologia 172: 97–110.

Palmer, C.M. 1969. A composite rating of algae tolerating organic pollution. Journal of Phycology 5:78–82.

Patrick, R. and C.W. Reimer. 1975. The Diatoms of the United States, Exclusive of Alaska and Hawaii. The Academy of Natural Sciences of Philadelphia, Philadelphia, PA.

Paulsen, S.G., A. Mayio, D.V. Peck, J.L. Stoddard, E. Tarquinio, S.M. Holdsworth, J. Van Sickle, L.L. Yuan, C.P. Hawkins, A.T. Herlihy, P.R. Kaufman, M.T. Barbour, D.P. Larsen, and A.R. Olsen. 2008. Condition of steam ecosystems in the US: an overview of the first national assessment. Journal of the North American Benthological Society 27:812–821.

Pennak, R.W. 1989. Fresh-Water Invertebrates of the United States: Protozoa to Mollusca. Third edition. John Wiley & Sons, Inc.

Relyea, C.D., G.W. Minshall, and R.J. Danehy. 2012. Development and validation of an aquatic fine sediment biotic index. Environmental Management 49:242–252.

Resh, V.H. and J.K. Jackson. 1993. Rapid assessment approaches to biomonitoring using benthic macroinvertebrates. Pages 195–233 in D.M. Rosenberg and V.H. Resh (Eds.), *Freshwater Biomonitoring and Benthic Macroinvertebrates*. Kluwer Academic Publishers, Norwell, MA.

Rinella, D.J. and D.L. Bogan. 2007. Development of macroinvertebrate and diatom biological assessment indices for Cook Inlet Basin streams. Prepared for Alaska Dept. of Environmental Conservation. Environment and Natural Resources Institute, University of Alaska Anchorage, Anchorage, AK.

Rinella, D.J., D.L. Bogan, and K. Kishaba. 2005. Development of a macroinvertebrate bioassessment index for Alexander Archipelago streams. Prepared for Alaska Dept. of Environmental Conservation. Environment and Natural Resources Institute, University of Alaska Anchorage, Anchorage, AK.

Rosenberg, D.M., V.H. Resh, and R.S. King. 2008. Use of Aquatic Insects in Biomonitoring. In: *An Introduction to Aquatic Insects of North America*, Fourth edition. Kendall/Hunt Publishing Company, Dubuque, Iowa. p. 123–137.

Stevens, D.L. and A.R. Olsen. 2004. Spatially balanced sampling of natural resources. Journal of the American Statistical Association 99:262–278.

Stevenson, R.J. and L.L. Bahls. 1999. Periphyton Protocols. Pages 6.1–6.22. In: M.T. Barbour, J. Gerritson, B.D. Snyder, and J.B. Stribling (Eds.), *Rapid Bioassessment Protocols for Use in Streams and Wadeable Rivers*. Second edition. US Environmental Protection Agency, Office of Water, Washington, DC. EPA 841-B-99-002.

Stewart, K.W. and M.W. Oswood. 2006. *The Stoneflies (Plecoptera) of Alaska and Western Canada*. The Caddis Press, Columbus, OH.

Thorpe, J.H. and A.P. Covich (editors). 2001. *Ecology and Classification of North American Freshwater Invertebrates*. Second edition. Academic Press.

Townsend C.R., S. Dolédec, R. Norris, K. Peacock, and C. Arbuckle. 2003. The influence of scale and geography on relationships between stream community composition and landscape variables: description and prediction. Freshwater Biology 48:768–785.

USEPA. 2002. Summary of Bioassessment Programs and Biocriteria Development for States, Tribes, Territories, and Interstate Commissions: Streams and Wadeable Rivers. U.S. Environmental Protection Agency, Office of Environmental Information and Office of Water, Washington, DC. EPA-822-R-02-048.

USEPA. 2004. Wadeable Streams Assessment: Field Operations Manual. U.S. Environmental Protection Agency, Office of Water and Office of Research and Development, Washington, DC. EPA841-B-04-004.

USEPA. 2014. An Assessment of Potential Mining Impacts on Salmon Ecosystems of Bristol Bay, Alaska. Region 10, Seattle, WA. EPA 910-R-14.001.

Van Dam, H., A. Mertens, and J. Sinkeldam. 1994. A coded checklist and ecological indicator values of freshwater diatoms from the Netherlands. Netherlands Journal of Aquatic Ecology 28:117–133.

Van Damme, P.A., C. Hamel, A. Ayala, and L. Bervoets. 2008. Macroinvertebrate community response to acid mine drainage in rivers of the high Andes (Bolivia). Environmental Pollution 156:1061–1068.

Wagener, S.M. and J.D. LaPerriere. 1985. Effects of placer mining on three invertebrate communities of interior Alaska streams. Freshwater Invertebrate Biology 4:208–214.

Wallace, J.B. and J.R. Webster. 1996. The role of macroinvertebrates in stream ecosystem function. Annual Review of Entomology 41:115–139.

Weiderholm, T. 1983. Chironomidae of the Holarctic Region: Keys and Diagnoses. Part 1, Larvae. Entomologica Scandinavica 19:1–457.

Woody, C.A., R. Shaftel, D. Rinella, and D. Bogan. 2014. Long-term monitoring plan for wadeable streams, Lime Hills Ecoregion, Kvichak and Nushagak watersheds. Center for Science in Public Participation and University of Alaska Anchorage. Report prepared for the Bristol Bay Heritage Land Trust. http://www.bristolbaylandtrust.org/reports/.

Wiggins, G.B. 1996. Larvae of the North American Caddisfly Genera (Trichoptera). Second edition. University of Toronto Press.

Zalack, J.T., N.J. Smucker, and M.L. Vis. 2010. Development of a diatom index of biotic integrity for acid mine drainage impacted streams. Ecological Indicators 10:287–295.

Zamzow, K.L. 2011. Investigations of surface water quality in the Nushagak, Kvichak, and Chulitna watersheds, Southwest Alaska 2009–2010. The Nature Conservancy, Anchorage AK.

# 21

# FRESHWATER NON-SALMON FISHES OF BRISTOL BAY

Carol Ann Woody, Center for Science in Public Participation (currently with the National Park Service)

## INTRODUCTION

Bristol Bay is world-renowned for supporting the world's largest most valuable wild salmon fisheries (see Rinella et al. this volume). In the watershed surrounding Bristol Bay there are a myriad of freshwater lakes, rivers, and streams where anadromous salmon spawn and other freshwater species live year round. These non-salmon freshwater fishes provide food security, economic stability, and world-class recreational fishing opportunities.

All Bristol Bay finfish species are customarily harvested or used for subsistence purposes according to the Alaska Board of Fisheries; the amount of non-salmon finfish considered necessary for subsistence use in the region was estimated to be 113,398 kg/year (250,000 lbs/yr) (Krieg et al. 2015). The 25 remote communities of Bristol Bay (Figure 21.1), which are not on any interstate system and are only accessible by plane or boat, rely on a diversity of wild renewable resources (e.g., fish, moose, caribou, marine mammals, berries, etc.) annually harvesting an average of about 149 kg/person (~329 lbs/person) or about 493 kg/household (1,087 lbs/household) (Krieg et al. 2015). Freshwater fish (non-salmon) comprise 9–11% of total wild harvests, with per capita use estimated at 21 kg/person (47 lbs/person) and range from 3.6–45.8 kg/person (8–101 lbs/person). Estimated 2008 costs to replace wild subsistence foods ranged from $4,851/household in Levelock to $14,973/household in Koliganek, two Kvichak Watershed villages (Krieg et al. 2015). In contrast to the salmon resource, which most remote communities can only intercept during upriver spawning runs, freshwater fish provide year-round harvest opportunities (Fall et al. 1996, Krieg et al. 2005, Hazell et al. 2015).

The majority of Bristol Bay households participate in non-salmon fish harvests, particularly outside of the two large regional centers, Dillingham and King Salmon-Naknek, where freshwater fish comprise less than 10% of the total subsistence harvest (Krieg 2005). Although harvests vary in composition among villages and years, freshwater species most frequently harvested for subsistence include: Arctic grayling (*Thymallus arcticus*); three char species, Dolly Varden (*Salvelinus malma*), Arctic char (*S. alpinus*), and lake trout (*S. namaycush*); rainbow trout (*Oncorhynchus mykiss*); various whitefish species, namely humpback whitefish (*Coregonus pidschian*) and round whitefish (*Prosopium cylindraceum*); and northern pike (*Esox lucius*) (Table 21.1). Alaska Peninsula villages (e.g., Egegik, Pilot Point, Ugashik, and Port Heiden) (Figure 21.1) harvest fewer freshwater fish and more anadromous smelt (spawn in freshwater, migrate to sea) during their upriver spawning runs (Fall 1996).

Recreational or sport fishing is the second most important economic sector in Bristol Bay—after commercial fishing. In 2007, southcentral area anglers—which includes Bristol Bay—spent an estimated $989 million, directly generating $240 million of income to 7,897 full and part-time jobs (Southwick et al. 2008). The Bristol Bay recreational fishing effort averaged 77,269 angler days (angler day = one angler fishing one day) during 2009–2014 and totaled 89,942 angler days in 2014 (Dye and Fo 2015); note these data include both freshwater non-salmon and salmon fishing efforts. The most recent per angler expenditure estimate shows Bristol Bay sport fishers spent an average of $282.75 per angler day of fishing (Southwick et al. 2008), which calculates roughly to $25.4 million dollars ($282.75 × 89,942 angler days). Duffield et al. (2007) estimated that $61 million was spent in Alaska during 2005 on Bristol Bay fishing trips.

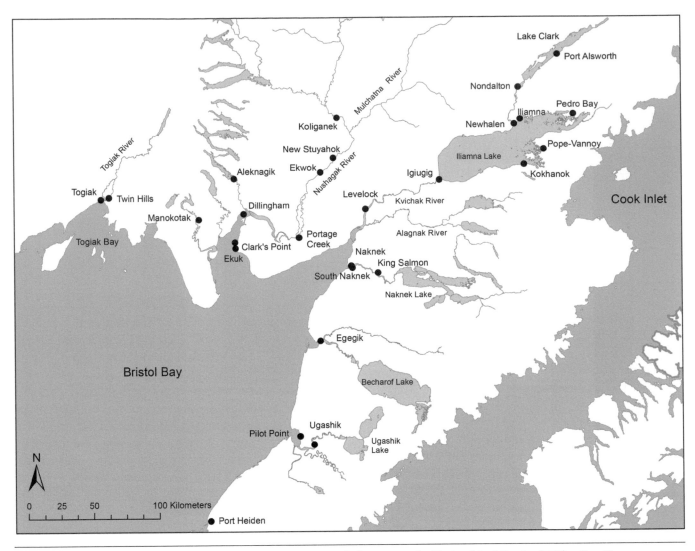

**Figure 21.1**    The communities of Bristol Bay located in southwest Alaska. *Map credit: National Park Service (NPS)— Dan Young.*

Sport fishers reported catching more than 350,000 fish during 2015 (Table 21.2), with Arctic grayling, char (Dolly Varden, Arctic char, lake trout), rainbow trout, and northern pike comprising the majority of catches (Table 21.2). It is important to note that only a small percent of these fish are actually harvested for food because many sport fishers practice *catch and release* fishing whereby captured fish are just viewed or photographed then returned to the water alive.

It's clear that freshwater fishes, especially salmonids (e.g., grayling, char, trout) and pike, are important to humans that subsist and recreate in Bristol Bay. It is also important to remember that the freshwater fish species that we tend to overlook or discount due to their small size, unappealing looks, or cryptic coloring or behavior, all play important roles in the greater Bristol Bay ecosystem. Most of the ubiquitous smaller fish species, such as sculpin (Figure 21.2), or stickleback may not be our preferred food source, but they are important prey for the larger fish that we humans prefer to catch, harvest, and eat.

**Table 21.1** Example of non-salmon subsistence fish harvests in the Bristol Bay region. Data are for five communities in the Kvichak River Drainage. Data from the Alaska Department of Fish and Game, Subsistence Division (see Krieg et al. 2005)

| Resource | mid-1980s | | | mid-1990s | | | 2002–2003 | | |
|---|---|---|---|---|---|---|---|---|---|
| | Total kg (lbs) | Per Capita kg (lbs) | % of total | Total kg (lbs) | Per Capita kg (lbs) | % of total | Total kg (lbs) | Per Capita kg (lbs) | % of total |
| Arctic grayling | 1,426 (3,143) | 1.5 (3.4) | 6.9 | 1,663 (3,666) | 1.8 (4.0) | 7.4 | 583 (1,286) | 0.8 (1.7) | 7.1 |
| Arctic char/Dolly Varden | 4,322 (9,528) | 4.7 (10.3) | 20.9 | 3,658 (8,065) | 4 (8.7) | 16.2 | 2,217 (4,887) | 2.9 (6.3) | 26.9 |
| Lake trout | 2,450 (5,401) | 2.6 (5.8) | 11.8 | 1,038 (2,289) | 1.1 (2.5) | 4.6 | 670 (1,476) | 0.9 (1.9) | 8.1 |
| Rainbow trout | 3,924 (8,651) | 4.3 (9.4) | 19.0 | 5,443 (11,999) | 5.9 (13.0) | 24.1 | 2,541 (5,602) | 3.3 (7.2) | 30.9 |
| Northern pike | 2,504 (5,521) | 2.7 (6.0) | 12.1 | 2,297 (5,063) | 2.5 (5.5) | 10.2 | 812 (1,789) | 1.0 (2.3) | 9.9 |
| Whitefish | 5,050 (11,133) | 5.4 (12.0) | 24.4 | 6,055 (13,349) | 6.6 (14.5) | 26.8 | 683 (1,505) | 0.9 (1.9) | 8.3 |
| Burbot | 595 (1,311) | 0.6 (1.4) | 2.9 | 99 (219) | 0.1 (0.2) | 0.4 | 29 (64) | 0.05 (0.1) | 0.4 |
| Smelt | 259 (570) | 0.3 (0.6) | 1.2 | 1,385 (3,054) | 1.5 (3.3) | 6.1 | 299 (658) | 0.4 (0.8) | 3.6 |
| Sucker | 165 (363) | 0.2 (0.4) | 0.8 | 934 (2,060) | 1 (2.2) | 4.1 | 402 (887) | 0.5 (1.1) | 4.9 |
| Totals | 20,695 (45,625) | 22.4 (49.3) | 100 | 22,575.3 (49,770) | 24.5 (53.9) | 99.9 | 8,235 (18,154) | 10.6 (23.3) | 100 |

**Table 21.2** Bristol Bay recreational catch of freshwater species during 2010–2015. Note only a small fraction of these fish were actually harvested for food because many recreational fishers practice catch and release fishing. Data courtesy of the Alaska Department of Fish and Game, Division of Sport fish, Dillingham, Alaska.

| Species | 2010 | 2011 | 2012 | 2013 | 2014 | 2015 | mean 2010–2015 |
|---|---|---|---|---|---|---|---|
| Arctic grayling | 42,471 | 35,913 | 36,569 | 55,361 | 38,090 | 74,760 | 47,197 |
| Dolly Varden/ Arctic char | 76,116 | 64,277 | 60,770 | 76,723 | 97,120 | 107,033 | 80,340 |
| Lake trout (lake char) | 9,664 | 9,295 | 1,659 | 5,357 | 4,995 | 7,014 | 6,331 |
| Rainbow trout | 193,229 | 155,765 | 128,574 | 153,146 | 182,736 | 191,034 | 167,414 |
| Northern pike | 6,608 | 3,857 | 4,927 | 8,016 | 4,787 | 9,112 | 6,218 |
| Total | 328,088 | 269,107 | 232,499 | 298,603 | 327,728 | 388,953 | 307,496 |

**Figure 21.2**   Two sculpin species commonly found in the freshwaters of Bristol Bay are the slimy sculpin and the coastrange sculpin (pictured here). Both species are small, well camouflaged, ubiquitous, and are important prey for many fish-eating species such as rainbow trout, common mergansers, and mink. *Photo credit: Carol Ann Woody.*

# BRISTOL BAY FRESHWATER FISH SPECIES

## Species Important to Subsistence and the Economy

Provided in the following text are basic information on the non-salmon freshwater fish species that are most important to the people of Bristol Bay, with brief reference to regional studies. The most important non-salmon species in Bristol Bay, relative to humans, are closely related to salmon and belong to the Salmonidae family, including: Arctic grayling; Dolly Varden char, Arctic char, and lake trout, also called lake char; rainbow trout; and whitefish, primarily humpback whitefish. Northern pike are also important to the people of Bristol Bay, but belong to the Esocidae family and will also be considered here.

All salmonids spawn in freshwater and some spend their entire life cycle there. Some remain in freshwater their entire life cycle but have the capacity for anadromy and can go to sea. For example, in Bristol Bay, Dolly Varden char originating in the Iliamna River of the Kvichak watershed do not migrate to sea (Jaecks 2010) but those originating from the Togiak National Wildlife Refuge (NWR) do (Lisac 2009).

Why do some salmonids within a species go to sea while others don't? Freshwater and anadromous life histories represent trade-offs between survival to first reproduction, and size and age at first reproduction. Within a species, freshwater residents generally mature at a smaller size, younger age and exhibit higher survival to reproduction and a greater probability of iteroparity (repeat spawning) compared to anadromous forms (Fleming and Reynolds 2004). Fish with potential for anadromy will likely go to sea if the benefits (e.g., increased growth, increased fecundity) outweigh the costs (e.g., higher mortality rates) (Gross 1987). Whether or not a salmonid goes to sea can be influenced by many factors including genetics, individual condition (e.g., lipid levels), and environmental factors (see review by Kendall et al. 2015).

Salmonids exhibit fidelity to feeding and freshwater natal habitats for spawning. Spawning site fidelity can help ensure that fish repeatedly spawn in habitats that maximize their reproductive success; spawning site fidelity can also reduce genetic exchange among populations. Reproductive isolation among populations in concert with differences in natural selection regimes among habitats (e.g., small, cold, shallow stream with lots of predators versus a large, warm, deep river with few predators), can contribute to differences evolving in a population's observable genetically based traits that effect reproductive success—such as age and size at maturity, spawn timing, among others (e.g., see Taylor 1991, Stearns 1992, Quinn 2005).

Salmonids are generally opportunistic feeders and will eat what is available, although they will migrate, sometimes great distances, to exploit a particularly rich food source. The hundreds of Bristol Bay salmon spawning habitats represent rich freshwater food sources, where salmon eggs, salmon carcasses, blowflies, and benthic insects stirred up from the substrate through the spawning act are available to hungry predators. Because salmon spawn time varies among habitats, freshwater fish can migrate to different smorgasbords and capitalize on the rich food subsidies.

Salmonids are native to cooler Northern Hemisphere waters, but introduced populations of some species, such as rainbow trout, occur widely. Many native wild fish populations are threatened, endangered, or extinct in the southern portions of their range (e.g., Hubbs and Lagler 1958, Dunham et al. 2008, NOAA 2017). Curious readers are encouraged to pursue further learning regarding Bristol Bay species habitats, habits, life cycles, and distributions from the references provided throughout this chapter as well as at the Alaska Department of Fish and Game (ADF&G) website: http://www.adfg.alaska.gov/index.cfm?adfg=species.main and at their fish distribution website: https://www.adfg.alaska.gov/sf/SARR/AWC/index.cfm?ADFG=main.interactive.

## Arctic Grayling

### Distinguishing Characteristics
Arctic grayling are easily distinguished by their large dorsal fin, which is often outlined in red (Figure 21.3), and their iridescent purple, blue, red, and sometimes golden coloration, which is particularly beautiful in males during spawning season.

### Distribution

Arctic grayling are a Holarctic species of salmonid that are widely distributed throughout freshwaters of Bristol Bay and mainland Alaska (ADF&G 2017). They occur on St. Lawrence and Nunivak Islands, both of which were once part of the Bering Land Bridge; they do not occur on the Aleutians. Grayling occur in northern Canada west of Hudson Bay and

**Figure 21.3** Arctic grayling male (top) and female (bottom); note the iridescent, sail-like dorsal fin. *Photo credit: U.S. Fish and Wildlife Service (USFWS).*

- At elevations ranging from 13–439 m (42.6–1,440.3 ft.) above mean sea level;
- Ranged in size from 49–8,049 ha (121–19,889 acres) (average 952 ha or 2,352 acres); and
- Had maximum depths ranging from 9–143 m (29–469 ft.) and mean depths ranging from 2.5–56.7 m (8.2–186 ft.).

Most of the lakes had a neutral pH (range 6.5–8.0), low conductivity averaging 42.2 micro siemens (μS; range 24–81 μS), and alkalinity averaging 34.8 mg/l (range 3–68 mg/l).

Russell (1980) surveyed lakes in the then proposed Lake Clark National Park and reported Arctic char in all surveyed lakes except those draining to Cook Inlet, the upper Stony River, and the upper Mulchatna River. He sampled 511 Arctic char from 10 lakes and found fish ranged in size from 32–580 mm FL (1.3–22.8 in.), and growth rates appeared exceptionally slow. The largest fish captured weighed 2,380 gs (5 lb 4 oz) and the maximum age observed was 15 years. He sampled 99 fish for age at maturity; fish matured as early as one, and most were mature by the age of six; average sizes reported for five-year-old fish from the different lakes ranged from 224–341 mm FL (8.8–13.4 in.).

Power et al. (2005) examined variation in fecundity for 32 North American populations of Arctic char across a latitudinal gradient for dwarf, normal, and anadromous populations and found fecundity declined in more northern populations. There was no indication of a latitudinal cline in egg size (e.g., larger eggs at higher latitudes or vice versa). The FL adjusted fecundity of Arctic char morphotypes ranged from: dwarf, 29–304 eggs (average = 115; CV = 53%); normal, 55–3,539 eggs (average 805; CV = 95%); and anadromous 1,000–10,692 eggs (average 4,180; CV = 37%) (Power et al. 2005). The consistent egg size among populations suggests that the fecundity declines that were observed in more northern populations were due to reduced growth afforded in the less productive, colder habitats.

## Lake Trout or Lake Char

### Distinguishing Characteristics

Lake trout are misnamed—they are not a trout but a char (*Salvelinus* spp.), and in the scientific literature are often referred to as lake char. Similar to other char, they have a dark fusiform body with light spots. (Figure 21.6) Their body is dark green to grey or brown colored with wormy or irregular light-colored spots on the back, fins, and body. Spots are never yellow or pink as in the other char and their tail is more deeply forked. The belly is white. Sometimes orange-red occurs on paired fins, especially in northern populations; the anterior edges of paired and anal fins sometimes have a white border.

### Distribution

The natural range of the lake trout correlates with the maximum limits of the Pleistocene glaciers (see Lindsey 1964). In Alaska, lake trout (char) are generally found in large glacially scoured lakes and their tributaries; they are not found in the Yukon-Kuskokwim drainages or in southeast Alaska (ADF&G 2017). Their range extends from Alaska across northern Canada, south to the Great Lakes basin and New England. They are widely introduced outside of their native range.

### Habitats and Life History

Lake trout appear limited to lake systems created by glacial scouring. They are found in many of Bristol Bay's large clear and glacial lakes and their tributaries. Studies in the Great Lakes (e.g., Behnke 1972, Chaverie et al. 2016) and Canada verify that like other char species, lake char populations within a particular lake system can look and behave differently from each other and the differences appear related to their use of different feeding, spawning, and/or habitat niches. Over time, phenotypic (observable) differences can evolve with up to four different morphs occurring within lakes based on diet (insects versus fish) and habitat use (e.g., shallow water versus deep); different morphs were also documented in a small lake (see Table 1 in Chaverie et al. 2016). Lake char studies are generally lacking in Bristol Bay, therefore it is unknown whether or not different forms or morphs occur.

Lake char can mature later than other salmonids in Bristol Bay. Russell (1980) surveyed 13 of the then proposed LACL lakes and found the youngest sexually mature lake char to be six years old (mean FL = 248 mm = 9.8 in.) and 74% of the breeding population to be comprised of fish that are 11–16 years old. The oldest lake char was 29 years old and measured 663 mm FL (26.1 in.) (Russell 1980); individuals can live to the age of 51 (Keyse et al. 2007). An age and size at sexual maturity study for fish in 11 Alaska lakes showed that 50% of females realized maturity at ages ranging from 4.9–14.3

**Figure 21.6**    Lake trout (char). *Photo credit: Paul Vecsei.*

years and sizes ranging from 348–481 mm FL (13.7–18.9 in); size at maturity was positively correlated to lake surface area (Burr 1993).

Lake char *home* to spawning sites (Binder et al. 2016) but do not build redds (nests) or establish spawning territories (see Esteve et al. 2005). Males arrive first on spawning grounds and wait for females; once the females arrive, often during high winds, fish move in groups over coarse gravel substrate. Males court females and attempt to spawn with them by nibbling their fins and pressing against their sides and quivering. Spawning occurs when females sink to the bottom, lift their tails, and release eggs; adjacent males concurrently release sperm then embryos fall into substrate crevices (Scott and Crossman 1973, Esteve et al. 2005). This act is repeated until the female releases all her eggs, sometimes over a wide area (Esteve et al. 2005). Mass spawning is documented with up to seven males and three females (Royce 1951). Spawning can be concentrated in a few nights or spread over two to three weeks, depending upon environmental conditions (see Esteve et al. 2005). Spawning can occur annually in more southern areas of their range and every other year in more northerly ones.

Although lake char are considered the least tolerant to salt water of the salmonids, Boulva and Simard (1968) reported them in coastal Arctic waters between 6% and 9% salinity. Lindsey (1964) pointed out that lake char had to swim through saltwater to colonize young islands of the Canadian Arctic Archipelago. In Bristol Bay, Rousenfel (1958) reported lake trout at the head of tidewater in the Nakenek River, but no anadromous populations or marine captures are thus far documented.

Anadromous populations are verified in the Canadian Arctic. Anadromous fish averaged 13 years old at first marine migration and fish that migrated annually to sea were generally in better condition compared to same-age lake-dwelling fish (Swanson et al. 2010). Growth rates for mass were significantly higher in the anadromous form than in the resident form, but no significant difference existed in absolute growth rates for length. The main advantage for these anadromous char is increased mass rather than length, which likely provides access to better spawning sites and mates, as well as increased fecundity (Gross 1987, McCart 1980).

## Rainbow Trout

### Distinguishing Characteristics

Adult rainbow trout (Figure 21.7) can generally be identified by a pink, or red stripe of color along the body midline; many black freckles on a light background covering the dorsal fin, body, and square tail; a lower jaw that doesn't extend past the eye, and an anal fin with 8–12 rays. Young rainbows have 8–13 parr marks on their sides and 5–10 parr marks between the dorsal fin and top of the head.

### Distribution

Rainbow trout are native to Pacific coast watersheds ranging from Mexico to the Kuskokwim River in Alaska, and to drainages of Russia's Kamchatka Peninsula (Morrow 1980, Mecklenburg et al. 2002); they are documented in all major Bristol Bay watersheds. The anadromous form Steelhead occurs in southeast Alaska to Cook Inlet and is present up to Port Mollar (see Anadromous Waters Catalog) (ADF&G 2017) but is not documented elsewhere in Bristol Bay and will not be discussed in detail here. The species has been widely introduced throughout Alaska and the world.

### Habitats, Life History

Rainbow trout are iteroparous (repeat) spring spawners. Similar to other salmonids, they exhibit complex life histories and may migrate to different habitats for spawning, feeding, and overwintering. Rainbow trout remain in fresh water their entire life whereas the anadromous Steelhead form migrate to sea to feed. Resident and anadromous forms can be sympatric (occur together), commonly interbreed, and their progeny can be resident or anadromous (Christie et al. 2011, Courter et al. 2013, Sloat and Reeves 2014). Factors influencing whether a rainbow trout will become anadromous include genetics, individual condition (e.g., lipid levels), and environmental factors (see review by Kendall et al. 2015). Frequency of freshwater versus anadromous rainbow trout was highest in cooler watersheds with higher summer flows, abundant food, and sufficient spawning habitat for smaller females (Kendall et al. 2015). Such conditions apparently

**Figure 21.7**   One of the world famous Naknek River rainbow trout. *Photo credit: Nanci Morris Lyon.*

increase rainbow trout growth opportunities and survival in fresh water, thereby reducing the benefits of a risky ocean migration. Stream or river size and lake presence influenced whether rainbow trout were more likely to become anadromous, with the freshwater resident form more common in watersheds with many small tributaries and/or large lakes (Kendall et al. 2015), such as occur throughout Bristol Bay.

Populations studied thus far in Bristol Bay are all freshwater rainbow trout that may migrate to tributaries, lake outlets, and/or main stem rivers, to spawn and/or feed (Russell 1977, Minard et al. 1992, Meka et al. 2003, Schwanke and Hubert 2003, Schwanke and Reed, in prep.). Bristol Bay rainbow trout generally migrate to spawning areas in late winter and early spring as ice thaws and break up begins (Russell 1977, Russell 1980, Dye 2008, Schwanke and Reed, in prep.). Females dig a redd (nest) in the gravel (March–July) with their tails, where gametes are deposited and buried. Incubation rates are generally faster than that of salmon, and rates vary depending on thermal regimes; at 5°C (41°F) embryos can hatch in 68 days and at 11°C (51.8°F) they can hatch in 28 days (see Quinn 2005). After hatching, alevins or hatchlings remain in the gravel absorbing their yolk sacs, usually emerging during July or August (Russell 1977, Johnson et al. 1994). Juveniles can grow quickly, and can double their length by fall (Russell 1977). Meanwhile, post-spawning fish often migrate to feeding areas to recuperate.

Bristol Bay rainbow trout will feed on fish, fish eggs, snails, aquatic and terrestrial insects, and small mammals. A Facebook post by the Togiak NWR (August 13, 2013) shows a rainbow trout that had about 16 shrews in its stomach. A 13-year study in the Wood River system of Bristol Bay found shrews were the dominant mammal observed every two to three years in both larger rainbow trout and Arctic grayling stomachs; notably, temporal and spatial correlations were observed in shrew consumption for both species among and within tributaries (Lisi et al. 2014).

When salmon return to spawn, rainbow trout can migrate to those spawning areas to take advantage of the reliable food source consisting of salmon eggs, carcasses, and associated insects (Brink 1995, Scheuerell et al. 2007). Russell (1977) observed that up- and downstream movements of rainbow trout in Lower Talarik Creek (Kvichak watershed) coincided with upstream spawning migrations of salmon and the later downstream movement of carcasses.

An 11-year study in two Wood River streams of Bristol Bay with a greater than 10-fold variation in annual spawning salmon density indicated that in poor salmon years rainbow trout and Arctic grayling exhibited significantly lower growth rates due to both decreased ration sizes and decreased proportions of salmon derived prey (Bentley et al. 2012). Rainbow trout were better at consuming energetically rich salmon eggs compared to grayling; even in low salmon density years, eggs occurred in > 65% of rainbow diets, but were rarely observed in grayling diets (Bentley et al. 2012). Although declines of > 50% in consumption rates of salmon-derived prey occurred in low salmon density years, rainbow trout partially compensated by increasing their intake of other prey. The relative impact on fish growth attributed to reduced spawning salmon densities differed between streams, with fish growth declining 46–68% in the less-productive stream compared to 26–34% in the more-productive stream. The study is important in that it shows that the relative impact of pulsed resources (e.g., spawning salmon) will be effected by the subsidy magnitude, the ability of fish to exploit the pulsed resource and the relative productivity of the water body (Bentley et al. 2012).

Radio tagging studies suggest that the extent of rainbow trout seasonal migrations to reproductive, feeding, and overwintering habitats can be greater in watersheds with large lakes than in those without lakes. Radio-tagged Alagnak River (Kvichak watershed) rainbow trout swam 4–35 km (2.5–21.7 miles) to reach spawning habitat, 4–72 km (2.5–44.7 miles) to reach summer feeding habitat, and 3–60 km (1.9–37.3 miles) to reach overwintering habitat (Meka et al. 2003). Naknek River rainbow trout moved greater than 50 km (31.1 miles) to summer feeding habitat (Burger and Gwartney 1986). In contrast, radio-tagged fish in the upper King Salmon and Kanektok rivers in southwest Alaska, both without lakes, moved less (Adams 1996, 1999).

The ages and sizes of Bristol Bay rainbow trout show wide variation among and within drainages, which is likely related to annual salmon subsidies, thermal regimes, and other factors. In the Nakenek River, fish at the age of 5 were 350–550 mm (13.8–21.7 in.) and 10-year-old fish were 550–700 mm (21.7–27.6 in.); the oldest observed fish was 13 (Minard and Dunaway 1991, Rifie 1994, Schwanke and Hubert 2003). Most fish were sexually mature at anywhere from age 5 to 7 and 551–650 mm FL (21.7–25.6 in.); minimum size at maturity was 424 mm FL (16.7 in.) (Schwanke and Hubert 2003). Mature Naknek River males were significantly longer (mean FL = 676 ± 2.9 mm; 26.6 ± 0.1 in.) than females (mean FL 617 ± 2.7 mm; 24.3 ± 0.1 in.) (Schwanke and Hubert 2003). Half of Wood River lake fish in Moose Creek greater than 376 mm FL (14.8 in.) were sexually mature (Dye 2008).

Recent studies of Lower Talarik Creek (Kvichak watershed) indicate reduced rainbow trout abundance, fewer mature, and fewer large fish compared to the 1970s. Stream surveys conducted in 1973 and 1974 counted 737 and 926 spawning rainbow trout (Russell 1977) in contrast, the highest count for spawning rainbow trout during 2009 to 2013 was 345 fish

(Schwanke and Reed, in prep.). The average length of mature fish as reported by Russell (1977) was at least 15 mm (0.6 in.) higher during 1974–1974 than during recent studies and cumulative length composition of fish indicates a higher proportion of large fish in the 1970s (Schwanke and Reed, in prep.).

Rainbow trout research conducted during spring 2004 on the Tazimina River in response to reports of decreased abundance and reduced fish size estimated 950 rainbow trout present in the Tazimina during April and May, with 16% sexually mature (Schwanke and Evans 2005). Fish averaged 12.1 inches long (FL) and ranged from 6.3 to 24 inches long (FL). Fall sampling that same year indicated that catch per unit effort was similar to a historic study, but fish were smaller.

Evidence of reduced resident freshwater fish size and abundance compared to the past from studies, sportfishers (Schwanke and Evans 2005, Schwanke and Reed, in prep.), and local subsistence fishers (Krieg et al. 2005) is likely linked to a number of factors including increased fishing pressure, climate changes, and the recent extended period of low spawning salmon abundance in the Kvichak watershed (Morstad and Brazil 2012). Such prolonged declines in a formerly abundant and reliable temporal food resource to the Kvichak's nutrient-poor waters could impact freshwater fish productivity (Bilby et al. 1996, Gresh et al. 2000, Bilby et al. 2001, Wipfli 2003, Hicks et al. 2005, Kline et al. 2007).

Schwanke and Reed (in prep.) found that immature Talarik Creek female lengths ranged from 330–545 mm (13.0–21.5 in.) (FL) and at sexual maturity, the same female lengths ranged from 398–611 mm (15.7–24.1 in.) (FL); immature male lengths ranged from 326–598 mm (12.8–23.5 in.) and mature fish recapture lengths (FL) ranged from 414–652 mm (16.3–25.7 in.) (FL).

## Humpback Whitefish

### Distinguishing Characteristics

Mature humpback whitefish are distinguished by their iridescent silver blue to creamy white color, lack of markings, an upper jaw that overhangs the lower, and in mature fish, a prominent nuchal hump (Figure 21.8). Compared to other salmonids, they have much larger scales, with more than 110 along the lateral line (Mecklenburg et al. 2002). They have a deeply forked tail, and are similar to other salmonids in their body form.

Humpback whitefish belong to what McPhail and Lindsey (1970) called the "*Coregonus clupeaformis* complex" of species, which includes Alaska whitefish (*C. nelsonii*) and lake whitefish (*C. clupeaformis*). The complex is roughly differentiated through ecotypes and modal gill raker counts. Humpback whitefish usually have 21–23 gill rakers on the first gill arch, but there is overlap in gill raker numbers (Alt 1979). Here, I treat the three as a single species.

### Distribution

Humpback whitefish are Holarctic and occur in the Arctic Ocean Basin from Sweden and Finland, to northeastern Siberia, Alaska, northern U.S., Canada, and the Sea of Okhotsk (Alt 1979, Morrow 1980, Mecklenburg et al. 2002).

### Habitats and Life History

Humpback whitefish occur in a wide diversity of Alaskan habitats including: glacial, clear, and floodplain lakes; slow moving, glacial, clear, and tannin stained rivers; shallow lakes with rooted plants, as well as deltas and estuaries with brackish waters. Early research in Alaska indicated abundant populations in tributaries of the Yukon, Kuskokwin, and Tanana rivers from the Canadian border to coastal streams (Alt 1979). Populations inhabiting large, deep lakes (Minchumina, Crosswind and Louise) reportedly remained year-round, while those in shallow lakes migrated out during freeze up. Later studies indicated lacustrine (lake use only), river resident, allacustrine (use of both lakes and rivers), and anadromous (use of freshwater and marine habitats) ecotypes of adult humpback whitefish in both Alaska and Canada (Alt 1979, Morrow 1980, Reist and Bond 1988, Fleming 1996). Humpback whitefish were first documented in the Kvichak watershed of Bristol Bay after a 1964 freshwater commercial fishery was attempted (Metsker 1967).

Humpback whitefish are long-lived (57 years) (Power 1978), iteroparous (repeat), broadcast spawners, and exhibit diverse life histories and extensive migratory patterns (Alt 1979, Reist and Bond 1988, Harper et al 2012, Brown 2013, Woody and Young 2007). They can exhibit distinct migrations from summer feeding to fall spawning areas and fidelity to feeding and spawning areas (Fleming 1996, Brown 2006, Dupuis and Brown 2013, Sutton 2014). Some migrate long distances to reach essential feeding and spawning habitats.

Tagging and otolith chemistry studies show some humpback whitefish may spend part of their life in brackish or marine waters and migrate long distances. Brown (2006) used otolith strontium distribution to document anadromy

**Figure 21.8**  Laura Sorensen of Dillingham holds a Lake Clark humpback whitefish for a photo before measuring and weighing it for a study. *Photo credit: C. Woody.*

in Yukon River humpback whitefish and determined anadromous individuals ranged as far as 1,700 km (1,056 miles) from the ocean. Harper et al. (2007) analyzed otoliths from humpback whitefish, captured in Whitefish Lake of the Kuskokwim River system and observed elevated strontium levels consistent with anadromy in 10 of 10 sampled fish. Radio tracking data from Harper et al. (2012) indicated that fish migrated more than 500–900 river km (310–559 miles) from multiple sites to spawning locations. Woody and Young (in prep.) examined humpback whitefish otoliths for potential anadromy in Lake Clark (n = 10) and Kaskanak River (n = 10) populations (Kvichak watershed) but did not find strong evidence of anadromy.

Humpback whitefish spawn during fall in lake (Anras et al. 1999) or river habitats (Alt 1979, Brown 2006, Harper et al. 2012, Young and Woody, in prep.) with rock, gravel, or sand substrates (McPhail and Lindsey 1970, Scott and Crossman 1973, Morrow 1980). Spawning in rivers is documented in waters from 1–3 m (3.2–9.8 ft.) deep, flows of 0.5 m/sec (1.6 ft./sec), and temperatures ranging from 0–3°C (32–37°F) (Kepler 1973, Alt 1979). Fish broadcast spawn their gametes, which sink to the bottom and lodge in spaces between substrates (Scott and Crossman 1973, Morrow 1980). Embryos develop over winter and hatch in spring. Fish may spawn annually (Brown 2006) or every one to three years (Reist and Bond 1988, Lambert and Dodson 1990).

A radio-tagging study of 216 humpback whitefish that were tagged at three sites (Chulitna Bay, Hardenburg Bay, and Nondalton Fish Camp) in Lake Clark during 2006–2008 showed that humpback whitefish moved throughout the Lake Clark/Newhalen drainage seasonally (Young and Woody, in prep). In early summer, tagged fish occupied shallow, near-shore habitats that were generally near the original tagging locations. When salmon arrived, fish tended to move to subsistence processing areas to capitalize on salmon carcasses that fishers returned to the water. During mid-August through September, some radio-tagged humpback whitefish from all tagged groups migrated to and congregated in Chulitna Bay prior to migrating up the Chulitna River for spawning, which later occurred between the Koksetna River and Lynx Creek confluences (Young and Woody, in prep.). Most fish remained on spawning grounds about 50 days, then migrated back to original tagging locations. The Chulitna River is a tannin stained, swift flowing river; average temperatures recorded in 2008 (n = 7) during spawning were 2.3°C (36.1°F) (Young and Woody, in prep.).

Studies in LACL indicate that juvenile humpback whitefish (from age 0 to 3) rear in vegetated, shallow areas (< 3m, < 9.8 ft.) of Chulitna Bay and Long Lake, whereas individuals > 4 years were captured across a wider range of habitat types including shallow (≤ 2 m, ≤ 6.6 ft.) small (< 6 km; 3.7 miles long) lakes with abundant aquatic plants (Long and Pickerel Lakes) and deep large glacial fjord lakes (Little Lake Clark) (Woody and Young, in prep.).

Humpback whitefish are omnivorous and similar to other salmonids in that their diet changes throughout their life cycle. Younger smaller fish eat zooplankton, whereas larger fish prey predominantly on snails, clams, midges, and caddis fly larvae (Metsker 1967, Russell 1980, Brown 2007). Their mollusk feeding also serves as an important dispersal mechanism for snails and clams, since some survive gut passage (Brown 2007). Humpback whitefish also feed on fish eggs, fish, and diverse aquatic insects, (Van Whye and Peck 1968, Woody and Young 2007, Claramunt et al. 2010).

Age and size at maturity of humpback whitefish varies by latitude, among watersheds, and by life history type. Alaskan humpback whitefish age at maturity ranges from 3 to 11 years (Alt and Kogl 1973, Alt 1979, Brown 2006, Harper et al. 2007, Woody and Young, in prep.) and size of maturity ranges from 227–400 mm FL (8.9–15.7 in.) (Alt 1979, Howland 2001, Woody and Young, in prep.). Fish from northern populations and from cooler, less-productive habitats mature later than fish from more southerly or warmer regions. Nonanadromous fish are reported from Lake Clark with males reproducing as early as the age of three (227 mm FL, 8.9 in.) and females at the age of four (372 mm FL; 14.6 in.) (Woody and Young, in prep.); in contrast, fish from the relatively cooler coastal Kuzitrin River matured at the age of six (both sexes), and samples from near the Arctic Ocean matured as late as the age of 11 (Alt 1979, Moulton et al. 1997, Harper et al. 2007). Half of Chatanika River males were mature at the age of six, and half of the females were mature at the age of seven (Sutton and Edenfield 2012). Kuskokwim River humpback whitefish were mature at a minimum length of 350 mm FL (13.8 in.) (Harper et al. 2007).

Clark and Bernard (1992) examined humpback whitefish fecundity in the Chatanika River and found that it ranged from 8,400 to 65,400 for fish ranging in size from 320–520 mm FL (13.8 in.). Dupuis and Sutton (2011) examined absolute fecundity for 60 humpback whitefish from the Chatanika River. Mean absolute fecundity was 45,000 eggs per female (range 11,747–108, 426 eggs per female) and was positively related to both FL ($r^2$ = 0.74) and wet weight ($r^2$ = 0.83); wet weight explained a greater proportion of the variance in absolute fecundity than did FL.

## Northern Pike

### Distinguishing Characteristics

Northern pike belong to the Esocidae family and are easily distinguished by a long slender body, dorsal and anal fins placed posteriorly on the body (to allow rapid acceleration; Hubbs and Lagler 2004); a wide alligator-like jaw lined with many small, sharp, recurved teeth that also line the upper mouth, tongue, and gill rakers (Figure 21.9); these teeth are constantly replaced (Scott and Crossman 1973). The cheeks are completely scaled. Their color varies depending on habitat and light levels, but resembles a hunter's camouflage, green to brown with lighter flanks and white spots. Fish from clear waters are generally light green while fish from tannin stained habitats can be much darker. The eyes are golden yellow and very active. The northern pike is a keystone predator that ambushes its prey relying on camouflage, vision, speed, and surprise.

### Distribution

Northern pike are circumpolar in North America and Eurasia (Mecklenburg et al. 2002). They are a common and abundant species and are found in 45% of the total freshwater area of North America (Carlander et al. 1978). Northern pike are native to Bristol Bay and the interior, both west and north of the Alaska Range (Mecklenburg et al. 2002); isolated

**Figure 21.9**   The author holding a 127 cm (50 inch) female northern pike from a small lake within Lake Clark National Park and Preserve. *Photo credit: Mike Booze.*

populations occur in the Ahrnklin River, these are believed to be relicts from the Pleistocene Epoch (Whooler et al. 2015). Introduced populations occur in the Susitna River, Anchorage area, and Kenai Peninsula (see: http://www.adfg.alaska .gov/index.cfm?adfg=invasivepike.main). They are widespread in Canada with the exception of British Columbia and the Maritimes (Scott and Cross 1973). Northern pike are native to the Great Lakes, upper Mississippi River drainage, and Atlantic basins to Pennsylvania, Missouri, Montana, and Nebraska (Page and Burr 1991, Holton and Johnson 1996). Introductions have occurred throughout the U.S. (see Bradford 2008). Temperature seems to limit their southern distribution.

## Habitats

Northern pike are best adapted to shallow (< 12 m), productive environments (Casselman and Lewis 1995). They generally inhabit low gradient rivers, sloughs, lakes, and ponds linked to wetlands and floodplains with abundant aquatic vegetation. In such areas, environmental characteristics can fluctuate widely with the seasons. Annual high water from snow melt creates extensive inundated areas during spring, then waters recede, revealing rich stands of submerged vegetation as summer begins; during autumn, rains again inundate floodplains as vegetation dies back and decays; winter brings a white blanket of snow and ice. These seasonal changes in typical pike habitat all coincide with wide variations in temperature, pH, dissolved oxygen, and nutrients.

Northern pike exhibit an exceptional ability to adapt to wide changes in abiotic conditions. Increased water temperatures over seven decades in eastern Ontario were correlated with increased northern pike abundance (Casselman and Dietriech 2003). Fish growth rate increased rapidly above 10°C and was highest between 19°C (for biomass) and 21°C (for length); upper lethal temperature was 29.4°C; pike tolerated temperatures as low as 0.1°C (Casselman 1978). Northern pike are very tolerant to varying levels of dissolved oxygen (DO)—surviving levels as low as 0.3 mg/l in shallow lakes and being captured alive from waters with DO concentrations as low as 0.04 mg/l; they actively seek higher $O_2$ at DO concentrations < 4 mg/l (see reviews by Casselman 1978, Casselman and Lewis 1996). Petit (1973) determined that northern pike ceased feeding at DO < 2 mg/l and fish maintained at a DO of 1 mg/l more than 20 hours did not survive. Critical DO concentration depends on temperature; the upper range of the lower incipient lethal oxygen concentration varies between 0.5 and 2.0 mg/l. Northern pike tolerate a wide range of pH; Margenau et al. (1998) notes their inhabiting lakes ranging from 6.1–8.6 pH, while Scott and Crossman (1973) refer to a population living in a Nebraska lake of pH 9.5. Inskip

(1982) mentions a population able to sustain itself at pH 5.0. While most populations of northern pike are confined to freshwaters, both anadromous and populations spawning in brackish water are documented in the Baltic Sea; pike are also reported in areas with salinities ranging from 7–25% (Engstedt et al. 2013). In Cook Inlet, Alaska, commercial fishermen report catching northern pike in commercial fishing nets indicating the species may disperse and perhaps colonize new areas via saltwater routes (Southcentral Alaska Northern Pike Control Committee 2006). Few fish species exhibit the ability to tolerate such wide fluctuations in this range of abiotic conditions.

## Life History

Northern pike are a spring spawning, iteroparous (repeat spawner) species with well-studied specific habitat preferences. They spawn during May and July in slow shallow waters over vegetation, after ice-off. Fish will migrate, even under ice, from overwintering habitats in lakes and rivers to wetlands, sloughs, and floodplains, as the melt begins and river waters rise (Cheney 1971, Chihuly 1979, Dye 2002). Optimal spawning substrate usually consists of flooded vegetation (sedges, grasses) in a shallow (10–70 cm) sheltered area (Casselman and Lewis 1995). Females deposit small batches of adhesive eggs over selected substrate and one to several males fertilize them; successive matings follow, sometimes over several days until fish are spent (Scott and Crossman 1973, Billard 1996). The total number and size of eggs a female produces is variable and influenced by female size, temperature, available food, fish density, and social interactions (Billard 1996). Fecundity increases with female size, small fish may have as few as 2,000 eggs whereas females greater than 14.5 kg (30 lb) may have as many as 600,000 eggs (Morrow 1980).

Early development is rapid for northern pike but is dependent on temperature, food availability, and access to suitable cover (Morrow 1980). Embryos generally hatch within a few weeks to a month depending on temperature. The 6–8 mm long larvae swim upward and attach to a piece of vegetation or other substrate by a sticky nasal gland and there they remain for one to two weeks, absorbing their yolk sacs and finishing their development (Bry 1996). When the nasal gland regresses, larvae swim to the surface, fill their swim bladders and begin swimming. At this active state they are more visible and vulnerable to predators and seek cover in shallows, with submerged vegetation interspersed with emergent and floating plants, which enhances survival (Casselman and Lewis 1996). Initially they feed on zooplankton, then aquatic insects, and then fish—cannibalistic behavior occurs among fry as small as 21 mm (Hunt and Carbine 1951) and as young as five weeks (Giles 1986). Once fry reach about 5 cm, fish become their primary prey (Scott and Crossman 1973). In Bristol Bay habitats, young-of-the-year northern pike are usually swimming and feeding by late June and early July (Chihuley 1979, Woody, pers. obs.).

North American pike growth rates decrease with increasing latitude—and temperature appears to be the main climate driver; a concordant increase in longevity with latitude is also observed (Rypel 2012). The age at which pike mature is dependent on temperature, food availability, and growth rate, and reflects size more than chronological age (Billard 1996). Age at maturity is also affected by exploitation (fishing), and can decline when larger fish are removed from the population (Diana 1996). In the Lower 48, some male northern pike can reach sexual maturity at the age of one (Diana 1983); in contrast, studies in the Wood River system of Bristol Bay indicate that northern pike begin to reach maturity around the age of three and that most fish are mature by the age of five and lengths of approximately 438–469 mm (Chihuly 1979). In Lake Clark, the smallest mature Chulitna River males observed measured 414 mm FL and the smallest mature females measured 508 mm FL (Dye 2002); females generally mature at older ages and larger sizes than males.

Movement studies of northern pike show a wide diversity of patterns among populations. Evidence that northern pike home to natal sites for spawning are revealed by otolith trace element (Engstedt et al. 2013), genetic, mark-recapture (Miller et al. 2011), and radio-telemetry studies (Chythlook and Burr 2002, Joy and Burr 2004, Vehanen 2006) although further research is needed. Northern pike radio telemetry studies also show they exhibit fidelity to overwintering habitats (Taube and Lubinski 1996, Scanlon 2009). Most movement studies of pike indicate they generally do not move far. A three year radio-telemetry study of northern pike in Long Lake (Lake Clark) showed that fish spawned in Long Lake in spring and the proportion of tagged fish remaining in Long Lake year-round ranged from 1.0 to 0.71; a few fish made forays, one down to Lake Clark, another about 32 km to Nikabuna Lake and a couple of fish apparently left the system. A Yukon River radio telemetry study indicated pike moved a maximum of 38.1 km upstream or 56.6 km downstream of their spawning drainage (Joy and Burr 2004) but fish returned annually to their spawning habitat.

A Missouri River tagging study documented one northern pike moving up to 322 km away, but the majority of tagged fish were recaptured within 32 km of their original tagging site (Moen and Henegar 1971).

The following citations provide additional information regarding northern pike: Inskip (1982), Harvey (2009), and Rypel (2012).

# RECOMMENDED RESEARCH

Human development and climate change (Hansen 2012, Rosenblat 2016, Hill et al. 2017, Runting 2017) will likely change freshwater fish habitat availability and suitability due to alterations in natural water flow and thermal regimes. This can impact *near-term* abundance, condition, and health of the region's fish, as well as their *long-term* likelihood of persistence. While it is critical for humans who rely on or manage Bristol Bay's aquatic resources to better understand and adapt to these climate-induced changes, doing so presents a wide range of research needs.

There is broad uncertainty regarding how melting glaciers, reduced snowpack, and altered rainfall patterns will impact seasonal water budgets; the timing, frequency and magnitude of floods; and freshwater thermal regimes (see Sections 4.6 and 5.6 of Reynolds and Wiggins 2012). Progress is being made to characterize these expected changes in freshwater systems at the regional-scale, but much work remains both for collaborative data collection efforts and associated data management and synthesis. Promising recent efforts along these lines are the cooperative freshwater temperature monitoring programs in the Kvichak and Nushagak watersheds lead by Mauger and Troll (2014), part of a broader effort advanced by the Western Alaska Landscape Conservation Cooperative (Reynolds et al. 2013), and the assessment of watershed characteristics controlling thermal regimes in the Wood River (Lisi et al. 2013, Schindler et al. 2015) and Togiak River systems. The efforts to project expected changes in the hydrologic cycle of the Upper Nushagak and Kvichak watersheds (Wobus et al. 2015) need to be repeated in the region's other major river systems, especially those whose current precipitation patterns differ from these systems. The projected changes in hydrologic cycle would then allow one to consider how current thermal refugia (e.g., spring-fed rivers, streams, and lakes) may be impacted by projected climate changes; this, of course, hinges on first identifying such refugia (e.g., Torgerson et al. 2012).

The uncertainties surrounding the region's physical responses are likely much less than those regarding the essential habitat requirements (spawning, feeding, rearing, overwintering) and tolerances of the freshwater fish discussed and not discussed here. Are thermal tolerances of species in Bristol Bay different than those characterized for these same species in Lower 48 studies? What is each species distribution relative to current thermal and flow regimes (as well as that of their major prey)? How might these distributions change under projected climate regimes? How much *plasticity*, or flexibility, do they have to survive in altered regimes? While there have been some investigations of this for salmon (e.g., Sparks 2016), little to no work has been conducted for freshwater species. Investigating *adaptive capacity*, however, requires first characterizing a species' basic life history details and essential habitat requirements. For most species covered in this chapter, extensive information exists regarding their essential habitat needs, mainly from other regions, but gaps exist. And do habitat suitability indices and habitat preference models developed in other regions translate to Bristol Bay species? Or are there important differences? Where are essential habitats and what abiotic and biotic characteristics define them—particularly relative to flow and temperature? Such knowledge is especially lacking for humpback whitefish.

The most pressing research need may be simply using the existing knowledge base to undertake adaptation planning for each of these species (e.g., Cross et al. 2012 and Stein et al. 2014). Such efforts would both clarify specific pathways of impact as well as priority information needs for reducing uncertainties.

# SUMMARY

The freshwater non-salmon fishes of Bristol Bay provide local food and economic security as well as rare world-class fishing opportunities in intact extraordinary landscapes. For resource managers and scientists, Bristol Bay also presents an opportunity to expand our knowledge regarding wild fish ecology in intact landscapes for application to complex fish restoration efforts in altered landscapes across the Lower 48 (and world).

The Bristol Bay freshwater (non-salmon) species that are most important to humans are closely related to salmon and are part of the Salmonidae family: Arctic grayling; char, Dolly Varden, Arctic char, and lake trout (also called lake char); rainbow trout, and whitefish, primarily humpback whitefish. The exception is the toothy, fish eating northern pike, which is in the Esocidae family. Although this chapter has a decidedly human bias and is focused on the species we like to catch and eat, it is crucial to remember that all species—even the diminutive, seemingly unimportant sculpin—play important roles in the freshwater food chain. Small fish feed larger fish and are particularly important when salmon are unavailable.

All species considered here spawn in freshwater (but see Baltic pike; Nilsson 2006) and within each species, some may spend their entire life cycle in freshwater. However, all species discussed in this chapter have documented anadromous populations and can go to sea; this tendency increases in the more northern latitudes. In Bristol Bay, only freshwater resident forms are documented for: Arctic grayling, Arctic char, lake trout, rainbow trout (but see Pt. Mollar ADF&G 2017), humpback whitefish, and northern pike, whereas anadromous forms are documented for Dolly Varden.

Freshwater and anadromous life histories represent trade-offs between survival to first reproduction and size and age at first reproduction. Within a species, freshwater residents generally mature at a smaller size, younger age, exhibit higher survival to reproduction and a greater probability of iteroparity (repeat spawning) compared to anadromous forms. Fish with potential for anadromy will likely go to sea if the benefits (e.g., increased growth, increased fecundity) outweigh the costs (e.g., higher mortality rates) (Gross 1987).

All species discussed in this chapter exhibit fidelity to freshwater natal (birth) habitats for spawning. Spawning site fidelity can help ensure that fish repeatedly spawn in habitats that maximize their reproductive success; spawning site fidelity can also reduce genetic exchange among populations and result in significant differences evolving among populations (biodiversity). This can be an important consideration relative to management and conservation.

All species considered here are generally opportunistic feeders, with some becoming primarily piscivorous (fish eating; e.g., lake trout, northern pike) as they mature. Evidence from many scientific studies and subsistence interviews indicate that Bristol Bay freshwater fish species will migrate, sometimes great distances, to salmon spawning habitats in order to exploit the rich indirect and direct temporal pulse of food resources available including the aquatic insects that salmon dislodge as they construct their redds (nests), salmon eggs, and post-spawn salmon carcasses.

# REFERENCES

Adams, F.J. 1996. Status of Rainbow Trout in the Kanektok River, Togiak National Wildlife Refuge, Alaska, 1993–1994. U.S. Fish and Wildlife Service, Alaska Fisheries Technical Report 39, King Salmon.

Adams, F.J. 1999. Status of Rainbow Trout in tributaries of the Upper King Salmon River, Becharof National Wildlife Refuge, Alaska, 1990–1992. U.S. Fish and Wildlife Service, Alaska Fisheries Technical Report 53, King Salmon.

ADF&G (Alaska Department of Fish and Game). 2017. The following site provides links to interactive maps, additional species information and contacts. http://www.adfg.alaska.gov/index.cfm?adfg=lands.main.

Alt, K. 1976. Inventory and cataloging of north slope waters. Alaska Department of Fish and Game, Federal Aid in Fish Restoration, Annual Performance Report, 1975–1976. Project F¥8, 17(G-I-0):129–150.

Alt, K. 1978. Inventory and cataloging of sport fish and sport fish waters of Western Alaska. Alaska Department of Fish and Game, Federal Aid in Fish Restoration, Annual Performance Report, 1977–1978. Project F-9-10, 19(G-I-P):36–60.

Alt, K.T. 1979. Contributions to the life history of the Humpback Whitefish in Alaska. Transactions of the American Fisheries Society 108(2):156–160.

Alt, K. 1980. Inventory and cataloging of sport fish and sport fish waters of Western Alaska. Alaska Department of Fish and Game, Federal Aid in Fish Restoration, Annual Performance Report, 1979–1980. Project F-9-12, 2l(G-I-P):32–59.

Alt, K. and Kogl. 1973. Notes on the whitefish of the Colville River. J. Fish. Res. Bd. CA. 30:554–556.

Anras, M.L.B., P.M. Cooley, R.A. Bodaly, L. Anras, and R.J.P. Fudge. 1999. Movement and habitat use by lake whitefish during spawning in a boreal lake: integrating acoustic telemetry and geographic information systems. Transactions of the American Fisheries Society 128(5):939–952.

Arbour, J.H., D.C. Hardie and J.A. Hutchings. 2011. Morphometric and genetic analyses of two sympatric morphs of Arctic char (*Salvelinus alpinus*) in the Canadian high Arctic. Can. J. Zool. 89: 19–30.

Armstrong, R.H. and J.E. Morrow. 1980. The Dolly Varden char, *Salvelinus malma*. Pages 99–140 inE.K. Balon (Ed.), *Charrs: Salmonid Fishes of the Genus Salvelinus*. Dr. W. Junk Publishers, The Hague, Netherlands.

Armstrong, R.H. 1986. A review of Grayling studies in Alaska, 1952–1982. Biological papers of the University of Alaska. Institute of Arctic Biology. 23:3–18.

Baccante, D. 2011. Further evidence of size gradient of Arctic Grayling (*Thymallys arcticus*) along stream length. Journal of Ecosystems and Management. 11:13–17.

Behnke, R.J. 1972. The systematics of salmonid fishes of recently deglaciated lakes. Journal of the Fisheries Research Board of Canada 29:639–671.

Beitinger, T.L., W.A. Bennett, and R.W. McCauley. 2000. Temperature tolerances of North American freshwater fishes exposed to dynamic changes in temperature. Environmental Biology of Fishes. 58:237.

Bendock, T. 1979. Inventory and cataloging of Arctic area waters. Alaska Department of Fish and Game, Federal Aid in Fish Restoration, Annual Performance Report, 1978–1979. Project F-9-11, 20(G-I-I). 64 p.

Bentley, K.T., D.E. Schindler, J.B. Armstrong, R. Zhang, C.P. Ruff, and P.J. Lisi. 2012. Foraging and growth responses of stream-dwelling fishes to inter-annual variation in a pulsed resource subsidy. Ecosphere 3(12):113. http://dx. doi.org/10.1890/ES12-00231.1.

Bilby, R.E., Fransen, B.R., and Bisson, P.A. 1996. Incorporation of nitrogen and carbon from spawning Coho Salmon into the trophic system of small streams: evidence from stable isotopes. Can. J. Fish. Aquat. Sci. 53(1): 164–173. doi:10. 1139/f95-159.

Bilby R.E., B.R. Fransen, J.K. Walter, J. Cederholm, and W.J. Scarlett. 2001. Preliminary evaluation of the use of nitrogen stable isotope ratios to establish escapement levels for Pacific salmon. Fisheries 26:6–14.

Binder, T.R. and seven coauthors. 2016. Spawning site fidelity of wild and hatchery Lake Trout (*Salvelinus namaycush*) in Northern Lake Huron. CJFAS. 73(1)18–34.

Binder, T.R., S.C. Riley, and 6 co-authors. 2016. Spawning site fidelity of wild and hatchery Lake Trout (*Salvelinus namaycush*) in Northern Lake Huron. CJFAS 73:18–34.

Brink, S.R. 1995. Summer habitat ecology of Rainbow Trout in the Middle Fork of the Gulkana River, Alaska. Master's thesis. University of Alaska, Fairbanks, AK.

Blackett, R.F. 1973. Fecundity of resident and anadromous Dolly Varden in Southeastern Alaska. J. Fish. Res. Board Can. 30:543–548.

Boulva, J.S. and A. Simard. 1968. Presence du *Salvelinus namaycush* (Pisces:Salmonidae) dans les eaux marines de l'Arctique occidental Canadien. J. Fish. Res. Board Can. 25:1501–1504.

Brown, R.J. 2006. Humpback Whitefish *Coregonus pidschian* of the upper Tanana River drainage. U. S. Fish and Wildlife Service, Alaska Fisheries Technical Report No. 90, Fairbanks, AK.

Brown, R. 2013. Seasonal migrations and essential habitats of Broad Whitefish, Humpback Whitefish, and Least Cisco in the Selawik River Delta, as Inferred from Radiotelemetry Data, 2004–2006. U.S. Fish and Wildlife Service, Alaska Fisheries Data Series Number 2013.

Brown, R.J., Lunderstadt, C., and Schulz, B. 2002. Movement Patterns of Radio-Tagged Adult Humpback Whitefish in the Upper Tanana River Drainage. U. S. Fish and Wildlife Service, Alaska Fisheries Data Series No. 2002–1, Fairbanks.

Brown, R.J., N. Bickford, and K. Severin. 2007. Otolith trace element chemistry as an indicator of anadromy in Yukon River drainage Coregonine fishes. Transactions of the American Fisheries Society 136:678–690.

Burger, C.V. and L.A. Gwartney. 1986. A radio tagging study of Naknek drainage Rainbow Trout. U. S. National Park Service, Alaska Regional Office, Anchorage, AK.

Burr, J.M. 1993. Maturity of Lake Trout from 11 lakes in Alaska. Northwest Science 67(2)78–87.

Buzby, K.M., and L.A. Deegan. 2000. Inter-annual fidelity to summer feeding sites in Arctic Grayling. Environmental Biology of Fishes. 59: 31–327.

Carlander, K.D., J.S. Campbell, and R.J. Muncy. 1978. Inventory of Percid and Esocid habitat in North America. Am. Fish. Soc. Spec. Publ. 11: 27–38.

Chavarie, L., K. Howland, P. Venturelli, B.C. Kissinger, R. Tallman, and W. Tonn. 2016. Life history variation among four shallow water morphotypes of Lake Trout from Great Bear Lake, Canada. Journal of Great Lakes Research 42:193–203.

Chihuly, M.B. 1979. Biology of the Northern Pike, *Esox lucius* Linnaeus, in the Wood River Lakes system of Alaska, with emphasis on Lake Aleknagik. University of Alaska, Fairbanks, AK.

Christie, M.R., Marine, M.L., and Blouin, M.S. 2011. Who are the missing parents? Grandparentage analysis identifies multiple sources of gene flow into a wild population. Mol. Ecol. 20: 1263–1276.

Clark, J.H. and D.R. Bernard. 1988. Fecundity of Humpback Whitefish and Least Cisco in the Chatanika River, Alaska. Transactions of the American Fisheries Society 121:268–273.

Chythlook, J. and J.M. Burr 2002. Seasonal movements and length composition of Northern Pike in the Dall River, 1999–2001. Alaska Department of Fish and Game, Fishery Data Series No. 02-07, Anchorage, AK.

Courter, I.I., D.B. Child, J.A. Hobbs, T.M. Garrison, J.J.G. Glessner, and S. Duery, S. 2013. Resident Rainbow Trout produce anadromous offspring in a large interior watershed. Can. J. Fish. Aquat. Sci. 70(5): 701–710.

Craig, P.C. and V.A. Poulin. 1975. Movements and growth of Arctic Grayling (*Thymallus Arcticus*) and Juvenile Arctic Char (*Salvelinus alpinus*) in a small Arctic stream, Alaska. J. Fish. Res. Board of Canada. 32:689–697.

Crane, P., M. Lisac, B. Spearman, E. Kretschmer, C. Lewis, S. Miller, and J. Wenburg. 2003. Development and application of microsatellites to population structure and mixed stock analyses of Dolly Varden from the Togiak River drainage. U.S. Fish and Wildlife Service, Office of Subsistence Management, Fisheries Resource Monitoring Program, Final Report of Study 00-011, Anchorage, AK.

Crane P.A., C.J. Lewis, E.J. Kretschmer, S.J. Miller, W.J.Spearman, A.L. DeCicco, M.J. Lisac, and J.K. Wenburg. 2004. Characterization and inheritance of seven microsatellite loci from Dolly Varden, *Salvelinus malma*, and cross-species amplification in Arctic char, *S. alpinus*. Conserv Genet 5:737–741.

Crane, P., T. Viavant, and J. Wenburg. 2005. Overwintering patterns of Dolly Varden *Salvelinus malma* in the Sagavanirktok River in the Alaskan North Slope inferred using mixed-stock analysis. U.S. Fish and Wildlife Service, Alaska Fisheries Technical Report No. 84, Anchorage, AK.

Cross, M.S., P.D. McCarthy, G. Garfin, D. Gori, and C.A.F. Enquist. 2012. Accelerating adaptation of natural resource management to address climate change. *Conservation Biology*, 27 (1): 4–13.

de Bruyn, M. and P. McCart. 1974. Life History of the Grayling *(Thymallus arcticus)* in Beaufort Sea drainages in the Yukon Territory. In: P.J. McCart (Ed.), Fisheries Research Associated with Proposed Gas Pipeline Routes in Alaska, Yukon and Northwest Territories. Canadian Arctic Study, Ltd., Calgary, Biological Report Series 15(2). 39 p.

DeCicco, A.L. 1992. Long-distance movements of anadromous Dolly Varden between Alaska and the U.S.S.R. Arctic Vol. 45. No. 2, p. 120–123.

DeCicco, A.L. and R.J. Brown. 2006. Direct validation of annual growth increments on sectioned otoliths from adult Arctic Grayling and a comparison of otolith and scale ages. North American Journal of Fisheries Management 26(3): 580–586.

Diana, J.S. 1983. Growth, maturation, and production of Northern Pike in three Michigan lakes, Transactions of the American Fisheries Society, 112: 38–46.

Dunham, J. and 12 coauthors. 2008. Evolution, ecology and conservation of Dolly Varden, White-Spotted Char and Bull Trout. Fisheries. 33(11)537–550.

Dupuis, A.W. and Sutton, T.M. 2011. Reproductive biology of female Humpback Whitefish *Coregonus pidschian* in the Chatanika River, Alaska. Journal of Applied Ichthyology 27:1365–1370.

Dye, J., M. Wallendorf, G.P. Naughton, and A.D. Gryska. 2002. Stock assessment of Northern Pike in Lake Aleknagik, 1998–1999; Alaska Department of Fish and Game, Fishery Data Series No. 02-14. Alaska Department of Fish and Game, Division of Sport Fish, Anchorage, AK.

Dye, J.E. 2008. Stock Assessment of Rainbow Trout in the Wood River Lakes System, 2003–2005. Alaska Department of Fish and Game, Fishery Data Series No. 08-50, Anchorage, AK.

Dye, J.E. and I.K. Fo. 2015. Recreational fisheries in the Bristol Bay management area, 2013–2015. Alaska Department of Fish and Game, Fishery Management Report No. 15-40, Anchorage, AK.

Eloranta, A.P. and 5 coauthors. 2015. Lake size and fish diversity determine resource use and trophic position of a top predator in high-latitude lakes. Ecology and Evolution 2015; 5(8):1664–1675.

Esteve, M. 2005. Observations of spawning behaviour in Salmoninae: *Salmo, Oncorhynchus*, and *Salvelinus*. Rev Fish Biol Fisher 15:1–21.

Everett, R.J., R.L. Wilmot, and C.C. Krueger. 1997. Population genetic structure of Dolly Varden from Beaufort Sea drainages of Northern Alaska, and Canada. American Fisheries Society Symposium 19:240–249.

Fall, J.A., D.L. Holen, B. Davis, T. Krieg, and D. Koster. 2006. Subsistence harvests and uses of wild resources in Iliamna, Newhalen, Nondalton, Pedro Bay, and Port Alsworth, Alaska, 2004. Alaska Department of Fish and Game Division of Subsistence, Technical Paper No. 302: Juneau. http://www.adfg.alaska.gov/techpap/tp302.pdf.

Fall, J.A., T.M. Krieg, and D.L. Holen. 2009. Overview of the subsistence fisheries of the Bristol Bay management area. Alaska Department of Fish and Game Division of Subsistence Special Publication No. BOF 2009-07: Anchorage, AK.

Fleming, D.F. 1996. Stock assessment and life history studies of whitefish in the Chatanika River during 1994 and 1995.

Fleming, I.A. and Reynolds, J.D. 2004. Salmonid breeding systems. In: *Evolution Illuminated: Salmon and Their Relatives*. A. Hendry and S.C. Stearns (Eds.), Oxford University Press, Oxford. p. 264–294.

Foote, C.F. and G.S. Brown. 1998. Ecological relationship between freshwater sculpins (genus *Cottus*) and beach-spawning Sockeye Salmon (*Oncorhynchus nerka*) in Iliamna Lake, Alaska. Canadian Journal of Fisheries and Aquatic Sciences, 1998, 55(6): 1524–1533.

Fraley, K.M., J.A. Falk, R. Yanusz, and S. Ivey. 2016. Seasonal movements and habitat use of potamodromous Rainbow Trout across a complex Alaska riverscape. Transactions of the American Fisheries Society. 145(5)1077–1092.

Giles, N., R.M. Wright, and M.E. Nord. (1986), Cannibalism in pike fry, *Esox lucius* L.: some experiments with fry densities. Journal of Fish Biology, 29:107–113.

Gresh T, Lichatowich J., and P. Schoonmaker. 2000. An estimation of historic and current levels of salmon production in the Northeast Pacific ecosystem: evidence of a nutrient deficit in the freshwater systems of the Pacific Northwest. Fisheries 25: 15–21.

Gross, M. 1987. Evolution of diadromy in fishes. Am. Fish. Soc. Symp. 1:14–25.

Gryska, A.D. 2015. Seasonal distributions of Arctic grayling in the Upper Delta River. Alaska Department of Fish and Game, Fishery Data Series No. 15-21, Anchorage, AK.

Hansen, J., M. Sato, and R. Ruedy. 2012. Perception of climate change. PNAS. Vol. 109 no. 37.

Harper K.C., Harris F., Brown R., Wyatt T., and D. Cannon. 2007. Stock Assessment of Broad Whitefish, Humpback Whitefish and Least Cisco in Whitefish Lake, Yukon Delta National Wildlife Refuge, Alaska, 2001–2003. Alaska Fisheries Technical Report No. 88, Kenai, Alaska: U.S. Fish and Wildlife Service. http://alaska.fws.gov/fisheries/fish/Technical_Reports/t_2005_88.pdf.

Harper K.C., F. Harris, S. Miller, and D. Orabutt. 2008. Migratory behavior of Broad and Humpback Whitefish in the Kuskokwim River, 2006. Alaska Fisheries Data Series 2007–11. Kenai, Alaska: U.S. Fish and Wildlife Service. Available: http://alaska.fws.gov/fisheries/fish/Data_Series/ d_2007_11.pdf (December 2011).

Harper K.C, F. Harris, S. Miller, and D. Orabutt. 2009. Migration timing and seasonal distribution of Broad Whitefish, Humpback Whitefish, and Least Cisco from Whitefish Lake and the Kuskokwim River, Alaska 2004 and 2005. Alaska Fisheries Technical Report No. 105. Kenai, Alaska: U.S. Fish and Wildlife Service.

Harper, K.C., F. Harris, S. J. Miller, J. M. Thalhauser, and S. D. Ayers. 2012. Life history traits of adult Broad Whitefish and Humpback Whitefish. Journal of Fish and Wildlife Management: June 2012, Vol. 3, No. 1, p. 56–75.

Hart, L.M, M.H. Bond, S.L. May-McNally, J.A. Miller and T.P. Quinn. 2015. Use of otolith microchemistry and stable isotopes to investigate the ecology and anadromous migrations of northern Dolly Varden from the Egegik River, Bristol Bay, Alaska. Environ. Biol. Fish. 98:1633–1643.

Harvey, B. 2009. A biological synopsis of Northern Pike (*Esox lucius*). Canadian Manuscript Report Fisheries Aquatic Sciences 2885: v + 31 p.

Hazell, S.M., C. Welch, J.T. Ream, S.S. Evans, T.M. Krieg, H.Z. Johnson, G. Zimpelman, and C. Carty. 2015. Whitefish and other non-salmon fish trends in Lake Clark and Iliamna Lake, Alaska, 2012 and 2013. Alaska Department of Fish and Game Division of Subsistence, Technical Paper No. 411. Anchorage, AK.

Heim, K.C., M.S. Whitman, and L.L. Moulton. 2016. Arctic Grayling (*Thymallus arcticus*) in saltwater: a response to Blair et al. (2016). Conserv. Physiol 4(1): cow055; doi:10.1093/conphys/cow055.

Hicks, B.J., M.S. Wipfli, D.W. Lang, and M.E. Lang. 2005. Marine-derived nitrogen and carbon in freshwater-riparian food webs of the Copper River Delta, Southcentral Alaska. Oecologia, 144:558–569.

Hill, E.A., R.J. Carr, and C.R. Stokes. 2017. A review of recent changes in major marine-terminating outlet glaciers in Northern Greenland. Frontiers in Earth Science. Vol. 4. Article No. UNSP 111.

Holton, G.D. and H.E. Johnson. 1996. A field guide to Montana fishes. Second edition. Montana Department of Fish, Wildlife and Parks. Helena, MT. 104 p.

Howland, K.L., W.M. Tonn, J.A. Babaluk, and R.F. Tallman. 2001. Identification of freshwater and anadromous Inconnu in the Mackenzie River system by analysis of otolith strontium. Transactions of the American Fisheries Society 130:725–741.

Hubbs, C.L. and K.F. Lagler. 1958. Fishes of the Great Lakes region. University of Michigan Press, Ann Arbor, MI.

Hubert, W.A., R.S. Helzner, L.A. Lee, and P.C. Nelson. 1985. Habitat suitability index models and instream flow suitability curves: Arctic Grayling riverine populations. U.S. Fish Wildl. Serv. Biol. Rept. 82(10.110). 34 p.

Hughes, N.F. 1992. Ranking of feeding positions by drift-feeding Arctic Grayling (*Thymallus arcticus*) in dominance hierarchies. Can. J. Fish. Aquat. Sci. 49:1994–1998.

Hughes, N. 1999. Population processes responsible for larger-fish-upstream distribution patterns of Arctic Grayling (*Thymallus arcticus*) in Interior Alaskan runoff rivers. Can. J. Fish. Aquat. Sci. 56:2292–2299.

Hughes, N. and J. Reynolds. 1994. Why do Arctic Grayling get bigger as you go upstream? Can. J. Fish. Aquat. Sci. 51:2154–2163.

Inskip, J.D. 1982. Habitat suitability index models. Northern Pike. US Dept. Interior, USFWS, FWS/OBS-82/10.17. 40 p.

Jaecks, T.A. 2010. Population dynamics and trophic ecology of Dolly Varden in the Iliamna River, Alaska: life history of freshwater fish relying on marine food subsidies. Master's thesis, University of Washington, Seattle, WA.

Jennings, G.B., K. Sundet, and A.E. Bingham. 2011. Estimates of participation, catch, and harvest in Alaska sport fisheries during 2009. Fishery Data Series No. 11-45. Alaska Department of Fish and Game, Division of Sport Fish, Anchorage, AK.

Jessop, C.S. and J.W. Lilley. 1975. An evaluation of the fish resources of the MacKenzie River Valley based on 1974 data. Environment Canada, Fisheries and Marine Service, Technical Report Series CEN/T-75-6, Winnipeg, Manitoba.

Johnsson B. and N. Johnsson. 2000. Polymorphism and speciation in Arctic char. Journal of Fish Biology (2001) 58, 605–638.

Johnson, S.W., J.F. Thedinga, and A.S. Feldhausen. 1994. Juvenile salmonid densities and habitat use in the main-stem Situk River, Alaska, and potential effects of glacial flooding. Northwest Science 68(4):284–293.

Joy, P. and J.M. Burr. 2004. Seasonal movements and length composition of Northern Pike in Old Lost Creek, 2001–2003; Alaska Department of Fish and Game, Fishery Data Series No. 04-17. Alaska Department of Fish and Game, Division of Sport Fish, Anchorage, AK.

Jung, T.S., A. Milani, O.E. Barker, and N.P. Millar. 2011. American Pygmy Shrew, *Sorex hoyi*, consumed by an Arctic Grayling, *Thymallus arcticus*. Canadian Field-Naturalist 125(3):255–256.

Kendall, N.W. and 8 coauthors. 2015 Anadromy and residency in Steelhead and Rainbow Trout (*Oncorhynchus mykiss*): a review of processes and patterns. Can. J. Fish. Aquat. Sci. 72:319–342.

Kepler, P. 1973. Population studies of Northern Pike and Whitefish in the Minto Flats complex with emphasis on the Chatanika River. Federal Aid in Fish Restoration, Annual Performance Report, 1972–1973, Project F-9-5, 14 (G-II-J). Alaska Department of Fish and Game, Division of Sport Fish, Juneau, AK.

Keyse, M.D., and coauthors. 2007. Effects of large Lake Trout (*Salvelinus namaycush*) on the dietary habits of small lake trout: a comparison of stable isotopes (delta N-15 and delta C-13) and stomach content analyses. Hydrobiologia 579:175–185.

Kreiner, R. 2006. An investigation of Arctic char and lake trout in Lake Clark National Park. M.S. Thesis. Univ. Idaho. 69 p.

Krieg, T., M. Chythlook, P. Coiley-Kenner, D. Holen, K. Kamletz, and H. Nicholson. 2005. Freshwater fish harvest and use in communities of the Kvichak Watershed, 2003. Alaska Department of Fish and Game Division of Subsistence, Technical Paper No. 297: Juneau. http://www.adfg.alaska.gov/techpap/tp297.pdf.

Krieg, T.M., D.L. Holen, and D. Koster. 2009. Subsistence harvests and uses of wild resources in Igiugig, Kokhanok, Koliganek, Levelock, and New Stuyahok, Alaska, 2005. Alaska Department of Fish and Game Division of Subsistence, Technical Paper No. 322: Dillingham. http://www.adfg.alaska.gov/techpap/TP322.pdf.

Lee, D.P. 1999. A review of the life history and biology of Northern Pike *Esox lucius* Linnaeus. California Department of Fish and Game.

Lambert, Y. and Dodson, J.J. 1990. Freshwater migration as a determinant factor in the somatic cost of reproduction of two anadromous Coregonines of James Bay. Canadian Journal of Fisheries and Aquatic Sciences 47: 318–334.

Lindsey, C.C. 1964. Problems in zoogeography of the Lake Trout, *Salvelinus namaycush*. J. Fish. Res. Board Can. 21: 977–994.

Lisac, M.J. 2006. Run timing seasonal distribution and biological characteristics of Dolly Varden in the Kanektok River, Togiak National Wildlife Refuge, 2002–2003. U.S. Fish and Wildlife Service. Alaska Fisheries Technical Report Number 94.

Lisac, M. 2009. Seasonal distribution and biological characteristics of Dolly Varden in the Goodnews River, 2005–2006. U.S. Fish and Wildlife Service. Alaska Fisheries Technical Report Number 103.

Lisac, M. 2012. Dolly Varden. Eddies. Reflections on fisheries conservation. U.S. Fish and Wildlife Service. Spring 2012. https://www.fws.gov/eddies/pdfs/EddiesSpring2012.pdf.

Lisac, Mark J. and Jennifer Gregory. 2016. Inventory of Arctic char *Salvelinus alpinus*, Togiak National Wildlife Refuge, Alaska. 2015 Progress Report. U.S. Fish and Wildlife Service. Dillingham, Alaska.

Lisac, M.J. and R.D. Nelle. 2000. Migratory behavior and seasonal distribution of Dolly Varden *Salvelinus malma* in the Togiak River watershed, Togiak National Wildlife Refuge. U.S. Fish and Wildlife Service, Final Report, Dillingham, AK.

Lisi, P.J., D.E. Schindler, K.T. Bently, and G.R. Pess. 2013. Association between geomorphic attributes of watersheds water temperature, and salmon spawning in Alaska streams. Geomorphology. 185:78–86.

Lisi, P.J., K.T. Bentley, J.B. Armstrong, and D.E. Schindler. 2014. Episodic predation of mammals by stream fishes in a boreal river basin. Ecology of Freshwater Fish. 23:622–630.

May-McNally, S.L., T.P. Quinn and E.B. Taylor. 2015. Low levels of hybridization between sympatric Arctic char (*Salvelinus alpinus*) and Dolly Varden Char (*Salvelinus malma*) highlights their genetic and spatial distinctiveness. Ecology and Evolution 5: 3031–3045.

Martin, N.V., 1957. Reproduction of Lake Trout in Algonquin Park, Ontario. Trans. Am. Fish. Soc. 86:231–244.

Mauger, S. and T. Troll. 2014. Implementation plan: Bristol Bay regional water temperature monitoring network. Cook Inletkeeper, Homer, AK and Bristol Bay Heritage Land Trust, Dillingham, AK. 21 p.

McCart, P.J. 1980. A review of the systematics and ecology of Arctic char, *Salvelinus alpinus*, in the Western Arctic. Can. Tech. Rep. Fish. Aquat. Sci. No. 935, Fisheries and Oceans Canada, Winnipeg, Manitoba.

McDermid J.L., J.D. Reist, and R.A. Bodaly. 2007. Phylogeography and postglacial dispersal of whitefish (*Coregonus clupeaformis* complex) in Northwestern North America. Advances in Limnology 60:91–109.

McPhail J.D. and C.C. Lindsey. 1970. Freshwater fishes of northern Canada and Alaska. Bulletin 173. Fisheries Research Board of Canada.

Mecklenburg, C.W., T.A. Mecklenburg, and L.K. Thorsteinson. 2002. Fishes of Alaska. American Fisheries Society. Bethesda, Maryland.

Meka, J.M., E.E. Knudsen, D.C. Douglas, and R.B. Benter. 2003. Variable migratory patterns of different adult Rainbow Trout life history types in a Southwest Alaska watershed. Transactions of the American Fisheries Society 132:717–732.

Merkowsky, J.J. 1989. Assessment of Arctic Grayling populations in Northern Saskatchewan. Department of Parks and Renewable Resources, Saskatchewan Fisheries Laboratory, Fisheries Technical Report 89-4, Saskatoon.

Metsker, H. 1967. Iliamna Lake watershed freshwater commercial fisheries investigation of 1964, Informational Leaflet 95. Alaska Department of Fish and Game, Division of Commercial Fisheries, Dillingham, AK.

Michigan Department of Natural Resources. 2016. http://www.michigan.gov/dnr/0,4570,7-153-10364_18958-53612—,00.html checked 26 Oct 2016.

Miller, L.M., L. Kallemeyn, and W. Senanan. 2001. Spawning-site and natal-site fidelity by Northern Pike in a large lake: mark-recapture and genetic evidence. Transactions of the American Fisheries Society 130: 307–316.

Minard, R.E. and D.O. Dunaway. 1991. Compilation of age, weight, and length statistics for Rainbow Trout samples collected in Southwest Alaska, 1954 through 1989. Fishery Data Series No. 91-62. Alaska Department of Fish and Game, Division of Sport Fish, Anchorage, AK.

Minard, R.E., M. Alexandersdottir, and S. Sonnichsen. 1992. Estimation of abundance, seasonal distribution, and size and age composition of Rainbow Trout in the Kvichak River, Alaska, 1986 to 1991. Fishery Data Series No. 92-51. Alaska Department of Fish and Game, Division of Sport Fish, Anchorage, AK.

Moen, T. and D. Henegar. 1971. Movement and recovery of tagged Northern Pike in Lake Oahe, South and North Dakota, 1964–68. Am. Fish. Soc. Spec. Publ. 8:85–93.

Morrow, J.E. 1980. *The Freshwater Fishes of Alaska*. Alaska Northwest Publishing Company, Anchorage. 248 p.

Morstad, S. and C.E. Brazil. 2012. Kvichak River Sockeye Salmon stock status and action plan, 2012, a report to the Alaska Board of Fisheries. Alaska Department of Fish and Game, Special Publication No. 12-19, Anchorage.

Moulton L.L., L.M. Philo, and J.C. George. 1997. Some reproductive characteristics of Least Ciscoes and Humpback Whitefish in Dease Inlet, Alaska. American Fisheries Society Symposium 19:119–126.

Nilsson, J. 2006. Predation of Northern Pike (*Esox lucius* L.) eggs: a possible cause of regionally poor recruitment in the Baltic Sea. Hydrobiologia 553:161–169.

NOAA (National Oceanic and Atmospheric Association). 2017. http://www.westcoast.fisheries.noaa.gov/protected_species/salmon _steelhead/salmon_and_steelhead_listings/salmon_and_steelhead_listings.html (Accessed Feb. 13, 2017).

Northcote, T.G. 1995. Comparative biology and management of Arctic and European Grayling (Salmonidae *Thymallus*). Reviews in Fish Biology and Fisheries. 5(2)141–194.

Ostberg, C.O., D. Pavlov, and L. Hauser. 2009. Evolutionary relationships among sympatric life history forms of Dolly Varden inhabiting the landlocked Kronotsky Lake, Kamchatka, and a neighboring anadromous population. Transactions of the American Fisheries Society 138:1–14.

Page, L.M. and B.M. Burr. 1991. *A Field Guide to Freshwater Fishes of North America North of Mexico*. The Peterson Field Guide Series, volume 42. Houghton Mifflin Company, Boston, MA.

Petit, G.D. 1973. Effects of dissolved oxygen on survival and behavior of selected fishes of Western Lake Erie. Bull. Ohio Biol. Surv. New Ser. 4: 1–76.

Phillips, R.B., L.I. Gudex, K.M. Westrich, and A.L. DeCicco. 1999. Combined phylogenetic analysis of ribosomal ITS1 sequences and new chromosome data supports three subgroups of Dolly Varden Char (*Salvelinus malma*). Canadian Journal of Fisheries and Aquatic Sciences 56: 1504–1511.

Plumb, M.P. 2006. Ecological factors influencing fish distribution in a large subarctic lake system. M.S. University of Alaska, Fairbanks.

Power, M., J.B. Dempson, J.D. Reist, C.J. Schwarz, and G. Power. 2005. Latitudinal variation in fecundity among Arctic char populations in Eastern North America. Journal of Fish Biology (2005) 67, 255–273.

Quinn, T.P. 2005. *The Behavior and Ecology of Pacific Salmon and Trout*. University of Washington Press. Seattle. WA. 378 p.

Reilly, J.R., C.A. Paszkowski and D.W. Coltman. 2014. Population genetics of Arctic Grayling distributed across large, unobstructed river systems. Transactions of the American Fisheries Society, 143:3, 802–816, DOI: 10.1080/00028487.2014.886620.

Reist, J.D. and W.A. Bond. 1988. Life history characteristics of migratory Coregonids of the lower Mackenzie River, Northwest Territories, Canada. Finnish Fisheries Research 9:133–144.

Reist, J.D., J.D. Johnson, and T.J. Carmichael. 1997. Variation and specific identity of char from Northwestern Arctic Canada and Alaska. American Fisheries Society Symposium 19:250–261.

Reist, J.D., G. Low, J.D. Johnson, and D. McDowell. 2002. Range extension of Bull Trout, *Salvelinus confluentus*, to the central Northwest Territories, with notes on identification and distribution of Dolly Varden, *Salvelinus malma*, in the Western Canadian Arctic. Arctic 55: 70–76.

Reynolds, J.B. 2000. Life history analysis of Togiak River char through otolith microchemistry. Final Report. Unit Cooperative Agreement 1434-HG-97-RU-01582. Research Work Order 91. University of Alaska, Alaska Cooperative Fish and Wildlife Research Unit, Fairbanks, AK.

Reynolds, J.H., and H.V. Wiggins (Eds.). 2012. Shared science needs: report from the Western Alaska Landscape Conservation Cooperative science workshop. Western Alaska Landscape Conservation Cooperative, Anchorage, AK. 142 pp. https://westernalaskalcc .org/science/SitePages/sciencewkshp.aspx.

Reynolds, J.H., K. Murphy, and C. Smith. 2013. Alaska stream and lake temperature monitoring workshop, Anchorage, Alaska, November 5–6, 2012. Western Alaska Landscape Conservation Cooperative. http://westernalaskalcc.org.

Rhydderch, J.G. 2001. Population structure and microphylogeographic patterns of Dolly Varden (*Salvelinus malma*) along the Yukon north slope. M.Sc. Thesis, University of Guelph, Guelph, ON. v + 128 p.

Riffe, R. 1994. Complilation of age, weight and length statistics for Rainbow Trout samples collected in Southwest Alaska. 1990–1993. ADF&G Fishery Data Series 94-17, Anchorage, AK.

Roach, S. 1998. Site Fidelity, dispersal, and movements of radio-implanted Northern Pike in Minto Lakes, 1995–1997. Alaska Department of Fish and Game, Fishery Manuscript No. 98-1. Alaska Department of Fish and Game, Division of Sport Fish, Anchorage, AK.

Rosenblatt, A.E. and O.J. Schmitz. 2016. Climate change, nutrition, and bottom-up and top-down food web processes. Trends in Ecology and Evolution. 31:965–975.

Rounsefell, G., 1958. Anadromy in North American Salmonidae. U.S. Fish Wildl. Serv. Fish. Bull. 58, 171–185.

Royce, W.F., 1951. Breeding habits of lake trout in New York. Fish. Bull. U.S. Fish. Wildl. Serv. 52(59):59–76.

Runting, R.K., B.A. Bryan, LE. Dee, et al. 2017 Incorporating climate change into ecosystem service assessments and decisions: a review. Global Change Biology. 23:28–41.

Russell, R. 1977. Completion report for rainbow trout life history studies in Lower Talarik Creek–Kvichak drainage study G-II, Job No. G-II-E, Vol. 18. Alaska Department of Fish and Game, Division of Sport Fish, Juneau, AK.

Russell, R. 1980. A fisheries inventory of waters in the Lake Clark National Monument area. Alaska Department of Fish and Game. 124 p.

Rypel, A.L. 2012. Meta-analysis of growth rates for a circumpolar fish, the Northern Pike (*Esox lucius*), with emphasis on effects of continent, climate and latitude. Ecology of Freshwater Fish. 21:521–532.

Scanlon, B.P. 2000. The ecology of the Arctic char (*Salvelinus alpinus*) and the Dolly Varden (*Salvelinus malma*) in the Becharof Lake drainage, Alaska. MS thesis, University of Alaska Fairbanks. 103 p.

Scanlon, B.P. 2009. Movements and fidelity of Northern Pike in the lower Innoko River drainage, 2002–2004. Alaska Department of Fish and Game, Fishery Data Series No. 09-45, Anchorage.

Schindler, D.E., R. Hilborn, B. Chasco, C.P. Boatright, T.P. Quinn, L.A. Rogers, and M.S. Webster. 2010. Population diversity and the portfolio effect in an exploited species. Nature 465:609–612.

Schindler, D., P. Lisi, P. Walsh, M. Lisac, B. Berkhahn, and C. LeClair. 2015. Watershed control of hydrologic sources and thermal conditions in Southwest Alaska: A framework for forecasting effects of changing climate. LCC project final report. https:// westernalaskalcc.org/projects/SitePages/WA2011_05.aspx.

Schwanke, C.J. and D.G. Evans. 2005. Stock assessment of the Rainbow Trout in the Tazimina River. Alaska Department of Fish and Game, Fishery Data Series No. 05-73, Anchorage, AK.

Schwanke, C.J. and W.A. Hubert. 2003. Structure, abundance and movements of an allacustrine population of Rainbow Trout in the Naknek River, Southwest Alaska. Northwest Science. 77(4)340–348.

Schwanke, C.J. and D. J. Reed. In Prep. Stock assessment of Lower Talarik Creek Rainbow Trout, 2009–2013. Alaska Department of Fish and Game, Fishery Data Series, Anchorage, AK.

Scheuerell, M.D., J.W. Moore, D.E. Schindler, and C.J. Harvey. 2007. Varying effects of anadromous Sockeye Salmon on the trophic ecology of two species of resident salmonids in Southwest Alaska. Freshwater Biology. 52:1944–1956.

Scott, W.B. and E.J. Crossman, 1973. Freshwater Fishes of Canada. Bull. Fish. Res. Board Can. 184:1–966.

Sloat, M.R. and Reeves, G.H. 2014. Individual condition, standard metabolic rate, and rearing temperature influence steelhead and Rainbow Trout (*Oncorhynchus mykiss*) life histories. Can. J. Fish. Aquat. Sci. 71(4):491–501.

Southwick Associates Inc. and W.J. Romberg, A.E. Bingham, G.B. Jennings, and R.A. Clark. 2008. Economic impacts and contributions of sportfishing in Alaska, 2007. Alaska Department of Fish and Game, Professional Paper No. 08-01, Anchorage, AK.

Southcentral Alaska Northern Pike Control Committee. 2006. Management plan for invasive Northern Pike in Alaska. Alaska Department of Fish and Game, Anchorage, AK. 58 p.

Sparks, M. 2016. Phenology and embryo survival in a changing climate: adaptive responses by Sockeye Salmon in a warming world. MS Thesis, UAF SFOS. https://scholarworks.alaska.edu/handle/11122/6854?show=full.

Stearns, S.T. 1992. *The Evolution of Life Histories*. Oxford University Press. Oxford. 248 p.

Stein, B.A., P. Glick, N. Edelson, and A. Staudt (Eds.). 2014. Climate-smart conservation: putting adaptation principles into practice. National Wildlife Federation, Washington, D.C. https://www.nwf.org/pdf/Climate-Smart-Conservation/NWF-Climate-Smart-Conservation_5-08-14.pdf.

Sutton, T.M. and L.E. Edenfield. 2012. Stock characteristics of Humpback Whitefish and Least Cisco in the Chatanika River, Alaska. Arctic. 65(1) 67–75.

Swanson, H.K., K.A. Kidd, J.A. Babaluk, R.J. Wastle, P.P. Yang, N.M. Halden, and J.D. Reist. 2010. Anadromy in Arctic Populations of Lake Trout (*Salvelinus namaycush*): otolith microchemistry, stable isotopes, and comparisons with Arctic char (*Salvelinus alpinus*) Can. J. Fish. Aquat. Sci. 67:842–853.

Tack, S.L. 1980. Migrations and distributions of Arctic Grayling *Thymallus arcticus* (Pallas) in Interior and Arctic Alaska. Alaska Department of Fish and Game, Federal Aid in Fish Restoration, Annual Performance Report, 1979–1980. Project F-9-12, 21 (R-1).

Taube, T.T. and B.R. Lubinski. 1996. Seasonal migrations of Northern Pike in the Kaiyuh Flats, Innoko National Wildlife Refuge, Alaska. Alaska Department of Fish and Game, Fishery Manuscript No. 96-04, Anchorage, AK.

Taylor, E.B. 1991. A review of local adaptation in salmonidae with particular reference to Pacific and Atlantic salmon. Aquaculture. 98:185–207.

Taylor, E.B. 2016. The Arctic char complex in North America revisited. Hydrobiologia. 783:283–293.

Taylor, E.B., E. Lowery, A. Lilliestrale, A. Elz. and T.P. Quinn. 2008. Genetic analysis of sympatric char populations in Western Alaska: Arctic char (*Salvelinus alpinus*) and Dolly Varden (*Salvelinus malma*) are not two sides of the same coin. Journal of Evolutionary Biology. 21:1609–1625.

Taylor, E.B. and S.L. May-McNally. 2015. Genetic analysis of Dolly Varden (*Salvelinus malma*) across Its North American range: evidence for a contact zone in Southcentral Alaska. Canadian Journal of Fisheries and Aquatic Sciences 72: 1048–1057. doi:10.1139/cjfas-2015-0003.

VanWyhe, G. 1964. Investigations of the Tanana River Grayling fisheries: migration study. Alaska Department of Fish and Game, Federal Aid in Fish Restoration. Annual Report of Progress, 1963–1964. Project F-5-R-5, 5(14-B):353–368.

Van Whye, G.L. and J.W. Peck. 1968. A limnological survey of Paxson and Summit Lakes in Interior Alaska. Informational Leaflet 124. Alaska Department of Fish and Game, Division of Sport Fish, Juneau, AK.

Vehanen, T., Hyvarinen, P., Johansson, K. and Laaksonen, T. 2006. Patterns of movement of adult Northern Pike (*Esox lucius* L.) in a regulated river. Ecology of Freshwater Fish 15:154–160.

Vecsei, P., K. Dunmall, and J. Reist. 2014. Guide to identifying salmons and chars in the Arctic. Fisheries and Oceans Canada. 24 pp.

Wipfli, M.S., Hudson, J.P., Caouette, J.P., and Chaloner, D.T. 2003. Marine subsidies in freshwater ecosystems: salmon carcasses increase the growth rates of stream-resident salmonids. Trans. Am. Fish. Soc. 132:371–381.

Wobus, C., R. Prucha, D. Albert, C. Woll, M. Loinaz, and R. Jones. 2015. Hydrologic alterations from climate change inform assessment of ecological risk to Pacific salmon in Bristol Bay, Alaska. PLoS ONE 10(12): e0143905. doi:10.1371/journal.pone.0143905.

Wojcik, F. 1954. Spawning habits of Grayling in interior Alaska. U. S. Fish and Wildlife Service and Alaska Game Commission, Federal Aid in Fish Restoration, Quarterly Progress Report. Project F-l-R-4, 4(1). 3 p.

Wojcik, F. 1955. Life history and management of the Grayling in interior Alaska. M.S. Thesis, University of Alaska, Fairbanks. 54 pp.

Woods, P.J., D. Young, S. Sku´lason, S.S. Snorrason, and T.P. Quinn. 2013. Resource polymorphism and diversity of Arctic char *Salvelinus alpinus* in a series of isolated lakes. Journal of Fish Biology. 82, 569–587.

Woody, C.A. and D.B. Young. 2007. Life history and essential habitats of Humpback Whitefish in Lake Clark National Park, Kvichak River watershed, Alaska. U.S. Fish and Wildlife Service Office of Subsistence Management, annual report for study FIS 05-403: Anchorage, AK.

Woody, C.A. and D.B. Young. In Prep. Contributions to life history studies of Humpback Whitefish. National Park Service. Anchorage, AK.

Wooler, M.J., B. Gaglioti, T.L. Fulton, A. Lopez, and B. Shapiro. 2015. Post-glacial dispersal patterns of Northern Pike inferred from an 8800 year old Pike (*Esox* cf. *lucius*) skull from interior Alaska. Quaternary Science Reviews. 120:118–125.

Young, D.B. and C.A. Woody. In Prep. Seasonal movements of Humpback Whitefish to essential habitats. National Park Service. Anchorage, AK.

**Appendix 21.1**  Non-salmon fish documented in the freshwaters of Bristol Bay; since surveys are limited, other species may occur. Some notes on the types of habitats in which Bristol Bay species are likely to occur are noted. Information compiled from Russell (1980), Mecklenburg et al. (2002), personal observations, and the Alaska Department of Fish and Game, Fish Resource Monitor mapping tools (http://www.adfg.alaska.gov/index.cfm?adfg=ffinventory.interactive).

| Scientific family/ Common family name | Common name | Scientific name | Principal migratory patterns | Habitats |
|---|---|---|---|---|
| Petromyzontidae/ lampreys | Arctic lamprey | *Lethenteron camtschaticum* | Anadromous | Young documented in silty mud in backwaters; adults spawn in habitats similar to those preferred by salmon with gravel substrates in riffles and side channels |
| | Alaskan brook lamprey | *Lethenteron alaskense* | Non-anadromous | Young documented in silty mud in back waters and low flow systems; spawn in cool clear stream of moderate flow in gravel |
| Catostomidae/ suckers | Longnose sucker | *Catostomus catostomus* | Non-anadromous | Found in cold and sometimes warm lakes, rivers, ponds, in both clear and glacial systems; spawns over gravel in riffles |
| Esocidae/pikes | Northern pike | *Esox lucius* | Non-anadromous | Prefer shallow (< 12m) slow-moving water or lake systems with abundant aquatic vegetation |
| Umbridae/ mudminnows | Alaska blackfish | *Dallia pectoralis* | Non-anadromous | Found in vegetated slow-moving waters, swamps, and lakes |
| Osmeridae/smelts | Rainbow smelt | *Osmerus mordax* | Anadromous | Spawn in coastal streams and rivers in spring |
| | Pond smelt | *Hypomesus olidus* | Non-anadromous | Documented in coastal lakes, ponds, and tributaries |
| Salmonidae/ salmonids | Humpback whitefish | *Coregonus pidschian* | Non-anadromous and potentially anadromous | Documented in clear and glacial lakes and tributaries; young rear in shallow, warmer, vegetated habitats; adults spawn in stained or clear rivers over gravel |
| | Least cisco | *Coregonus sardinella* | Non-anadromous and anadromous | Occur in glacial and clear lakes and tributaries; documented as spawning in a tannin-stained, large river (Chulitna) |
| | Pygmy whitefish | *Prosopium coulterii* | Non-anadromous | Documented in a few upland lakes |
| | Round whitefish | *Prosopium cylindraceum* | Non-anadromous | Documented in larger upland streams as well as glacial and clear lakes |
| | Rainbow trout | *Oncorhynchus mykiss* | Non-anadromous in Bristol Bay | Occur in streams, rivers, lakes, and ponds; appear to prefer clear waters |
| | Arctic char | *Salvelinus alpinus* | Non-anadromous | Occur in upland lakes |
| | Dolly Varden | *Salvelinus malma* | Non-anadromous and anadromous | Dwarf forms documented in upland headwater tributaries, anadromous forms documented in rivers and selected lakes |
| | Lake trout | *Salvelinus namaycush* | Non-anadromous | Common in large upland lakes; not present in Wood River watershed |
| | Arctic grayling | *Thymallus arcticus* | Non-anadromous | Abundant/widespread in clear cold tributaries, often in eddies and near tributary outlets, lakes, and ponds |
| Gadidae/cods | Burbot | *Lota lota* | Non-anadromous | Documented in glacial and clear lakes as well as deep, sluggish, or still waters |

| Scientific family/ Common family name | Common name | Scientific name | Principal migratory patterns | Habitats |
|---|---|---|---|---|
| Gasterosteidae/ sticklebacks | Threespine stickleback | *Gasterosteus aculeatus* | Non-anadromous and anadromous | Locally abundant in slow moving tributaries, lakes, ponds, and quiet weedy ponds; spawning often associated with aquatic vegetation |
| | Ninespine stickleback | *Pungitius pungitius* | Non-anadromous | Documented in cool, quiet waters and shallow vegetated habitats; may migrate to deep waters for winter |
| Cottidae/sculpins | Coastrange sculpin | *Cottus aleuticus* | Non-anadromous | Ubiquitous in large clear and glacial lakes along coarse gravel beaches, slower shallower river reaches, small tributaries |
| | Slimy sculpin | *Cottus cognatus* | Non-anadromous | Headwaters to large lowland tributaries in coarse gravel and moderate to swift currents; found in small ponds and large lakes in waters ranging from gin-clear to turbid |

# 22

# DIVERSITY IN BRISTOL BAY SOCKEYE SALMON AND THEIR HABITAT: IMPLICATIONS FOR FISHERIES AND WILDLIFE

**Daniel E. Schindler, Lisa W. Seeb, and James E. Seeb**
**University of Washington**

## INTRODUCTION

Sockeye salmon (*Oncorhynchus nerka*) are the numerically dominant salmon species in Bristol Bay, owing to the prominence of large lakes connected via short rivers to a productive ocean. Sockeye salmon colonized this landscape following the most recent deglaciation and have proliferated into most freshwater habitats that are accessible from the marine environment. Over the last two decades, an average of about 40 million sockeye salmon have returned each year to spawn in Bristol Bay rivers and lakes, with the highest returns exceeding 60 million (Baker et al. 2009). Equally spectacular is the wide diversity of biological features observed among sockeye salmon populations that appear to be local adaptations to habitat characteristics such as water temperature, gravel size, and hydrology that vary considerably across the region and even within the same river basin. The variation in habitat conditions within the region, combined with the associated diversity of biological traits of sockeye salmon, produce a complex pattern of population dynamics among the individual stocks that comprise the stock complex of Bristol Bay sockeye salmon.

For decades scientists have recognized that not all populations boom and bust in the same years, thereby compensating for one another as their abundances wax and wane. Current theory along with supporting field and genetic evidence suggest that habitat variation maintains genetic diversity across Bristol Bay sockeye salmon, thereby producing the complexity in dynamics that stabilize the overall returns to watersheds of this region. Emerging discoveries in salmon genetics are offering new approaches for conserving biodiversity in sockeye salmon while maintaining productive, industrial-scale fisheries. Salmon landscapes in other areas of North America—where most salmon returning to rivers were released from hatcheries and where watersheds have been homogenized through development—have lost much of their population diversity, which is associated with less stability in returns among years (Griffiths et al. 2014).

Here we provide a brief overview of the state of knowledge about habitat complexity and biological diversity within Bristol Bay sockeye salmon. We will also describe how emerging genetics and genomics techniques are providing new insights into the evolutionary processes that generate this biodiversity and provide powerful new tools for fishery management. Finally, we discuss how this biodiversity within sockeye salmon contributes to sustainability of fisheries, enhances the benefits of marine-derived resources to watersheds, and how these benefits may be vulnerable to human activities that will change the character of habitat throughout the region.

# VARIATION IN SOCKEYE SALMON HABITAT THROUGHOUT BRISTOL BAY

Habitat used by sockeye salmon over their life cycle can be viewed as structured within a hierarchy, where populations are segregated among individual tributaries, which coalesce into the major rivers, all draining to the ocean, a habitat shared by all populations. At the broadest scale, sockeye salmon spend about half of their lives in the subarctic domain of the North Pacific Ocean where they achieve most of their growth before using a combination of cues from the Earth's magnetic fields, light, and olfaction to navigate back to the tributary where they hatched (Quinn 2005). As fish mature, they migrate east into Bristol Bay where they have already begun to segregate into their respective stocks returning to the eight major rivers (Igushik, Wood, Nushagak, Naknek, Kvichak, Alagnak, Egegik, Ugashik, and several smaller rivers in the Togiak region) (Dann et al. 2013). Fish then pass through the mouths of the major rivers (where some are captured by fisheries) and proceed on their spawning migrations into fresh water. Within each of the river basins, fish return to spawn in individual natal tributaries, rivers, or along lakeshores with appropriate gravel size and water flows to incubate eggs (hereafter called *beaches*). Habitat conditions change along the hierarchy, ranging from fine spatial scales associated with spawning locations, to intermediate sized tributaries and lakes also used for juvenile rearing and migration, and eventually to the major rivers flowing to the ocean.

Habitat features encountered by sockeye salmon during the freshwater components of their life cycle are formed by interactions between geomorphic and hydrologic processes. Tectonics and volcanism produced the mountains of Bristol Bay, and glaciation and flowing water have sculpted the landscape now inhabited by sockeye salmon. The mountain ranges of Bristol Bay were formed over eons, the most recent glacier recession was nearly complete 15,000 years ago (Kaufman and Manley 2004), and hydrologic erosion is a continuous force shaping aquatic habitats. Sea level rise and isostatic rebound following the last major deglaciation continue to change the distribution of estuaries throughout the region. Glaciers were particularly important for forming sockeye salmon habitat in Bristol Bay as they smoothed the landscape and produced a heavy overburden of gravels covering the watersheds, as well as carved most of the large lakes that provide the nurseries for most sockeye salmon in this region. Flowing water has eroded glacial deposits to produce the remarkable diversity in river geomorphic conditions that salmon encounter among Bristol Bay watersheds. Further, geomorphic conditions ultimately control other environmental features of salmon habitat, notably water temperature regimes and chemistry. For example, streams draining steep watersheds tend to be cooler than those draining flat watersheds (Lisi et al. 2014). Streams draining steep watersheds tend to have little dissolved organic matter compared to streams in flat watersheds which can have high concentrations of humic substances which give them that characteristic tea-colored water (Jankowski et al 2014). Though there are limits to the habitat conditions where sockeye salmon will spawn (Pess et al. 2014), sockeye salmon have successfully established in a wide variety of physical and chemical conditions. Thus, throughout Bristol Bay, sockeye salmon habitat is comprised of a remarkable assortment of conditions characterized by stream size and gradient, gravel size, water temperature regime, water chemistry, and the range of biological conditions that are structured by these features.

Stanford et al. (2005) described flood plain rivers as *shifting habitat mosaics* where a wide range of habitat conditions exist across the river basin, characterized by the successional and erosional processes that play out across the landscape. This model is an excellent description of sockeye salmon habitat in Bristol Bay because within any of its watersheds, there exists a broad range of habitat conditions characterized by variation in water source, water temperature and chemistry, stream gradient, gravel size, and many other features. What is equally important, however, is that this mosaic is continuously changing as erosion and succession continue to alter the configuration and distribution of salmon habitat; the evolution of life-history variation in sockeye salmon presumably is tracking this shifting mosaic of habitat conditions. Furthermore, the environmental conditions that salmon experience in any specific habitat are an expression of how the overriding climate regime is filtered through local habitat conditions. For example, one might expect that all streams would warm up during periods of hot summer weather. However, the magnitude of the response in stream temperatures to warmer air temperatures depends on the geomorphic features of watersheds and on seasonal timing. Lisi et al. (2015) showed that streams draining flat, low-elevation watersheds warmed during periods of hot weather, and that this effect was most pronounced near the summer solstice when days were longest and inputs of solar radiation were highest. Unexpectedly, however, water temperatures in small streams draining steep watersheds were remarkably buffered during heat waves as warmer temperatures increased snowmelt inputs to streams. This effect was pronounced during the early summer when snowpack at high elevations was still substantial, and declined later in the season when snowpack was virtually gone. Further, this effect of snow on stream temperatures appeared to be diminished following winters with little snow accumulation (Lisi et al. 2015).

A second example of how geomorphic characteristics of watersheds interact with overriding meteorological conditions to affect the quality and distribution of salmon habitat derives from studies of juvenile coho salmon in a small floodplain stream (Armstrong and Schindler 2013). In Bear Creek, a tributary of the Wood River system, water temperature drops abruptly as the stream flows out of a meadow complex near its headwaters. The drop in water temperature is a result of water inputs from groundwater springs that are insulated from solar radiation. Armstrong et al. (2013) showed that juvenile coho salmon perform daily movements between the cold lower reaches of Bear Creek where they feed on the eggs of spawning sockeye salmon and the warm upper reaches of the stream where their digestive efficiency is higher and they can accumulate growth from the egg resource. However, during years with high rainfall, stream flow increases to the point where the longitudinal variation in stream temperature is essentially washed out and the entire stream is cold. One would think, therefore, that the loss of warm water temperatures that enable juvenile coho salmon to digest energy-rich sockeye salmon eggs would inhibit their growth in high rainfall years. On the contrary—under rainy conditions, Bear Creek floods off-channel habitat that becomes substantially warmer than the main channel habitat. Juvenile coho salmon then begin moving between the main channel where they can consume sockeye salmon eggs and off-channel habitat where they can digest in warm water. Thus, the landscape of thermal conditions in coho salmon habitat shifts as hydrology and geomorphology interact to form a shifting mosaic of habitat conditions for juvenile fish. While not understood in specific detail, it is clear that all species of salmon throughout Bristol Bay have evolved genetically based adaptations to both the average habitat conditions experienced in specific locations used for various aspects of their life cycle, and also mechanisms to cope with the continuously changing landscape used for spawning, rearing, and migration in watersheds.

## THE GENETIC BASIS OF DIVERSITY IN BRISTOL BAY SOCKEYE SALMON

Anadromous sockeye salmon have the greatest homing fidelity of all of the salmonids, generally subdividing into discrete subpopulations along comparatively fine-scale geographical or temporal gradients. These discrete subpopulations, across the entire species range, are categorized primarily into two ecotypes based upon juvenile life history: lake-type and sea-type (Wood et al. 2008). Juveniles of lake-type sockeye salmon spend one or two years rearing in nursery lakes before migrating to sea. Wood et al. (2008) use the term sea-type, but we will refer to this ecotype as sea/river to better describe their riverine life history. Sea/river-type sockeye salmon usually spawn in larger rivers, and juveniles rear in side channels and sloughs before migrating seaward after a few weeks or months after they emerge from gravels. Sea/river-type sockeye salmon demonstrate weaker population structure than do the lake-type; sea/river-type are hypothesized to stray more than lake-type—reflecting the dynamic nature of floodplain habitats that support these populations (McPhee et al. 2009, Dann et al. 2013a). Sea/river-type sockeye salmon serve as colonists, radiating the species into novel ecosystems.

Early sockeye salmon colonists radiated into diverse habitats in Bristol Bay, spawning in large rivers, smaller streams, and on lake beaches. Juveniles initially may have followed different strategies, either rearing in lakes for one or a few years or migrating to saltwater immediately. Maturing adults that initially returned to watersheds likely had little specificity to particular locations in watersheds. But the variation in environmental conditions present in the large systems, such as different temperatures, gradients, cobble size, or pathogen load, provided the basis for a natural feedback mechanism to drive genetic diversification. Adults that homed to their natal habitat and produced juveniles that in turn homed to that habitat were able to adapt and be increasingly successful (Lin et al. 2008). The result, after thousands of years of adaptation, is the portfolio of discrete lake-type populations today that spawn in cold and warm tributaries, warmer-water lake outlets, rivers, and beaches. Sea/river-type populations are less differentiated; populations from different but adjacent drainages in Bristol Bay tend to cluster closely on genetic trees (Dann et al. 2013) (Figure 22.1).

Finding specific genes or gene frequency differences to identify and describe discrete populations is a primary goal of genetic conservation practice. Discrete populations accumulate genetic differences primarily from two mechanisms: natural selection on adaptively important traits or genetic drift. Adaptively important traits may be controlled by one, dozens, or possibly hundreds of genes. Natural selection shapes the frequency distribution of advantageous alleles at these gene loci to reflect local habitat conditions experienced by a population. Alternatively, genetic drift acts to change allele frequencies based on population size and random processes; random changes are accelerated in small populations, especially those that experience sharp reductions in number (for example, due to phenomena such as epizootics, floods, or droughts). The identification of the specific gene responsible for an adaptively important trait is not always possible given technological limitations (but see Larson et al. 2015). Even so, current technologies enable the identification of gene markers that distinguish populations; these markers may signal adaptively important genes, be closely linked to adaptively important genes on the genome, or signal random responses to demographic changes.

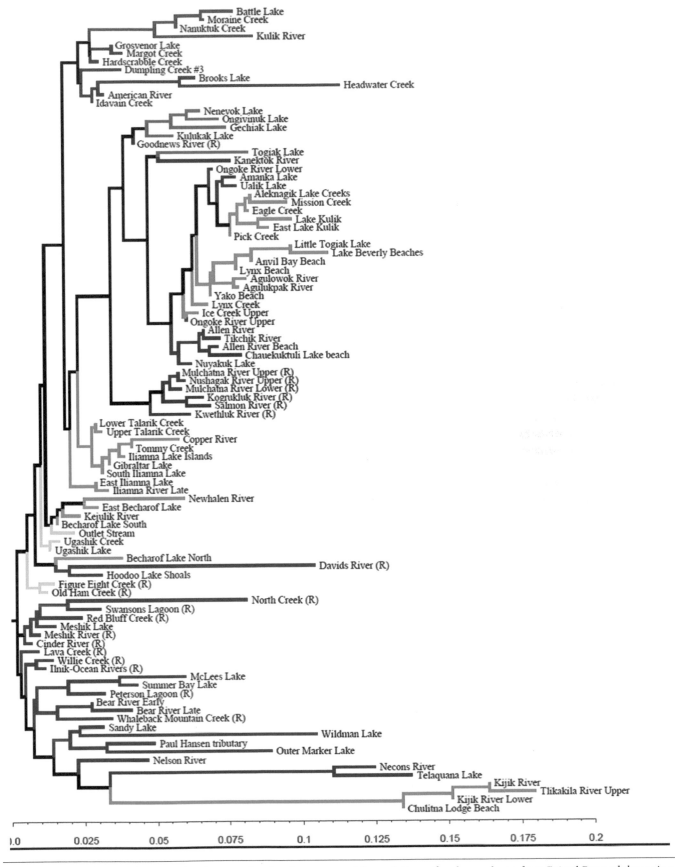

**Figure 22.1** Neighbor-joining tree showing genetic relatedness among 96 populations of sockeye salmon from Bristol Bay and the region. Genetic relatedness based on pairwise $F_{ST}$ values. Colors code for stock affiliations and river-type populations are denoted by an (R). Consensus nodes based on bootstrapping are noted as * (50%–90%) or ** (90%–100%). *Reprinted from Dann et al. (2013).*

During the last 40 years conservation practitioners have developed collections of DNA markers that may be quantified to reveal population structure, identify distribution and migration of Bristol Bay sockeye salmon on the high seas, and optimize harvest of specific populations in commercial fisheries.

## Genetics Tools for Studying Sockeye Salmon

The type of genetic markers used in fisheries genetics have changed immensely since the first applications to Bristol Bay sockeye salmon in the 1980s (Wilmot et al. 1989). That early study, based on variants at 50 protein-coding loci (allozymes) from 11 populations, found low divergence within Bristol Bay sockeye salmon with few variable loci and only 2% of the total variation due to differences among populations. Typical of that era, a single collection of samples was used to represent an entire river basin. These collections were often from mixtures of migrating fish, and data sets were not able to describe the ecotypic and temporal diversity within the drainages.

Protein markers were largely replaced by microsatellites (sections of DNA that have multiple repeat sequences that often show high levels of variation among populations) for studies of population structure in the mid 1990s. Microsatellites provided a higher level of polymorphism, ability to conduct nonlethal sampling, and simpler genotyping procedures (Nelson and Beacham 1999, Shaklee et al. 1999, Habicht et al. 2007).

Using microsatellites, researchers began exploring both genetic diversity among ecotypes and fine-scale diversity within and among Bristol Bay drainages. For example, Habicht et al. (2004) compared genetic diversity among ecotypes to assess demographic history and levels of gene flow among populations. Within the Kvichak River drainage, they found that tributary populations showed a strong signal of reduced effective population size compared to beach populations. They also found that populations located above migration obstacles showed signals consistent with reduction in historical population sizes that were associated with demographic bottlenecks. Ramstad and colleagues (Ramstad et al. 2004, Ramstad et al. 2007) described population bottlenecks in Lake Clark and described significant divergence of Lake Clark fish from downstream populations in Six Mile Lake and Lake Iliamna. These types of studies were important in describing the genetic diversity and diversification across the complex landscape of Bristol Bay. The inferences drawn from these studies reflect not only connectivity and barriers to migration but the glacial history and the time since colonization of individual populations returning to specific spawning locations.

In the mid-2000s, simpler and more easily automated DNA markers in the form of single-nucleotide polymorphisms (SNPs; variation in the DNA sequence at single base-pairs) (Smith et al. 2005) largely replaced microsatellites for studies of salmon in Bristol Bay. SNP technology also allowed development of assays for expressed sequence tags (EST) and major histocompatibility complex genes (MHC) (Elfstrom et al. 2006). MHC genes have been shown to be under both balancing and directional selection in sockeye salmon (Miller et al. 2001, Gomez-Uchida et al. 2012) and assist greatly in distinguishing among sockeye salmon ecotypes and populations occupying different habitats. MHC ultimately has become one of the most important markers in characterizing adaptive variation and distinguishing among populations and ecotypes of Bristol Bay (Figure 22.2) (Habicht et al. 2010, McGlauflin et al. 2011, Gomez-Uchida et al. 2012, Larson et al. 2014).

Technologies and platforms to conduct high-throughput genotyping of panels of SNPs have also rapidly expanded in the last decade; these advances provide researchers with a variety of options depending on the number of individuals and SNPs to be genotyped. Low-density arrays that facilitate the analysis of thousands or tens of thousands of individuals for panels of 96 SNPs have been widely adopted (Seeb et al. 2009, Dann et al. 2013b). These arrays provide an efficient automated approach in a cost effective manner. Thus, by 2010, genetic information derived from SNP data was routinely used in a variety of management and research activities at costs, efficiencies, and the power to discriminate among populations that were not fathomable a decade earlier. These applications not only improve the efficiency of harvest management to enable sustainable harvest rates on productive stocks while protecting stocks of concern (Dann et al. 2013), but also provide a means to understand the radiation of sockeye salmon throughout their range, and to protect this diversity through conservation actions.

## Hierarchical Structure of Bristol Bay Sockeye Salmon and Its Relationship to the Pacific Rim

An increasingly comprehensive understanding of the population structure of sockeye salmon from Bristol Bay has emerged, based on multiple marker types reflecting both neutral and adaptive variation (Figure 22.1). The neutral variation reflects processes such as gene flow, age of the population, and effective size, while adaptive variation reflects forces of selection that produce local adaptions among populations.

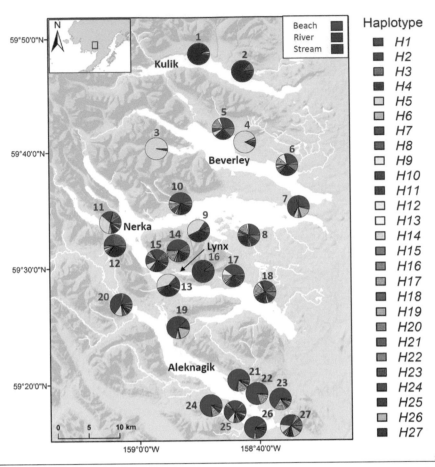

**Figure 22.2** Pie diagrams of frequencies of 27 MHC haplotypes from populations of sockeye salmon from the Wood River watershed. Populations are labeled by color numbers; the color of each number corresponds to the spawning ecotype, either beach, river, or stream. Nursery lakes are labeled in black. MHC has become one of the most important markers in distinguishing among populations and ecotypes of Bristol Bay. *Reprinted from Larson et al. (2014).*

Habicht et al. (2010) provided the first comprehensive evaluation of population genetic structure in Bristol Bay sockeye salmon using 45 SNPs, including two highly variable MHC SNPs; their dataset also included populations from across the species' range. The patterns of genetic similarity between populations detected by Habicht et al. (2010) and expanded by Dann et al. (2012a) were of finer scale but generally consistent with those patterns revealed by similar studies of sockeye salmon from throughout the Pacific Rim (Seeb et al. 2000, Beacham et al. 2006). These studies support a model of population structure based on the rearing in a network of nursery lakes (Wood et al. 1994, Seeb et al. 2000, Habicht et al. 2007) with populations within drainages and regions more similar to each other than to populations from other drainages. Exceptions to this nursery-lake structure are the river/sea-type populations of sockeye salmon that rear in off-channel sloughs or in estuarine habitats; these populations may be more genetically similar to one another, even between adjacent watersheds, than they are to lake-type populations within their watershed (Gustafson and Winans 1999, Wood et al. 2008).

Current genetic data shows a clear picture of the diversification of populations of sockeye salmon across the Bristol Bay landscape. Much of this variation is produced by variation in genes that support specific biological functions in salmon and not simply by random genetic drift. The inference we can make, therefore, is that populations returning to spawn in different locations are subjected to different environmental conditions that affect their ecology and evolutionary trajectories. Thus, the combination of continuously changing habitat variation and the biological diversity within sockeye salmon that it supports should be viewed as an integrated system that represents a unique resource with distinct intrinsic value.

## Key Features of Biological Variation in Bristol Bay Sockeye Salmon

Sockeye salmon have developed a broad range of local adaptations, in many biological characters, to the variety of habitat conditions they experience throughout their residence in freshwater ecosystems of Bristol Bay; much of this biological variation was described by Quinn (2005). For example, the dominant age at migration from freshwater habitat is correlated with growth potential in fresh water (Quinn et al. 2009), and the dominant age at maturation after achieving growth in the sea assorts with spawning habitat conditions [i.e., spawners tend to be older in populations that spawn in deep water where sexual selection favors large body size (Quinn et al. 2001)]. Similarly, size-at-age of adult fish also co-varies with the dimensions of spawning habitat whereby populations spawning in deep water are relatively large for their age compared to fish in small streams where size-selective predation and stranding make large body size maladaptive (Quinn et al. 2001). Seasonal spawn timing also varies considerably among populations, with populations occupying cold streams spawning earlier in the season than those that occupy warmer habitat (Brannon 1987, Lisi et al. 2013). The mechanisms for this pattern of spawn timing are not particularly well understood, but it appears to be a result of thermal constraints on embryo incubation rates (longer in cold environments; Brannon 1987) and temperature controlled risk of disease or parasitism (Larson et al. 2016). Other good examples of local environmental controls on biological characteristics include: egg sizes scaling with the dominant gravel sizes (Quinn et al. 1995), and rates of post-spawning senescence scaling with stream size, due to its control on salmon vulnerability to predation by bears during spawning (Carlson et al. 2007). A glacial ecotype was documented in Lake Clark National Park where males spawning in highly turbid habitats had larger snouts, which serve as weapons during breeding competition (Quinn and Foote 1994), and females exhibited reduced spawning coloration, making them less visible to visual predators (Ramstad et al. 2010).

Thus, sockeye salmon within Bristol Bay express a remarkable diversity of biological attributes that appear to be evolutionary adaptations to the specific habitat conditions they encounter across their life cycle. While there has been considerable progress made toward understanding the association between biological attributes of sockeye salmon and habitat features in freshwater systems, there is no reason to believe that similar population-specific adaptions do not also exist during the marine phase of their lives. For example, Johnson and Schindler (2013) showed that the carbon and nitrogen stable isotope ratios in adult sockeye salmon from different populations within the Wood River were consistent with different marine foraging strategies needed to achieve different body sizes at maturity (i.e., fish from large-bodied populations fed at a higher trophic position than those from small-bodied populations). Seeb et al. (2011b) used genetic markers to estimate the origins of juveniles collected on marine surveys in late summer and found population-specific migration routes in the eastern Bering Sea. Other population-specific biological strategies during the marine phase of sockeye salmon life may remain obscure due to the difficulty of studying fish in the ocean. Nevertheless, the important point relevant to this chapter is that sockeye salmon throughout Bristol Bay exhibit a broad array of biological characteristics that appear to be local evolutionary adaptations to specific habitat conditions that not only generate the observed genetic diversity but also the diversity in population dynamics among stocks in the region.

# IMPORTANCE OF SOCKEYE SALMON DIVERSITY FOR FISHERIES AND ECOLOGICAL FUNCTIONS

The biological diversification of sockeye salmon across Bristol Bay (i.e., their *biocomplexity*; Hilborn et al. 2003) has fascinated biologists interested in understanding evolutionary processes within individual species under the selective regimes encountered in different habitats. However, there is also a growing appreciation of the functional effects of biocomplexity on ecosystem processes and services provided to people. If Bristol Bay sockeye salmon were characterized by a homogenous set of life-history strategies among populations, the roles of these fish in watersheds and the value to fisheries would be vastly different than they are. Here we highlight some of the more obvious ecological effects of these diversity effects.

## Population Dynamics and Synchrony

The number of sockeye salmon that have returned to Bristol Bay watersheds over the last century has varied considerably. Two prominent components of this variability are the large interannual swings in abundance owing to the population cycles in Kvichak River sockeye salmon during much of the twentieth century (Eggers and Rogers 1987), and the large step up in production in response to a change in the Pacific Decadal Oscillation in the mid-1970s. However, all rivers have shown changes in the number of sockeye salmon produced at both interannual and decadal scales causing substantial swings in fisheries catches from year to year and also from decade to decade.

Data documenting the long-term dynamics of Bristol Bay sockeye salmon demonstrate clearly that not all populations march to the beat of the same drum—there is considerable variation in the dynamics among different populations within rivers, and the river stock complexes across Bristol Bay also show little coordinated variation through time. Current understanding of these diverse responses suggests that they are produced through complex interactions between local habitat features and the overriding climate changes that are producing novel environments that the populations are exposed to at different stages of their lives (Hilborn et al. 2003, Rogers and Schindler 2013). Further, these population-specific environmental conditions interact with the genetically determined biological attributes of fish to produce the dynamics that characterize each population.

The observation that individual streams have unique population dynamics has been recognized for decades. W.F. Thompson (1962), the founding principal investigator of the Fisheries Research Institute at the University of Washington (FRI), wrote the following upon reflection of what FRI had learned after the first decade of study of Bristol Bay sockeye salmon from 1945–1958.

> "Numerous small, independent, spawning populations exist in the lakes and streams. During seaward migration, the young form successively more complex mixtures and then segregate progressively as they return to spawn in their home streams, where interbreeding occurs within each local stock. When thus segregated, the summed mortality of the life of the generation is most evident in the variability of these units. Each mixture during the homeward migration of the adults must then have an abundance, which is based upon the varying productivity of its components, hence in any given generation, it must be much more stable in numbers than the individual components."

In recent years a number of investigators have more explicitly quantified the extent to which different populations covary through time (Stewart et al. 2003, Rogers and Schindler 2008, 2011, Moore et al. 2010, Schindler et al. 2010, Quinn et al. 2012). The important result derived from this body of work is that there is only modest synchronization among populations of sockeye salmon in Bristol Bay, expressed across a wide range of spatial scales. While populations inhabiting streams that drain to a common nursery lake tend to vary more coherently than populations draining to different lakes (Rogers and Schindler 2009), even neighboring populations often have very different population dynamics.

Rogers et al. (2013) used paleolimnological data to reconstruct five centuries of population dynamics of sockeye salmon from the major nursery lakes of western Alaska to show that this response of diversity in population dynamics was also evident over century-long time scales. Even over the course of the last 500 years in Bristol Bay, there was no overriding pattern of variation in sockeye salmon production; most stocks appeared to show unique dynamics, often characterized by century-long fluctuations.

## Portfolio Effects in Bristol Bay Sockeye Salmon

The total number of sockeye salmon that return to Bristol Bay watersheds each year is less variable than it would be if there was no biocomplexity in the ecosystem because stocks returning to different rivers and streams do not all have the same population dynamics among years. Because some populations peak in years when other populations are below average, the aggregate response is less variable than the individual populations that comprise the overall stock complex (Schindler et al. 2010).

Because the population structure of Bristol Bay sockeye salmon is hierarchical (i.e., the overall return is represented by fish from the major rivers, each of which is composed of dozens of individual populations, and a variety of life-history strategies within populations), the variability of the components decreases at increasingly coarser scales of the hierarchy (Figure 22.3) (Schindler et al. 2010). The dominant cohorts within each stream are approximately 25% more variable (as measured by the inter-annual coefficient of variation) than individual populations; individual populations are approximately 25% more variable than the returns to major rivers; and returns to rivers are about 40% more variable than the total return to all of Bristol Bay. This reduction of variability at increasingly coarse scales of aggregation results from statistical averaging across the dynamics expressed at finer scales of the biocomplexity hierarchy. Thus, if Bristol Bay were composed of a single population of sockeye salmon, characterized by a single life-history strategy inhabiting the same habitat conditions everywhere, the annual returns could be 220% more variable from year to year than they currently are (Schindler et al. 2010). This stabilizing effect of sockeye salmon biodiversity on fisheries, wildlife, and ecosystem properties likely has distinct economic and ecological value that we are only beginning to understand.

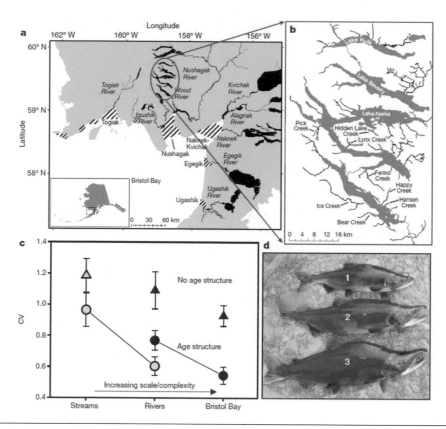

**Figure 22.3** Illustration of portfolio effect in the variability of sockeye salmon returns to Bristol Bay rivers and streams. (a) Map of major sockeye salmon rivers, nursery lakes, and associated commercial fishing districts for sockeye salmon in Bristol Bay, Alaska. (b) Map of sockeye salmon habitat within the Wood River drainage with spawning streams shown. Labeled streams denote those for which we have long-term data (from 1962–2007) on the annual abundance of spawning sockeye salmon. (c) Inter-annual variability, expressed as the coefficient of variation (CV), in the number of returning sockeye salmon at three levels of spatial aggregation and two levels of life-history aggregation. Grey symbols are for eight populations within the Wood River and black symbols are for the major rivers of Bristol Bay, or for the Bristol Bay aggregate (far right, 1958–2008). Circles show the variability in sockeye salmon returns at each level of spatial scale with the full complement of age structure. Triangles show observed variability in individual dominant age classes at each level of the spatial hierarchy. Taken together, these patterns show that returns of Bristol Bay sockeye salmon increase in stability when aggregated across more of the region, and across more of the life-history diversity in sockeye salmon. (d) Photos of male sockeye salmon from a population in the Wood River showing individuals that have spent one to three years at sea before returning to spawn. Each of these individuals, though spawning in the same year, were themselves spawned during different brood years and would likely have migrated to the ocean in different years. *Reprinted from Schindler et al. (2010).*

Schindler and Rogers (2009) first used the term *portfolio effect* to describe this effect of complexity on the variability in sockeye salmon returns to Bristol Bay. In essence, what W.F. Thompson was describing in terms of the compensatory dynamics among populations of sockeye salmon populations was a portfolio effect in the stock complexes of Bristol Bay sockeye salmon. His early concerns were with fisheries overexploiting small populations that co-migrated with larger components. Interestingly, this management concern persists today in both Bristol Bay fisheries and in fisheries globally.

While the benefits of portfolio effects to fisheries have become increasingly clear, both within Bristol Bay and throughout much of the range of Pacific salmon (Griffiths et al. 2014), these effects also likely extend to the wide array of wildlife that rely on salmon as a critical seasonal source of food. Both predators of spawning salmon and scavengers of post-spawn salmon depend heavily on this superabundant and rich food for achieving much of their annual growth (Schindler et al. 2003). A wide variety of mammals, birds, fishes, and invertebrates consume sockeye salmon on the spawning grounds, and in Bristol Bay many of these species achieve most of the annual growth by consuming salmon resources during the short period of time they are active on the spawning grounds. Thus, it is reasonable to assume that a vast array of other species within Bristol Bay watersheds also benefit from the reliability in salmon returns generated by population diversity, much like fisheries do. At present, the wildlife benefits of salmon biodiversity are not particularly well understood, but we summarize some noteworthy examples here.

Species that have limited mobility relative to the spatial variation in salmon populations (e.g., benthic insects) cannot benefit from the portfolio effects from salmon (i.e., because they cannot consume salmon from multiple populations within the same year). One situation where consumers located in a single location would benefit from population diversity would be those inhabiting migratory corridors such as rivers. In these situations, predators could consume different populations passing by a single location (e.g., McGlauflin et al. 2011), thereby benefiting from the stabilizing effects of temporal diversity as individual salmon migrate past them—either as adults on their way to the spawning grounds or as juveniles leaving for the ocean. However, small organisms with limited mobility (e.g., aquatic insects) certainly can benefit from the portfolio effects that are produced from within-population life-history diversity, likely experiencing much less year-to-year variability in the influx of marine-derived resources than if there was no life-history diversity within populations.

For organisms that are mobile enough to exploit multiple populations of sockeye salmon within the same year, among-population portfolio effects have the potential to stabilize the year-to-year variability in the abundance of this resource to them. Large mammals, birds, and resident fishes are all mobile enough to move across the landscape to capitalize on populations that are particularly abundant in any year. For example, individually tagged rainbow trout and Arctic grayling can easily achieve movements extending tens of kilometers within a single summer and are not associated with their own spawning migrations (Meka et al. 2003, Bentley et al. 2015). Although it is not clear that these movements are specifically directed toward exploiting individual populations of spawning sockeye salmon, this level of mobility clearly affords these fish the possibility of consuming resources from multiple populations within the same year, thereby avoiding being fully dependent on the production from any single salmon population. It is clear that birds such as eagles and gulls also aggregate at sites with especially high returns in any given year. We can assume that large mobile mammals such as brown bears and wolves also benefit from portfolio effects at the watershed scale by exploiting populations that are particularly abundant and avoiding sites with small returns, though these movements and the benefits they confer to consumers are only now being quantified.

Predators and scavengers of salmon resources also benefit from population diversity in sockeye salmon that is produced through the expression of variation in spawn timing among populations. Given that most predators and scavengers can only access adult salmon during or shortly after the spawning season, and that individual sockeye salmon populations are typically active on their spawning habitat for a month or less (Schindler et al. 2010), some consumers will be distinctly constrained by the duration of time in which they have access to salmon within a given year. However, because spawn timing varies considerably across watersheds according to geomorphic and thermal conditions in individual spawning locations (Lisi et al. 2013), sockeye salmon are actively spawning for up to three months at the river-basin scale in Bristol Bay rivers (Schindler et al. 2010). Thus, consumers that can move to track the timing of salmon spawning across the landscape benefit from a distinctly longer salmon feeding season than if this life-history variation did not exist (Ruff et al. 2010, Schindler et al. 2013). How sensitive predators and scavengers are to the loss of spawn-timing variation at the landscape scale has not been quantified; however, a variety of mechanisms—ranging from climate change to landscape development—could shorten the duration of spawn timing across watersheds and, therefore, also shorten the seasonal duration that consumers have access to salmon resources.

# EMERGING GENETICS TOOLS FOR FISHERIES MANAGEMENT AND CONSERVATION OF SOCKEYE SALMON BIODIVERSITY

Previously, we discussed that the diversity of habitat and its associated biological and genetic diversity within Bristol Bay sockeye salmon benefit not only the fisheries that exploit this species but also the wide range of wildlife that depend on sockeye salmon across the region. An important challenge to management is to conserve the genetic resources that provide resilience to these ecosystem processes and services to people, while simultaneously supporting large-scale fisheries. The genetic population structure of sockeye salmon is better understood than that of many other intensely managed animals and can be used as a rich source of information to inform fisheries management in real-time during the harvest in Bristol Bay. Homing of individuals from populations with diverse life histories, along with concomitant genetic drift and adaptive evolution, provide a wealth of signals detectable with genotyping—a method commonly referred to as genetic stock identification (GSI). GSI has been used successfully to identify the composition of mixed aggregations of Bristol Bay sockeye salmon across their entire migration route and life history, including outmigration and marine phases (Habicht et al. 2010, Seeb et al. 2011b) and as returning adults at test fishery sites (Dann et al. 2013). Additionally, the comprehensive Western Alaska Salmon Stock Identification Program (WASSIP) (http://www.adfg.alaska.gov/index.cfm?adfg=wassip.main)

detailed the harvest patterns throughout western Alaska of sockeye salmon originating from Bristol Bay (see upcoming text) (Dann et al. 2012b, Habicht et al. 2012).

The Port Moller test fishery, which intercepts fish as they migrate eastward into Bristol Bay along the north side of the Alaska Peninsula, historically provided managers with a crude index of the numbers of sockeye salmon that would enter Bristol Bay fisheries about a week later (Flynn and Hilborn 2004). GSI techniques were incorporated into the Port Moller test fishery beginning in 2006 (Dann et al. 2013). The authors used 38 SNPs, assayed in test fish, to provide real-time, stock-specific abundance data and assess spatial and temporal trends in stock composition in the Port Moller test fishery for the years 2006–2010. They reliably and repeatedly detected abundances that were not anticipated based upon preseason forecasts—fishing effort was shifted accordingly, and the fleet and local economies benefited while the risk of overharvesting weak stocks was minimized. Genetic monitoring of the Port Moller test fishery on a real-time basis is ongoing, although the number of SNPs used has been refined through time.

While informative, the Dann et al. (2013) panel of 38 SNPs did not provide the finer-scale accuracy that the Alaska Department of Fish and Game sought for the large WASSIP study. Additional SNPs were developed (Storer et al. 2012), and a panel of 96 SNPs was chosen for the baseline for spawning populations (Dann et al. 2012a, Dann et al. 2012c) to differentiate nine subregional groups within the Bristol Bay region (Togiak, Igushik, Wood, Nushagak, Kvichak, Alagnak, Naknek, Egegik, and Ugashik Rivers). Samples from over 80,000 sockeye salmon from 217 separate strata across years and regions were part of the WASSIP study. The majority of catches within Bristol Bay consisted of local stocks originating from Bristol Bay drainages. However, Bristol Bay stocks were widely represented within the Alaska Peninsula Area fisheries (Dann et al. 2012b). These estimated stock proportions were expanded to harvest rates across western Alaska by Habicht et al. (2012) to provide a comprehensive analysis of the stock-specific harvest of Bristol Bay sockeye salmon. Equally important, these studies provide a detailed analysis of migration routes and timing.

New genome-wide approaches have emerged in recent years. The ability to obtain millions of DNA sequences across genomes using next generation sequencing based methods has revolutionized SNP discovery in nonmodel species. Rather than survey the entire genome, researchers commonly use approaches to subsample the genome to reduce complexity by constructing reduced representation libraries (RRL). By creating an RRL for a small set of ascertainment individuals, thousands of putative SNPs can be detected; DNA primer sequences can then be developed for specified SNPs to further validate amplification and polymorphism (Seeb et al. 2011a, Helyar et al. 2012, Candy et al. 2015). The new genomics era promises increased understanding of the relationship between biological processes and adaptation in Bristol Bay populations; concomitant with this increased understanding will come higher resolution GSI and increased power to manage biological diversity in Bristol Bay sockeye salmon.

## SUMMARY AND FUTURE RESEARCH

Rivers draining the watersheds of Bristol Bay are characterized by remarkable physical complexity that has been produced by the interacting forces of tectonics, glaciation, and erosion. The physical characteristics of salmon habitat throughout this region continue to evolve as rivers wander across unconstrained floodplains due to the lack of human infrastructure such as roads, canals, and dams that impede floodplain evolution. Since the last deglaciation, sockeye salmon have radiated into the remarkably rich assortment of populations and life-history strategies that characterize the current stock complex. New genetics and genomics approaches are providing new insights into the evolutionary processes that generated this spectacular diversity and powerful tools that can be used in management and conservation of sockeye salmon across this region and elsewhere.

Recent and rapid advancements in DNA technology signal future opportunities to further refine genetic tools for managing sockeye salmon fisheries. Genome-wide resources are becoming available, better enabling the identification of genes that control adaptively important traits. These genes will enable higher resolution of stock structure that will, in turn, provide higher resolution of the mixture composition of the Port Moller test fishery and in the commercial fisheries. Likewise, higher resolution of historical fisheries through analysis of archived scales will enable an improved understanding of the relationship between harvest and other natural demographic processes.

The primary message of this chapter is that the genetic and life-history variation within Bristol Bay sockeye salmon is a critical and unique resource of this region. Not only is this within-species biodiversity a spectacular component of the natural heritage of Bristol Bay, it also provides a growing list of tangible benefits to people and to wildlife. It is clear that subsistence, recreational, and commercial fisheries all benefit from the reliability in salmon returns that is supported by the combined geomorphic and biological complexity that characterize Bristol Bay's salmon ecosystems. However, this complexity is also likely to be very important for maintaining resilience of sockeye salmon as climate change alters the character of ecosystems. We suspect that habitat complexity and genetic diversity within sockeye salmon makes this species and the human enterprises it supports far less vulnerable to global change than if this complexity did not exist. An important research goal is to understand how ongoing global and regional changes to climate and watersheds will alter the complexity of salmon resources and their habitat, and how management might reduce the erosion of complexity that is currently intact across this region.

Diversity in Bristol Bay sockeye salmon is potentially threatened, directly and indirectly, by a variety of human activities. Climate change and associated effects on thermal regimes and hydrology (Wobus et al. 2015) will likely alter the selective forces that have produced diversity in sockeye salmon. Reduced winter snow accumulation may be particularly important, as it serves to generate much of the within-watershed thermal variation that appears to be a major selective force among populations (Lisi et al. 2013, Larson et al. 2016). Selective harvest by commercial fisheries has the potential to change the distribution of population representation in the stock complex of Bristol Bay sockeye salmon and increases extinction risk of very small populations that may represent an important component of diversity. However, current management that attempts to distribute harvest across the range of run timing appears to minimize these risks. Emerging genetics tools are enabling real-time monitoring of stocks of concern that will improve the ability of managers to protect weak stocks and monitor relative vulnerability of ecotypes while maintaining productive fisheries on abundant stocks (Dann et al. 2013). Future research should continue to focus on developing ways to assess the selective nature of fisheries to reduce impacts of heavy exploitation on genetic diversity across the region.

Last, development of watersheds poses a serious threat to Bristol Bay sockeye salmon because it will likely erode the habitat heterogeneity that generates genetic variation in sockeye salmon. Construction of infrastructure such as roads will impede the disturbance and erosional processes that generate and maintain habitat variation that salmon benefit from over ecological and evolutionary time scales, from headwaters to estuaries. Similarly, flooding events pose serious risk to human infrastructure and therefore become a focus for civil engineering projects to protect rural communities. However, flood suppression and flow regulation will change the magnitude and frequency of the disturbance processes that maintain productive salmon habitat. Changing climate is expected to alter the precipitation, and therefore, flood regimes that will pose new challenges to human infrastructure. Unfortunately, engineered solutions to minimize risks to infrastructure have the potential to eliminate the physical processes that maintain complexity in salmon habitat. Thus, there are serious needs for developing planning scenarios that will enable economic opportunities for Bristol Bay communities, but in ways that will not pose substantial risks to the resilience of salmon ecosystems—particularly as ongoing climate change may alter the distribution of productive salmon habitat across the Bristol Bay landscape.

In other parts of the range of sockeye salmon, it has been assumed that habitat loss can be mitigated with habitat enhancement and hatcheries. This assumption has proved wrong as about 30% of population diversity in Pacific salmon has been lost in the Lower 48 states, with about half of the population diversity of sockeye salmon and Chinook salmon lost—probably because they are particularly vulnerable during their prolonged freshwater life history (Gustafson et al. 2007). Associated with this loss of biodiversity and habitat, and an increased reliance on hatchery production, has been a distinct weakening of the portfolio effects that make Bristol Bay fisheries so reliable from year to year (Moore et al. 2010, Griffiths et al. 2014). Similar erosion of habitat and genetic complexity of Bristol Bay salmon ecosystems would likely also reduce the resilience of fish stocks to climate change and the reliability of fisheries.

# REFERENCES

Armstrong, J.B. and D.E. Schindler. 2013. Going with the flow: spatial distributions of juvenile coho salmon track an annually shifting mosaic of water temperature. Ecosystems 16(8):1429–1441.

Baker, T.T., L.F. Fair, F.W. West, G.B. Buck, X. Zhang, S. Fleishman, and J. Erickson. 2009. Review of salmon escapement goals in Bristol Bay, Alaska, 2009. Alaska Department of Fish and Game, Fishery Management Series No. 09-05, Anchorage, AK.

Beacham, T.D., B. McIntosh, C. MacConnachie, K.M. Miller, and R.E. Withler. 2006. Pacific rim population structure of sockeye salmon as determined from microsatellite analysis. Transactions of the American Fisheries Society 135:174–187.

Bentley, K.T., D.E. Schindler, J.B. Armstrong, T.J. Cline, and G.T. Brooks. 2015. Summer movements of stream-dwelling salmonids throughout a network of lake tributaries. PLoS ONE. 10: e0136985. doi: 10.1371/journal.pone.0136985.

Candy, J.R., N.R. Campbell, M.H. Grinnell, T.D. Beacham, W.A. Larson, and S.R. Narum. 2015. Population differentiation determined from putative neutral and divergent adaptive genetic markers in eulachon (*Thaleichthys pacificus*, Osmeridae), an anadromous Pacific smelt. Mol Ecol Resour:n/a-n/a.

Carlson, S.M., R. Hilborn, A.P. Hendry, and T.P. Quinn. 2007. Predation by bears drives senescence in natural populations of salmon. PLoS ONE 2(12): e1286. doi: 10.1371/journal.pone.0001286.

Dann, T.H., C. Habicht, T.T. Baker, and J.E. Seeb. 2013. Exploiting genetic diversity to balance conservation and harvest of migratory salmon. Canadian Journal of Fisheries and Aquatic Sciences 70(5): 785–793.

Dann, T.H., C. Habicht, J.R. Jasper, E.K.C. Fox, H.A. Hoyt, H.L. Liller, E.S. Lardizabal, P.A. Kuriscak, Z.D. Grauvogel, and W.D. Templin. 2012a. Sockeye salmon baseline for the western Alaska salmon stock identification project. Alaska Department of Fish and Game, Special Publication No. 12-12, Anchorage. http://www.adfg.alaska.gov/FedAidpdfs/SP12-12.pdf.

Dann, T.H., J.R. Jasper, H.A. Hoyt, H. Hildebrand, and C. Habicht. 2012c. Western Alaska salmon stock identification program technical document 6: Selection of the 96 SNP marker set for sockeye salmon. Alaska Department of Fish and Game, Division of Commercial Fisheries, Regional Information Report 5J12-11, Anchorage, AK.

Eggers, D.M. and D.E. Rogers. 1987. The cycle of runs of sockeye salmon (Oncorhynchus nerka) to the Kvichak River, Bristol Bay, Alaska: Cyclic dominance or depensatory fishing? Can. Spec. Publ. Fish. Aquat. Sci. 96: 343–366.

Elfstrom, C.M., C.T. Smith, and J.E. Seeb. 2006. Thirty-two single nucleotide polymorphism markers for high-throughput genotyping of sockeye salmon. Molecular Ecology Notes 6:1255–1259.

Everett, M.V., M.R. Miller, and J.E. Seeb. 2012. Meiotic maps of sockeye salmon derived from massively parallel DNA sequencing. BMC Genomics 13: Article Number 521, DOI 10.1186/1471-2164-13-521.

Flynn, L. and R. Hilborn. 2004. Test fishery indices for sockeye salmon (Oncorhynchus nerka) as affected by age composition and environmental variables. Canadian Journal of Fisheries and Aquatic Sciences 61:80–92.

Gomez-Uchida, D., James E. Seeb, C. Habicht, and Lisa W. Seeb. 2012. Allele frequency stability in large, wild exploited populations over multiple generations: Insights from Alaska sockeye salmon (Oncorhynchus nerka). Canadian Journal of Fisheries and Aquatic Sciences 69:916–929.

Griffiths, J.R., D.E. Schindler, J.B. Armstrong, M.D. Scheuerell, D.C. Whited, R.A. Clark, R. Hilborn, C.A. Holt, S.T. Lindley, J.A. Stanford, and E.C. Volk. 2014. Performance of salmon fishery portfolios across Western North America. Journal of Applied Ecology 51(6) 1554–1563.

Gustafson, R.G., R.S. Waples, J.M. Myers, L.A. Weitkamp, G.J. Bryant, O.W. Johnson, and J.J. Hard. 2007. Pacific salmon extinctions: quantifying lost and remaining diversity. Conservation Biology 21(4):1009–1020.

Gustafson, R.G., and G.A. Winans. 1999. Distribution and population genetic structure of river- and sea-type sockeye salmon in western North America. Ecology of Freshwater Fish 8:181–193.

Habicht, C., A.R. Munro, T.H. Dann, D.M. Eggers, W.D. Templin, M.J. Witteveen, T.T. Baker, K.G. Howard, J.R. Jasper, S.D. Rogers Olive, H.L. Liller, E.L. Chenoweth, and E.C. Volk. 2012. Harvest and harvest rates of sockeye salmon stocks in fisheries of the western Alaska salmon stock identification program (WASSIP), 2006–2008. Alaska Department of Fish and Game, Special Publication No. 12-24, Anchorage. http://www.adfg.alaska.gov/FedAidpdfs/SP12-24.pdf.

Habicht, C., J. Olsen, L. Fair, and J. Seeb. 2004. Smaller effective population sizes evidenced by loss of microsatellite alleles in tributary-spawning populations of sockeye salmon from the Kvichak River, Alaska drainage. Environmental Biology of Fishes 69:51–62.

Habicht, C., L.W. Seeb, K.W. Myers, E.V. Farley, and J.E. Seeb. 2010. Summer-fall distribution of stocks of immature sockeye salmon in the Bering Sea as revealed by single-nucleotide polymorphisms (SNPs). Transactions of the American Fisheries Society 139:1171–1191.

Habicht, C., L.W. Seeb, and J.E. Seeb. 2007. Genetic and ecological divergence defines population structure of sockeye salmon populations returning to Bristol Bay, Alaska, and provides a tool for admixture analysis. Transactions of the American Fisheries Society 136:82–94.

Helyar, S.J., M.T. Limborg, D. Bekkevold, M. Babbucci, J. van Houdt, G.E. Maes, L. Bargelloni, R.O. Nielsen, M.I. Taylor, R. Ogden, A. Cariani, G.R. Carvalho, F. Panitz, and F. Consortium. 2012. SNP discovery using next generation transcriptomic sequencing in Atlantic herring (Clupea harengus). PloS One 7.

Johnson, S.P. and D.E. Schindler. 2012. Marine trophic diversity in an anadromous fish is linked to its life-history variation in fresh water. Biology Letters 9, 20120824. http://dx.doi.org/10.1098/rsbl.2012.0824.

Kaufman, D.S. and W.F. Manley. 2004. Pleistocene maximum and late Wisconsinan glacier extents across Alaska, U.S.A. Pages 9–28. In: J. Ehlers and P.L. Gibbard (Eds.), Quaternary Glaciations Extent and Chronology Part II: North America.

Larson, W.A., J.E. Seeb, T.H. Dann, D.E. Schindler, and L.W. Seeb. 2014. Signals of heterogeneous selection at an MHC locus in geographically proximate ecotypes of sockeye salmon. Molecular Ecology 23(22):5448–5461.

Larson, W.A., G.J. McKinney, M.T. Limborg, M.V. Everett, L.W. Seeb, and J.E. Seeb. 2015. Identification of multiple QTL hotspots in sockeye salmon (Oncorhynchus nerka) using genotyping-by-sequencing and a dense linkage map. Journal of Heredity 107: 122–133.

Larson, W.R., P.J. Lisi, J.E. Seeb, L.W. Seeb, and D.E. Schindler 2016. Major histocompatibility complex diversity is positively associated with stream water temperatures in proximate populations of sockeye salmon. Journal of Evolutionary Biology 29: 1846–1859.

Lin, J., T.P. Quinn, R. Hilborn, and L. Hauser. 2008. Fine-scale differentiation between sockeye salmon ecotypes and the effect of phenotype on straying. Heredity 101:341–350.

Lisi, P.J., D.E. Schindler, K.T. Bentley, and G.R. Pess. 2013. Association between geomorphic attributes of watersheds, water temperature, and salmon spawn timing in Alaskan streams. Geomorphology 185:78–86.

Lisi, P.J., D.E. Schindler, T.J. Cline, M.D. Scheuerell, and P.B. Walsh. 2015. Watershed geomorphology and snowmelt control stream thermal sensitivity to air temperature. Geophysical Research Letters 42(9):3380–3388.

McGlauflin, M.T., D.E. Schindler, L.W. Seeb, C.T. Smith, C. Habicht, and J.E. Seeb. 2011. Spawning habitat and geography influence population structure and juvenile migration timing of sockeye salmon in the Wood River lakes, Alaska. Transactions of the American Fisheries Society 140(3):763–782.

Meka, J.M., E.E Knudsen, D.C. Douglas, and R.B. Benter. 2003. Variable migration patterns of different adult rainbow trout life history types in a wouthwest Alaska watershed. Transactions of the American Fisheries Society 132:717–723.

Miller, K.M., K.H. Kaukinen, T.D. Beacham, and R.E. Withler. 2001. Geographic heterogeneity in natural selection on an MHC locus in sockeye salmon. Genetica 111:237–257.

Moore, J.W., M. McClure, L.A. Rogers, and D.E. Schindler. 2010. Synchronization and portfolio performance of threatened salmon. Conservation Letters 3:340–348.

Nelson, R.J. and T.D. Beacham. 1999. Isolation and cross species amplification of microsatellite loci useful for study of Pacific salmon. Animal Genetics 30:228–229.

Quinn, T.P. The Behavior and Ecology of Pacific Salmon & Trout. 2005. American Fisheries Society, Bethesda, MD.

Quinn, T.P., H.B. Rich, Jr., D. Gosse, and N. Schtickzelle. 2012. Population dynamics and asynchrony at fine spatial scales: a case history of sockeye salmon (*Oncorhynchus nerka*) population structure in Alaska, USA. Canadian Journal of Aquatic and Fishery Sciences 69:297–306.

Quinn, T.P., K. Doctor, N. Kendall, and H.B. Rich Jr. 2009. Anadromy and the life history of salmonid fishes: nature, nurture, and the hand of man. American Fisheries Society Symposium 69, 23–42. Bethesda, MD.

Quinn, T.P., A.P. Hendry and L.A. Wetzel. 1995. The influence of life history trade-offs and the size of incubation gravels on egg size variation in sockeye salmon (*Oncorhynchus nerka*). Oikos 74:425–438.

Quinn, T.P., L.A. Wetzel, S. Bishop, K. Overberg, and D.E. Rogers. 2001. Influence of breeding habitat on bear predation, and age at maturity and sexual dimorphism of sockeye salmon populations. Canadian Journal of Zoology 79:1782–1793.

Pess, G.R., T.P. Quinn, D.E. Schindler, and M.C. Liermann. 2014. Freshwater habitat associations between pink (*Oncorhynchus gorbuscha*), chum (*O. keta*) and Chinook Salmon (*O. tshawytscha*) in a watershed dominated by sockeye salmon (*O. nerka*) abundance. Ecology of Freshwater Fish 23(3):360–372.

Ramstad, K.M., C.A. Woody, and F.W. Allendorf. 2010. Recent local adaptation of sockeye salmon to glacial spawning habitats. Evolutionary Ecology 24(2):391–411.

Ramstad, K.M., C.A. Woody, C. Habicht, G.K. Sage, J.E. Seeb, and F.W. Allendorf. 2007. Concordance of nuclear and mitochondrial DNA markers in detecting a founder event in Lake Clark sockeye salmon. Sockeye Salmon Evolution, Ecology, and Management 54:31–50.

Ramstad, K.M., C.A. Woody, G.K. Sage, and F.W. Allendorf. 2004. Founding events influence genetic population structure of sockeye salmon (*Oncorhynchus nerka*) in Lake Clark, Alaska. Molecular Ecology 13:277–290.

Rogers, L.A. and D.E. Schindler. 2008. Asynchrony in population dynamics of sockeye salmon in southwest Alaska. Oikos 117(10):1578–1586.

Rogers, L.A. and D.E. Schindler. 2011. Scale and the detection of climatic influences on the productivity of salmon populations. Global Change Biology 17:2546–2558.

Rogers, L.A., D.E. Schindler, P.J. Lisi, G.W. Holtgrieve, P.R. Leavitt, L. Bunting, B.P. Finney, D.T. Selbie, G. Chen, I. Gregory-Eaves, M.J. Lisac, and P.B. Walsh. 2013. Centennial-scale fluctuations and regional complexity characterize Pacific salmon population dynamics over the past five centuries. Proceedings of the National Academy of Sciences, USA 110(5):1750–1755.

Ruff, C.P., D.E. Schindler, J.B. Armstrong, K.T. Bentley, G.T. Brooks, G.W. Holtgrieve, M.T. McGlauflin, C.E. Torgersen, and J.E. Seeb. 2011. Temperature-associated population diversity in salmon confers benefits to mobile consumers. Ecology 92(11) 2073–2084.

Schindler, D.E., J.B. Armstrong, K.T. Bentley, K. Jankowski, P.J. Lisi, and L.X. Payne. 2013. Riding the crimson tide: mobile terrestrial consumers track phenological variation in spawning of an anadromous fish. Biology Letters 9(3), article number 20130048, DOI 10.1098/rsbl.2013.0048.

Schindler, D.E., X. Augerot, E. Fleishman, N.J. Mantua, B. Riddell, M. Ruckelshaus, J. Seeb, and M. Webster. 2008. Climate change, ecosystem impacts, and management for Pacific salmon. Fisheries 33(10) 502–506.

Schindler, D.E., R. Hilborn, B. Chasco, C.P. Boatright, T.P. Quinn, L.A. Rogers, and M.S. Webster. 2010. Population diversity and the portfolio effect in an exploited species. Nature 465(7298):609–612.

Schindler, D.E. and L.A. Rogers. 2009. Responses of Pacific salmon populations to climate variation in freshwater ecosystems. p. 1127–1142. In: C.C. Krueger and C.E. Zimmerman (Eds.), *Pacific Salmon: Ecology and Management of Western Alaska's Populations*. American Fisheries Society, Symposium 70, Bethesda MD.

Seeb, L.W., A. Antonovich, A.A. Banks, T.D. Beacham, A.R. Bellinger, S.M. Blankenship, A.R. Campbell, N.A. Decovich, J.C. Garza, C.M. Guthrie, T.A. Lundrigan, P. Moran, S.R. Narum, J.J. Stephenson, K.J. Supernault, D.J. Teel, W.D. Templin, J.K. Wenburg, S.E. Young, and C.T. Smith. 2007. Development of a standardized DNA database for Chinook salmon. Fisheries 32:540–552.

Seeb, L.W., C. Habicht, W.D. Templin., K.E. Tarbox, R.Z. Davis., L.K. Brannian, and J.E. Seeb. 2000. Genetic diversity of sockeye salmon of Cook Inlet, Alaska, and its application to management of populations affected by the *Exxon Valdez* oil spill. Transactions of the American Fisheries Society 129:1223–1249.

Seeb, J.E., C.E. Pascal, E.D. Grau, L.W. Seeb, W.D. Templin, T. Harkins, and S.B. Roberts. 2011a. Transcriptome sequencing and high-resolution melt analysis advance single nucleotide polymorphism discovery in duplicated salmonids. Molecular Ecology Resources 11:335–348.

Seeb, J.E., C E. Pascal, R. Ramakrishnan, and L.W. Seeb. 2009. SNP genotyping by the 5′-nuclease reaction: advances in high throughput genotyping with non-model organisms. Pages 277–292. In: A. Komar (Ed.), *Methods in Molecular Biology, Single Nucleotide Polymorphisms*, 2nd Edition. Humana Press.

Seeb, L.W., J.E. Seeb, C. Habicht, E.V. Farley, and F.M. Utter. 2011. Single-nucleotide polymorphic genotypes reveal patterns of early juvenile migration of sockeye salmon in the Eastern Bering Sea. Transactions of the American Fisheries Society 140(3):734–748.

Seeb, L.W., J.E. Seeb, C. Habicht, E.V. Farley, and F.M. Utter. 2011b. Single-nucleotide polymorphic genotypes reveal patterns of early juvenile migration of sockeye salmon in the Eastern Bering Sea. Transactions of the American Fisheries Society 140:734–748.

Shaklee, J.B., T.D. Beacham, L. Seeb, and B.A. White. 1999. Managing fisheries using genetic data: case studies from four species of Pacific salmon. Fisheries Research 43:45–78.

Smith, C.T., W.D. Templin, J.E. Seeb, and L.W. Seeb. 2005. Single nucleotide polymorphisms (SNPs) provide rapid and accurate estimates of the proportions of U.S. and Canadian Chinook salmon caught in Yukon River fisheries. North American Journal of Fisheries Management 25:944–953.

Stephenson, J., M. Campbell, J. Hess, C. Kozfkay, A. Matala, M. McPhee, P. Moran, S. Narum, M. Paquin, O. Schlei, M. Small, D. Van Doornik, and J. Wenburg. 2009. A centralized model for creating shared, standardized, microsatellite data that simplifies interlaboratory collaboration. Conservation Genetics 10:1145–1149.

Stewart, I.J., R. Hilborn, and T.P. Quinn. 2003. Coherence of observed adult sockeye salmon abundance within and among spawning habitats of the Kvichak River watershed. Alaska Fishery Research Bulletin 10(1):28–41.

Storer, C.G., C.E. Pascal, S.B. Roberts, W.D. Templin, L.W. Seeb, and J.E. Seeb. 2012. Rank and order: evaluating the performance of SNPs for individual assignment in a non-model organism. PLoS ONE 7:e49018.

Thompson, W.G. 1962. The research program of the Fisheries Research Institute in Bristol Bay, 1945–1958. Page 3–36. In: T.W.Y. Koo (Ed.), *Studies of Alaska Red Salmon*. University of Washington Press, Seattle, WA.

Wilmot, R.L., R.J. Everett, and W.A. Gellman. 1989. Genetic stock identification of sockeye and chum salmon from Bristol Bay, Alaska, U. S. Dep. Commer., NOAA, OCSEAP Final Rep. 63 (1989):553–599.

Wobus, C., R. Prucha, D. Albert, C. Woll, M. Loinaz, and R. Jones. 2015. Hydrologic alterations from climate change inform assessment of ecological risk to Pacific salmon in Bristol Bay, Alaska. PLoS ONE 10(2): e0143905.

Wood, C.C., J.W. Bickham, R.J. Nelson, C.J. Foote, and J.C. Patton. 2008. Recurrent evolution of life history ecotypes in sockeye salmon: implications for conservation and future evolution. Evolutionary Applications 1:207–221.

Wood, C.C., B.E. Riddell, D.T. Rutherford, and R.E. Withler. 1994. Biochemical genetic survey of sockeye salmon (*Oncorhynchus nerka*) in Canada. Canadian Journal of Fisheries & Aquatic Sciences 51:114–131.

# 23

# FRESHWATER SEALS OF ILIAMNA LAKE

Jennifer Burns, Ph.D., University of Alaska Anchorage
David Withrow, National Oceanic and Atmospheric Administration
James M. Van Lanen, Alaska Department of Fish & Game

## INTRODUCTION

Worldwide, there are six freshwater, lake-dwelling seal populations, of which four are likely descended from ringed seals that were isolated during the last ice age. These four populations are all found in the northern hemisphere; in Lake Baikal (~85,000 individuals, *Phoca sibirica*) and Lake Ladoga (~3,000 seals, *P. hispida ladogensis*) in Russia, Lake Saimaa (~270 seals, *P. h. saimensis*) in Finland, and in eastern Europe in the brackish (salinity ranges from < 1 to ~13 ppt, Dumont 1998) Caspian Sea (~105,000 seals, *P. caspica*) (Everitt and Braham 1980, Smith et al. 1994, Smith et al. 1996, Rice 1998). The fifth population of freshwater seals (150–600 individuals) is found within the Lacs des Loups Marins area of the Ungava Peninsula, in northern Québec, Canada (COSEWIC 2007) and is recognized as a subspecies (*P.v. mellonae*) of harbor seals (Smith et al. 1994), while the sixth population is in Iliamna Lake, Alaska (Figure 23.1). All freshwater seal populations are relatively small, and concerns about population sustainability due to anthropogenic and climate effects

**Figure 23.1**  View of the many islands within Iliamna Lake taken during an aerial survey for harbor seals. *Photo credit: David Withrow.*

493

have resulted in protective measures for seals in the Caspian Sea, Lake Baikal, and Lake Saimaa. In Alaska, concerns about the seals within Iliamna Lake have centered on the potential impact of natural resource development and climate change (CBD 2012), and have been magnified by a general lack of scientific knowledge about the seals' population biology or ecology.

The lack of historical scientific data on the Iliamna Lake seals, however, does not equate with an absence of knowledge, as the Bristol Bay region in which Iliamna Lake is located has a rich cultural history including Yup'ik Eskimos, Aleuts, Athabascan Indians, and Inupiaq people (Borras and Knott, Chapter 1 of this volume), and these cultures maintain a rich and detailed oral history and traditional knowledge of the natural world with which they interact. Indeed, the local and traditional knowledge about seal biology and ecology represents a valuable resource and many insights presented in this chapter come from interviews conducted with local residents during 2011 and 2012, which were part of a community-based research project (Burns et al. 2012, Burns et al. 2013).

The Iliamna Lake region was originally settled 7,600 to 4,000 years ago by people of the Ocean Bay Tradition, with two other coastal marine cultures, the Arctic Small Tool Tradition [~3,000 years before present (ybp)] and the Norton Culture (2,500 to 1,500 ybp) (Reger 2005) also present. The name *Iliamna* is derived from the Dena'ina Athabascan name *Nila Vena*, which means *island's lake* in reflection of the many small islands within the lake (Kari and Kari 1982). Historically, major village sites have been located on the eastern end of the lake, with the oldest, Pedro Bay, having been settled ~4,500 ybp and occupied by multiple cultures since (Reger 2005). By comparison, Old Iliamna Village was established in 1838 by the Dena'ina near a Russian post (est. 1821) (Ellanna 1986), but was moved approximately 65 km (40 miles) to the southwest in 1935 to the existing site on the western shore of Iliamna Lake. Currently, the native villages of Iliamna (IL), Newhalen (NH), Pedro Bay (PB), Kokhanok (KK), and Iguigig (IG) are found on the shoreline of the lake, with Levelock (LV) along the shores of the Kvichak River (Figure 23.2). Residents of all of these villages contributed to the information

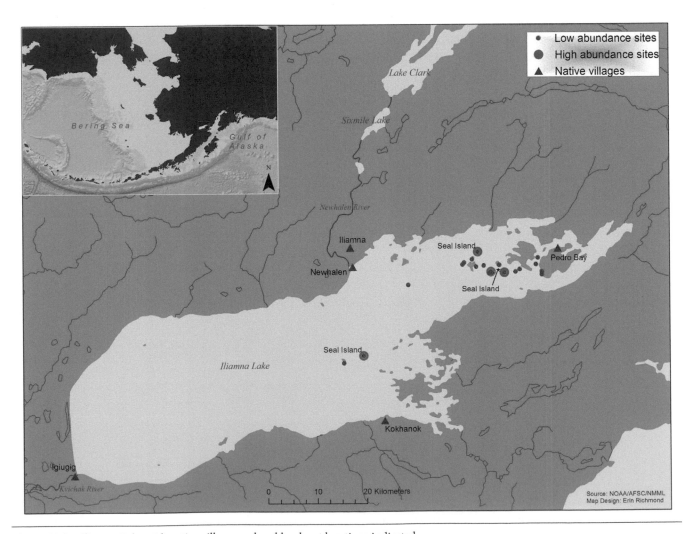

**Figure 23.2**    Iliamna Lake with native villages and seal haul-out locations indicated.

presented here during the course of their participation in two research projects funded by the North Pacific Research Board that integrated scientific surveys with traditional ecological knowledge (Burns et al. 2013).

## THE ORIGIN OF SEALS WITHIN ILIAMNA LAKE

Iliamna Lake is Alaska's largest lake and the eighth largest in the U.S., measuring roughly 130 km (80 miles) long by 40 km (25 miles) wide, for a total area of 3,000 square km (1,150 square miles). It lies approximately 362 km (225 miles) southwest of Anchorage and just south of Lake Clark National Park and Preserve (Winfree, Chapter 4 of this volume). Fed by numerous creeks and rivers, Iliamna Lake drains westward via the Kvichak River approximately 110 km (70 miles) into Bristol Bay (Figure 23.2), and, as indicated before, is home to a unique population of freshwater seals. However, it remains unclear when seals first began occupying Iliamna Lake. Harbor and spotted seals are abundant along Bristol Bay and the sandbars off the mouth of the Kvichak River (Allen and Angliss 2012), and Iliamna residents have observed and utilized the lake seals as long as they can recall. When questioned about the origin of the seals in the lake, native elders and hunters indicated that seals had been in the lake for a long time, and had moved to the lake following the salmon resources. For example, a Newhalen hunter stated, "*I never heard the elders say that there was any point* [in time] *that there were no seals*," while a second said, "*These seals have been here since way before my time, and I think way before my grandparents' time . . . they were here before we* [humans] *were*" (NH, male, 50s). Similarly, an Iguigig hunter said, "*Seals got here by following the fish, thousands of years ago, and they stayed here because of the fish population*," (IG, male, 70s) while a Kokhanok hunter said "*I imagine the seals, at one time, they came from the saltwater up here and they just stayed*," (KK, male, 50s). Of ethnographic interest are the reflections of an Iliamna elder recounting stories about the seals' origins that he had heard from his elders in the past: "*There are some stories; seals came from the ocean under the cracks from the ocean floor to the lake, from outside of this area. Legend says that seals came from under the mountains. There is a big crack somewhere*," (IL, male, 60s). Certainly, studies of harbor seal populations throughout the Pacific northwest (Brown and Mate 1983, Lowry et al. 2001) have indicated that journeys of 110–160 km (70–100 miles) up river are well within the scope of normal seal behavior, and that harbor seals can, and do, follow salmon upriver. Thus, both scientific studies and traditional knowledge suggest that originally, some seals likely followed migrating salmon—a favorite prey item—upstream during their spawning run into Iliamna Lake, and chose to remain within the lake.

## WRITTEN HISTORY OF SEALS IN THE LAKE

While traditional ecological knowledge suggests that seals have been in the lake for hundreds, if not thousands of years, the first written record of the seals' presence within the lake comes from Russian explorers who first travelled to the region in the late 1700s (Michael 1967). Two rival Russian companies (Shelikov and Lebedev-Lastochkin) competed for control of trade within the Cook Inlet area and the latter plundered the Iliamna and Nushagak villages in 1792, where they then established a small trading post (VanStone and Townsend 1970). In retaliation for this and other mistreatments, villagers twice destroyed the Russian trading post on Iliamna Lake, but by the early 1810s, relations had stabilized, as reflected in an 1818 journal entry by the Russian explorer Petr Korsakovskiy that also includes the earliest known account of seal hunting in the lake: "*. . . we said farewell to the Indians and set out on the return journey* [from Lake Clarke] *to Lake Iliamna . . . Toward evening we came upon our baydarkas, got into them, and set out along the* [Newhalen River] *to the lake. Patyukov killed a harbor seal. We went ashore, pitched our tent, and settled down for the night.*" Additional information on the presence and native use of the seals in the lake can be found in Lemuel E. Bonham's journal, who in 1901 remarked that "*among these islands* [in the NE portion of Iliamna Lake] *are plenty of hair seals. Perhaps the only place on earth where seals are found in fresh water.*" He also described how natives hunted seals in winter by spearing them through a small hole in the ice made next to a seal breathing hole (Branson 2007). After this report, sporadic Iliamna Lake seal reports occur in the literature, but the first recorded abundance survey was not until 1984 (Withrow and Yano 2010) and despite increased attention recently (Hauser et al. 2008, Withrow et al. 2011, Burns et al. 2012, Burns et al. 2013), information on the ecology and behavior of Iliamna Lake seals remains scarce as compared to that available for marine seal populations.

## TAXONOMY AND PHYLOGENY

Indeed, until 2013, even the species identity of the freshwater seals in Iliamna Lake was ambiguous. Bristol Bay is home to both harbor (*Phoca vitulina*) and spotted seal (*Phoca largha*) populations, and the native name for the seals within Iliamna Lake translates to spotted seal, but this is due to the seals' pelage pattern, and does not refer to *Phoca largha* specifically. It is difficult to visually distinguish between the two species as adults are similar in size and appearance (Burns et al. 1984, Lowry et al. 1998), in part because spotted seals (Figure 23.3A) and harbor seals (Figure 23.3B) share a recent common ancestor (~1.1 million years ago) (Higdon et al. 2007). In general, spotted seals are thought to be more *ice-associated* than harbor seals, but Bristol Bay is seasonally ice covered and inhabited by both species (Lowry et al. 1998, Kelly et al. 2010, Allen and Angliss 2012), thus removing this behavioral cue.

Recent genetic analysis of tissue samples recovered from 12 seals harvested in Iliamna Lake during fall and spring of five different years over a 16-year period (1996 to 2012) has helped shed light on the species identity of Iliamna Lake's freshwater seals. All 12 seals were categorized as harbor seals using multi-locus genotypes, based on nine microsatellite loci that had previously been screened in both spotted (n = 202) and North Pacific harbor (n = 684) seals (Burns et al. 2013). In addition, only one mitochondrial DNA (mtDNA) haplotype, Pvit-Hap#7, was present—an unusually low level of mtDNA diversity when judged against similar sampling regimes for this species elsewhere in Alaska. This single haplotype occurs at a frequency of 21% of harbor seals sampled in eastern Bristol Bay (n = 16/76) and in 22% of the harbor seals from across the entire Bay (n = 24/109) (O'Corey-Crowe et al. 2003). If immigration and reproductive mixing from the Bristol Bay harbor seal population into a resident Iliamna Lake harbor seal population was ongoing, there is a near 80%

**Figures 23.3A and 23.3B**    A single spotted seal (A) and two harbor seals (B) on ice in Alaska. *Photo credit: David Withrow.*

likelihood that mtDNA haplotypes other than Hap#7 would have been introduced. The rate of such immigration would, of course, determine the frequency of such haplotypes in the lake population and thus, also in a sample of this population. That none of these other haplotypes have been recorded as yet in Iliamna suggests that immigration of harbor seals, if it occurs, is not substantial (Burns et al. 2013). In addition, genetic differentiation and homogeneity tests suggest that the population of harbor seals in Iliamna has differentiated from that in Bristol Bay with restricted male and female-mediated dispersal between the two. However, sample size was small and kinship among samples tested remains unknown, so these results must be interpreted with caution (Burns et al. 2013). In general, these findings confirm the oral tradition that the seals in the lake originated from Bristol Bay, and highlight the value of including local and traditional knowledge about the seals when assessing their behavior and ecology.

## DESCRIPTION

In general size, shape, and appearance, the freshwater seals in Iliamna Lake appear similar to the harbor seals within Bristol Bay. There is little scientific documentation on how harbor seals of Iliamna Lake differ from their marine counterparts, and ethnographic interviews with Alaska Native hunters and elders living in the Iliamna Lake region are equivocal. While some residents reported difficulty in being able to tell the difference between saltwater and freshwater seals, others indicate that the seals in the lake "*look totally different*" (KK, male, 50s) and are larger and fatter than nearby marine harbor seals. However, the scant empirical data on seal size from harvested animals does not indicate that seals within the lake differ from those within Bristol Bay (Burns et al. 2013). In addition, while many local residents state that the freshwater seals' coats are less "*greyish*" (KK, male, 30s), exhibit more distinct color patterns, and have a softer, less coarse fur coat than saltwater seals (Burns et al. 2013), variation in pelage patterns are documented across the harbor seal range (Kelly 1981), as have tactile differences between pelts from fresh and saltwater populations (COSEWIC 2007).

## DISTRIBUTION AND ABUNDANCE

Within Iliamna Lake, seals are primarily seen hauled out on islands at the eastern end of the lake, are rarely seen outside this region, and have never been seen hauled out anywhere along the lake's vast shoreline, or the shore of any of the rivers that feed the lake. For example, an Iliamna respondent who has flown the air routes between Iliamna, Kokhanok, and Igiugig almost daily for over 40 years said, "*Very, very seldom do you ever see them* [seals] *anywhere past* [southwest] *a direct line between Kokhanok and Newhalen*" (IL, male, 60s). However, local residents report seals in the Kvichak River, as well as up rivers and at river mouths throughout the lake, especially when salmon are in the area and when set net fishing occurs. Systematic aerial surveys of the entire lake identified 24 different sites where seals haul out (Figure 23.2), but the majority of all sightings (92%) occur on four low sandy islands at the eastern end of the lake. While most seals haulouts are unnamed by both locals and USGS maps, there are several sites referred to by local residents as *Seal Island*. However, the location referred to as Seal Island by Kokhanok residents differs from that named by Iliamna and Newhalen residents, and that on the USGS map.

Data on Iliamna Lake seal abundance have been collected by biologists since 1984 when a boat-based survey recorded 77 seals hauled out in the lake (T.C. Kline, pers. comm.). In 1991, University of Alaska Fairbanks (UAF) researchers Mathisen and Kline, under contract to the National Marine Mammal Laboratory (NMML), conducted the first complete aerial lake seal count, and recorded 137 seals hauled out. The Alaska Department of Fish and Game (ADF&G) and NMML intermittently conducted aerial surveys between 1998 and 2005; and from 2008 through 2014, the NMML and the University of Alaska Anchorage (UAA) surveyed Iliamna Lake multiple times a year. Surveys were timed so that one was conducted while the lake was mostly frozen (late March/early April), one during pupping (mid-July), and one to three during the August molt, when the greatest number of seals typically haul out on shore and are visible to aerial observers (Figure 23.4). These unadjusted seal counts[1] vary quite extensively, but the highest counts in August each year have averaged in the low to mid 200s (Figure 23.5), while the highest unadjusted count, recorded in 2014, indicated 356 seals hauled out. This number is nearly identical to the high count in 2008 of 357 from surveys conducted by Alaska Biological Research, Inc. (ABR), and 321 counted by the ADF&G in 1998. These data suggest that the population of seals in

---

[1] Raw aerial survey counts have not been corrected to account for weather, visibility, time of day, lake water height, and seals in the water and not hauled out and therefore not observable, etc. These corrections are necessary prior to estimating population size.

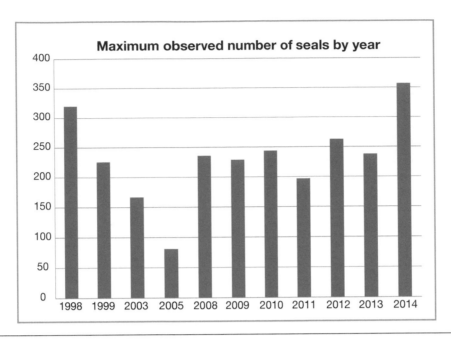

**Figure 23.4**    The maximum number of seals observed in each year during aerial surveys conducted between 1998 and 2014.

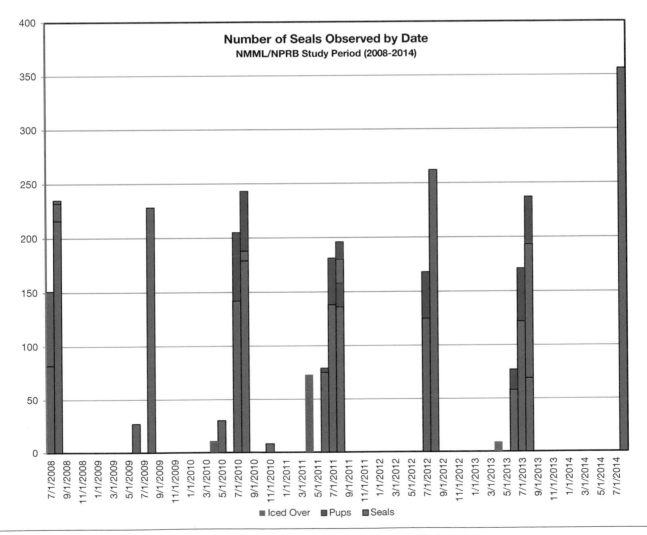

**Figure 23.5**    The number of seals, including pups, observed by date from 2008 through 2014.

Iliamna Lake has remained low but stable over the past two decades. An analysis recently completed by NMML (Boveng et al. in prep.) estimates the total number of seals in the lake to be around 400, with an intrinsic growth rate of 5% per year. This rate is similar to the average annual harvest as estimated from household surveys (ADF&G 2013) and also suggests that the current population has remained stable for (at least) the last decade or two.

# SEASONAL VARIATION IN ABUNDANCE AND BEHAVIOR

## Winter

The vast majority of the surface of Iliamna Lake freezes annually during the winter, although the date of freezing onset and the maximum extent of ice formation varies (Lindsey et al. 2011). While harbor seals can utilize ice for haul out and pupping, particularly in glacial areas, throughout most of their range they never encounter ice and so are not considered an ice associated seal like bearded (*Erignathus barbatus*), ribbon (*Histriophoca fasciata*), ringed (*Phoca hispida*), or spotted seals. Still, local residents report that harbor seals can be found in Iliamna Lake throughout the winter (Burns et al. 2013). During the ice covered period, generally late winter to early spring, seals are seldom seen. During this time, the lake can be completely covered with ice (up to 18 inches thick) and ice covers most of the streams that flow into the lake, as well as much of the Kvichak River, which flows from the lake to Bristol Bay. Small areas of water may remain open, particularly in the eastern portion of the lake where it is deepest and the seals are known to occur. These areas may be kept open by water circulation patterns, springs, hydrothermal activity, bottom topography, or combinations of these and other, unknown factors. Cracks also form in the ice, primarily in the late spring when the ice thins and winds push and pull the ice apart. Seals use these cracks (Figure 23.6), especially on sunny days, to haul out. During an aerial survey on April 14, 2011, we observed 73 seals hauled out along cracks and open areas, however it is not uncommon to complete an aerial survey without sighting a single seal hauled out. Late winter/early spring subsistence hunting by Alaska Natives occurs along these cracks when the seals haul out. Cracks can freeze and thaw several times during the season. Holes are not uncommon in these areas, and local observers suggest that they can be maintained or enhanced by the seals using their teeth or claws (Burns et al. 2013).

But where are the seals when there are no apparent openings? Seals may access air spaces for breathing and dry platforms for hauling out by exploiting air pockets that develop along shorelines when the water level drops after lake ice forms. The water level in the lake continues to drop during the fall and winter (Burns et al. 2013; T. Quinn, University of Washington, unpub. data). The ice becomes propped up on its shore-ward edge and an air-filled gap forms where the lake surface has fallen away from the ice (Figure 23.7). Similar gaps can also form when pressure ridges form (i.e., the ice becomes broken and jumbled against the rocky reef areas, small islets, and along the shore) from the pressure exerted

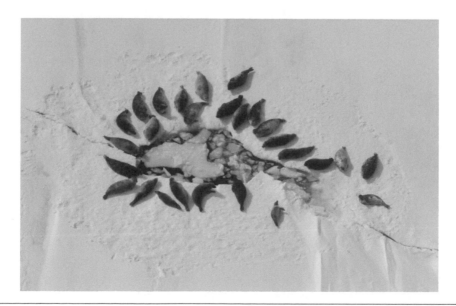

**Figure 23.6** Aerial view of seals hauled out along a pressure crack on Iliamna Lake during the late winter/early spring. *Photo credit: David Withrow.*

**Figure 23.7**    Diagram of seal hauled out under lake ice in an air gap that formed between the lake ice and shoreline due to ice surging along the shore and lake levels dropping. *Drawing credit: Jessica Crance.*

by winds on the overall ice surface. Local residents report hearing seals in such spaces during the winter. Use of such under-ice spaces is a likely explanation for low seal counts obtained during surveys when the lake is ice covered (Burns et al. 2013). This strategy is also thought to be used by the freshwater harbor seals found in Lacs des Loups Marin, Quebec (Smith and Horonowitsch 1987), not only for overwintering shelter but also for birth lairs (DFO 2009). In Iliamna Lake, it's unclear to what extent the seals use this strategy, but seems that the use of under-ice lairs could provide not only air for the seals to breath, but a haulout that is sheltered from terrestrial predators such as bears and wolves.

## Spring

While the occurrence of pressure cracks and open water locations varies in response to weather conditions, as does the time of ice break-up, prominent pressure cracks in the lake ice normally begin emerging some time in March, followed by break up beginning in early to mid-May. As the air temperatures begin to rise in early spring (mid-March to April), and the ice starts to break up, seals begin to appear hauled out on the lake ice. Respondents explained that, following a long winter of residing under the ice, the seals are eager for an opportunity to haul out and will thus emerge on the ice as soon as the first pressure cracks open in order to "*start basking in the sun*," (LV, male, 60s). An Iliamna elder said, "*When it starts turning warm here you'll see seals all over the place. They haul out in 20–30 to a bunch*" (IL, male, 60s). While some respondents reported that more seals can be seen on the lake during early spring than any other time of the year, they also indicated that it is at this time of year that some seals begin traveling extensively. Respondents explained that, as the lake transitions to becoming completely ice free, seals begin to consistently haul out on islands and sandbars between Kokhanok and Newhalen, but that seals remain scarce farther southwest, and are only rarely observed near the lake outlet to the Kvichak River. A Pedro Bay hunter said that the seals are somewhat lethargic, "*tame*" (PB, male, 60s), and easy to hunt at this time due to low energy from surviving a long winter.

## Summer

While the seals are primarily found in the northeast end of the lake during spring, respondents explained that, beginning in mid- to late June when the lake is ice free, greater numbers of seals are known to migrate down the lake to intercept salmon entering the lake. An Iliamna elder said that seals could be expected in the Igiugig area "*as the salmon come in, anywhere from late June to the first week or so of July*" (IL, male, 60s). Respondents explained that salmon entering the lake move quickly and that around the first of July most of the seals have moved back up the lake to the mouth of rivers and streams to feed on salmon entering rivers and streams to spawn. For example, an Iliamna respondent said: "*They* [seals] *follow the fish. As soon as the salmon come in the lake they're everywhere. They really spread fast. It only takes a couple of days for the salmon to travel and they start congregating at the mouths of every one of these streams, and there are a lot of middle-of-the-lake spawners and beach spawners and stream spawners*" (IL, male, 60s). Respondents also reported that seals will travel some distance up rivers and creeks, following the fish. Indeed, respondents from Igiugig and Levelock reported that seals could be observed in the Kvichak River feeding on salmon or hauled out on sandbars inside the Kvichak River during the summer when salmon are running. Whether seals seen near the headwaters of the Kvichak River originate from Iliamna Lake or have moved up the river from Bristol Bay is not clear; similarly, seals seen closer to the mouth of the river might have originated from the lake or the much larger Bristol Bay population. Within Iliamna Lake, river and creek mouths are utilized not only by salmon, but also by other fish species such as rainbow trout, Dolly Varden, and grayling; and respondents consistently reported these to be the primary feeding locations for the seals within Iliamna Lake.

## Fall

Throughout summer and into fall and early winter seals continue to haul out on sandbars, spits, and the beaches and rocks of islands used during the summer months. In addition, as the season shifts to fall, the mouths of rivers and creeks within Iliamna Lake remain active with seals feeding as spawning salmon continue amassing in those locations. "*In the fall time this whole lake is just full of salmon, so wherever there is any major spawning grounds you'll find them* [seals]—*especially the mouths of creeks*," (KK, male, 30s) said a Kokhanok hunter. For local residents, the heavy reliance of seals on salmon is reflected in the taste of the seal meat, as local hunters report that it develops a strong fishy taste at the time, which is undesirable. As temperature and light levels drop, Iliamna lake freezes up in a process where the ice surges and retreats in sheets. Ice formation and melting can occur several times before the lake is completely covered in ice, with the deeper northeastern portion of the lake freezing last, although this may not occur until early January.

## REPRODUCTION

Throughout most of Alaska, female harbor seals give birth to their pups in May and June, and aerial surveys to document reproduction are typically flown during this time. However, prior to 2009, no pups had ever been sighted during scientific surveys of the harbor seals within Iliamna Lake. The question of whether female seals give birth within the lake is of importance, because in the absence of local reproduction, all mortality within the lake would need to be balanced by the immigration of seals from Bristol Bay or else the population would decline. Thus, information on reproduction was a critical focus when conducting ethnographic interviews. However, responses to initial survey questions about reproductive timing were varied, with some interviewees stating that pups could be seen year-round, while others stated that pups were born during the winter, and yet others stating that pupping did not occur in the lake at all. One elder reported that "*things happen around here about one month later than everywhere else*" (KK, female, 50s). As a result of this insight, in 2009 aerial surveys aimed at detecting pups were shifted to July and August, and pups were successfully identified[2] in photographs taken between mid-July and the first week of August. Since initially documented, the peak number of pups observed has varied between a low of 43 and a high count of 63 (see Figure 23.5); much of this variance can likely be attributed to survey timing, which in some years may have missed the peak pup production. This is because while newborn pups are about half the length of their mother at birth, they grow rapidly due to the high fat milk (Bowen et al. 1992), and soon can no longer be reliably distinguished from other immature seals. Pups are weaned at approximately four weeks of age (Bowen et al. 1992); thus, peak counts in August include a mix of all age classes. Although pupping in mid-July is three to four weeks later than most of the Alaska harbor seals, harbor seal pups on Otter Island in the Bering Sea's Pribilof Islands have been observed suckling in mid-July, and pups have been observed in Bristol Bay in July (Jemison et al. 2006). Local residents generally describe the pups seen in the lake as having a light grey or spotted coat and being born on the offshore islands, which fits with what is known about the reproductive biology of harbor seals—that they give birth on land to pups with an adult-like pelage (Figure 23.8), rather than the soft, light lanugo pelage that is characteristic of spotted and other ice-associated seals.

---

[2] In order for a small seal to be categorized as a pup during an aerial survey, it had to be smaller than the female, in a nursing position, or within one body length of the larger adult.

**Figure 23.8**    An Iliamna Lake seal pup hauled out on the beach of a small unnamed island in late June 2009. *Photo credit: Ron Aaberg, Pedro Bay.*

## SEAL DIET AND FORAGING ECOLOGY

Fish are the primary source of food for the seals within Iliamna Lake. A Newhalen elder said, *"They subsist entirely on fish"* (NH, male, 60s). Salmon are particularly important, especially sockeye salmon (*Oncorhynchus nerka*). Chinook salmon (*O. tshawytscha*), coho salmon (*O. kisutch*), pink salmon (*O. gorbuscha*), and chum salmon (*O. keta*) are also potential sources of food for the seals within Iliamna Lake, and salmon carcasses that show signs of being depredated by seals are commonly found (Hauser et al. 2008). Still, salmon do not make up the entirety of the seal diet, as respondents explained that the seals within Iliamna Lake feed on both salmon and nonsalmon fish, even during the summer. Once the salmon runs are completed, seals feed on a wide variety of freshwater fish species (Bond and Becker 1963, Hauser et al. 2008). Seasonal changes in the availability and abundance of forage species is reflected in seal movements. For example, seal activity in Iliamna Lake increases during summer and fall when the salmon enter the lake, as an Iliamna elder said, once the salmon runs begin *"the seals are all over, eating salmon"* (IL, male, 60s). Indeed, many respondents expressed a belief that the abundance of salmon in Iliamna Lake is the primary reason for the origin of the population and the continuing presence of seals. For example, when asked about traditional stories of seal origins in the lake an Igiugig elder said, *"I guess they just follow the salmon up, the food they use to survive. All* [the seals] *have to follow the salmon up and then when the salmon quit spawning, they find the lake trout, and rainbows, and grayling, and whitefish,"* (IG, male, 70s). As salmon abundance declines, respondents say that seals begin working their way back toward the northeastern portion of the lake for overwintering and for feeding on nonsalmon fish, especially lake trout. *"They change their locations depending on where fish are,"* (PB, female, 60s) said a Pedro Bay elder.

Additional insight into the foraging ecology of Iliamna Lake seals can be gleaned from analysis of stomach contents and tissue samples. Stomach contents represent the diet over the past day or so, and the limited analysis of seal stomach contents (n = 8) supports the previous remarks, in that stomachs contained a mix of salmonid and freshwater species (Burns et al. 2013). A longer-term picture of the diet (weeks to months to years) has been obtained through biochemical analysis (stable isotope values) of different tissues. The isotopic signatures of different tissues are thought to reflect the signature of the source molecules (dietary or endogenous), as well as tissue specific fractionation processes (discriminate factors) and turnover times (Newsome et al. 2010, Germain et al. 2012). Using this approach, we characterized the isotopic signature of soft tissues (muscle, organs), which integrated diet over the past several months, and found them to be depleted in both $\delta^{15}N$ and $\delta^{13}C$, as compared to seals from marine environments. When compared to potential prey items, these results suggest that seals harvested in Iliamna Lake had been feeding primarily on freshwater fish during the several months prior to harvest. In addition, isotopic analysis of whiskers, which represent the diet at the time the whisker was biosynthesized (Hirons et al. 2001, Greaves et al. 2004, Zhao and Schell 2004) suggest that salmon are seasonally important, but that freshwater fish provide an increasing fraction of nutrients as the winter progresses (Burns et al. 2013). Finally, an examination of the naturally occurring isotopes of strontium (Sr) and oxygen (O) also suggests that seals that are harvested within the lake are year-round lake residents. Figure 23.9 shows the $^{87}Sr/^{86}Sr$ ratio profile in the dentine of a canine tooth of four Iliamna seals and one marine seal. The lower levels in the entire dentine record from pulp to enamel edge for the harbor seals from Iliamna Lake as compared to both marine water and seals from Bristol Bay indicate that the Iliamna Lake seals were lifelong residents in the lake, and suggests that the seals only began to utilize the seasonally abundant adult salmon later in life (Burns et al., unpub. data). This conclusion is also suggested by the much lower oxygen isotope values in the apex, middle, and enamel-dentine interface of teeth from the lake seals, as compared to that of a seal from Bristol Bay (Figure 23.10). Thus all dietary analyses performed support the conclusion that the seals are life-long residents within Iliamna Lake and that they subsist primarily on a diet of freshwater fishes, with seasonal inclusion of salmon potentially more important for older animals.

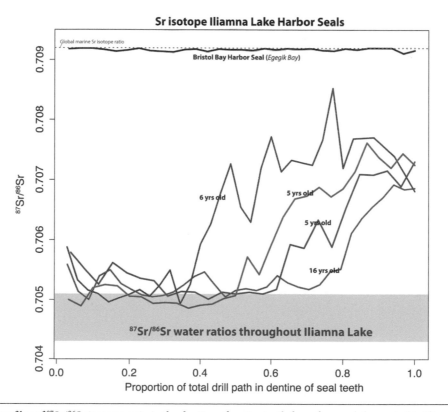

**Figure 23.9**   Lifelong profiles of $^{87}Sr/^{86}Sr$ isotope ratio in the dentine of canine teeth from four seals harvested in Iliamna Lake, one seal from Bristol Bay, and both the lake and marine waters. Samples from when the seal were youngest are at the start of the drill path, and age increases to the right, with the end of the drill path representing the diet of the seals near the time of their death at the ages indicated.

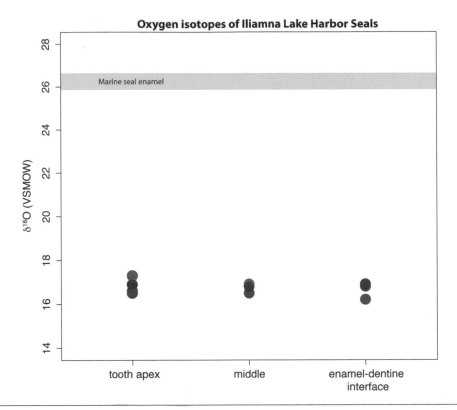

**Figure 23.10** The oxygen isotope ($d^{18}O$) values of enamel measured at the apex, middle, and enamel-dentine interface of the four seals harvested in Iliamna Lake and one seal (mean value ± 1S.D.) from Bristol Bay (same as in Figure 23.9).

## PREDATION AND THREATS

While seals within Iliamna Lake are not exposed to any aquatic predators, they are preyed on by wolves (*Canis lupus*) and wolverines (*Gulo gulo*) inhabiting the watershed. Both of these predators have been observed by lake residents following pressure cracks on the ice in search of hauled out seals. Respondents said that once a wolf or wolverine locates seals hauled out on the ice, they will stalk the seals at the haulout location. A Pope-Vannoy hunter and two different Kokhanok hunters reported observations of seal-kill sites near open-water in the Tommy Point area. A Kokhanok hunter explained that when a group of seals hauled out on the ice becomes startled, they will all rush to the hole and try to jump back in the water, but that the seals get tangled up and several of them usually get stalled and that this is how predators can successfully obtain a seal trapped on the ice. An Iliamna elder reported that during the winter of 2010–2011, a wolf pack killed seals in an area near Pedro Bay. The elder said that the wolves had only recently "*learned the pressure cracks*" and that, as far as he knew, wolves preying on seals is "*a new behavior*" (IL, male, 60s). A Kokhanok hunter reported that he recently had observed a wolverine travelling alongside the windward side of a pressure crack near Tommy Point. The respondent believed that the wolverine was seal hunting. The respondent also said that, due to this seal-stalking behavior, wolverines had been occasionally harvested opportunistically by local hunters while seal hunting. In addition, while local residents did not include brown bears (*Ursus arctos*) as depredating seals, bears are occasionally seen on or near seal haulout sites (Burns et al. 2013).

Bears, coyotes, and foxes are all known predators of seals in other areas of their range (Steiger et al. 1989, Moss 1992), and predation risk is hypothesized as the underlying cause for the seals' use of offshore islands and sandbars as haulout locations rather than coastal shorelines (DaSilva and Terhune 1988, Savarese 2004, Savarese and Burns 2010). Respondents did not report observing seals ever being hauled out on the shoreline of the lake. "*You'll see them cruising around in front of the streams and rivers on the shoreline there but I've never seen a seal actually on the beach on any shoreline other than one of the islands,*" (IL, male, 60s) said an Iliamna respondent. The respondent suggested one reason for this is due to the threat of predation on the main shoreline. "*They're smart. They're on these islands for a very good reason. There's not going to be bears or wolves*" (IL, male, 60s).

Other concerns about the health of the seal population in Iliamna Lake stem from the potential impact of climate change on the abundance and availability of food resources to the seals, and the extent and seasonality of the ice cover within the lake (CBD 2012). In addition, there is potential for resource development within Iliamna Lake's watersheds to impact seal populations, either directly through disturbance and exposure to contaminants, or indirectly through impacts on other ecosystem processes (CBD 2012). However, as yet there have been no studies focused on either of these issues; thus, the magnitude of these potential impacts remains unresolved.

## SUBSISTENCE HUNTING AND TRADITIONAL USE OF LAKE SEALS

Local knowledge asserts that seals have been hunted in Iliamna Lake for a long time: "*As long as people have been here* [inhabiting the Iliamna Lake area], *people went out with the skin boats, they hunted them* [seals]. *During winters, they walked on the ice to hunt seals,*" (IL, male, 60s) said an Iliamna elder. Historical documentation confirms local historical knowledge from at least the nineteenth century (see previous text and VanStone 1988). Indeed, many respondents from the study communities recalled their grandparents and parents hunting seals in the lake from the late nineteenth century through the twentieth century. "*I've always known seals to be in the lake, always. I remember the old people talking about their hunts,*" (IG, male, 70s) recalled an Igiugig elder.

L.E. Bonham, an early Bristol Bay fisherman and explorer, in a letter to his parents dated March 10, 1901, stated that "*. . . they* [natives] *hunt them only when the lake is frozen over, and the weather is very cold. The seals live under the ice, and have places at the islands where they come at frequent intervals to breathe—a small hole in the ice. The Natives find these holes—they cut a hole in the ice about a foot across, and two or three feet from the hole where the seal comes to breathe. The Indian sits by this hole he has cut in the ice with a spear. So arranged that when it is plunged into the seal, the end is detached to which is fastened a strong rope. When the seal comes to breathe the Native sees him through this hole he had cut in the ice and throws the spear into him . . .*" (an excerpt from Branson 2007).

Elders reported that prior to the use of motor boats, much of the seal hunting that occurred in the northeastern portion of the lake was done in the winter on the ice. An Iliamna elder said, "*We know where the pressure cracks are. So we just walk and make sure we are downwind from them* [seals], *and just walk to where the pressure cracks are going to open up . . . and you can get to a high point on the lake so people can look and people just walked up and had a little sled they towed, and some a kind of camouflage, white canvas, or clothes to help camouflage them*" (IL, male, 60s). Bonham's account of 1900 also suggests that it was traditional for seal hunters to hunt seals in Iliamna Lake during winter once the lake had iced over. During this time, seals were hunted with toggled harpoons (Branson 2007).

More recently, motor boats, snow machines, and ATVs share equal importance for seal hunting transport. An Igiugig hunter explained that today, hunters target seals in Iliamna Lake "*by boat in the summer and snow machine or 4-wheeler in the winter*" (IG, male, 20s). Airplanes are another type of modern transport that has been used for seal hunting. While he no longer does so, a Pedro Bay elder said that he used to search for seals by air and, once hauled-out seals were located, would land on the ice at a distance and stalk the seals. While one Igiugig elder reported that he continues to carry a harpoon in his boat in case it becomes useful for retrieving a seal. Today seals are hunted with small-caliber rifles, such as .22, .22-250, .243, .270, and even sometimes with 12 gauge shotguns.

However, reliance on seals differs among communities along the lake, with most of the hunting taking place in the northeast portion of the lake and residents along the Kvichak River only pursuing seals more opportunistically. As a Levelock elder explained, "*We get them* [seals] *if we see em', but we never depended on them. We do put away what we catch for the winter, usually about 2–3* [seals per family]. *Going up to Igiugig* [in a boat on the Kvichak River] *and 'look out here, there's a seal!' and then we'd stick with it 'til we get it. I grew up in the time when you hunt what you need for food, there were not many stores. You see a seal, that's 'game.' And mom would take care of the skin and make something out of it*" (LV, male, 60s). The elder said that, historically, Levelock hunters were far more dependent upon moose and caribou than they were on seals and that no one from the community ever hunted in Iliamna Lake because it was too far away. However, some Elders said that local residents no longer hunt for seals as actively as the previous generations did. "*Nobody hardly hunt* [seals] *no more now,*" (IG, female, 80s) said an Igiugig elder.

Seal products are traditional foods that are highly relished by the elders, and much of the seal distribution goes to elders in the communities who have retired from hunting activities. "*I give a lot of it* [a seal] *away to people who really want it. Especially the elders, they really prize seal oil,*" (PV, female, 50s) explained a Pope-Vannoy hunter when describing what she does with the meat and fat from seals she harvests in Iliamna Lake. Similarly, another Kokhanok hunter said, "*I bring the whole seal back to the village, cut it up, hand out the meat, everybody, elders first and then whoever wants it"* (KK, Male, 20s). Another Kokhanok hunter, describing how he goes about distributing meat and fat from seals he harvests, said, *"I'll go to certain houses; I'll start with the elders and then other people that I know who like seal meat"* (KK, male, 30s). Seal skins are also highly valued by Iliamna Lake residents, as they are a traditional fabric used for making clothing, parkas, hats, boots, pants, and gloves. An Iliamna elder recalled a family history of utilizing the skins of the seals within Iliamna Lake: *"I don't think we've ever let a seal skin go to waste. We always use them. My mom, that's how she made our clothes when we were growing up. She made coats, and mukluks, and mitts with them for us kids"* (IL, male, 60s). In addition, a hunter's first seal harvest is an event worthy of celebration for many community members. For the first harvest, the hunter must give all of the harvest away in order to demonstrate that they can provide for the community first. An Iliamna elder told about sharing the yield from his grandson's first seal harvest: *"This summer my grandson, fifteen-years-old, got his first seal. I am very proud and happy! We give all of it away to relatives; to his godfather, godmother, and older people…just give it all away, that is our custom . . . Here, young man just gives it all away . . . we have a dinner, and then he* [the hunter] *tells about the hunt"* (IL, male, 60s).

Traditional hunting practices are closely tied to conservation measures, with the primary conservation mechanism adherence to a rule to only harvest what the community needs and to thus produce no waste. Sharing networks within the communities facilitate the distribution of harvested seal parts among the community and respondents reported that seal harvests are limited to ensuring that households that use seals have it available. Respondents reported recognition that seals are a limited resource and explained that, once the households have seal oil for the winter, hunting will normally be discontinued until the following spring. A Kokhanok hunter explained that if a hunter harvests a seal and shares it with other households, then usually the people from those households that received seal will not hunt that year. Another conservation measure employed by seal hunters is careful selection of the seal to be harvested. Hunters reported that they are careful not to harvest females with pups or potentially pregnant females during the springtime. These conservation measures are exemplified by a Newhalen hunter, who said, *"I take the old people's advice: 'take only one seal a year and try not to shoot the females. One a year. Just enough for seal oil and to share the meat too, if people want the meat or fat"* (NH, male, 50s). All in all, community respondents placed a large emphasis on the importance of both monitoring and protecting the population of seals in Iliamna Lake for the future.

# RECOMMENDED RESEARCH

The information presented here represents an integrated effort to improve our knowledge about the ecology of the unique seals that inhabit Alaska's largest lake. Aerial surveys have documented the abundance and location of seals that were hauled out during periods when the lake was open water and ice covered, while interviews with village residents have captured traditional knowledge about seal behavior and their importance within the local cultures, and tissue samples provided by local hunters have improved our understanding of the stock structure, diet, and lifetime habitats of these seals. While research efforts over the past decade have substantially improved our understanding, significant work remains to be done. Indeed, in response to local concerns about the potential impact of pending management action, and in recognition of the outstanding need for additional information about the status of the Iliamna Lake harbor seal population, we recommend that the following research and outreach efforts be continued: (1) aerial surveys to document seasonal patterns of abundance and location of seal haul-out areas; (2) collection and analysis of tissue samples provided by participating hunters to provide additional information on genetic structure, dietary habits, and contaminant loads; and (3) information exchange between local residents and groups involved in the management and protection of the seal population.

In addition, local communities have expressed interest in developing a research plan that leads toward a capture and tagging program that will directly monitor seal movements in, and possibly out of, the lake. Such an effort might include seal capture and tagging operations conducted by a team consisting of federal, state, and university personnel, with the assistance and advice of local residents. This type of program would support the development of spatially and temporally accurate correction factors that could be used to transform haul-out counts to total abundance, and so address the question of if and how seal population numbers have changed over time. Meaningful collaboration and coordination in research and management decisions with local users of the resource is a core tenet of co-management programs (Marine Mammal Commission 2008); is a stated policy of the U.S. Department of Commerce, Marine Mammal Protection Act, Section 119 of the Marine Mammal Protection Act Amendments of 1994 (Public Law 103-238); and is a goal that remains critical to expanding our knowledge about Iliamna Lake seals.

# SUMMARY

The seals in Iliamna Lake are indeed a special group of animals. They are one of only six lake-dwelling seal species in the world and one of two freshwater seal populations in North America. Written records of seals in Iliamna Lake go back to the early 1800s, but ethnographic stories passed down for generations suggest that the seals have inhabited Iliamna Lake at least since humans have occupied the area (4,000 to 6,000 years) and likely much longer. Recent genetic analysis confirms the seals within Iliamna Lake are harbor seals that originated from Bristol Bay, while the lack of any variation in haplotypes suggests low genetic diversity within the population, but further study is needed. Within Iliamna Lake, the harbor seals primarily haul out on four islands at the eastern end of the lake, except when foraging for salmon in rivers during the summer and fall months. To date, it remains unclear the degree to which individual seals migrate between Bristol Bay and Iliamna Lake, although dietary evidence suggests that it is rare, and the ethnographic data strongly supports the hypothesis that the seals are year-round residents within the lake. Both aerial surveys and local interviews demonstrate that the seals are present in the lake year-round, and confirm the presence of a breeding population, with pups produced in mid-July—slightly later than in other regions of Alaska. Peak numbers typically occur in mid-August during the annual molt, while fewer seals are observed in winter. This may be because, when the lake is covered by substantial amounts of ice, seals may haul out underneath the ice in air pockets formed by dropping lake levels and/or ice ridging. Despite seasonal variation in apparent numbers, a recent analysis estimates that the population size has held relatively steady at approximately 400 seals for the last two decades. The seals are an important subsistence resource, providing oil, meat, and fur to village residents and the annual harvest is likely close to the population's intrinsic growth rate (~5%). Our understanding of the behavior and ecology of this unique population of harbor seals is still limited, and there is ample opportunity for new research and insights.

# REFERENCES

Allen, B.M. and R.P. Angliss. 2012. Alaska Marine Mammal Stock Assessments, 2011. NOAA Technical Memorandum NMFS-AFSC-234, National Marine Fisheries Service: 288.

Bond, C.E. and C.D. Becker. 1963. Key to the fishes of the Kvichak River System. Fisheries Research Institute, University of Washington, School of Fisheries and Aquatic Sciences Publication Office, Circular #189.

Bowen, W.D., O.T. Oftedal, and D.J. Boness. 1992. Mass and energy transfer during lactation in a small phocid, the harbor seal (*Phoca vitulina*). Physiol. Zool. 65:844–866.

Boveng, P.L., J.M. Ver Hoef, D.E. Withrow, and J. M. London. In prep. A Bayesian analysis of abundance, trend and population viability for harbor seals in Iliamna Lake, Alaska. Unpublished manuscript. NOAA Alaska Fisheries Science Center, Seattle, WA.

Branson, J.B. 2007. The canneries, cabins, and caches of Bristol Bay, Alaska. United States Department of the Interior.

Brown, R.F. and B.R. Mate. 1983. Abundance, movements, and feeding habits of harbor seals (*Phoca vitulina*) at Netarts and Tillamook Bays, Oregon. Fish Bull 81:291–301.

Burns, J.M., H. Alderman, T. Askoak, and D. Withrow. 2012. Local and Scientific Understanding of the Iliamna Lake Freshwater Seals. In: C. Carothers et al. (eds) *Fishing People of the North: Cultures, Economies, and Management Responding to Change*, Alaska SeaGrant, p. 211–228.

Burns, J.J., F.H. Fay, and G.A. Fedoseev. 1984. Craniological analysis of harbor and spotted seals of the North Pacific region. In: F. H. Fay and G. A. Fedoseev (eds) *Soviet-American cooperative research on marine mammals*, Vol 1. Pinnipeds. U.S. Dept. Commerce, NOAA Tech Rept NMFS 12, Seattle, WA, p. 5–16.

Burns, J.M., J. Van Lanen, D. Withrow, D. Holen, T. Askoak, H. Aderman, G. O'Corey-Crowe, G. Zimpelman, and B. Jones. 2013. Integrating local traditional knowledge and subsistence use patterns with aerial surveys to improve scientific and local understanding of the Iliamna Lake seals. Report to the North Pacific Research Board Final Report 1116. 189 p.

CBD (Center for Biological Diversity). 2012. Petition to list Iliamna Lake seal, a distinct population segment of Pacific harbor seal (*Phoca vitulina richardsi*) under the Endangered Species Act. 77 p. Center for Biological Diversity, San Francisco, CA.

Clementz, M.T. and P.L. Koch, 2001, Differentiating Aquatic Mammal Habitat and Foraging Ecology with Stable Isotopes in Tooth Enamel Oecologia, 129:461–472.

COSEWIC. 2007. COSEWIC assessment and update status report on the harbour seal Atlantic and Eastern Arctic subspecies *Phoca vitulina concolor* and Lacs des Loups Marins subspecies *Phoca vitulina mellonae* in Canada. Committee on the Status of Endangered Wildlife in Canada. 40 p.

da Silva, J. and J.M. Terhune. 1988. Harbour seal grouping as an anti-predator strategy. Anim. Behav. 36: 1309–1316.

Department of Fisheries and Oceans. 2009. Recovery potential assessment for freshwater harbour seal, *Phoca vitulina mellonae*, (Lac des Loups marins Designated Unit (DU)). DFO, Canadian Science Advisory Secretariat, Science Advisory Report 2008/062.

Dumont, H. J. 1998. "The Caspian Lake: History, Biota, Structure, and Function," Limnol. Oceanogr. 43:44–52.

Ellanna, L.J. 1986. Lake Clark Sociocultural Study: Phase 1. U.S. National Park Service, Lake Clark National Park and Preserve, Alaska.

Germain, L.R., P.L. Koch, J.T. Harvey, M.D. McCarthy. 2012. Nitrogen isotope fractionation in amino acids from harbor seals: implications for compound-specific trophic position calculations. Mar. Ecol. Prog. Ser. 482:265–277.

Greaves, D.K., M.O. Hammill, J.D. Eddington, D. Pettipas, and J.F. Schreer. 2004. Growth rate and shedding of vibrissae in the gray seal, *Halichoerus grypus*: a cautionary note for stable isotope analysis. Mar. Mamm. Sci. 20:296–304.

Hauser, D.W., C.S. Allen, H.B. Rich Jr., and T.P. Quinn. 2008. Resident harbor seals (*Phoca vitulina*) in Iliamna Lake, Alaska: Summer diet and partial consumption of adult sockeye salmon (*Oncorhynchus nerka*). Aquat. Mammal. 34:303–309.

Higdon, J.W., O.R.P Bininda-Emonds, R.M.D. Beck, and S.H. Ferguson. 2007. Phylogeny and divergence of the pinnipeds (Carnivora: Mammalia) assessed using a multigene dataset. BMC Evol. Biol. 2007, **7**:216 doi:10.1186/1471-2148-7-216.

Hirons, A.C., D.M. Schell, and B.P. Finney. 2001. Temporal records of delta $^{13}$C and delta $^{15}$N in North Pacific pinnipeds: inferences regarding environmental change and diet. Oecol. 1129:591–601.

Jemison, L.A., G.W. Pendleton, C.A. Wilson, and R.J. Small. 2006. Long-term trends in harbor seal numbers at Tugidak Island and Nanvak Bay, Alaska. Mar. Mamm. Sci. 22:339–360.

Kari, J., and P.R. Kari. 1982. Tanaina county = Dena'ina Ełnena. Alaska Native Language Center, University of Alaska Fairbanks. 109 p.

Kelly, B.P. 1981. Pelage polymorphism in pacific harbor seals. Can. J. Zool. 59:1212–1219.

Kelly, B.P., J.L. Bengtson, P.L. Boveng, M.F. Cameron, S.P. Dahle, J.K. Jansen, E.A. Logerwell, J.E. Overland, C.L. Sabine, G.T. Waring, and J.M. Wilder. 2010. Status review of the spotted seal (*Phoca largha*). U.S. Department of Commerce, NOAA Technical Memorandum NMFS-AFSC-200. 153 p.

Knapp, G., M. Guettabi, and S. Goldsmith. 2013. The economic importance of the Bristol Bay salmon industry. Report prepared for the Bristol Bay Regional Seafood Development Association. http://www.iser.uaa.alaska.edu/Publications/2013_04-TheEconomic ImportanceOfTheBristolBaySalmonIndustry.pdf.

Lindsay, C., P. Spenser, and B. Hill. 2011. Interannual variability of lake ice phenology in southwest Alaska: Integrating remote sensing and climate data. The 2011 George Wright Society Conference on Parks, Protected Areas, and Cultural Sites. March 14–18, New Orleans, LA.

Lowry, L.F., K.J. Frost, R. Davis, D.P. DeMaster, and R.S. Suydam. 1998. Movements and behavior of satellite-tagged spotted seals (*Phoca largha*) in the Bering and Chukchi Seas.

Lowry L.F., K.J. Frost, J.M. Ver Hoef, and R.L. DeLong. 2001. Movements of satellite-tagged subadult and adult harbor seals in Prince William Sound, Alaska. Mar Mamm Sci 17: 835–861.

Marine Mammal Commission, 2008. Report of the Marine Mammal Commission Review of co-management efforts in Alaska, 6-8 Feb 2008, Anchorage, AK. Available online at www.mmc.gov/pdf/mmc_comgmt.pdf.

Michael, H.N. 1967. Lieutenant Zagoskin's travels in Russian America, 1942–1844: the first ethnographic and geographic investigations in the Yukon and Kuskokwim valleys of Alaska. University of Toronto Press, Toronto.

Morris, J.M. 1986. Subsistence Production and Exchange in the Iliamna Lake Region, Southwest Alaska, 1982–1983. Technical Paper #136. Division of Subsistence, Alaska Division of Fish and Game, Juneau, Alaska. 185 p.

Moss, J.M. 1992. Environmental and biological factors that influence harbor seal haulout behavior in Washington, and their consequences for the design of population surveys. University of Washington, MSc. Thesis, 178 p.

Newsome, S.D., M.T. Clementz, and P.L. Koch. 2010. Using stable isotope biogeochemistry to study marine mammal ecology. Mar. Mamm. Sci. 26: 509–572.

O'Corey-Crowe, G.M., K.K. Martien, and B.L. Taylor. 2003. The analysis of population genetic structure in Alaska harbor seals, *Phoca vitulina*, as a framework for the identification of management stocks. National Marine Fisheries Service, Southwest Fisheries Science Center Administrative Report LJ-03-08, 54 p.

Reger, D.R. 2005. Prehistory at the Pedro Bay Site. Pedro Bay Village Council, Pedro Bay, AK.

Savarese, D.M. 2004. Seasonal trends in harbor seal abundance at the terminus of the Bering Glacier in southcentral Alaska. MSc. Thesis, University of Alaska Anchorage, p. 1–59.

Savarese, D.M. and J.M. Burns. 2010. Harbor seal (*Phoca vitulina richardii*) use of the Bering Glacier habitat: implications for management. In E. Josberger and R. A. Shuchman (Eds.), *Bering Glacier and its Environment. Bering Glacier: Interdisciplinary Studies of Earth's Largest Temperate Surging Glacier.* GSA Special Paper 462:181–192.

Smith,T.G. and G. Horonowitsch. 1987. Harbour seals in the Lacs des Loups Marins and eastern Hudson Bay drainage. Can. Tech. Rep. Fish. Aquat. Sci. 1536:1–17.

Smith, R.J., D.M. Lavigne, and W.R. Leonard. 1994. Subspecific status of the freshwater harbor seal (*Phoca vitulina mellonae*): a reassessment. Mar. Mamm. Sci. 10:105–110.

Steiger G.H., J. Calambokidis, J.C. Cubbage, D.E. Skilling, A.W. Smith, and D.H. Gribble. 1989. Coyote (*Canis latrans*) predation and scavenging on harbor seal (*Phoca vitulina*) pups. J Wild Dis 25:319–328.

Townsend, J.B. 1965. Ethnohistory and culture change of the Iliamna Tanaina. Ph. D. Thesis, University of California Los Angeles

VanStone, J.W. 1988. Russian exploration in Southwest Alaska: the travel journals of Petr Korsakovskiy (1818) and Ivan Ya, Vasilev (1829), edited with an introduction by J.W. VanStone, translated by D.H. Kraus. University of Alaska Press; Volume IV of Historical Translation Series, 120 p.

VanStone, J.W. and Townsend, J.B. 1970. Kijik: an historic Tanaina Indian settlement. Chicago Field Museum of Natural History.

Withrow, D.E. and Yano, K.M. 2010. Freshwater Harbor Seals of Lake Iliamna, Alaska: Updated Counts and Research Coordination for 2010. Alaska Marine Science Symposium, Anchorage, AK. Jan 2010.

Withrow, D., K. Yano, J. Burns, C. Gomez, and T. Askoak. 2011. Freshwater Harbor Seals of Lake Iliamna, Alaska: Do they pup and over-winter in the lake? Poster Presented at the 2011 Alaska Marine Science Symposium, Anchorage, AK, Jan 17–20, 2011.

Zhao, L. and Schell, D.M. 2004. Stable isotope ratios in harbor seal *Phoca vitulina* vibrissae: effects of growth patterns on ecological records. Mar. Ecol. Prog. Ser. 281:267–273.

# SECTION VI

## NON-BIOLOGICAL RESOURCES OF BRISTOL BAY

# 24

# NORTH ALEUTIAN BASIN (BRISTOL BAY) OIL AND GAS POTENTIAL

**Kirk W. Sherwood and Michael T. Lu**
**Bureau of Ocean Energy Management, U.S. Department of the Interior**

## INTRODUCTION

Reconnaissance seismic data gathered by the U.S. Geological Survey in the 1970s (Cooper et al. 1979, Kirschner 1988) revealed several geologic basins beneath the broad, flat Bering Sea continental shelf. These geologic basins, highlighted in the map of Figure 24.1, are isolated from each other beneath a thin (< 3,000 feet) carapace of strata that mantles the entire Bering Sea continental shelf.

The term *geologic basin* refers to an area of significant subsidence of the earth's crust. Geologic basins are filled with sedimentary strata that can exceed 20,000 feet in thickness and in the deepest parts are subjected to high temperatures. Geologic basins are instrumental to the creation of oil and natural gas. Oil and gas are largely derived from the heating of organic matter that originated from photosynthesis of plant material. In the case of oil, the linkage to plant material is evidenced by molecules known as biomarkers that persist in the oil but that originated in plant material created by photosynthesis. Strata in the deep parts of geologic basins can be subjected to temperatures sufficient to convert organic matter to oil and/or natural gas, depending upon the composition of the original organic matter. Once generated, the oil or gas can escape to the surface as natural seeps but some is usually trapped within closed volumes of porous rocks like sandstones and can be extracted by conventional production wells.

The Bureau of Ocean Energy Management (BOEM), an agency within the U.S. Department of the Interior (DOI), periodically assesses the potential volumes of undiscovered recoverable oil and gas beneath the outer continental shelves (OCS)[1] of the U.S. These assessments are the basis for planning oil and gas leasing programs that may lead to exploration drilling, discoveries, and commercial development by private firms. Generally, geologic basins that are forecast to contain oil or gas in quantities sufficient to justify economic development are nominated for lease sales. However, remote, gas-prone basins in Alaska offer poor chances for commercial development because of the high cost of development in the context of low natural gas prices and are typically excluded from further consideration early in the planning process. The oil and gas assessments are periodically updated to account for new geologic information or changing economic factors like market prices and development costs.

---

[1]The *continental shelf* in a geographic sense is the submerged extension of the adjoining continental land area. The *shelf* inclines gently offshore to a shelf break at the top of the *continental slope* that inclines more steeply toward the floor of the flanking deep oceanic basin. In most cases, the continental shelf is a direct extension of the adjoining land mass and both are quite different in their shared subsurface geology from the deep ocean basins that flank the continents. In many cases, most of the offshore production of oil and gas is confined largely to the continental shelf areas, with some notable exceptions like the deepwater Gulf of Mexico. The *Outer Continental Shelf* (hereafter, OCS) is an area defined in U.S. law (the 1953 Outer Continental Shelf Lands Act and succeeding amendments; see BOEM 2016a) that captures the continental shelf, but that extends from 3 nautical miles offshore—the limit of State waters in most cases—to include deepwater areas ranging from 200 to 300 miles offshore (Dellagiarino 1986). The OCS in most cases lies within the U.S. Exclusive Economic Zone (hereafter, EEZ) that extends 200 nautical miles (230 statute miles) seaward from the legal coastline of the U.S. and overseas territories and possessions (Dellagiarino 1986). The subject of this chapter—the North Aleutian Basin Planning Area—is entirely contained within both the U.S. OCS and the U.S. EEZ.

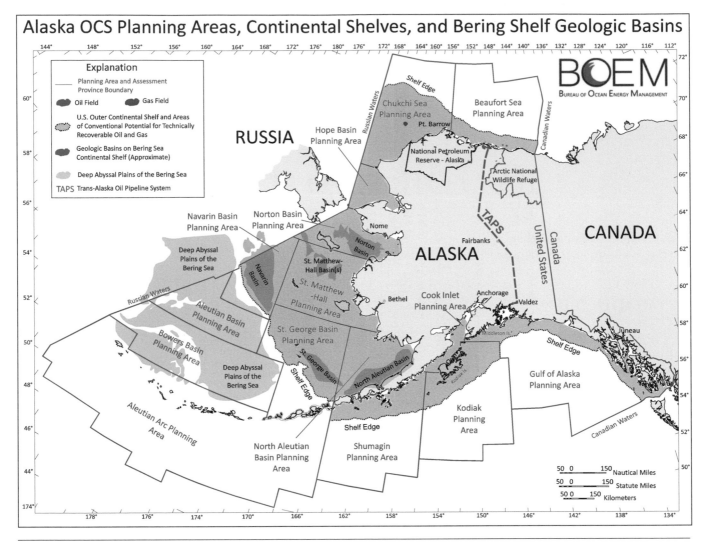

**Figure 24.1**   Location map for Alaska outer continental shelf planning areas and geologic basins of the Bering Sea continental shelf. *Adapted from Sherwood et al. (2006).*

The assessments and regulatory activities on the OCS of the U.S. are organized around *planning areas* adopted as management units by the U.S. DOI. The central feature acknowledged within each planning area is a geologic basin and the planning areas are purposely designed to circumscribe the basin.

The Navarin Basin, Norton Basin, and St. George Basin Planning Areas of the Bering Sea continental shelf incurred leasing and exploratory drilling in the 1980s that failed to discover any significant quantities of oil or gas. More far-reaching, the geologic data obtained from the exploration wells condemned these basins as unpromising for further exploration. The last well that was drilled into these basins was completed in 1985. No plans presently exist for future leasing or exploration in the Navarin Basin, Norton Basin, or St. George Basin Planning Areas.

The North Aleutian Basin that underlies Bristol Bay is considered the most promising of the Bering Sea shelf basins, as reflected in the oil and gas endowments reported in the ranked list in Table 24.1. A single stratigraphic test or COST[2] well drilled in 1983 to a depth of 17,155 feet below Kelly bushing (bkb)[3] found thick, highly permeable sandstones that

---

[2]COST: continental offshore stratigraphic test.

[3]bkb: "below Kelly bushing." The Kelly bushing is located on the drill floor where drill pipe enters the top of the well and is the zero point for measurements of depth for drilling operations and wireline measurements (*logs*) of geophysical properties of rocks in the well. For the COST well, the Kelly bushing was located 85 feet above mean sea level on the drill floor of the semisubmersible drilling vessel (Turner et al. 1988).

**Table 24.1** Comparison of the risked (downgraded to acknowledge chance for existence or success), technically recoverable (no economic thresholds or constraints) oil, gas, and aggregate (barrels of oil-equivalent (BOE) summation of oil + gas) endowments of Bering continental shelf OCS Planning Areas, with numbers of wells

**Undiscovered Technically Recoverable Oil and Gas for OCS Planning Areas of the Bering Sea Continental Shelf**

*Resource Endowment with No Consideration of Economics, 2016 Resource Assessment*

| OCS Planning Area | Oil (BSTB) | | | Gas (TSCF) | | | BOE (BSTB-e) | | | Gas BOE/Total BOE at Means (fraction) | Wells Drilled | |
|---|---|---|---|---|---|---|---|---|---|---|---|---|
| | Minimum (95% Chance) | Mean | Maximum (5% Chance) | Minimum (95% Chance) | Mean | Maximum (5% Chance) | Minimum (95% Chance) | Mean | Maximum (5% Chance) | | Stratigraphic Test | Exploration |
| North Aleutian Basin | 0.12 | 0.75 | 1.82 | 1.47 | 8.62 | 17.37 | 0.38 | 2.29 | 4.91 | 0.67 | 1 | 0 |
| St. George Basin | 0.00 | 0.21 | 0.57 | 0.00 | 2.80 | 6.69 | 0.00 | 0.71 | 1.76 | 0.70 | 2 | 10 |
| Norton Basin | 0.00 | 0.06 | 0.17 | 0.00 | 3.06 | 9.65 | 0.00 | 0.60 | 1.89 | 0.90 | 2 | 6 |
| Navarin Basin | 0.00 | 0.13 | 0.42 | 0.00 | 1.22 | 3.67 | 0.00 | 0.35 | 1.07 | 0.63 | 1 | 8 |
| St. Matthew-Hall Basin | Negligible Technically Recoverable Oil and Gas Resources | | | | | | | | | | 0 | 0 |

*The Aleutian (volcanic) Arc and the deep-water Aleutian and Bowers Basins west of the Bering Sea continental shelf edge are assessed as offering negligible potential for undiscovered oil and gas recoverable by conventional or traditional development methods.

BSTB, billions of stock-tank barrels of oil; TSCF, trillions of standard cubic feet of gas; BOE (BSTB-e), oil plus gas in energy-equivalent billions of barrels (1 barrel of oil = 5,620 cubic feet of gas); Gas BOE/Total BOE at Means (%), percentage of overall energy endowment represented by natural gas at mean values. Minimum (95% Chance) refers to the quantity corresponding to a 95% chance (to meet or exceed) in the probability distribution for technically recoverable resources. Maximum (5% Chance) refers to the quantity corresponding to a 5% chance (to meet or exceed) in the probability distribution for technically recoverable resources.

can be traced with seismic data into large traps that exceed 100,000 acres in potential productive area.[4] The chemistry of the organic matter in rocks recovered by the North Aleutian Shelf COST well suggests that natural gas will be the dominant hydrocarbon resource. The deepest parts of the North Aleutian Basin have been exposed to elevated temperatures sufficient to convert organic matter into gas—and possibly oil—that could readily migrate to prominent traps imaged in seismic data. Although the stratigraphic test well did not encounter significant quantities of oil or gas, the geologic data offer some promise for significant pools of natural gas. The North Aleutian Basin thus remains attractive for exploration.

The potential size of the largest gas pool forecast by the most recent assessment of the North Aleutian Basin is 2.9 trillion cubic feet—approximately the size of the largest commercial gas field in Cook Inlet (Kenai gas field, > 2.46 trillion cubic feet produced as of May 1, 2016). A gas field of this size, if demonstrated by drilling to actually exist, forms a potential candidate for commercial development. Development economics for the North Aleutian Basin are favored by direct access to Pacific Rim markets via a liquefied natural gas (LNG) export model. However, current prices for natural gas delivered as LNG to Pacific Rim markets are unfavorable to development.

## LOCATION OF THE NORTH ALEUTIAN BASIN PLANNING AREA AND THE NORTH ALEUTIAN GEOLOGIC BASIN

The North Aleutian Basin Planning Area encompasses an area of 52,230 square miles and includes most of the southeastern part of the Bering Sea continental shelf, as outlined in Figures 24.1, 24.2, and 24.3. The North Aleutian Basin proper (where strata exceed 3,000 feet in total thickness) is about 17,500 square miles in area and underlies the northern coastal plain of the Alaska Peninsula and the southern waters of Bristol Bay. The North Aleutian Basin is sometimes alternatively referred to as the *Bristol Bay Basin*. On the west, the shallowest strata[5] that cap the North Aleutian Geologic Basin fill extend into the St. George Basin. Water depths range from 15 to 700 feet, and in the area of the most important prospects, the water depths are approximately 300 feet.

## LEASING AND EXPLORATION

The prospects in the central part of the North Aleutian Basin have long been the focus of exploration interest and twenty-three blocks[6] were leased for total high bids of $95.4 million in OCS Lease Sale 92 in 1988. The leases issued in Lease Sale 92 are located in Figure 24.2 and partly identify the locations of the prospects that attracted the greatest bidding interest. All of the 1988 leases were returned to the U.S. government in 1995 and are no longer active. In this assessment, as well as in past assessments, most of the hypothetical, undiscovered oil and gas resources of the North Aleutian Basin OCS Planning Area are associated with the prospects in the central part of the North Aleutian Basin beneath and near the Lease Sale 92 leases.

On October 26, 2005, the State of Alaska received 37 bids on 37 tracts[7] in the Port Moller area (Figure 24.2). High bids totaled $1.149 million and all bids were submitted by either Shell Offshore Inc. (33 tracts, $0.842 million) or Hewitt Mineral Corp. (4 tracts, $0.307 million). The locations of the 2005 leases are shown in Figure 24.2. The 2005 leases cover onshore areas where thrust-faulted and folded Mesozoic and Cenozoic rocks are exposed at the surface as well as a known gas seep (Wilson et al. 1995, Decker at al. 2005). On February 28, 2007, a second State lease sale was held and attracted a single Hewitt Mineral Corp. bid on a single block adjacent to the blocks leased in 2005. State of Alaska lease sales in 2008 and 2009 did not attract bids on any Alaska Peninsula blocks. In February 2009, Shell relinquished the 33 leases acquired in 2005 (Bailey 2009); and in October 2010, Hewitt relinquished their five leases (Lidji 2010). State of Alaska lease sales in 2010, 2011, 2012, and 2013 did not attract bids on any Alaska Peninsula blocks (AK DOG 2015a). By early 2014, no

---

[4]One hundred thousand acres is a large prospect, on the areal scale of the 2.9-billion-barrel Kuparuk field (170,000 acres) and the 13.7-billion-barrel Prudhoe Bay field (150,000 acres) in northern Alaska (MMS, 2006d). In terms of recoverable reserves, Prudhoe Bay field is the largest oil field ever discovered in North America.

[5]Strata: a layer of rock formed from deposition of sediments (sand, gravel, mud, shell fragments, etc.).

[6]Blocks offered in the North Aleutian Lease Sale 92 were typically 4,800 m (or 15,748 ft) along an edge and 2,304 hectares (or 5,693 acres) in area.

[7]Tracts offered in the State of Alaska lease sales on the Alaska Peninsula are one-quarter of a township (the latter comprised of 36 sections) or a group of nine one-mile-wide square sections, each tract aggregating approximately nine square miles or 5,760 acres in area.

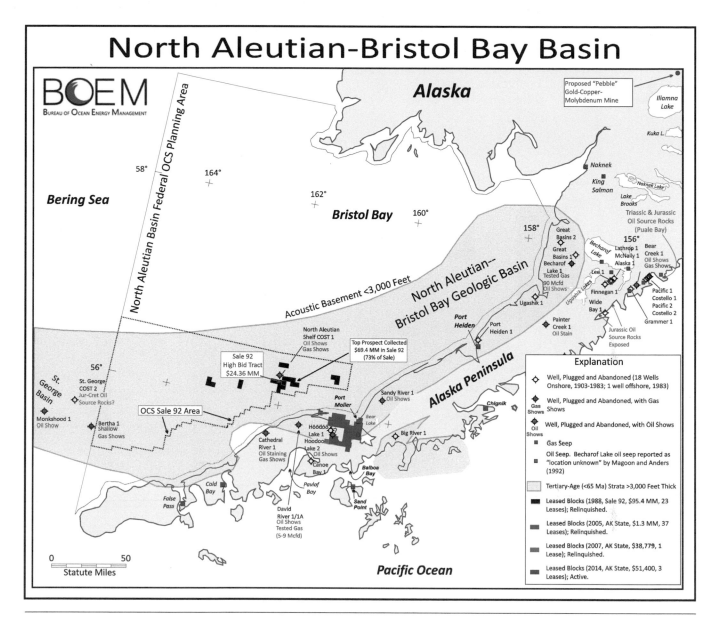

**Figure 24.2** Regional map for the North Aleutian Basin, with well control, Lease Sale 92 outline and issued leases (relinquished in 1995 buy-back), State of Alaska leases near Port Moller (2005, 2007, and 2014 sales), and regional distribution of Tertiary-age sedimentary rock. *Adapted from Sherwood et al. (2006).*

active leases remained on the Alaska Peninsula, but a lease sale held on May 7, 2014 drew bids from Novus Terra Ltd. and Auxillium Alaska Inc. on three tracts in State of Alaska waters north of the area of the 2005 and 2007 leasing (AK DOG 2014) (Figure 24.2). State lease offerings in May 2015 and May 2016 on the Alaska Peninsula did not receive any bids (AK DOG 2016).

Onshore, nine exploration wells have tested fold and thrust-fault structures[8] along the southern limit of the North Aleutian Basin. Several wells in OCS waters to the west in the St. George Basin tested strata age-equivalent to the North Aleutian Basin fill. Offshore, the principal point of geological control for the Cenozoic-age fill in the North Aleutian Basin is the North Aleutian Shelf COST 1 well (hereafter the COST well) drilled in 1983 by an industry consortium led by ARCO Exploration Company. None of these wells encountered any sizeable accumulations of oil or gas, although several

[8]Fold and thrust-fault structures: contortions in rock layers formed by horizontal compression and shortening of the earth's crust, generally associated with the rise of mountains.

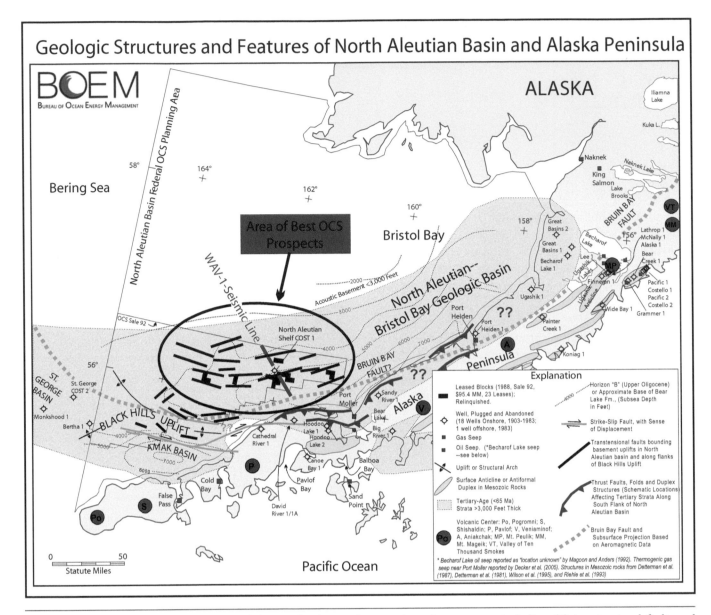

**Figure 24.3**   Principal geologic structures of the North Aleutian Basin and contiguous areas, including: (1) transtensional faults and basement uplifts in western parts of the basin; (2) wrench-fault structures along the Black Hills uplift; and (3) fold/thrust belts onshore along the southeast margin of the basin. The fold/thrust structures do not appear to extend into the Federal OCS (> 3 n. miles or 3.45 st. miles from shoreline). *Adapted from Sherwood et al. (2006).*

wells detected oil and gas shows[9] and two onshore wells tested minor gas pools at rates up to 90 thousands of standard[10] cubic feet per day (abbreviated as 90 MSCF/d).

The North Aleutian and Amak Basins (located in Figure 24.3) are covered by gridded (one- to five-mile spacing between lines in the high-potential areas) two-dimensional seismic data, mostly acquired in the period from 1975 to 1988 (Sherwood et al. 2006). Seismic data gathered to date within the North Aleutian Basin OCS Planning Area consists of 61,438 line miles of conventional, two-dimensional, common-depth-point data and 3,234 line miles of shallow-penetrating,

---

[9]Shows are small quantities of oil or natural gas detected as staining or fluorescence of rock samples from wells, or the presence of hydrocarbon compounds in drilling fluids detected by chemical analyses.

[10]Standard refers to volumes measured at 60°F and 1 atmosphere (14.73 psia).

high-resolution data. Airborne magnetic data in the area consists of 9,596 line miles, while airborne gravity data consists of 6,400 line miles.

## HISTORY OF MORATORIA ON LEASING OR EXPLORATION IN THE NORTH ALEUTIAN BASIN PLANNING AREA

In October 1989, the North Aleutian Basin Planning Area was placed under a Congressional moratorium that prohibited U.S. DOI expenditures in support of any petroleum leasing or development activities in the North Aleutian Basin (as well as the Atlantic, Pacific, and eastern Gulf of Mexico OCS Planning Areas). Bristol Bay is the center of a very important commercial salmon fishery. Enactment of the North Aleutian moratorium was in reaction to widespread demands for fisheries protection in Bristol Bay by Alaska Native organizations, Alaska Native villages, local fishing interests, and the State of Alaska following the March 1989 *Exxon Valdez* grounding and oil spill in Prince William Sound, Alaska. This moratorium was extended by Congress several times during the 1990s and reinforced by a *Presidential Withdrawal* in 1990 by President George H.W. Bush. OCS leases issued in the 1988 OCS Lease Sale 92 in the North Aleutian Basin were returned to the Federal government in a 1995 buy-back agreement. On June 12, 1998, President William J. Clinton issued a *Presidential Withdrawal* on North Aleutian Basin (as well as the Atlantic, Pacific, and eastern Gulf of Mexico continental shelves) until June 30, 2012.

In the 2004 fiscal year, Congressional budget bill for the DOI, the language forbidding funding of oil and gas activities (i.e., the *Congressional moratorium*) in the North Aleutian Basin OCS Planning Area, was deleted. This action effectively ended the Congressional moratorium on oil and gas activities in the Planning Area. The Congressional moratorium on all affected OCS areas ended on October 1, 2008, when Congress failed to renew the ban. No OCS areas are currently subject to *annual* or *Congressional* moratoria (BOEM, 2015).

The Minerals Management Service [(MMS)—predecessor to BOEM)] proposal for the 2007–2012 leasing program (MMS 2010) initiated the environmental studies and public process that could lead to future lease sales. The draft plan proposed North Aleutian Basin lease sales for years 2010 and 2012, contingent upon public review, public support, favorable resolution of environmental issues, and cancellation of the then standing *Presidential Withdrawal*.

The *Presidential Withdrawal* was amended on January 9, 2007, to delete references to the North Aleutian Basin Planning Area, thereby opening the path to consideration of a leasing program. The *Presidential Withdrawal* on all OCS areas (except for all marine sanctuaries) was finally lifted by President George W. Bush through a presidential memorandum published on July 14, 2008 (MMS 2008b, The White House 2008).

The 2007–2012 OCS Oil and Gas Final Proposed Program (MMS 2007) proposed lease sale 214 for year 2011 in the same area as the 1988 Lease Sale 92. A *Call for Information and Nominations* (Call) and *Notice of Interest* (NOI) to prepare an *Environmental Impact Statement* (EIS) for possible sale 214 was published in the Federal Register on April 8, 2008 (MMS 2008a). However, on March 31, 2010, the U.S. DOI withdrew the North Aleutian Basin from the lease sale planning process (Bailey 2010). On December 16, 2014, President Barack Obama enacted a new *Presidential Withdrawal* of the Federal waters of Bristol Bay from consideration for oil and gas leasing (The White House 2014).

## GEOLOGIC SETTING OF NORTH ALEUTIAN BASIN

The North Aleutian Basin is one of several basins of primarily Cenozoic[11] age[12] that dot the Bering Sea shelf, as shown in Figure 24.1. The basin is roughly 100 miles wide and 400 miles in length, and the sedimentary strata exceed 20,000 feet in thickness in its deepest parts. The North Aleutian Basin extends onshore beneath the lowlands along the north shore of the Alaska Peninsula, where it has been penetrated by several wells (Figure 24.2). At its west end, a series of arches isolate the basin from the similar St. George and Amak Basins (Figure 24.3), where the thicknesses of Cenozoic-age strata reach 40,000 feet and 12,500 feet, respectively (Comer et al. 1987). Along the southern margin of the North Aleutian Basin, the Cenozoic-age basin fill is deformed in fold and thrust-fault structures like those widely exposed on the Alaska Peninsula.

---

[11]Cenozoic or *new or recent life*: extends from the present to 66 million years ago, characterized by the proliferation of mammals.

[12]Geological ages: all attributions of absolute ages (millions of years) to geological areas, periods, and epochs are based upon the time scale published by Walker et al. (2012).

In contrast, the interior of the North Aleutian Basin is dominated by uplifted fault blocks with intervening down-dropped blocks in a pattern that suggests crustal extension.

Most mapped prospects in the North Aleutian Basin proper are simple domes draped over the crests of fault-bounded basement[13] uplifts [Figure 24.3 and WAV-1 (See page 539 for more details on these WAV items and how to download them)]. In the central part of the North Aleutian Basin 65 miles northwest of Port Moller, these domes are surrounded by deeply buried sedimentary strata between uplifted fault blocks where oil and gas may have been generated (WAV-1). Oil and gas generated in these basin deeps may have migrated upward along faults or laterally along the base of a regional (impermeable) shale seal and thence into sandstone reservoirs draped over the basement uplifts. In the most recent assessment, as well as in past assessments, most of the hypothetical, undiscovered oil and gas resources are associated with the simple domes draped over basement uplifts in the central part of the basin, highlighted as the *area of best OCS prospects* in Figure 24.3.

## NATURE OF BASEMENT ROCKS BENEATH THE NORTH ALEUTIAN BASIN

In the western Alaska Range (northwest of the study area), the Bruin Bay fault (located in Figure 24.3) forms a regional contact between Mesozoic-age[14] *granitic* volcano-plutonic[15] rocks on the north, and Mesozoic sedimentary rocks on the south (Magoon et al. 1976). The volcano-plutonic rocks on the north are the roots of a Mesozoic-age magmatic arc,[16] and range in radiometric[17] ages from 179 to 107 million years ago (Magoon et al. 1976). The Mesozoic-age sedimentary rocks (here, ~220 to 100 million years ago) (Figure 24.4) south of the Bruin Bay fault represent a Mesozoic-age basin that flanked the contemporary magmatic arc to the north (Bally and Snelson 1980).

Southwest of Becharof Lake, the Bruin Bay fault passes beneath volcanic rocks and glacial sediments of Quaternary[18] age of the Bristol Bay Lowlands. Farther southwest, the location of any extension of the Bruin Bay fault is not exposed and its location becomes a matter for speculation on the basis of remote sensing data.

A regional map (Figure 24.5) indicating the intensity of the magnetic field[19] for the southern Bering Sea shelf west of 162° west longitude (Childs et al. 1981) shows two principal magnetic domains with very different field characters. In the north, the map shows high-frequency, high-intensity magnetic anomalies. In the south, the map shows low-frequency, low-intensity magnetic anomalies. The two magnetic domains are clearly separated by a sharp, west-trending line that may represent the subsurface extension of the Bruin Bay fault into the southern Bering Sea.

Sherwood et al. (2006) argue that the northern magnetic domain in the mapping by Childs et al. is the southwestward and offshore extension of the Mesozoic-age magmatic arc terrane[20] that rises to surface exposures north of the Bruin Bay fault in the western Alaska Range. Similarly, the southern magnetic domain is regarded as the southwestward offshore extension of the folded Mesozoic-age sedimentary terrane exposed south of the Bruin Bay fault on the Alaska Peninsula.

---

[13]Basement: refers to the substrate beneath the geologic basin—often crystalline rocks lacking stratification or layering.

[14]Mesozoic or *middle life*: a period of geologic time ranging between 66 and 252 million years ago and including the age of the proliferation of dinosaurs.

[15]Volcano-plutonic: originating from molten fluids (*magma*) that crystallize into mineral aggregates upon cooling. If the molten material reaches the surface, it is extruded as a volcano. Molten material that fails to rise to the surface but that is injected into shallow rocks and then crystallizes is referred to as *plutonic*.

[16]Magmatic arc: belt of volcanoes and underlying chambers of molten rocks (*magma*) near an oceanic trench where oceanic crust is forced to descend beneath a separate crustal slab, often part of a continent. The descent of the oceanic crust and the sediments resting upon it causes it to melt. The molten rock rises to create the volcanoes at the land surface.

[17]Radiometric age: determined by measuring the relative quantities of unstable isotopes and *daughter* isotopes that are the product of radio-genic decay (e.g., potassium to argon, uranium to thorium, uranium to lead, rubidium to strontium, etc.).

[18]Quaternary or *fourth era*: extends from the present to 2.6 million years ago and is characterized by a number of expansions and contractions of continental glaciers, as well as the proliferation of human life.

[19]Magnetic intensity: caused by the interaction of the earth's natural magnetic field and the magnetic properties of subsurface rocks. Metallic deposits in the earth cause disruptions of the magnetic intensity at the surface and are thus easily identified by magnetic surveys. Higher magnetic intensity is generally associated with crystalline rocks such as granitic intrusions and volcanic rocks.

[20]Terrane: a geologic term for a fragment of the earth's crust that is very distinct in age, makeup, or origins from adjacent fragments. Terranes are often joined along faults and have traveled great distances along faults to their present locations where the crustal fragments have collected and welded into an amalgamation.

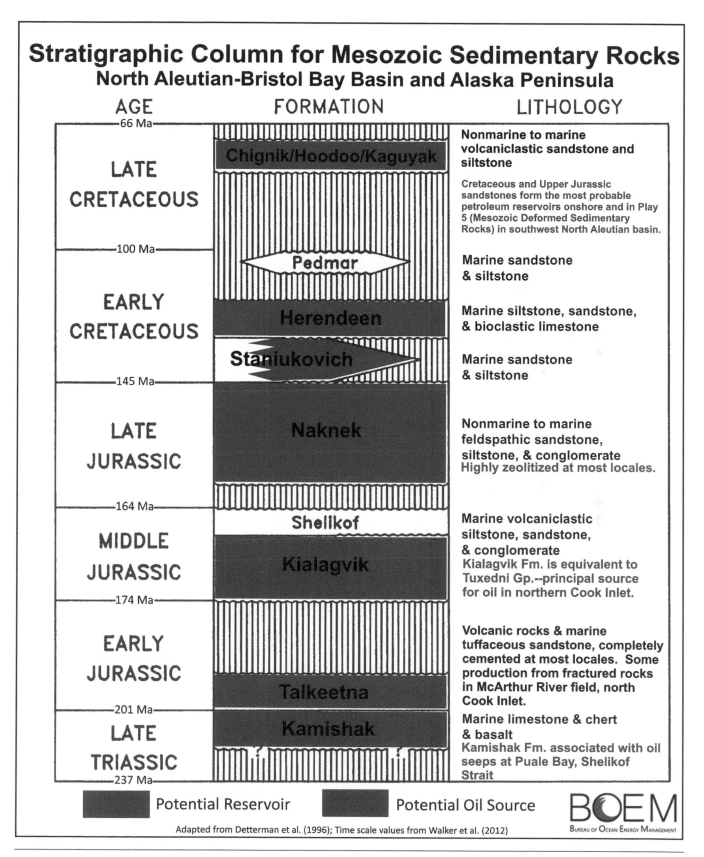

**Figure 24.4**  Mesozoic stratigraphy, Alaska Peninsula, and substrate beneath the southwest part of the North Aleutian Basin. *Adapted from Sherwood et al. (2006).*

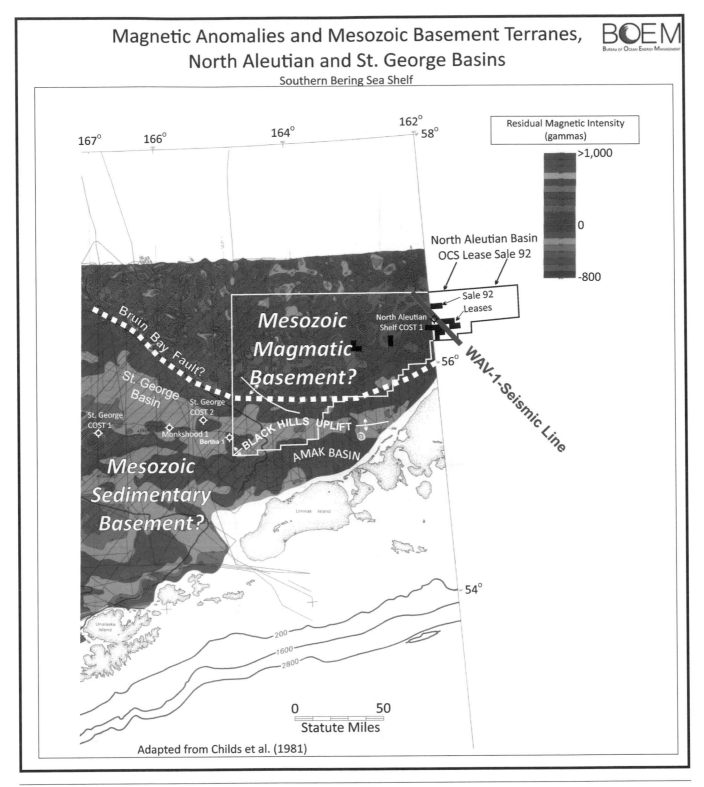

**Figure 24.5**    Magnetic intensity map with offshore speculative extrapolation of the Bruin Bay fault. *Adapted from Childs et al. (1981).*

A speculative course of the Bruin Bay fault through the North Aleutian Basin is located as thick dashed lines in Figures 24.3 and 24.5.

    The best oil and gas prospects in the OCS part of the North Aleutian Basin (prospects partly located by Lease Sale 92 leases) occupy the part of the basin that overlies the inferred Mesozoic-age magmatic-arc terrane north of the Bruin Bay fault (Figure 24.3). It is very unlikely that oil or gas in any quantity can be sourced out of the Mesozoic magmatic-arc rocks interpreted to floor the part of the North Aleutian Basin north of the Bruin Bay fault. This interpretation implies that any hydrocarbon charge to prospects in this area must be sourced from Cenozoic-age rocks of the North Aleutian Basin fill rather than from underlying magmatic-arc rocks. In northern Cook Inlet, oil was generated in an underlying Mesozoic-age basin and has migrated upward to fill shallower traps in Cenozoic-age rocks. A Cook Inlet analogy for formation of oil and gas fields probably cannot be extended to this part of the North Aleutian Basin.

## PROSPECTS—POTENTIAL OIL AND GAS TRAPS

Most mapped prospects in the North Aleutian Basin proper are simple domes[21] draped over the crests of fault-bounded basement uplifts. In the *area of best OCS prospects* (Figure 24.3), these domes range up to 93,000 acres in capture or closure area (area enclosed by the depth contour below which fluids can escape from the trap[22]). The domes in the *area of best offshore prospects* are surrounded by deep parts of the basin where the lowermost strata have been heated to temperatures sufficient for conversion of organic matter to oil or gas. The domes and the intervening deeps are imaged by the seismic data presented as WAV-1 and in the schematic shown later in Figure 24.8A.

The data gathered from the North Aleutian Shelf COST well provides estimates of the temperature history, or thermal exposure experienced by the basin fill and reveals whether a rock may (or may not) have generated oil or gas. The aggregate thermal exposure of rocks is estimated for rock samples obtained from wells by measurements of the ability of certain types of coaly[23] organic matter to reflect light. The reflectance of light (*vitrinite reflectance*) is commonly used to assess whether buried rocks have been heated sufficiently to generate and expel hydrocarbons. Thresholds for oil and gas generation corresponding to certain values for vitrinite reflectance (reported as "%Ro") are posted on the seismic-reflection profile in the graphic at WAV-1, and show that the thermal exposures of the deep-buried rocks in the basin deeps between uplifts exceed Ro = 0.5% and thus are adequate for oil and gas generation given appropriate organic compositions. Any oil and gas generated in these basin deeps (likely, but not known to have occurred) could have migrated upward into the domed sandstone strata that overlie the adjoining basement uplifts.

The dominant geological feature of the southwest part of the North Aleutian Basin OCS Planning Area is the Black Hills uplift (Figure 24.3). On the Black Hills uplift, broad domes in Cenozoic-age strata range up to 133,000 acres in area. Over the Black Hills uplift, the Cenozoic-age sequence is thin (2,000 to 5,000 feet) and thermally immature. However, the area is underlain by Mesozoic-age strata that might form a source for oil as observed for correlative Mesozoic-age strata in Cook Inlet. Areas where buried strata have been heated sufficiently for gas generation (pink areas) or oil generation (green areas), given appropriate compositions of organic matter, are shown in Figure 24.6.

## MESOZOIC-AGE STRATIGRAPHY AND RESERVOIR AND HYDROCARBON-SOURCE FORMATIONS

A stratigraphic chart for the Mesozoic-age rocks of the Alaska Peninsula is presented in Figure 24.4. Sandstone- and conglomerate-dominated units that might form reservoirs for oil and gas include the Talkeetna, Naknek, Staniukovich, Herendeen, and Hoodoo (or equivalent Chignik and Kaguyak) Formations[24] and are highlighted in yellow in Figure 24.4. Because the Mesozoic-age basin in which these strata were deposited is located adjacent to an extinct volcanic arc, most clastic sediments are rich in volcanic- and plutonic-rock fragments. Upon burial and heating, the volcanic-rock fragments readily degraded into laumontite and other minerals that also precipitated in intergranular pores and fully occluded the original intergranular porosity. Generally, absent an upsurge in volcanic activity, younger sandstone formations in the Mesozoic-age column contain less volcanic-rock fragments because of erosion of older clastic formations into new sediments and winnowing of volcanic-rock fragments made susceptible to disintegration by chemical and physical

---

[21]Dome: an uplift of strata that is semicircular in map view.

[22]Trap spill level: A petroleum trap is like an inverted cup that captures migrating oil or gas as it seeks to rise by buoyancy above the more dense water that saturates most subsurface rock. If the *cup* is completely filled, additional oil or gas will escape the trap at the spill level or the lip of the inverted *cup*.

[23]*Coaly* refers to organic matter from the cellulosic material (stems, leaves, wood fragments) associated with land plants.

[24]Formations: geologic formations are typically named after the geographic localities where the formation rocks are widely exposed in outcroppings at the land surface. Formations that are the same age as formations exposed in different geographic locales are said to be *equivalent*.

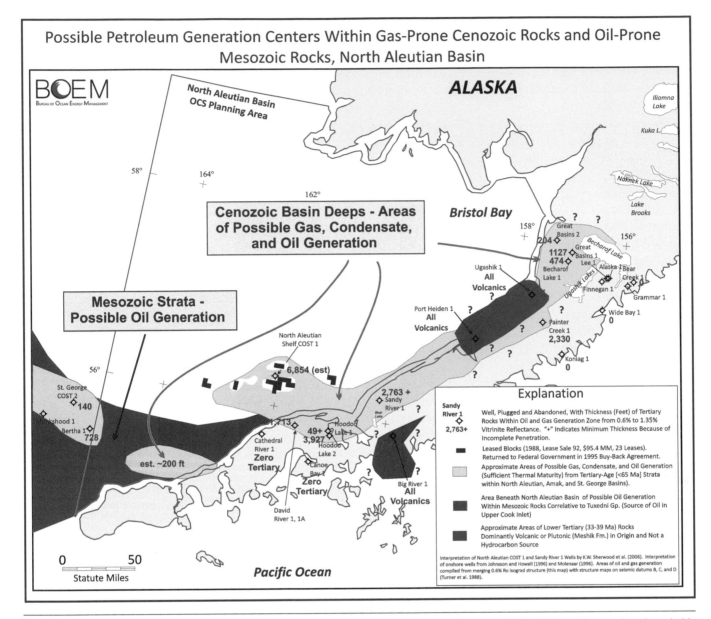

**Figure 24.6**    Isopach map for thickness of Cenozoic-age rocks within oil generation zone (0.6% to 1.35% vitrinite reflectance) with probable areas of thermal maturity sufficient for gas and oil generation within and beneath the North Aleutian Basin. *Adapted and revised from Sherwood et al. (2006).*

weathering. Younger formations have also been less deeply buried because of their position near the top of the Mesozoic-age basin fill and have presumably suffered less porosity destruction related to compaction and temperature-driven pore-filling chemical precipitates. Nonetheless, because of pervasive pore-filling cements, no Mesozoic-age sandstone-bearing formations have been found to contain significant quantities of oil and gas among the wells drilled on the Alaska Peninsula.

Potential sources for oil are identified in the Kialagvik and Kamishak Formations that are exposed on the Alaska Peninsula. These formations are highlighted in green in the stratigraphic column in Figure 24.4. These formations host oil seeps at exposures and minor amounts of oil are found in these formations in a number of wells (localities mapped in Figure 24.2). These potential oil-source formations probably extend beneath the North Aleutian Basin fill south of the Bruin Bay fault (Figures 24.3 and 24.5).

# CENOZOIC STRATIGRAPHY, RESERVOIR FORMATIONS, AND PLAY SEQUENCES

The most complete source of stratigraphic information for the Cenozoic-aged North Aleutian Basin fill is the COST well. The stratigraphy encountered by the COST well is presented alongside well-log geophysical data and interpretations of sedimentary environment in the stratigraphic column accessible as WAV-2 (download this item at www.jrosspub.com/wav). The *Paleo-Bathymetry* classifications, obtained from studies of microfossils in well samples and plotted on the left in WAV-2, show that, in general, the lowermost (and older) strata were deposited in nonmarine or continental settings when the basin began to subside, but transitioned into a marine shelf settings for the shallower strata that cap the basin fill.

The COST well reached total depth at 17,155 feet md bkb in rocks dated by fossil pollen grains as Eocene (33.9–56.0 millions of years) in age. Radiometric ages on volcanic rocks near the base of the well yielded ages from 31.6 to 47.1 millions of years. The basin floor was not penetrated, and interpretations of reflection-seismic data suggest that an additional 3,000 feet of strata lie between the base of the well and the Mesozoic-age rocks beneath the basin floor.

The most attractive reservoir sandstones in the basin fill—those with the highest preserved porosity[25] (up to 40%) and permeability[26] (> 1,000 millidarcies)—are found between 3,000 and 5,700 feet in the COST well. These sandstones are associated with the Miocene-age[27] Bear Lake Formation and the upper 800 feet of the Oligocene-age[28] Stepovak Formation. The seismic profile in WAV-1 shows that this sandstone sequence (highlighted in yellow) can be traced away from the COST well to domed uplifts over fault-bounded basement uplifts where the largest potential traps are found. Any exploration drilling for oil and gas in the North Aleutian Basin will target the Bear Lake/Stepovak sandstones in the potential traps overlying the basement uplifts.

The Eocene-age[29] Tolstoi Formation fills the basin deeps between the basement uplifts imaged in the seismic profile in WAV-1. The formation contains many coal beds and sandstone is abundant in the part of the formation encountered below 15,500 ft, as shown in the stratigraphic column of WAV-2. The sandstones do not form attractive reservoirs for oil and gas because diagenesis[30] has softened the volcanic-rock sedimentary particles and the sandstone grain framework has collapsed, destroying most porosity and severely degrading permeability. In the COST well, coal beds are common below 8,000 feet in the Tolstoi Formation. These coal beds form a potential source for natural gas. Shale beds are also rich in organic carbon, primarily coaly particles and laminations, and also form a potential source for natural gas.

# THERMAL MATURITY OF NORTH ALEUTIAN BASIN FILL

Figure 24.6 maps the areal extent and shows the thickness values for Cenozoic-age strata that lie within the zone of thermal maturity commonly associated with oil and early gas generation (bracketed by vitrinite reflectance values ranging from 0.6% to 1.35%). Areas highlighted in pink in Figure 24.6 are potential generation *kitchens* for gas, condensate, and (possibly, but unlikely) oil. The belt of potential generation (pink) in the basin follows the northeast-trending basin center between Becharof Lake and the area of the Sale 92 leases. The belt of potential generation is segmented by the presence of a massive 33–39-million-year-old volcanic center[31] penetrated at the Port Heiden 1 and Ugashik 1 wells (shown in red, Figure 24.6).

---

[25]Porosity: fraction or percent of bulk rock volume attributed to open pores between mineral grains. Most commercial oil and gas reservoirs range between 10% and 25% in pore volume.

[26]Permeability: ability of a rock to transmit fluids through interconnected pores. Permeability is measured in laboratories by measuring the transmission rate of fluid forced through the rock samples. Most commercial oil and gas reservoirs are associated with permeability in the range of 10 millidarcies ($10^{-3}$ Darcy or "mD") to 1,000 mD.

[27]Miocene: 5.3 to 23.0 millions of years in age (Ma).

[28]Oligocene: 23.0 to 33.9 millions of years in age (Ma).

[29]Eocene: 33.0 to 56.0 millions of years in age (Ma).

[30]Diagenesis: chemical transformation of the mineral makeup of sedimentary particles and formation of pore-filling chemical deposits—generally acting to reduce porosity.

[31]Volcanic center: ancient analog to the large, active volcanoes now found along the Alaska Peninsula—examples including Mount Veniaminov, Aniakchak Crater, and the Katmai complex (Valley of 10,000 Smokes).

# ORGANIC COMPOSITIONS AND SOURCE ROCK POTENTIAL

We recognize two potential sources of thermogenic petroleum that might generate and expel oil or gas to prospects in the North Aleutian Basin: (1) Mesozoic-age sedimentary rocks that underlie the southwest part of the basin and the Black Hills uplift; and (2) Cenozoic-age rocks that comprise the basin fill.

## Mesozoic-Age Source Rocks

The principal point of geologic information for the source rock potential of Mesozoic rocks in the North Aleutian Basin OCS Planning Area is the Cathedral River 1 well (Figures 24.2 and 24.3). Some indicators for source rock potential of the Mesozoic rocks in this well, as suggested by pyrolysis[32] data, are illustrated by Sherwood et al. (2006).

For most of the Mesozoic sequence penetrated by the Cathedral River 1 well, the rocks are rated as poor to fair sources for gas. However, shale and tuffaceous limestone in the interval from 8,700 to 9,300 feet in the Kialagvik Formation appear to rate as *good* potential sources for oil and wet gas. In addition, cherty shale and marlstones in the interval from 12,000 to 12,700 feet in the Talkeetna Formation also appear to form *good* sources for oil and wet gas. Unfortunately, the pyrolysis data are inconclusive because the well samples may have been contaminated by drilling mud additives (Peters 1986, Sherwood et al. 2006).

Thermal maturity data indicate that the rocks in the Cathedral River 1 well were previously much more deeply buried (Sherwood et al. 2006). Because the well is located on the Black Hills uplift where Mesozoic strata are unconformably overlain by Eocene-age rocks (north flank), it appears that the thermal maturation occurred during a time of deep burial that antedated the deposition of the overlying Cenozoic-age reservoir formations and associated traps. The Mesozoic-age oil sources in this area may have generated and expelled their oil long before the deposition of Cenozoic-age reservoir sandstones or the formation of drape domes in Cenozoic-age strata atop the Black Hills uplift.

## Cenozoic-Age Source Rocks

The principal point of geologic information for the source rock potential of the Cenozoic-age fill for the North Aleutian Basin is the COST well. A graphical summary of a large body of source rock geochemical information is presented by Sherwood et al. (2006), but two indicators for source potential in the COST well, total organic carbon (TOC) and hydrogen index[33] (HI), are profiled in a graphic available as WAV-3 (download this item at www.jrosspub.com/wav). TOC data classify COST well rock samples as ranging from *poor* to *very good*. HI values suggest mostly gas sources, with some sample intervals where HI > 300 that might be capable of generating oil. Sample descriptions reveal that most samples with TOC values exceeding 1.0% include coal, as either discrete fragments in cuttings or coaly laminations in core samples. In the several depth intervals below 8,000 feet in the graphic at WAV-3 where HI > 300, nearly all of the elevated HI values are associated with COST well samples described as containing coal. The key question then is whether or not these high-HI coals, or possibly non-coal rocks mixed with coal material in samples, are legitimately capable of generating oil. Sherwood et al. (2006) summarize direct chemical determinations for H/C atomic ratios for coal-bearing (sidewall and conventional core) samples and conclude that the coaly intervals with HI > 300 represent *false-positives* and are actually hydrogen-poor and offer little potential for generation of liquid hydrocarbons.

Sherwood et al. (2006), Robertson Research (1983), and Turner et al. (1988) unanimously conclude that the Cenozoic-age basin fill penetrated by the North Aleutian Shelf COST 1 well is dominated by gas-prone organic matter. This gas-prone organic matter occurs in coal beds or is dispersed as finely divided material in clastic rocks and forms poor to very good sources for gas, with minor potential for condensates and light oil.

---

[32]Pyrolysis: laboratory study of rocks in which samples are heated in a closed chamber and measurements of the quantities of released and evolved hydrocarbons are obtained. The laboratory experiments mimic the effect of heating rocks through burial beneath younger strata.

[33]The hydrogen index is a proxy for the hydrogen richness of the organic material in the rock samples. Oil is a hydrogen-rich substance and is generated by hydrogen-rich organic material. The HI is determined by heating a rock sample and measuring the weight of escaping hydrocarbons or "S2" in the temperature interval 400–470°C. The weight of escaping hydrocarbons is divided by the weight of TOC, and the HI is calculated as 100*S2 (milligrams)/TOC (grams). Oil-source rocks typically offer HI values in excess of 300, while rocks with HI values less than 150 are primarily gas sources. The hydrogen index is a less expensive and fast alternative to chemical determinations of the ratio of atomic hydrogen to atomic carbon.

# OIL AND GAS OCCURRENCES AND BIOMARKER CORRELATIONS

Gas was recovered by three flow tests at rates summing to 90 thousand cubic feet per day (90 MSCF/d) from three zones in the Tolstoi Formation in the Becharof Lake 1 well. Gas was also recovered in flow tests from two intervals in the Tolstoi Formation at rates of 5 to 10 MSCF/d (with 300–400 barrels of water per day) in the David River 1/1A well. Minor amounts of oil and gas were noted elsewhere in wells offshore (St. George and North Aleutian Basins) and in wells on the Alaska Peninsula (annotated on the map in Figure 24.2). In all cases, the quantities of natural gas encountered in these wells are too small to justify economic development.

Gas seeps are observed as gas *chimneys*[34] in some OCS seismic profiles over the Black Hills uplift. Onshore, oil and gas seeps are known primarily from the area near the east end of Becharof Lake, where they are observed along the axes of exposed anticlines in Mesozoic-age rocks or along important faults. Gas seeps there are composed mostly of carbon dioxide ($CO_2$) and are probably the result of magmatic intrusions that intrude and heat limestone[35] in the subsurface. Decker at al. (2005), however, report a gas seep near Port Moller (Figures 24.2 and 24.3) that consists of 91% methane, 7% nitrogen, and 2% $CO_2$. Natural gas recovered from the Becharof Lake 1 well consists of 87.5% methane, 4.7% ethane, 2.3% propane, 0.8% butane, 1.0% hydrogen, and 3.7% other gases (AOGCC 1985, DST data).

Oil seeps emanating from Mesozoic-age rocks along the crests of exposed anticlines attracted the earliest exploration drilling to the Alaska Peninsula in the early 1900s. Geochemical studies of these seep oils suggest that they were sourced from the Kamishak Formation (Figure 24.4) as known from exposures at Puale Bay (locale posted in Figure 24.2), with possible contribution from the Kialagvik Formation (Magoon and Anders 1992).

Modest oil and gas shows[36] and elevated gas *wetness* values[37] of 30% to 95% were noted below 15,450 feet bkb in the North Aleutian Shelf COST 1 well. These observations may signal the presence of migrated oil and unseen oil-source rocks in or beneath the North Aleutian Basin. It therefore became important to determine the source of this oil in order to properly characterize the resource forecast. The possible petroleum liquids raised two questions:

1. Was the oil generated in small or insignificant quantities by coaly organic matter like that dominating the Cenozoic-age rocks penetrated by the COST well? or,
2. Did the oil originate from quantitatively significant oil sources within unrecognized Mesozoic-age rocks beneath the basin floor?

To try to identify the source of the oil below 15,450 feet bkb in the COST well, we conducted extraction and biomarker studies[38] on rock samples[39] from the interval with high gas wetness values. The two extractions obtained in this study supplement the data that were previously obtained by Robertson Research (1983). The extract data are discussed by Sherwood et al. (2006), who conclude that the minor amounts of oil in the interval from 15,450 to 16,800 feet bkb in the North Aleutian Shelf COST 1 well originated from nonmarine Cenozoic-age rocks rather than marine Mesozoic-age rocks. This interpretation extends from the following observations reported in the extract geochemical data (the evidence is summarized in detail by Sherwood et al. 2006):

1. Low sulfur content suggests Cenozoic-age sources.
2. Isoprenoid ratios are terrestrial (as opposed to marine).
3. Pristane/phytane ratios are terrestrial (as opposed to marine).

---

[34]Gas chimney: chimney-shaped vertical zones of disruption of sonic-wave reflections observed in seismic data and caused by the presence of gas venting from surface accumulations to surface seeps.

[35]Decarbonation of limestone: $CaCO_3$ + Heat → $CaO$ (*lime*) + $CO_2$.

[36]Show: small amounts of oil or gas detected in drilling fluids circulated to the surface with rock drill cuttings in the course of drilling a well. The rock cuttings are also tested for the presence of oil or gas as part of the screening for *shows*.

[37]Wetness calculated as: 100*(ethane+propane+butane)/(methane+ethane+propane+butane). High wetness values suggest the presence of petroleum liquids.

[38]Extraction: the rock sample is ground and then immersed in a liquid solvent to capture the liquid hydrocarbon compounds containing more than 15 carbon atoms. Mass spectrometry is used to identify and assay the component compounds. *Biomarkers* are relics of the living plant material that contributed to the organic matter that was incorporated into the sediments, survived the conversion to petroleum, and help correlate the petroleum to the formation from which it was derived. Similar studies and correlations help fingerprint the sources of illegal discharges of petroleum into the environment.

[39]Archived at the Alaska (State) Geologic Materials Center (http://dggs.alaska.gov/gmc/index.php).

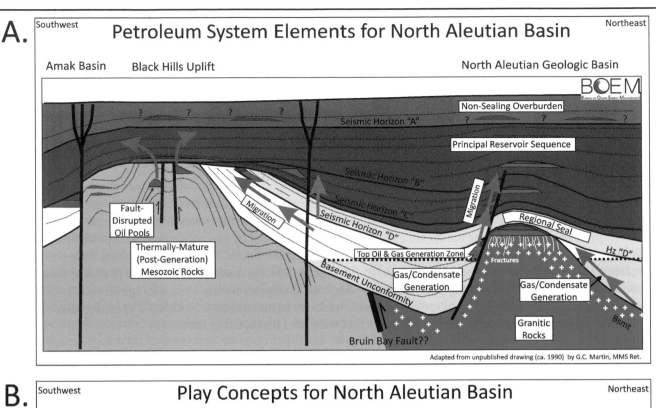

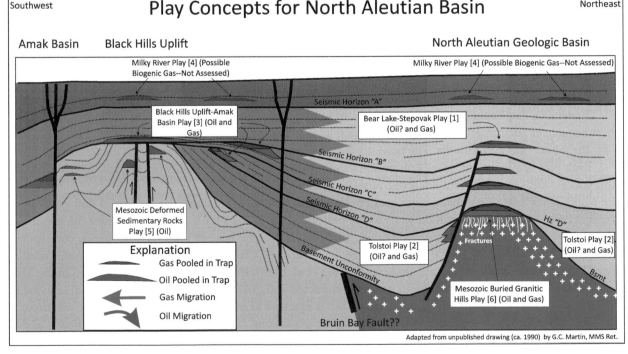

**Figure 24.8** Schematic cross sections illustrating petroleum system elements and play concepts for North Aleutian Basin OCS Planning Area. **A)** Petroleum system elements, including regional reservoir sequence floored by a regional seal and underlain by deep gas/condensate "kitchens" in basin deeps flanking uplifts. Petroleum generated in "kitchens" migrates to traps in shallow reservoir formations draped over basement uplifts via faults that pierce the regional seal. The Black Hills uplift may be reached by long-distance lateral migration of petroleum across highly faulted areas. Fault disruption of Mesozoic oil pools beneath the Black Hills uplift may release oil into overlying strata. Arrows show hypothetical migration paths for gas (red) and oil (green). **B)** Six oil and gas plays defined for North Aleutian Basin OCS Planning Area, separated on the basis of reservoir formation, structural style, and access to petroleum sources.

Late Oligocene to Early Pliocene (~28 to 4.5 millions of years, respectively) in the COST well (shown in WAV-1 and WAV-2). The Bear Lake-Stepovak sequence includes the upper part of the Stepovak Formation, the entire Bear Lake Formation, and the lower part of the Milky River Formation. The Bear Lake-Stepovak sequence at the North Aleutian Shelf COST 1 well is rich in thick (up to 277 feet), porous (up to 40+% porosity), and permeable (up to 7,722 mD) sandstone beds which comprise 61% of the sequence (Sherwood et al. 2006). The sequence contains a total of 3,120 net feet of sandstones in beds greater than 10 feet thick (an assumed practical minimum thickness for productive reservoirs) and 1,443 net feet in beds over 100 feet thick. Because of the ample, thick sandstones and the excellent preservation of porosity and permeability, the Bear Lake-Stepovak sequence is the most attractive reservoir target for exploration-well drilling in the North Aleutian Basin. The Bear Lake-Stepovak sequence is highlighted in yellow in the seismic profile of WAV-1.

The Bear Lake-Stepovak play sequence can be traced in seismic data southwest from the COST well to the Black Hills uplift, where it is draped over fault-bounded uplifts cored by Mesozoic sedimentary rocks. But the Black Hills uplift is treated as a separate play (Play 3) because it has access to hypothetical oil-prone sources in the underlying Mesozoic assemblage, shown in blue in Figure 24.8B. The Black Hills uplift has limited access (large distances across highly faulted areas) to the deeply buried gas-prone Cenozoic-age sources that are hypothesized to charge the Bear Lake-Stepovak traps of Play 1.

The Tolstoi sequence is the basis for Play 2 and ranges in age from Early Eocene to Early Oligocene (~56 to 28 millions of years, respectively) in the COST well. The Tolstoi sequence, as shown in the stratigraphic column of WAV-2, includes the Tolstoi Formation and the lower, shaly part of the Stepovak Formation (7,900–10,380 feet bkb), bracketed by seismic marker "C" and a Mesozoic-age basement not penetrated by the well, but below the floor of the basin fill interpreted in seismic data in WAV-1. The Tolstoi sequence in the COST well is characterized by sparse (10% to 30% of interval), thin (maximum = 57 feet) sandstones. A shale sequence between 9,555 and 10,380 feet bkb in the lower part of the Stepovak Formation in the COST well in WAV-2 forms a regional seal above seismic marker "D", which corresponds to an unconformity that truncates the crests of uplifted fault blocks, as illustrated in the seismic profile in WAV-1. Below seismic marker "D" at 10,380 feet, the Tolstoi sandstones sampled by the COST well are largely impermeable because diagenesis (to clay and zeolites) has softened the volcanic-rock framework grains and the sandstone grain framework has collapsed, as described in petrographic studies by AGAT (1983, p. 2) and Turner et al. (1988).

The substrates of Mesozoic rocks beneath the Cenozoic-age basin and the Black Hills uplift are divided into a southern province of deformed sedimentary rocks (Play 5) and a northern province of *granitic*[48] magmatic-arc (extinct) rocks (Play 6). The Mesozoic deformed sedimentary rocks on the south include rocks that correlate to regional oil sources and that may host oil-charged reservoirs. The granitic Mesozoic province on the north forms the cores of basement uplifts and if properly fractured, might form a reservoir for hydrocarbons, probably thermogenic gas sourced from coal beds in the Tolstoi Formation that fills the basin deeps between uplifted fault blocks. The possible reservoirs formed by these potentially-fractured rocks are set aside as oil and gas Play 6. Although unusual, fractured granitic rocks serving as oil and gas reservoirs are important exploration targets in the Bohai (North China) Basin and offshore Vietnam (Guangming and Quanheng 1982, Areshev et al. 1992). The largest oil field in Vietnam (Bach Ho/White Tiger) is lodged in a fractured granite reservoir with estimated reserves of 1.4 billion stock-tank barrels (STB) and has produced oil at field rates up to 280,000 STB/day (Hung and Le 2004, Brown 2005).

## OIL AND GAS RESOURCES OF THE NORTH ALEUTIAN BASIN

The 2006–2016 assessments of the North Aleutian Basin OCS Planning Area identified six exploration plays. Five of the six plays are assessed with technically recoverable oil and gas while one play (Play 4) is assessed with negligible potential (Table 24.2). Technically recoverable oil and gas represents any hydrocarbons that could be recovered through conventional development technologies without any consideration of extraction costs, market price, or economic thresholds. The *technically recoverable* resource can be thought of as the overall endowment. Only a fraction of the overall endowment can be economically recovered (i.e., at a profit) at any given set of cost and price assumptions. In very remote basins with small forecast endowments of oil and gas resources, no profitable development may be possible at any market price.

---

[48]Granite: rock crystallized from liquid (magma), light in color, and consisting of quartz, feldspar, and minor iron-bearing minerals. Commonly found beneath extinct volcanoes that were once active in magmatic arcs—like the modern Alaska Peninsula and Aleutian Island chain of active volcanoes.

**Table 24.2** 2016 assessment of the risked (downgraded to acknowledge chance for existence or success) technically recoverable (no economic thresholds or constraints) oil, condensate, solution gas, free gas, total gas, total liquids, and aggregate (BOE) endowments for six exploration plays in the North Aleutian Basin OCS Planning Area

**2016 Assessment Results for Technically Recoverable Oil and Gas, North Aleutian Basin OCS Planning Area**

*Risked, Undiscovered, Technically Recoverable (No Economic Thresholds) Oil and Gas Resources, 2016 Assessment*

| Play Number | Play Name | BOE Resources (MMSTB-e)* | | | Oil Resources (MMSTB) | | | Gas-Condensate Liquid Resources (MMSTB) | | |
|---|---|---|---|---|---|---|---|---|---|---|
| | | Minimum (95% Chance) | Mean | Maximum (5% Chance) | Minimum (95% Chance) | Mean | Maximum (5% Chance) | Minimum (95% Chance) | Mean | Maximum (5% Chance) |
| 1 | Bear Lake-Stepovak (Oligocene-Miocene) | 0 | 1,400 | 3,749 | 0 | 271 | 828 | 0 | 136 | 349 |
| 2 | Tolstoi (Eocene-Oligocene) | 91 | 568 | 1,293 | 9 | 62 | 139 | 10 | 61 | 141 |
| 3 | Black Hills Uplift-Amak Basin (Eocene-Miocene) | 0 | 210 | 1,077 | 0 | 149 | 706 | 0 | 6 | 38 |
| 4 | Milky River Biogenic Gas (Plio-Pleistocene) | Play 4 Assessed with Negligible Resources | | | | | | | | |
| 5 | Mesozoic Deformed Sedimentary Rocks (Triassic-Cretaceous) | 0 | 41 | 197 | 0 | 38 | 183 | 0 | 0 | 0 |
| 6 | Mesozoic Buried Granitic Hills (Jurassic-Cretaceous Magmatic Rocks) | 0 | 67 | 330 | 0 | 26 | 93 | 0 | 5 | 29 |
| Sum of All Plays | | 380 | 2,290 | 4,910 | 9 | 545 | 1,948 | 10 | 208 | 556 |

\* BOE, total energy content of oil and gas, in MMSTB-e or millions of barrels-equivalent (assumes 5,620 cubic feet of gas per barrel of oil, energy-equivalent); MMSTB, millions of stock-tank barrels of oil or liquids; TSCF, trillions of standard cubic feet of natural gas. Stock-tank or standard conditions are set at 60°F and 1 atmosphere (14.73 psia) for purposes of measurements of volumes. Some columns may not sum precisely because of rounding and use of alternative aggregation methods.

\*\* Free gas, occurring as gas caps associated with oil and as oil-free gas pools (non-associated gas).

Generally, at low market prices or high development costs, only the very largest oil or gas pools can be realized at a profit and the economically recoverable resource is a small fraction of the overall endowment. At high prices and/or low development costs, the smaller oil or gas pools become profitable, and a larger fraction of the overall technically recoverable endowment becomes economically recoverable.

As shown in Table 24.2, 61% of mean aggregate BOE resources (calculated as 100\*1400/2290 from Column 2) of the hypothetical, risked, undiscovered, technically recoverable oil and gas resources in the basin are associated with Play 1—the *Bear Lake-Stepovak* play. The resource dominance of the *Bear Lake-Stepovak* play reflects the following important characteristics: (1) numerous traps covering large areas of potential production (up to 93,000 acres); (2) thick, high-porosity potential sandstone reservoirs; and (3) proximity to anticipated paths for migration of thermogenic oil and gas.

**Table 24.2** con't

| Free Gas Resources (TSCF)** | | | Solution Gas Resources (TSCF) | | | Total Liquid Resources (MMSTB)* | | | Total Gas Resources (TSCF)* | | |
|---|---|---|---|---|---|---|---|---|---|---|---|
| Minimum (95% Chance) | Mean | Maximum (5% Chance) | Minimum (95% Chance) | Mean | Maximum (5% Chance) | Minimum (95% Chance) | Mean | Maximum (5% Chance) | Minimum (95% Chance) | Mean | Maximum (5% Chance) |
| 0.000 | 5.473 | 14.131 | 0.000 | 0.113 | 0.330 | 0 | 410 | 1,720 | 0.000 | 5.590 | 11.440 |
| 0.401 | 2.476 | 5.640 | 0.003 | 0.025 | 0.053 | 20 | 120 | 160 | 0.400 | 2.500 | 6.380 |
| 0.000 | 0.249 | 1.588 | 0.000 | 0.063 | 0.289 | 0 | 150 | 850 | 0.000 | 0.310 | 1.310 |
| Play 4 Assessed with Negligible Resources | | | | | | | | | | | |
| 0.000 | 0.000 | 0.000 | 0.000 | 0.017 | 0.079 | 0 | 40 | 190 | 0.000 | 0.020 | 0.070 |
| 0.000 | 0.195 | 1.128 | 0.000 | 0.010 | 0.041 | 0 | 30 | 240 | 0.000 | 0.210 | 0.520 |
| **0.401** | **8.393** | **22.487** | **0.003** | **0.229** | **0.791** | **120** | **750** | **1,820** | **1.470** | **8.620** | **17.370** |

The aggregate risked, technically recoverable, undiscovered hydrocarbon endowment of all of the plays in the North Aleutian Basin OCS Planning Area ranges from a minimum of 380 million (MM) STB-e[49] (corresponding to a 95% chance to meet or exceed) to a maximum of 4,910 MMSTB-e (5% chance), with a mean value or expectation of 2,290 MMSTB-e, as shown in Table 24.2. At the mean, the (risked, undiscovered) total gas (sum of free gas and solution gas in oil) resources total 8.620 TSCF[50] but could range up to a maximum (5% chance) potential of 17.370 TSCF. Total mean risked, undiscovered liquid resources (sum of oil resources and gas-condensate liquid resources) is estimated at

---

[49]STB-e: oil and gas summed as energy-equivalent stock-tank (surface) barrels, assuming that 1 STB = 5,620 standard cubic feet of gas. MM signifies *millions*.

[50]The reported quantities in the tables carry significant digits sufficient to *not* report zero resources in instances where rounding would report zero for very small quantities. The significant digits are not intended to convey accuracy or precision of measurement. Rather, they represent a reporting convention for reported results that is designed to avoid substituting zero values for small quantities.

750 MMSTB, but could range up to a maximum potential of 1,820 MMSTB. The planning area is assessed as overall gas-prone, with 67% of the undiscovered mean technically recoverable resources consisting of natural gas.[51]

The wide ranges in the resource assessments of individual plays and the overall planning area as reported in Table 24.2 reflect the high degree of uncertainty in assessment work in sparsely drilled frontier basins like the North Aleutian Basin. Important volumetric parameters like reservoir porosity, the thickness of reservoir formation occupied by hydrocarbons, hydrocarbon fill fractions for prospects, and factors governing reservoir productivity (permeability, fluid viscosity, volume changes in fluids when lifted to the surface, etc.) are poorly constrained by available data and must be estimated or extrapolated from distant points of control. The estimates are constructed as ranges described for computational purposes as probability distributions. The wide ranges in the distributions acknowledge uncertainty by providing the full spectrum of possibility with the mean value for the distribution representing the expectation or *most likely* value.

The five quantified plays in the North Aleutian Basin Planning Area are estimated to contain a maximum of 119 undiscovered pools—ranging five orders of magnitude in mean sizes from 1,838 STB-e to 826,550,000 STB-e. Most of these pools are far too small for any consideration of commercial development.

The sizes of the natural gas and liquids resources estimated for the top ten pools in the *area of best OCS prospects* (Figure 24.3) are shown in Table 24.3. At the mean, the largest pool in the North Aleutian Basin Planning Area is forecast to contain 2,914.5 BSCF of natural gas, modestly larger than the largest commercial gas field in the analogous Cook Inlet geologic basin (Kenai gas field, 2,458 BSCF EUR[52]; AK DOG 2008).

## SCENARIOS FOR DEVELOPMENT AND ECONOMIC RESULTS FOR NORTH ALEUTIAN BASIN GAS DEVELOPMENT

Water depths in the North Aleutian Basin Planning Area range from 50 to 700 feet, although in the *area of best OCS prospects* (Figure 24.3) the water depths are approximately 300 feet. Sea ice typically encroaches into this area during winter months, with maximum encroachment in recent years in the month of March. Marine operations can be conducted year-round, although frequent storms during fall and winter will delay activities. Exploration drilling could use jack-up rigs in water depths less than 150 feet and floating platforms (drill-ships, semisubmersibles) in deeper water. A marine shore-base for these activities would likely be an expansion of facilities at Dutch Harbor and an air support base would probably be located at the 10,000-foot paved runway at Cold Bay on the Alaska Peninsula.

Past (2006–2011) development scenarios assumed the production of both gas and oil. Gas constitutes the majority of the resource potential and was forecast to drive initial development. Condensate (petroleum liquids dissolved in gas) will be recovered mostly by producing gas reservoirs and recovering the liquids at surface separators. Small oil pools, if present, could be developed as satellites or by wells reaching from gas-production platforms. Production operations in past scenarios were assumed to utilize gravity-based structures in water depths < 300 feet and floating platforms (ship-shapes or semisubmersibles) in deeper water. Subsea wells and production equipment were employed to tap the distal parts of large pools and for smaller satellite pools. Maximum feasible tie-back distances from subsea templates to central production platforms were assumed to be < 50 miles for oil production and < 100 miles for gas production. Past development scenarios assumed that gas and oil produced by offshore facilities were carried by trenched subsea pipelines to a landfall on the north side of the Alaska Peninsula and then overland to a port in a deep-water bay on the Pacific side. Here, the gas was chilled for conversion to LNG, and then transported via LNG tankers to markets on the U.S. West Coast (Sherwood et al. 2006).

Given the long distances to potential gas markets in East Asia or the U.S. West Coast, LNG is clearly the most efficient strategy for delivering North Aleutian Basin gas to market. Because this basin is credited with a relatively high gas resource potential (8.620 (TSCF) mean technically recoverable gas) and is within 30 statute miles of land, the BOEM[53] assumed for the 2006 and 2011 assessments that an onshore LNG facility and marine terminal would be constructed on the Pacific side of the Alaska Peninsula. Typically, the high cost for LNG facilities including a marine loading terminal

---

[51]Calculated as $(8.620*10^{12} / (5,620*10^6)) / 2,290 = 1,533.8/2,290 = 0.6698$; assumes that 1 STB is energy-equivalent to 5,620 SCF of natural gas.

[52]EUR = estimated ultimate recovery or original economically recoverable reserves. As of May 1, 2016, the Kenai gas field had produced 2,464 BSCF of natural gas (AOGCC 2016) and is now over 100% depleted based upon the 2008 EUR. As of May 1, 2016, only two of nine field reservoirs continue to produce gas, at an aggregate rate of 1,341,115,000 SCF/month.

[53]And predecessor agency Minerals Management Service.

**Table 24.3** Ranked listing of total gas and total liquids (at means) for the ten largest pools forecast by the 2016 assessment for the North Aleutian Basin OCS Planning Area

**North Aleutian Basin OCS Planning Area, Top Ten Individual Pool Volumes in Area of Best Prospects**

Assessment Results, 2016

| Play Number | Play Name | Pool Rank (BOE) in Play | Liquid Resources (Oil + Condensate, Mean MMSTB)* | Total Gas Resources (Free Gas + Solution Gas, Mean BSCF)* | Fraction of Resource as Natural Gas (Energy-Equivalent Basis**) | Fraction of Liquids as Condensate at Mean Values |
|---|---|---|---|---|---|---|
| 1 | Bear Lake/Stepovak (Miocene/Oligocene) | 1 | 307.9 | 2,914.5 | 0.63 | 0.23 |
| 1 | Bear Lake/Stepovak (Miocene/Oligocene) | 2 | 78.3 | 1,683.9 | 0.79 | 0.53 |
| 1 | Bear Lake/Stepovak (Miocene/Oligocene) | 3 | 65.0 | 1,013.2 | 0.73 | 0.38 |
| 2 | Tolstoi Fm. (Eocene/Paleocene) | 1 | 45.1 | 915.5 | 0.78 | 0.50 |
| 1 | Bear Lake/Stepovak (Miocene/Oligocene) | 4 | 36.5 | 774.7 | 0.79 | 0.52 |
| 6 | Mesozoic Basement – Buried "Granite Hills" | 1 | 50.2 | 546.9 | 0.66 | 0.26 |
| 1 | Bear Lake/Stepovak (Miocene/Oligocene) | 5 | 28.3 | 569.6 | 0.78 | 0.49 |
| 2 | Tolstoi Fm. (Eocene/Paleocene) | 2 | 25.5 | 551.4 | 0.79 | 0.53 |
| 1 | Bear Lake/Stepovak (Miocene/Oligocene) | 6 | 20.3 | 445.1 | 0.80 | 0.54 |
| 2 | Tolstoi Fm. (Eocene/Paleocene) | 3 | 18.5 | 400.4 | 0.79 | 0.53 |

*Conditional (unrisked), technically recoverable pool resources, in millions of stock-tank barrels (MMSTB) of liquids (oil and condensate) and billions of standard cubic feet of natural gas (free gas and solution gas). "Stock-Tank" barrels or "Standard" cubic feet refer to volumes measured at surface or standard conditions (60°F and 14.73 psia or 1 atmosphere).

**Energy-Equivalent Basis = total recoverable hydrocarbon energy at Stock-Tank or surface conditions (60°F, 14.73 psia or 1 atmosphere), expressed in barrels-of-oil-equivalent, where 1 barrel of oil = 5,620 cubic feet of natural gas.

and LNG ships would require a minimum reserve base of approximately five TSCF with co-produced liquids. However, recent advances in technology suggest an alternative development scenario that employs floating liquefied natural gas (FLNG) production facilities.

In an FLNG development scenario, all of the production equipment and storage tanks necessary for production are hosted by a single large offshore floating vessel. For the 2016 development scenario, it is assumed that a large FLNG vessel would be anchored in Bristol Bay near any developed gas fields and that no pipeline or shore-based export facility would be required. In this scenario, LNG created and stored at the FLNG facility is assumed to be offloaded to specialized tankers and conveyed to receiving terminals in markets on the U.S. West Coast or East Asia. Light oil and condensate is loaded on tankers and transported to Nikiski (Cook Inlet) for processing and local consumption, or alternatively conveyed via tankers to Valdez, Alaska and then transshipped to the U.S. West Coast by the TAPS[54] oil tanker fleet. The average oil tariff (600 miles to Cook Inlet) is estimated at $7.47/STB ($2015 U.S.) and the average gas tariff (2,800 miles to East Asia) is estimated at $9.31/MSCF ($2015 U.S.), with an uncertainty range of approximately ±25%.

Selected results of the 2006, 2011, and 2016 economic assessments are assembled in Table 24.4. Each of the assessments was constructed around a wide range of assumed market prices and gas-discount factors,[55] but results for only a few price points are reported in Table 24.4. For the 2006 and 2011 assessments reported in Table 24.4, a single gas price representative of the prices that typified the targeted markets at each of those times was chosen. For the 2016 economic assessment, the forecast economic resources for three price cases are reported.

The $7.12/MSCF[56] ($2015 U.S.) case assumes a gas price approximate to that expected at a destination on the U.S. West Coast where the gas would compete with abundant shale gas produced in the central part of the U.S. Additionally, in the early months of 2015, Japan-landed LNG prices have fallen from ~$15/MSCF to ~$7.50/MSCF (FERC, 2015; Reuters, 2015), so the $7.12/MSCF price case is also relevant to recent gas prices in the East Asia market. For the $7.12/MSCF case in either of these export market scenarios, the mean, risked, economically recoverable gas resource is very small—only 0.320 TSCF—indicating a small probability for commercial success.

Table 24.4 also shows the 2016 results for gas at two additional price cases—$10.68/MSCF and $17.79/MSCF—a range of prices more consistent with those obtained in the period 2012–2014 at East Asia export destinations (e.g., $13.13/MSCF to $17.59/MSCF) (EIA 2015). The assumption of the high prices of the 2012–2014 East Asia import markets predicts much higher realizations of economic resources, 0.798 to 1.655 TSCF (mean to maximum) for the $10.68/MSCF case and a remarkable 6.640 to 13.805 TSCF (mean to maximum) for the $17.79/MSCF case.

The results reported in Table 24.4 are not intended to forecast the future course of gas values in these markets, but are shown to illustrate the sensitivity of resources recovered by commercial development projects to the choice of export market and anticipated prices.

## SUMMARY

The North Aleutian Basin underlies approximately the southern one-third of the North Aleutian Basin OCS Planning Area and extends onshore beneath the Alaska Peninsula. In 1983, one stratigraphic test well was drilled by a petroleum industry consortium into the OCS North Aleutian Basin to obtain geologic data. The well was located so as to avoid penetration of any oil or gas accumulations. Although small amounts of hydrocarbons were detected, the well did not encounter any significant quantities of oil or gas. However, the data obtained have been shown to be extremely valuable for subsequent oil and gas assessments conducted by the Bureau of Ocean Energy Management.

From 1988 OCS Lease Sale 92, twenty-three oil and gas leases were issued for $95 million in the North Aleutian Basin Planning Area. The leases were returned to the U.S. government in 1995. Although additional lease sales have been contemplated in subsequent years, the area has been placed off-limits to oil and gas leasing by Congressional Moratoria and Presidential Withdrawals.

---

[54]TAPS: Trans-Alaska Pipeline System, delivering North Slope crude oil to tankers at Valdez, Alaska for transshipment to the U.S. West Coast.

[55]The relative value of gas to oil on an energy basis, assuming that one STB of oil is equal to 5,620 cubic feet of gas. The *gas price discount*, as a percent, is calculated as 100*(gas price$ per MSCF*5.62)/(oil price$ per STB).

[56]Gas prices are also commonly reported for quantities of energy units such as British thermal units or *Btu*. In most cases, depending upon the composition of the gas, a thousand cubic feet of natural gas is approximately equivalent to a million Btus, such that $/MSCF ≈ $/MMBtu. All prices are in U.S. dollars.

**Table 24.4** Synopsis of selected results from the 2006, 2011, and 2016 economic assessments for the North Aleutian Basin Planning Area. Three gas-price cases for the 2016 assessment illustrate the sensitivity of realized resources to market prices

**Assessment Results, 2006–2016, North Aleutian Basin OCS Planning Area, Reflecting Changes in Gas Market Conditions**

| Year, Resource Type, Economic Conditions | BOE Resources (MMSTB-e) | | | Total Liquid Resources (MMSTB) | | | Total Gas Resources (TSCF) | | |
|---|---|---|---|---|---|---|---|---|---|
| | Minimum (95% Chance) | Mean | Maximum (5% Chance) | Minimum (95% Chance) | Mean | Maximum (5% Chance) | Minimum (95% Chance) | Mean | Maximum (5% Chance) |
| 2006 Undiscovered Technically Recoverable Oil and Gas; No Economic Conditions.[1] | 91 | 2,287 | 6,647 | 19 | 753 | 2,505 | 0.404 | 8.622 | 23.278 |
| 2016 Undiscovered Technically Recoverable Oil and Gas; No Economic Conditions.[1] | 380 | 2,290 | 4,910 | 120 | 750 | 1,820 | 1.470 | 8.620 | 17.370 |
| 2006 Undiscovered Economically Recoverable Oil ($60/STB and Gas ($9.07/MSCF).[2] | 73 | 2,068 | 6,145 | 17 | 706 | 2,388 | 0.317 | 7.653 | 21.117 |
| 2011 Undiscovered Economically Recoverable Oil ($110/STB and Gas ($7.83/MSCF).[3] | 8 | 913 | 3,019 | 7 | 561 | 2,002 | 0.004 | 1.978 | 5.718 |
| 2016 Undiscovered Economically Recoverable Oil ($100/STB and Gas ($7.12/MSCF). U.S. West Coast LNG and 2015 East Asia LNG Markets. | 69 | 552 | 1,464 | 64 | 495 | 1,322 | 0.030 | 0.320 | 0.799 |
| 2016 Undiscovered Economically Recoverable Oil ($100/STB and Gas ($10.68/MSCF). 2012–2014 East Asia LNG Market. | 72 | 662 | 1,664 | 66 | 520 | 1,369 | 0.035 | 0.798 | 1.655 |
| 2016 Undiscovered Economically Recoverable Oil ($100/STB and Gas ($17.79/MSCF). 2012–2014 East Asia LNG Market. | 222 | 1,866 | 4,155 | 89 | 684 | 1,699 | 0.752 | 6.640 | 13.805 |

Data Sources: [1] MMS (2006b) and BOEM (2016); [2] MMS (2006c); [3] BOEM (2014).

BOE, total energy content of oil and gas, in MMSTB-e or millions of barrels-equivalent (assumes 5,620 cubic feet of gas per barrel of oil, energy-equivalent); MMSTB, millions of stock-tank barrels of oil or liquids; TSCF, trillions of standard cubic feet of natural gas. Stock-tank or standard conditions are set at 60°F and 1 atmosphere (14.73 psia) for purposes of measurements of volumes. All volumes are "risked" (downgraded by the chance for existence or success).

The 2006 gas price ($9.07) represents an 85% "gas-price discount" relative to the oil price ($60) and reflects high gas value at that time in the U.S. Midwest gas market. The "gas price discount" as a percent is calculated as 100*(gas price$ per MSCF*5.62)/(oil price$ per STB). The 2011 and the 2016 $7.12/MSCF gas prices are only 40% of the energy-equivalent oil prices and reflect the robust supply of shale gas entering the U.S. market over that period of time. The 2016 gas-price scenarios of $10.68/MSCF and $17.79/MSCF (60% and 100% gas-price-discounts) are more representative of LNG delivered to Japan from 2012 to 2014 ($13.13 to $17.59/MSCF; EIA, 2015). However, in the spring of 2015, landed LNG prices in Japan fell to ~$7.50/MSCF (FERC, 2015; Reuters, 2015). The consequences of such market volatility on the economically recoverable gas resources in the North Aleutian basin are aptly illustrated by the three 2016 assessment price cases presented here.

The North Aleutian Basin is filled with strata of Cenozoic age that reach thicknesses in excess of 20,000 feet. Most of these strata were penetrated by the 1983 stratigraphic test which reached a total depth of 17,155 feet. The Cenozoic-age strata appear (from magnetic data) to overlap two very different substrates beneath the floor of the geologic basin: 1) an extinct Mesozoic-age magmatic arc consisting of granitic and volcanic rocks on the north; and 2) a Mesozoic-age sedimentary basin on the south. The Mesozoic-age sedimentary basin can be traced into Cook Inlet, where it hosts the oil-source rocks that generated the oil reserves of Cook Inlet. However, the most attractive prospects in the North Aleutian Basin are located where the basin floor rests upon the extinct magmatic arc on the north and are isolated from the Mesozoic-age potential oil-source rocks that underlie southern parts of the basin.

Geochemistry data for rock samples recovered by the stratigraphic test well indicate that the composition of the organic material in the basin fill is most likely to generate gas. The lower half of the basin fill has been buried to sufficient depths to achieve the temperature thresholds for conversion to natural gas. The possible generation of significant volumes of gas, and possibly oil, began between 27.1 Ma and 38.5 Ma. Generation may continue at present as younger strata are buried and reach the necessary time-temperature thresholds for generation.

The most attractive prospects formed as subtle domes draped over uplifted fault blocks of granitic rocks. These potential traps began to form 28.3 Ma and continued to grow up to 4.5 Ma. Trap formation coincided with the time of possible generation of gas and oil. If gas and oil generation actually occurred in significant volumes, the relative timing indicates ample opportunities for migrating hydrocarbons to have invaded and "charged" prospects. The key prospects are forecast to most likely contain natural gas of thermogenic and biogenic origins.

Six exploration plays are identified in the North Aleutian Basin and these form the framework for the resource assessment. The prospects in each play share common reservoirs, hydrocarbon sources, and potential traps. Input data for play volumetric parameters (porosity, prospect areas, etc.) are framed as wide-ranging probability distributions that reflect uncertainty, and as a result, the forecasts for resource volumes range widely. The reporting convention for BOEM assessments ranges the forecasts from a minimum (corresponding to a 95% probability or chance to meet or exceed) to a maximum (5% chance) about a mean as the preferred value of central tendency.

The technically recoverable (no consideration of economic factors) gas resources for the North Aleutian Basin are estimated to range between 1.470 TSCF and 17.370 TSCF, with a mean forecast of 8.620 TSCF. Petroleum liquids (free oil and condensate recovered from gas) comprise only 33% of basin resources on an energy basis and are estimated to range between 120 MMSTB and 1,820 MMSTB, with a mean forecast of 750 MMSTB.

Natural gas is the dominant expected resource in the North Aleutian Basin. Gas economics control the amount of the technically recoverable resource endowment that might be expected to enter the world market. North Aleutian Basin gas is most likely to reach distant markets as LNG. The 2006 and 2011 economic assessments assumed that natural gas production would be piped from platforms in the Bering Sea across the Alaska Peninsula to a future port and LNG processing plant on the Pacific coast for tanker transport to the U.S. West Coast. The 2016 economic assessment instead assumes a lower-cost alternative that invokes the use of new FLNG technology, in which production, conversion, storage, and transfer to tankers are all carried out at the OCS field site, thus avoiding the high cost of constructing a pipeline, port, and onshore LNG processing plant. Three price scenarios from a 2016 assessment for export of North Aleutian Basin gas illustrate economic sensitivity to market. A gas price of $7.12/MSCF ($2015 U.S.) can be considered to approximately represent current (2016[57]) prices realized on both the U.S. West Coast and East Asia (i.e., Japan) and yields an economically-recoverable gas resource of 0.320 TSCF or ~4% of the technically recoverable gas endowment. LNG landed in Japan from 2012–2014 garnered prices ranging from $13/MSCF to nearly $18/MSCF ($U.S.). This market is approximately represented by the $10.68/MSCF and $17.79/MSCF ($2015 U.S.) price cases in the 2016 economic assessment. The $10.68/MSCF ($2015 U.S.) price case yields a mean economically recoverable gas resource of 0.798 TSCF or ~9% of the technically recoverable gas endowment. At $17.79/MSCF ($2015 U.S.), the economically recoverable gas resource rises dramatically to 6.640 TSCF or ~77% of the technically recoverable gas endowment. These price cases do not represent gas price forecasts, but instead are chosen as historically observed examples to illustrate the sensitivity of resource realizations to prices in the export market.

---

[57]Spot prices for LNG landed in Japan in February 2015 were $7.60/MMbtu (Reuters, 2015). Spot prices for LNG cargoes landed in Japan in early June 2016 fell to $4.90/MMbtu (Reuters 2016) (1 MMbtu = 1 million British Thermal Units ≈ 1 MSCF).

## ONLINE WAV™ ILLUSTRATIONS

The following are descriptions of illustrations that can be found in the WAV (Web Added Value) section of the publisher's website at www.jrosspub.com/wav. Download and use these items as you progress through the text to enhance interaction with the figures.

> WAV-1. Seismic profile through basement uplifts in western North Aleutian Basin, showing Tertiary sedimentary sequences and basin structures. Profiles adapted from Turner et al. (1988) and Sherwood et al. (2006). The location of the profile is shown in Figures 24.3 and 24.5.
>
> WAV-2. Wellbore stratigraphy, North Aleutian Shelf COST 1 stratigraphic test well, drilled by an industry consortium in 1983. Adapted from Sherwood et al. (2006).
>
> WAV-3. Generation potential (total organic carbon), hydrocarbon type (hydrogen index) indicators, thermal maturity (vitrinite reflectance), stratigraphy, and organization of play sequences for Cenozoic-age rocks in the North Aleutian Shelf COST 1 well. Adapted from Sherwood et al. (2006).
>
> WAV-4. Burial history model for North Aleutian COST 1 well, showing high probability for hydrocarbons reaching sandstones in prospects as suggested by overlap of two *critical events*, primarily: (1) the time of oil and gas generation (38.5 Ma to the present); and (2) the formation of traps in Bear Lake-Stepovak reservoir sequence (28.3 Ma to 4.5 Ma). Diagram simplified after Sherwood et al. (2006). TTI thresholds and correlations to vitrinite reflectance (Ro) from Waples (1980). Software: *Lopatin-From Here to Maturity*, ver. 1.0, 1985, by Platte River Associates, Inc., and D. Waples. Geologic time scale from Walker et al. (2012). TTI forecasts and vitrinite reflectance (Ro) data do not correspond below 15,000 feet bkb, suggesting that the geothermal gradient in the past was lower than observed at present in the North Aleutian COST 1 well.

## REFERENCES

AGAT (Consultants, Inc.). 1983. Arco North Aleutian Shelf C.O.S.T. #1 well, Bristol Basin, Alaska, Section I, final integration of lithologic and reservoir quality analysis of core, core samples, sidewall core samples and cuttings samples in the interval 2,000–17,155 ft (T.D.): Report for ARCO Exploration Company and industry consortium, October 1983.

AK DOG (State of Alaska, Department of Natural Resources, Division of Oil and Gas). 2008. 2007 Annual Report: January 2008. Available: http://dog.dnr.alaska.gov/Publications/AnnualReports.htm.

AK DOG (State of Alaska, Department of Natural Resources, Division of Oil and Gas). 2014. 2014 Alaska Peninsula Areawide Lease Sale Results: Documents Archived in Zip File. Available: http://dog.dnr.alaska.gov/Leasing/Documents/SaleResults/AKPeninsula/LeaseSaleResults-AKPeninsula2014.zip.

AK DOG (State of Alaska, Department of Natural Resources, Division of Oil and Gas). 2015. 2014 Annual Report: January 2015. Available: http://dog.dnr.alaska.gov/Publications/AnnualReports.htm.

AK DOG (State of Alaska, Department of Natural Resources, Division of Oil and Gas). 2016. Lease Sale Results, Alaska Peninsula Areawide Lease Sale Results (2016–2005). Available: http://dog.dnr.alaska.gov/Leasing/SaleResults.htm#AKPeninsula.

AOGCC (Alaska Oil and Gas Conservation Commission). 1985. Well completion report for Amoco Becharof Lake No. 1 well: Public well data file at Alaska Oil and Gas Conservation Commission, 333 W. 7th Ave., Anchorage, AK.

AOGCC (Alaska Oil and Gas Conservation Commission). 2015. 2015–Present, Oil and Gas Pools—Statistics Pages: Continuously updated by Alaska Oil and Gas Conservation Commission, Anchorage, AK. Available: http://doa.alaska.gov/ogc/annual/current/annindex_current.html.

AOGCC (Alaska Oil and Gas Conservation Commission). 2016. April 2016 Production: Excel spreadsheet "Apr16prod.xlsx", worksheet tab "Gas Production Summary By Activ", Kenai Cum Gas, Field Total. Available: http://doa.alaska.gov/ogc/production/pindex.html.

Areshev, E.G., T.L. Dong, N.T. San, and O.A. Shnip. 1992. Reservoirs in fractured basement on the continental shelf of southern Vietnam: Journal of Petroleum Geology 15(4):451–464.

Bailey, A. 2009. The Explorers 2009: Alaska Peninsula & North Aleutian basin: Petroleum News, 14(46), week of November 15, 2009. Available: http://www.petroleumnews.com/pnads/33872555.shtml.

Bally, A.W. and S. Snelson. 1980. Realms of Subsidence. Pages 9–94 in A.D. Miall, editor, Facts and Principals of World Petroleum Occurrence. Canadian Society of Petroleum Geologists Memoir 6.

Baseline DGSI. 2003. Characterization of hydrocarbons extracted from two intervals in the North Aleutian COST 1 well: Report 03-529-A prepared by Baseline DGSI, 8701 New Trails Drive, The Woodlands, Texas for U.S. Minerals Management Service, Anchorage, Alaska. Available to public as Alaska Geologic Materials Center Report DGGS GMC 309, State of Alaska Geological

Materials Center, 3651 Penland Parkway, Anchorage, AK. CD-ROM and printed copy, 47 p. Also available digitally as Appendix 3 in report published on CD by Sherwood et al. (2006; cited elsewhere here). Printed report (only) available: http://dggs.alaska .gov/pubs/id/19160.

BOEM (Bureau of Ocean Energy Management). 2010. North Aleutian Basin Lease Sale 214. Available: http://www.boem.gov/Oil-and -Gas-Energy-Program/Leasing/Regional-Leasing/Alaska-Region/Alaska-Lease-Sales/Sale-214/Index.aspx.

BOEM (Bureau of Ocean Energy Management). 2014. Assessment of Undiscovered Technically Recoverable Oil and Gas Resources of the Nation's Outer Continental Shelf, 2011 (Includes 2014 Atlantic Update): Page 4, Table 2. Available: http://www.boem.gov/ Oil-and-Gas-Energy-Program/Resource-Evaluation/Resource-Assessment/2011-RA-Assessments.aspx or http://www.boem .gov/2011-National-Assessment-Factsheet/.

BOEM (Bureau of Ocean Energy Management). 2015. Areas Under Moratoria. Available: http://www.boem.gov/Areas-Under- Moratoria/.

BOEM (Bureau of Ocean Energy Management). 2016a. OCS Lands Act History. Available: http://www.boem.gov/OCS-Lands -Act-History/.

BOEM (Bureau of Ocean Energy Management). 2016b. Assessment of Undiscovered Technically Recoverable Oil and Gas Resources of the Nation's Outer Continental Shelf, 2016. Available: http://www.boem.gov/Resource-Assessment/.

Brown, D. 2005. Vietnam Finds Oil in the Basement: American Association of Petroleum Geologists Explorer 26(2):8–11.

Burk, C.A. 1965. Geology of the Alaska Peninsula—Island Arc and Continental Margin: Geological Society of America Memoir 99, 250 p.

Childs, J.R., A.K. Cooper, and A.W. Wright. 1981. Residual magnetic map of Umnak Plateau region, southwestern Bering Sea: U.S. Geological Survey Geophysical Investigations Map GP-939, scale 1:1,000,000. Available: http://dggs.alaska.gov/pubs/id/24004.

Claypool, G.E., C.N. Threlkeld, and L.B. Magoon. 1980. Biogenic and thermogenic origins of natural gas in Cook Inlet basin, Alaska: American Association of Petroleum Geologists Bulletin 64(8):1131–1139.

Comer, C.D., B.M. Herman, and S.A. Zerwick. 2015. Geologic Report for the St. George Basin Planning Area, Bering Sea, Alaska: Minerals Management Service, OCS Report MMS 87-0030, 84 p. Available: http://www.boem.gov/Alaska-Reports-1980/.

Cooper, A.K., M.S. Marlow, A.W. Parker, and J.R. Childs. 1979. Structure-Contour map on acoustic basement in the Bering Sea: U.S. Geological Survey Miscellaneous Field Studies Map MF-1165, 1 Sheet, scale 1:2,500,000. Available: http://dggs.alaska.gov/pubs/ id/23599.

Decker, P.L., E.S. Finzel, K.D. Ridgway, R.R. Reifenstuhl, and R.B. Blodgett. 2005. Preliminary summary of the 2005 field season: Port Moller, Herendeen Bay, and Dillingham areas, Bristol Bay Basin, Alaska Peninsula: Alaska Division of Geological & Geophysi-cal Surveys Preliminary Interpretive Report 2005-7, 55 p., 2 sheets. doi:10.14509/7190. Available: http://dggs.alaska.gov/pubs/ id/7190.

Dellagiarino, G. 1986. Offshore resource evaluation program: background and functions: U.S. Department of the Interior, Minerals Management Service, Offshore Resource Evaluation Division, OCS Report MMS-85-0091, 42 p.

Detterman, R.L. 1990. Correlation of exploratory wells, Alaska Peninsula: U.S. Geological Survey Open file Report OF 90-279, 2 plates. Available: http://dggs.alaska.gov/pubs/id/11783.

Detterman, R.L., J.E. Case, S.W. Miller, F.J. Wilson, and M.E. Yount. 1996. Stratigraphic framework of the Alaska Peninsula: U.S. Geo-logical survey Bulletin 1969-A, 74 p. Available: http://dggs.alaska.gov/pubs/id/3754.

Detterman, R.L., J.E. Case, F.J. Wilson, and M.E. Yount. 1987. Geologic map of the Ugashik, Bristol Bay, and western part of Karluk quadrangles, Alaska: U.S. Geological Survey Miscellaneous Investigations Map I-1685, 1:250,000. Available: http://dggs.alaska .gov/pubs/id/12927.

Detterman, R.L., T.P. Miller, M.E. Yount, and F.J. Wilson. 1981. Geologic map of the Chignik and Sutwick Island quadrangles, Alaska: U.S. Geological Survey Miscellaneous Investigations Map I-1229, 1:250,000. Available: http://dggs.alaska.gov/pubs/id/12940.

EIA (U.S. Energy Information Administration). 2015. Price of Liquefied U.S. Natural Gas Exports to Japan, 1989–2014. Available: http://www.eia.gov/dnav/ng/hist/n9133ja3m.htm.

Exlog (Exploration Logging Inc. of U.S.A.). 1983. Geochemical final well report, ARCO Exploration Company, North Aleutian Shelf C.O.S.T. Well No. 1, September 1982–January 1983: Report to ARCO Exploration Company and Industry Consortium by Exlog. Available digitally as Appendices 2 and 4 in report published on CD by Sherwood et al. (2006; cited elsewhere here) and in public well file, Bureau of Ocean Energy Management, 3801 Centerpoint Drive, Suite 500, Anchorage, AK. 40 p.

FERC (Federal Energy Regulatory Commission). 2015. World LNG Estimated June 2015 Landed Prices. Available: http://www.ferc .gov/market-oversight/mkt-gas/overview/ngas-ovr-lng-wld-pr-est.pdf.

Guangming, Z. and Z. Quanheng. 1982. Buried-Hill oil and gas pools in the North China Basin: Pages 317–335 in M.T. Halbouty, editor. The Deliberate Search for the Subtle Trap. American Association of Petroleum Geologists Memoir 32.

Hung, N.D. and H.V. Le. 2004. Petroleum geology of Cuu Long basin—offshore Vietnam: American Association of Petroleum Ge-ologists Search and Discovery Article #10062 (2004). Available: http://www.searchanddiscovery.com/documents/2004/hung/ images/hung.pdf.

Hunt, J.M. 1979. Petroleum Geochemistry and Geology: W.H. Freeman and Company, San Francisco, 617 p.

Johnsson, M.J. and D.G. Howell. 1996. Thermal maturity of sedimentary basins in Alaska—an overview: Pages 1–10 in M.J. Johnsson and D.G. Howell, editors. Thermal Evolution of Sedimentary Basins in Alaska. U.S. Geological Survey Bulletin 2142. Available: http://dggs.alaska.gov/pubs/id/4331.

Kirschner, C.E. 1988. Map showing sedimentary basins of onshore and continental shelf areas, Alaska: U.S. Geological Survey Miscellaneous Investigations Series I-1873, scale 1:2,500,000, 1 Sheet. Available: http://dggs.alaska.gov/pubs/id/12926. (June 2015).

Lidji, E. 2010. Hewitt drops N. Aleutian leases; Armstrong adds North Slope acres: Petroleum News 15(46), week of November 14, 2010. Available: http://www.petroleumnews.com/pnads/311186021.shtml.

Lopatin, N.V. 1971. Temperature and geologic time as factors in coalification (in Russian): Akad. Nauk. SSSR Isv. Ser. Geol. 3: 95–106.

Magoon, L.B. and D.E. Anders. 1992. Oil-to-source-rock correlation using carbon-isotopic data and biological marker compounds, Cook Inlet-Alaska Peninsula, Alaska: Pages 241–274 in J.M. Moldowan, P. Albrecht, and R.P. Philip, editors. Biological Markers in Sediments and Petroleum. Prentice Hall, NJ.

Magoon, L.B., W.L. Adkison, and R.M. Egbert. 1976. Map showing geology, wildcat wells, Cenozoic plant fossil localities, K-Ar age dates, and petroleum operations, Cook Inlet area, Alaska: U.S. Geological Survey Miscellaneous Investigations Series, Map I-1019, 3 sheets incl. map, 1:250,000. Available: http://dggs.alaska.gov/pubs/id/12849.

MMS (Minerals Management Service). 2006a. North Aleutian Basin Planning Area (Alaska)—Province Summary: Page 12, Table 3. Available: http://www.boem.gov/uploadedFiles/BOEM/About_BOEM/BOEM_Regions/Alaska_Region/Resource_Evaluation/2006-Assessment-Files/North%20Aleutian%20Basin%20Province%20Summary-2006%20Assessment.pdf.

MMS (Minerals Management Service). 2006b. 2006 assessment, economically-recoverable oil and gas at $60/STB and $9.07/MSCF. Available: http://www.boem.gov/uploadedFiles/BOEM/About_BOEM/BOEM_Regions/Alaska_Region/Resource_Evaluation/Undiscovered-Oil-and-Gas-Resources-Alaska-2006.pdf.

MMS (Minerals Management Service). 2006c. Chukchi Sea Planning Area—Province Summary, 2006 Oil and Gas Assessment. Available: http://www.boem.gov/About-BOEM/BOEM-Regions/Alaska-Region/Resource-Evaluation/2006-assessment-AK.aspx, "Chukchi Sea Province Summary".

MMS (Minerals Management Service). 2007. Proposed Final Program Outer Continental Shelf Oil and Gas Leasing Program 2007–2012, April 2007, 146 p. Available: http://www.boem.gov/Oil-and-Gas-Energy-Program/Leasing/Five-Year-Program/MMSProposedFinalProgram2007-2012-pdf.aspx.

MMS (Minerals Management Service). 2008a. Call for Information and Nominations (Call) and Notice of Interest (NOI) to Prepare an Environmental Impact Statement (EIS): published in Federal Register, v. 73, no. 68, Tuesday, April 8, 2008, Notices:19,905–19,098.

MMS (Minerals Management Service). 2008b. Annual Moratoria (also referred to as Congressional Moratoria). Not available: http://www.mms.gov/5-year/moratoria.htm.

MMS (Minerals Management Service). 2010. Revised OCS Oil & Gas Leasing Program for 2007–2012: December 1, 2010, 250 p. Available: http://www.boem.gov/Oil-and-Gas-Energy-Program/Leasing/Five-Year-Program/2007-2012-5-Year-Program.aspx.

Molenaar, C.M. 1996. Thermal maturity patterns and geothermal gradients on the Alaska Peninsula: Pages 11–20 in M.J. Johnsson and D.G. Howell, editors. Thermal Evolution of Sedimentary Basins in Alaska. U.S. Geological Survey Bulletin 2142. Available: http://dggs.alaska.gov/pubs/id/4332.

Peters, K.E. 1986. Guidelines for evaluating petroleum source rock using programmed pyrolysis: American Association of Petroleum Geologists Bulletin 70(3):318–329.

Reuters. 2015. Japan Feb LNG spot price falls a quarter to $7.60/mmBtu: Reuters online, Tuesday, March 10, 2015, 12:20 am EDT. Available: http://www.reuters.com/article/2015/03/10/lng-japan-spot-idUSL4N0WC1JL20150310.

Reuters. 2016. Japan May LNG spot price falls to lowest in more than 2 years: Reuters online, Thursday, June 9, 2016, 12:57 am EDT. Available: http://www.reuters.com/article/us-lng-japan-spot-idUSKCN0YV0BL.

Riehle, J.R., R.L. Detterman, M.E. Yount, and J.W. Miller. 1993. Geologic map of the Mount Katmai quadrangle and adjacent parts of the Naknek and Afognak quadrangles, Alaska: U.S. Geological Survey Miscellaneous Investigations Map I-2204, 1:250,000. Available: http://dggs.alaska.gov/pubs/id/12845.

Robertson Research (Inc.). 1983. Geochemical analysis of the North Aleutian Shelf, COST No. 1 well, Alaska: Report to ARCO Exploration Company and Industry Consortium by W.G. Dow, Robertson Research (U.S.) Inc., 16730 Hedgecroft, Suite 306, Houston, TX 77060-3697. Available, Appendices 2 and 5: http://www.boem.gov/uploadedFiles/BOEM/About_BOEM/BOEM_Regions/Alaska_Region/Resource_Evaluation/North-Aleutian-Basin-Assessment-Charts-and-Tables.zip.

Sherwood, K.W., J. Larson, C.D. Comer, J.D. Craig, and C. Reitmeier. 2006. North Aleutian Basin OCS Planning Area, assessment of undiscovered technically-recoverable oil and gas as of 2006: February 2006, 138 p. with 4 plates. Available: http://www.boem.gov/About-BOEM/BOEM-Regions/Alaska-Region/Resource-Evaluation/Resource-Evaluation-Report.aspx. (2006 Oil and Gas Assessment of North Aleutian Basin Planning Area).

Teledyne Isotopes. 1983. K-Ar determinations for 3 samples from the North Aleutian shelf COST 1 well: Report to ARCO Exploration Company prepared by Teledyne Isotopes, 50 Van Buren Ave., Westwood, NJ 07675, April 11, 1983. Available: Bureau of Ocean Energy Management, 3801 Centerpoint Drive, Suite 500, Anchorage, AK.

The White House. 2008. President Bush discusses Outer Continental Shelf exploration: Presidential announcement of memorandum lifting the executive prohibition on exploration in the U.S. Outer Continental Shelf, Office of the Press Secretary, July 14, 2008. Memorandum on Modification of the Withdrawal Areas of the United States Outer Continental Shelf from Leasing Disposition. Weekly Compilation of Presidential Documents, vol. 44 (July 14, 2008), p. 986. Available: https://www.gpo.gov/fdsys/pkg/WCPD-2008-07-21/pdf/WCPD-2008-07-21-Pg986-2.pdf.

The White House. 2014. President Obama protects Alaska's Bristol Bay from future oil and gas drilling: The White House, Office of the Press Secretary, Statements and Releases, December 16, 2014. Available: https://www.whitehouse.gov/the-press-office/2014/12/16/president-obama-protects-alaska-s-bristol-bay-future-oil-and-gas-drillin.

Turner, R.F., C.M. McCarthy, M.B Lynch, P.J. Hoose, G.C. Martin, J.A. Larson, T.O. Flett, K.W. Sherwood, and A.J. Adams. 1988. Geological and operational summary, North Aleutian shelf COST No. 1 well, Bering Sea, Alaska: Minerals Management Service OCS Report MMS 88-0089, 266 p. Available: http://www.boem.gov/Alaska-Reports-1980/.

Walker, J.D., J.W. Geissman, S.A. Bowring, and L.E. Babcock. (compilers). 2012. Geologic time scale v. 4.0: Geological Society of America, doi: 10.1130/2012.CTS004R3C. Available: http://www.geosociety.org/science/timescale/timescl.pdf.

Waples, D.W. 1980. Time and temperature in petroleum formation—application of Lopatin's method to petroleum exploration: American Association of Petroleum Geologists Bulletin 64(6):916–926.

White, D.A. 1993. Geologic risking guide for prospects and plays: American Association of Petroleum Geologists Bulletin 77(12): 2048–2061.

Wilson, F.H., R.L. Detterman, J.W. Miller, and J.E. Case. 1995. Geologic map of the Port Moller, Stepovak Bay, and Simeonof Island quadrangles, Alaska Peninsula, Alaska: U.S. Geological Survey Miscellaneous Investigations Map I-2272, 1:250,000. Available: http://dggs.alaska.gov/pubs/id/12905.

# 25

# MINERAL RESOURCES OF THE BRISTOL BAY WATERSHED AND THEIR ENVIRONMENTAL CHARACTERISTICS

**Robert R. Seal, II, US Geological Survey**

## INTRODUCTION

Alaska truly is the *last frontier* in the United States when it comes to mineral resource exploration and development because of its limited accessibility. The Bristol Bay watershed is one of its least explored areas because the only access is by air or water and roads are sparse. To improve decision making regarding land management in southwestern Alaska, the mineral resource potential was evaluated by the U.S. Geological Survey (USGS) for the U.S. Bureau of Land Management (Schmidt et al. 2007). The assessment methodology was based on geologic settings conducive to the formation of specific mineral deposit types in a broad sense. This qualitative assessment considered the mineral resource potential of over a dozen base-metal [copper (Cu), lead (Pb), zinc (Zn), nickel (Ni), molybdenum (Mo)] and precious-metal (gold (Au), silver (Ag), platinum-group metals) deposit types. The geologic setting within the Bristol Bay watershed, or more specifically the Nushagak and Kvichak watersheds, is permissive for the occurrence of several mineral deposit types that are amenable for large-scale development. Of these deposit types, porphyry copper deposits (e.g., Pebble) (Lang et al. 2013) and intrusion-related gold deposits (e.g., Shotgun) (Rombach and Newberry 2001) are probably the most important on the basis of the current maturity of exploration activities in the watershed by the mining industry.

Despite its remoteness, significant mineral resources have been identified in the Bristol Bay watershed. The most notable is the Pebble porphyry Cu-Mo-Au deposit, which ranks among the world's largest deposits of this type. It sits astride the watershed divide between the Nushagak and Kvichak river systems near Iliamna Lake (Figure 25.1). The geologically similar, but significantly less-explored Humble and Big Chunk porphyry copper prospects are within the Nushagak watershed (Figure 25.1). A more complete understanding of the economic potential of these deposits and their geologic similarity to Pebble will rely upon additional exploration. The H and D Block prospects, west of Pebble, potentially represent additional porphyry copper exploration targets in the watershed (see Figure 25.1).

The Shotgun deposit in the northern part of the Nushagak watershed, approximately 177 km (110 miles) northwest of Pebble, is currently the most notable example of an intrusion-related gold deposit within the Bristol Bay watershed (Figure 25.1). An additional deposit type that has received historical interest is the placer gold deposit. Placer deposits are heavy mineral deposits in unconsolidated or poorly consolidated gravels that form from the weathering of gold-bearing rocks.

Because of the Pebble prospect, and to a lesser extent the Shotgun and other gold prospects, mining has been identified as a potential source of future large-scale development in the Bristol Bay watershed. This possibility prompted the Bristol Bay watershed assessment conducted by the U.S. Environmental Protection Agency (USEPA), which was released in 2014 (USEPA 2014).

The goal of this chapter is to summarize, largely on the basis of literature review, the geologic and environmental characteristics of the two most important mineral deposit types within the Bristol Bay watershed. The geologic characteristics of mineral deposits are paramount to determining their geochemical signatures in the environment. The geologic characteristics of different types of mineral deposits provide guidelines for exploration, as well as guidelines for predicting

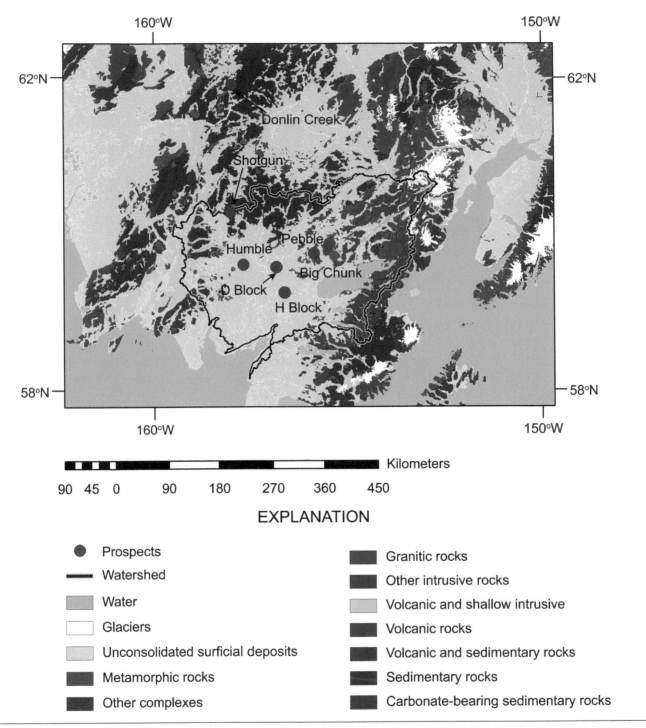

**Figure 25.1**   Generalized geologic map of the Bristol Bay watershed showing the general locations of the Pebble, Humble, Big Chunk, and other prospects. *Adapted from Wilson et al. (2015). Map credit: Ruth Schulte (USGS).*

potential environmental impacts from mining. Examples of distinctive characteristics of different types of mineral deposits include:

1. Ore minerals and associated alteration assemblages;
2. Geochemical associations of elements, including the commodities being sought;
3. The grade and tonnage of the deposit;
4. The likely mining and ore-processing methods used;

5. The environmental attributes of the deposit, such as acid-generating and acid-neutralizing potentials of geologic materials that may become mine waste; and

6. The susceptibility of the surrounding ecosystem to various stressors related to the deposit and its mining, among other features.

This geologically based approach for understanding the environmental characteristics of a mineral deposit is known as a geoenvironmental mineral deposit model, and it will form the basis of this chapter (Seal and Hammarstrom 2003).

The environmental features of a mine—in essence its environmental footprint—arise from three main factors: the geology of the deposit; the local setting (climate, hydrology); and the mining and ore-processing approaches, including waste management. The geology of a deposit, which influences many of the chemical risks associated with a specific mine, is predetermined for a given ore deposit. Much of the local setting is also predetermined, but various aspects of the local hydrologic setting can be managed or otherwise influenced during and after mining. In contrast, most aspects of mining, ore processing, and waste management have a number of options that can be considered and planned before, during, and after mining.

## GEOLOGY OF THE BRISTOL BAY WATERSHED

The Bristol Bay watershed is characterized by a complex geologic history. The history, going back at least 100 million years, has been dominated by northward movement and thrusting (subduction) of the oceanic crust beneath the Alaskan continental landmass, which continues today. In fact, it is this continued motion of the Pacific plate against the North American plate that is the cause of most of the earthquakes in Alaska (Page et al. 1991). The northward subduction of oceanic crust led to the accretion of island land masses (or terranes) to the Alaskan mainland. The divide between the Nushagak and Kvichak watersheds is near the geologic boundary between the Peninsular Terrane to the southeast and the Kahiltna Terrane to the northwest (Decker et al. 1994, Nokleberg et al. 1994).

The Peninsular Terrane consists of Permian limestone, Triassic limestone, chert and volcanic rocks, Jurassic volcanic and intrusive rocks, and Jurassic to Cretaceous sedimentary rocks. The Pebble deposit and the Humble and Big Chunk prospects are located within the southern Kahiltna Terrane (Figure 25.1). The southern Kahiltna Terrane consists of a deformed sequence of Triassic to Jurassic volcanic rocks (basalt, andesite, and tuff), chert, and minor limestone of the Chilikadrotna Greenstone, which is overlain by the Jurassic to Cretaceous Koksetna River sequence of sandstones, siltstone, and shales (Wallace et al. 1989). The area was intruded by deep-seated Cretaceous to Tertiary igneous intrusions, which include those associated with the Pebble deposit. The area also was partially covered by Tertiary to Quaternary volcanic rocks and varying thicknesses of glacial deposits (Detterman and Reed 1980, Bouley et al. 1995).

The Shotgun gold deposit is hosted by the Upper Cretaceous Kuskokwim Group—a sequence of marine sedimentary rocks—and is associated with slightly younger granitic porphyry rocks (Rombach and Newberry 2001). The deposit is located at the southwestern edge of the Kuskokwin Mineral Belt, which includes the Donlin Creek, Nixon Fork, and Vinasale deposits, all in the Kuskokwim River watershed.

## MINERAL RESOURCE POTENTIAL OF THE NUSHAGAK AND KVICHAK WATERSHEDS

The geology of the Nushagak and Kvichak watersheds has characteristics that indicate that the region is favorable for several different mineral-deposit types (Schmidt et al. 2007). These deposit types include porphyry copper deposits, copper and iron skarn (limestone replacement) deposits, intrusion-related gold deposits, tin greisen deposits, epithermal gold-silver vein deposits, hot spring mercury deposits, placer gold deposits, and sand and gravel deposits (Table 25.1). Of these deposit types, porphyry copper deposits and intrusion-related gold deposits are represented by current prospects that could prompt large-scale development. Copper skarn deposits hold less potential, in the absence of infrastructure from other mine development in the region, because of their typical smaller size (John et al. 2010). Placer gold deposits are a deposit type of widespread, but smaller scale importance. Significant exploration activity associated with porphyry copper deposits is currently being done at the Pebble prospect, and to a lesser extent the Humble and Big Chunk prospects. Several other porphyry copper prospects are immediately adjacent to Pebble, including the H Block and D Block prospects. Exploration has also been done in the watershed at several gold properties, most notably the Shotgun deposit.

**Table 25.1** Primary deposit types with significant resource potential for large-scale mining in the Nushagak and Kvichak watersheds

| Deposit type | Commodities | Examples | References |
|---|---|---|---|
| Porphyry copper | Cu, Mo, Au, Ag | Pebble, Big Chunk, Kijik River | Schmidt et al. (2007); Bouley et al. (1995) |
| Intrusion-related gold | Au, Ag | Shotgun/Winchester, Kisa, Bonanza Hills | Schmidt et al. (2007); Rombach and Newberry (2001) |
| Copper(-iron-gold) skarn | Cu, Au, Fe | Kasna Creek, Lake Clark Cu, Iliamna Fe, Lake Clark | Schmidt et al. (2007); Newberry et al. (1997) |

The Pebble deposit is the most advanced among the mining prospects in the Bristol Bay watershed in terms of exploration and progress toward the submission of mine permit applications. Therefore, the potential for large-scale mining development within the watershed in the near future is greatest for porphyry copper deposits. Intrusion-related gold deposits, such as Shotgun, and Donlin Creek, within the nearby Kuskokwim River watershed, also hold potential for future development. Placer gold deposits can form in areas of bedrock sources of gold, such as around intrusion-related gold deposits or a gold-bearing porphyry copper deposit. However, significant placer gold deposits have yet to be identified in the Nushagak and Kvichak watersheds (Nokleberg et al. 1987). Therefore, the remainder of this chapter will focus on porphyry copper deposits and intrusion-related gold deposits.

# GENERAL CHARACTERISTICS OF IMPORTANT MINERAL DEPOSIT TYPES

## Porphyry Copper Deposits

Porphyry copper deposits are found around the world, most commonly in areas with active or ancient volcanism (Figure 25.2). Southern Alaska is part of the Circum-Pacific zone known as the *Ring of Fire* where earthquakes and volcanoes are localized, and other geologic events such as geothermal activity and the formation of hydrothermal ore deposits occur. Porphyry copper deposits currently supply approximately 65% of the world's copper and together with porphyry molybdenum deposits, account for over 95% of the molybdenum production (Sinclair 2007, John et al. 2010, Singer and Menzie 2010). The economic viability of porphyry copper deposits is dictated by the economy of scale—they typically are low grade (average 0.44% copper in 2008), large tonnage deposits (typically hundreds of millions to billions of metric tons of

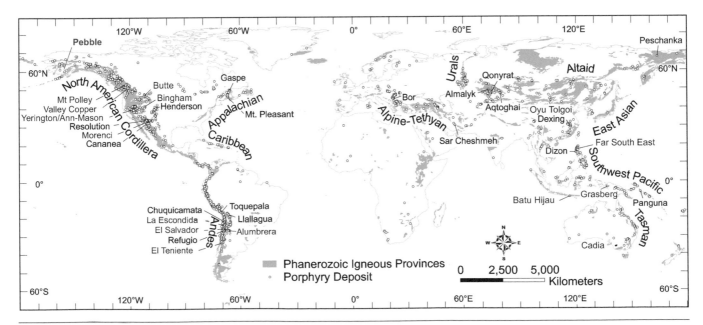

**Figure 25.2** Map showing location of Phanerozoic porphyry deposits with representative deposits labeled. *Map credit: modified from Seedorff et al. (2005) and John et al. (2010).*

ore) that are developed by bulk mining techniques (John et al. 2010). Copper is the primary commodity for which these deposits are sought, but they may also contain significant amounts of byproduct commodities such as Mo, Au, and Ag, in addition to critical elements such as tellurium (Te), selenium (Se), rhenium (Re), and platinum-group metals. Because of their large size, their mine lives typically span decades. The geologic characteristics of porphyry copper deposits recently have been reviewed by John et al. (2010), Sinclair (2007), and Seedorff et al. (2005).

Primary (hypogene) ore minerals found in porphyry copper deposits are disseminated throughout hydrothermally altered igneous rocks in narrow quartz veinlets. The porphyry intrusions were typically emplaced at shallow levels in the crust. They form when a crystallizing magma expels high temperature metal-rich (hydrothermal) fluids during the late stages of crystallization. A combination of cooling and reactions with the surrounding rocks leads to the precipitation of metals as sulfide minerals. Mineralization commonly occurs both within the associated intrusions and in the surrounding country rocks. The primary minerals fill veins, veinlets, stockworks, and breccias. Pyrite ($FeS_2$) is generally the most abundant sulfide mineral. The main copper-sulfide ore minerals are chalcopyrite ($CuFeS_2$) and bornite ($Cu_5FeS_4$). A number of other minor copper sulfide minerals are commonly found; most notable from an environmental perspective is the arsenic (As)-bearing mineral enargite ($Cu_3AsS_4$). Molybdenite ($MoS_2$) is the main molybdenum mineral. Gold in porphyry copper deposits can be associated with bornite, chalcopyrite, and pyrite; the gold may occur as a trace element within these sulfide minerals or as micrometer-scale grains of native gold (Kesler et al. 2002). Hydrothermal mineralization produces hydrothermal alteration haloes that are much larger than the actual ore deposit. The classic alteration zonation includes a potassium feldspar-biotite rich core, surrounded by a muscovite/illite sericitic (phyllic) alteration zone, which is surrounded by a clay-rich (argillic) alteration zone and finally by a chlorite-epidote rich propylitic zone (Figure 25.3) (Lowell and Gilbert 1970). The ore zones generally coincide with the potassic and sericitic alteration zones.

Supergene (weathering) processes, which occur long after the initial hydrothermal mineralizing events, can lead to zones of supergene enrichment near the tops of these deposits (John et al. 2010). The supergene enrichment zones can be either oxide- or sulfide-dominated depending on the prevailing oxidation state at the site of formation, the depth of the water table, and local climate. Mined material from the oxide enrichment zone is amenable to a heap-leaching method of ore processing known as *solvent extraction and electrowinning* (SX-EW) (Jergensen 1999). The supergene enrichment zone at Pebble is poorly developed, in part due to recent glaciation, and is dominated by the secondary copper sulfide minerals covellite (CuS), digenite ($Cu_{1-x}S$), and chalcocite ($Cu_2S$) (Bouley et al. 1995). Therefore, this approach to ore-processing is not likely to occur in the Bristol Bay watershed.

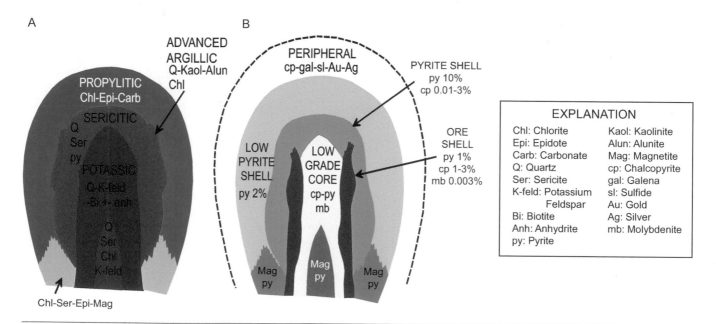

**Figure 25.3** Idealized vertical cross section through a porphyry copper deposit showing the relationship of the ore zone to various alteration types. The total depth covered by this generic cross section can exceed 1,000 m in typical deposits. A—Distribution of alteration types; B—Distribution of ore mineral assemblages. The causative intrusion corresponds to the potassic alteration zone. *Illustration credit: from John et al. (2010) and modified from Lowell and Guilbert (1970).*

Porphyry copper deposits can be divided into three subtypes on the basis of Au (g/t)/Mo (%) ratios: porphyry Cu, porphyry Cu-Mo, and porphyry Cu-Au deposits, where Cu-Au deposits have Au/Mo ratios greater than or equal to 30, Cu-Mo deposits have Au/Mo ratios less than or equal to three, and Cu deposits are all other deposits not within these bounds (Sinclair 2007, Singer et al. 2008). On the basis of these criteria, the Pebble deposit would be classified as a porphyry Cu deposit.

The United States consumed 1,730,000 metric tons of copper in 2015, of which 30% was imported, chiefly from Chile, Canada, and Peru. In the same year, the U.S. consumed 48,000 metric tons of molybdenum, and was a net exporter. In 2015, the U.S. consumed 380 metric tons of gold of which 33% was imported, primarily from Canada, Mexico, Peru, and Chile. These commodities serve myriad uses (USGS 2016). Copper is used primarily in building construction—wiring and pipes (49%), electric and electronic products (20%), vehicles (12%), consumer products (10%), and industrial machinery and equipment (9%). Molybdenum is primarily used as a steel alloy (75%). Gold is used mainly for jewelry (69%), and electrical and electronic products (9%). Silver is used for a variety of applications including industrial and medical uses, electronics, coins, and silverware, along with photography (albeit a declining application). Rhenium is principally used as an alloy in turbine engines (70%) and for petroleum refining (20%). Tellurium is primarily used as an alloy with steel, iron, and lead, but increasingly is being used in photovoltaic cells. Platinum-group metals [platinum (Pt), palladium (Pd), rhodium (Rh), ruthenium (Ru), iridium (Ir), and osmium (Os)] are predominantly used in vehicle catalytic converters, as catalysts for chemical manufacturing, in electronics, and in emerging applications to fuel cells.

The grade and tonnage of porphyry copper deposits vary widely (Singer et al. 2008). Summary statistics compiled for 256 porphyry copper deposits are presented in Figure 25.4. The Pebble deposit contains 10.8 billion metric tons of ore: grading 0.34% Cu, 0.023% Mo, and 0.31 g/t (0.01 oz/t) Au (more information can be found at *http://www.pebblepartnership.com/*). These statistics place Pebble in the upper 5th percentile for metric tons of ore, the lower 50th percentile for Cu grade, the upper 10th percentile for Mo grade, and the upper 10th percentile for Au grade for porphyry copper deposits globally. The amount of metal contained in the Pebble deposit corresponds to a 20-year supply of Cu for the United States, a 133-year supply of Mo, and a 22-year supply of Au, based on 2015 consumption statistics (Table 25.2). From the perspective of future discoveries in the watershed, it is therefore highly unlikely that new deposits will approach the size of Pebble, but instead will be considerably smaller, closer to the median size for this deposit type (Singer et al. 2005).

The Pebble deposit is the only porphyry copper deposit in the watershed with a detailed, published description of its geology (Bouley et al. 1995, Kelley et al. 2010, Lang et al. 2013). The deposit may be viewed as consisting of two contiguous ore bodies: Pebble West and Pebble East, with the buried Pebble East having the higher ore grades. Pebble West was discovered in 1989 at the surface; and delineation drilling in 2005 resulted in discovery of Pebble East beneath a 300 to 600 m thick cover of Tertiary volcanic rocks. The deposit has been explored extensively with more than 1,150 drill holes that total greater than 289,250 m (949,000 feet) (Northern Dynasty Minerals 2011).

The oldest rocks in the vicinity of the deposit are Jurassic to Cretaceous (approximately 150 million years old, or *mega-annum*, Ma) clastic sedimentary rocks (i.e., mudstone, siltstone, and sandstone), which were intruded by dominantly granitic plutons from 100 to 90 Ma. Granodiorite stocks and sills, spatially and genetically related to the Cu-Au-Mo mineralization, were intruded at 90 Ma (Kelley et al. 2010). Intrusion of these granodiorite bodies resulted in hydrothermal activity that produced the mineralization and associated alteration of the intrusions and surrounding rocks. The Pebble West deposit extends from the surface to a depth of about 500 m and encompasses roughly six square kilometers on the surface. Pebble East is covered by a wedge of post-mineralization Tertiary volcanic rocks that exceeds 600 m in thickness toward the east. Early copper mineralization was dominated by pyrite, chalcopyrite, and gold, which was overprinted by pyrite, bornite, digenite, covellite, and minor enargite, followed by quartz-molybdenite veinlets (Bouley et al. 1995, Kelley et al. 2010, Lang et al. 2013).

Geologic information on the Humble prospect (also known as Kemuk) is limited to the details found on the Millrock Resources, Inc. website (*http://www.millrockresources.com/projects/humble/*). The prospect is covered by glacio-fluvial gravels and sands from 30 m to greater than 140 m thick. The site was identified on the basis of the presence of an airborne geophysical (magnetic) anomaly and the presence of igneous rocks similar to those found at Pebble. The Humble Oil Company drilled the property in 1958 and 1959 as an iron prospect. No mention is made of Cu-Au-Mo mineralization from the 1950s drilling, and there are no recent data available. Information on the Big Chunk Super Project is limited to details on the Liberty Star Uranium and Metals Corporation website (*http://www.libertystaruranium.com/www/projects/big-chunk-super-project*).

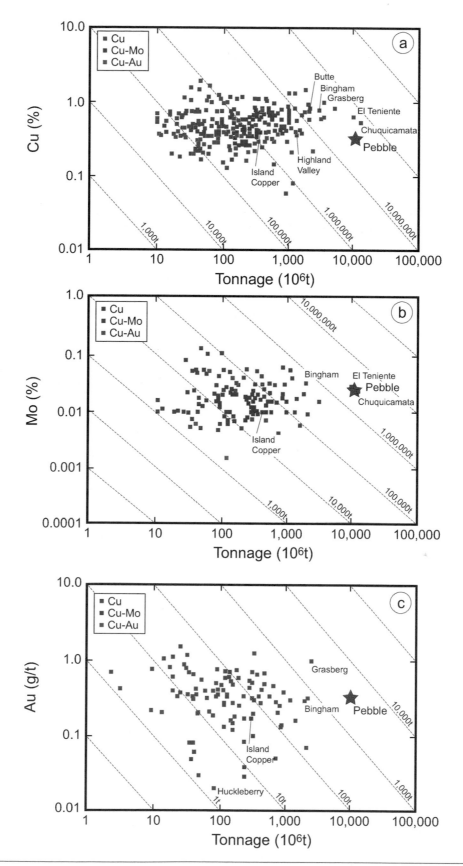

**Figure 25.4**   Grade-tonnage characteristics of the Pebble deposit compared to other porphyry-type deposits. A—Copper; B—Molybdenum; C—Gold. The Pebble deposit is shown as the yellow star. Selected, noteworthy deposits are labeled. Pebble is classified as a porphyry Cu deposit (red squares). The dashed diagonal lines represent the total contained metal. *Illustration credit: modified from Sinclair (2007).*

**Table 25.2** Annual consumption of copper, molybdenum, and gold compared to the Pebble and Shotgun deposits

| Commodity | U.S. Annual Consumption (2015)[1] | Resource | Years of Consumption |
|---|---|---|---|
| Pebble Copper (metric tons)[2] | 1,800,000 | 36,636,364 | 20 |
| Pebble Molybdenum (metric tons)[2] | 19,000 | 2,531,818 | 133 |
| Pebble Gold (metric tons)[2] | 150 | 3,337 | 22 |
| Shotgun Gold (metric tons)[3] | 150 | 30 | 0.2 |
| Donlin Creek Gold (metric tons)[4] | 150 | 790 | 5.3 |

Sources: [1] USGS (2016); [2] PLP (0.3% Cu cut-off grade), includes measured, indicated, and inferred resources (*http://www.pebblepartnership.com/*); [3] Rombach and Newberry (2001); [4] Goldfarb et al. (2004).

## Intrusion-Related Gold Deposits

Intrusion-related gold deposits are a recently recognized class that bear many similarities to the more common *orogenic* gold deposits (Figure 25.5), also known as low-sulfide quartz vein gold deposits. Orogenic gold deposits are found around the world, most commonly in areas that experienced ancient mountain building, such as the Mother Lode Belt in California, western Australia, and east-central Canada, among other places. Intrusion-related gold deposits have a more restricted distribution, typically being found at the boundary between stable cratons (continental masses) and the tectonically active margins of ancient continents (Figure 25.6). Gold is commonly produced from numerous deposit types, some of which produce gold as a primary commodity, whereas others host it as a byproduct commodity. Thus, intrusion-related gold deposits fail to have the same global dominance in supply for gold as porphyry copper deposits have for copper. The economic viability of intrusion-related gold deposits is based on their gold and silver contents. Gold grades typically range between 0.7 and 1.5 g/t (0.02 and 0.05 oz/ton), but cutoff grades can be as low as 0.4 g/t (0.01oz/ton). Ore tonnages range between 10 and 300 million metric tons (Mt) (Hart 2007). Deposits with more equivocal origins due to a

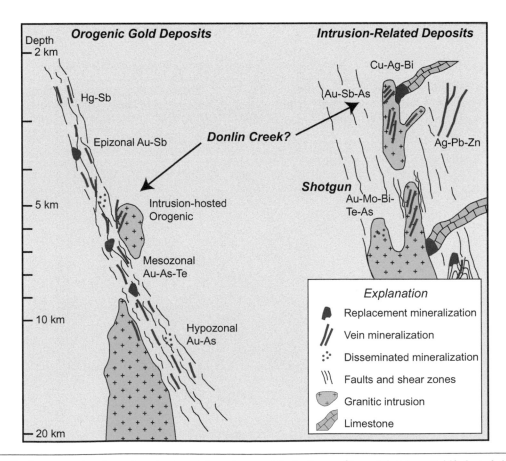

**Figure 25.5**　Idealized cross section through various gold deposit types. *Illustration credit: modified from Goldfarb et al. (2005).*

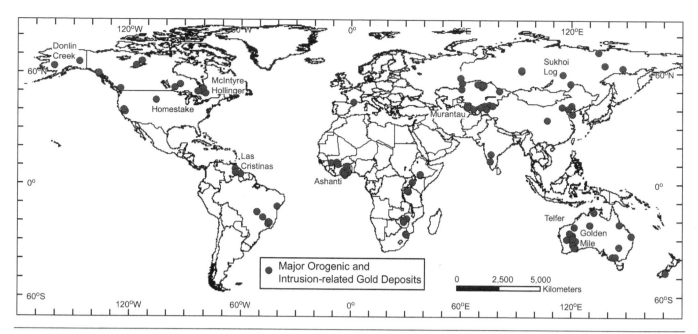

**Figure 25.6**  Map showing location of major intrusion-related and orogenic gold deposits. Deposits containing more than 1,000 metric tons of gold are labeled. Note Donlin Creek contains slightly less than 1,000 metric tons, but is the largest included because of its discussion in this chapter. *Map credit: modified from Goldfarb et al. (2005).*

less clear link to a causative intrusion, such as Donlin Creek (2.24 g/t (0.07 oz/ton), 540 Mt) and Pogo (12.5 g/t (0.4 oz/ton), 12.3 Mt), both in Alaska, have grades and tonnages that range to higher values (Goldfarb et al. 2005, Hart 2007). The Shotgun deposit in the Bristol Bay watershed contains 32.2 Mt of ore with an average gold grade of 0.93 g/t (0.03 oz/ton) (Rombach and Newberry 2001). In terms of total contained gold, Donlin Creek falls near the upper end of the spectrum, whereas Shotgun is found near the lower end (Figure 25.7). Because of the size of these deposits, mine lives can span one or more decades. The geologic characteristics of intrusion-related gold deposits have been reviewed by Thompson et al. (1999), Groves et al. (2003), Goldfarb et al. (2005), and Hart (2007).

Intrusion-related gold deposits are closely associated with, and hosted by, granitic intrusions thought to be the source of the ore-forming hydrothermal fluids (Thompson et al. 1999, Goldfarb et al. 2005, Hart 2007). Mineralization typically

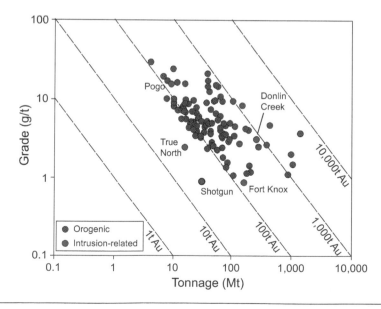

**Figure 25.7**  Grade-tonnage characteristics of gold deposits. *Illustration credit: modified from Goldfarb et al. (2005).*

and will eventually require disposal (Lindsay et al. 2015). Most tailings will end up in a tailings storage facility, in essence a large impoundment, but there are several options as to how this is done. Conventional tailings disposal involves pumping tailings as a slurry to dammed impoundments. The dams are typically earthen, or constructed of rock, or tailings. The dam levels are raised as the impoundments fill, and water is decanted and recycled in the mill as solids settle. Several strategies for raising the dam level are used, referred to as *upstream*, *downstream*, and *center-line* construction. Upstream dams have the highest failure rate (Rico et al. 2008). Abnormally high rain events are the most common causes of failure, followed by seismic liquefaction, and undetermined causes (Rico et al. 2008). More recent advances in tailings disposal include paste or thickened disposal and dry stacking, both of which remove significant amounts of water from the tailings (Edraki et al. 2014). The lower water content increases the stability of the tailings pile, which reduces its footprint on the landscape. It also exposes less water to sources of contaminants. However, paste and dry stack disposal is more expensive than conventional disposal. For some designs of underground mines, tailings can be disposed underground in the mine workings. The crushing of rock doubles by volume of the crushed material, so only about half of the tailings can be returned underground. The other half will require management on the surface.

A number of options are available for the location of tailings storage facilities because they can be pumped great distances as a slurry. For example, at the Los Pelambres porphyry copper mine near the crest of the Andes Mountains in Chile, tailings are currently pumped over 50 km to a suitable site. The ability to pump tailings great distances provides an opportunity to select disposal sites to reduce ecological or human health risks and to better address seismic risks. Subaqueous disposal, either in constructed tailings ponds, in lakes or on the seafloor have been used as well. The choice of best tailings disposal strategy for a given mine depends upon the geologic characteristics of the waste material and local (e.g., topographic, geographic, hydrologic, climatic, ecological) conditions.

The choice of tailings disposal method and the location of tailings storage facilities influences seismic risks related to tailings dam failures. Conventional disposal of tailings as a slurry, using *upstream* strategies for raising tailings dams has the greatest risk for failure by seismic liquefaction (Rico et al. 2008). Seismic risks in Alaska are related to the proximity of a site to the *subduction* zone where the Pacific Plate of dense oceanic crust is thrust northward, beneath the less dense North American Plate of continental crust along the Aleutian Trench. The Aleutian Trench is part of the Pacific *Ring of Fire*, which is the locus of significant earthquake and volcanic activity where crustal plates collide. The Bristol Bay watershed is in an area that could be subjected to strong ground shaking, with seismic risks generally decreasing going from southeast to northwest in the watershed (Wesson et al. 2007).

For deposits that are mined by open pit methods where the floor of the pit reaches below the water table, a pit lake is typically part of the post-closure landscape because back-filling the pit is usually cost prohibitive. Water quality in pit lakes from porphyry copper deposits spans a considerable range. For example, pH has been documented ranging from 2.6 to 8.2, copper concentrations from 0.027 to 89 mg/L, and sulfate from 258 to 8100 mg/L among various examples of porphyry copper deposit pit lakes (Shevenell et al. 1999, Bowell 2002, Castendyk et al. 2015a, b).

The range of values is influenced by the local geology, mining history, hydrologic setting, and climate setting (Castendyk et al. 2015a). The lowest pH values and highest concentrations in copper and sulfate are found in the Berkeley Pit in the Butte mining district in Montana. Open pit mining at Butte followed a long history of underground mining and consolidated many underground mines into a single hydrologic system (Castendyk et al. 2015b). As such, the anomalous water chemistry in the Berkeley Pit can, at least in part, be attributed to the access of water and oxygen to fractured rock granted by the extensive underground mine workings.

---

*Environmental Chemistry Concentrations*: A number of potentially toxic elements can be toxic in trace quantities, commonly in the parts per million (ppm) or parts per billion (ppb) range. Strictly speaking, when expressing concentrations as ppm or ppb, units of mass per mass should be used. For example, 1 milligram (1 1/1,000th of a gram) per 1 kilogram (1,000 grams) (1 mg/kg) is the same as 1 ppm. Likewise, 1 microgram (1/1,000,000th of a gram) per 1 kilogram (1 µg/kg) is the same as 1 ppb. One liter (L) of pure water has a mass of 1 kilogram at room temperature. However, when solids are dissolved in water, the mass will increase. This difference means that 1 liter of water will have a mass greater than 1 kilogram. At typical environmental concentrations, the difference is trivial, which means that 1 milligram per liter (mg/L) in water is approximately equal to 1 part per million. Likewise, 1 microgram per liter (µg/L) is approximately equal to 1 part per billion.

# ENVIRONMENTAL CHARACTERISTICS OF PORPHYRY COPPER DEPOSITS AND INTRUSION-RELATED GOLD DEPOSITS

## Overview

Porphyry copper deposits and intrusion-related gold deposits share some environmental characteristics, but they also have many important differences. Both deposit types can pose geochemical risks to aquatic and terrestrial ecosystems, and to human health. The risks can range from small to large and depend on a variety of factors. Factors that influence the environmental characteristics of mineral deposits range from geologic setting (both local and regional), hydrologic setting, climatic settings, and mining methods, to ore beneficiation methods. The sources of the risk can be considered in the broad categories of acid-generating potential; trace element mobility; and mining, ore processing, and waste management practices. The significance of these sources of risk will vary from deposit to deposit, but some generalizations can be made about these deposit types.

## Acid-Generating Potential

The pH of the water or the acid-generating potential of solid mine waste can be considered *master variables* for aqueous risks. Metals and other cations are more soluble at low pH than at neutral or high pH. Conversely, elements that tend to combine with oxygen when they dissolve, termed *oxyanions*, such as arsenic, selenium, chromium, vanadium, and uranium, are generally more soluble at high pH. Therefore, the acid-generating or acid-neutralizing potentials of the waste rock, tailings, and mine walls are of prime importance in identifying the potential environmental risks associated with mining and ore beneficiation. A staged approach is commonly used to evaluate the acid-generating potential of various types of mine waste (Price 2009, INAP 2011). *Static* tests are used to analyze larger numbers of samples from outcrop, exploratory drill cores, or material used for metallurgical testing. Static tests consist of a single analysis or combination of analyses to get a result. Static tests are in contrast to *kinetic* tests where a single sample is leached numerous times over the course of weeks or even years to predict long-term behavior.

An example of a static test is *acid-base accounting*, which is used to evaluate the maximum acid-generating potential of a sample. In acid-base accounting, the acid-generating potential (AP) of a sample of rock or mine waste is estimated on the basis of the sulfide (reduced)-sulfur content of a sample, which is generally equated with the pyrite content of the sample, and the acid-neutralizing potential (NP) is estimated on the basis of the carbonate content of the sample. The concept is explained by two competing reactions that describe the oxidation of pyrite ($FeS_2$) to form acid ($H^+$):

$$FeS_2 + 3.75\ O_2 + 3.5\ H_2O = Fe(OH)_3 + 2\ SO_4^{2-} + 4\ H^+$$

and the ability of the mineral calcite ($CaCO_3$) to neutralize acid:

$$CaCO_3 + 2\ H^+ = Ca^{2+} + H_2O + CO_2$$

Both AP and NP values are expressed in terms of kilograms of calcium carbonate per metric ton of sample (kg $CaCO_3$/t), so that the competing values can be directly compared to one another. If the AP and NP are equal (NP/AP = 1), the sample is theoretically *net* neutral. An NP/AP ratio of less than 1 would be considered to be *net* acid because of the excess in theoretical acid-generating potential. In theory any sample with an NP/AP greater than 1 should neutralize acid. In practice, an NP/AP greater than 2 is considered to be *not potentially acid generating* (NAG), those between 1 and 2 are considered to have uncertain acid-generating potential, and those less than 1 are considered to be potentially acid generating (PAG). Note that the requirement that non-PAG material have an NP/AP value greater than 2 represents twice the amount of alkalinity needed for net neutrality under equilibrium conditions. In practice, reaction rates are important, which is why an NP/AP greater than 2 is desirable.

The rocks associated with porphyry copper deposits, in general, tend to straddle the boundary between having net acid-generating potential and not having net acid-generating potential. This aspect is illustrated well by data in Figure 25.8A. The distribution of acid-generating and nonacid-generating material correlates with the idealized cross section of porphyry copper deposits shown in Figure 25.3B. The AP values for porphyry copper deposits dominantly reflect the distribution of pyrite. The alteration cross section (Figure 25.3A) provides the best conceptual framework for acid-neutralizing potential. The hydrothermal process that leads to the formation of argillic and sericitic tend to destroy acid-neutralizing potential. The propylitic alteration zone is the only volume of rock where carbonate minerals are expected in minor amounts, unless a porphyry copper deposit happens to have formed in the vicinity of limestones or dolomites, which is not the case with the Pebble deposit.

During mining of porphyry copper deposits, a variety of waste rocks and mill tailings with differing acid-generating and acid-neutralizing potentials may be encountered or produced. The low NP/AP ratio, largely barren pyrite shell likely represents waste rock that may need to be removed to access the ore (Figure 25.3B). The boundary between the ore shell and the pyrite shell is cryptic and typically is defined operationally on the basis of a cut-off copper grade. The intrusions that produce porphyry copper deposits can intrude any rock type. Therefore, the acid-generating potential of the country rock of undiscovered deposits cannot be predicted reliably. Likewise, geologic events following ore formation could juxtapose a variety of rock types against an ore deposit, which can have a range of acid-generating potentials. In the case of Pebble, subsequent volcanic activity after mineralization covered the eastern part of the deposit with material that has limited acid-generating potential (Kelley et al. 2010, Pebble Partnership 2011, Lang et al. 2013).

In the specific case of Pebble, the volcanic rocks overlaying Pebble East have limited amounts of pyrite and are generally classified as NAG material, which would not require special handling to mitigate acidic drainage (Pebble Partnership 2011). In fact, this material could be used for a variety of construction projects on site (e.g., road fill, tailings dam construction). In contrast, the pre-Tertiary rocks at Pebble are generally classified as PAG, with some samples having uncertain potential for generating acid and fewer classified as NAG.

The acid-base accounting characteristics of intrusion-related gold deposits are similar to those of porphyry copper deposits. No information is available for the acid-base accounting characteristics of the Shotgun deposit. However, data are available for the nearby geologically similar Donlin Creek deposit just north of the Bristol Bay watershed (U.S. Army Corps of Engineers 2015). Data are provided for ore and waste rock anticipated during mining for mill tailings samples derived from metallurgical testing. Compared to Pebble, the AP shows a more limited range for Donlin Creek, but the NP shows a slightly larger range (Figure 25.8B). The NP/AP ratios for ore and waste rock span the range from PAG to NAG. The tailings samples from Donlin Creek at have NP/AP ratios greater than 1, and half of those are above 2.

## Trace Element Mobility

At mine sites, trace elements may pose environmental risks either dissolved in water or as particulates. In water, trace elements may threaten aquatic organisms in surface water or they may pose threats to humans through drinking water supplies (surface water or groundwater). Particulates, such as mill tailings or ore concentrates, pose risks to stream, river, and lakes sediments, which mainly target aquatic organisms, or as windblown dusts that can contaminate soils. The solubility of a trace element in water ultimately determines if it is likely to be an aqueous or particulate risk, or both.

The geochemical characteristics of waste-rock dump drainage from porphyry deposits, in general, have been investigated by several studies. Day and Rees (2006) conducted a study of dump seepage associated with several operating or recently closed porphyry copper and porphyry molybdenum mines in British Columbia, many of which are located in the Fraser River watershed. These deposits fell into two groups: those that produced low pH drainage and those that did not. The lowest pH values reached a value of 2 and the highest was pH 8.5. The concentrations of sulfate and metals were negatively correlated with pH. They found that the most important dissolved constituents were sulfate (< 30,000 mg/L), Al (< 1,000 mg/L), Mn (< 100 mg/L), Cu (< 1000 mg/L), and Zn (< 100 mg/L).

In the vicinity of the proposed Pebble mine, the best insights into leachate chemistry from waste rock and mill tailings come from the humidity-cell tests conducted by the Pebble Limited Partnership (Pebble Partnership 2011). Humidity-cell tests are long-term laboratory leaching tests meant to simulate the weathering of mine waste at the Earth's surface. The procedure involves a weekly cycle of weathering and leaching with durations lasting from a minimum of 20 weeks but can last for many years (Price 2009).

The results of the Pebble Partnership (2011) humidity-cell tests are summarized in Table 25.3 and presents the mean composition of leachate from a number of individual tests divided into three groups: hydrothermally altered pre-Tertiary rocks (undifferentiated) from Pebble West, hydrothermally altered pre-Tertiary rocks (undifferentiated) from Pebble East, and mill tailings from metallurgical testing. The table also includes average data for the supernatant associated with mill tailings. The supernatant is the water decanted from the tailings during disposal. At most mines, the solution is pumped back into the mill for additional ore processing. It may represent an important part of the water budget of an active mine, but has little relevance after decommissioning of a tailings storage facility.

The results from Pebble include an extensive list of parameters considered, but only a subset that is relevant to the current discussion is highlighted in the table (Pebble Partnership 2011). The present discussion focuses on pH, sulfate, Cu, Mo, As, and Zn because of the geologic context, their concentrations, and likely environmental concern. The pH of

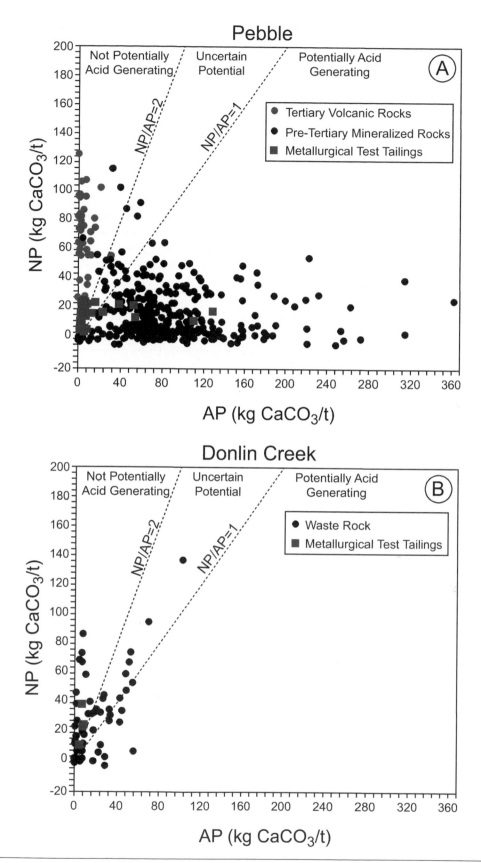

**Figure 25.8** Plot of acid-neutralizing potential (NP) and acid-generating potential (AP) for rocks in the vicinity of: A—the Pebble porphyry copper deposit. Data credit: Pebble Limited Partnership (2011); and B—the Donlin Creek gold deposit. Data credit: U.S. Army Corps of Engineers (2015).

**Table 25.3** Summary of average geochemical results from mean humidity-cell tests on waste-rock and mill tailings samples, and the supernatant from metallurgical testing from the Pebble Partnership (2011) compared to environmental guidelines

| Parameter | Units | Pebble West Pre-Tertiary Waste Rock | Pebble East Pre-Tertiary Waste Rock | Tailings Humidity Cell | Supernatant | Environmental Guidelines | |
|---|---|---|---|---|---|---|---|
| | | | | | | Drinking Water[1] | Acute Aquatic Health[2] |
| pH | S.U. | 6.6 | 4.8 | 7.8 | 7.9 | | 6.5–9 |
| Alkalinity | mg/L CaCO$_3$ | 18.5 | 9.9 | 59.7 | 74.8 | | |
| Hardness | mg/L CaCO$_3$ | 59.2 | 21.9 | 66.8 | 322.8 | | |
| SO$_4$ | mg/L | 60.8 | 51.9 | 17.4 | 318.7 | 250 | |
| Al | mg/L | 0.32 | 0.38 | 0.02 | 0.07 | 200 | 750 |
| As | µg/L | 1.5 | 8.0 | 5.5 | 17.2 | 10 | 340 |
| Ca | mg/L | 12.7 | 6.3 | 22.6 | 116.0 | | |
| Cd | µg/L | 0.4 | 3.2 | 0.01 | < 0.01 | 5 | 1.8 |
| Cu | µg/L | 1,599 | 1,416 | 5.3 | 7.8 | 1,300 | 13 |
| Fe | µg/L | 1,671 | 10,195 | 30.0 | 20.0 | 300 | |
| K | mg/L | 1.41 | 0.96 | 4.02 | 25.95 | | |
| Mg | mg/L | 6.69 | 1.50 | 2.55 | 8.00 | | |
| Mn | µg/L | 729 | 339 | 44.1 | 71.9 | 50 | |
| Mo | µg/L | 1.8 | 4.3 | 33.5 | 69.7 | | |
| Na | mg/L | 2.05 | 2.07 | 2.10 | 43.78 | | |
| Pb | µg/L | 0.2 | 0.4 | 0.06 | 0.23 | 15 | 65 |
| Zn | µg/L | 55.6 | 478.6 | 3.2 | 04.3 | 5,000 | 120 |

[1] USEPA (2009a)

[2] USEPA (2009b)

a solution controls the solubility of most elements. Sulfate, in part, reflects the extent of acid drainage formation. Copper is the most likely inorganic ecologic stressor expected at the site, especially for aquatic organisms, because of its abundance in the deposit, and the inability of ore-processing to be 100% effective in recovering it. Zinc commonly occurs in base-metal hydrothermal systems, but typically not in economic concentrations in porphyry copper deposits. Arsenic and molybdenum are oxyanion species, which behave differently from cations; arsenic is a potentially significant stressor, especially with respect to drinking-water contamination; and molybdenum is an important ore constituent with less potential to be an environmental stressor.

The leachates from the pre-Tertiary rocks are characterized by neutral to acidic pH values. As expected from the role of pyrite oxidation in acid generation, the samples that generated the lowest pH values had the higher sulfate concentrations and lower alkalinity values. For example, the mean pH for humidity-cell leachates for Pebble East was 4.8 ± 1.9 compared to 6.6 ± 1.7 for Pebble West, presumably reflecting the higher grade and pyrite content of Pebble East and the corresponding greater ability to generate acid. The pH of the samples correlated negatively with the alkalinity of the leachates. Copper concentrations generally correlate with sulfate concentrations and low pH, as would be expected from the higher solubility of metals with acidic pH conditions. The mean concentrations of copper in humidity-cell leachates from both Pebble West and Pebble East were high compared to other metals and exceeded 1 mg/L. The mean zinc concentration reached 0.5 mg/L. In contrast, the highest mean molybdenum concentration was less than 0.005 mg/L and the highest mean arsenic concentration was 0.008 mg/L. At an operating mine, the drainage from waste-rock piles will be a mixture of direct leachates from the waste rock and local ambient surface water and precipitation. The relative proportion of these sources will depend upon local climatic conditions, the natural topography, alterations to the natural topography made during mine construction, and engineering controls put in place during mine construction to manage surface water, which makes prediction of drainage compositions difficult.

A greater number of environmental concerns are associated with tailings due to their finer grain size compared to waste rock. Like waste rock, tailings can weather and the associated leachate can contaminate surface water and groundwater (Stollenwerk 1994, Brown et al. 1998, Khorasanipour et al. 2011). Because of the sand- to silt-size grains, tailings are prone to be transported by waters, especially in the case of tailings dam failure, and by wind. Thus, they present additional potential risks to aquatic organisms through sediment contamination. Insights into aquatic concerns associated with tailings can be found in case studies from mines. The geochemical characteristics of tailings seepage have been investigated by several studies. Smuda et al. (2008) investigated the geochemical environment associated with tailings at the El Teniente porphyry copper deposit, Chile. They found a range of values for various water-quality parameters associated with the tailings pond. These parameters included pH (7.2–10.2), sulfate (1,556–5,574 mg/L), Fe (1.44–8.59 mg/L), Al (below detection–0.886 mg/L), Mn (0.001–20.1 mg/L), Ni (0.008–0.393 mg/L), Cu (0.003–0.250 mg/L), Zn (0.007–130 mg/L), Mo (0.033–13.2 mg/L), and As (below detection–0.345 mg/L). Khorasanipour et al. (2011) studied the geochemical environment associated with tailings at the Sarcheshmeh mine, Iran. They too found a range of values for water-quality parameters such as pH (3.6–7.9), sulfate (1,348–4,479 mg/L), Fe (< 0.01–19.3 mg/L), Al (< 0.5–154 mg/L), Mn (5.6–73.7 mg/L), Ni (0.088–1.74 mg/L), Cu (< 0.002–149.9 mg/L), Zn (0.094–20.3 mg/L), Mo (0.027–2.9 mg/L), and As (< 0.005–0.04 mg/L).

As with the waste rock at Pebble, the best insights into the potential behavior of mill tailings come from the humidity-cell tests being conducted by the Pebble Limited Partnership (Pebble Partnership 2011). Humidity-cell tests represent one of the best, albeit imperfect, predictors of long-term weathering of tailings in an aerobic environment (Price 2009). The test conditions are most representative of unsaturated tailings exposed at the surface of a pile, which may not necessarily be representative of conditions chosen if Pebble ever becomes a mine. The geochemical environment found at depth in the saturated zone is typically quite different (Blowes et al. 2003). As for the waste-rock samples, the following discussion focuses on pH, sulfate, copper, zinc, molybdenum, and arsenic.

The mean humidity-cell results for tailings samples had pH values ranging between 7 and 8.5 in experiments lasting up to five years (Table 25.3). As with the waste-rock samples, individual humidity-cell tests for tailings can show a range of leachate concentrations that vary over the course of the experiment. In general, the concentrations of dissolved constituents are most erratic and highest in the initial flush covering the first few one-week cycles in humidity-cell tests; several weeks after the start of the experiments, the concentrations of dissolved constituents tends to be less erratic. The average release values used in Table 25.3 obscure this variability. Sulfate concentrations are generally below 40 mg/L after the initial flush of soluble sulfate salts. The mean sulfate release concentration was $17.4 \pm 8.0$ mg/L. The mean Cu ($5.3 \pm 2.2$ µg/L) and Zn ($3.2 \pm 1.7$ µg/L) concentrations were less than those from the waste-rock samples, whereas the Mo ($33.5 \pm 23.7$ µg/L) and As ($5.5 \pm 8.4$ µg/L) concentrations were higher (Table 25.3). The supernatant solution has the highest copper concentration and water hardness of the three end members.

Gold deposits may or may not have acid drainage depending upon the pyrite content of the waste material (either waste rock or mill tailings) and the carbonate mineral content. In the case of Shotgun, sulfide minerals typically constitute less than 1% of the ore and carbonate minerals are also present in minor amounts; in addition to pyrite, arsenopyrite is notably present at Shotgun (Rombach and Newberry 2001). At the nearby Donlin Creek deposit, sulfide minerals are slightly more abundant and the carbonate minerals dolomite and ankerite are similarly abundant. In addition to pyrite, arsenopyrite, stibnite, and cinnabar are found at Donlin Creek in trace amounts (Goldfarb et al. 2004). Drainage from abandoned mines that developed orogenic gold deposits generally have pH values that are near neutral, but range from slightly acidic to slightly alkaline (pH 4.5 to 8.5) (Seal and Hammarstrom 2003). Particularly at the lower pH values, elevated concentrations of Cu (< 2 mg/L), Zn (< 3 mg/L), and Cd (< 0.1 mg/L) can be commonly expected in mine drainage from waste rock or tailings piles at abandoned mines. At all pH values, As (< 0.2 mg/L) and Sb (< 0.02 mg/L) can reach elevated levels at abandoned mines (Seal and Hammarstrom 2003, Craw et al. 2015). For deposits containing the mercury mineral cinnabar, mercury can leach into mine drainage. However, a more important risk related to mercury is during cyanide leaching of gold. Mercury can complex with cyanide and end up in the doré (unrefined gold-silver alloy) prior to smelting. Retorting prior to smelting can effectively capture the mercury, thus keeping it from being emitted to the atmosphere (Marsden and House 2006).

Information about trace element mobility related to the Shotgun deposit is lacking. Instead, information about trace element mobility from the environmental impact statement (U.S. Army Corps of Engineers 2015) for the nearby Donlin Creek deposit in the Kuskokwin River watershed to the north can be used to get general insights about trace element mobility for this deposit type. For most site waters estimated from humidity-cell tests for waste rock at Donlin Creek, the pH values were near neutral (5.8–6.5), alkalinity concentrations were moderate to high (11–530 mg/L CaCO₃), and

Decker, J., S.C. Bergman, R.B. Blodgett, S.E. Box, T.K. Bundtzen, J.G. Clough, W.L. Coonard, W.G. Gilbert, M.L. Miller, J.M. Murphy, M.S. Robinson, and W.K. Wallace. 1994. Geology of Southwestern Alaska., In: G. Plafker and H.C. Berg (Eds.), The Geology of Alaska: Boulder, Colorado, Geological Society of America, The Geology of North America, v. G-1, p. 285–310.

Detterman, R.L. and B.L. Reed. 1980. Stratigraphy, Structure, and Economic Geology of the Iliamna Quadrangle, Alaska: U.S. Geological Survey Bulletin 1368-B, 86 p.

Edraki, M., T. Baumgartl, E. Manlapig, D. Bradshaw, D.M. Franks, and C.J. Moran. 2014. Designing Mine Tailings for Better Environmental, Social and Economic Outcomes: a Review of Alternative Approaches; Journal of Cleaner Production, v. 84, p. 411–429.

Fuerstenau, M.C., G. Jameson, and R.-H. Yoon (Eds.), 2007. Froth Flotation: A Century of Innovation: Society of Mining, Metallurgy, and Exploration, Littleton, CO, 891 p.

Goldfarb, R.J., R. Ayuso, M.L. Miller, S.W. Ebert, E.E. Marsh, S.A. Petsel, L.D. Miller, L.D. Bradley, C. Johnson, and W. McClelland. 2004. The Late Cretaceous Donlin Creek Gold Deposit, Southwestern Alaska: Controls on Epizonal Ore Formation: Economic Geology, v. 99, p. 643–671.

Goldfarb, R.J., T. Baker, B. Dubé, D.I. Groves, C.J.R. Hart, and P. Gosselin. 2005. Distribution, Character, and Genesis of Gold Deposits in Metamorphic Terranes: Economic Geology 100th Anniversary Volume, p. 407–450.

Groves, D.I., R.J. Goldfarb, F. Robert, C.J.R. and Hart. 2003. Gold Deposits in Metamorphic Belts: Overview of Current Understanding, Outstanding Problems, Future Research, and Exploration Significance: Economic Geology, v. 98, p. 1–29.

Hart, C.J.R. 2007. Reduced Intrusion-Related Gold Systems. In: W.D. Goodfellow (Ed.), Mineral Deposits of Canada: A Synthesis of Major Deposit Types, District Metallogeny, the Evolution of Geological Provinces, and Exploration Methods: Geological Association of Canada, Mineral Deposits Division, Special Publication No. 5, p. 95–112.

International Network for Acid Prevention (INAP), 2011, Global Acid Rock Drainage Guide (*http://www.gardguide.com*, accessed 11/10/2011).

Jergensen, G.V., II (Ed.), 1999, Copper Leaching, Solvent Extraction, and Electrowinning Technology: Society of Mining, Metallurgy, and Exploration, Littleton, CO, 296 p.

John, D.A., R.A. Ayuso, M.D. Barton, R.J. Blakely, R.J. Bodnar, J.H. Dilles, F. Gray, F.T. Graybeal, J.C. Mars, D.K. McPhee, R.R. Seal, R.D. Taylor, and P.G. Vikre. 2010. Porphyry Copper Deposit Model, Chap. B *of* Mineral deposit Models for Resource Assessment: U.S. Geological Survey Scientific Investigations Report 2010–5070–B, 169 p.

Kelley, K.D., J. Lang, and R.G. Eppinger. 2010. Exploration Geochemistry at the Giant Pebble Porphyry Cu-Au-Mo Deposit, Alaska: SEG Newsletter, no. 80, p. 1–23.

Kesler, S.E., S.L. Chryssoulis, and G. Simon. 2002. Gold in Porphyry Copper Deposits: Its Abundance and Fate: Ore Geology Reviews, v. 21, p. 103–124.

Khorasanipour, M., M.H. Tangestani, R. Naseh, and H. Hajmohammadi. 2011. Hydrochemistry, Mineralogy and Chemical Fractionation of Mine and Processing Wastes Associated with Porphyry Copper Mines: a Case Study of the Sarcheshmeh Mine, SE Iran: Applied Geochemistry, v. 26, p. 714–730.

Lang, J.R., M.J. Gregory, C.M. Rebagliati, J.G. Payne, J.L. Oliver, and K. Roberts. 2013. Geology and Magmatic-Hydrothermal Evolution of the Giant Pebble Porphyry Copper-Gold-Molybdenum Deposit, Southwest Alaska: Economic Geology, v. 108, p. 437–462.

Laubscher, D.H. 1994. Cave Mining—the State of the Art: Journal of the South African Institute of Mining and Metallurgy, v. 94, p. 279–293.

Lindsay, M.B.J., M.C. Moncur, J.G. Bain, J.L. Jambor, C.J. Ptacek, and D.W. Blowes. 2015. Geochemical and Mineralogical Aspects of Sulfide Mine Tailings: Applied Geochemistry, v. 57, p. 157–177.

Lowell, J.D., and J.M. Guilbert. 1970. Lateral and Vertical Alteration-Mineralization Zoning in Porphyry Ore Deposits: Economic Geology, v. 65, p. 373–408.

Marsden, J.O., and C.I. House. 2006. The Chemistry of Gold Extraction, 2nd Edition: Society of Mining, Metallurgy, and Exploration, Littleton, CO, 651 p.

Newberry, R.J., G.L. Allegro, S.E. Cutler, J.H. Hagen-Levelle, D.D. Adams, L.C. Nicholson, T.B. Weglarz, A.A. Bakke, K.H. Clautice, G.A. Coulter, M.J. Ford, G.L. Myers, and D.J. Szumigala. 1997. Skarn Deposits of Alaska. In: R.J. Goldfarb and L.D. Miller. (Eds.), *Mineral Deposits of Alaska*. Economic Geology Monograph Series Volume 9, p. 355–395.

Nokleberg, W.J., T.K. Bundtzen, H.C. Berg, D.A. Brew, D. Grybeck, M.S. Robinson, T.E. Smith, and W. Yeend. 1987. Significant Metalliferous Lode Deposits and Placer Districts of Alaska: U.S. Geological Survey Bulletin 1786, 104 p.

Nokleberg, W.J., G. Plafker, and F.H. Wilson. 1994. Geology of South-Central Alaska., In: G. Plafker, and H.C. Berg (Eds.), *The Geology of Alaska*: Boulder, Colorado, Geological Society of America, The Geology of North America, v. G-1, p. 311–366.

Northern Dynasty Minerals, 2011, Preliminary Assessment of the Pebble Project: unpublished report, Wardrop, February 17, 2011, 579 p.

Page, R.A., N.N. Biswas, J.C. Lahr, and H. Pulpan. 1991. Seismicity of Continental Alaska. In: D.B. Slemmons, E.R. Engdahl, M.D. Zoback, and D.D. Blackwell (Eds.), *Neotectonics of North America*: Boulder, Colorado, Geological Society of America, Decade Map, v. l, p. 47–68.

Pebble Partnership, 2011, Pebble Project Environmental Baseline Document 2004 through 2008, unpublished report, 30, 378 p.

Porter, K.E., and D.I. Bleiwas. 2003. Physical Aspects of Waste Storage from a Hypothetical Open Pit Porphyry Copper Operation: U.S. Geological Survey Open-File Report 03-143, 63 p.

Price, W.A., 2009, Prediction Manual for Drainage Chemistry from Sulphidic Geologic Materials: MEND Report 1.20.1, 579 p.

Rico, M., G. Benito, A.R. Salgueeiro, A. Díez-Herrero, and H.G. Pereira. 2008. Reported Tailings Dam Failures: A Review of the European Incidents in the Worldwide Context: Journal of Hazardous Materials, v. 152, p. 846–852.

Rombach, C.S., and R.J. Newberry. 2001. Shotgun Deposit: Granite Porphyry-Hosted Gold-Arsenic Mineralization in Southwestern Alaska, USA: Mineralium Deposita, v. 36, p. 607–621.

Schmidt, J.M., T.D. Light, L.J. Drew, F.H. Wilson, M.L. Miller, and R.W. Saltus. 2007. Undiscovered Locatable Mineral Resources in the Bay Resource Management Plan Area, Southwestern Alaska: A Probabilistic Assessment: U.S. Geological Survey Scientific Investigations Report 2007-5039, 50 p.

Seal, R.R., II, and J.M. Hammarstrom. 2003. Geoenvironmental Models of Mineral Deposits: Examples from Massive Sulfide and Gold Deposits: Environmental Aspects of Mine Wastes, J.L. Jambor, D.W. Blowes, and A.I.M. Ritchie (Eds.), Mineralogical Association of Canada Short Series, v. 31, p. 11–50.

Seedorff, E., J.H. Dilles, J.M. Proffett, Jr., M.T. Einaudi, L. Zurcher, W.J.A. Stavast, D.A. Johnson, and M.D. Barton. 2005. Porphyry Deposits: Characteristics and Origin of Hypogene Features: Economic Geology 100th Anniversary Volume, p. 251–298.

Shevenell, L.A., K.A. Connors, and C.D. Henry. 1999. Controls on Pit Lake Water Quality at Sixteen Open-Pit Mines in Nevada. Appl. Geochem., v. 14, p. 669–687.

Sinclair, W.D., 2007, Porphyry Deposits. In: W.D. Goodfellow (Ed.), *Mineral Deposits of Canada*: A Synthesis of Major Deposit-Types, District Metallogeny, the Evolution of Geological Provinces, and Exploration Methods: Geological Association of Canada, Mineral Deposits Division, Special Publication No. 5, p. 223–243.

Singer, D.A., V.I. Berger, W.D. Menzie, and B.R. Berger. 2005. Porphyry Copper Deposit Density: Economic Geology, v. 100, p. 491–514.

Singer, D.A., V.I. Berger, and B.C. Moring. 2008. Porphyry Copper Deposits of the World: Database and Grade and Tonnage Models, 2008: U.S. Geological Survey Open-File Report 2008-1155, 45 p.

Singer, D.A. and W.D. Menzie. 2010. *Quantitative Mineral Resource Assessments—An Integrated Approach*: New York, Oxford University Press, 219 p.

Smuda, J., B. Dold, J.E. Spangenberg, and H.-R. Pfeifer. 2008. Geochemistry and Stable Isotope Composition of Fresh Alkaline Porphyry Copper Tailings; Implications on Sources and Mobility of Elements during Transport and Early Stages of Deposition: Chemical Geology, v. 256, p. 62–76.

Stollenwerk, K.G., 1994. Geochemical Interactions between Constituents in Acidic Groundwater and Alluvium in an Aquifer near Globe, Arizona. Applied Geochemistry, v. 9, p. 353–369.

Tarkian, M., and B. Stribrny. 1999. Platinum-Groups Elements in Porphyry Copper Deposits: a Reconnaissance Study: Mineralogy and Petrology, v. 65, p. 161–183.

Thompson, J.F.H., R.H. Sillitoe, T. Baker, J.R. Lang, and J.K. Mortensen. 1999. Intrusion-Related Gold Deposits Associated with Tungsten-Tin Provinces: Mineralium Deposita, v. 34, p. 323–334.

U.S. Army Corps of Engineers, 2015, Donlin Gold Project: Draft Environmental Impact Statement, November 2015.

USEPA, 2003, EPA and Hardrock Mining—A source book for industry in the Northwest and Alaska: Seattle, Wash., U.S. Environmental Protection Agency, 57 p., accessed October 27, 2016, at http://yosemite.epa.gov/R10/WATER.NSF/Sole+Source+Aquifers/hardrockmining.

USEPA, 2009a, National Primary Drinking Water Regulations: EPA 816-F-09-0004, accessed November 16, 2016, at: https://www.epa.gov/ground-water-and-drinking-water/table-regulated-drinking-water-contaminants.

USEPA, 2009b, National Recommended Water Quality Criteria, accessed November 16, 2016, at https://www.epa.gov/wqc/national-recommended-water-quality-criteria-aquatic-life-criteria-table.

USEPA, 2014, An Assessment of Potential Mining Impacts on Salmon Ecosystems of Bristol Bay, Alaska: EPA 910-R-14-001A, January 2014, 1402 p.; accessed November 16, 2016, at https://www.epa.gov/bristolbay.

USGS, 2016, Mineral Commodity Summaries 2016, 201 p. (http://minerals.usgs.gov/minerals/pubs/mcs/2016/mcs2016.pdf).

Wallace, W.K., C.L. Hanks, and J.F. Rogers. 1989. The Southern Kahiltna Terrane: Implications for the tectonic Evolution of Southwestern Alaska: Geological Society of America Bulletin, v. 101, p. 1389–1407.

Wesson, R.L., O.S. Boyd, C.S. Mueller, C.G. Bufe, A.D. Frankel, and M.D. Petersen. 2007. Revision of Time-Independent Probabilistic Seismic Hazard Maps for Alaska: U.S. Geological Survey Open-File Report 2007-1043, 33 p.

Wilson, F.H., C.P. Hults, C.G. Mull, and S.M. Karl. 2015. Geologic map of Alaska: U.S. Geological Survey Scientific Investigations Map 3340, pamphlet 196 p., 2 sheets, scale 1:1,584,000, http://dx.doi.org/10.3133/sim3340.

# 26

# RENEWABLE ENERGY RESOURCES

**Tom Marsik, University of Alaska Fairbanks Bristol Bay Campus**

## INTRODUCTION

Renewable energy resources are natural resources that can be replenished on a human timescale. Bristol Bay Alaska is a region full of natural resources, some of which can be used to supply renewable energy. This chapter deals specifically with energy that can be harnessed from river, wind, solar, biomass, geothermal, and tidal resources. While other renewable energy resources exist, such as ocean thermal gradients, these are not included in this chapter because they have minimal relevance to Bristol Bay.

While the energy resources discussed in this chapter are renewable, the devices needed for the conversion of these resources into a useful form are not; for example wind turbines used in the utilization of wind energy require copper and other finite resources. Therefore, energy efficiency, which can significantly reduce the need for renewable or nonrenewable resources, is also included in this chapter, despite the fact that energy efficiency itself is not exactly a natural resource.

Historically, renewable energy in Bristol Bay has been used on a small scale, such as wood stoves for space heating, small wind turbines to generate electricity, and sail boats for transportation. While renewable energy resources are considered abundant in Bristol Bay, the feasibility of harvesting these resources on a larger scale is a different issue. Bristol Bay, together with some other areas of rural Alaska, serve as a globally significant test fields for new technologies and approaches. Because of long distances between communities, these communities are typically not electrically connected and each of them has their own small grid, also called a micro-grid. Each micro-grid is served by at least one power plant that is typically diesel fueled. The amount of diesel burned can be reduced by integrating renewable energy sources into the micro-grid and supplying a portion of the electrical energy from these renewable sources. The electrical micro-grids with integrated renewable energy are often viewed as a possible model for less-developed parts of the world.

Renewable energy resources in Bristol Bay are diverse and vary greatly. Therefore, no universal approach to harvest this energy exists, and specific options need to be evaluated for individual situations. Some very basic evaluations are provided in this chapter. For the purposes of economic feasibility, costs of electricity for communities within the Bristol Bay area for the fiscal year 2014 are used (see Table 26.1). The second column in Table 26.1 shows the cost to the customer within the communities of Bristol Bay, in terms of dollars per kilowatt-hour ($/kWh). This cost includes not only the fuel cost, but also the fixed costs (e.g., maintenance of power lines, labor, insurance, etc.) distributed among the kWh sold. Renewable energy and energy efficiency measures only reduce fuel costs, not the fixed costs; this means that the direct societal benefits are represented by the fuel cost, not the total cost to the customer. Therefore, the fuel cost in $/kWh, listed in the last column of Table 26.1, is the cost that will be used for the economic analysis in this chapter.

While quantifying savings in external costs would be beyond the scope of this chapter, it is important to keep in mind that there are significant societal benefits associated with reduced use of nonrenewable resources. External costs refer to the costs that are not paid directly by the energy user, but are paid by the society as a whole. Examples of external costs include clean-ups of oil spills, treatments of respiratory diseases associated with pollution from burning fossil fuels, relocating communities due to climate change impacts, and many others. Energy efficiency and renewable energy measures significantly reduce these external costs.

The following sections describe the individual renewable energy resources (including energy efficiency) from the perspectives of resource magnitude, resource stability and predictability, technologies for the utilization of the resource and related economics, and other aspects. They also show examples of related projects in Bristol Bay. The purpose of this chapter is to give an overall view of the renewable energy resources within Bristol Bay—not to provide final answers

The previous example portrays only one out of many energy efficiency improvements being performed in Bristol Bay. Besides energy efficiency projects funded by private sources (and thus rarely reported), the following are programs with external funding (through the Alaska Housing Finance Corporation and Alaska Energy Authority) that have been utilized in Bristol Bay: Weatherization, Home Energy Rebate, Commercial Building Energy Audit, Village Energy Efficiency, and the Energy Efficiency and Conservation Block Grant Program (The Energy Efficiency Partnership 2015). Additional state programs that have not yet been used in the region exist as well: Alaska Energy Efficiency Revolving Loan and the Alternative Energy and Conservation Loan Program.

Of global significance is the fact that Bristol Bay has what is officially the tightest residential building in the world, as recognized by the World Record Academy. In addition to air tightness, the building has other energy efficiency features, and saves approximately 95% of heat energy compared to an average house in rural Alaska (Marsik 2012). The building is located in Dillingham.

# RIVER HYDRO

## Magnitude of Resource

Bristol Bay has lakes and rivers that could (if magnitude of resource was the only factor) support many large capacity hydroelectric projects (Robert W. Retherford Associates 1979). A map of investigated hydroelectric projects is shown in Figure 26.1.

## Stability and Predictability

The stability and predictability of hydro systems depends on whether it is a storage, run-of-river, or in-stream (hydrokinetic) alternative. A storage system has a dam (or in some situations a natural feature that serves as a dam) that allows substantial storage of water, the outflow of which can be well controlled, based on the electricity demand. Run-of-river systems have a dam with limited storage or no dam at all (instead, a natural feature, such as a water fall, can be used), and the electricity generation pattern (with respect to the maximum generation capacity) is mainly determined by the natural water flow pattern. Hydrokinetic systems have no dam and consist of one or more in-stream turbines placed in the natural water current. The term *hydrokinetic* refers to the fact that these systems harvest the kinetic energy of the flowing water, unlike the previous two systems, which harvest the potential energy of the water on the higher side of the dam (or natural feature).

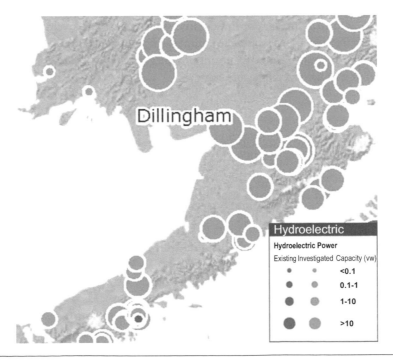

**Figure 26.1**   Map of investigated hydroelectric projects. *Source credit: the Renewable Energy Atlas of Alaska (AEA and REAP 2013).*

The storage system alternative provides a very stable energy source and can provide a good control on an annual scale, depending on the storage volume. The run-of-river alternative, due to limited storage capacity, is stable and controllable on a short-term scale, but not on an annual scale (spring water runoff cannot be stored for later use). The hydrokinetic alternative is not controllable and is the least stable option, but due to the nature of river flow (where changes are slow), the stability is still significantly better than for wind or solar. River hydrokinetic is also more stable than tidal hydrokinetic, but the fluctuations of rivers are less predictable than tides. With respect to predicting an annual energy production, some uncertainty exists for all three alternatives—storage, run-of-river, and hydrokinetic.

## Technologies and Economics

The economics of all three alternatives—storage, run-of-river, and hydrokinetic—strongly depend on the local conditions and the proximity of the resource to the point of use or existing power lines. While the economics of storage and run-of-river systems can be evaluated if the local conditions are known, the economics of hydrokinetic systems cannot be evaluated because that technology does not exist yet in a commercial, reliable form for Alaskan conditions (see note regarding this emerging technology in the following section).

## Other Aspects

Hydrokinetic systems under current consideration have essentially no impact on the water flow patterns. Run-of-river systems result in slightly modified streams where water flows are similar to the baseline hydrological conditions. Storage systems result in stream flows that are significantly altered from the baseline hydrological conditions. Therefore, storage systems typically have the highest impacts on downstream aquatic ecosystems. While the interactions between in-stream turbines and aquatic life have not been completely evaluated yet, the dams associated with storage and run-of-river systems can have a significant impact on aquatic life.

Another important factor to mention is the fact that storage and run-of-river hydro are very mature technologies compared to some other renewable energy technologies, such as wind. Therefore, storage and run-of-river systems have a higher potential for reliable operation. The hydrokinetic device for Alaskan conditions, on the other hand, is an emerging technology in its infancy (significantly less developed than wind for example, let alone storage or run-of-river systems). Debris management (including ice) is one of the unresolved issues that researchers are working on (ACEP 2015).

There are many other factors that need to be considered when evaluating river hydro as an alternative for electricity generation. Important factors to consider include: land ownership and public use issues, permitting from the Federal Energy Regulatory Commission (FERC) and other agencies, aesthetics, accumulation of silt behind dams (resulting in a reduced lifetime of the system), risk of dam failure, acceptance of the project by local culture, and other factors.

## Examples of Use

The only operating hydro project in Bristol Bay is the Tazimina River hydroelectric plant. It is a run-of-river system that uses a natural feature (Tazimina Falls) as opposed to a man-made dam. A portion of the water from above the falls is diverted into an underground powerhouse and then released back into the river near the base of the falls. This 824 kW hydroelectric plant serves the communities of Iliamna, Newhalen, and Nondalton (Wikimapia 2012). In fiscal year 2014, the Tazimina hydro plant generated 4,006,061 kWh, which means about 98% of the Iliamna, Newhalen, and Nondalton electric needs came from renewable sources (AEA 2015b).

At Chignik Lagoon, the 177 kW run-of-river Packer's Creek hydroelectric project is under construction. It is expected to start operating in August 2015 (AEA 2015a). Several other hydro projects are in preconstruction stages. This includes a 150 kW run-of-river Knutson Creek project at Pedro Bay, and the restoration and upgrade of an antiquated 60 kW hydroelectric system on Indian Creek at Chignik (SWAMC et al. 2013).

Two experimental hydrokinetic systems were temporarily installed during the summer of 2014 in the Kvichak River near the outlet from Lake Iliamna at the community of Igiugig. The first one is a RivGen® 25 kW unit developed by the Ocean Renewable Power Company (ORPC). The system successfully delivered electricity to an on-shore station and environmental monitoring of the device showed that the system had no known negative impacts on fish and other aquatic life (ORPC 2014). The device was removed from the Kvichak River in September 2014, to be modified for a planned 2015 redeployment (Gray Stassel Engineering, Inc. 2015). The second system tested during the summer of 2014 is a 5 kW Cyclo-Turbine™ unit developed by Boeschma Research, Inc. (BRI). The system successfully delivered energy into Igiugig's electrical grid before a mechanical failure occurred and the Cyclo-Turbine™ was removed from the river (Gray Stassel Engineering, Inc. 2015).

# WIND

## Magnitude of Resource

Figure 26.2 maps the wind resource for the Bristol Bay area and illustrates that the wind quantity varies widely. As seen in the figure, there are areas with a Class 7 (or *superb*) wind resource; however, the issue is that these areas are often very distant from the potential point of use and existing electrical grids or on top of steep terrain that creates logistical and cost barriers for successful projects. Overall, wind is a resource worthy of attention in the Bristol Bay region and was once used as one of the primary sources of electricity (before inexpensive diesel). Long-term wind studies using meteorological towers with anemometers have been done for several locations in Bristol Bay (AEA 2014b).

## Stability and Predictability

Wind is an unstable energy resource. Even though the wind speed fluctuations on a semi-long-term scale (hours to days) can be partially predicted using weather forecasting methods, the fluctuations on a short-term scale (seconds to minutes) are highly unpredictable. This variability and low predictability of wind pose challenges for its use in the generation of electricity. One possible solution is the integration of wind with more controllable power sources, such as diesel generators or hydro systems. With such installations, though, there can be issues with power quality, especially with high penetration systems, where a large portion of the total energy is supplied from wind. These power quality issues can be solved, or at least reduced; for example, by using short-term energy storage devices, such as flywheels or super-capacitors. Other solutions include sophisticated control systems, secondary loads (e.g., used for space heating during periods of excess wind), and wind curtailment (e.g., changing the pitch of blades to intentionally generate less power). Combinations of these strategies are often used, as opposed to a single strategy.

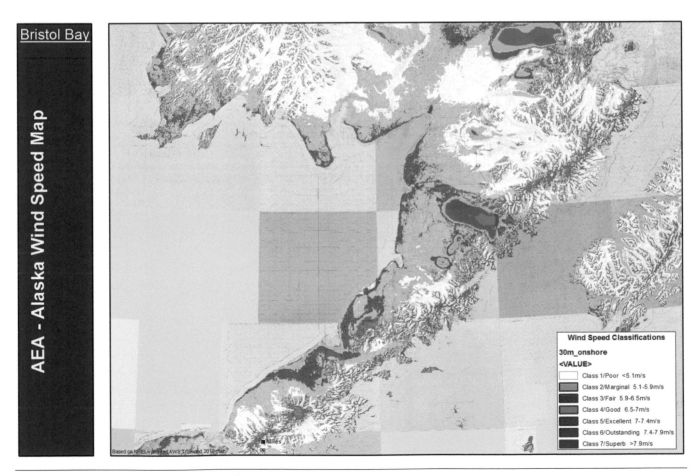

**Figure 26.2**    Wind resource map. *Source credit: AEA.*

Another solution is the combination of the wind system with a long-term energy storage system, such as a large battery facility or a pumped hydro system. A battery facility is a building (or a part of a building) containing many batteries connected together. It can be very expensive, partly due to the hazardous nature of common battery types requiring special measures (electrolyte spill containment, etc.) and because of intensive maintenance that is due to the limited lifetime of batteries. A pumped hydro system is a system where water is pumped uphill during periods of excess energy (e.g., during periods of too much wind). This water is stored in a pool at a higher elevation for later use when there is energy demand (e.g., when there is not enough wind); then the water is allowed to flow back down through a turbine to generate the needed energy. A pumped hydro system can be expensive, depending on natural topographical features that are available for operating two pools of water at different elevations. Another problem with pumped hydro systems can be their impact on fish and other life forms.

Instability, due to fluctuations in wind speed, is one of the biggest drawbacks of wind compared to some other sources (e.g., diesel and hydro). A chart summarizing basic solutions to the instability problem is shown in Figure 26.3.

## Technologies and Economics

The economics of a wind project depend on the size of wind turbines used, magnitude of wind resource, cost of electricity for the system into which wind is being integrated (typically diesel generators), and many other factors. Economies of scale have a significant effect for two main reasons: (1) in terms of dollars per one watt of installed capacity, larger wind turbines are, in general, cheaper than smaller wind turbines; and (2) larger wind turbines are typically higher where wind speed is higher, which, in general, results in a larger capacity factor.

The following is an example of some broad economic feasibility calculations for Dillingham (Bristol Bay's largest community). A long-term wind study using a 30-meter meteorological tower with anemometers was done in Dillingham at the Kanakanak site (near the hospital). The study indicated a Class 3 (or *fair*) wind resource (Vaught 2007a). As a part of the study, simulations were done for several wind turbine models and the average capacity factor was about 26%, which translates to about 2300 kWh produced annually from 1 kW of installed capacity. Dillingham produces electricity from diesel generators, with the fuel cost of about $0.24/kWh—which means that 2300 kWh produced from wind would save about $550 annually in fuel costs. The average cost per 1 kW of installed wind capacity in rural Alaska is about $9600

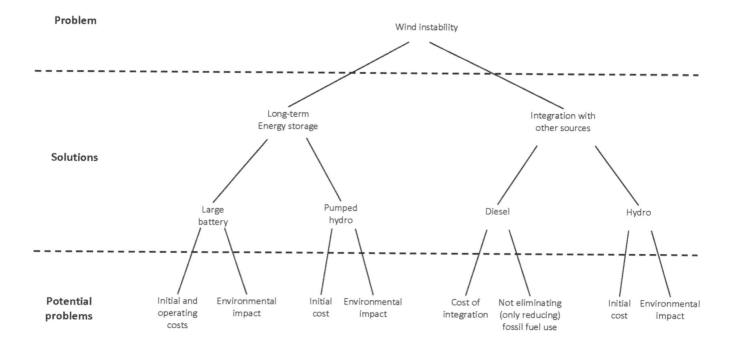

Note: See main text for more details, such as secondary loads, flywheels, and other technologies used in combination with the solutions shown in the figure.

**Figure 26.3**   Basic solutions to the instability problem of wind.

(Fay et al. 2010). With $9600 of capital cost and $550 annual savings, the simple payback period would be about 17 years. This is a very simplified analysis that ignores operation and maintenance costs and other factors (such as the fact that larger wind turbines would have a smaller cost per installed kW). However, one can use this analysis as a baseline for more detailed studies.

## Other Aspects

There are many other factors that need to be considered when evaluating wind as an alternative for electricity generation. Important factors to consider include: effect on birds and bats, land ownership, permitting, aesthetics, shadow flicker for turbines close to communities, technology readiness for northern climates (icing issues, etc.), skills of local workforce, soils and their effect on foundation costs (can be significant), and acceptance of the technology by local culture.

Another thing to take into account is the fact that in some communities, the current diesel generation systems operate in winter as combined heat and power plants. It means that the byproduct heat from the diesel generators is utilized for heating buildings. If some diesel-generated electricity is replaced with wind-generated electricity, the byproduct heat available from the diesel-generation system will be reduced and the needed heat might have to be supplied from other sources, such as heating oil. Some of this heat can come from excess wind energy in higher penetration systems.

## Example of Use

The only utility scale wind system in Bristol Bay is a 180 kW wind farm in Kokhanok, a community with a Class 6 (or *outstanding*) wind (Vaught 2007b). It was installed in 2010 (SWAMC et al. 2013) and consists of two 90 kW Vestas V-17 turbines on 85 foot lattice towers, one grid forming inverter, battery storage, a synchronous condenser, secondary load, and controls (Lyons 2013). In fiscal year 2013, the wind system produced about 29% of Kokhanok's total electrical needs (AEA 2014a). This number dropped to about 7% in fiscal year 2014 (AEA 2015b) due to technical issues. While the primary purpose of the wind system is to provide electricity, excess energy generated during high-wind/low-load situations is used to heat the school and reduce its use of heating oil.

# SOLAR

## Magnitude of Resource

Figure 26.4 maps the solar resource for the Bristol Bay area. In general, the solar resource in Bristol Bay is relatively poor.

## Stability and Predictability

Solar radiation is a seasonal, intermittent resource with limited predictability. When using solar to generate electricity, basic solutions to these problems are similar to those discussed in the previous section on wind. When using solar to produce heat, typical solutions to these problems are by using a thermal storage system and by combining the solar system with another heating system that has a controllable energy source, such as heating oil.

## Technologies and Economics

### Passive Solar Thermal

*Passive solar* refers mainly to the solar energy entering a building through windows. Since most buildings have windows anyway (for day lighting, egress, etc.), passive solar can be a very economical renewable energy source. Suitable orientation of a building and suitable distribution of windows (typically with the majority of them on the south-facing wall) might represent no added construction cost and increased passive solar gain, and thus decreased needs to provide heat from other sources. However, adding windows specifically for the purpose of the passive solar gain that are not needed for other purposes is typically not advantageous in Bristol Bay because the additional heat loss through the windows typically outweighs the passive solar gain. This problem can be overcome by advanced technologies, such as insulated shutters (to be closed at night), but these technologies have not been shown to be economical in Bristol Bay.

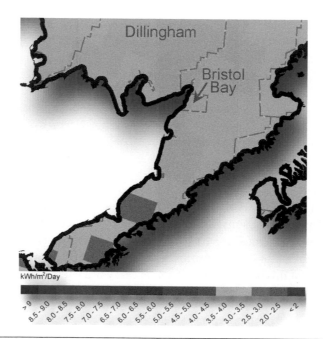

**Figure 26.4 Solar resource map.** *Source credit: National Renewable Energy Lab.*

## Active Solar Thermal

*Active solar thermal* refers to the use of solar thermal panels and an active system (typically a pump) to transfer the heat from the panels into a building through a moving fluid. While using an active solar thermal system for space heating is challenging because the resource is needed the most (in winter) when it is available the least, an active solar thermal system is often considered for hot water needs because hot water is needed all year round (including seasons when the solar resource is well available). Economic estimates are often done based on performance data collected on existing systems in similar conditions; however, the challenge with active solar thermal systems is that heat is difficult/expensive to measure and thus many systems operate without collecting performance data. Some systems use cheaper data collection, which often results in very inaccurate results (ACEP 2013). The problem is that active solar thermal systems operate with a relatively low delta T (supply temperature minus return temperature of the heat transfer fluid) and a small error in the temperature output of the sensors represents a large error in the calculated heat. The only solar hot water system in Bristol Bay that has a fairly accurate data collection system was evaluated to have the simple payback period of about 30 years (UAF BBC 2013).

## Solar Photovoltaic

Solar photovoltaic (PV) refers to technology for the direct conversion of solar energy into electrical energy. The following paragraph is an example of some broad economic feasibility calculations for Dillingham (Bristol Bay's largest community).

The National Renewable Energy Laboratory (NREL) published solar radiation data for various locations for a typical meteorological year (TMY). The published data are results of measurements combined with modeling. TMY data, version 3 (TMY3) includes solar radiation data for Dillingham. The following calculations use a vertical south-facing orientation of PV panels as a baseline. Even though vertical orientation does not maximize the energy production potential, it provides a relatively maintenance free operation because of low susceptibility to snow and dust accumulation. It also helps reduce wind loading if installed on a south-facing wall of a building. Using TMY3 data for Dillingham, the average insolation on a vertical south-facing surface is about 725 Btu/ft$^2$-day, assuming a full exposure with no obstructions. After a unit conversion (1 Btu/ft$^2$-day = 0.1314 W/m$^2$), this average insolation is 95 W/m$^2$, which is 9.5% of the standard insolation used for rating the power output of PV panels (1000 W/m$^2$). It means that vertical south-facing PV panels in

Dillingham would operate with about a 9.5% capacity factor because the power output of PV panels is approximately proportional to the insolation. A 9.5% capacity factor translates to about 832 kWh produced annually from 1 kW of installed capacity. Dillingham produces electricity from diesel generators, with the fuel cost of about $0.24/kWh, which means that 832 kWh produced from solar would save about $200 annually in fuel costs. The average cost per 1 kW of installed PV capacity for rural Alaska projects proposed in Round 7 (2013) of the Alaska Energy Authority Renewable Energy Grant Program was $6200 (AEA 2014c), which is the value used for the analysis herein. With $6200 of capital cost and $200 annual savings, the simple payback period would be about 31 years. This is a very simplified analysis that ignores maintenance costs and other factors; however, one can use this analysis as a baseline for more detailed studies.

## Other Aspects

There are many other factors that need to be considered when evaluating solar as an alternative for electricity generation. Important factors to consider are, for example: space needs and land ownership, environmental impact of manufacturing (high energy needs, pollution), environmental impact of the disposal of the panels at the end of their useful life, aesthetics, loss of by-product heat from current diesel generators, and acceptance of the project by local culture.

## Examples of Use

In 2014, an 80 kW solar PV system was installed at the Naknek High School and it is considered the largest PV system in Alaska. It is comprised of three separate arrays—a 30 kW ground mount, a 15 kW wall mount, and a 35 kW roof mount. Each array is optimized for different seasons. No performance data is available because the system is not operating yet as of the writing of this chapter, but it is expected to produce about 73,000 kWh annually when in operation (Malony 2014).

# BIOMASS

## Magnitude of Resource

Figure 26.5 maps the biomass resource for the Bristol Bay area. The Bristol Bay biomass resource includes wood, fish byproducts, and municipal waste. Woody biomass is a relatively scarce resource. It is mainly in the form of trees and can be used, for example, as fuel for wood stoves. Municipal waste, even if an appropriate technology for waste-to-energy conversion on a small scale existed, would only be able to provide a very small fraction of the needed energy (Marsik 2009). Bristol Bay has several major fish processing plants that generate fish byproduct waste.

## Stability and Predictability

Biomass energy is considered demand energy, which means it is typically available as needed. For the fish byproduct waste obtained during the fishing season, the storage time might be limited by certain issues (e.g., oxidation and biological growth).

## Technologies and Economics

### Woody Biomass

Woody biomass can be used for heating or for electricity generation. Because woody biomass in Bristol Bay is relatively scarce, and likely can be considered only as a supplemental fuel source covering a fraction of energy needs, it is technologically easier and cheaper to use the available biomass to substitute some of the diesel used for heating, as opposed to trying to substitute the diesel used for electricity generation. Technologies that use biomass for heating (such as wood stoves, or wood-fired boilers) are much more mature than technologies that use biomass for electricity production. The following is an example of the economics of using wood stoves.

In Dillingham (Bristol Bay's largest community), a cord of split spruce costs about $300 and when dry has about 16,000,000 Btu. Assuming a relatively new wood stove with 75% efficiency, a cord of dry spruce will provide about 12,000,000 Btu of useful heat. In comparison, a gallon of #1 heating oil costs around $5, has about 135,000 Btu, and if burned in a relatively new heating appliance with 85% efficiency, it provides about 115,000 Btu of useful heat. If 115,000 Btu of useful heat were to be provided from dry spruce, the numbers previously mentioned would translate to a cost of

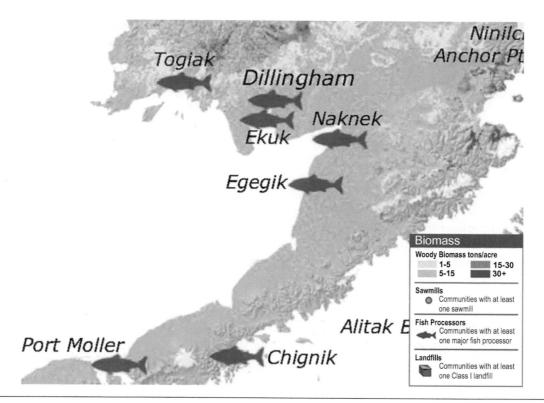

**Figure 26.5**  Biomass resource map. *Source credit: Renewable Energy Atlas of Alaska (AEA and REAP 2013).*

about $2.90. It means that heating with wood, from the economic perspective, is about the same as heating with $2.90 per gallon heating oil. While this is cheaper than heating with oil (oil costs about $5 per gallon), the financial savings need to be weighed against other factors, such as labor involved in handling wood, capital cost of a wood stove, air pollution, and sustainability of local forests.

## Municipal Waste

Several challenges exist when trying to use municipal waste for heating. One of them is that the incineration would have to happen relatively close to facilities that can use the heat; however, from the air pollution perspective, it is desirable to have the incineration far away from other facilities. Another challenge is that most small-scale incineration systems suitable for Bristol Bay communities need diesel to operate, which is the opposite of what is desired (using municipal waste to reduce diesel consumption). It is even more challenging to try to use municipal waste for electricity generation. No mature technologies were found that would be suitable for the small-scale systems that would be required in Bristol Bay.

A special category of using waste for energy is the use of waste motor oil (which does not fall under the stricter definition of biomass, but does fall under the broader definition of biomass that includes waste). Waste motor oil is successfully used in Bristol Bay for heating using waste oil burners. The economics are favorable—mainly thanks to two aspects: (1) the fuel source is free; (2) utilizing the waste oil reduces costs that would otherwise be incurred for the proper disposal of waste oil as hazardous waste.

## Fish Byproducts

A lot of the waste from fish processing ends up being discarded into the ocean, despite that fact that it contains large quantities of fish oil. This is because the equipment and process for recovering the fish oil are expensive. However, harvesting the fish oil becomes economical when fishmeal is being produced from the fish waste. This is because extracting the fish oil is a process closely affiliated with the fishmeal manufacturing process. Even when fish oil is extracted, it does not always end up being used for energy because other possible uses exist (e.g., further processing into a pharmaceutical supplement). When the fish oil is used for energy, it is often used in the boilers needed for the fishmeal manufacturing process, but sometimes it is also used in diesel generators to produce electricity. Because there is no operating fishmeal plant in Bristol Bay as of writing this chapter, no fish oil is being harvested from fish waste in Bristol Bay. Trident Seafoods is planning to start operating its new fishmeal plant in Naknek in late April 2015, but no plans of using the fish oil for energy were announced (Colton 2015).

## Other Aspects

There are many other factors that need to be considered when evaluating biomass as an energy source. Important factors to consider include: land ownership, air pollution, resource sustainability, space for biomass storage, odors, labor, and acceptance of the project by local culture.

## Examples of Use

Due to the relatively limited resource, the examples of use are limited also. Woody biomass is used for heating in communities that have at least some forests close by, such as Dillingham, New Stuyahok, or Igiugig. Special types of waste products (such as shipping pallets or waste motor oil) are also being used, even though the quantities are very limited. There is no known use of fish byproduct as a direct source of energy as of writing this chapter.

# GEOTHERMAL

## Magnitude of Resource

While most of Bristol Bay does not have any known shallow geothermal resource, the east boundary of Bristol Bay is located along the Ring of Fire, a very rich geothermal area. Figure 26.6 maps associated volcanoes, springs, and wells. While this is an extraordinary renewable energy resource, the problem is that it is very far away from main population centers.

## Stability and Predictability

If developed and operated correctly, geothermal energy is a relatively stable, long-term resource. An important problem to mention is that in almost all geothermal installations in the world, heat is being removed faster than it is replaced, which means the available output power is gradually dropping (Boyle 2004).

## Technologies and Economics

Geothermal energy can be used for heating and for electricity production. Geothermal energy systems are normally economical where a shallow geothermal resource exists in the proximity of the point of energy use or distribution line, and where energy need is high enough to justify resource explorations and development. Where a shallow geothermal resource does not exist, an enhanced geothermal system (EGS) can be considered to transfer heat from great depths to the surface; however, EGS is not a mature technology and only few pilot projects exist in the world. Usable heat exists within drillable depths in most areas of the earth, the question that needs to be answered, though, is how economical it would be to transfer the heat to the surface. Deep explorations need to be done to answer that question, which can be very expensive.

Geothermal energy can also be used for cooling, using technologies called absorption chillers and adsorption chillers. In theory, this could be used, for example, for ice production for the fishing industry.

## Other Aspects

There are many other factors that need to be considered when evaluating geothermal energy as an alternative energy source. Important factors to consider include: potential for simultaneous use for several purposes (electricity generation, heating, food production in greenhouses, etc.), land ownership, and acceptance of the project by local culture.

## Examples of Use

One example of an attempt to use geothermal energy is in Naknek/King Salmon, an area served by the Naknek Electric Association (NEA). The NEA area does not have a known shallow geothermal resource, but the NEA attempted to

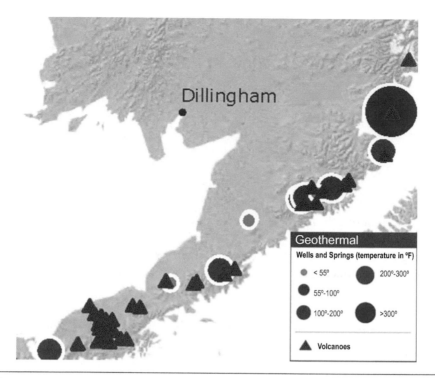

**Figure 26.6 Geothermal resource map.** *Source credit: Renewable Energy Atlas of Alaska (AEA and REAP 2013).*

develop an EGS system. It began drilling the first exploratory geothermal well on August 16, 2009. In December 2009, at a depth of 11,218 feet, a drilling bit broke. A sidetrack well was attempted, but the first attempt failed. The second side-track well was successfully drilled to 11,387 feet. In April 2010, flow testing of the well was attempted, but the well refused to flow under its own pressure due to the drilling mud clogging the well. A large air compressor was utilized to try to clean out the mud, but without significant success. Other unexpected circumstances occurred. In September 2010, with a debt of approximately $40 million associated with the geothermal project, the NEA filed a reorganization plan in U.S. Bankruptcy Court in Anchorage. In the fall of 2012, the NEA stopped efforts to confirm the viability of the geothermal resource (LeRoy 2013).

# TIDAL

## Magnitude of Resource

Figure 26.7 maps the tidal current speed for the Bristol Bay area and demonstrates that the Nushagak Bay has extraordinary tidal currents.

## Stability and Predictability

Tidal currents are an intermittent resource as the currents stop and change direction with every tide. However, tidal currents occur regularly, four times a day (with various magnitude), and can be predicted years in advance. The predictability of tidal energy makes it easier to integrate it with other systems, such as diesel generators.

## Technologies and Economics

Tidal energy can be harvested using barrages or using hydrokinetic devices. Barrages are dams built across suitable estuaries; and they rely on the same technology as conventional river hydro systems that harvest the potential energy associated with different levels of water on either side of the dam. Because of their enormous impact on the ecosystem, barrages are unlikely to be ever developed in Bristol Bay, and therefore, are not further discussed in this chapter. Hydrokinetic devices are in-stream turbines that harvest the kinetic energy of the flowing water. Because no dam is needed,

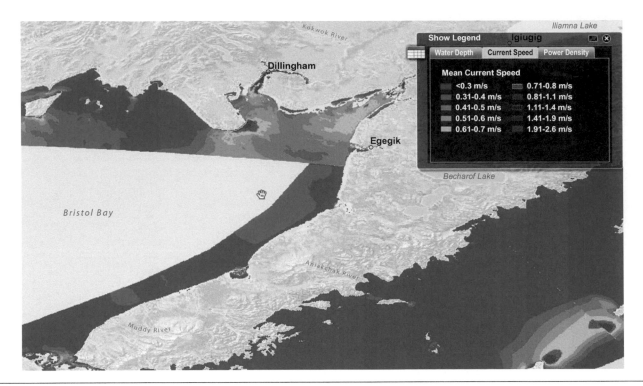

**Figure 26.7**   Tidal resource map. *Source credit: Assessment of Energy Production Potential from Tidal Streams in the United States (Georgia Institute of Technology 2011).*

their impact on the ecosystem is significantly lower than barrages. Insufficient tidal current data coupled with the fact that no mature technology exists for harvesting tidal energy in Alaskan conditions pose high challenges to performing an economic estimate.

## Other Aspects

A very important factor to consider with hydrokinetic devices is that it is a new technology and has been mostly used in pilot projects. The world's first commercial-scale tidal generator, a 1.2 MW underwater turbine, was installed in Northern Ireland in 2008 (McKimm 2012).

There are many environmental and technical challenges related to the deployment of tidal hydrokinetic devices in Alaska that still need to be overcome. The main ones include the impact on aquatic life and resistance of the technology to debris and ice.

## Examples of Use

Tidal energy is not used anywhere in Bristol Bay. University of Alaska Fairbanks Bristol Bay Campus (UAF BBC) has been collecting tidal current data in the Nushagak Bay using an Acoustic Doppler Current Profiler (ADCP).

## RECOMMENDED RESEARCH

Successful implementation of a sustainable energy project depends on the characteristics of the energy resources, available technology for resource utilization, knowledge of the performance characteristics of given technology, and other factors. The recommendations of the author are to continue data collection and modeling related to characterizing renewable energy resources in Bristol Bay, continue data collection on the installed sustainable energy systems, and continue developing technology for the utilization of these resources. This includes energy efficiency technology. Broader recommendations include developing energy policy supporting sustainable energy, developing educational programs for energy users, and social research to encourage habitual changes to conserve energy.

# SUMMARY

Renewable energy resources can be considered abundant in Bristol Bay; however, their utilization poses numerous challenges. One of them is the transfer of the energy from the source to the point of use, which often involves long distances, steep terrain, or other obstacles. Another challenge is that commercially available technology doesn't exist for the utilization of some of these resources in local conditions (cold climate; small communities with small energy load). Another issue is high installation costs of renewable energy systems due to the need to transport the material, and sometimes also heavy equipment and skilled labor, to Bristol Bay. The intermittent nature of some renewable energy resources is also a problem. Because of these and other challenges, renewable energy has not been much utilized in the region. While accurate data on heat energy is not being collected, electrical energy data shows that only about 7% of the electrical energy in Bristol Bay comes from renewable sources (mostly hydro) and the remaining amount (about 93%) comes from diesel (AEA 2015b).

The challenges summarized in this chapter also present opportunities. Bristol Bay, together with some other parts of rural Alaska, serve as globally significant test fields for new technologies and approaches. Micro-grids with integrated renewable energy are often viewed as possible models for less developed parts of the world with similar challenges (e.g., small communities and long distances).

The most utilized resource—and the one with the highest potential in the near-term future—is energy efficiency. Energy efficiency, in general, provides the highest reduction in fossil fuel used per dollar invested. However, while implementing energy efficiency measures, research and pilot projects in renewable energy should continue, so that the renewable energy technology is in a better position to supply the reduced amount of energy after the cost-effective energy efficiency measures have been implemented.

# ACKNOWLEDGMENTS

This work was done at the University of Alaska Fairbanks Bristol Bay Campus and was supported by the National Science Foundation and the National Institute of Food and Agriculture, U.S. Department of Agriculture. The author would also like to thank the following organizations and individuals for their help with this chapter: Alaska Center for Energy and Power: Christopher Pike; Alaska Energy Authority: David Lockard, Richard Stromberg, Katie Conway, Sean Skaling, Rebecca Garrett; Bristol Bay Borough School District: Bill Hill; Bristol Bay Native Association: Lawrence Sorensen; Center for Science in Public Participation: Carol Ann Woody; Crimp Energy Consulting: Peter Crimp; Iliamna Newhalen Nondalton Electric Cooperative: George Hornberger; Naknek Electric Association: Donna Vukich; Native Council of Port Heiden: John Christensen Jr.; Renewable Energy Alaska Project: Chris Rose, Shaina Kilcoyne; University of Alaska Fairbanks: Gabriel Dunham, Kristin Donaldson, Todd Radenbaugh.

# REFERENCES

ACEP (Alaska Center for Energy and Power). 2013. An investigation of solar thermal technology in arctic environments: a project by Kotzebue Electric Association. Available: http://energy-alaska.wdfiles.com/local—files/feasibility-of-solar-hot-water-systems/KEA%20Solar%20Thermal%20Final%201-31-13.pdf.

ACEP. 2015. Alaska Hydrokinetic Energy Research Center. Available: http://acep.uaf.edu/programs/alaska-hydrokinetic-energy-research-center.aspx.

AEA (Alaska Energy Authority). 2014a. Power cost equalization program: statistical data by community, July 1, 2012 to June 30, 2013. Available: http://www.akenergyauthority.org/Programs/PCE.

AEA. 2014b. Wind energy analysis data. Available: http://www.akenergyauthority.org/Programs/AEEE/Wind/analysisdata.

AEA. 2014c. Alaska renewable energy fund round 7—economic summaries. Available: ftp://ftp.aidea.org/REFund7/2_Project_Specific_Docs/economic_analysis_summaries/REF7_EconomicSummaries.pdf.

AEA. 2015a. Alaska renewable rnergy fund status report, rounds I-VII. Available: ftp://www.aidea.org/REFund/Round%208/documents/REFStatusAppendix01302015.pdf.

AEA. 2015b. Power cost equalization program: statistical data by community, July 1, 2013 to June 30, 2014. Available: http://www.akenergyauthority.org/Programs/PCE.

AEA and REAP (Renewable Energy Alaska Project). 2013. Renewable energy atlas of Alaska: a guide to Alaska's clean, local, and inexhaustible energy resources. Available: http://alaskarenewableenergy.org/wp-content/uploads/2009/04/2013-RE-Atlas-of-Alaska-FINAL.pdf.

Boyle, G. 2004. Renewable Energy—Power for a Sustainable Future. Second Edition. Oxford University Press Inc., New York, NY.

Colton, H. 2015. Fishmeal plant to go online soon in Naknek. Bristol Bay Times. 36(17):1, 15. Available: http://www.thebristol baytimes.com/article/1517fishmeal_plant_to_go_online_soon_in_naknek.

Fay, G., T. Schwoerer, and K. Keith. 2010. Alaska isolated wind-diesel systems: performance and economic analysis. Prepared for AEA. Available: http://www.iser.uaa.alaska.edu/Publications/wind_diesel10022010.pdf.

Georgia Institute of Technology. 2011. Assessment of energy production potential from tidal streams in the United States. Available: http://www.tidalstreampower.gatech.edu/.

Gray Stassel Engineering, Inc. 2015. Kvichak River RISEC 2014 Project Summary Report. Available: ftp://ftp.aidea.org/RPSU/Igiugig/RISEC_Deploy/2014%20Summary%20Report%20to%20AEA.pdf.

Helliwell, J., R. Layard, and J. Sachs. 2013. World happiness report. Available: http://unsdsn.org/resources/publications/world-happiness-report-2013/.

Lee, M. and G. Reinhardt. 2003. Eskimo architecture: dwelling and structure in the early historic period. University of Alaska Press, Fairbanks, AK.

LeRoy, E. 2013. Second amended disclosure statement for plan of reorganization. In the United States Bankruptcy Court for the district of Alaska. Available: http://bankrupt.com/misc/NAKNEK_ELECTRIC_ds_2amended.pdf.

Lyons, J. 2013. Kokhanok wind-diesel system update. Available: http://www.uaf.edu/files/acep/2013_REC_Integration%20of%20Renewable%20Energy%20in%20Kokhanok_John%20Lyons.pdf.

Malony, L. 2014. Infrastructure builds wrap up. Alaska Business Monthly. December 2014: 46-47. Available at: http://www.akbizmag-digital.com/akbizmag/december_2014?pg=46#pg46.

Marsik, T. 2009. Basic study of renewable energy alternatives for electricity generation in Dillingham/Aleknagik region, prepared for Nushagak Electric and Telephone Cooperative, Inc. Available: http://www.agnewbeck.com/pdf/bristolbay/Dillingham_Comp_Plan/Report_NushagakAlternatives.pdf.

Marsik, T. 2012. Net zero energy ready home in Dillingham, Alaska. Alaska Building Science News. Available: http://www.uaf.edu/files/ces/newsletters/ABSN/63_2012-Summer.pdf.

McKimm, M. 2012. Strangford Lough generator given all-clear. BBC News. Available: http://www.bbc.com/news/uk-northern-ireland-16595752.

ORPC (Ocean Renewable Power Company). 2014. ORPC's RivGen® power system demonstration project a major success. Available: http://www.orpc.co/newsevents_pressrelease.aspx?id=pOH6DBxeTtg%3d.

Ristinen, R. and J. Kraushaar. 2006. *Energy and the Environment*. Second edition. John Wiley & Sons, Inc., Hoboken, New Jersey, USA.

Robert W. Retherford Associates. 1979. Bristol Bay energy and electric power potential. U.S. Department of Energy 85-79AP 0002.000. Available: http://akenergyinventory.org/hyd/SSH-1979-0075.pdf.

SWAMC (Southwest Alaska Municipal Conference), Bristol Bay Native Association (BBNA), and Information Insights. 2013. Bristol Bay Regional Energy Plan, Phase I: Preliminary Planning, Resource Inventory& Data Collection. Available: http://bristolbay energy.org/documents/.

The Energy Efficiency Partnership. 2013. Alaska energy efficiency. Available: http://www.akenergyefficiency.org/.

The Energy Efficiency Partnership. 2015. Energy projects coordination matrix. Available: http://www.akenergyefficiency.org/energy-projects-coordination-matrix/.

UAF BBC. 2013. U.S. Fish and Wildlife Service/UAF Bristol Bay Campus solar thermal. Available: http://energy-alaska.wikidot.com/usfws-uaf-bristol-bay-campus-solar-thermal.

Vaught, D. 2007a. Dillingham, Alaska wind resource report—Kanakanak site, prepared for AEA. Available: http://www.akenergy authority.org/Content/Programs/AEEE/Wind/WindResourceAssessment/Dillingham-KanakanakWindResourceReport 3-31-07.pdf.

Vaught, D. 2007b. Kokhanok, Alaska wind resource report, prepared for AEA. Available: http://www.akenergyauthority.org/Content/Programs/AEEE/Wind/WindResourceAssessment/KokhanokWindResourceReport3-31-07.pdf.

Wikimapia. 2012. Tazimina River hydroelectric plant. Available: http://wikimapia.org/22619226/Tazimina-River-Hydroelectric-Plant.

# INDEX